Acta Numerica 2001

Acta

Numerica

2001

Volume 10

CAMBRIDGE
UNIVERSITY PRESS

CAMBRIDGE UNIVERSITY PRESS
Cambridge, New York, Melbourne, Madrid, Cape Town, Singapore,
São Paulo, Delhi, Dubai, Tokyo, Mexico City

Cambridge University Press
The Edinburgh Building, Cambridge CB2 8RU, UK

Published in the United States of America by Cambridge University Press, New York

www.cambridge.org
Information on this title: www.cambridge.org/9780521157698

© Cambridge University Press 2001

First published 2001
First paperback edition 2010

A catalogue record for this publication is available from the British Library

ISBN 978-0-521-80312-0 Hardback
ISBN 978-0-521-15769-8 Paperback

Contents

Acta Numerica (2001), pp. 1–102

An optimal control approach to *a posteriori* error estimation in finite element methods

Roland Becker and Rolf Rannacher

Institut für Angewandte Mathematik
Universität Heidelberg
INF 293/294, D-69120 Heidelberg, Germany
http://gaia.iwr.uni-heidelberg.de
E-mail: Roland.Becker@iwr.uni-heidelberg.de
Rolf.Rannacher@iwr.uni-heidelberg.de

This article surveys a general approach to error control and adaptive mesh design in Galerkin finite element methods that is based on duality principles as used in optimal control. Most of the existing work on *a posteriori* error analysis deals with error estimation in global norms like the 'energy norm' or the L^2 norm, involving usually unknown 'stability constants'. However, in most applications, the error in a global norm does not provide useful bounds for the errors in the quantities of real physical interest. Further, their sensitivity to local error sources is not properly represented by global stability constants. These deficiencies are overcome by employing duality techniques, as is common in *a priori* error analysis of finite element methods, and replacing the global stability constants by computationally obtained local sensitivity factors. Combining this with Galerkin orthogonality, *a posteriori* estimates can be derived directly for the error in the target quantity. In these estimates local residuals of the computed solution are multiplied by weights which measure the dependence of the error on the local residuals. Those, in turn, can be controlled by locally refining or coarsening the computational mesh. The weights are obtained by approximately solving a linear adjoint problem. The resulting *a posteriori* error estimates provide the basis of a feedback process for successively constructing economical meshes and corresponding error bounds tailored to the particular goal of the computation. This approach, called the 'dual-weighted-residual method', is introduced initially within an abstract functional analytic setting, and is then developed in detail for several model situations featuring the characteristic properties of elliptic, parabolic and hyperbolic problems. After having discussed the basic properties of duality-based adaptivity, we demonstrate the potential of this approach by presenting a selection of results obtained for practical test cases. These include problems from viscous fluid flow, chemically reactive flow, elasto-plasticity, radiative transfer, and optimal control. Throughout the paper, open theoretical and practical problems are stated together with references to the relevant literature.

CONTENTS

1. Introduction

Solving complex systems of partial differential equations by discretization methods may be considered in the context of 'model reduction': a conceptually *infinite-dimensional* model is approximated by a *finite-dimensional* one. Here, the quality of the approximation depends on the proper choice of the discretization parameters, for example, the mesh width, the polynomial degree of the trial functions, and the size of certain stabilization parameters. As the result of the computation, we obtain an approximation to the desired output quantity of the simulation and, besides that, certain accuracy indicators, such as local cell residuals. Controlling the error in such an approximation of a continuous model of a physical system requires us to determine the influence factors of the local error indicators on the target quantity. Such a sensitivity analysis with respect to local perturbations of the model is common in optimal control theory and naturally introduces the concept of an 'adjoint' (or 'dual') problem.

For illustration, consider a *continuous* model governed by a linear differential operator A and a force term f, and a related discrete model depending on a discretization parameter $h \in \mathbb{R}_+$:

$$Au = f, \qquad A_h u_h = f_h. \tag{1.1}$$

In designing this discretization, we have to detect the interplay of the various error propagation effects in order to achieve

    (i) *a posteriori error control,* that is, control of the error in quantities of physical interest such as stress values, mean fluxes, or drag and lift coefficients, and

(ii) *solution-adapted meshing*, that is, design of economical meshes for computing these quantities with optimal efficiency.

Traditionally, *a posteriori* error estimation in Galerkin finite element methods is done with respect to a natural 'energy norm' $\|\cdot\|_E$ induced by the underlying differential operator. This results in estimates of the form

$$\|u - u_h\|_E \leq c_s \|\rho(u_h)\|_E^*, \tag{1.2}$$

with a suitable dual norm $\|\cdot\|_E^*$ and the computable 'residual' $\rho(u_h) = f - Au_h$, which is well defined in the context of a Galerkin finite element method. Now, the main goal is to localize the residual norm in order to make it computable as a sum of cell-wise contributions. This approach was initiated by the pioneering work of Babuška and Rheinboldt (1978a, 1978b) and was then further developed by Ladeveze and Leguillon (1983), Bank and Weiser (1985), and Babuška and Miller (1987), to mention only a few of the most influential papers. For discussions and further references, we refer to the surveys by Verfürth (1996) and Ainsworth and Oden (1997a). In this article, 'energy error' estimation will be addressed only when it is important for our own topic.

Energy error estimation seems rather generic as it is directly based on the variational formulation of the problem and allows us to exploit its natural coercivity properties. However, in most applications the error in the energy norm does not provide useful bounds on the errors in the quantities of real physical interest. A more versatile method for *a posteriori* error estimation with respect to relevant error measures is obtained by using duality arguments as is common from the *a priori* error analysis of finite element methods (the so-called 'Aubin–Nitsche trick'). Let $J(u)$ be a quantity of physical interest derived from the solution u by applying a functional $J(\cdot)$. The goal is to control the error $J(u) - J(u_h)$ in terms of local residuals $\rho_K(u_h)$ computable on each of the mesh cells K. An example is control of the local total error $e_K = (u - u_h)_{|K}$. By superposition, e_K splits into two components, the locally produced 'truncation error' and the globally transported 'pollution error', $e_K^{tot} = e_K^{loc} + e_K^{trans}$, assuming for simplicity that the underlying problem is linear. In view of the error equation $A(u-u_h) = \rho(u_h)$, the effect of the cell residual ρ_K on the error $e_{K'}$, at another cell K', is governed by the Green function of the continuous problem. In practice it is mostly impossible to determine the complex error interaction by analytical means; rather, it has to be detected by computation. This leads to 'weighted' *a posteriori* error estimates

$$|J(u) - J(u_h)| \approx \langle \rho(u_h), \omega_h(z) \rangle, \tag{1.3}$$

where the sensitivity factor $\omega_h(z)$ is obtained by approximately solving an 'adjoint problem' $A^*z = j$, with j a density function associated with $J(\cdot)$.

The adjoint solution z may be viewed as a generalized Green's function with respect to the output functional $J(\cdot)$, and accordingly the weight $\omega_h(z)$ describes the effect of local variations of the residual $\rho(u_h)$ on the error quantity $J(u) - J(u_h)$, for instance as the consequence of mesh adaptation. This approach to *a posteriori* error estimation is called the 'dual-weighted residual method' (or, for short, 'DWR method'). On the basis of *a posteriori* error estimates like (1.3), we can design a feedback process in which error estimation and mesh adaptation go hand-in-hand, leading to economical discretization for computing the quantities of interest. This approach is particularly designed for achieving high accuracy at minimum computational cost. The additional work required by the evaluation of the error bounds is usually acceptable since, particularly in nonlinear cases, it amounts to only a moderate fraction of the total costs.

The use of duality arguments in *a posteriori* error estimation goes back to ideas of Babuška and Miller (1984a, 1984b, 1984c) in the context of post-processing of 'quantities of physical interest' in elliptic model problems. It has since been systematically pursued by Eriksson and Johnson (1988, 1991) and their collaborators for more general situations (see also Johnson (1994) and the survey paper by Eriksson, Estep, Hansbo and Johnson (1995)). Here, stability constants for adjoint problems are mostly derived by analytical arguments. In Becker and Rannacher (1996a, 1996b) this approach is further developed into a computation-based feedback method, the 'DWR method', for error control and mesh optimization in computing local quantities of interest (for surveys and applications to problems in mechanics, physics and chemistry see Rannacher (1998a, 2000)). More recently, following ideas by Babuška and Miller (1984a, 1984b, 1984c), related techniques based on 'energy-norm' error estimation have been proposed by Machiels, Patera and Peraire (1998) and Oden and Prudhomme (1999).

We illustrate the ideas underlying the DWR method within an abstract setting. Let $A(\cdot, \cdot)$ be a bilinear form and $F(\cdot)$ a linear functional defined on some function space V, such that the given equation $Au = f$ has the variational formulation

$$A(u, \varphi) = F(\varphi), \quad \text{for all } \varphi \in V. \tag{1.4}$$

For a *finite-dimensional* subspace $V_h \subset V$, with a mesh-size parameter $h \in \mathbb{R}_+$, the Galerkin approximation $u_h \in V_h$ is determined as the solution of the discrete equation

$$A(u_h, \varphi_h) = F(\varphi_h), \quad \text{for all } \varphi_h \in V_h. \tag{1.5}$$

Then, the value $J(u_h)$ is an approximation for the target quantity $J(u)$. In practice, the existence of a solution $u \in V$ of the variational equation (1.4) has to be guaranteed by separate arguments outside the present general frame. Here, we only assume that the bilinear form $A(\cdot, \cdot)$ is sufficiently regular on V and V_h, such that this solution as well as its approximation $u_h \in V_h$ is uniquely determined.

Our approach to estimating the functional error $J(u) - J(u_h)$ is based on embedding the given problem into the framework of optimal control. To this end, we consider the following trivial constraint optimization problem for $u \in V$:

$$J(u) = \min!, \quad A(u, \varphi) = F(\varphi), \quad \text{for all } \varphi \in V, \tag{1.6}$$

which is equivalent to evaluating $J(u)$ from the solution of $Au = f$. Introducing the corresponding Lagrangian $L(u, z) := J(u) + F(z) - A(u, z)$, with the 'adjoint' variable $z \in V$, the minimal solution u is characterized as the first component of a stationary point of $L(u, z)$. That is determined by the Euler–Lagrange system consisting of (1.4) and the adjoint problem

$$A(\varphi, z) = J(\varphi), \quad \text{for all } \varphi \in V. \tag{1.7}$$

Again, the actual existence of the adjoint solution $z \in V$ satisfying (1.7) is guaranteed in concrete situations by separate arguments. By construction, the solutions u and z are mutually adjoint to each other in the sense that $J(u) = A(u, z) = F(z)$. Hence, for computing $J(u)$, one could equally well try to compute $F(z)$. The Euler–Lagrange system is approximated by the Galerkin method in V_h resulting in the discrete equations (1.5) for $u_h \in V_h$ and

$$A(\varphi_h, z_h) = J(\varphi_h), \quad \text{for all } \varphi_h \in V_h,$$

for $z_h \in V_h$. For both errors $e := x - x_h$ and $e^* := z - z_h$, we have the Galerkin orthogonality property $A(e, \cdot) = 0 = A(\cdot, e^*)$ on V_h. Therefore, the mutual duality of u and z carries over to the corresponding errors,

$$J(e) = A(e, z) = A(e, e^*) = A(u, e^*) = F(e^*).$$

We introduce the notation $\rho(u_h, \cdot) := F(\cdot) - A(u_h, \cdot)$ and $\rho^*(z_h, \cdot) := J(\cdot) - A(\cdot, z_h)$ for the residuals of u_h and z_h, respectively. Then, again using Galerkin orthogonality, we have $\rho(u_h, z - \varphi_h) = A(e, e^*) = \rho^*(z_h, u - \varphi_h)$, with arbitrary $\varphi_h \in V_h$. From this, we see that

$$J(e) = \min_{\varphi_h \in V_h} \rho(u_h, z - \varphi_h) = \min_{\varphi_h \in V_h} \rho^*(z_h, u - \varphi_h) = F(e^*). \tag{1.8}$$

These error representations are to be evaluated computationally and serve as a basis for adaptive control of the discretization in computing $J(u)$. This duality of the primal error e and the adjoint error e^* is characteristic of the Galerkin method and is also found in nonlinear problems. The method for a posteriori error estimation described so far will be applied below in several concrete situations of 'conforming' Galerkin finite element discretizations of elliptic problems, also including an optimal control problem for which (1.6) becomes nontrivial.

However, in many applications the Galerkin approximation of the given problem $Lu = f$ is based on modified variational formulations which are mesh-dependent. This is the case, for example, in the 'discontinuous' Galerkin method for transport-oriented or time-dependent problems and when standard Galerkin schemes are stabilized by introducing least-squares or streamline diffusion terms. In order to cover discretizations of this type too, the above framework has to be properly extended. In this context, the DWR method is mainly used as a formal guideline for deriving useful *a posteriori* error estimators. Because of the complexity of the concrete setting, this derivation may occasionally lack full mathematical rigour but eventually finds its justification by computational success.

The contents of this article are as follows. The concept underlying the DWR method will be introduced in Section 2 within an abstract setting. This serves as a basis for the application of the method to various rather different situations of variational problems. This will be made concrete in Section 3 for some linear model problems. In the same context, in Section 4 we will discuss some other approaches to functional-oriented error estimation which are based on 'energy norm' error estimates. The details of the practical realization of the DWR method will be addressed in Section 5. Though the derivation of the abstract *a posteriori* error estimates is equally simple for linear and nonlinear problems, in the latter case its practical evaluation poses particular difficulties. These will be discussed in Section 6. Section 7 deals with the application to Galerkin methods for time-dependent problems, particularly the heat equation and the acoustic wave equation.

The remaining sections are devoted to the application of the DWR method to various practical problems. In Section 8, we show an application to viscous incompressible fluid flow. Here, the main aspect is the interaction of different physical mechanisms with numerical stabilization, and its effect on the accuracy in computing a local quantity, in this case the 'drag coefficient'. The complexity of this physical model will be further increased in Section 9 by including compressibility effects due to temperature changes and chemical reactions. Section 10 contains an example from elasto-plasticity which involves a quasi-linear operator with nonlinearity not everywhere differentiable. This shows that the restrictive smoothness assumptions made in Section 2 are not really necessary for the method to work. Next, Section 11 describes the application of the DWR method to a nonstandard integro-differential equation governing the transfer of energy by radiation and scattering. This is an example in which, because of the high dimensionality of the problem, adaptive mesh design is indispensable. In Section 12, we present an application to a simple optimal control problem with linear state equation and quadratic cost functional. In this example the abstract framework of the DWR method introduced in Section 2 shows its full power.

Finally, Section 13 concludes the paper by presenting open problems and an outlook to future developments.

2. A paradigm for *a posterori* error control

We will develop the DWR method initially within an abstract functional analytic setting. The starting point is a general (though almost trivial) result on *a posteriori* error estimation in the Galerkin approximation of variational problems. From this, we then derive more detailed error representations for variational equations which yield residual-based error estimators with respect to functionals of the solution.

2.1. Galerkin approximation of stationary points

We begin with a general result for *a posteriori* error estimation in the Galerkin approximation of variational problems. The key point is that, here, the error is naturally measured in terms of the generating 'energy functional'. Let X be a function space and $L(\cdot)$ a differentiable functional on X. Its derivatives are denoted by $L'(\cdot;\cdot)$, $L''(\cdot;\cdot,\cdot)$ and $L'''(\cdot;\cdot,\cdot,\cdot)$. Here, we use the convention that in semilinear forms like $L'(\cdot;\cdot)$ the form is linear with respect to all arguments on the right of the semicolon. We seek a stationary point x of $L(\cdot)$ on X, that is,

$$L'(x;y) = 0, \quad \text{for all } y \in X. \tag{2.1}$$

This equation is approximated by a Galerkin method using a finite-dimensional subspace $X_h \subset X$, where $h \in \mathbb{R}_+$ is a discretization parameter. The discrete problem seeks $x_h \in X_h$ satisfying

$$L'(x_h; y_h) = 0, \quad \text{for all } y_h \in X_h. \tag{2.2}$$

To estimate the error $e := x - x_h$, we write

$$L(x) - L(x_h) = \tfrac{1}{2}L'(x_h; e) + \tfrac{1}{2}L'(x; e) + \tfrac{1}{2}\int_0^1 L''(x_h + se; e, e)\, ds. \tag{2.3}$$

The second derivative $L''(x;\cdot,\cdot)$ is symmetric and, for convex $L(\cdot)$, we have $L''(x_h + se; e, e) \geq 0$. From this, we obtain

$$0 \leq L(x_h) - L(x) \leq -\tfrac{1}{2}L'(x_h; e), \tag{2.4}$$

and, using definition (2.2) of x_h,

$$0 \leq L(x_h) - L(x) \leq -\tfrac{1}{2}\min_{y_h \in X_h} L'(x_h; x - y_h). \tag{2.5}$$

The term $L'(x_h; x - y_h)$ represents a residual involving the weight factor $\omega := x - y_h$. Below, we will derive explicit representations for such terms depending on the concrete form of the underlying problem. This gives an *a posteriori* bound for the error measured in terms of the energy functional $L(\cdot)$. Now, we consider the general case when the functional $L(\cdot)$ is not necessarily convex.

Proposition 2.1. For the Galerkin approximation of the variational problem (2.1), we have the *a posteriori* error representation

$$L(x) - L(x_h) = \tfrac{1}{2} \min_{y_h \in X_h} L'(x_h; x - y_h) + R. \tag{2.6}$$

The remainder term is given by

$$R := \tfrac{1}{2} \int_0^1 L'''(x_h + se; e, e, e)\, s(s-1)\, \mathrm{d}s$$

and vanishes if the functional $L(\cdot)$ is quadratic.

Proof. We introduce the notation

$$L'(\overline{x x_h}; y) := \int_0^1 L'(x_h + se; y)\, \mathrm{d}s.$$

and we begin with the trivial identity

$$L(x) - L(x_h) = L'(\overline{x x_h}; e) + \tfrac{1}{2}L'(x_h; e) - \tfrac{1}{2}L'(x_h; e) - \tfrac{1}{2}L'(x; e),$$

where we use the fact that $L'(x; e) = 0$. From this we obtain, by definition of x_h,

$$L(x) - L(x_h) = \tfrac{1}{2}L'(x_h; x - y_h) + L'(\overline{x x_h}; e) - \tfrac{1}{2}L'(x_h; e) - \tfrac{1}{2}L'(x; e),$$

with an arbitrary $y_h \in X_h$. The last two terms on the right are just the approximation of the second one by the trapezoidal rule. Recalling the corresponding remainder term

$$R = \tfrac{1}{2} \int_0^1 L'''(x_h + se; e, e, e)\, s(s-1)\, \mathrm{d}s,$$

we obtain the desired error representation (2.6). □

2.2. Galerkin approximation of variational equations

Now, we present an approach to *a posteriori* error estimation for the standard Galerkin approximation of variational equations based on the general result of Proposition 2.1. Let $A(\cdot; \cdot)$ be a differentiable semilinear form and $F(\cdot)$ a linear functional defined on some function space V. We seek a solution $u \in V$ to the variational equation

$$A(u; \varphi) = F(\varphi), \quad \text{for all } \varphi \in V. \tag{2.7}$$

For a finite-dimensional subspace $V_h \subset V$, again parametrized by $h \in \mathbb{R}_+$, the corresponding Galerkin approximation $u_h \in V_h$ is determined by

$$A(u_h; \varphi_h) = F(\varphi_h), \quad \text{for all } \varphi_h \in V_h. \tag{2.8}$$

We assume that equations (2.7) and (2.8) possess (locally) unique solutions. Then, we have the Galerkin orthogonality relation

$$A(u; \varphi_h) - A(u_h; \varphi_h) = 0, \quad \varphi_h \in V_h. \tag{2.9}$$

The conventional approach to *a posteriori* error estimation in this situation is based on assumed coercivity properties of $A(\cdot; \cdot)$, namely

$$\|u - v\|_E \leq c_s \sup_{z \in V, \|z\|=1} |A(u; z) - A(v; z)|, \quad u, v \in V, \tag{2.10}$$

with some generic 'energy norm' $\| \cdot \|_E$ on V. Then, using Galerkin orthogonality, we obtain a first *a posteriori* estimate for the error $e = u - u_h$:

$$\|e\|_E \leq c_s \sup_{z \in V, \|z\|=1} \left\{ \min_{\varphi_h \in V_h} |\rho(u_h, z - \varphi_h)| \right\}, \tag{2.11}$$

where the residual $\rho(u_h; \cdot)$ is defined by

$$\rho(u_h; \varphi) := F(\varphi) - A(u_h; \varphi) \quad \varphi \in V. \tag{2.12}$$

We note that, by construction, $\rho(u_h; \cdot)$ vanishes on V_h.

Now, we consider again the case that *a priori* known coercivity properties of the form $A(\cdot; \cdot)$ are not available. Furthermore, we consider the more general situation of computing an approximation to $J(u)$, with a given differentiable functional $J(\cdot)$. We want to embed this situation into the general setting of Proposition 2.1. To this end, we note that the task of computing $J(u)$ from the solution of (2.7) can be equivalently formulated as solving a (trivial) constrained optimization problem for $u \in V$:

$$J(u) = \min!, \quad A(u; \varphi) = F(\varphi), \quad \text{for all } \varphi \in V. \tag{2.13}$$

Minima u correspond to stationary points $\{u, z\} \in V \times V$ of the Lagrangian

$$L(u; z) := J(u) + F(z) - A(u; z), \tag{2.14}$$

with the adjoint variable $z \in V$. Hence, we seek solutions $\{u, z\} \in V \times V$ to the Euler–Lagrange system

$$\begin{aligned} A(u; \varphi) &= F(\varphi), \quad \text{for all } \varphi \in V, \\ A'(u; \varphi, z) &= J'(u; \varphi), \quad \text{for all } \varphi \in V. \end{aligned} \tag{2.15}$$

Observe that the first equation of this system is just the given variational equation (2.7). In order to obtain a discretization of (2.15), we solve, in addition to (2.8), the following discrete adjoint equation:

$$A'(u_h; \varphi_h, z_h) = J'(u_h; \varphi_h), \quad \varphi_h \in V_h. \tag{2.16}$$

Again, we suppose that equations (2.15) and (2.16) possess unique solutions. To the solution $z_h \in V_h$ we associate the 'adjoint error' e^* and the 'adjoint residual'

$$\rho^*(z_h; \varphi) := J'(u_h; \varphi) - A'(u_h; \varphi, z_h), \quad \varphi \in V. \tag{2.17}$$

Proposition 2.2. For the Galerkin approximation of the Euler–Lagrange system (2.15), we have the *a posteriori* error representation

$$J(u) - J(u_h) = \tfrac{1}{2} \min_{\varphi_h \in V_h} \rho(u_h; z - \varphi_h) + \tfrac{1}{2} \min_{\varphi_h \in V_h} \rho^*(z_h; u - \varphi_h) + R \tag{2.18}$$

with the residual $\rho(u_h; \cdot)$ defined in (2.12) and the adjoint residual $\rho^*(z_h; \cdot)$ defined in (2.17). The remainder term is given by

$$R := \tfrac{1}{2} \int_0^1 \{ J'''(u_h + se; e, e, e) - A'''(u_h + se; e, e, e, z_h + se^*) \\ - 3A''(u_h + se; e, e, e^*) \} \, s(s-1) \, ds \tag{2.19}$$

and vanishes if $A(\cdot; \cdot)$ is linear and if $J(\cdot)$ is quadratic.

Proof. At the solutions $\{u, z\} \in V \times V$ and $\{u_h, z_h\} \in V_h \times V_h$,

$$L(u; z) - L(u_h; z_h) = J(u) - J(u_h).$$

Hence, Proposition 2.1, applied to the Lagrangian $L(\cdot; \cdot)$ with $X := V \times V$, $x = \{u, z\}$, $x_h = \{u_h, z_h\}$, etc., yields a representation for the error $J(u) - J(u_h)$ in terms of the residuals $\rho(u_h; \cdot)$ and $\rho^*(z_h; \cdot)$. The remainder term has the form

$$\tfrac{1}{2} \int_0^1 L'''(x_h + se; \cdot, \cdot, \cdot) \, s(s-1) \, ds.$$

Notice that $L(u; z)$ is linear in z. Consequently, the third derivative of $L(\cdot; \cdot)$ consists of only three terms, namely,

$$J'''(u_h + se; e, e, e) - A'''(u_h + se; e, e, e, z_h + se^*) - 3A''(u_h + se; e, e, e^*).$$

This completes the proof. □

The remainder term R in (2.19) is cubic in the errors e and e^* and may usually be neglected. The evaluation of the resulting error estimator

$$\eta_\omega(u_h, z_h) := \tfrac{1}{2} \min_{\varphi_h \in V_h} \rho(u_h; z - \varphi_h) + \tfrac{1}{2} \min_{\varphi_h \in V_h} \rho^*(z_h; u - \varphi_h)$$

requires approximations to the exact primal and dual solutions u and z. The error representation (2.18) is the nonlinear analogue of the (trivial) representation in the linear case derived in the introduction, as will be seen later. The relation between the primal and dual residuals is given in the next proposition.

Proposition 2.3. With the notation of Proposition 2.2,

$$\min_{\varphi_h \in V_h} \rho^*(z_h, u - \varphi_h) = \min_{\varphi_h \in V_h} \rho(u_h, z - \varphi_h) + \Delta\rho, \tag{2.20}$$

with

$$\Delta\rho = \int_0^1 \left\{ A''(u_h + see; e, e, z_h + se^*) - J''(u_h + se; e, e) \right\} ds. \qquad (2.21)$$

As a consequence, we have the following simplified error representation:

$$J(u) - J(u_h) = \min_{\varphi_h \in V_h} \rho(u_h, z - \varphi_h) + R, \qquad (2.22)$$

with a quadratic remainder

$$R = \int_0^1 \left\{ A''(u_h + se; e, e, z) - J''(u_h + se; e, e) \right\} s\, ds.$$

Proof. Let us define

$$g(s) := J'(u_h + se; u - \varphi_h) - A'(u_h + se; u - \varphi_h, z_h + se^*).$$

Then we have

$$g(1) = J'(u; u - \varphi_h) - A'(u; u - \varphi_h, z) = 0,$$

by the definition of z. Further,

$$g(0) = J'(u_h; u - \varphi_h) - A'(u_h; u - \varphi_h, z_h) = \rho^*(z_h, u - \varphi_h).$$

The derivative of g is given by

$$\begin{aligned} g'(s) \;=\; & J''(u_h + se; e, u - \varphi_h) - A''(u_h + se; e, u - \varphi_h, z_h + se^*) \\ & - A'(u_h + se; u - \varphi_h, e^*). \end{aligned}$$

Therefore, we find that

$$\begin{aligned} \rho^*(z_h, u - \varphi_h) \;=\; & \int_0^1 \left\{ A''(u_h + se; e, u - \varphi_h, z_h + se^*) \right. \\ & \left. - J''(u_h + se; e, u - \varphi_h) \right\} ds \\ & + \int_0^1 A'(u_h + se; u - \varphi_h, e^*)\, ds. \end{aligned}$$

Notice that the last term is just the primal residual, where we can substitute e^* by $z - \varphi_h$ with an arbitrary $\varphi_h \in V_h$.

The simplified error representation (2.22) can be derived by comparison of the remainder terms in (2.19) and (2.21) and some tedious computation. However, we prefer to give a direct argument. Integration by parts yields

$$R = -\int_0^1 \left\{ A'(u_h + se; e, z) - J'(u_h + se; e) \right\} ds + A'(u; e, z) - J'(u; e),$$

and the last term vanishes by definition of z. Therefore, we have

$$R = J(u) - J(u_h) + \rho(u_h, z).$$

Noticing that $\rho(u_h, z) = \rho(u_h, z - \varphi_h)$, for $\varphi_h \in V_h$, completes the proof. $\square$

Remark 2.1. We note that the results of Propositions 2.2 and 2.3 hold true in the following more general situation. Let V and W be Banach spaces. We consider a Lagrangian $L : V \times W \to \mathbb{R}$ defined as in (2.14). The corresponding stationary point $\{u, z\}$ is approximated by a Galerkin method using subspaces $V_h \subset V$ and $W_h \subset W$. Then, for the corresponding discrete solution $\{u_h, z_h\}$, we have

$$J(u) - J(u_h) = \tfrac{1}{2} \min_{\varphi_h \in W_h} \rho(u_h; z - \varphi_h) + \tfrac{1}{2} \min_{\varphi_h \in V_h} \rho^*(z_h; u - \varphi_h) + R,$$

where R has the same form as in (2.19). This generalization allows us to include Petrov–Galerkin schemes into the general framework.

In the case of a *linear* functional $J(\cdot)$ and *linear* variational equation, Proposition 2.3 states that the two residuals coincide. In general, the difference is quadratic in e and can be supposed to be relatively small in many practical situation. In general, the difference $\Delta \rho$ can be used as an indicator of the influence of the nonlinearity on the error.

In the following proposition, we derive an alternative representation where no remainder occurs.

Proposition 2.4. For the Galerkin approximation (2.8) we have the *a posteriori* error representation

$$J(u) - J(u_h) = \min_{\varphi_h \in V_h} \rho(u_h; z - \varphi_h), \qquad (2.23)$$

where z is defined by the adjoint problem

$$\int_0^1 A'(u_h + se; \varphi, z) \, ds = \int_0^1 J'(u_h + se; \varphi) \, ds, \quad \text{for all } \varphi \in V. \qquad (2.24)$$

Proof. By elementary calculus,

$$A(u; \varphi) - A(u_h; \varphi) = \int_0^1 A'(u_h + se; e, \varphi) \, ds = A'(\overline{uu_h}; e, \varphi),$$

and analogously for $J(u) - J(u_h)$. Taking $\varphi = e$ in the adjoint problem (2.24), we obtain

$$A'(\overline{uu_h}; e, z) = J'(\overline{uu_h}; e),$$

and combining this with the previous identities,

$$J(u) - J(u_h) = A'(\overline{uu_h}; e, z) = A(u; z) - A(u_h; z).$$

Now, using the Galerkin orthogonality (2.9) yields

$$J(u) - J(u_h) = A(u; z - \varphi_h) - A(u_h; z - \varphi_h) = F(z - \varphi_h) - A(u_h; z - \varphi_h),$$

with an arbitrary $\varphi_h \in V_h$. This implies (2.23). $\qquad \square$

Remark 2.2.   We note that the result of Proposition 2.4 can also be inferred from the general Proposition 2.1. To this end, we construct an artificial optimization problem which is equivalent to the present situation.
Let $\hat{u}$ and $\hat{u}_h$ be two fixed elements of V and set $\hat{e} := \hat{u} - \hat{u}_h$. The linear
Lagrangian

$$\tilde{L}(u; z) := \int_0^1 J'(\hat{u}_h + s\hat{e})(\hat{u} - u))\,ds + \int_0^1 A'(\hat{u}_h + s\hat{e}; \hat{u} - u, z)\,ds$$

has stationary point $\{\tilde{u}, \tilde{z}\}$ defined by

$$A(\tilde{u}; \varphi) = A(\hat{u}; \varphi), \quad \text{for all } \varphi \in V,$$

and

$$A'(\overline{\hat{u}\hat{u}_h}; \varphi, \tilde{z}) = J'(\overline{\hat{u}\hat{u}_h}; \varphi), \quad \text{for all } \varphi \in V.$$

Let $\{\tilde{u}_h, \tilde{z}_h\} \in V_h \times V_h$ be the Galerkin approximation to these equations.
If we now choose $\hat{u} = u$ and $\hat{u}_h = u_h$, we see that $\tilde{u} = u$ and $\tilde{z} = z$, and the
two previous equations reduce to the continuous equations (2.7) and (2.24).
Furthermore, their discrete analogues generate equation (2.8) for u_h and

$$A'(\overline{uu_h}; \varphi, z_h) = J'(\overline{uu_h}; \varphi_h), \quad \text{for all } \varphi_h \in V_h$$

for $z_h \in V_h$. Now applying the general result of Proposition 2.1, we find that

$$J(u) - J(u_h) = \tfrac{1}{2}\tilde{\rho}(u_h; z - \varphi_h) + \tfrac{1}{2}\tilde{\rho}^*(z_h; u - \varphi_h),$$

since the remainder vanishes because of the linearity of $\tilde{L}(\cdot; \cdot)$. By definition,
we have that $\tilde{\rho}(u_h; \varphi) = \rho(u_h; \varphi)$ and $\tilde{\rho}^*(z_h; u) = J(u) - J(u_h)$, which
completes the argument.

The result of Proposition 2.4 cannot be used directly in practice, since the
underlying adjoint problem (2.24) involves the unknown solution u. However, in order to study the effect of the nonlinearity, one might be willing
to approximate (2.24) by means of some approximation $\hat{u}_h$ obtained in a
possibly richer space $\hat{V}_h \subset V$, that is,

$$A'(\overline{\hat{u}_h u_h}; \varphi, z) = J'(\overline{\hat{u}_h u_h}; \varphi), \quad \text{for all } \varphi \in V. \tag{2.25}$$

For the simplest choice $\hat{u}_h = u_h$, (2.25) reduces to

$$A'(u_h; \varphi, z) = J'(u_h; \varphi), \quad \text{for all } \varphi \in V, \tag{2.26}$$

and we are in the situation of Proposition 2.2. In the following, we provide
a theoretical result on the effect of using the approximate adjoint problem
(2.25).

Proposition 2.5.   Suppose that $\hat{u}_h \in \hat{V}_h$ is an improved approximation
to the solution u with corresponding error $\hat{e} := u - \hat{u}_h$. Let $z \in V$ be

the solution of the linearized adjoint problem (2.25). Then, for the Galerkin scheme (2.8), we have the error identity

$$J(u) - J(u_h) = \min_{\varphi_h \in V_h} \rho(u_h; z - \varphi_h) + R \qquad (2.27)$$

with a remainder bounded by

$$|R| \le \tfrac{1}{2} \max_{\xi \in \overline{u \hat{u}_h u_h}} |J''(\xi; \hat{e}, e) - A''(\xi; \hat{e}, e, z)|,$$

where $\overline{u \hat{u}_h u_h}$ denotes the 'triangle' in V spanned by the three points $\{u, \hat{u}_h, u_h\}$. The remainder term vanishes if $A(\cdot; \cdot)$ and $J(\cdot)$ are linear.

Proof. We set $\hat{e}_h := \hat{u}_h - u_h$. Then, again by elementary calculus,

$$J(u) - J(u_h) = J'(\overline{\hat{u}_h u_h}; e) + \int_0^1 \{J'(u_h + se; e) - J'(u_h + s\hat{e}_h; e)\} \, ds,$$

$$A(u; z) - A(u_h; z) = A'(\overline{\hat{u}_h u_h}; e, z)$$

$$+ \int_0^1 \{A'(u_h + se; e, z) - A'(u_h + s\hat{e}_h; e, z)\} \, ds.$$

Taking $\varphi = e$ in the adjoint problem (2.25), we conclude from the previous identities that

$$J(u) - J(u_h) = A(u; z) - A(u_h; z) + R(\overline{u \hat{u}_h u_h}; \hat{e}, e, z),$$

with the remainder term

$$R := \int_0^1 \{J'(u_h + se; e) - J'(u_h + s\hat{e}_h; e)\} \, ds$$

$$- \int_0^1 \{A'(u_h + se; e, z) - A'(u_h + s\hat{e}_h; e, z)\} \, ds.$$

Observing that $u_h + se - u_h + s\hat{e}_h = s\hat{e}$, the first integral on the right is rewritten as

$$\int_0^1 \{J'(u_h + se; e) - J'(u_h + s\hat{e}_h; e)\} \, ds$$

$$= -\int_0^1 \left\{ \int_0^1 J''(u_h + s\hat{e}_h + ts\hat{e}; s\hat{e}, e) \, dt \right\} ds,$$

and analogously for the second one. We note that $u_h + s\hat{e}_h + ts\hat{e} = (1-s)u_h + tsu + s(1-t)\hat{u}_h$ is a convex combination of the three points. Hence,

$$|R| \le \tfrac{1}{2} \max_{\xi \in \overline{u \hat{u}_h u_h}} |J''(\xi; \hat{e}, e) - A''(\xi; \hat{e}, e, z)|,$$

as asserted. Now, using the Galerkin orthogonality (2.9) again yields

$$J(u) - J(u_h) = \rho(u_h; z - \varphi_h) + R,$$

with an arbitrary $\varphi_h \in V_h$. From this, we conclude (2.27). Clearly, the remainder term vanishes if $A(\cdot;\cdot)$ and $J(\cdot)$ are linear. $\qquad\square$

Assuming that $\hat{e}$ is much smaller than e, the remainder term $R(\overline{uu_h}; \hat{e}, e, z)$ may be neglected in comparison with the residual term

$$\hat{\eta}_\omega(u_h) := \min_{\varphi_h \in V_h} \rho(u_h; z - \varphi_h).$$

This result provides the basis for successive improvement of the reliability of the error estimator $\hat{\eta}_\omega(u_h)$ by using improved approximations to $\hat{u}_h$ in the linearization. This concept will be discussed in more detail below for a concrete situation.

2.3. Extension to nonstandard Galerkin methods

In many concrete situations the Galerkin approximation is based on a modified variational formulation which is mesh-dependent. This is the case, for example, in the 'discontinuous' Galerkin method for transport-oriented or time-dependent problems (see Section 7) and also when a standard Galerkin scheme is stabilized by introducing least-squares or streamline diffusion terms (see Sections 3, 8, 9, and 11). In order to cover discretizations of this type too, the framework developed so far has to be properly extended.

Let $\hat{V}$ be some larger space containing the solution space V, and, in particular, the solution $u \in V$, of the problem

$$A(u; \varphi) = F(\varphi), \quad \text{for all } \varphi \in V, \tag{2.28}$$

as well as the finite-dimensional space V_h. In most concrete situations, we simply set $\hat{V} := V \oplus V_h$. On $\hat{V}$, we use an h-dependent 'stabilization form' $S_h(\cdot;\cdot)$ in defining

$$A_h(\cdot;\cdot) := A(\cdot;\cdot) + S_h(\cdot;\cdot),$$

and consider the approximate problem

$$A_h(u_h; \varphi_h) = F(\varphi_h), \quad \text{for all } \varphi_h \in V_h, \tag{2.29}$$

We assume again that the modified form $A_h(\cdot;\cdot)$ is regular on V_h, such that the finite-dimensional problem (2.29) has a unique solution $u_h \in V_h$. The residual of this solution $u_h \in V_h$ is defined by $\rho_h(u_h, \cdot) := F(\cdot) - A_h(u_h; \cdot)$. Further, this discretization is required to be 'consistent' with the original problem in the sense that the solution $u \in V$ of (2.28) automatically satisfies (2.29),

$$A_h(u; \varphi) = F(\varphi), \quad \varphi \in V.$$

This immediately implies Galerkin orthogonality in the form

$$\rho_h(u_h; \varphi_h) = A_h(u; \varphi_h) - A_h(u_h; \varphi_h) = 0, \quad \varphi_h \in V_h. \tag{2.30}$$

This consistency condition may have to be relaxed in some concrete situations introducing additional perturbation terms in the resulting *a posteriori* error estimates. For controlling the error $J(u) - J(u_h)$, we consider the h-dependent adjoint problem

$$\tilde{A}'_h(u; \varphi, z) := A'(u; \varphi, z) + \tilde{S}'_h(u; \varphi, z) = J'(u; \varphi), \quad \text{for all } \varphi \in \hat{V}, \quad (2.31)$$

where $\tilde{S}'_h(u; \cdot, \cdot)$ may consist of only a definite part of $S'_h(u; \cdot, \cdot)$ chosen such that (2.31) can be guaranteed to possess a unique solution $z \in \hat{V}$. Clearly, by its construction this adjoint solution z usually also depends on h.

Remark 2.3. The previous construction requires some comment. For example, in the case of a discontinuous Galerkin scheme the adjoint solution z usually exists in the original solution space V and is independent of h. On the other hand, in the presence of least-squares stabilization the adjoint solution may exist only in a weak sense in $\hat{V}$ and its limit behaviour for $h \to 0$ is not clear. However, this may not be too critical, since in the practical evaluation of the residual $\rho_h(u_h; z - \varphi_h)$ the weights $z - \varphi_h$ are, after all, approximated using the solution $z_h \in V_h$ of the discretized adjoint problem

$$\tilde{A}'_h(u_h; \varphi_h, z_h) = J'(u_h; \varphi_h), \quad \text{for all } \varphi_h \in V_h. \quad (2.32)$$

This problem, in turn, can often be interpreted as stabilized approximation of a formal adjoint problem in terms of the given differential operator.

In the following, we derive an analogue of the error representation in Proposition 2.3 for the present situation.

Proposition 2.6. With the above notation, we have the error representation

$$J(u) - J(u_h) = \min_{\varphi_h \in V_h} \rho_h(u_h, z - \varphi_h) + R_h, \quad (2.33)$$

with the remainder term

$$R_h = (\tilde{S}'_h - S'_h)(u; e, z) + \int_0^1 \{A''_h(u_h + se; e, e, z) - J''(u_h + se; e, e)\} s \, ds.$$

Proof. We note that

$$J(u) - J(u_h) = J'(u; e) - \int_0^1 J''(u_h + se; e, e) s \, ds.$$

Taking $\varphi := e$ in (2.31) and using the result in the previous equation gives us

$$J(u) - J(u_h) = \tilde{A}'_h(u; e, z) - \int_0^1 J''(u_h + se; e, e) s \, ds$$

$$= A'_h(u; e, z) + (\tilde{S}'_h - S'_h)(u; e, z) - \int_0^1 J''(u_h + se; e, e) s \, ds.$$

We combine this with the relation

$$A'_h(u; e, z) = \int_0^1 A'_h(u_h + se; e, z) \, ds + \int_0^1 A''_h(u_h + se; e, e, z) s \, ds$$

$$= \rho_h(u_h; z) + \int_0^1 A''_h(u_h + se; e, e, z) s \, ds,$$

and use the Galerkin orthogonality (2.30) to obtain

$$J(u) - J(u_h) = \rho_h(u_h; z - \varphi_h) + (\tilde{S}'_h - S'_h)(u; e, z)$$

$$+ \int_0^1 \{A''_h(u_h + se; e, e, z) - J''(u_h + se; e, e)\} s \, ds.$$

with an arbitrary $\varphi_h \in V_h$. This completes the proof. □

In concrete situations the first part of the remainder term, $(\tilde{S}'_h - S'_h)(u; e, z)$, contains small parameters and can be neglected in comparison with the residual term $\rho_h(u_h; z - \varphi_h)$.

Notes and references
Global energy and L^2-norm error estimates for nonlinear problems exploiting coercivity properties or assuming bounds for Frechét derivatives have been derived, for instance, by Caloz and Rappaz (1997), Medina, Picasso and Rappaz (1996), and Verfürth (1993, 1994, 2000). Duality arguments as described above are very common in the *a priori* error analysis of Galerkin finite element methods (see, *e.g.*, Ciarlet (1978) and Brenner and Scott (1994)). Their use for *a posteriori* error estimation particularly in nonlinear problems was suggested by Johnson (1994), Eriksson and Johnson (1995*b*), and Johnson and Rannacher (1994) (see also Eriksson *et al.* (1995)). The presentation based on optimal control principles uses ideas from Rannacher (200x) and Becker, Kapp and Rannacher (2000*b*).

3. The dual-weighted residual method

Having prepared the general frame for *a posteriori* error estimation in Section 2, we now discuss the application of these concepts to concrete situations. For illustration, we first consider some simple model situations: the Poisson equation, an eigenvalue problem of the Laplace operator and a linear transport equation. This setting is also used to introduce the basic facts about Galerkin finite element approximation, as they will be used throughout the paper. Below, we will need some notation from the theory of function spaces. Readers who are familiar with the usual Lebesgue- and Sobolev-space notation may want to skip this and continue with the next subsection on finite element approximation.

Notation for function spaces

For a domain $Q \subset \mathbb{R}^d$, we let $L^2(Q)$ denote the Lebesgue space of square-integrable functions on Q, which is a Hilbert space with the scalar product and norm

$$(v, w)_Q = \int_Q v\,w\,\mathrm{d}x \quad \text{and} \quad \|v\|_Q = \left(\int_Q |v|^2\,\mathrm{d}x \right)^{1/2}.$$

Analogously, $L^2(\partial Q)$ denotes the space of square-integrable functions defined on the boundary ∂Q equipped with the scalar product and norm

$$(v, w)_{\partial Q} = \int_{\partial Q} v\,w\,\mathrm{d}s \quad \text{and} \quad \|v\|_{\partial Q} = \left(\int_{\partial Q} |v|^2\,\mathrm{d}s \right)^{1/2}.$$

The Sobolev spaces $H^1(Q)$ and $H^2(Q)$ consist of the functions $v \in L^2(Q)$ that possess first- and second-order (distributional) derivatives $\nabla v \in L^2(Q)^d$ and $\nabla^2 v \in L^2(Q)^{d \times d}$, respectively. For functions in these spaces, we use the seminorms

$$\|\nabla v\|_Q = \left(\int_Q |\nabla v|^2\,\mathrm{d}x \right)^{1/2}, \quad \|\nabla^2 v\|_Q = \left(\int_Q |\nabla^2 v|^2\,\mathrm{d}x \right)^{1/2}.$$

The space $H^1(Q)$ can be embedded in the space $L^2(\partial Q)$, such that for each $v \in H^1(Q)$ there exists a trace $v_{|\partial Q} \in L^2(\partial Q)$. Further, the functions in the subspace $H_0^1(Q) \subset H^1(Q)$ are characterized by the property $v_{|\partial Q} = 0$. By the Poincaré inequality,

$$\|v\|_Q \leq c \|\nabla v\|_Q, \quad v \in H_0^1(Q), \tag{3.1}$$

the H^1-seminorm $\|\nabla v\|_Q$ is a norm on the subspace $H_0^1(Q)$. If the set Q is the set Ω on which the differential equation is posed, we usually omit the subscript Ω in the notation of norms and scalar products, for instance, $\|v\| = \|v\|_\Omega$. All the above notation will be synonymously used also for vector- or matrix-valued functions $v : \Omega \to \mathbb{R}^d$ or $\mathbb{R}^{d \times d}$.

3.1. Finite element discretization of the Poisson equation

We begin with the model problem

$$-\Delta u = f \quad \text{in } \Omega, \quad u = 0 \quad \text{on } \partial\Omega, \tag{3.2}$$

posed on a polygonal domain $\Omega \subset \mathbb{R}^2$. The natural solution space for this boundary value problem is the Sobolev space $V = H_0^1(\Omega)$. The variational formulation of (3.2) seeks $u \in V$ such that

$$A(u, \varphi) = (f, \varphi), \quad \text{for all } \varphi \in V, \tag{3.3}$$

where $A(u, \varphi) := (\nabla u, \nabla \varphi)$. The finite element approximation of (3.3) uses finite-dimensional subspaces

$$V_h = \{v \in V : v_{|K} \in P(K),\ K \in \mathbb{T}_h\},$$

defined on decompositions $\mathbb{T}_h$ of Ω into triangles or quadrilaterals K (*cells*) of width $h_K = \operatorname{diam}(K)$; we write $h = \max_{K \in \mathbb{T}_h} h_K$ for the *global* mesh width. Here, $P(K)$ denotes a suitable space of polynomial-like functions defined on the cell $K \in \mathbb{T}_h$. In the numerical results discussed below, we have mostly used 'bilinear' finite elements on quadrilateral meshes in which case $P(K) = \tilde{Q}_1(K)$ consists of shape functions obtained via a bilinear transformation from the space of bilinear functions $Q_1(\hat{K}) = \operatorname{span}\{1, x_1, x_2, x_1 x_2\}$ on the reference cell $\hat{K} = [0, 1]^2$. Local mesh refinement or coarsening is realized by using *hanging nodes*, in such a way that global conformity is preserved, that is, $V_h \subset V$. For technical details of finite element spaces, the reader may consult the standard literature, for instance Ciarlet (1978) or Brenner and Scott (1994), and, particularly for the treatment of hanging nodes, Carey and Oden (1984). Now, approximations $u_h \in V_h$ are determined by

$$A(u_h, \varphi_h) = (f, \varphi_h), \quad \text{for all } \varphi_h \in V_h. \tag{3.4}$$

The corresponding residual and error of $u_h \in V_h$ are $\rho(u_h, \cdot) := (f, \cdot) - A(u_h, \cdot)$ and $e = u - u_h$, respectively. By construction, we have 'Galerkin orthogonality'

$$\rho(u_h, \varphi_h) = A(e, \varphi_h) = 0, \quad \varphi_h \in V_h. \tag{3.5}$$

In order to convert (3.4) into an algebraic equation, one uses a 'nodal basis' $\{\varphi_h^i\}_{i=1,\dots,n}$ ($n := \dim V_h$) of the finite element space V_h. The coefficient vector $x_h = (x_i)_{i=1}^n$ in the expansion $u_h = \sum_{i=1}^n x^i \varphi_h^i$ is determined by the linear system

$$A_h\, x_h = b_h, \tag{3.6}$$

with the 'stiffness matrix' $A = (a_{ij})_{i,j=1}^n$ and the 'load vector' $b_h = (b_i)_{i=1}^n$ defined by $a_{ij} = A(\varphi_h^j, \varphi_h^i)$ and $b_i = (f, \varphi_h^i)$.

3.1.1. A priori error analysis
We begin with a brief review of the a priori error analysis for the scheme (3.4). We let $i_h u \in V_h$ denote the natural 'nodal interpolation' of $u \in C(\bar{\Omega})$ satisfying $i_h u(P) = u(P)$ at all nodal points P. We have

$$\|\nabla(u - i_h u)\|_K \le c_i h_K \|\nabla^2 u\|_K \tag{3.7}$$

(see, *e.g.*, Ciarlet (1978) and Brenner and Scott (1994)) and, furthermore,

$$\|u - i_h u\|_K + h_K^{1/2} \|u - i_h u\|_{\partial K} \le c_i h_K^2 \|\nabla^2 u\|_K, \tag{3.8}$$

for certain positive 'interpolation constants' c_i.

(i) By the projection property of the Galerkin finite element scheme the interpolation estimate (3.7) directly implies the 'energy error estimate'

$$\|\nabla e\| = \inf_{\varphi_h \in V_h} \|\nabla(u - \varphi_h)\| \leq c_i h^2 \|\nabla^2 u\|. \qquad (3.9)$$

(ii) Further, employing a duality argument ('Aubin–Nitsche trick'),

$$-\Delta z = \|e\|^{-1} e \quad \text{in } \Omega, \quad z = 0 \text{ on } \partial\Omega, \qquad (3.10)$$

we obtain

$$\|e\| = (e, -\Delta z) = A(e, z) = A(e, z - i_h z) \leq c_i c_s h \|\nabla e\|, \qquad (3.11)$$

where the constant c_s is defined by the *a priori* bound $\|\nabla^2 z\| \leq c_s$. In view of the energy error estimate (3.9), this implies the 'L^2-error estimate'

$$\|e\| \leq c_i^2 c_s h^2 \|\nabla^2 u\|. \qquad (3.12)$$

(iii) On the basis of the global error estimates (3.9) and (3.12), we can also obtain error bounds for various other quantities, *e.g.*, point values or boundary moments (in $\mathbb{R}^2$):

$$|\nabla e(P)| \approx h^{-1} \|\nabla e\|, \qquad |(e, \psi)_{\partial\Omega}| \approx h^{-1/2} \|e\|.$$

All these estimates are only suboptimal in h. This tells us that estimating the error in functional output usually requires special effort and cannot simply be reduced to the standard energy or L^2-error estimates.

3.1.2. A posteriori error analysis

Next, we derive *a posteriori* error estimates. Suppose that the error is to be controlled with respect to some (linear) 'error functional' $J(\cdot)$ defined on V. Let $z \in V$ be the solution of the corresponding adjoint problem

$$A(\varphi, z) = J(\varphi), \quad \text{for all } \varphi \in V, \qquad (3.13)$$

and $z_h \in V_h$ its finite element approximation defined by

$$A(\varphi_h, z_h) = J(\varphi_h), \quad \text{for all } \varphi_h \in V_h. \qquad (3.14)$$

The 'adjoint' error is denoted by $e^* := z - z_h$. Applying the abstract *a posteriori* error representation of Proposition 2.3 to this situation yields

$$J(e) = \rho(u_h, z - \varphi_h), \qquad (3.15)$$

for arbitrary $\varphi_h \in V_h$. Now, the residual expression for u_h has to be further developed. By cell-wise integration by parts, we obtain

$$\rho(u_h, z - \varphi_h) = \sum_{K \in \mathbb{T}_h} \{(f + \Delta u_h, z - \varphi_h)_K - (n \cdot \nabla u_h, z - \varphi_h)_{\partial K}\}. \qquad (3.16)$$

At this point, we have assumed that the domain Ω is polygonal (or polyhedral) in order to ease the approximation of the boundary $\partial\Omega$. In the presence of curved parts of $\partial\Omega$ the formula (3.16) contains additional terms representing the error caused by the polygonal approximation of the boundary (see, *e.g.*, Becker and Rannacher (1996*b*)). In the tests below, these terms are suppressed since they are usually negligibly small. The representation (3.16) can be rewritten by combining the contributions from cell edges in the following form:

$$J(e) = \sum_{K \in \mathbb{T}_h} \{ (R(u_h), z - \varphi_h)_K - (r(u_h), z - \varphi_h)_{\partial K} \}, \tag{3.17}$$

with the cell and edge residuals $R(u_h)$ and $r(u_h)$, respectively, defined by

$$R(u_h)_{|K} = f + \Delta u_h, \qquad r(u_h)_{|\Gamma} := \begin{cases} \frac{1}{2} n \cdot [\nabla u_h], & \text{if } \Gamma \subset \partial K \backslash \partial\Omega, \\ 0, & \text{if } \Gamma \subset \partial\Omega, \end{cases}$$

where $[\nabla u_h]$ denotes the jump of ∇u_h across the inter-element edges Γ, and $z - \varphi_{h|\Gamma} - 0$ on $\Gamma \subset \partial\Omega$ is observed. From the error identity (3.17), we can infer the following result.

Proposition 3.1. For the finite element approximation of the Poisson equation (3.2), we have the *a posteriori* error estimate

$$|J(e)| \leq \eta_\omega(u_h) := \sum_{K \in \mathbb{T}_h} \rho_K \omega_K, \tag{3.18}$$

where the cell residuals ρ_K and weights ω_K are given by

$$\rho_K := \|R(u_h)\|_K + h_K^{-1/2} \|r(u_h)\|_{\partial K},$$

$$\omega_K := \|z - \varphi_h\|_K + h_K^{1/2} \|z - \varphi_h\|_{\partial K},$$

with a suitable approximation $\varphi_h \in V_h$ to z.

Remark 3.1. We give an interpretation of the *a posteriori* error estimate (3.18). On each cell, we have an 'equation residual' $R(u_h) := f + \Delta u_h$ and a 'flux residual' $r(u_h) := n \cdot [\nabla u_h]$, the latter one expressing smoothness of the discrete solution. Both residuals can easily be evaluated. They are multiplied by the weighting function $z - \varphi_h$, which provides quantitative information about the impact of these cell residuals on the error $J(e)$ in the target quantity. In this sense $z - \varphi_h$ may be viewed as *sensitivity factors* as in optimal control problems. Setting $\varphi_h = i_h z$, we have

$$\frac{\partial J(e)}{\partial \rho_K} \approx \omega_K \approx c_i h_K^2 \|\nabla^2 z\|_K.$$

Accordingly, the influence factors have the behaviour $h_K^2 \|\nabla^2 z\|_K$, characteristic of the Galerkin finite element approximation ('projection method').

This allows us to exploit the 'Galerkin orthogonality' in deriving the local weights ω_K. We note that in a standard finite difference or finite volume discretization of (3.2), lacking the full Galerkin orthogonality property, the corresponding *influence factors* would behave like $\omega_K \approx h_K \|\nabla z\|_K$. This reflects the lack of accuracy of these methods in computing the target quantity.

Remark 3.2. We have developed the error estimate (3.18) into a rather compact form in order to make its structure transparent. For practical use in mesh adaptation, we suggest avoiding taking norms but, rather, directly evaluating the error representation in the form

$$\eta_\omega(u_h) = \sum_{K \in \mathbb{T}_h} \eta_K = \sum_{K \in \mathbb{T}_h} |(R(u_h), z - \varphi_h)_K - (r(u_h), z - \varphi_h)_{\partial K}|. \quad (3.19)$$

Corresponding strategies will be discussed in Section 5.

Remark 3.3. The evaluation of the *a posteriori* error bound (3.18) requires us to provide sufficiently accurate approximations to the adjoint solution z. Techniques for generating these will be discussed in Section 5. In Proposition 3.1, we have only used the residual term corresponding to u_h. In view of the identities (3.15), we could also have used the relation

$$J(e) = \min_{\varphi_h \in V_h} \rho^*(z_h, u - \varphi_h) = (f, e^*),$$

resulting in the *a posteriori* error bound

$$|(f, e^*)| \le \eta_\omega^*(z_h) := \sum_{K \in \mathbb{T}_h} \rho_K^* \omega_K^*,$$

with the residual terms and weights, defined analogously,

$$\rho_K^* := \|R^*(z_h)\|_K + h_K^{-1/2}\|r^*(z_h)\|_{\partial K},$$
$$\omega_K^* := \|u - \varphi_h\|_K + h_K^{1/2}\|u - \varphi_h\|_{\partial K},$$

with a suitable approximation $\varphi_h \in V_h$ to u. Obviously, the evaluation of this error bound requires us to compute sufficiently accurate approximations to the solution u. Which approach to the estimation of $|J(e)|$ is used, the one based on u_h or the alternative one based on z_h, may depend on how expensive the construction of more accurate approximations to z or u will be. Roughly speaking, the underlying principle is as follows. Computing a more accurate approximation to z yields an accurate estimate for $J(e)$, but then an even better approximation to $J(u)$ should be obtained by evaluating (f, z). However, to ensure this improved approximation accuracy by an *a posteriori* error estimate, in turn, requires an even better approximation to u. Which way to proceed in practice depends on the different regularity properties of u and z. In the examples presented below, particularly the nonlinear ones, we have chosen to base error estimation as well as mesh adaptation on an *a posteriori* error estimate in terms of the approximate solution u_h.

Remark 3.4. For the first-degree finite elements considered here, the contribution from the edge residual $r(u_h)$ dominates that from the equation residual $R(u_h)$ and the latter can therefore be neglected in most cases (see Carstensen and Verfürth (1999) and Becker and Rannacher (1996*b*)). Further, in the case of a curved boundary or nonzero boundary data, the error estimator $\eta_\omega(u_h)$ in (3.18) contains additional terms measuring the effects of the approximation along the boundary. In regular cases these terms are usually smaller than the edge terms and will be suppressed in the following for simplicity (for a rigorous *a posteriori* error analysis of boundary approximation see Dörfler and Rumpf (1998)).

Remark 3.5. The *a posteriori* error estimate (3.18) directly extends to more general diffusion operators with variable coefficients,

$$Lu := -\nabla \cdot \{a\nabla u\} + bu = f \quad \text{in } \Omega,$$

and mixed boundary conditions, $\partial\Omega = \Gamma_D \cup \Gamma_N$,

$$u_{|\Gamma_D} = 0, \quad n \cdot \{a\nabla u_h\}_{|\Gamma_N} = g.$$

The only difference is in the definition of the equation and edge residuals, which take the form $R(u_h)_{|K} = f + \nabla \cdot \{a\nabla u_h\} - bu_h$ and

$$r(u_h)_{|\Gamma} := \begin{cases} \frac{1}{2} n \cdot [a\nabla u_h], & \text{if } \Gamma \subset \partial K \backslash \partial\Omega, \\ 0, & \text{if } \Gamma \subset \Gamma_D, \\ n \cdot \{a\nabla u_h\} - g, & \text{if } \Gamma \subset \Gamma_N, \end{cases}$$

respectively. The incorporation of non-homogeneous Dirichlet boundary conditions will be discussed below in the context of concrete examples.

3.1.3. Examples
We want to illustrate some features of the DWR method by two simple examples.

Example 1: Computation of mean boundary stress. The first example is meant as an illustrative exercise. For problem (3.2) posed on the unit circle $\Omega := \{x \in \mathbb{R}^2 : |x| < 1\}$, we consider the functional

$$J(u) = \int_{\partial\Omega} \partial_n u \, ds \quad \left(= -\int_\Omega f \, dx \right),$$

and ask the question: What is an optimal mesh-size distribution for computing $J(u)$? The corresponding adjoint problem

$$a(\varphi, z) = (1, \partial_n \varphi)_{\partial\Omega}, \quad \text{for all } \varphi \in V \cap C^1(\bar{\Omega})$$

has a measure solution with density of the form $z \equiv -1$ in Ω, $z = 0$ on $\partial\Omega$. In order to avoid the use of measures, we consider the regularized functional

$$J_\varepsilon(\varphi) = |S_\varepsilon|^{-1} \int_{S_\varepsilon} \partial_n \varphi \, dx = \int_{\partial\Omega} \partial_n \varphi \, ds + \mathcal{O}(\varepsilon), \quad \varepsilon = \mathrm{TOL},$$

where $S_\varepsilon = \{x \in \Omega : \mathrm{dist}\{x, \partial\Omega\} < \varepsilon\}$. Then, the adjoint solution is given by

$$z_\varepsilon = \begin{cases} -1, & \text{in } \Omega \setminus S_\varepsilon, \\ -\varepsilon^{-1}\mathrm{dist}\{x, \partial\Omega\}, & \text{in } S_\varepsilon. \end{cases}$$

This implies that

$$J_\varepsilon(e) \leq c_i \sum_{K \in \mathbb{T}_h, K \cap S_\varepsilon \neq \emptyset} \rho_K \, h_K^2 \|\nabla^2 z_\varepsilon\|_K,$$

that is, there is no contribution to the error from cells in the interior of Ω. Hence, whatever the form of the right-hand side f, the optimal strategy is to refine the elements adjacent to the boundary and to leave the others unchanged. This requires, of course, that the function f is integrated exactly in computing the components of the load vector.

Example 2: Computation of point stresses. Next, we consider the Poisson problem (3.2) on the square domain $\Omega := (-1, 1)^2$. For a smooth solution, we want to compute $J(u) = \partial_1 u(0)$. By *a priori* analysis, we know that, for our first-degree finite element approximation on quasi-uniform meshes, we have

$$|\partial_1 e(0)| = \mathcal{O}(h) \quad (h \to 0), \tag{3.20}$$

provided that the solution has bounded second derivatives $\nabla^2 u$ (see Rannacher and Scott (1982)). Hence, in order to achieve a solution accuracy TOL, we need (in two dimensions) a mesh with $N \approx h^{-2} \approx \mathrm{TOL}^{-2}$ cells. We will examine what we can achieve on meshes constructed on the basis of the weighted *a posteriori* error estimate (3.18). Also in the present case, the adjoint solution does not exist in the sense of $H_0^1(\Omega)$, such that for practical use we have to regularize the functional, for instance taking

$$J_\varepsilon(u) = |B_\varepsilon|^{-1} \int_{B_\varepsilon} \partial_1 u \, dx = \partial_1 u(0) + \mathcal{O}(\varepsilon^2),$$

where $B_\varepsilon = \{x \in \Omega, |x| < \varepsilon\}$, and $\varepsilon = \mathrm{TOL}$. The corresponding adjoint solution z behaves like $|\nabla^2 z(x)| \approx d(x)^{-3}$, where $d(x) = |x| + \varepsilon$. Using this *a priori* information and the local interpolation estimate (3.8), we can bound the weights ω_K in the *a posteriori* error estimate (3.18) as follows:

$$\omega_K \leq c_i h_K^2 \|\nabla^2 z\|_K \leq \tilde{c}_i h_K^3 d_K^{-3},$$

Table 3.1. Computing $\partial_1 u(0)$ using the weighted
estimator $\eta_\omega(u_h)$ $(L = \#$ levels)

TOL	N	L	$|J_\varepsilon(e)|$	$\eta_\omega(u_h)$
4^{-3}	940	9	4.10e-1	1.42e-2
4^{-4}	4912	12	4.14e-3	3.50e-3
4^{-5}	20980	15	2.27e-4	9.25e-4
4^{-6}	86740	17	5.82e-5	2.38e-4

Fig. 3.1. Refined mesh and approximate adjoint solution for computing
$\partial_1 u(0)$ using the *a posteriori* error estimator $\eta_\varepsilon(u_h)$, with $\varepsilon = \text{TOL} = 4^{-4}$

where $d_K := \min_{x \in K} |d(x)|$. This leads us to the *a posteriori* error estimate

$$|J_\varepsilon(e)| \leq \eta_\omega(u_h) := \tilde{c}_i \sum_{K \in \mathbb{T}_h} \rho_K h_K^3 d_K^{-3}, \qquad (3.21)$$

which can be used for successive mesh refinement by different strategies
which will be discussed in detail in Section 5, below. Here, we follow the so-
called 'error-balancing strategy', by which the mesh is adapted such that,
for the given tolerance TOL, the 'error indicators' $\eta_K := \rho_K h_K^3 d_K^{-3}$ are
equilibrated (as well as possible) over the mesh $\mathbb{T}_h$ according to

$$\eta_K \approx \frac{\text{TOL}}{N}, \quad N = \#\{K \in \mathbb{T}_h\}. \qquad (3.22)$$

Results obtained in this way for a sequence of error tolerances TOL are shown
in Table 3.1, while Figure 3.1 displays the balanced mesh for $\text{TOL} = 4^{-4}$
and the approximation to the adjoint solution z_ε, computed on this mesh

(see Becker and Rannacher (1996b)). In order to interpret Table 3.1, we supply the following argument. Localization of the a priori error estimate (3.20) suggests that

$$\rho_K = \mathcal{O}(h_K) \quad (\text{TOL} \to 0),$$

a (heuristic) assumption which will be addressed in some more detail in Section 5, below. Under this assumption the above error estimate becomes

$$|J_\varepsilon(e)| \approx \tilde{c}_i \sum_{K \in \mathbb{T}_h} h_K^4 d_K^{-3}. \tag{3.23}$$

Let us assume that the mesh has been adapted according to (3.22):

$$\eta_K \approx \frac{h_K^4}{d_K^3} \approx \frac{\text{TOL}}{N}.$$

From this, we derive

$$h_K^2 \approx d_K^{3/2} \left(\frac{\text{TOL}}{N} \right)^{1/2},$$

and consequently,

$$N = \sum_{K \in \mathbb{T}_h} h_K^2 h_K^{-2} = \left(\frac{N}{\text{TOL}} \right)^{1/2} \sum_{K \in \mathbb{T}_h} h_K^2 d_K^{-3/2} \approx \left(\frac{N}{\text{TOL}} \right)^{1/2}.$$

This implies that $N \approx \text{TOL}^{-1}$, which is better than the $N \approx \text{TOL}^{-2}$ that could be achieved on uniformly refined meshes. This predicted asymptotic behaviour is well confirmed by the results shown in Table 3.1. We emphasize that in this example strong 'mesh refinement' occurs, though the solution is smooth. In fact, this phenomenon should rather be interpreted as 'mesh coarsening' away from the point of evaluation.

Remark 3.6. In the two preceding examples the adjoint solution z could be described analytically, leading to a priori bounds for the weights ω_K in the a posteriori error estimates. In practice, this will usually not be the case and information about z has to be provided numerically by solving the adjoint problem. Strategies for the computational evaluation of the a posteriori error estimates will be discussed in Section 5, below.

3.2. Approximation of a symmetric eigenvalue problem

We present a simple application of the optimal control approach laid out in Proposition 2.1. Consider the first eigenvalue problem of the Laplacian,

$$-\Delta u = \lambda u \quad \text{in } \Omega, \quad u = 0 \text{ on } \partial\Omega, \tag{3.24}$$

on a bounded domain $\Omega \subset \mathbf{R}^d$. Computing the smallest eigenvalue and the corresponding eigenfunction of this problem may be written in variational form: we seek $u \in V := H_0^1(\Omega)$ such that

$$A(u, u) = \min_{v \in V} \left\{ A(v, v) : \|v\|^2 = 1 \right\}, \tag{3.25}$$

where again $A(u, \varphi) = (\nabla u, \nabla \varphi)$. We call problem (3.24) 'H^2-regular' if eigenfunctions $u \in V$, $\|u\| = 1$, are also in $H^2(\Omega)$ and satisfy the a priori bound

$$\|\nabla^2 u\| \leq c_s \|\Delta u\| = c_s \lambda. \tag{3.26}$$

The constrained optimization problem (3.25) is solved by the Lagrangian approach. Introducing the Lagrangian functional defined on pairs $x = \{u, \lambda\} \in X := V \times \mathbf{R}$,

$$L(x) := A(u, u) - \lambda \{\|u\|^2 - 1\},$$

the solution $x = \{u, \lambda\} \in X$ of (3.24) is a stationary point of $L(\cdot)$,

$$L'(x; y) = 0, \quad \text{for all } y = \{\varphi, \mu\} \in X, \tag{3.27}$$

with the derivative functional

$$L'(x; y) := 2A(u, \varphi) - 2\lambda(u, \varphi) + \mu \{\|u\|^2 - 1\}.$$

For discretizing this problem, we choose finite element subspaces $V_h \subset V$ as described above and set $X_h := V_h \times \mathbf{R}$. The approximations $x_h = \{u_h, \lambda_h\} \in X_h$ are determined by the finite-dimensional system

$$L'(x_h; y_h) = 0, \quad \text{for all } y_h \in X_h, \tag{3.28}$$

or, in explicit form,

$$A(u_h, \varphi_h) = \lambda_h(u_h, \varphi_h), \quad \text{for all } \varphi_h \in V_h, \quad \|u_h\|^2 = 1. \tag{3.29}$$

Using Proposition 2.1, we will deduce the following result.

Proposition 3.2. Suppose that, for certain solutions u and u_h of the eigenvalue problems (3.24) and (3.29), we have

$$\|u - u_h\| \leq 2, \tag{3.30}$$

and that the problem is H^2-regular. Then we have the a posteriori error estimate

$$0 \leq \lambda_h - \lambda \leq 2 c_i c_s \lambda_h \left(\sum_{K \in \mathbb{T}_h} h_K^4 \rho_K^2 \right)^{1/2}, \tag{3.31}$$

with the cell-wise residual terms

$$\rho_K := \|R(u_h, \lambda_h)\|_K + h_K^{-1/2} \|r(u_h)\|_{\partial K},$$

where the cell residual is $R(u_h, \lambda_h) := \Delta u_h + \lambda_h u_h$, and the edge residual $r(u_h)$ is defined as before.

Proof. **(i)** For any solution $x = \{u, \lambda\} \in X$ of (3.27) and any solution $x_h = \{u_h, \lambda_h\} \in X_h$ of (3.28),

$$\lambda = J(u) := A(u, u), \qquad \lambda_h = J(u_h) := A(u_h, u_h),$$

and, moreover,

$$0 \le J(u_h) - J(u) = L(x_h) - L(x).$$

From Proposition 2.2, we have for the error $e := x - x_h$ the identity

$$L(x_h) - L(x) = -\tfrac{1}{2} \min_{y_h \in X_h} L'(x_h; x - y_h) - R,$$

with the remainder

$$R = \tfrac{1}{2} \int_0^1 L'''(x_h + se; e, e, e)\, s(s-1)\, \mathrm{d}s.$$

The residual term on the right can be expressed in the form

$$\tfrac{1}{2} L'(x_h; x - y_h) = A(u_h, u - \varphi_h) - \lambda_h(u_h, u - \varphi_h) =: \rho(u_h, \lambda_h; u - \varphi_h).$$

Further, observing that the Lagrangian functional is quadratic in u and linear in λ, the third derivative $L'''(\cdot; \cdot, \cdot, \cdot)$ consists of only three nonzero terms:

$$L'''_{uu\lambda} = L'''_{u\lambda u} = L'''_{\lambda uu} = -2\mu(\psi, \varphi).$$

Accordingly, the remainder term becomes

$$R(x, x_h; e, e) = \tfrac{1}{4}(\lambda_h - \lambda)\|u - u_h\|^2.$$

This implies that

$$0 \le \lambda_h - \lambda = \rho(u_h, \lambda_h; u - \varphi_h) + \tfrac{1}{4}(\lambda_h - \lambda)\|u - u_h\|^2,$$

and consequently, in view of assumption (3.30),

$$0 < \lambda_h - \lambda \le 2 \min_{\varphi_h \in H_h} |\rho(u_h, \lambda_h; u - \varphi_h)|. \tag{3.32}$$

(ii) Splitting the integrals on the right in (3.32) into their contributions from the cells $K \in \mathbb{T}_h$ and integrating cell-wise by parts, we obtain similarly

$$|\rho(u_h, \lambda_h; u - \varphi_h)| = \left| \sum_{K \in \mathbb{T}_h} \{(R(u_h, \lambda_h), u - \varphi_h)_K - (r(u_h), u - \varphi_h)_{\partial K}\} \right|$$

$$\le \sum_{K \in \mathbb{T}_h} \rho_K \omega_K,$$

with the residual terms

$$\rho_K := \|R(u_h, \lambda_h)\|_K + h_K^{-1/2}\|r(u_h)\|_{\partial K},$$

and the weights

$$\omega_K := \|u - \varphi_h\|_K + h_K^{1/2}\|u - \varphi_h\|_{\partial K}.$$

(iii) By the interpolation estimate (3.7), we have

$$\min_{\varphi_h \in V_h} \left(\sum_{K \in \mathbb{T}_h} h_K^{-4} \omega_K^2 \right)^{1/2} \leq c_i \|\nabla^2 u\|. \tag{3.33}$$

This implies for the term in (3.32) that

$$\min_{\varphi_h \in V_h} |A(u_h, u-\varphi_h) - \lambda_h(u_h, u-\varphi_h)| \leq c_i \left(\sum_{K \in \mathbb{T}_h} h_K^4 \rho_K^2 \right)^{1/2} \|\nabla^2 u\|.$$

Hence, by the H^2-regularity (3.26) and observing $\lambda \leq \lambda_h$, it follows that

$$\lambda_h - \lambda \leq 2c_i c_s \lambda_h \left(\sum_{K \in \mathbb{T}_h} h_K^4 \rho_K^2 \right)^{1/2},$$

which completes the proof. ☐

Notes and references
Estimate (3.31) was derived by Larson (2000), by direct computation. Nystedt (1995) gives the alternative error bound for the eigenvalues

$$\lambda_h - \lambda \leq c\lambda_h \sum_{K \in \mathbb{T}_h} h_K^2 \rho_K^2, \tag{3.34}$$

which reflects the dependence of the eigenvalue error on the square of the energy norm error of the eigenfunctions known from *a priori* analysis. The approach described above has been extended to *non-selfadjoint* eigenvalue problems in Heuveline and Rannacher (2001*a*). Its application for eigenvalue problems in hydrodynamic stability theory is developed in Heuveline and Rannacher (2001*b*).

3.3. A linear transport problem

The previous examples illustrate the DWR method applied to elliptic problems in which error propagation is isotropic. Next, we consider the other extreme of uni-directional error transport, as present in linear transport problems of the form

$$\beta \cdot \nabla u = f \quad \text{in } \Omega, \quad u = g \text{ on } \Gamma_-. \tag{3.35}$$

Here, $\Gamma_- = \{x \in \partial\Omega, n \cdot \beta < 0\}$ is the 'inflow boundary' and $\Gamma_+ = \partial\Omega \setminus \Gamma_-$ the 'outflow boundary'. The natural solution space is $V := \{v \in L^2(\Omega) : \beta \cdot \nabla v \in L^2(\Omega)\}$. This problem is discretized by the Galerkin finite element with streamline diffusion stabilization called SDFEM (see Hughes and Brooks (1982) and Johnson (1987)). The discretization is based on first-degree polynomials on regular quadrilateral meshes $\mathbb{T}_h = \{K\}$ as described above:

$$V_h = \{v \in H^1(\Omega),\ v|_K \in \tilde{Q}_1(K),\ K \in \mathbb{T}_h\} \subset V.$$

The discrete solution $u_h \in V_h$ is defined by

$$(\beta \cdot \nabla u_h, \Phi_h) - (\beta_n u_h, \varphi_h)_- = (f, \Phi_h) - (\beta_n g, \varphi_h)_-, \quad \text{for all } \varphi_h \in V_h, \tag{3.36}$$

where $\Phi_h := \varphi_h + \delta\beta \cdot \nabla\varphi_h$, and $\beta_n = \beta \cdot n$, while the parameter function δ is determined locally by $\delta_K = \kappa h_K$ (assuming that $|\beta| = 1$). In the formulation (3.36) the inflow boundary condition is imposed in the weak sense. The right-hand and left-hand side of (3.36) define a bilinear form $A_h(\cdot,\cdot)$ and a linear form $F_h(\cdot)$, respectively. Using this notation, (3.36) may be written as

$$A_h(u_h, \varphi_h) = F_h(\varphi_h), \quad \text{for all } \varphi_h \in V_h. \tag{3.37}$$

The corresponding residual is $\rho_h(u_h, \cdot) := F_h(\cdot) - A_h(u_h, \cdot)$.

Let $J(\cdot)$ be a given functional defined on V for controlling the error $e = u - u_h$. Following our general approach, we consider the corresponding adjoint problem

$$A_h(\varphi, z) = J(\varphi), \quad \text{for all } \varphi \in V, \tag{3.38}$$

which is a (generalized) transport problem with transport in the negative β-direction. We note that here we use the stabilized bilinear form $A_h(\cdot,\cdot)$ in the duality argument, in order to achieve an optimal treatment of the stabilization terms. The existence of the adjoint solution $z \in V$ follows by standard variational arguments. Then, the general result of Proposition 2.6 yields the error representation

$$J(e) = \rho_h(u_h, z - \varphi_h) = (\beta \cdot \nabla e, z - \varphi_h + \delta\beta \cdot \nabla(z - \varphi_h)) - (\beta_n e, z - \varphi_h)_-,$$

for arbitrary $\varphi_h \in V_h$. This results in the following a posteriori error estimate.

Proposition 3.3. (Rannacher 1998b) For the approximation of the linear transport equation by the SDFEM, we have the a posteriori error estimate

$$|J(e)| \le \eta_\omega(u_h) := \sum_{K \in \mathbb{T}_h} \{\rho_K^1 \omega_K^1 + \rho_K^2 \omega_K^2\}, \tag{3.39}$$

where the residuals and weights are defined by

$$\rho_K^1 := \|f - \beta \cdot \nabla u_h\|_K, \quad \omega_K^1 := \|z - \varphi_h\|_K + \delta_K \|\beta \cdot \nabla(z - \varphi_h)\|_K,$$

$$\rho_K^2 := h_K^{-1/2} \|\beta_n(u_h - g)\|_{\partial K \cap \Gamma_-}, \quad \omega_K^2 := h_K^{1/2} \|z - \varphi_h\|_{\partial K \cap \Gamma_-},$$

with a suitable approximation $\varphi_h \in V_h$ to z.

This *a posteriori* error bound explicitly contains the mesh size h_K and the stabilization parameter δ_K as well. This gives us the possibility of simultaneously adapting both parameters, which may be particularly advantageous for capturing sharp layers in the solution.

Examples. We want to illustrate the error estimator (3.39) by two thought experiments. Let $\beta \equiv$ const and $f = 0$. First, we take the functional $J(e) := (1, \beta_n e)_+$. The corresponding adjoint solution is $z \equiv 1$, so that $J(e) = 0$. This reflects the global conservation property of the SDFEM. Next, we set

$$J(e) := (1, e) + (1, \delta\beta_n e)_+.$$

The corresponding adjoint problem reads

$$(-\beta \cdot \nabla z, \varphi - \delta\beta \cdot \nabla\varphi) + (\beta_n z, \varphi)_+ = (1, \varphi) + (1, \delta\beta_n\varphi)_+.$$

Assuming that $\delta \equiv$ const, this adjoint problem has the same solution as

$$-\beta \cdot \nabla z = 1 \quad \text{in } \Omega, \quad z = \delta \text{ on } \Gamma_-.$$

Consequently, z is linear almost everywhere, that is, the weights in the *a posteriori* bound (3.39) are nonzero only along the lines of discontinuity. Therefore, the mesh refinement will be restricted to these critical regions although the cell residuals $\rho_K(u_h)$ may be nonzero everywhere.

Numerical test (from Hartmann (200x))
We consider the model problem (3.35) on the unit square $\Omega = (0, 1) \times (0, 1) \subset \mathbb{R}^2$ with the right-hand side $f \equiv 0$, the (constant) transport coefficient $\beta = (-1, -0.5)^T$, and the inflow data $g = 1$ on Γ_1 and $g = 0$ elsewhere. The quantity to be computed is part of the outflow, as shown in Figure 3.2:

$$J(u) := \int_{\Gamma_2} \beta \cdot nu \, ds.$$

The corresponding meshes and the solutions u_h and z_h are shown in Figure 3.2. There is no mesh refinement along the upper line of discontinuity of the solution since here the cell residuals of the adjoint solution z_h are almost zero (z is constant).

Notes and references
Here we have considered only the adaptive SDFEM for simple *linear and scalar* transport problems (for more examples see Houston, Rannacher and Süli (2000)). Analogous results have also been obtained by the discontinuous Galerkin finite element 'dG(r) method' (for a comparison see Hartmann (2000)). Earlier work on the adaptive SDFEM for linear as well as nonlinear problems (*e.g.*, Burgers' equation and Euler equations) using duality

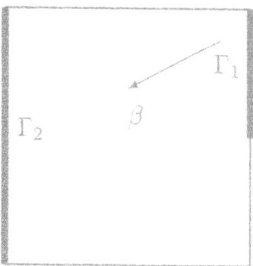  

Fig. 3.2. Configuration and results for the model transport problem (3.35): primal solution (centre) and dual solution (right) on an adaptively refined mesh

and 'stability constants' is Johnson (1990), Hansbo and Johnson (1991), Eriksson and Johnson (1993), Eriksson, Johnson and Larsson (1998), and Johnson and Szepessy (1995). The DWR approach in combination with the SDFEM and the dG(r) method for nonlinear problems has been considered by Führer (1996), Führer and Rannacher (1997), and Hartmann (200x). An application to transport equations with stiff source terms is described in Hebeker and Rannacher (1999).

4. Norm-based error estimation

In this section, we discuss techniques for deriving *a posteriori* estimates for the error in global norms and also with respect to local output quantities which use the concept of 'energy error estimation'. For simplicity, we continue using the Poisson problem as our prototype.

4.1. A posteriori error bounds in global norms

By the same type of argument as used in Proposition 3.1, we can also derive the traditional global error estimates in the energy and the L^2 norm.

(i) **Energy error bound.** First, we use the functional

$$J(\varphi) = \|\nabla e\|^{-1}(\nabla e, \nabla \varphi)$$

in the adjoint problem. Its solution $z \in V$ satisfies $\|\nabla z\| \leq 1$. We obtain the estimate

$$\|\nabla e\| \leq \sum_{K \in \mathbb{T}_h} \rho_K \, \omega_K \leq \left(\sum_{K \in \mathbb{T}_h} h_K^2 \rho_K^2 \right)^{1/2} \left(\sum_{K \in \mathbb{T}_h} h_K^{-2} \omega_K^2 \right)^{1/2},$$

with residual terms and weights as defined above. Now, we use an extension of the interpolation estimate (3.7),

$$\left(\sum_{K \in \mathbb{T}_h} \{ h_K^{-2} \| z - \tilde{\imath}_h z \|_K^2 + h_K^{-1} \| z - \tilde{\imath}_h z \|_{\partial K}^2 \} \right)^{1/2} \leq \tilde{c}_i \, \| \nabla z \|, \qquad (4.1)$$

where $\tilde{\imath}_h z \in V_h$ is a modified nodal interpolation which is defined and stable on $H^1(\Omega)$ (for such a construction see, e.g., Brenner and Scott (1994)). This gives us

$$\| \nabla e \| \leq \tilde{c}_i \left(\sum_{K \in \mathbb{T}_h} h_K^2 \rho_K^2 \right)^{1/2} \| \nabla z \|.$$

Finally, observing the *a priori* bound for $\| \nabla z \|$, we conclude the *a posteriori* *energy error* estimate

$$\| \nabla e \| \leq \eta_E(u_h) = \tilde{c}_i \left(\sum_{K \in \mathbb{T}_h} h_K^2 \rho_K^2 \right)^{1/2}. \qquad (4.2)$$

(ii) **L^2-norm error bound.** Next, we use the functional

$$J(\varphi) = \| e \|^{-1} (e, \varphi)$$

in the adjoint problem. If the (polygonal) domain Ω is convex, the dual solution $z \in V$ is in $H^2(\Omega)$ and admits the *a priori* bound $c_s := \| \nabla^2 z \| \leq 1$. As in (i), this yields the error estimate

$$\| e \| \leq \left(\sum_{K \in \mathbb{T}_h} h_K^4 \rho_K^2 \right)^{1/2} \left(\sum_{K \in \mathbb{T}_h} h_K^{-4} \omega_K^2 \right)^{1/2}.$$

Now we use the stronger version of the interpolation estimate (4.1),

$$\left(\sum_{K \in \mathbb{T}_h} \{ h_K^{-4} \| z - i_h z \|_K^2 + h_K^{-3} \| z - i_h z \|_{\partial K}^2 \} \right)^{1/2} \leq c_i \, \| \nabla^2 z \|, \qquad (4.3)$$

to obtain

$$\| e \| \leq c_i \left(\sum_{K \in \mathbb{T}_h} h_K^4 \rho_K^2 \right)^{1/2} \| \nabla^2 z \|.$$

Finally, observing the *a priori* bound for $\| \nabla^2 z \|$, we conclude the L^2-norm *a posteriori* error estimate

$$\| e \| \leq \eta_{L^2}(u_h) = c_i c_s \left(\sum_{K \in \mathbb{T}_h} h_K^4 \rho_K^2 \right)^{1/2}. \qquad (4.4)$$

In this estimate, the information about the global error sensitivities contained in the dual solution is condensed into just one 'stability constant'

$c_s = \|\nabla^2 z\|$. This is appropriate in estimating the global L^2 error in the case with constant diffusion coefficient. For more general situations with strongly varying coefficients, it may be advisable rather to follow the DWR approach in which the information from the dual solution is kept within the a posteriori error estimator as weights:

$$\|e\| \leq c_i \sum_{K \in \mathbb{T}_h} h_K^2 \rho_K \|\nabla^2 z\|_K. \tag{4.5}$$

Below, we will present an example to support this claim.

4.2. Application for L^2-error estimation

We consider the finite element approximation of a diffusion equation with variable coefficient,

$$-\nabla \cdot \{a\nabla u\} = f \quad \text{in } \Omega, \quad u = 0 \quad \text{on } \partial\Omega. \tag{4.6}$$

The error is to be estimated in the L^2 norm. By this, we want to demonstrate that the use of 'weighted' a posteriori error estimates as proposed by the DWR method (using the approximations described above) can also be beneficial for estimating the error in a global norm. The adjoint problem for estimating the L^2 error is

$$-\nabla \cdot \{a\nabla z\} = \|e\|^{-1} e \quad \text{in } \Omega, \quad z = 0 \quad \text{on } \partial\Omega. \tag{4.7}$$

The a posteriori error estimates (4.4) and (4.5) have natural generalizations to the present situation, that is, the usual L^2-error estimate

$$\|e\| \leq \eta_{L^2}(u_h) := c_i c_s \left(\sum_{K \in \mathbb{T}_h} h_K^4 \rho_K^2 \right)^{1/2}, \quad c_s := \|\nabla^2 z\|, \tag{4.8}$$

and the 'weighted' L^2-error estimate

$$\|e\| \leq \eta_{L^2}^\omega(u_h) := c_i \sum_{K \in \mathbb{T}_h} h_K^2 \rho_K \omega_K, \quad \omega_K := \|\nabla^2 z\|_K, \tag{4.9}$$

with the cell residuals $\rho_K := \|R(u_h)\|_K + h_K^{-1/2} \|r(u_h)\|_{\partial K}$ as defined in Remark 3.5. Both error estimators are evaluated by replacing the second derivatives of the dual solution z by second-order difference quotients of an approximation $z_h \in V_h$, as described above:

$$\omega_K \approx \tilde{\omega}_K := \|\nabla_h^2 z_h\|_K, \quad c_s \approx \tilde{c}_s := \left(\sum_{K \in \mathbb{T}_h} \|\nabla_h^2 z_h\|_K^2 \right)^{1/2}. \tag{4.10}$$

The interpolation constant is set to $c_i = 0.2$. The functional $J(\cdot)$ is evaluated by replacing the unknown solution u by a patch-wise higher-order interpolation $I_h^{(2)} u_h$ of the computed approximation u_h, that is, $e \approx I_h^{(2)} u_h - u_h$.

The quality of the resulting two *a posteriori* error estimators $\tilde{\eta}_{L^2}(u_h)$ and $\tilde{\eta}^\omega_{L^2}(u_h)$ for mesh adaptation will be compared below for a representative model case.

Numerical test
We choose the square domain $\Omega = (-1,1) \times (-1,1)$ and the nonconstant coefficient function $a(x) = 0.1 + e^{3(x_1+x_2)}$ with right-hand side $f \equiv 0.1$. A reference solution is generated by a computation on a very fine mesh. The meshes are refined according to the 'error-balancing strategy' by balancing (as well as possible) the cell-wise indicators $\eta_K := h_K^4 \rho_K^2$ or $\eta_K := h_K^2 \rho_K \tilde{\omega}_K$ over the mesh $\mathbb{T}_h$. Figure 4.1 shows error plots and meshes obtained by the two different strategies.

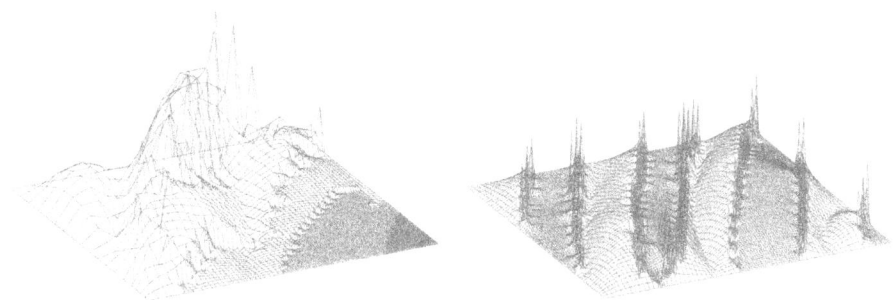

Fig. 4.1. Errors obtained by η_{L^2} (left, scaled by $1/30$) and $\eta^\omega_{L^2}$ (right, scaled by $1/10$) on meshes with $N \sim 10000$ cells

We see that carrying the weights $\|\nabla^2 z\|_K$ within the estimator yields a distribution of mesh cells that is much better adapted to the solution structure determined by the nonconstant coefficient $a(x)$. Condensing all the weighting terms into just one global 'stability constant' $c_s \approx \|\nabla^2 z\|_\Omega$ loses this detail information. Further, the results in Table 4.1 show that the error bounds resulting from the 'weighted' *a posteriori* error estimates are more accurate.

4.3. A posteriori error estimates for functional output based on energy error estimates

Several other approaches to 'goal-oriented' *a posteriori* error control have been proposed in the literature which, like the DWR method, use duality arguments for error representation but employ global 'energy norm' error estimates. We continue to use the notation of the preceding section particularly that of the model problem (3.2). Let $J(\cdot)$ again be a (linear) error

Table 4.1. Results obtained by the global L^2-error estimator $\tilde{\eta}_{L_2}$ (top) compared to the weighted estimator $\tilde{\eta}_{L_2}^\omega$ (bottom) ($I_{\text{eff}} := \eta(u_h)/\|e\|$, $L = \#$ refinement levels)

TOL	N	L	$\|e\|$	$\tilde{\eta}_{L^2}$	I_{eff}	$\tilde{c}_s$
4^{-2}	2836	9	6.40e-2	2.32e-1	3.62	3.02
4^{-3}	5884	10	2.13e-2	1.21e-1	5.68	3.26
4^{-4}	15736	11	7.36e-3	4.76e-2	6.46	3.55
4^{-5}	23380	11	5.59e-3	3.12e-2	5.58	3.39

TOL	N	L	$\|e\|$	$\tilde{\eta}_{L^2}^\omega$	I_{eff}	$\tilde{c}_s$
4^{-4}	220	5	6.77e-2	5.24e-2	0.77	....
4^{-5}	592	6	2.21e-2	2.59e-2	1.17	....
4^{-6}	892	6	1.19e-2	1.54e-2	1.29	....
4^{-7}	2368	7	5.11e-3	7.17e-3	1.40	....

functional, $z \in V$ the associated dual solution and $z_h \in V_h$ its finite element projection. For the errors $e := u - u_h$ and $e^* := z - z_h$, by Galerkin orthogonality, we have

$$J(e) = A(e, z) = A(e, e^*) = A(u, e^*) = (f, e^*). \qquad (4.11)$$

We recall the definition of the corresponding residual functionals

$$\rho(u_h, \varphi) := (f, \varphi) - A(u_h, \varphi), \qquad \rho^*(z_h, \varphi) := J(\varphi) - A(\varphi, z_h).$$

Clearly, these residuals vanish for $\varphi_h \in V_h$. Next, a suitable enlarged space $\hat{V}_h \supset V_h$ is introduced, for instance, by adding certain higher-order polynomials in each cell (see again Ainsworth and Oden (1993, 1997a)). The functions in $\hat{V}_h$ are allowed to be discontinuous. We assume that the residual functionals ρ and ρ^* possess proper extensions $\hat{R}$ and $\hat{R}^*$ to $\hat{V}_h$ (involving, for example, local stress averaging). The natural 'cell-wise' defined extension of the bilinear form $A(\cdot, \cdot)$ is denoted by $A_h(\cdot, \cdot)$. With this notation approximations $\hat{e}$ and $\hat{e}^* \in \hat{V}_h$ are generated for the errors e and e^* by solving the defect equations

$$\begin{aligned}
A_h(\hat{e}, \hat{\varphi}_h) &= \hat{\rho}(u_h, \hat{\varphi}_h), \quad \text{for all } \hat{\varphi}_h \in \hat{V}_h, \\
A_h(\hat{\varphi}_h, \hat{e}) &= \hat{\rho}^*(z_h, \hat{\varphi}_h), \quad \text{for all } \hat{\varphi}_h \in \hat{V}_h.
\end{aligned} \qquad (4.12)$$

Owing to the allowed discontinuity of functions in $\hat{V}_h$, solving (4.12) usually reduces to solving local defect problems separately on each mesh cell $K \in \mathbb{T}_h$. Now, the assumption is made that the space $\hat{V}_h$ is sufficiently large that

$$e \in \hat{V}_h. \tag{4.13}$$

Of course, in general, this condition cannot be satisfied exactly by a finite-dimensional space $\hat{V}_h$. Therefore, condition (4.13) renders all the resulting conclusions heuristic. Nevertheless, assuming (4.13) to be true, we can take e as a test function in (4.12), obtaining

$$A_h(\hat{e}, e) = \hat{\rho}(u_h, e) = \rho(u_h, e) = \|\nabla e\|^2,$$

$$A_h(\hat{e}^*, e) = \hat{\rho}^*(z_h, e) = \rho^*(z_h, e) = J(e).$$

This trivially implies that $\|\nabla e\| \leq \|\nabla \hat{e}\|_h$, and consequently,

$$|J(e)| = |A_h(e, \hat{e}^*)| \leq \|\nabla e\| \, \|\nabla \hat{e}^*\| \leq \|\nabla \hat{e}\| \, \|\nabla \hat{e}^*\|. \tag{4.14}$$

By an elementary argument, this error bound can be improved to

$$|J(u) - \hat{J}(u_h)| \leq \tfrac{1}{2}\|\nabla \hat{e}\| \, \|\nabla \hat{e}^*\|, \tag{4.15}$$

where $\hat{J}(u_h) := J(u_h) + \tfrac{1}{2}A_h(\hat{e}, \hat{e}^*)$ (Machiels et $al.$ 1998). The error bound (4.14) requires precise energy-norm estimates for e as well as e^*, which are difficult to achieve for singular z, such as, for example, in point error estimation. Further, (4.14) does not directly yield criteria for local mesh refinement. In order to obtain those, one may return to the very basic error representation (4.11) which directly implies the error bound

$$|J(e)| \leq \sum_{K \in \mathbb{T}_h} \|\nabla e\|_K \|\nabla e^*\|_K.$$

This is then replaced heuristically by

$$|J(e)| \approx \sum_{K \in \mathbb{T}_h} \|\nabla \hat{e}\|_K \|\nabla \hat{e}^*\|_K. \tag{4.16}$$

This last step lacks justification (and may even lead to a systematically wrong result) since the local total error $\|\nabla e^*\|_K$ is not necessarily bounded by the locally obtained approximation $\|\nabla \hat{e}^*\|_K$ due to possible error pollution. This defect may not be critical in generating good meshes for elliptic model problems but can be disastrous in situations where strong error transport occurs. The DWR method avoids this problem by correctly replacing the terms $\|\nabla e\|_K\|\nabla e^*\|_K$ by $\rho_K(u_h)\omega_K(z)$ where the local residual $\rho_K(u_h)$ can be evaluated exactly and the local weights $\omega_K(z) \sim \|z - i_h z\|_K$ do not suffer from error pollution provided, of course, that z is computed with sufficient accuracy. Alternatively, it has been suggested that we convert the globally multiplicative a $posteriori$ estimate (4.14) into a locally additive one similar to (4.20):

$$|J(e)| \leq \sum_{K \in \mathbb{T}_h} \{\theta \|\nabla \hat{e}\|_K^2 + \theta^{-1} \|\nabla \hat{e}^*\|_K^2\}, \tag{4.17}$$

for some $\theta > 0$. Again, because of the necessarily global choice of θ, this procedure does not allow us to balance the cell-wise contributions of $\hat{e}$ and $\hat{e}^*$ properly, which is critical in the case of strongly differing behaviour of u and z.

For completeness, we briefly sketch a variant of the above approach, which requires upper *and* lower bounds for the energy-norm error and fully exploits the fact that the energy bilinear form is a scalar product. Suppose that we have available approximations $\hat{e}$ and $\check{e}$ for the error e which yield upper as well as lower bounds for the energy error:

$$\|\nabla \underline{e}\| \leq \|\nabla e\| \leq \|\nabla \hat{e}\|. \tag{4.18}$$

Starting from the error representation (4.11), the idea is to use the parallelogram identity for scalar products to obtain

$$J(e) = A(\theta e, \theta^{-1} e^*) = \tfrac{1}{4} \|\nabla(\theta e + \theta^{-1} e^*)\|^2 - \tfrac{1}{4} \|\nabla(\theta e - \theta^{-1} e^*)\|^2,$$

with a suitable balancing parameter $\theta > 0$. In view of (4.18), this implies the following a *posteriori* error estimate:

$$\check{\eta}_E(u_h, z_h) \leq J(e) \leq \hat{\eta}_E(u_h, z_h), \tag{4.19}$$

with the lower and upper bounds given by

$$\check{\eta}_E(u_h, z_h) := \tfrac{1}{4} \|\nabla(\theta \check{e} + \theta^{-1} \check{e}^*)\|^2 - \tfrac{1}{4} \|\nabla(\theta \hat{e} - \theta^{-1} \hat{e}^*)\|^2,$$

$$\hat{\eta}_E(u_h, z_h) := \tfrac{1}{4} \|\nabla(\theta \hat{e} + \theta^{-1} \hat{e}^*)\|^2 - \tfrac{1}{4} \|\nabla(\theta \check{e} - \theta^{-1} \check{e}^*)\|^2.$$

Again it is not clear how to organize effective local mesh adaptation on the basis of a global estimate of the form (4.19). The simple idea of 'localizing' the error bound $\hat{\eta}_E(u_h, z_h)$ according to

$$\hat{\eta}_E(u_h, z_h) \leq \tfrac{1}{4} \sum_{K \in \mathbb{T}_h} \left| \|\nabla(\theta \hat{e} + \theta^{-1} \hat{e}^*)\|_K^2 - \tfrac{1}{4} \|\nabla(\theta \check{e} - \theta^{-1} \check{e}^*)\|_K^2 \right| \tag{4.20}$$

again suffers from the fact that the parameter θ is to be chosen globally. Hence, the multiplicative local interaction of e and e^*, which is the essential feature of the DWR method, would be lost here.

Finally, we note that the argument leading to the a *posteriori* error bound (4.14) can be generalized to nonsymmetric problems. If the symmetric energy form $A(\cdot, \cdot)$ is replaced by a nonsymmetric bilinear form, for example,

$A(\cdot,\cdot) := (\nabla\cdot,\nabla\cdot) + (b\cdot\nabla\cdot,\cdot)$, the error estimate (4.14) contains an additional constant measuring the deviation of $A(\cdot,\cdot)$ from symmetry:

$$|J(e)| \le \frac{1+\lambda\mu}{2}\,\|\nabla\hat{e}\|\,\|\nabla\hat{e}^*\|, \tag{4.21}$$

where

$$\lambda := \inf_{v\in V}\left\{\sup_{\varphi\in V}\frac{A(v,\varphi)}{\|\nabla v\|\|\nabla\varphi\|}\right\}, \qquad \mu := \inf_{v\in V}\left\{\sup_{\varphi\in V}\frac{A(\varphi,v)}{\|\nabla\varphi\|\|\nabla v\|}\right\}.$$

In practice, it would be very difficult to obtain realistic bounds for the *global* stability parameters λ and μ. The DWR method is designed to avoid this critical difficulty simply by numerically solving the associated adjoint problem with higher accuracy and thereby implicitly detecting the *local* sensitivities of the problem.

Notes and references
The approaches described above were originally proposed in a series of papers by Babuška and Miller (1984a, 1984b, 1984c) as a post-processing tool for achieving improved accuracy. Recently, they have been extensively elaborated for several model applications in Paraschivoiu and Patera (1998), Peraire and Patera (1998), and Oden and Prudhomme (1999).

5. Evaluation of estimates and mesh adaptation

In this section, we discuss some practical aspects of the DWR method. The first question is how to evaluate the *a posteriori* error representations and estimates derived in Section 3 in practice. This particularly involves the approximation of the 'adjoint solution'. The second question concerns the use of these estimates in the course of successive mesh adaptation. Finally, in the solution of nonlinear problems we have to combine error estimation and mesh adaptation with nonlinear iteration and linearization. The latter aspect will be discussed in Section 6.

5.1. Evaluation of a posteriori error bounds

First, we will discuss the evaluation of the error representation (3.17) or the resulting error bound (3.18), again using the Poisson problem as our prototype (see Becker and Rannacher (1996b)). There are essentially two different approaches. In the first one the dual solution z is replaced by some approximation $\hat{z}_h$ obtained from solving the adjoint problem by a higher-order method or on a finer mesh. In the second one the quantity $z - \varphi_h$ is estimated using the finite element projection $z_h \in V_h$ of z on the same

Table 5.1. Efficiency of the *weighted* error indic-
ators for computing the point-error $J(e) = |e(0)|$
on uniformly refined meshes

L	N	$I_{\text{eff}}(\eta_\omega^1)$	$I_{\text{eff}}(\eta_\omega^2)$	$I_{\text{eff}}(\eta_\omega^3)$
1	4^3	6.667	12.82	2.066
2	4^4	1.253	13.51	45.45
3	4^5	1.052	3.105	9.345
4	4^6	1.007	2.053	7.042
5	4^7	1.003	1.886	6.667

mesh. As our quality measure, we use the over-estimation factor

$$I_{\text{eff}} := \left| \frac{\eta(u_h)}{J(e)} \right|,$$

referred to as the (reciprocal) 'effectivity index' (see Babuška and Rheinboldt
(1978*b*)). Results of tests with these methods are reported in Table 5.1.

(i) Global higher-order approximation. Let $\hat{z}_h \in \hat{V}_h$ be an approximation to z obtained by a higher-order method or on a finer mesh $\hat{\mathbb{T}}_h$ with
corresponding finite element space $\hat{V}_h$:

$$a(\hat{\varphi}_h, \hat{z}_h) = J(\hat{\varphi}_h), \quad \text{for all } \hat{\varphi}_h \in \hat{V}_h.$$

The additional error caused by this approximation can be described by writing the error representation (3.17) in a symmetric form. To this end, we use
Galerkin orthogonality of $e := u - u_h$ and $\hat{e}^* := z - \hat{z}_h$ to obtain

$$J(e) = a(e, \hat{z}_h - \varphi_h) + a(u - \hat{\varphi}_h, \hat{e}^*),$$

with arbitrary $\varphi_h \in V_h$ and $\hat{\varphi}_h \in \hat{V}_h$. This can be rewritten as

$$
\begin{aligned}
J(e) = \ & \sum_{K \in \mathbb{T}_h} \left\{ (R(u_h), \hat{z}_h - \varphi_h)_K - (r(u_h), \hat{z}_h - \varphi_h)_{\partial K} \right\} \\
& + \sum_{K \in \hat{\mathbb{T}}_h} \left\{ (u - \hat{\varphi}_h, R^*(\hat{z}_h))_K - (u - \hat{\varphi}_h, r^*(\hat{z}_h))_{\partial K} \right\},
\end{aligned}
\tag{5.1}
$$

with the cell and edge residuals $R(u_h)$ and $r(u_h)$ of u_h, and $R^*(\hat{z}_h)$ and
$r^*(\hat{z}_h)$ of $\hat{z}_h$, respectively, defined analogously as before. The first sum on
the right can easily be evaluated, while the evaluation of the second one
would require generation of an improved approximation to u. Therefore,
the simplest approach is to assume that this second sum is small compared

to the first and neglect it. The resulting error estimator reads

$$\eta_\omega^1(u_h) := \left| \sum_{K \in \mathbb{T}_h} \left\{ (R(u_h), \hat{z}_h - i_h\hat{z}_h)_K - (r(u_h), \hat{z}_h - i_h\hat{z}_h)_{\partial K} \right\} \right|,$$

with the nodal interpolation $i_h \hat{z}_h \in V_h$. Note that we cannot simplify this approach by using $\hat{z}_h = z_h$ since then the whole error estimator would vanish. We have tested this approach using bi-quadratic approximation for $\hat{z}_h$ on the primal mesh $\mathbb{T}_h$. In the model case of the Poisson equation (3.2), it is seen by the numerical results as well as by theoretical analysis that, in this case, $\hat{\eta}_\omega(u_h)$ is asymptotically sharp, that is, $\lim_{TOL \to 0} I_{eff} = 1$ (see Table 5.1). However, approximating z by a higher-order finite element scheme does not seem very economical in estimating the error in the low-order scheme. A more practical alternative would be to compute an approximation to z on a sufficiently fine 'super-mesh'. Then, increasing the accuracy, arbitrarily close approximations to $J(e)$ could be obtained.

(ii) **Local higher-order approximation.**   In most practical cases it suffices to use the finite element projection $z_h \in V_h$ of z computed on the current mesh. It is assumed that an improved approximation can be obtained by patch-wise higher-order interpolation $i_h^+ z_h$. This construction requires special care on elements with hanging nodes, in order to preserve the higher-order accuracy of the interpolation process. The resulting global error estimator reads

$$\eta_\omega^2(u_h) := \left| \sum_{K \in \mathbb{T}_h} \left\{ (R(u_h), i_h^+ z_h - z_h)_K - (r(u_h), i_h^+ z_h - z_h)_{\partial K} \right\} \right|.$$

For bi-quadratic interpolation, we observe that $\limsup_{TOL \to 0} I_{eff} \approx 1.2-3$.

(iii) **Approximation by difference quotients.**   We use the cell-wise interpolation estimate (3.7) to estimate the weights ω_K by

$$\omega_K = \|z - i_h z\|_K + h_K^{1/2} \|z - i_h z\|_{\partial K} \le c_i h_K^2 \|\nabla^2 z\|_K.$$

The second derivative $\nabla^2 z$ is then replaced by a suitable second-order difference quotient $\nabla_h^2 z_h$. A useful substitute is the approximation

$$\|\nabla^2 z\|_K \approx h_K^{-1/2} \|r(z_h)\|_{\partial K},$$

which corresponds to the edge residual $\rho_{\partial K}$ of u_h, the evaluation of which has to be implemented anyway. This yields the heuristic error estimator

$$\eta_\omega^3(u_h) := c_i \sum_{K \in \mathbb{T}_h} h_K^{-1/2} \rho_K \|r(z_h)\|_{\partial K}.$$

In this case, we observe that $\limsup_{TOL \to 0} I_{eff} \approx 2-10$ depending on the problem. The estimator $\eta_\omega^3(u_h)$ involves an interpolation constant c_i for

which sharp bounds are difficult to get. Therefore it is not suited to error
estimation, but by localization it yields rather effective mesh refinement
indicators $\eta_K := \rho_K \|r(z_h)\|_{\partial K}$.

The effectivity indices observed for these strategies in computing the
point-value $u(0)$ (on uniformly refined meshes) are listed in Table 5.1 (from
Becker and Rannacher (1996b)). These results indicate that, even for the
simplest model situations, an asymptotic effectivity index $I_{\text{eff}} = 1$ is achiev-
able only at the expense of high extra cost, for instance by approximating
the dual solution using a higher-order method.

5.2. Strategies for mesh adaptation

Now, we discuss some popular strategies for mesh adaptation based on an
a posteriori error estimate of the form

$$|J(e)| \leq \eta(u_h) := \sum_{K \in \mathbb{T}_h} \eta_K, \tag{5.2}$$

with certain cell-error indicators $\eta_K = \eta_K(u_h)$ obtained on the current mesh
$\mathbb{T}_h$. Suppose that a tolerance TOL for the error $J(e)$ or a maximum number
$N_{\max}$ of mesh cells has been prescribed. Starting from an approximate
solution $u_h \in V_h$ obtained on the current mesh $\mathbb{T}_h$, the mesh adaptation
may be organized by one of the following strategies.

Error-balancing strategy. Cycle through the mesh and equilibrate the
local error indicators according to $\eta_K \approx \text{TOL}/N$ with $N := \#\{K \in$
$\mathbb{T}_h\}$. This leads eventually to $\eta(u_h) \approx \text{TOL}$. Since N changes when
the mesh is locally refined or coarsened, this strategy requires iteration
with respect to N on each mesh level.

Fixed mesh (or error) fraction strategy. Order cells according to the
size of η_K and refine a certain percentage (say 20%) of cells with
largest η_K (or those which make up 70% of the estimator value $\eta(u_h)$)
and coarsen those cells with smallest η_K. By this strategy, one may
achieve a prescribed rate of increase of N or keep it constant within a
nonstationary computation, if that is desirable.

Mesh-optimization strategy. Use a (proposed) representation

$$\eta(u_h) := \sum_{K \in \mathbb{T}_h} \eta_K \approx \int_\Omega h(x)^2 \Phi(x) \, dx \tag{5.3}$$

for directly generating a formula for an optimal mesh-size distribution:

$$h_{\text{opt}}(x) = \left(\frac{W}{N_{\max}} \right)^{1/2} \Phi(x)^{-1/4}, \quad W := \int_\Omega \Phi(x)^{1/2} \, dx. \tag{5.4}$$

Here, we think of a smoothly distributed mesh-size function $h(x)$ such that $h_{|K} \approx h_K$. The existence of such a representation with an h-independent error density function $\Phi(x) \approx |\nabla^2 u(x)| \, |\nabla^2 z(x)|$ can be rigorously justified only under very restrictive conditions, but is generally supported by computational experience (Becker and Rannacher 1996b). Even for rather 'irregular' functionals $J(\cdot)$ the quantity W in (5.4) is bounded. For example, the evaluation of $J(u) = \partial_i u(P)$ for smooth u in two dimensions leads to $\Phi(x) \approx |x - P|^{-3}$ and, consequently, $W < \infty$.

Problem 5.1. Establish an asymptotic error representation of the form (5.3) on adaptively refined meshes. The essential step for this is the bound

$$\limsup_{N \to \infty} \|\nabla_h^2 u_h\|_\infty \leq c \, \|\nabla^2 u\|_\infty, \tag{5.5}$$

for second-order difference quotients $\nabla_h^2 u_h$ of first-degree finite elements.

Problem 5.2. The explicit formula for $h_{\mathrm{opt}}(x)$ has to be used with care in designing a mesh. Its derivation implicitly assumes that it actually corresponds to a *scalar* mesh-size function of an isotropic mesh such that $h_{\mathrm{opt}|K} \approx h_K$. However, this condition is not incorporated into the formulation of the mesh-optimization problem. The treatment of anisotropic meshes containing stretched cells requires a more involved concept of mesh description and optimization.

6. Algorithmic aspects for nonlinear problems

In this section, we discuss the practical realization of the DWR method for nonlinear problems. This concerns the iterative treatment of the nonlinearity and particularly the evaluation of the a *posteriori* error estimators by linearization.

6.1. The nested solution approach

In the following, we briefly describe how the concepts of adaptivity laid out so far can be combined with an iterative solution process, for instance a Newton-type method, for general nonlinear problems as in Section 2:

$$A(u; \varphi) = F(\varphi), \quad \text{for all } \varphi \in V. \tag{6.1}$$

This inherits the basic structure of the solution processes employed in all the nonlinear applications which will be presented below. Let a desired error tolerance TOL or a maximum mesh complexity $N_{\max}$ be given. Starting from a coarse initial mesh $\mathbb{T}_0$, a hierarchy of successively refined meshes $\mathbb{T}_i$, $i \geq 1$, and corresponding finite element spaces V_i, is generated by the following algorithm.

Adaptive solution algorithm.

(0) Initialization $i = 0$: Compute an initial approximation $u_0 \in V_0$.

(1) Defect correction iteration: For $i \geq 1$, start with $u_i^{(0)} := u_{i-1} \in V_i$.

(2) Iteration step: For $j \geq 0$ evaluate the defect

$$(d_i^{(j)}, \varphi) := F(\varphi) - A(u_i^{(j)}; \varphi), \quad \varphi \in V_i. \tag{6.2}$$

Pick a suitable approximation $\tilde{A}'(u_i^{(j)}; \cdot, \cdot)$ to the derivative $A'(u_i^{(j)}; \cdot, \cdot)$ (with good stability and solubility properties) and compute a correction $v_i^{(j)} \in V_i$ from the linear equation

$$\tilde{A}'(u_i^{(j)}; v_i^{(j)}, \varphi) = (d_i^{(j)}, \varphi), \quad \text{for all } \varphi \in V_i. \tag{6.3}$$

For that, Krylov space and multigrid methods are employed using the hierarchy of already constructed meshes $\{\mathbb{T}_i, \dots, \mathbb{T}_0\}$ (see Becker and Braack (2000)). Then, update $u_i^{(j+1)} = u_i^{(j)} + \lambda_i v_i^{(j)}$, with some relaxation parameter $\lambda_i \in (0, 1]$, set $j := j + 1$ and go back to (2). This process is repeated until a limit $\tilde{u}_i \in V_i$, is reached with a prescribed accuracy.

(3) Error estimation: Accept $\tilde{u}_i = u_i$ as the solution on mesh $\mathbb{T}_i$, solve the discrete linearized adjoint problem

$$z_i \in V_i: \quad A'(u_i; \varphi, z_i) = J'(u_i; \varphi), \quad \text{for all } \varphi \in V_i,$$

and evaluate the *a posteriori* error estimate

$$|J(u) - J(u_i)| \approx \eta_\omega(u_i). \tag{6.4}$$

For controlling the reliability of this bound, that is, the accuracy of the linearization and the determination of the approximate adjoint solutions z_i, one may use the algorithm described below. If the error estimator $\eta_\omega(u_i)$ is detected to be reliable and $\eta(u_i) \leq \text{TOL}$ or $N_i \geq N_{\max}$, then stop. Otherwise mesh adaptation yields the new mesh $\mathbb{T}_{i+1}$. Then, set $i := i+1$ and go to (1).

Remark 6.1. The defect-correction iteration (1)–(3) has to be terminated when one is close enough to the solution $u_i \in V_i$. This can be controlled by monitoring the algebraic residual $\|d_i^{(j)}\|$, which can additionally be weighted cell-wise by the current approximation z_i to the adjoint solution. However, the development of a fully satisfactory stopping criterion on rigorous grounds has not yet been accomplished. The situation is better for the iterative solution of the linear defect equations (6.3). Here, the multigrid method is used which inherits projection properties similar to those of the underlying finite element discretization. This can be exploited for deriving weighted *a posteriori* error estimates which simultaneously control the discretization and the iteration error. The application of this concept has been described in Becker, Johnson and Rannacher (1995) for the Poisson model problem (3.2), in Becker (1998a) for the Stokes problem in primal-mixed formulation, and in Larson and Niklasson (1999) for semilinear elliptic problems. In both papers only energy-norm and L^2-norm error estimates have been given. However, the extension of this argument for also incorporating estimation of error functionals is straightforward.

Remark 6.2.   We note that the evaluation of the *a posteriori* error estimate (6.4) involves only the solution of *linearized* problems. Hence, the whole error estimation may amount to only a relatively small fraction of the total cost for the solution process. This has to be compared to the usually much higher cost when working on non-optimized meshes.

6.2. Approximation of the 'exact' adjoint problem

The use of the error estimator $\eta_\omega(u_i)$ in (6.4) for mesh adaptation relies on the assumption that the remainder terms in the 'exact' error representations (2.18) or (2.27) are small compared to the residual terms and may therefore be neglected. This may even be justified for the simplest possible linearization using $\hat{u}_h = u_h$ and

$$J(u) - J(u_h) = \min_{\varphi_h \in V_h} \rho(u_h; z - \varphi_h) + R(\overline{uu_h}; e, e, z), \qquad (6.5)$$

provided the mesh is sufficiently fine, depending on the stability properties of the continuous problem. However, there are several aspects which require special thought.

- The *a posteriori* error estimate (6.4) may lead to meshes that are rather coarse in areas of less importance for the computation of $J(u)$. This means that global error quantities like the energy-norm or the L^2-norm error may not be so small. In turn, the linearization may be less accurate globally.

- In situations when the problem is close to being singular (for example, close to a bifurcation point), the second derivative $A''(\overline{uu_h}; \cdot, \cdot, \cdot)$ may be very large, such that the remainder term may not be small enough to be neglected.

The estimation of the remainder term requires bounds for global integral expressions of the error $e = u - u_h$ in which the adjoint solution appears as a weight, for example $\|e^2 z\|_{L^1}$. It seems natural that the control of the linearization also reflects the sensitivity of the problem, that is, the dependence of the target quantity on the local residuals in terms of the adjoint solution z. Now, we want to discuss the possibility of making the *a posteriori* error representation (6.5) reliable for estimating the error $J(u)-J(u_h)$ on meshes for which the remainder term cannot be guaranteed to be negligible. This must also include the generation of approximations $\hat{z}_h$ to the exact solution of the perturbed adjoint problem. Given a mesh $\mathbb{T}_h$, an associated finite element space V_h and an approximating solution $u_h \in V_h$, we have to approximate the solution $z \in V$ of the 'exact' adjoint problem (see Proposition 2.4)

$$A'(\overline{uu_h}; \varphi, z) = J'(\overline{uu_h}; \varphi), \quad \text{for all } \varphi \in V. \tag{6.6}$$

In Proposition 2.5, we have analysed the effect of linearizing (6.6) by replacing u by an improved approximation $\hat{u}_h \in \hat{V}_h$ in a richer finite element space $\hat{V}_h \supset V_h$:

$$A'(\overline{\hat{u}_h u_h}; \varphi, z) = J'(\overline{\hat{u}_h u_h}; \varphi), \quad \text{for all } \varphi \in V. \tag{6.7}$$

Then, the most straightforward approximation would be to solve this linearized adjoint problem on the same super-mesh for $\hat{z}_h \in \hat{V}_h$:

$$A'(\overline{\hat{u}_h u_h}; \hat{\varphi}_h, \hat{z}_h) = J'(\overline{\hat{u}_h u_h}; \hat{\varphi}_h), \quad \text{for all } \hat{\varphi}_h \in \hat{V}_h. \tag{6.8}$$

Of course, for solving (6.7), one could also use an even richer space than $\hat{V}_h$, depending on the particular properties of the adjoint problem. This variant will not be further discussed here. The resulting *a posteriori* error estimators may then be evaluated by one of the techniques described in Section 5. Involving more and more accurate approximations $\hat{u}_h$ and $\hat{z}_h$, the resulting error estimator on the current space V_h tends to become sharp. This approximation process can be realized in form of a feedback algorithm. For sequences of increasingly enriched spaces $V_i := V_{h_i}$ and $\hat{V}_{i+j} := \hat{V}_{h_{i+j}} \supset V_i$ $(i, j = 0, 1, \ldots)$, we write $u_i := u_{h_i}$, $z_i := z_{h_i}$, and analogously $\hat{u}_{i+j} := \hat{u}_{h_{i+j}}$, $\hat{z}_{i+j} := \hat{z}_{h_{i+j}}$. The resulting approximate error estimators are accordingly denoted by $\hat{\eta}_\omega^j(u_i)$.

Adaptive algorithm controlling linearization and discretization.

(0) Choose appropriate parameters $\varepsilon > 0$ and $0 < \varepsilon_0 \leq \gamma\varepsilon$. Then, the algorithm aims at achieving effectivity index $I_{\text{eff}} = 1 + \varepsilon$, while $\gamma \ll 1$ is a safety factor for the stopping criterion.

(1) Start from coarse meshes $\mathbb{T}_0$ and $\hat{\mathbb{T}}_0 := \mathbb{T}_0$ and compute the corresponding solutions u_0 and $\hat{z}_0$. Evaluate the approximate error estimator $\eta_\omega^0(u_0)$ and construct the new mesh $\mathbb{T}_1$.

(2) For $i \geq 1$, compute $u_i \in V_i$. Set $j = 0$.

(3) For $j \geq 0$, solve the adjoint problem with the bilinear form

$$A'(\overline{\hat{u}_{i+j} u_i}; \cdot, \cdot)$$

on the mesh $\hat{\mathbb{T}}_{i+j}$ and evaluate the resulting error bound $\hat{\eta}_\omega^j(u_i)$. If

$$\left|\eta_\omega^j(u_i) - \eta_\omega^{j-1}(u_i)\right| > \varepsilon\, \eta_\omega^{j-1}(u_i),$$

then set $j := j+1$ and proceed.

(4) For $i \geq 0$, if $|\hat{\eta}_\omega^0(u_i) - \hat{\eta}_\omega^j(u_i)| \leq \varepsilon_0\, \hat{\eta}_\omega^j(u_i)$, then neglect the effect of linearization on the error estimator and continue with the mesh adaptation process using the error estimators $\eta_\omega(u_i) := \hat{\eta}_\omega^0(u_i)$. Further, if $\eta_\omega(u_i) \leq \text{TOL}$ or $N_i \geq N_{\max}$, then stop; otherwise go to (3).

6.2.1. A semilinear example

We consider a semilinear problem of the form

$$L(u) := -\Delta u + s(u) = f \text{ in } \Omega, \quad u = 0 \text{ on } \partial\Omega, \tag{6.9}$$

defined on a bounded domain $\Omega \subset \mathbb{R}^2$. The nonlinearity $s(u)$ is assumed to have at most polynomial growth. Then, the corresponding semilinear form

$$A(u; \varphi) := (\nabla u, \nabla\varphi) + (s(u), \varphi)$$

is considered on the function space $V := H_0^1(\Omega)$. Its derivatives are

$$A'(u; \psi, \varphi) = (\nabla\psi, \nabla\varphi) + (s'(u)\psi, \varphi)$$
$$A''(u; \xi, \psi, \varphi) = (s''(u)\xi\psi, \varphi).$$

The variational formulation of (6.9) seeks $u \in V$ such that

$$A(u; \varphi) = (f, \varphi), \quad \text{for all } \varphi \in V. \tag{6.10}$$

The Galerkin approximation of (6.9) uses finite element spaces $V_h \subset V$ as defined in Section 3. The discrete solutions $u_h \in V_h$ are determined by

$$A(u_h; \varphi_h) = (f, \varphi_h), \quad \text{for all } \varphi_h \in V_h. \tag{6.11}$$

In this case the general a posteriori error representation (2.27) results in the error estimate

$$|J(u) - J(u_h)| \leq \sum_{K \in \mathbb{T}_h} \rho_K\, \omega_K + |R(\overline{u\hat{u}_h u_h}; \hat{e}, e, z)|, \tag{6.12}$$

where the weights ω_K have the same form as in Proposition 3.1 and the residual terms ρ_K are defined by

$$\rho_K := \|R(u_h)\|_K + \|r(u_h)\|_{\partial K}$$

with $R(u_h)_{|K} := f + \Delta u_h - s(u_h)$, and $r(u_h)$ as defined above. The remainder term can be bounded by

$$|R(\overline{u\hat{u}_h u_h}; \hat{e}, e, z)| \leq \tfrac{1}{2}\max_{v \in \overline{u\hat{u}_h u_h}} \|s''(v)\|_{L^\infty} \|\hat{e}ez\|_{L^1}. \tag{6.13}$$

Numerical test (from Vexler (2000))

To illustrate the adaptive algorithm described above, we consider the non-linear model problem

$$-\Delta(u-\hat{u}) + \lambda\big((u-\hat{u})^3 - (u-\hat{u})\big) = 0, \quad \text{in } \Omega, \quad u = 0, \quad \text{on } \partial\Omega, \qquad (6.14)$$

on the two-dimensional domain $\Omega := (0,1)^2$, with parameter $\lambda \geq 0$. The equilibrium solution is given by $\hat{u}(x_1, x_2) = 256x_1(1-x_1)x_2(1-x_2)$. Bifurcation from $\hat{u}$ occurs at the first critical value $\lambda_1^* = 2\pi^2 = 19.7392\ldots$. We want to compute the function value $u(0.75, 0.75)$ with a reliable error bound, for the only slightly subcritical value $\lambda = 19$. We will see that by the above algorithm on a fixed mesh the effectivity index of the approximate error estimator can actually be driven to $I_{\text{eff}} = 1$. However, in the present almost singular situation, this requires an unacceptable amount of additional computational work on finer meshes. Therefore, we restrict our goal to achieving an effectivity of reasonable size $I_{\text{eff}} \leq 2$. The stopping parameters are chosen as $\varepsilon = 0.6$, $\varepsilon_0 = 0.06$. For this example, the described algorithm for controlling the linearization and discretization in the adjoint problem seems to work quite well. It achieves acceptable effectivity and the stopping criterion for the inner loops limits the amount of additional work.

Table 6.1. Control of linearization: numbers of cells N of the primary mesh $\mathbb{T}_i$ and $\hat{N}$ of the secondary mesh $\hat{\mathbb{T}}_i$

N	9	14	41	119	393	1367	5233	20338
$\hat{N}$	322	2256	1670	2247	2184	1367	5233	20338
I_{eff}	0.585	1.354	1.493	1.554	1.577	1.798	1.421	1.369

7. Realization for time-dependent problems

In the preceding sections, we have discussed the DWR method and some of its practical aspects for simple stationary diffusion and transport model problems. Now we want to extend this concept of adaptivity to nonstationary problems including parabolic as well as hyperbolic cases. We will see that in all these situations the general pattern of the DWR method remains the same: *variational formulation*; *error representation via duality argument*; *Galerkin orthogonality*; *approximate computation of adjoint solution*. This also applies to time-dependent problems provided that the discretization is organized as a Galerkin method in space and time. The following examples are used to discuss some aspects of the DWR method which are specific to the different types of differential equations. For simplicity, we restrict the analysis here to *linear* problems. The treatment of *nonlinear time-dependent* problems by the DWR method is the subject of ongoing research.

7.1. Parabolic model problem: heat equation

We consider the heat-conduction problem

$$
\begin{aligned}
\partial_t u - \nabla \cdot (a \nabla u) &= f && \text{in } Q_T, \\
u_{|t=0} &= u^0 && \text{on } \Omega, \\
u_{|\partial \Omega} &= 0 && \text{on } I,
\end{aligned}
\tag{7.1}
$$

on a space-time region $Q_T := \Omega \times I$, where $\Omega \subset \mathbb{R}^d$, $d \geq 1$, and $I = (0, T)$; the coefficient a, the source term f and the initial value u^0 are assumed to be smooth. This model is used to describe diffusive transport of energy or certain species concentrations. The corresponding strong solution is unique in the space

$$
V := H^1(0, T; L^2(\Omega)) \cap L^2(0, T; H_0^1(\Omega)).
$$

The discretization of problem (7.1) is by a Galerkin finite element discretization simultaneously in space and time. We split the time interval $(0, T)$ into subintervals $I_n = (t_{n-1}, t_n]$, where

$$
0 = t_0 < \cdots < t_n < \cdots < t_N = T, \qquad k_n := t_n - t_{n-1}.
$$

At each time level t_n, let $\mathbb{T}_h^n$ be a regular finite element mesh as defined above with local mesh width $h_K = \operatorname{diam}(K)$, and let $V_h^n \subset H_0^1(\Omega)$ be the corresponding finite element subspace with d-linear shape functions. Extending the spatial mesh to the corresponding space-time slab $\Omega \times I_n$, we obtain a global space-time mesh consisting of $(d+1)$-dimensional cubes $Q_K^n := K \times I_n$. On this mesh, we define the global finite element space

$$
\begin{aligned}
V_{h,k} := \big\{ v \in L^\infty(0, T; H_0^1(\Omega)) : \\
v(\cdot, t)_{|Q_K^n} \in \tilde{Q}_1(K),\ v(x, \cdot)_{|Q_K^n} \in P_r(I_n) \text{ for all } Q_K^n \big\},
\end{aligned}
$$

with some $r \geq 0$. For this setting, we introduce the space $\hat{V} := V \oplus V_{h,k}$. For functions from this space, we use the notation

$$
v^{n+} := \lim_{t \to t_n+0} v(t), \quad v^{n-} := \lim_{t \to t_n-0} v(t), \quad [v]^n := v^{n+} - v^{n-}.
$$

The discretization of problem (7.1) is based on a variational formulation which allows the use of discontinuous functions in time. This method, called the 'dG(r) method' (*discontinuous* Galerkin method in time), determines approximations $u_h \in V_{h,k}$ by

$$
A_k(u_h, \varphi_h) = F(\varphi_h), \quad \text{for all } \varphi_h \in V_{h,k},
\tag{7.2}
$$

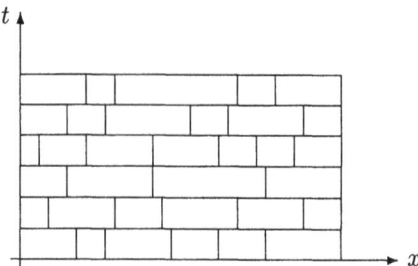

Fig. 7.1. Sketch of a space-time mesh with hanging nodes

with the bilinear form

$$A_k(v, \varphi) := \sum_{n=1}^{N} \int_{I_n} \left\{ (\partial_t v, \varphi) + (a\nabla v, \nabla \varphi) \right\} dt$$

$$+ \sum_{n=2}^{N} ([v]^{n-1}, \varphi^{(n-1)+}) + (v^{0+}, \varphi^{0+}),$$

and the linear functional

$$F(\varphi) := (f, \varphi) + (u^0, \varphi^{0+}),$$

which are both well defined on $\hat{V}$. We note that the continuous solution u also satisfies equation (7.2), which again implies Galerkin orthogonality for the error $e := u - u_h \in \hat{V}$ with respect to the bilinear form $A(\cdot, \cdot)$. Since the test functions $\varphi_h \in V_{h,k}$ may be discontinuous at times t_n, the global system (7.2) decouples and can be written in the form of a time-stepping scheme,

$$\int_{I_n} \left\{ (\partial_t u_h, \varphi_h) + (a\nabla u_h, \nabla \varphi_h) \right\} dt + ([u_h]^{n-1}, \varphi_h^{(n-1)+}) = \int_{I_n} (f, \varphi_h) \, dt,$$

for all $\varphi_h \in V_h^n$, $n = 1, \ldots, N$. In the following, we consider for simplicity only the lowest-order case $r = 0$, the 'dG(0) method', which is equivalent to a variant of the backward Euler scheme. We concentrate on control of the spatial L^2-error $\|e^{N-}\|$ at the end-time $T = t_N$. To this end, we use the adjoint problem

$$\partial_t z - a\Delta z = 0 \quad \text{in } \Omega \times I,$$
$$z_{|t=T} = \|e^{N-}\|^{-1} e^{N-} \quad \text{in } \Omega, \quad z_{|\partial\Omega} = 0 \quad \text{on } I, \tag{7.3}$$

the strong solution of which automatically satisfies

$$A_k(\varphi, z) = J(\varphi) := \|e^{N-}\|^{-1} (\varphi^{N-}, e^{N-}), \quad \text{for all } \varphi \in \hat{V}.$$

Notice that, in this special situation, the adjoint solution z does not depend on the parameter k. Then, from the general results of Proposition 2.6, we obtain the error representation

$$J(e) = F(z - \varphi_h) - A_k(u_h, z - \varphi_h), \tag{7.4}$$

for any $\varphi_h \in V_{h,k}$. By elementary reordering of terms in (7.4), we infer the estimate

$$\|e^{N-}\| \le \sum_{n=1}^{N} \sum_{K \in \mathbb{T}_h^n} \left| (R(u_h), z - \varphi_h)_{K \times I_n} - (r(u_h), z - \varphi_h)_{\partial K \times I_n} \right.$$
$$\left. - ([u_h]^{n-1}, (z - \varphi_h)^{(n-1)+})_K \right|, \tag{7.5}$$

with the cell and edge residuals

$$R(u_h)|_{K \times I_m} := f + \nabla \cdot (a \nabla u_h) - \partial_t u_h,$$

$$r(u_h)|_{\Gamma \times I_m} := \begin{cases} \frac{1}{2} n \cdot [\nabla u_h], & \text{if } \Gamma \subset \partial K \setminus \partial \Omega, \\ 0, & \text{if } \Gamma \subset \partial \Omega, \end{cases}$$

and an arbitrary approximation $\varphi_h \in V_{h,k}$. From this, we infer the following result.

Proposition 7.1. (Hartmann 1998) For the dG(0) finite element method applied to the heat equation (7.1), we have the *a posteriori* error estimate

$$\|e^{N+}\| \le \eta_\omega(u_h) := \sum_{n=1}^{N} \sum_{K \in \mathbb{T}_h^n} \{\rho_K^{n,1} \omega_K^{n,1} + \rho_K^{n,2} \omega_K^{n,2}\}, \tag{7.6}$$

where the cell-wise residuals and weights are defined by

$$\rho_K^{n,1} := \|R(u_h)\|_{K \times I_n} + h_K^{-1/2} \|r(u_h)\|_{\partial K \times I_n},$$
$$\omega_K^{n,1} := \|z - \varphi_h\|_{K \times I_n} + h_K^{1/2} \|z - \varphi_h\|_{\partial K \times I_n},$$
$$\rho_K^{n,2} := k_n^{-1/2} \|[u_h]^{n-1}\|_K, \qquad \omega_K^{n,2} := k_n^{1/2} \|(z - \varphi_h)^{(n-1)+}\|_K,$$

with a suitable approximation $\varphi_h \in V_{h,k}$ to the adjoint solution z.

Remark 7.1.   The statement of Proposition 7.1 extends to higher-order dG(r) methods and also to the related cG(r) methods (*continuous* Galerkin methods). In this case the flexibility in choosing the approximation φ_h in (7.5) can be used to replace the residual terms by

$$\rho_K^{n,1} := \|R(u_h) - \overline{R(u_h)}\|_{K \times I_n} + h_K^{-1/2} \|r(u_h) - \overline{r(u_h)}\|_{\partial K \times I_n},$$

for certain time-averages indicated by over-lines.

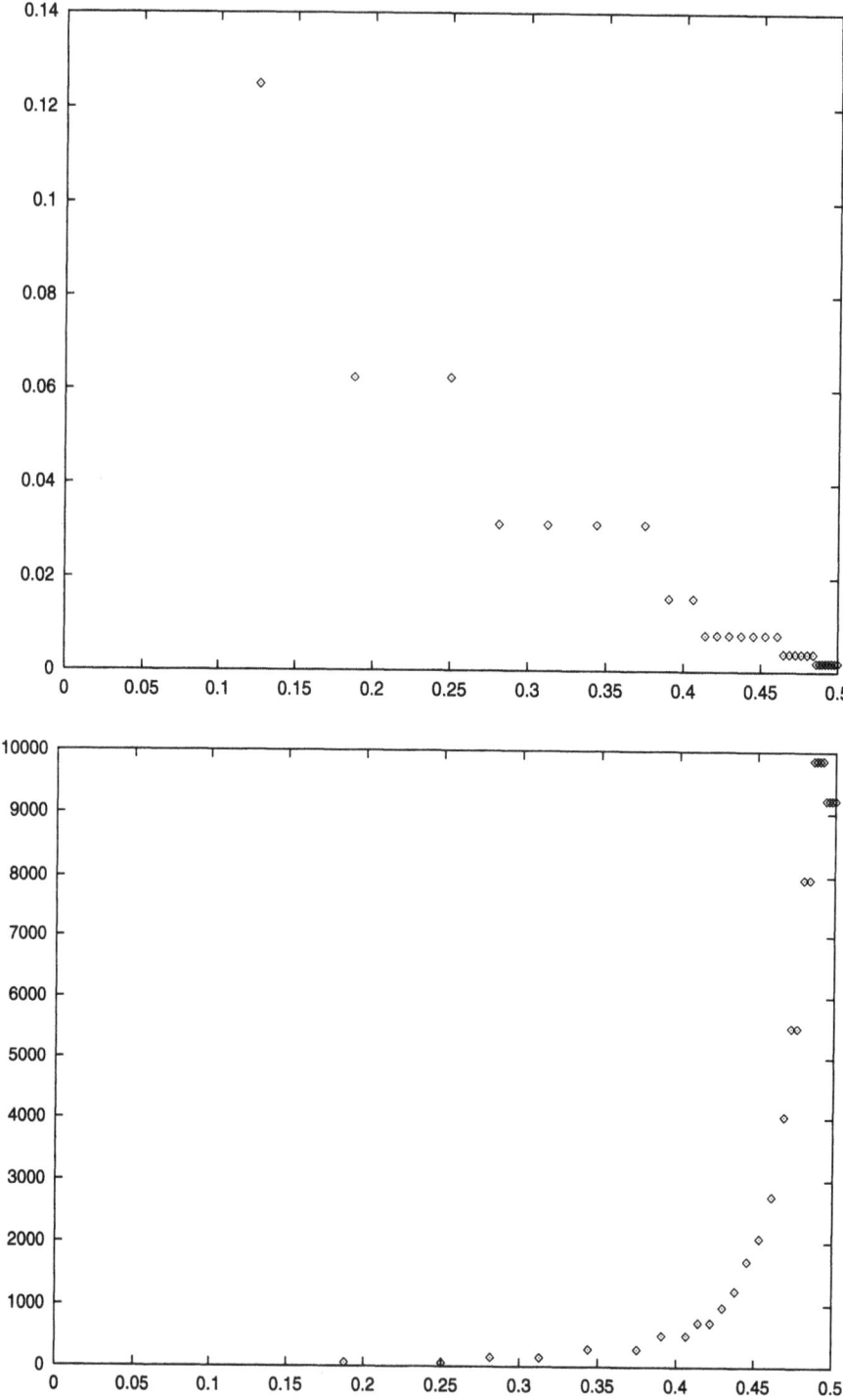

Fig. 7.2. Development of the time-step size (above) and the number N of mesh cells (below) by the DWR method over the time interval $I = [0, 0.5]$

Numerical test (from Hartmann (1998))

The performance of the error estimator (7.6) is illustrated by a simple test in two space dimensions where the (known) exact solution represents a smooth rotating bump with a suitably adjusted force f on the unit square. The computation is carried over the time interval $0 \leq t \leq T = 0.5$. In Figure 7.2, we show the development of the time-step size k and the spatial mesh complexity N, while Table 7.1 contains the corresponding results. Figure 7.3 shows a sequence of adapted meshes at successive times obtained by controlling the spatial L^2 error at the end-time $t_N = 0.5$. We clearly see the localizing effect of the weights in the error estimator, which suppress the influence of the residuals during the initial period.

Table 7.1. Results by simultaneous adaptation of spatial and time discretization by the DWR method ($M = \#$ time-steps, $N = \#$ mesh-cells)

M	N	$J(e)$	$\eta_\omega(u_h)$	I_{eff}
8	256	4.19e-2	3.04e-1	7.27
21	256	1.20e-2	7.14e-2	5.59
46	760	2.91e-3	6.60e-3	2.27
81	3472	7.92e-4	1.08e-3	1.36
119	9919	3.99e-4	6.64e-4	1.67

Notes and references

The traditional approach to *a posteriori* error estimation for parabolic problems combines 'energy norm' error estimates for spatial discretization (like the ones described in Section 4) with heuristic truncation error estimation (see, *e.g.*, Ziukas and Wiberg (1998) and Lang (1999)). The resulting error *indicators*, on principle, cannot reflect the true behaviour of the global error in time since the possible time-accumulation of the local truncation errors is not taken into account. Space-time duality arguments for error estimation in dG-finite element approximation of parabolic problems have been intensively used in a sequence of papers by Eriksson, Johnson and Larsson (Eriksson and Johnson 1991, 1995a, 1995b, 1995c, Eriksson *et al.* 1998) and by Estep and Larsson (1993). The same approach can also be used in the context of ordinary differential equations (see Estep (1995), Estep and French (1994) and Böttcher and Rannacher (1996)).

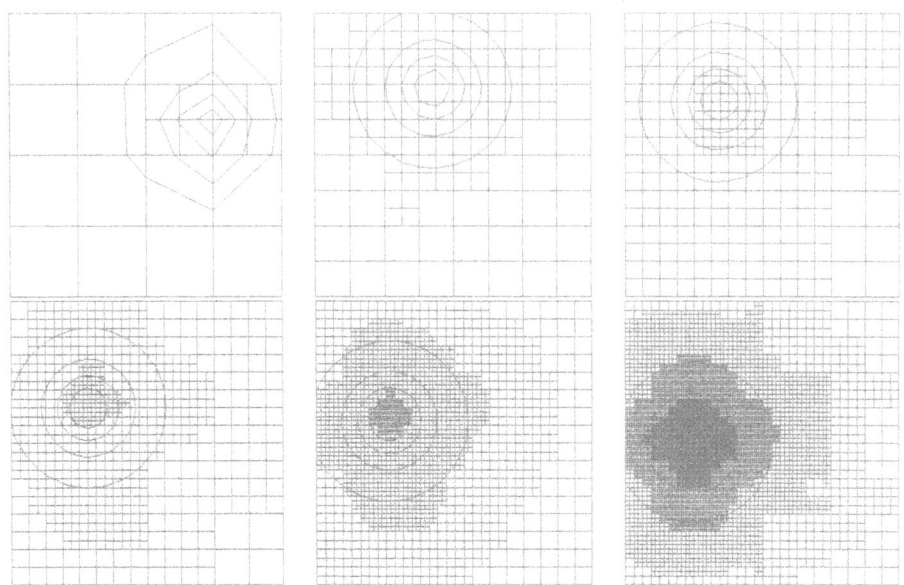

Fig. 7.3. Sequence of refined meshes for controlling the end-time
error $\|e^{N-}\|$ shown at six consecutive times $t_n = 0.125, \ldots, 0.5$

7.2. Hyperbolic model problem: acoustic wave equation

We consider the acoustic wave equation

$$
\begin{aligned}
\partial_t^2 w - \nabla\cdot\{a\nabla w\} &= 0 &&\text{in } Q_T, \\
w_{|t=0} = w^0, \quad \partial_t w_{|t=0} &= v^0 &&\text{on } \Omega, \\
n\cdot a\nabla w_{|\partial\Omega} &= 0 &&\text{on } I,
\end{aligned}
\tag{7.7}
$$

on a space-time region $Q_T := \Omega \times I$, where $\Omega \subset \mathbb{R}^d$, $d \geq 1$, and $I = (0,T)$; the elastic coefficient a may vary in space. This equation frequently occurs in the simulation of acoustic waves in gaseous or fluid media, seismics, and electrodynamics. We approximate problem (7.7) by a 'velocity-displacement' formulation which is obtained by introducing a new velocity variable $v := \partial_t w$. Then, the pair $u = \{w, v\}$ satisfies the system of equations

$$
\partial_t w - v = 0, \quad \partial_t v - \nabla\cdot\{a\nabla w\} = 0,
\tag{7.8}
$$

with the natural solution space

$$
V := [H^1(0,T; L^2(\Omega)) \cap L^2(0,T; H^1(\Omega))] \times H^1(0,T; L^2(\Omega)).
$$

We derive a Galerkin discretization in space-time of (7.7) based on the formulation (7.8). To this end, the time interval $I = (0,T)$ is again decomposed into subintervals $I_n = (t_{n-1}, t_n]$ with length $k_n = t_n - t_{n-1}$

and at each time level t_n a quadrilateral mesh $\mathbb{T}_h^n$ on Ω is chosen. On each time slab $Q^n := \Omega \times I_n$, we define intermediate meshes $\bar{\mathbb{T}}_h^n$ which are composed of the mutually finest cells of the neighbouring meshes $\mathbb{T}_h^{n-1}$ and $\mathbb{T}_h^n$, and obtain a decomposition of the time slab into space-time cubes $Q_K^n = K \times I_n, K \in \bar{\mathbb{T}}_h^n$. This construction is used in order to allow continuity in time of the trial functions when the meshes change with time. The discrete 'trial spaces' $V_{h,k}$ in space-time domain consist of functions that are $(d+1)$-linear on each space-time cell Q_K^n and globally *continuous* on Q_T. This prescription requires the use of 'hanging nodes' if the spatial mesh changes across a time level t_n. The corresponding discrete 'test spaces' $W_{h,k}$ consist of functions that are constant in time on each cell Q_K^n, while they are d-linear in space and globally continuous on Ω. On these spaces, we introduce the bilinear form

$$A(u, \varphi) := (\partial_t w, \xi)_{Q_T} - (v, \xi)_{Q_T} + (w(0), \xi(0))$$
$$+ (\partial_t v, \psi)_{Q_T} + (a\nabla w, \nabla\psi)_{Q_T} + (v(0), \psi(0)),$$

and the linear functional

$$F(\varphi) = (w^0, \xi(0)) + (v^0, \psi(0)).$$

The Galerkin approximation of (7.7) seeks $u_h = \{w_h, v_h\} \in V_{h,k}$ satisfying

$$A(u_h, \varphi_h) = F(\varphi_h), \quad \text{for all } \varphi_h = \{\psi_h, \xi_h\} \in W_{h,k}. \tag{7.9}$$

For more details, we refer to Bangerth (1998) and Bangerth and Rannacher (1999). The scheme (7.9) is a 'Petrov–Galerkin' method. Since the solution $u = \{w, v\}$ also satisfies (7.9), we again have Galerkin orthogonality for the error $e := \{e^w, e^v\}$:

$$A(e, \varphi_h) = 0, \quad \varphi_h \in V_{h,k}. \tag{7.10}$$

This time-discretization scheme is called the 'cG(1) method' (*continuous* Galerkin method) in contrast to the dG method used in the preceding section. We note that, from this scheme, we can recover the standard Crank–Nicolson scheme in time (combined with a spatial finite element method):

$$(w^n - w^{n-1}, \varphi) - \tfrac{1}{2} k_n (v^n + v^{n-1}, \varphi) = 0,$$
$$(v^n - v^{n-1}, \psi) + \tfrac{1}{2} k_n (a\nabla(w^n + w^{n-1}), \nabla\psi) = 0. \tag{7.11}$$

The system (7.11) splits into two equations, a discrete Helmholtz equation and a discrete L^2-projection.

In order to embed the present situation into the general framework laid out in Section 2, we introduce the spaces

$$\hat{V} := V \oplus V_{h,k}, \quad \hat{W} := V \oplus W_{h,k}.$$

We want to control the error in terms of a functional of the form

$$J(e) := (j, e^w)_{Q_T},$$

with some density function $j(x, t)$. To this end, we again use a duality argument in space-time employing the time-reversed wave equation

$$\partial_t^2 z^w - \nabla \cdot \{a \nabla z^w\} = j \quad \text{in } Q_T,$$
$$z^w|_{t=T} = 0, \quad -\partial_t z^w|_{t=T} = 0 \quad \text{on } \Omega, \tag{7.12}$$
$$n \cdot a \nabla z^w|_{\partial \Omega} = 0 \quad \text{on } I.$$

Its strong solution $z = \{-\partial_t z^w, z^w\} \in \hat{W}$ satisfies the variational equation

$$A(\varphi, z) = J(\varphi), \quad \text{for all } \varphi \in \hat{V}. \tag{7.13}$$

Then, from the general results of Proposition 2.3, we obtain the error identity

$$J(e) = F(z - \varphi_h) - A(w_h, z - \varphi_h), \tag{7.14}$$

for arbitrary $\varphi_h = \{\varphi_h^w, \varphi_h^v\} \in W_{h,k}$. Recalling the definition of the bilinear form $A(\cdot, \cdot)$, we obtain

$$|(j, w)_{Q_T}| \leq \sum_{n=1}^{N} \sum_{K \in \mathbb{T}_h^n} \left| (R^u(u_h), \partial_t z^w - \varphi_h^v)_{K \times I_n} \right.$$
$$\left. - (R^v(u_h), z^w - \varphi_h^w)_{K \times I_n} - (r(u_h), z^w - \varphi_h^w)_{\partial K \times I_n} \right|,$$

with the cell residuals

$$R^w(u_h)_{|K} := \partial_t w_h - v_h, \qquad R^v(u_h)_{|K} := \partial_t v_h - \nabla \cdot \{a \nabla w_h\},$$

and the edge residuals

$$r(w_h)_{|\Gamma \times I_m} := \begin{cases} \frac{1}{2} n \cdot [\nabla w_h], & \text{if } \Gamma \subset \partial K \backslash \partial \Omega, \\ 0, & \text{if } \Gamma \subset \partial \Omega. \end{cases}$$

From this, we infer the following result.

Proposition 7.2. (Bangerth 1998, Bangerth and Rannacher 1999)
For the cG(1) finite element method applied to the acoustic wave equation (7.7), we have the *a posteriori* error estimate

$$|J(e^N)| \leq \eta_\omega(u_h) := \sum_{n=1}^{N} \sum_{K \in \mathbb{T}_h^n} \{\rho_K^{n,1} \omega_K^{n,1} + \rho_K^{n,2} \omega_K^{n,2}\}, \tag{7.15}$$

where the cell-wise residuals and weights are defined by

$$\rho_K^{n,1} := \|R_1(u_h)\|_{K \times I_n} + h_K^{-1/2}\|r(u_h)\|_{\partial K \times I_n},$$

$$\omega_K^{n,1} := \|\partial_t z^w - \varphi_h^v\|_{K \times I_n} + h_K^{1/2}\|z^w - \varphi_h^w\|_{\partial K \times I_n},$$

$$\rho_K^{n,2} := \|R_2(u_h)\|_{K \times I_n}, \qquad \omega_K^{n,2} := \|z^w - \varphi_h^w\|_{K \times I_n},$$

with a suitable approximation $\{\varphi_h^w, \varphi_h^v\} \in V_{h,k}$ to the adjoint solution $\{z^w, z^v\}$.

Below, we will compare the error estimator (7.15) with a simple heuristic 'energy error' indicator measuring the spatial smoothness of the computed solution w_h:

$$\eta_E(u_h) := \left(\sum_{n=1}^{N} \sum_{K \in \mathbb{T}_h^n} \rho_K^{n,3}(u_h)^2 \right)^{1/2}. \tag{7.16}$$

Numerical test (from Bangerth (1998))
The error estimator (7.15) is illustrated by a simple test: the propagation of an outward travelling wave on $\Omega = (-1,1)^2$ with a strongly heterogeneous coefficient. Layout of the domain and structure of the coefficient are shown in Figure 7.4. Boundary and initial conditions were chosen as follows:

$$n \cdot \{a\nabla u\} = 0 \quad \text{on } y = 1, \qquad w = 0 \quad \text{on } \partial\Omega \backslash \{y = 1\},$$

$$w_0 = 0, \quad v_0 = \theta(s - r) \exp(-|x|^2/s^2)(1 - |x|^2/s^2),$$

with $s = 0.02$ and $\theta(\cdot)$ the jump function. The region of origin of the wave field is significantly smaller than shown in Figure 7.4. Notice that the lowest

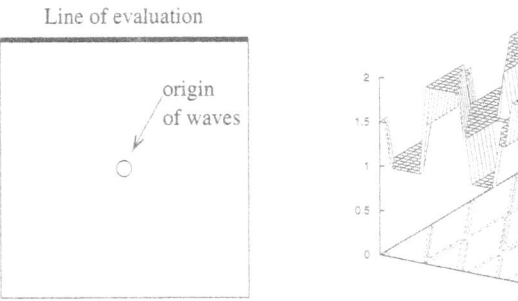

Fig. 7.4. Layout of the domain (left) and structure of the coefficient $a(x)$ (right)

Table 7.2. Results obtained by adaptation of spatial
discretization using the DWR method (reference value
$J(w) \approx$ -4.515e-6, M =# time-steps, N =# mesh-cells)

Weighted estimator		Heuristic indicator	
$N \times M$	$J(w_h)$	$N \times M$	$J(w_h)$
327,789	-2.085e-6	327,789	-2.085e-6
920,380	-4.630e-6	920,380	-4.630e-6
2,403,759	-4.286e-6	2,403,759	-4.286e-6
1,918,696	-4.177e-6	5,640,223	-4.385e-6
2,975,119	-4.438e-6	10,189,837	-4.463e-6
6,203,497	-4.524e-6	17,912,981	-4.521e-6
		41,991,779	-4.517e-6

frequency in this initial wave field has wavelength $\lambda = 4s$; hence taking the
common minimum ten grid points per wavelength would yield $62,500$ cells
for the largest wavelength. Uniform grids quickly get to their limits in such
cases. If we consider this example as a model of propagation of seismic
waves in a faulted region of rock, then we would be interested in recording
seismograms at the surface, here chosen as the top line Γ of the domain. A
corresponding functional output is

$$J(w) = \int_0^T \int_\Gamma w(x,t) \, \omega(\xi,t) \, d\xi \, dt,$$

with a weight factor $\omega(\xi,t) = \sin(3\pi\xi) \sin(5\pi t/T)$, and end-time $T = 2$.
The frequency of oscillation of this weight is chosen to match the frequencies
in the wave field to obtain good resolution of changes.

In Figure 7.5, we show the grids resulting from refinement by the weighted
error estimator (7.15) compared with the energy error indicator (7.16). Both
resolve the wave field quite well, including reflections from discontinuities in
the coefficient. The first additionally takes into account that the lower parts
of the domain lie outside the domain of influence of the target functional if we
truncate the time domain at $T = 2$; this domain of influence constricts to the
top as we approach the final time, as is reflected by the produced grids. The
meshes obtained in this way are obviously much more economical, without
degrading the accuracy in approximating the quantity of interest (for more
examples see Bangerth and Rannacher (1999)).

Remark 7.2. The evaluation of the *a posteriori* error estimate (7.15) requires a careful approximation of the adjoint solution z. Therefore we have
used a higher-order method (bi-quadratic elements) for solving the space-

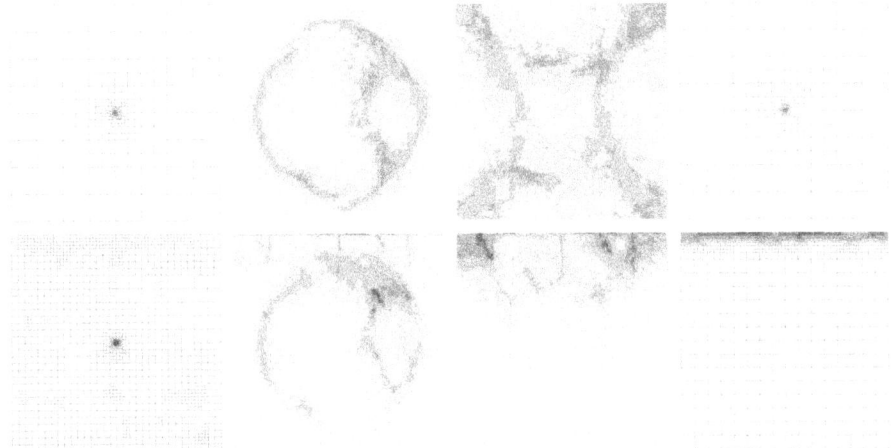

Fig. 7.5. Grids produced by the energy error indicator (top row) and by the dual-weighted estimator (bottom row) at times $t = 0, \frac{2}{3}, \frac{4}{3}, 2$

time adjoint problem, though this does not seem feasible for complex higher-dimensional problems.

Notes and references

The *a priori* error analysis of dG methods for the wave equation using space-time duality arguments has been initiated by Johnson (1993). Results obtained by heuristic ZZ-type indicators are discussed, for instance in Ziukas and Wiberg (1998). The extension of the adaptive cG method to the *elastic wave equation* is described by Bangerth (2000).

8. Application to incompressible viscous flow

In this section, we present an application of the DWR method to computing incompressible fluid flow governed by the classical Navier–Stokes equations. The purpose is to demonstrate that, by solving the global adjoint problem, one can actually capture the complex interaction of all physical mechanisms such that meshes are generated that are significantly more economical than those obtained by simple *ad hoc* adaptation. In this application (and in that presented in Section 9, below), we reach a point at which the concept of the DWR method laid out in Section 2 has to be taken only as a formal guideline for deriving meaningful *a posteriori* error estimates. A mathematically fully rigorous argument is not possible because of the complexity of the problem and the discretization procedure.

8.1. The incompressible Navier–Stokes equations

We consider viscous incompressible Newtonian fluid flow modelled by the Navier–Stokes equations

$$-\nu\Delta v + v\cdot\nabla v + \nabla p = f, \qquad \nabla\cdot v = 0, \tag{8.1}$$

for the velocity v and the pressure p in a bounded domain $\Omega \subset \mathbb{R}^2$. Here, $\nu > 0$ is the normalized viscosity (density $\rho \equiv 1$), and the volume force is assumed as $f \equiv 0$. At the boundary $\partial\Omega$, the usual no-slip condition is imposed along rigid walls together with suitable inflow and 'free-stream' outflow conditions,

$$v|_{\Gamma_{\text{rigid}}} = 0, \quad v|_{\Gamma_{\text{in}}} = \hat{v}, \quad \nu\partial_n v - pn|_{\Gamma_{\text{out}}} = 0.$$

As a concrete example, below, we will consider flow around a cylinder in a channel.

The variational formulation of (8.1) uses the function spaces

$$\hat{V} := L\times\hat{H}, \quad V := L\times H \subset \hat{V},$$

where $L := L^2(\Omega)$, $\hat{H} := H^1(\Omega)^2$, and $H := \{v \in H^1(\Omega)^2 : v|_{\Gamma_{\text{in}}\cup\Gamma_{\text{rigid}}} = 0\}$. For pairs $u = \{p, v\}$, $\varphi = \{q, w\} \in \hat{V}$, we define the semilinear form

$$A(u; \varphi) := \nu(\nabla v, \nabla w) + (v\cdot\nabla v, w) - (p, \nabla\cdot w) + (q, \nabla\cdot v)$$

and seek $u = \{p, v\} \in V + \{0, \hat{v}\}$, such that

$$A(u; \varphi) = (f, w), \quad \text{for all } \varphi = \{q, w\} \in V. \tag{8.2}$$

We assume that this problem possesses a (locally) unique solution that is stable, that is, the Fréchet derivative $A'(u; \cdot, \cdot)$ is coercive in the strong sense,

$$A'(u; \varphi, \varphi) \geq \gamma\|\varphi\|^2, \quad \varphi \in V.$$

For discretizing this problem, we use a finite element method based on the quadrilateral Q_1/Q_1-Stokes element with globally continuous (piecewise isoparametric) bilinear shape functions for both unknowns, pressure and velocity. As described before, we allow 'hanging' nodes while the corresponding unknowns are eliminated by linear interpolation. The corresponding finite element subspaces are denoted by

$$L_h \subset L, \quad \hat{H}_h \subset \hat{H}, \quad H_h \subset H, \quad \hat{V}_h := L_h\times\hat{H}_h, \quad V_h := L_h\times H_h,$$

and $\hat{v}_h \in \hat{H}_h$ is a suitable interpolation of the boundary function $\hat{v}$. This construction is oriented by the situation of a polygonal domain Ω for which the boundary $\partial\Omega$ is exactly matched by the mesh domain $\Omega_h := \cup\{K \in \mathbb{T}_h\}$. In the case of a general curved boundary (as in the flow example, below) some standard modifications are necessary, the description of which is omitted here.

In order to obtain a stable discretization of (8.2) in these spaces with 'equal-order interpolation' of pressure and velocity, we use the least-squares technique proposed by Hughes, Franca and Balestra (1986). Following Hughes and Brooks (1982), a similar approach is employed for stabilizing the convective term. The Navier–Stokes system can be written in vector form for the unknown $u = \{p, v\} \in \hat{V}$ as

$$A(u) := \begin{bmatrix} -\nu\Delta v + v\cdot\nabla v + \nabla p \\ \nabla\cdot v \end{bmatrix} = \begin{bmatrix} f \\ 0 \end{bmatrix} =: F.$$

Then, the strong solution $u = \{p, v\}$ of (8.1) satisfies $A(u) = F$ in the L^2-sense. With the operator $A(\cdot)$, we associate a derivative $A'(u)$ at u and an approximation $S(u)$ which act on $\varphi = \{q, w\} \in \hat{V}$ via

$$A'(u)\varphi := \begin{bmatrix} -\nu\Delta w + v\cdot\nabla w + w\cdot\nabla v + \nabla q \\ \nabla\cdot w \end{bmatrix}, \qquad S(u)\varphi := \begin{bmatrix} v\cdot\nabla w + \nabla q \\ 0 \end{bmatrix}.$$

With this notation, we introduce the stabilized form

$$A_h(u; \varphi) := A(u; \varphi) + (A(u) - F, S(u)\varphi)_h,$$

with the mesh-dependent inner product and norm

$$(v, w)_h := \sum_{K \in \mathbb{T}_h} \delta_K (v, w)_K, \qquad \|v\|_h = (v, v)_h^{1/2}.$$

The stabilization parameter is chosen according to

$$\delta_K = \alpha \big(\nu h_K^{-2}, \beta |v_h|_{K;\infty} h_K^{-1} \big)^{-1}, \qquad \delta := \max_{K \in \mathbb{T}_h} \delta_K, \tag{8.3}$$

with the heuristic choice $\alpha = \frac{1}{12}$, $\beta = \frac{1}{6}$. Now, in the discrete problems, we seek $\{p_h, v_h\} \in V_h + \hat{u}_h$, such that

$$A_h(u_h; \varphi_h) = F(\varphi_h), \quad \text{for all } \varphi_h \in V_h. \tag{8.4}$$

This approximation is fully consistent in the sense that the solution $u = \{p, v\}$ also satisfies (8.4). This again implies Galerkin orthogonality for the error $e = \{e^p, e^v\} := \{p - p_h, v - v_h\}$ with respect to the form $A_h(\cdot; \cdot)$:

$$A_h(u; \varphi_h) - A_h(u_h; \varphi_h) = 0, \quad \varphi_h \in V_h. \tag{8.5}$$

We note that, instead of S, one can also use $-S^*$ in the stabilized form $A_h(\cdot; \cdot)$ (see the compressible flow example below).

We now turn to the question of a posteriori error estimation in the scheme (8.4). Let the goal of the computation again be the evaluation of a quantity $J(u)$ expressed in terms of a linear functional $J(\cdot)$ on $\hat{V}$ (for simplicity). Here, we think of local quantities such as point values of pressure, drag and lift coefficients or averages of vorticity localized to certain recirculation zones. In order to control the error $J(u) - J(u_h)$, we consider the h-dependent adjoint problem

$$\tilde{A}'_h(u; \varphi, z) := A'(u; \varphi, z) + (S(u)\varphi, S(u)z)_h = J(\varphi), \quad \text{for all } \varphi \in \hat{V}, \quad (8.6)$$

with the approximate derivative $S(u)$ defined above. Since the stabilizing bilinear form $(S(u)\cdot, S(u)\cdot)_h$ is coercive, the strong coercivity of $A'(u; \cdot, \cdot)$ also implies the (unique) solvability of the adjoint problem (8.6). With this notation, we have the following result.

Proposition 8.1. For a (linear) functional $J(\cdot)$ let $z = \{z^p, z^v\} \in V$ be the solution of the linearized adjoint problem (8.6). Then we have the a *posteriori* error estimate

$$|J(e)| = \eta_\omega(u_h) := \sum_{K \in \mathbb{T}_h} \left\{ \sum_{\alpha \in \{p,v\}} \rho_K^\alpha \omega_K^\alpha + \cdots \right\} + R_h, \quad (8.7)$$

where the residual terms and weights are given by

$$\rho_K^p := \|R^p(u_h)\|_K, \quad \omega_K^p := \|z^p - \varphi_h^p\|_K,$$

$$\rho_K^v := \|R^v(u_h)\|_K + h_K^{-1/2}\|r^v(u_h)\|_{\partial K},$$

$$\omega_K^v := \|z^v - \varphi_h^v\|_K + h_K^{1/2}\|z^v - \varphi_h^v\|_{\partial K} + \delta_K \|v_h \cdot \nabla(z^v - \varphi_h^v) + \nabla(z^p - \varphi_h^p)\|_K,$$

with suitable approximations $\{\varphi_h^p, \varphi_h^v\} \in V_h$ to $\{z^p, z^v\}$. The dots '...' in (8.7) stand for additional terms measuring the errors in approximating the inflow data and the curved cylinder boundary. The remainder term can be bounded by

$$|R_h| \leq \|e^v\| \|\nabla e^v\| \|z^v\|_\infty + \delta\, C(u, e, z), \quad (8.8)$$

with a function $C(u, e, z)$ linear in e. For simplicity, the explicit dependence of the stabilization parameter δ on the solution is neglected.

Proof. We apply the abstract result of Proposition 2.6. Neglecting the errors due to the approximation of the inflow boundary data $\hat{v}$ as well as the curved cylinder boundary S, we have to evaluate the residual $\rho(u_h; \varphi) := F(\varphi) - A_h(u_h, \varphi)$ for $\varphi = \{q, w\} \in V$. By definition,

$$\rho(u_h; \varphi) = (f, w) - \nu(\nabla v_h, \nabla w) - (v_h \cdot \nabla v_h, w) + (p_h, \nabla \cdot w)$$
$$- (q, \nabla \cdot u_h) - (A(u_h) - F, S(u_h)\varphi)_h.$$

Splitting the integrals into their contributions from each cell $K \in \mathbb{T}_h$ and integrating cell-wise by parts yields, analogously to deriving (3.18),

$$\rho(u_h; \varphi) = \sum_{K \in \mathbb{T}_h} \{(R^v(u_h), w)_K + (r^v(u_h), w)_{\partial K} + (q, R^p(u_h))_K$$
$$+ \delta_K(R^v(u_h), v_h \cdot \nabla w + \nabla q)_K\}.$$

From this, we obtain the error estimate (8.7) if we set $\varphi := z - z_h$. It remains to estimate the remainder term R_h which, observing that $A(u) - F = 0$, in the present situation has the form

$$R_h = (S(u)e, S(u)z)_h) - (A'(u)e, S(u)z)_h + \int_0^1 A_h''(u_h + se; e, e, z)s\, ds.$$

We recall that, for arguments $u = \{v, p\}, \varphi = \{q, w\}$ and $z = \{z^p, z^v\}$,

$$A'(u)\varphi := \begin{bmatrix} -\nu\Delta w + v\cdot\nabla w + w\cdot\nabla v + \nabla q \\ \nabla\cdot w \end{bmatrix}, \qquad S(u)\varphi := \begin{bmatrix} v\cdot\nabla w + \nabla q \\ 0 \end{bmatrix},$$

$$A''(u)\varphi z := \begin{bmatrix} w\cdot\nabla z^v + z^v\cdot\nabla w \\ 0 \end{bmatrix}, \qquad S'(u)\varphi z := \begin{bmatrix} w\cdot\nabla z^v \\ 0 \end{bmatrix}.$$

Hence, the first two terms in the remainder R_h can be written in the form

$$(S(u)e, S(u)z)_h) - (A'(u)e, S(u)z)_h = -(-\nu\Delta e^v + e^v\cdot\nabla v, v\cdot\nabla z^v + \nabla z^p)_h.$$

Further, the first derivative of $A_h(\cdot;\cdot)$ has, for the arguments $u = \{p, v\}, \varphi = \{q, w\} \in V$ and $z = \{z^p, z^v\} \in V$, the explicit form

$$A_h'(u; \varphi, z) = (A'(u)\varphi, z) + (A'(u)\varphi, S(u)z)_h + (A(u), S'(u)\varphi z)_h,$$

with $A'(u)\varphi$ and $S(u)z$ defined as above. Since $S''(u) \equiv 0$, the second derivative of $A_h(\cdot;\cdot)$ has, for arguments $u_s = \{p_s, v_s\}, e = \{e^p, e^v\}$ and $z = \{z^p, z^v\}$, the form

$$A_h''(u_s; e, e, z) = 2(e^v\cdot\nabla e^v, z^v) + 2(e^v\cdot\nabla e^v, v_s\cdot\nabla z^v + \nabla z^p)_h$$
$$+ 2\left(-\nu\Delta e^v + v_s\cdot\nabla e^v + e^v\cdot\nabla v_s + \nabla e^p, e^v\cdot\nabla z^v\right)_h.$$

From this, we easily infer the proposed bound for the remainder term.   □

We will compare the weighted error estimator $\eta_\omega(u_h)$ of Proposition 8.1 with several heuristically based error indicators. In all cases the mesh refinement is driven by local 'error indicators' $\eta_K = \eta_K(v, p)$ which are cheaply obtained from the computed solution $\{p_h, v_h\}$. Examples of common *heuristic* indicators are:

Vorticity. $\eta_K := \|\nabla \times v_h\|_K$,

Pressure-gradient. $\eta_K := \|\nabla p_h\|_K$,

'Energy error'. $\eta_K := \|R^p(u_h)\|_K + \|R^v(u_h)\|_K + h_K^{1/2}\|r^v(u_h)\|_{\partial K}$,
    with the cell and edge residuals defined by

$$R^p(u_h)_{|K} := \nabla\cdot v_h,$$

$$R^v(u_h)_{|K} := -\nu\Delta v_h + v_h\cdot\nabla v_h + \nabla p_h,$$

$$r^v(u_h)_{|\Gamma} := \begin{cases} -\frac{1}{2}[\nu\partial_n v_h - p_h n], & \text{if } \Gamma \not\subset \partial\Omega, \\ -(\nu\partial_n v_h - p_h n), & \text{if } \Gamma \subset \Gamma_{\text{out}}, \\ 0, & \text{if } \Gamma \subset \Gamma_{\text{rigid}} \cup \Gamma_{\text{in}}. \end{cases}$$

The vorticity and the pressure-gradient indicators measure the 'smoothness' of the computed solution $\{v_h, p_h\}$ while the 'energy error' indicator additionally contains information concerning local conservation of mass and momentum. However, neither of them contains information about the effect of changing the local mesh size on the error in the target quantity. This is only achieved by the 'weighted' indicators derived from $\eta_\omega(u_h)$.

Numerical results (from Becker, Braack and Rannacher (2000a))
We consider the flow around the cross section of a cylinder with surface S in a channel with a narrowed outlet (see Figure 8.1). Here, a quantity of physical interest is, for example, the 'drag coefficient' defined by

$$J(v, p) := \frac{2}{U^2 D} \int_S n \cdot \sigma(v, p) \psi \, ds, \tag{8.9}$$

where $\psi := (0, 1)^T$. Here, $\sigma(v, p) = \frac{1}{2}\nu(\nabla v + \nabla v^T) + pI$ is the stress acting on the cylinder, D is its diameter, and $U := \max |\hat{v}|$ is the reference inflow velocity. In this example, the Reynolds number is $\mathrm{Re} = UD/\nu = 50$, such that the flow is stationary.

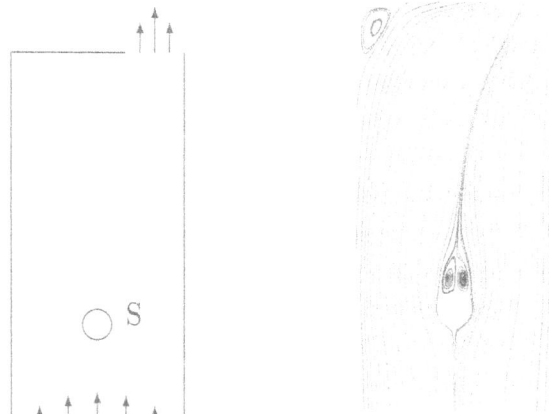

Fig. 8.1. Configuration and streamline plot of the 'flow around a cylinder'

In Figure 8.2, we compare the results for approximating the drag coefficient $J(v, p)$ on meshes which are constructed by using the different error indicators with that obtained on uniformly refined meshes. It turns out that all the above indicators perform equally weakly. A better result is obtained by using the 'weighted indicators' derived from (8.7). We clearly see the advantage of this error estimator which reflects the sensitivity of the error $J(v, p) - J(v_h, p_h)$ with respect to changes of the cell-wise residuals under mesh refinement.

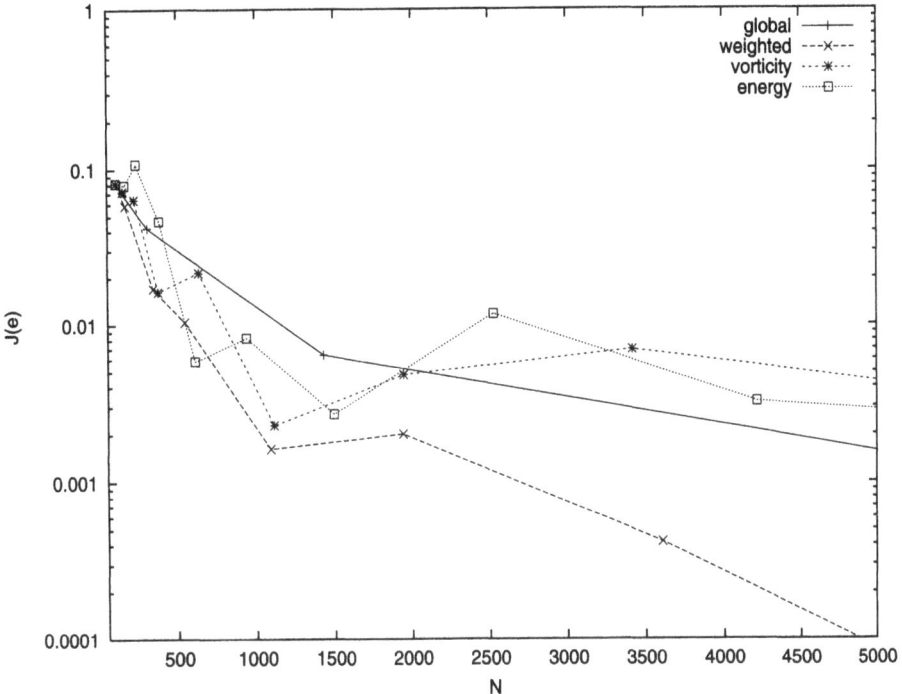

Fig. 8.2. Mesh efficiency obtained by global uniform refinement ('global' +),
the weighted indicator ('weighted' ×), the vorticity indicator ('vorticity' *),
and the energy indicator ('energy' □)

In the estimate (8.7) the additional terms representing the errors in approximating the inflow data and the curved boundary component S are
neglected; they can be expected to be small compared to the other residual
terms. The bounds for the dual solution $\{z^p, z^v\}$ are obtained computationally using patch-wise biquadratic interpolations as discussed in Section 5.
This avoids the use of interpolation constants. The quantitative results of
Figure 8.2 and the corresponding meshes shown in Figure 8.3 confirm the
superiority of the weighted error indicator in computing local quantities.

Remark 8.1. In our computation the drag coefficient has actually been
evaluated from the formula

$$J(v,p) := \frac{2}{\hat{H}^2 D} \int_\Omega \left\{ \sigma(p,v)\nabla\bar{\psi} + \nabla\cdot\sigma(p,v)\bar{\psi} \right\} dx, \qquad (8.10)$$

where $\bar{\psi}$ is an extension of the directional vector $\psi := (0,1)^T$ from S to Ω
with support along S. By integration by parts, one sees that this definition
is independent of the choice of $\bar{\psi}$ and that it is equivalent to the original one
(8.9) as a contour integral. However, on the discrete level the two formulations differ. In fact, computational experience shows that formula (8.10)

Fig. 8.3. Meshes with about 5000 cells obtained by the vorticity indicator (left), the 'energy' indicator (middle), and the weighted indicator (right)

yields significantly more accurate approximations of the drag coefficient (see Giles, Larsson, Levenstam and Süli (1997) and Becker (1998b)).

Notes and references

Energy norm-type *a posteriori* error estimates for finite element approximations of the Stokes and Navier–Stokes equations have been derived by Verfürth (1993), Bernardi, Bonnon, Langouët and Métivet (1995), Oden, Wu and Ainsworth (1993) and Ainsworth and Oden (1997b). An error analysis based on duality arguments for estimating the L^2 error was developed by Johnson and Rannacher (1994) and Johnson, Rannacher and Boman (1995). Another variant of this technique also including functional error estimation along the lines described in Section 4 can be found in Machiels *et al.* (1998) and Machiels, Peraire and Patera (1999). In Becker and Rannacher (1996a) and Becker (1995, 1998b), the application of the DWR approach to drag and lift computation was developed for the 'cylinder flow' benchmark described in Schäfer and Turek (1996).

9. Application to thermal and reactive flow

Now the complexity of the preceding example (incompressible Navier–Stokes equations) is further increased by including compressibility effects due to energy transfer and heat-release by chemical reactions. This constitutes the most complex situation the DWR method has been applied to yet. Though the model involves various interacting physical mechanisms (diffusion, transport and reaction) and several physical quantities of different sizes (density, velocity, temperature, chemical species), the general approach works surprisingly well. It not only detects the quantitative dependencies of the target quantities on the local cell residuals but, by careful evaluation of the adjoint solution, also the interaction of the different solution components.

9.1. Low-Mach-number heat-driven flow

We concentrate on so-called 'low-Mach-number' gas flows where density variations are mainly due to temperature gradients. Such conditions often occur in chemically reactive flows and are characterized by hydrodynamically incompressible behaviour. Below, we will consider 'heat-driven' natural convection in a cavity as a typical example.

The underlying mathematical model is the full set of the (stationary) compressible Navier–Stokes equations in the so-called 'low-Mach-number approximation' due to the low speed of the resulting flow. Accordingly, the total pressure is split like $P(x) = P_{th} + p(x)$ into a thermodynamic part P_{th}, which is constant in space and used in the gas law, and a hydrodynamic part $p(x) \ll P_{th}$ used in the momentum equation. Then, the governing system of conservation equations can be written in the following form:

$$
\begin{aligned}
\nabla \cdot v - T^{-1} v \cdot \nabla T &= 0, \\
\rho v \cdot \nabla v + \nabla \cdot \tau + \nabla p &= (\rho - \rho_0)\, g, \\
\rho v \cdot \nabla T - \nabla \cdot (\kappa \nabla T) &= f_T,
\end{aligned}
\tag{9.1}
$$

supplemented by the law of an ideal gas $\rho = P_{th}/RT$. The stress tensor is given by $\tau = -\mu \{ \nabla v + (\nabla v)^T - \frac{2}{3}(\nabla \cdot v) I \}$.

In the model case considered below, there are no heat sources (e.g., due to chemical reactions), that is, $f_T = 0$. The boundary conditions are no-slip for the velocity along the whole boundary, $v_{|\partial\Omega} = 0$, Neumann boundary conditions for the temperature along adiabatic walls, $\partial_n T_{|\Gamma_N} = 0$, and Dirichlet conditions for the temperature along the heated or cooled walls, $T_{|\Gamma_D} = \hat{T}$.

The variational formulation of (9.1) uses the following semilinear form defined for triples $u = \{p, v, T\}$, $\varphi = \{\xi, \psi, \theta\}$:

$$
\begin{aligned}
A(u; \varphi) :=\ & (\nabla \cdot v - T^{-1} v \cdot \nabla T, \xi) + (\rho v \cdot \nabla v, \psi) - (\tau, \nabla \psi) - (p, \nabla \cdot \psi) \\
& - (p, \nabla \cdot \psi) - (\rho g, \psi) + (\rho v \cdot \nabla T, \theta) + (\kappa \nabla T, \nabla \theta).
\end{aligned}
$$

Further, we define the functional

$$
F(\varphi) := -(\rho_0 g, \psi).
$$

The natural solution spaces are

$$
\hat{V} = L^2(\Omega)/\mathbb{R} \times H^1(\Omega)^2 \times H^1(\Omega), \quad V := L^2(\Omega)/\mathbb{R} \times H_0^1(\Omega)^2 \times H_0^1(\Gamma_D; \Omega),
$$

where $H_0^1(\Gamma_D; \Omega) := \{\theta \in H^1(\Omega),\ \theta = 0 \text{ on } \Gamma_D\}$. With this notation, the variational form of (9.1) seeks $u = \{p, v, T\} \in V + \hat{u}$, with $\hat{u} := \{0, 0, \hat{T}\}$, satisfying

$$A(u; \varphi) = F(\varphi), \quad \text{for all } \varphi \in V, \tag{9.2}$$

where ρ is considered as a (nonlinear) coefficient determined by the temperature through the equation of state $\rho = P_{\text{th}}/RT$. For more details on the derivation of this model, we refer to Braack and Rannacher (1999), and the literature cited therein. Here, we assume that (9.2) possesses a (unique) solution $u \in \hat{V}$, which is stable in the sense that the corresponding Fréchet derivative $A'(u; \cdot, \cdot)$ is strongly coercive.

The discretization of the system (9.2) again uses the continuous Q_1-finite element for all unknowns and employs least-squares stabilization for the velocity-pressure coupling as well as for the transport terms. We do not explicitly state the corresponding discrete equations since they have an analogous structure, as already seen in the preceding example of the incompressible Navier–Stokes equations. The derivation of the related adjoint problem and the resulting a *posteriori* error estimates follows the same line of argument. For economy reasons, we do not use the full Jacobian of the coupled system in setting up the adjoint problem, but only include its dominant parts. The same simplification is used in the nonlinear iteration process. For details, we refer to Braack and Rannacher (1999) and Becker, Braack and Rannacher (2000a). Below, we again use the mesh-dependent inner product and norm

$$(v, \psi)_h := \sum_{K \in \mathbb{T}_h} \delta_K (v, \psi)_K, \quad \|v\|_h = (v, v)_h^{1/2},$$

with some stabilization parameters δ_K. The discrete problems seek $u_h = \{p_h, v_h, T_h\} \in V_h + \hat{u}_h$, satisfying

$$A_h(u_h; \varphi_h) = F(\varphi_h), \quad \text{for all } \varphi_h \in V_h, \tag{9.3}$$

with the stabilized form

$$A_h(u_h; \varphi_h) := A(u_h; \varphi_h) + (A(u_h) - F, S(u_h)\varphi_h)_h.$$

Here, the operator $A(u_h)$ is the generator of the form $A(u_h; \cdot)$, and the operator $S(u_h)$ in the stabilization term is chosen according to

$$S(u_h) := \begin{bmatrix} 0 & \text{div} & 0 \\ \nabla & \rho_h v_h \cdot \nabla + \nabla \cdot \mu \nabla & 0 \\ -T_h^{-1} v_h \cdot \nabla & 0 & \rho_h v_h \cdot \nabla + \nabla \cdot \kappa \nabla \end{bmatrix}.$$

As on the continuous level, the discrete density is determined by the temperature through the equation of state $\rho_h := P_{\text{th}}/RT_h$. We introduce the following notation for the cell residuals of the solution $u_h = \{p_h, v_h, T_h\}$

of (9.3):

$$R^p(u_h)_{|K} := \nabla \cdot v_h - T_h^{-1} v_h \cdot \nabla T_h,$$

$$R^v(u_h)_{|K} := \rho_h v_h \cdot \nabla v_h - \nabla \cdot (\mu \nabla v_h) + \nabla p_h + (\rho_0 - \rho_h)g,$$

$$R^T(u_h)_{|K} := \rho_h v_h \cdot \nabla T_h - \nabla \cdot (\kappa \nabla T_h) - f_T.$$

Further, we define the edge residuals

$$r^v(u_h)_{|\Gamma} := \begin{cases} -\frac{1}{2}[\nu \partial_n v_h - p_h n], & \text{if } \Gamma \not\subset \partial\Omega, \\ -(\nu \partial_n v_h - p_h n), & \text{if } \Gamma \subset \Gamma_{\text{out}}, \\ 0, & \text{if } \Gamma \subset \Gamma_{\text{rigid}} \cup \Gamma_{\text{in}}. \end{cases}$$

$$r^T(u_h)_{|\Gamma} := \begin{cases} -\frac{1}{2}[\kappa \partial_n T_h], & \text{if } \Gamma \not\subset \partial\Omega, \\ -\kappa \partial_n T_h, & \text{if } \Gamma \subset \Gamma_N, \\ 0, & \text{if } \Gamma \subset \Gamma_D. \end{cases}$$

with $[\cdot]$ again denoting the jump across an interior edge Γ. These quantities will be needed below for defining the *a posteriori* error estimator. Using this notation the stabilizing part in $A_h(\cdot;\cdot)$ can be written in the form

$$\begin{aligned}
(A(u_h), S(u_h)\varphi)_h &= (R^p(u_h), \nabla \cdot \psi)_h \\
&+ (R^v(u_h), \nabla \xi + \rho_h v_h \cdot \nabla \psi + \nabla \cdot (\mu \nabla \psi))_h \\
&+ (R^T(u_h), \rho_h v_h \cdot \nabla \theta + \nabla \cdot (\kappa \nabla \theta) - T_h^{-1} v_h \cdot \nabla \xi)_h,
\end{aligned}$$

for $\varphi = \{\xi, \psi, \theta\}$. These terms comprise stabilization of the stiff velocity-pressure coupling in the low-Mach-number case and stabilization of transport in the momentum and energy equation case as well as the enforcement of mass conservation. The parameters $\delta_K = \{\delta_K^p, \delta_K^v, \delta_K^T\}$ may be chosen differently in the three equations following rules like (8.3). The stability of this discretization has been investigated in Braack (1998).

Now, we turn to the question of *a posteriori* error estimation in the scheme (9.3). Let $J(\cdot)$ again be a (for simplicity) linear functional defined on $\hat{V}$ for evaluating the error $e = \{e^p, e^v, e^T\}$. As for the previous section on incompressible flow, the adjoint problem is again set up using a reduced Jacobian in order to guarantee existence of the adjoint solution. Accordingly, we consider the h-dependent adjoint problem

$$\tilde{A}_h'(u; \varphi, z) := A'(u; \varphi, z) + (S(u)\varphi, S(u)z)_h = J(\varphi), \quad \text{for all } \varphi \in \hat{V}. \quad (9.4)$$

Then, from Proposition 2.6, we obtain for the present situation the following.

Proposition 9.1. Let $z = \{r, w, S\} \in V$ be the solution of the linearized adjoint problem (9.4). Then, we have the *a posteriori* error estimate

$$|J(e)| \approx \eta_\omega(u_h) := \sum_{K \in \mathbb{T}_h} \left\{ \sum_{\alpha \in \{p,v,T\}} \rho_K^\alpha \omega_K^\alpha \right\} + R_h, \quad (9.5)$$

where the residual terms and weights are defined by

$$\rho_K^p := \|R^p(u_h)\|_K,$$

$$\omega_K^p := \|r - r_h\|_K + \delta_K^p \|S^p(u_h)(r - r_h)\|_K,$$

$$\rho_K^v := \|R^v(u_h)\|_K + h_K^{-1/2} \|r^v(u_h)\|_{\partial K},$$

$$\omega_K^v := \|w - w_h\|_K + h_K^{1/2} \|w - w_h\|_{\partial K} + \delta_K^v \|S^v(u_h)(w - w_h)\|_K,$$

$$\rho_K^T := \|R^T(u_h)\|_K + h_K^{-1/2} \|r^T(u_h)\|_{\partial K},$$

$$\omega_K^T := \|S - S_h\|_K + h_K^{1/2} \|S - S_h\|_{\partial K} + \delta_K^T \|S^T(u_h)(S - S_h^T)\|_K,$$

with a suitable approximation $\{r_h, w_h, S_h\} \in V_h$ to z. The remainder term R_h can be bounded as in Proposition 8.1 for incompressible flow.

The details of the proof are omitted. The argument is similar to that used in the proof of Proposition 8.1 for incompressible flow. This analysis assumes for simplicity that viscosity μ and heat conductivity κ are determined by the reference temperature T_0, i.e., these quantities are not included in the linearization process. This is justified because of their relatively small variation with T. Furthermore, the explicit dependence of the stabilization parameters δ_K on the discrete solution u_h is neglected.

The weights in the a *posteriori* error bound (9.5) are again evaluated by solving the adjoint problem numerically on the current mesh and approximating the exact adjoint solution z by patch-wise higher-order interpolation of its computed approximation $\tilde{z}_h$, as described above. This technique shows sufficient robustness and does not require the determination of any interpolation constant. For illustration, we state the strong form of the adjoint problem (suppressing terms related to the least-squares stabilization) used in our test computations:

$$\nabla \cdot w = j^p,$$

$$-\rho(v \cdot \nabla)w - \nabla \cdot \mu \nabla w - \rho \nabla r = j^v,$$

$$T^{-1}v \cdot \nabla r + T^{-2}v \cdot \nabla T \cdot r - \rho v \cdot \nabla S - c_p^{-1} \nabla \cdot (\lambda \nabla S) - (Df^T)_T S = j^T,$$

where $\{j^p, j^v, j^T\}$ is a suitable function representation of the error functional $J(\cdot)$. This system is closed by appropriate boundary conditions.

Numerical results (from Becker, Braack and Rannacher (2000a))
The first example (without chemistry) is a 2D benchmark 'heat-driven cavity' (for details see Figure 9.1 and Quere and Paillere (2000) or Becker *et al.* (2000a)). Here, the flow is confined to a square box with side length $L = 1$ and is driven by a temperature difference $T_h - T_c = 2\varepsilon T_0$ between the left ('hot') and the right ('cold') wall under the action of gravity g in the y-direction.

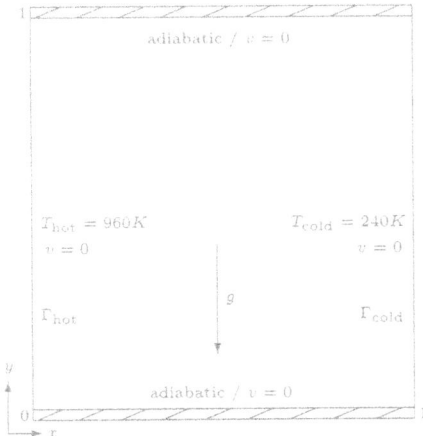 

Fig. 9.1. Configuration of the heat-driven cavity problem and plot of computed temperature isolines

For the viscosity μ, we use Sutherland's law,

$$\mu(T) \;=\; \mu^* \left(\frac{T}{T^*}\right)^{1/3} \frac{T^* + S}{T + S},$$

with the Prandtl number $\mathrm{Pr} = 0.71$, $T^* = 273\,K$, $\mu^* = 1.68 \cdot 10^{-5} \mathrm{kg/ms}$, and $S := 110.5\,K$. Further, the heat conductivity is $\kappa(T) = \mu(T)/\mathrm{Pr}$. In the stationary case the thermodynamic pressure is defined by

$$P_{\mathrm{th}} \;=\; P_0 \left(\int_\Omega T_0^{-1}\,\mathrm{d}x\right)\left(\int_\Omega T^{-1}\,\mathrm{d}x\right)^{-1},$$

where $T_0 = 600\,K$ is a reference temperature and $P_0 = 101,325\,\mathrm{Pa}$. Accordingly, the Rayleigh number is determined by

$$\mathrm{Ra} \;=\; \mathrm{Pr}\,g \left(\frac{\rho_0 L}{\mu_0}\right)^2 \frac{T_h - T_c}{T_0} \approx 10^6, \qquad \varepsilon \;=\; \frac{T_h - T_c}{T_h + T_c} = 0.6,$$

where $\mu_0 := \mu(T_0)$, $\rho_0 := P_0/RT_0$, and $R = 287\,\mathrm{J/kgK}$.

In this benchmark, one of the quantities to be computed is the average Nusselt number along the cold wall defined by

$$J(T) := c \int_{\Gamma_{\mathrm{cold}}} \kappa \partial_n T\,\mathrm{d}s, \qquad c := \frac{\mathrm{Pr}}{2\mu_0 T_0 \varepsilon}.$$

Table 9.1. Results of computing the Nusselt number by using the 'energy error' indicator (left) and the weighted error indicator (right)

N	$\langle Nu \rangle_c$	$J(e)$	N	$\langle Nu \rangle_c$	$J(e)$	η_ω	I_{eff}
524	-9.09552	4.1e-1	523	-8.86487	1.8e-1	3.4e-1	1.9
945	-8.67201	1.5e-2	945	-8.71941	3.3e-2	1.5e-1	4.5
1708	-8.49286	1.9e-1	1717	-8.66898	1.8e-2	6.6e-2	3.7
3108	-8.58359	1.0e-1	5530	-8.67477	1.2e-2	2.1e-2	1.8
5656	-8.59982	8.7e-2	9728	-8.68364	3.0e-3	1.1e-2	3.7
18204	-8.64775	3.9e-2	17319	-8.68744	8.5e-4	6.3e-3	7.4
32676	-8.66867	1.8e-2	31466	-8.68653	6.9e-5	3.7e-3	53.6
58678	-8.67791	8.7e-3	56077	-8.68653	6.7e-5	2.1e-3	31.3
79292	-8.67922	7.4e-3	78854	-8.68675	1.5e-4	1.5e-3	10.0

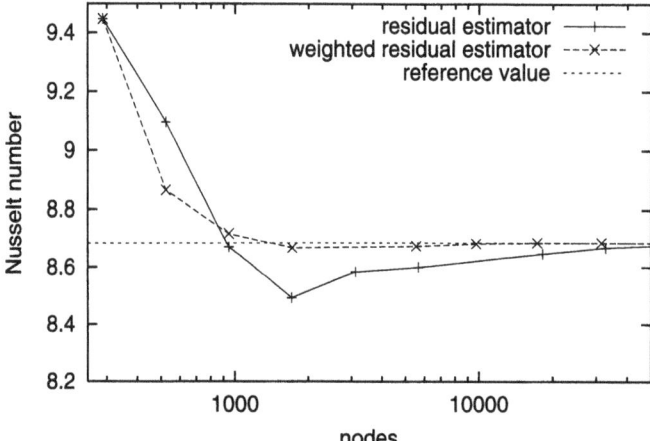

Fig. 9.2. Results of computing the Nusselt number by using a heuristic 'residual indicator' (symbol '+') and the 'weighted residual estimator' (symbol '×')

The results shown in Table 9.1 and Figure 9.2 demonstrate the mesh efficiency of the 'weighted' error indicator compared to a heuristic 'energy error' indicator similar to the one discussed in Section 8 for the incompressible flow case. The superiority of the 'weighted' indicator is particularly evident if higher solution accuracy is required; see Table 9.1. As usual in problems of such complex structure, the *a posteriori* bound $\eta(u_h)$ tends to

Fig. 9.3. Sequences of refined meshes obtained by the 'energy error' indicator (upper row, $N = 524, 5656, 58678$) and the weighted error estimator (lower row, $N = 523, 5530, 56077$)

over-estimate the true error but it appears to be 'reliable'. A collection of refined meshes produced by the two error indicators is shown in Figure 9.3.

Remark 9.1. The most important feature of the *a posteriori* error estimate (9.11) is that the local cell residuals related to the various physical effects governing flow and transfer of temperature are systematically weighted according to their impact on the error quantity to be controlled. For illustration, let us consider control of the mean velocity

$$J(v) = |\Omega|^{-1} \int_\Omega v \, dx.$$

Then, in the adjoint problem the right-hand sides j^p and j^T vanish, but because of the coupling of the variables all components of the adjoint solution $z = \{r, w, S\}$ will be nonzero. Consequently, the error term to be controlled is also affected by the cell residuals of the mass-balance and the energy equation. This sensitivity is quantitatively represented by the weights involving r and S.

9.2. Chemically reactive flow in a 'flow reactor'

Next, we extend the 'low-Mach-number' flow model (9.1) by including chemical reactions. In this case the equations of mass, momentum and energy conservation are supplemented by the equations of species mass conservation:

$$\nabla \cdot v - T^{-1}v \cdot \nabla T - M^{-1}v \cdot \nabla M = 0, \tag{9.6}$$

$$(\rho v \cdot \nabla)v + \nabla \cdot \tau + \nabla p = \rho f_e, \tag{9.7}$$

$$\rho v \cdot \nabla T - c_p^{-1}\nabla \cdot (\lambda \nabla T) = c_p^{-1} f_t(T, w), \tag{9.8}$$

$$\rho v \cdot \nabla w_i - \nabla \cdot (\rho D_i \nabla w_i) = f_i(T, w), \quad i = 1, \dots, n. \tag{9.9}$$

The gas law takes the form

$$\rho = \frac{P_* M}{RT}, \tag{9.10}$$

with the mean molar mass $M := (\sum_{i=1}^{n} w_i/M_i)^{-1}$ and the species mole masses M_i (see, e.g., Braack and Rannacher (1999)).

Owing to exponential dependence on temperature (Arrhenius' law) and polynomial dependence on w, the source terms $f_i(T, w)$ are highly nonlinear. In general, these zero-order terms lead to a coupling between all chemical species mass fractions. For robustness the resulting system of equations is to be solved by an implicit and fully coupled process that uses strongly adapted meshes. The discretization of the full flow system again uses continuous Q_1-finite elements for all unknowns and employs least-squares stabilization for the velocity-pressure coupling as well as for all the transport terms. We do not state the corresponding discrete equations since they have the same structure as seen before. The derivation of the related (linearized) adjoint problem corresponding to some (linear) functional $J(\cdot)$ and the resulting a *posteriori* error estimates follows the same line of argument. Again, we use a simplified Jacobian of the coupled system in setting up the adjoint problem (for details see Braack and Rannacher (1999). The resulting a *posteriori* error estimator has the same structure as that in Proposition 9.1, only that additional terms occur due to the balance equations for the chemical species:

$$|J(e)| \approx \eta_\omega(u_h) := \sum_{K \in \mathbb{T}_h} \left\{ \sum_{\alpha \in \{p,v,T,w_i\}} \rho_K^\alpha \omega_K^\alpha \right\}, \tag{9.11}$$

with the chemistry-related additional residual terms and weights

$$\rho_K^{w_i} := \|R^{w_i}(u_h)\|_K + h_K^{-1/2}\|r^{w_i}(u_h)\|_{\partial K}, \quad i = 1, \dots, n,$$

$$\omega_K^{w_i} := \|v_i - v_{i,h}\|_K + h_K^{1/2}\|v_i - v_{i,h}\|_{\partial K} + \delta_K^{w_i}\|S^{w_i}(u_h)(v_i - v_{i,h})\|_K.$$

The weights in the a *posteriori* error bound (9.11) are again computed in the same way as described above.

Numerical results (from Waguet (2000))

As a practically relevant example, we consider a chemical flow reactor (see Figure 9.4) for determining the reaction velocity of the heterogeneous relaxation of vibrationally excited hydrogen and its energy transfer in collisions with deuterium ('slow' chemistry),

$$H_2^{(1)} \rightarrow_{\text{wall}} H_2^{(0)}, \qquad H_2^{(1)} + D_2^{(0)} \rightarrow H_2^{(0)} + D_2^{(1)}.$$

The quantity to be computed is the CARS signal (coherent anti-Stokes Raman spectroscopy)

$$J(c) = \kappa \int_{-R}^{R} \sigma(s) c(r - s)^2 \, \mathrm{d}s,$$

where $c(r)$ is, for example, the concentration of $D_2^{(1)}$ along the line of the laser measurement (see Segatz *et al.* (1996)).

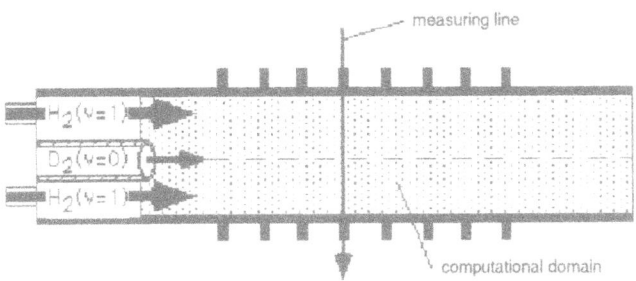

Fig. 9.4. Configuration of flow reactor

Table 9.2 and Figure 9.5 (overleaf) show results obtained by the DWR method for computing the mass fraction of $D_2^{(1)}$ and $D_2^{(0)}$. The comparison is against computations on heuristically refined tensor-product meshes. We observe higher accuracy on the systematically adapted meshes: in particular, monotone convergence of the quantities is achieved.

Notes and references

More complex combustion processes (gas-phase reactions: 'fast' chemistry) like the flame-sheet model of the Bunsen burner (laminar methane flame), the 3-species ozone recombination, or a model with detailed chemistry of (laminar) combustion in a methane burner involving 17 species and 88 reactions have been treated using the DWR method by Braack (1998); see also Becker, Braack, Rannacher and Waguet (1999), Braack and Rannacher (1999), and Becker *et al.* (2000a).

Table 9.2. Some results of simulation for the relaxation experiment on hand-adapted (left) and on automatic-adapted (right) meshes

Heuristic refinement				Adaptive refinement			
L	N	$D_2^{(\nu=0)}$	$D_2^{(\nu=1)}$	L	N	$D_2^{(\nu=0)}$	$D_2^{(\nu=1)}$
2	481	0.742228	0.002541	2	244	0.738019	0.004020
3	1793	0.780133	0.002531	3	446	0.745037	0.002600
4	1923	0.782913	0.002729	4	860	0.756651	0.002010
5	2378	0.785116	0.001713	5	1723	0.780573	0.001390
6	3380	0.791734	0.001162	6	3427	0.785881	0.001130
7	5374	0.791627	0.001436	7	7053	0.799748	0.001090

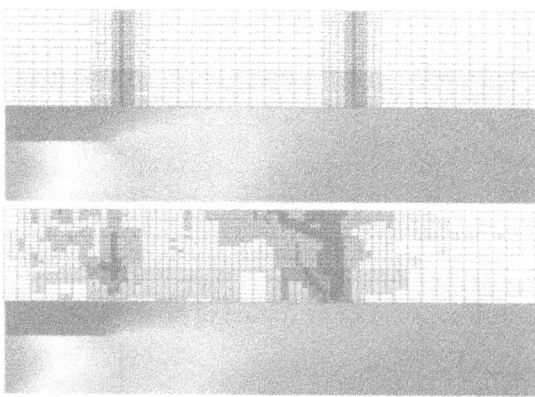

Fig. 9.5. Mass fraction of $D_2^{(1)}$ in the flow re-actor computed on a tensor-product mesh (top) and on a locally adapted mesh (bottom)

10. Application to elasto-plasticity

We present an example of the application of the DWR method in solid mechanics, particularly in elasto-plasticity theory. The problem to be solved is strongly nonlinear with a nonlinearity not everywhere differentiable. Nevertheless, the straightforward application of the DWR method to this situation gives satisfactory results. This indicates that the method is more robust in practice than one might expect from theory.

10.1. The Hencky model of perfect plasticity

The fundamental problem in the static deformation theory of linear-elastic perfect plastic material (the so-called *Hencky* model) reads

$$\nabla \cdot \sigma = -f, \quad \varepsilon(u) = A{:}\sigma + \lambda \quad \text{in } \Omega,$$
$$\lambda{:}(\tau - \sigma) \le 0, \quad \text{for all } \tau \in \Sigma := \{\tau, \mathcal{F}(\tau) \le 0\}, \tag{10.1}$$
$$u = 0 \text{ on } \Gamma_D, \quad \sigma \cdot n = g \text{ on } \Gamma_N,$$

where σ and u are the stress tensor and displacement vector, respectively, while λ denotes the plastic growth. This system describes the deformation of an elasto-plastic body occupying a bounded domain $\Omega \subset \mathbb{R}^d$ ($d = 2$ or 3) which is fixed along a part Γ_D of its boundary $\partial\Omega$, under the action of a body force with density f and a surface traction g along $\Gamma_N = \partial\Omega \backslash \Gamma_D$. The displacement u is assumed to be small in order to neglect geometric nonlinear effects, so that the strain tensor can be written in the form $\varepsilon(u) = \frac{1}{2}(\nabla u + \nabla u^T)$. We assume a linear-elastic isotropic material law, that is, the material tensor A is given in the form $A = C^{-1}$, where

$$\sigma = C\varepsilon := 2\mu\varepsilon^D + \kappa \operatorname{tr}(\varepsilon)I,$$

with material-dependent constants $\mu > 0$ and $\kappa > 0$. The plastic behaviour follows the von Mises flow rule

$$\mathcal{F}(\sigma) := |\sigma^D| - \sigma_0 \le 0,$$

with some $\sigma_0 > 0$. Here, $\varepsilon^D := \varepsilon - \frac{1}{d}\operatorname{tr}(\varepsilon)I$ and $\sigma^D := \sigma - \frac{1}{d}\operatorname{tr}(\sigma)I$ are the deviatoric parts of ε and σ, respectively. For setting up the *primal* variational formulation of problem (10.1), we introduce the Hilbert space $V := \{u \in H^1(\Omega)^d, u_{|\Gamma_D} = 0\}$. Further, we define the nonlinear material function

$$C(\varepsilon) := \begin{cases} C\varepsilon, & \text{if } |2\mu\varepsilon^D| \le \sigma_0, \\ \dfrac{\sigma_0}{|\varepsilon^D|}\varepsilon^D + \kappa \operatorname{tr}(\varepsilon)I, & \text{if } |2\mu\varepsilon^D| > \sigma_0. \end{cases}$$

The tensor-function $C(\cdot)$ is globally only Lipschitz and not differentiable along the yield surface $\{\tau \in \mathbb{R}^{d \times d}_{\text{sym}}, |2\mu\tau^D| = \sigma_0\}$. Below, we will use the piecewise differential $C'(\tau)$ defined by

$$C'(\tau)\varepsilon := \begin{cases} C\varepsilon, & \text{if } |2\mu\tau^D| \le \sigma_0, \\ \dfrac{\sigma_0}{|\tau^D|}\Big\{I - \dfrac{(\tau^D)^T\tau^D}{|\tau^D|^2}\Big\}\varepsilon^D + \kappa \operatorname{tr}(\varepsilon)I, & \text{if } |2\mu\tau^D| > \sigma_0. \end{cases}$$

Then, we seek a displacement $u \in V$, satisfying

$$A(u; \varphi) = F(\varphi), \quad \text{for all } \varphi \in V, \tag{10.2}$$

with the semilinear and linear forms

$$A(u; \varphi) := (C(\varepsilon(u)), \varepsilon(\varphi)), \qquad F(\varphi) := (f, \varphi) + (g, \varphi)_{\Gamma_N}.$$

The finite element approximation of problem (10.2) seeks $u_h \in V_h$ such that

$$A(u_h; \varphi_h) = F(\varphi_h), \quad \text{for all } \varphi_h \in V_h, \tag{10.3}$$

where V_h is the finite element space of bilinear shape functions as described above. Having computed the displacement u_h, we obtain a corresponding stress by $\sigma_h := C(\varepsilon(u_h))$. Details of the solution process can be found in Rannacher and Suttmeier (1999). In practice, the exact evaluation of $C(\cdot)$ is difficult, so the formal scheme (10.2) requires some modification. This may be accomplished by storing information only at Gaussian points (2×2-Gauss points for Q_1-elements) at which $C(\cdot)$ is evaluated exactly. However, in the following, it is always assumed that the discretization is realized in its ideal form without numerical integration. The nonlinear algebraic problem (10.3) is approximated by a modified Newton iteration as described in Section 6. The linear subproblems are solved by the CR method with multigrid acceleration (see Suttmeier (1996) for more details of the algebraic solution techniques).

Now, we turn to the *a posteriori* error analysis. Given a (linear) functional $J(\cdot)$ on V, we consider the corresponding linearized adjoint problem

$$A'(u; \varphi, z) = J(\varphi), \quad \text{for all } \varphi \in V, \tag{10.4}$$

where in the tangent bilinear form

$$A'(u; \varphi, z) = (C'(\varepsilon(u))\varepsilon(\varphi), \varepsilon(z)),$$

the matrix operator $C'(\varepsilon(u_h)$ is defined as above. As before, we define the equation residual

$$R_{|K} := f + \nabla \cdot \sigma_h, \quad \sigma_h := C(\varepsilon(u_h)),$$

and the edge residuals

$$r_{|\Gamma} := \begin{cases} \frac{1}{2} n \cdot [\sigma_h], & \text{if } \Gamma \subset \partial K \backslash \partial \Omega, \\ 0, & \text{if } \Gamma \subset \Gamma_D, \\ n \cdot \sigma_h - g, & \text{if } \Gamma \subset \Gamma_N. \end{cases}$$

With this notation, we obtain from Proposition 2.3 the following result.

Proposition 10.1. (Rannacher and Suttmeier 1998) For the approximation of the Hencky model (10.1) by the finite element scheme (10.3), we have the following a *posteriori* estimate for the error $e := u - u_h$:

$$|J(e)| \approx \eta_\omega(u_h) := \sum_{K \in \mathbb{T}} \rho_K \omega_K + R, \tag{10.5}$$

modulo a remainder term R that is formally quadratic in e, where the local residual terms and weights are defined by

$$\rho_K := \|R(u_h)\|_K + h_K^{-1/2}\|r(u_h)\|_{\partial K}, \qquad \omega_K := \|z - \varphi_h\|_K + h_K^{1/2}\|z - \varphi_h\|_{\partial K},$$

with a suitable approximation $\varphi_h \in V_h$ to the adjoint solution z.

Proof. From Proposition 2.3, we have the abstract error representation

$$J(e) = A(x_h; z - y_h) + R,$$

with a remainder term R, which is not further considered here. Using the definition of the semilinear form $A(\cdot; \cdot)$, this can be written in the concrete form

$$J(e) \approx \sum_{K \in \mathbb{T}_h} \left\{ (R(u_h), z - z_h)_K - (r(u_h), z - z_h)_{\partial K} \right\},$$

from which we immediately conclude the asserted estimate.   □

The *a posteriori* error estimator (10.5) may be evaluated by either one of the strategies described above. For simplicity, the linearization of the adjoint problem uses a decomposition of the mesh domain $\Omega_h = \cup\{K \in \mathbb{T}_h\}$ into its 'elastic' component Ω_h^e and 'plastic' component Ω_h^p defined by

$$\Omega_h^e := \cup\{K \in \mathbb{T}_h, |2\mu\varepsilon^D|_K| \le \sigma_0\}, \qquad \Omega_h^p := \Omega \setminus \Omega_h^e.$$

Then, we define on each cells $T \in \mathbb{T}_h$:

$$C_h'(\tau)(\varepsilon)_{|K} := \begin{cases} C\varepsilon, & \text{if } K \subset \Omega_h^e, \\ \dfrac{\sigma_0}{|\tau^D|}\left\{ I - \dfrac{(\tau^D)^K \tau^D}{|\tau^D|^2} \right\}\varepsilon^D + \kappa\,\mathrm{tr}(\varepsilon)I, & \text{if } K \subset \Omega_h^p. \end{cases}$$

Using this notation, the adjoint solution z is approximated by the solution $\tilde{z}_h \in V_h$ of the discretized adjoint problem

$$(\varepsilon(\varphi_h), C_h'(\varepsilon(u_h))^*\varepsilon(\tilde{z}_h)) = J(\varphi_h), \quad \text{for all } \varphi_h \in V_h, \tag{10.6}$$

The evaluation of the coefficient $C_h'(\varepsilon(u_h))^*$ on cells in the elastic–plastic transition zone is usually done by simple numerical integration. This may appear to be a rather crude approximation, but it works in practice. The reason may be that the critical situation only occurs in cells intersecting the elastic–plastic transition zone, which is a lower-dimensional surface. The weights ω_K may then again be approximated as described in Section 5. We emphasize that the computation of the adjoint solution requires us to solve only *linear* problems and normally only amounts to a small fraction of the total cost within a Newton iteration for the nonlinear problem. We will compare the weighted error estimator $\eta_\omega(u_h)$ with two heuristic indicators for the stress error $e_\sigma := \sigma - \sigma_h$.

- The heuristic ZZ-error indicator of Zienkiewicz and Zhu (1987) (see, e.g., Ainsworth and Oden (1997b)) uses the idea of higher-order stress recovery by local averaging,

$$\|e_\sigma\| \approx \eta_{ZZ}(u_h) := \left(\sum_{K \in \mathbb{T}_h} \|M_h \sigma_h - \sigma_h\|_K^2 \right)^{1/2}, \qquad (10.7)$$

where $M_h \sigma_h$ is a local (super-convergent) approximation of σ.

- The heuristic energy error estimator of Johnson and Hansbo (1992a, 1992b) and Hansbo (200x) is based on the decomposition $\Omega = \Omega_h^e \cup \Omega_h^p$:

$$\|e_\sigma\| \approx \eta_E(u_h) := c_i \left(\sum_{K \in \mathbb{T}_h} \eta_K^2 \right)^{1/2}, \qquad (10.8)$$

where

$$\eta_K^2 := \begin{cases} h_K^2 \max_K |R(u_h)|^2, & \text{if } K \in \Omega_h^e, \\ h_K^{-1} \max_K |C\varepsilon(u_h) - M_h C\varepsilon(u_h)| \int_K |R(u_h)|\, dx, & \text{if } K \in \Omega_h^p. \end{cases}$$

Numerical results (from Rannacher and Suttmeier (1998))
The approach described above is applied for a typical model problem in elasto-plasticity employing the two-dimensional 'plane strain' model (for more details and further examples see Rannacher and Suttmeier (1997)). A square disc with a crack is subjected to a constant boundary traction acting on the upper boundary (see Figure 10.1). Along the right-hand side and the lower boundary the disc is fixed and the remaining part of the boundary (including the crack) is left free. This problem is interesting as its solution develops a singularity at the tip of the crack where a strong stress concentration occurs which causes plastification locally. The material parameters chosen are those commonly used for aluminium: $\mu = 80193.80\,\text{Nmm}^{-2}$, $\kappa = 164206\,\text{Nmm}^{-2}$, and $\sigma_0 = \sqrt{2/3}\,450$.
We want to compute the mean normal stress over the clamped boundary,

$$J(\sigma) = \int_{\Gamma_u} n^T \sigma n\, ds. \qquad (10.9)$$

Since this functional is irregular, it is regularized as

$$J_\varepsilon(\tau) := \frac{1}{|\Gamma_\varepsilon|} \int_{\Gamma_\varepsilon} n^T \tau n\, dx, \quad \Gamma_\varepsilon = \{x \in \mathbb{R}^2, \text{dist}(x, \Gamma_D) < \tfrac{1}{2}\varepsilon\}, \qquad (10.10)$$

with $\varepsilon = \text{TOL}$. A reference solution σ_{ref} is computed on a fine mesh with about 200,000 cells for determining the relative error and the corresponding

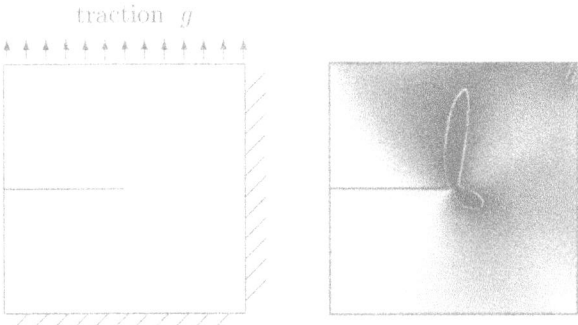

Fig. 10.1. Geometry of the test problem 'square disc with crack' and plot of $|\sigma^D|$ (plastic regions black) computed on a mesh with $N \approx 64,000$ cells

Table 10.1. Results obtained by the weighted *a posteriori* error estimator

N	$J(\sigma_h)$	E_{rel}	I_{eff}
1000	2.2224e+02	1.5757e-02	1.7
2000	2.2344e+02	1.0430e-02	2.1
4000	2.2405e+02	7.7387e-03	1.7
8000	2.2475e+02	4.6647e-03	1.6
16000	2.2532e+02	2.1360e-03	1.8
∞	2.2580e+02		

'effectivity index' (used here for expressing over-estimation) defined by

$$E_{\mathrm{rel}} := \left| \frac{J_\varepsilon(\sigma_h - \sigma_{\mathrm{ref}})}{J_\varepsilon(\sigma_{\mathrm{ref}})} \right|, \quad I_{\mathrm{eff}} := \left| \frac{\eta_\omega(u_h, \sigma_h)}{J_\varepsilon(\sigma_h - \sigma_{\mathrm{ref}})} \right|.$$

In this case the weighted error estimator turns out to be rather sharp even on relatively coarse meshes; see Table 10.1. This indicates that the strategy of evaluating the weights ω_K computationally also works for the present nonlinear problem with nonsmooth nonlinearity. Further, as in the linear case, we obtain more economical meshes than by the other heuristic error estimators (Figure 10.2).

Remark 10.1. In the plastic region the material behaviour will be almost incompressible, which causes stability problems of the discretization based on the formulation (10.2). In order to cope with this problem, a stabilized

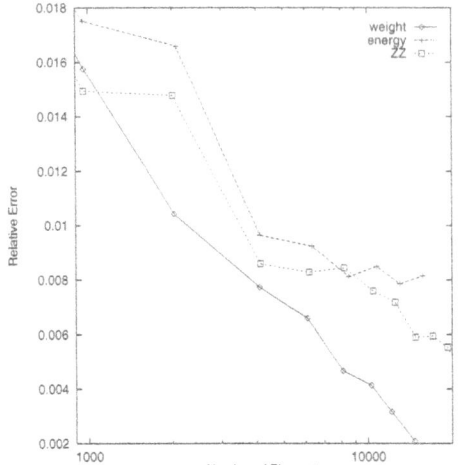

Fig. 10.2. Relative error for $J(\sigma)$ on grids based on the different estimators (left) and structure of optimized grid with $N \approx 8100$ (right)

finite element discretization may be employed by introducing an auxiliary 'pressure' variable (see Suttmeier (1998)).

Notes and references

The adaptive finite element method described above for solving the *stationary* Hencky problem in perfect plasticity has been extended in Rannacher and Suttmeier (1999, 200x) to the *quasi-stationary* Prandtl–Reuss model. Here, the time/load stepping is done by the backward Euler scheme and it is shown that the resulting incremental errors in each time/load step can accumulate at most linearly. The incorporation of the elasto-plasticity problem into the general framework of the DWR method relies on its reformulation as a nonlinear variational *equation*. For the application of duality techniques for *a posteriori* error estimation in the direct approximation of variational *inequalities*, we refer to Johnson (1992) and particularly to the recent papers of Blum and Suttmeier (1999, 2000) and Suttmeier (2001), where a variant of the DWR method is used.

11. Application to radiative transfer

In this section, we consider an example from astrophysics. The continuum model for the transfer of light including scattering is an integro-differential equation for the intensity as a function of time, frequency, space and direction. In astrophysical applications particular difficulties arise by the strong heterogeneity of the coefficients in the model. This type of problem poses highest requirements on numerical simulation, and adaptive discretization is mandatory for achieving accurate results. This example is intended to demonstrate that the DWR method can directly be applied to finite element discretization of this nonstandard model.

11.1. The monochromatic radiative transfer equation

The emission of light of a certain wavelength from a cosmic source is described by the (monochromatic) 'radiative transfer equation'

$$\theta \cdot \nabla_x u + (\kappa + \sigma)u - \sigma \int_{S_2} P(\theta, \theta')u \, d\theta' = B \quad \text{in } \Omega \times S_2, \qquad (11.1)$$

for the radiation intensity $u = u(x, \theta)$. Here, $x \in \Omega \subset \mathbb{R}^3$ is a bounded domain and $\theta \in S_2$ the unit-sphere in $\mathbb{R}^3$. The usual boundary condition is $u = 0$ at the 'inflow' boundary $\Gamma_{in,\theta} = \{x \in \partial\Omega, n\cdot\theta < 0\}$. The absorption and scattering coefficients $\kappa(x) > 0$, $\sigma(x) > 0$, the redistribution kernel $P(\theta, \theta') \geq 0$, and the source term B (Planck function) are given. In applications these functions exhibit strong variations (several orders of magnitude) in space, which requires the use of locally refined meshes. Further, because of these coefficient irregularities, quantitative a priori information about the properties of the radiation operator and the solution's regularity are not available (see Kanschat (1996, 1998)).

The starting point of a numerical solution of the radiative transfer problem (11.1) is again its variational formulation. To this end, we introduce the solution space

$$V := \{v \in L^2(\Omega \times S_2), \ \theta \cdot \nabla_x u \in L^2(\Omega \times S_2)\}$$

and the operators

$$T_\theta u := \theta \cdot \nabla_x u, \qquad \Sigma u := \sigma u - \sigma \int_{S_2} P(\theta, \theta')u \, d\theta'.$$

Then, we seek $u \in V$ satisfying

$$A(u, \varphi) = F(\varphi), \quad \text{for all } \varphi \in V, \qquad (11.2)$$

with the linear forms

$$A(u, \varphi) := (T_\theta u + \kappa u + \Sigma u, \varphi)_{\Omega \times S_2} + (\theta \cdot n u, \varphi)_{\Gamma_{in,\theta}}, \qquad F(\varphi) := (B, \varphi)_{\Omega \times S_2}.$$

Notice that, in this formulation, the 'inflow' boundary conditions are incorporated weakly. The existence of unique solutions of (11.2) can be inferred by abstract functional analytic arguments. The discretization of this problem uses standard (continuous) Q_1-finite elements in $x \in \Omega$, on meshes $\mathbb{T}_h = \{K\}$ with local width h_K, and (discontinuous) P_0-finite elements in $\theta \in S_2$, on meshes $\mathbb{D}_k = \{\Delta\}$ of uniform width k_Δ. Again, streamline diffusion as described in Section 3.3 is employed in order to stabilize the transport term $\theta \cdot \nabla_x u$. The discretization of the ordinate space S_2 is equivalent to using the midpoint quadrature formula on the integral operator Σ. The x-mesh is adaptively refined, while the θ-mesh is kept uniform (suggested by *a priori* error analysis).

Let $V_h \subset V$ be the finite element subspaces. The Galerkin finite element approximations $u_h \in V_h$ are defined by

$$A_h(u_h, \varphi_h) = F_h(\varphi_h), \quad \text{for all } \varphi_h \in V_h, \tag{11.3}$$

with the stabilized forms

$$A_h(u_h, \varphi_h) := A(u_h, \varphi_h + \delta T_\theta \varphi_h), \quad F(\varphi_h) := F(\varphi_h + \delta T_\theta \varphi_h),$$

and some mesh-dependent parameter $\delta \sim h \min\{\kappa^{-1}, \sigma^{-1}\}$. Clearly (11.3) is also satisfied by the exact solution u, such that Galerkin orthogonality again holds with respect to the form $A_h(\cdot, \cdot)$.

The refinement process is organized as described before on the basis of a weighted *a posteriori* error estimate for the target quantity $J(u)$. To this end, we consider the following adjoint problem for $z \in V$:

$$A_h(\varphi, z) = J(\varphi), \quad \text{for all } \varphi \in V, \tag{11.4}$$

which represents a modified radiative transport problem with source term $J(\cdot)$. The solvability of (11.4) follows from the general theory as for the original problem (11.2). From Proposition 2.6, we then obtain the following result.

Proposition 11.1. (Kanschat 1996) Let z be the solution of the adjoint problem (11.4). Then, for the finite element approximation (11.3) of the radiative transfer problem (11.1), we have the following *a posteriori* estimate:

$$|J(e)| \le \eta_\omega(u_h) := \sum_{\Delta \in \mathbb{D}_k} \sum_{K \in \mathbb{T}_h} \{\rho_K^1 \omega_K^1 + \rho_K^2 \omega_K^2\}, \tag{11.5}$$

where the residual terms and weights are defined by

$$\rho_K^1 := \|B - (T_\theta + \kappa I - \Sigma)u_h\|_{K \times \Delta},$$
$$\omega_K^1 := \|z - \varphi_h\|_{K \times \Delta} + \delta_K \|T_\theta(z - \varphi_h)\|_{K \times \Delta},$$
$$\rho_K^2 := h_K^{-1/2}\|\theta \cdot n u_h\|_{\partial K \cap \Gamma_{in,\theta}},$$
$$\omega_K^2 := h_K^{1/2}\|z - \varphi_h\|_{\partial K \cap \Gamma_{in,\theta}},$$

with a suitable approximation $\varphi_h \in V_h$ to z.

The 'weighted' estimator estimator η_ω will be compared to a heuristic L^2-error indicator of the form

$$\|e\|_{\Omega \times S_2} \le c_s c_i \left(\sum_{\Delta \in \mathbb{D}_k} \sum_{K \in \mathbb{T}_h} (h_K^2 + k_\Delta^2) \, \rho_K^2 \right)^{1/2}, \tag{11.6}$$

where the stability constant c_s is either computed by solving numerically the adjoint problem corresponding to the source term $\|e\|_{\Omega \times S_2}^{-1} e$, or simply set to $c_s = 1$. The 'interpolation constant' c_i is usually of moderate size $c_i \sim 0.1 - 1$. The inclusion of the scaling factor $h_K^2 + k_\Delta^2$ is only based on heuristic grounds since useful analytical *a priori* estimates for the adjoint problem (11.4) are not available.

Numerical results (from Kanschat (1996))
We consider a prototypical example from astrophysics. A satellite-based observer measures the light (at a fixed wavelength) emitted from a cosmic source hidden in a dust cloud. A sketch of this situation is shown in Figure 11.1. The measurement is compared with results of a (two-dimensional) simulation which assumes certain properties of the coefficients in the underlying radiative transfer model (11.1). Because of the distance to the source, only the mean value of the intensity emitted in the direction θ_{obs} can be measured. Thus, the quantity to be computed is

$$J(u) = \int_{\{n \cdot \theta_{\text{obs}} \ge 0\}} u(x, \theta_{\text{obs}}) \, ds,$$

where $\{n \cdot \theta_{\text{obs}} \ge 0\}$ is the outflow boundary of the computational domain $\Omega \times S_1$ ($\Omega \subset \mathbb{R}^2$ a square) containing the radiating object.

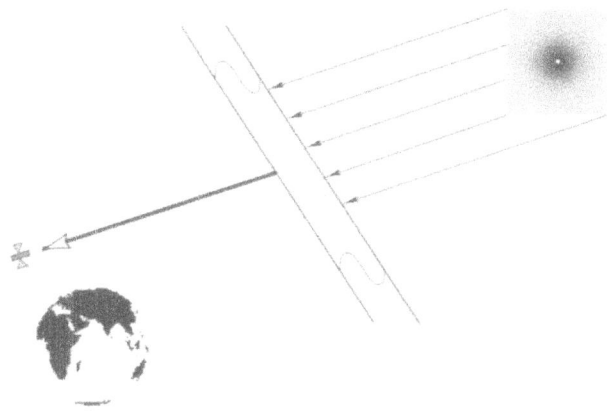

Fig. 11.1. Observer configuration of radiation emission

Table 11.1. Results obtained by the (heuristic) L^2-error indicator and the weighted error estimator, the total number of unknowns being $N_{\text{tot}} = N_x \cdot 32$

	L^2-indicator		Weighted estimator			
L	N_x	$J(u_h)$	N_x	$J(u_h)$	$\eta_\omega(u_h)$	$\eta_\omega(u_h)/J(e)$
2	1105	0.210	1146	0.429	1.0804	8.62
3	2169	0.311	2264	0.461	0.7398	7.11
4	4329	0.405	4506	0.508	0.2861	3.94
5	8582	0.460	9018	0.555	0.1375	3.33
6	17202	0.488	18857	0.584	0.0526	2.39
7	34562	0.537	39571	0.599	0.0211	1.76
8	68066	0.551	82494	0.608	0.0084	1.41
∞				0.618		

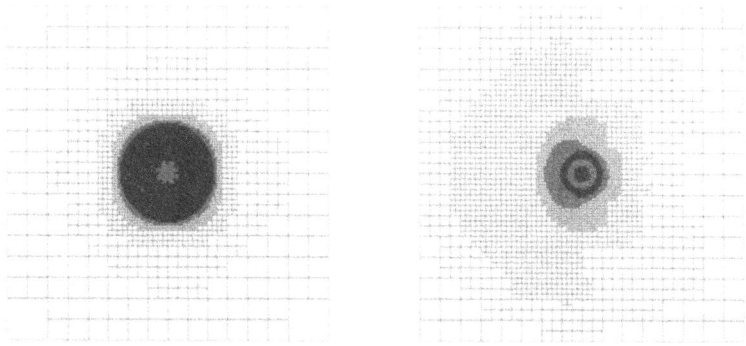

Fig. 11.2. Optimized meshes generated by the (heuristic) L^2-error indicator (left) and the weighted error estimator (right). The observer direction is to the left

The results in Table 11.1 demonstrate the superiority of the weighted error estimator over the heuristic global L^2-error indicator. The effect of the presence of the weights on the mesh refinement is shown in Figure 11.2.

Notes and references
Spatially three-dimensional radiative transfer problems in astrophysics can be solved with satisfactory accuracy *only* by systematically exploiting mesh adaptation and parallel computers. The application of the DWR method in

3D and the parallel implementation of the fully adaptive solution method has been accomplished by Kanschat (1996, 2000). A rigorous *a posteriori* error analysis can be found in Führer and Kanschat (1997), while the particular aspects of streamline diffusion in the context of radiative transport have been discussed by Kanschat (1998). For realistic astrophysical applications using the DWR see Wehrse, Meinkoehn and Kanschat (1999).

12. Application to optimal control

As the last example, we demonstrate the application of the DWR method in optimal control. This is a situation which is naturally suited to 'goal-oriented' *a posteriori* error estimation.

12.1. A linear model problem

We have chosen a very simple model problem with linear state equation in order to make the particular features of our approach clear. The state equation is given in terms of the mixed boundary value problem of the Helmholtz operator

$$- \Delta u + u = f \quad \text{on } \Omega,$$

$$\partial_n u = q \text{ on } \Gamma_C, \quad \partial_n u = 0 \text{ on } \partial\Omega \setminus \Gamma_C, \tag{12.1}$$

defined on a bounded domain $\Omega \subset \mathbb{R}^2$ with boundary $\partial\Omega$. The control q acts on the boundary component Γ_C, while the observations $u_{|\Gamma_O}$ are taken on a component Γ_O; see Figure 12.1, on page 90. The cost functional is defined by

$$J(u, q) = \tfrac{1}{2}\|u - u_O\|_{\Gamma_O}^2 + \tfrac{1}{2}\alpha\|q\|_{\Gamma_C}^2, \tag{12.2}$$

with a prescribed function u_O and a parameter $\alpha > 0$. We want to apply the general formalism of Section 2 to the Galerkin finite element approximation of this problem. This may be considered within the context of 'model reduction' in optimal control theory. First, we have to prepare the corresponding functional analytic setting. The functional of interest is the Lagrangian functional of the optimal control problem,

$$L(x) = J(u, q) + (\nabla u, \nabla z) + (u - f, z) - (q, z)_{\Gamma_C},$$

defined for triples $x = \{u, z, q\}$ in the Hilbert space $X := V \times V \times Q$, where $V := H^1(\Omega)$ and $Q := L^2(\Gamma_C)$. The Euler–Lagrange equations for stationary points $x = \{u, z, q\} \in X$ of $L(\cdot)$ are

$$L'(x; y) = 0, \quad \text{for all } y \in X, \tag{12.3}$$

or, written in explicit form,

$$
\begin{aligned}
(\varphi^u, u - u_O)_{\Gamma_O} + (\nabla\varphi^u, \nabla z) + (\varphi^u, z) &= 0, \quad \text{for all } \varphi^u \in V, \\
(\nabla u, \nabla \varphi^z) + (u - f, \varphi^z) - (q, \varphi^z)_{\Gamma_C} &= 0, \quad \text{for all } \varphi^z \in V, \\
(z - \alpha q, \varphi^q)_{\Gamma_C} &= 0, \quad \text{for all } \varphi^q \in Q.
\end{aligned}
\tag{12.4}
$$

The corresponding discrete approximations $x_h = \{u_h, z_h\, q_h\}$ are determined in the finite element space $X_h = V_h \times V_h \times Q_h \subset V$ by

$$
\begin{aligned}
(\psi_h, u_h - u_O)_{\Gamma_O} + (\nabla\psi_h, \nabla z_h) + (\psi_h, z_h) &= 0, \quad \text{for all } \psi_h \in V_h, \\
(\nabla u_h, \nabla \pi_h) + (u_h - f, \pi_h) - (q_h, \pi_h)_{\Gamma_C} &= 0, \quad \text{for all } \pi_h \in V_h, \\
(z_h - \alpha q_h, \chi_h)_{\Gamma_C} &= 0, \quad \text{for all } \chi_h \in Q_h.
\end{aligned}
\tag{12.5}
$$

Here, the trial spaces V_h for the state and co-state variables are as defined above in Section 3 (isoparametric bilinear shape functions), and the spaces Q_h for the controls consist of traces on Γ_C of V_h-functions, for simplicity. Following the general formalism, we seek to estimate the error $e = \{e^u, e^z, e^q\}$ with respect to the Lagrangian functional $L(\cdot)$. Proposition 2.1 yields the following abstract *a posteriori* estimate for the error in the Lagrangian:

$$
|L(x) - L(x_h)| \leq \eta(x_h) := \inf_{y_h \in X_h} \tfrac{1}{2}|L'(x_h; x - y_h)|.
\tag{12.6}
$$

In the present case, the cost functional is quadratic and the state equation linear, so that the remainder term in (2.6) vanishes. Since $\{u, z, q\}$ and $\{u_h, z_h, q_h\}$ satisfy (12.4) and (12.5), respectively,

$$
\begin{aligned}
L(x) - L(x_h) &= J(u, q) + (\nabla u, \nabla z) + (u - f, z) - (q, z)_{\Gamma_C} \\
&\quad - J(u_h, q_h) - (\nabla u_h, \nabla z_h) - (u_h - f, z_h) + (q_h, z_h)_{\Gamma_C} \\
&= J(u, q) - J(u_h, q_h).
\end{aligned}
$$

Hence, error control with respect to the Lagrangian functional $L(\cdot)$ and the cost functional $J(\cdot)$ are equivalent. Now, evaluation of the abstract error bound (12.6) again employs splitting the integrals into the contributions by the single cells, cell-wise integration by parts and Hölder's inequality (for the detailed argument see Kapp (2000) and Becker, Kapp and Rannacher (2000c)). For the following analysis, we introduce the cell residuals

$$
\begin{aligned}
R^u(x_h)_{|K} &:= f + \Delta u_h - u_h, \\
R^z(x_h)_{|K} &:= \Delta z_h - z_h,
\end{aligned}
$$

and edge residuals

$$r^u(x_h)|_\Gamma := \begin{cases} \frac{1}{2}n \cdot [\nabla u_h], & \text{if } \Gamma \not\subset \partial\Omega, \\ \partial_n u_h, & \text{if } \Gamma \subset \partial\Omega \backslash \Gamma_C, \\ \partial_n u_h - q_h, & \text{if } \Gamma \subset \Gamma_C, \end{cases}$$

$$r^z(x_h)|_\Gamma := \begin{cases} \frac{1}{2}n \cdot [\nabla z_h], & \text{if } \Gamma \not\subset \partial\Omega, \\ \partial_n z_h, & \text{if } \Gamma \subset \partial\Omega \backslash \Gamma_O, \\ \partial_n z_h + u_h - u_O, & \text{if } \Gamma \subset \Gamma_O, \end{cases}$$

$$r^q(x_h)|_\Gamma := \begin{cases} z_h - \alpha q_h, & \text{if } \Gamma \subset \Gamma_C, \\ 0, & \text{if } \Gamma \not\subset \Gamma_C, \end{cases}$$

associated with the solution $x_h = \{u_h, z_h, q_h\}$ of the system (12.5), where the notation $[\cdot]$ has its usual meaning. With this notation, we obtain from the abstract a *posteriori* error estimate (12.6) the following result for the present situation.

Proposition 12.1. For the finite element discretization of the system (12.5), we have the a *posteriori* error estimate

$$|J(u, q) - J(u_h, q_h)| \leq \eta_\omega(x_h)$$

$$:= \sum_{K \in \mathbb{T}_h} \{\rho_K^u \omega_K^z + \rho_K^z \omega_K^u + \rho_K^q \omega_K^q\}, \qquad (12.7)$$

where the cell-wise residuals and weights are defined by

$$\rho_K^z := \|R^z(x_h)\|_K + h_K^{-1/2}\|r^z(x_h)\|_{\partial K},$$

$$\omega_K^u := \|u - i_h u\|_K + h_K^{1/2}\|u - i_h u\|_{\partial K},$$

$$\rho_K^u := \|R^u(x_h)\|_K + h_K^{-1/2}\|r^u(x_h)\|_{\partial K},$$

$$\omega_K^z := \|z - i_h z\|_K + h_K^{1/2}\|z - i_h z\|_{\partial K},$$

$$\rho_K^q := h_K^{-1/2}\|r^q(x_h)\|_{\partial K},$$

$$\omega_K^q := h_K^{1/2}\|q - i_h q\|_{\partial K},$$

with suitable approximations $\{i_h u, i_h z, i_h q\} \in V_h \times V_h \times Q_h$.

Notice that the a *posteriori* error estimate (12.7) has very particular features. Its evaluation does not require the additional solution of an 'adjoint problem': the weights are rather generically obtained from the solution itself. The residuals of the state equation are weighted by the adjoint variable and, in turn, those of the adjoint equation by the primal variable. In this way, the particular sensitivities inherent to the optimization problem are reflected by the error estimator.

We will compare the performance of the weighted error estimator (12.7) with a more traditional error indicator. Control of the error in the 'energy norm' of the Euler–Lagrange equations (12.4) alone leads to the *a posteriori* error indicator

$$\eta_E(u_h) := c_i \left(\sum_{K \in \mathbb{T}_h} h_K^2 \left\{ (\rho_K^u)^2 + (\rho_K^z)^2 + (\rho_K^q)^2 \right\} \right)^{1/2}, \tag{12.8}$$

with the residual terms as defined above. This *ad hoc* criterion aims at satisfying the state equation uniformly with good accuracy. However, this concept seems questionable since it does not take into account the sensitivity of the cost functional with respect to the discretization. Capturing these dependencies is the particular feature of the DWR method.

Numerical results (from Becker, Kapp and Rannacher (2000c))
We consider the configuration as shown in Figure 12.1 with a T-shaped domain Ω of width one. The control acts along the lower boundary Γ_C, whereas the observations are taken along the (longer) upper boundary Γ_O. The cost functional is chosen as in (12.2) with $u_O \equiv 1$ and $\alpha = 1$, that is, the stabilization term constitutes a part of the cost functional.

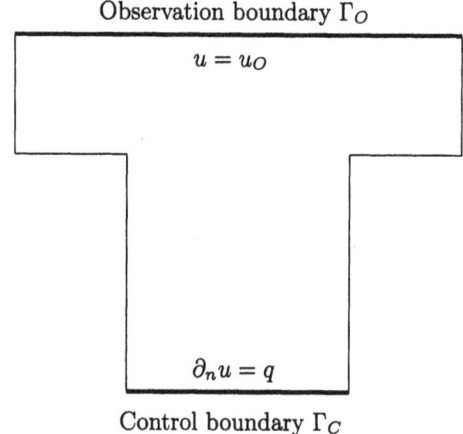

Fig. 12.1. Configuration of the boundary control
model problem

Figure 12.2 shows the quality of the weighted error estimator (12.7) for quantitative error control. The relative error E_{rel} and the effectivity index I_{eff} are defined as before. The reference value is obtained on a mesh with more than $200{,}000$ elements. We compare the weighted error estimator

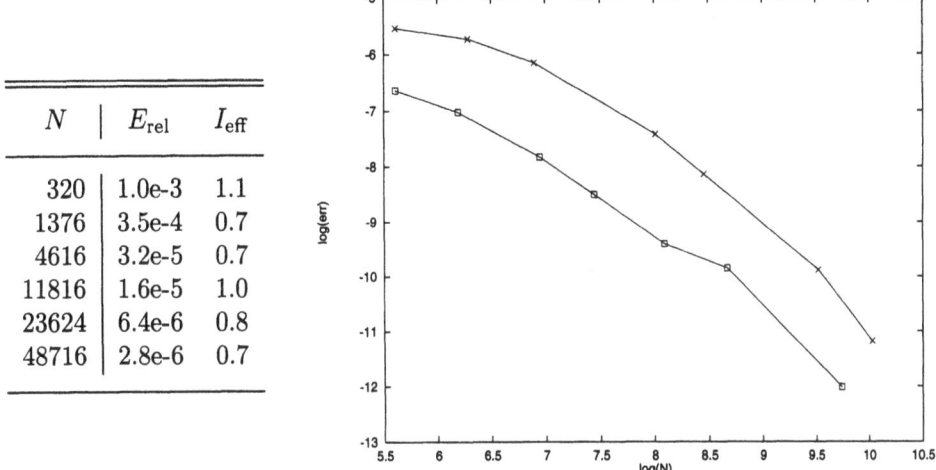

N	E_{rel}	I_{eff}
320	1.0e-3	1.1
1376	3.5e-4	0.7
4616	3.2e-5	0.7
11816	1.6e-5	1.0
23624	6.4e-6	0.8
48716	2.8e-6	0.7

Fig. 12.2. Effectivity of the weighted error estimator (left), and comparison of the efficiency of the meshes generated by the two estimators, '×' error values by the energy estimator, '□' error values by the weighted estimator (log−log scale)

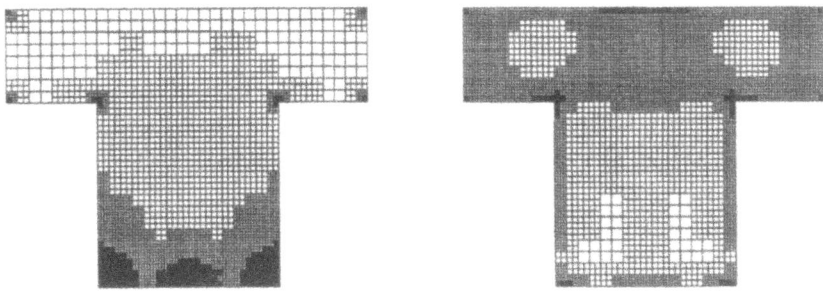

Fig. 12.3. Comparison between meshes obtained by the energy error estimator (left) and the weighted error estimator (right); $N \sim 5,000$ cells in both cases

with the 'energy error' estimator (12.8) for the state equation. Figure 12.3 shows meshes generated by the two estimators.

The difference in the meshes can be explained as follows. Obviously, the energy error estimator observes the irregularities introduced on the control boundary by the jump in the non-homogeneous Neumann condition, but it tends to over-refine in this region and to under-refine at the observation boundary. The weighted error estimator observes the needs of the optimization process by distributing the cells more evenly.

Notes and references

Here we have only presented results for optimal control problems with *linear* state equation. More realistic examples with nonlinear state equation (simplified Ginzburg–Landau model of superconductivity and Navier–Stokes equations in fluid mechanics) have been treated in Becker *et al.* (2000*c*) and Kapp (2000). The minimization of drag in a viscous flow is the subject of Becker (2001, 200x) and Becker, Heuveline and Rannacher (2001*b*).

13. Conclusion and outlook

In this article, we have presented a general approach to error control and mesh adaptation (the DWR method) for finite element approximations of variational problems. By employing optimal-control principles, *a posteriori* error estimates have been derived for the approximation of functionals of the solution related to quantities of physical interest. These error bounds are evaluated by numerically solving linearized 'adjoint problems'. In this context, we have discussed several theoretical and practical aspects of *a posteriori* error estimation and mesh adaptation.

The performance of the DWR method has been demonstrated for several linear and nonlinear model problems mainly from fluid and solid mechanics. In these tests the *dual-weighted* error estimators prove to be asymptotically correct and provide the basis of constructing economical meshes. Effectivity comparisons have been made with some of the traditional heuristic refinement indicators (*e.g.*, ZZ and energy error indicators). The evaluation of the 'weighted' error estimates requires us to solve a linear 'adjoint problem' on each mesh level. In the nonlinear case, this amounts to about the equivalent of *one* extra step within the Newton iteration on this mesh level. In this case, the extra work for mesh adaptation usually makes up 5–25% of the total work on the *optimized* mesh. However, the implementation may appear difficult when existing software components like mesh generators, multigrid solvers, *etc.*, cannot be directly used.

There are several open problems in the theoretical foundation of the DWR method, as well as in its practical realization, that need further investigation.

Theoretical foundation. The strategies for mesh adaptation are largely based on heuristic grounds. One hard open problem is the rigorous proof of the convergence of local residual terms and weights to certain 'limits' for TOL $\rightarrow 0$ or $N \rightarrow \infty$ as was (heuristically) assumed in (5.5). Further, the extension of the DWR concept for generating solution-adapted *anisotropic* meshes, either by simple cell stretching or by more sophisticated mesh reorientation, is still to be developed.

Control of linearization. In stiff nonlinear problems, *e.g.*, in the neighbourhood of bifurcation points, the linearization in the DWR method involves risks. We have shown that the effect of this linearization can, in principle, be controlled by *a posteriori* strategies. However, the realization of this approach for practical problems is still a critical open problem.

hp-**finite element method.** In this article the DWR method has been developed only for low-order finite element approximation (piece-wise linear or bilinear shape functions). The extension to higher-order approximation does not pose problems in principle. However, designing effective criteria for simultaneous adaptation of mesh size h and polynomial degree p is still not a fully satisfactorily solved problem, even for 'energy error' control. For the DWR method this problem is largely open.

Time-dependent problems. Multidimensional time-dependent problems constitute a major challenge for the DWR method. Rigorous error control in the space-time frame requires us to solve a space-time adjoint problem. Especially for nonlinear problems, this may be prohibitive with respect to storage space and computing time. The question is how to exploit the option of solving on coarser meshes only and that of data compression. The realization of these concepts for practical problems beyond simple model situations is still in an immature state.

3D problems. The use of the DWR method for spatially 3D problems, though theoretically straightforward, requires special care in setting up the data structure for mesh organization and assembling of the adjoint problem. In particular, the realization for 3D flow problems is a demanding task which has not yet been accomplished.

Finite volume methods. Residual-based methods for *a posteriori* error estimation like the DWR method rely on the variational formulation and the Galerkin orthogonality property of the finite element scheme. This allows us to locally extract additional powers of the mesh size, leading to sensitivity factors of the form $h_K^2 \|\nabla^2 z\|_K$. Other 'non-variational' discretizations such as the finite volume methods usually have a different error behaviour, governed by sensitivity factors like $h_K \|\nabla z\|_K$. This may be seen by re-interpreting these discretizations as perturbed finite element schemes obtained by evaluating local integrals by special low-order quadrature rules. It would be desirable to develop a rigorous *a posteriori* error analysis for finite volume schemes following the DWR approach.

Model adaptivity. The concept of *a posteriori* error control for single quantities of interest via duality may also be applicable to other situations when a full model (such as a differential equation) is reduced by projection to a subproblem (such as a finite element model). Model reduction within scales of hierarchical sub-models is a recent development in structural as well as in fluid mechanics (see Babuška and Schwab (1996), Stein and Ohnimus (1997), and Hughes, Feijoo, Mazzei and Quincy (1998)). Quantitative error control in these techniques by computational means like the DWR method seems to be a promising idea.

Adaptivity in optimal control. Most numerical simulation is eventually optimization. Complex multidimensional optimal control and parameter identification problems constitute highly demanding computational tasks. Goal-oriented model reduction by adaptive discretization has high potential to manage large-scale optimization problems in structural and fluid mechanics, such as, for example, minimization of drag, or control of flow-induced structural vibrations. The use of mesh adaptation techniques such as the DWR method for this purpose has just begun.

These are only some of the immediately obvious questions and directions of possible developments. In particular, the practical realization of the DWR method involves many other difficulties varying with the particular features of the problem to be solved, for instance, multi-target mesh adaptation. However, we think the effort invested is worthwhile, even for very complex problems, since, once the method works, the possible gain in accuracy and solution efficiency can be significant.

Acknowledgements

The authors wish to thank several members of the Numerical Analysis Group at their institute for important contributions to this survey article, particularly W. Bangerth, M. Braack, R. Hartmann, G. Kanschat, H. Kapp, F.-T. Suttmeier, C. Waguet, and B. Vexler. They have provided various results for particular cases and much of the computational material. Their comments and proof-reading have greatly helped the presentation.

We emphasize that the whole project would not have been realized without the availability of flexible software for carrying out the numerous calculations involving local mesh adaptation and fast multigrid solution. We credit G. Kanschat, F.-T. Suttmeier and the first author with the development of the finite element C++ package DEAL (Becker, Kanschat and Suttmeier 2001a), which in collaboration with W. Bangerth has since grown into the new package DEAL II (Bangerth, Hartmann and Kanschat 2001). This software was used for the computations reported in Sections 3–7 and 10–12. Most of the flow examples in Sections 8 and 9 were computed by the C++ code GASCOIGNE by M. Braack and the first author (Braack and Becker 2001).

Last, but not least, we acknowledge the financial support of the Deutsche Forschungsgemeinschaft (DFG) via SFB 359 'Reaktive Strömungen, Diffusion und Transport'.

REFERENCES

M. Ainsworth and J. T. Oden (1993), 'A unified approach to a posteriori error estimation using element residual methods', *Numer. Math.* **65**, 23–50.

M. Ainsworth and J. T. Oden (1997a), 'A posteriori error estimation in finite element analysis', *Comput. Meth. Appl. Mech. Eng.* **142**, 1–88.

M. Ainsworth and J. T. Oden (1997b), 'A posteriori error estimators for the Stokes and Oseen equations', *SIAM J. Numer. Anal.* **34**, 228–245.

I. Babuška and A. D. Miller (1984a), 'The post-processing approach in the finite element method, I: Calculations of displacements, stresses and other higher derivatives of the displacements', *Int. J. Numer. Meth. Eng.* **20**, 1085–1109.

I. Babuška and A. D. Miller (1984b), 'The post-processing approach in the finite element method, II: The calculation of stress intensity factors', *Int. J. Numer. Meth. Eng.* **20**, 1111–1129.

I. Babuška and A. D. Miller (1984c), 'The post-processing approach in the finite element method, III: A posteriori error estimation and adaptive mesh selection', *Int. J. Numer. Meth. Eng.* **20**, 2311–2324.

I. Babuška and A. D. Miller (1987), 'A feedback finite element method with a posteriori error estimation', *Comput. Meth. Appl. Mech. Eng.* **61**, 1–40.

I. Babuška and W. C. Rheinboldt (1978a), 'Error estimates for adaptive finite element computations', *SIAM J. Numer. Anal.* **15**, 736–754.

I. Babuška and W. C. Rheinboldt (1978b), 'A posteriori error estimates for the finite element method', *Int. J. Numer. Meth. Eng.* **12**, 1597–1615.

I. Babuška and C. Schwab (1996), 'A posteriori error estimation for hierarchic models of elliptic boundary value problems on thin domains', *SIAM J. Numer. Anal.* **33**, 221–246.

W. Bangerth (1998), 'Finite element approximation of the acoustic wave equation: Error control and mesh adaptation', Diplomarbeit, Institut für Angewandte Mathematik, Universität Heidelberg.

W. Bangerth (2000), Mesh adaptivity and error control for a finite element approximation of the elastic wave equation, in *Fifth International Conference on Mathematical and Numerical Aspects of Wave Propagation* (Waves2000) (A. Bermúdez et al., eds), SIAM, pp. 725–729.

W. Bangerth and R. Rannacher (1999), 'Finite element approximation of the acoustic wave equation: Error control and mesh adaptation', *East–West J. Numer. Math.* **7**, 263–282.

W. Bangerth, R. Hartmann and G. Kanschat (2001) deal.II: Differential Equations Analysis Library, Technical Reference. Available from:
http://gaia.iwr.uni-heidelberg.de/~deal/

R. E. Bank and A. Weiser (1985), 'Some a posteriori error estimators for elliptic partial differential equations', *Math. Comp.* **44**, 283–301.

R. Becker (1995), 'An adaptive finite element method for the incompressible Navier–Stokes equations on time-dependent domains', Doktorarbeit, Institut für Angewandte Mathematik, Universität Heidelberg.

R. Becker (1998*a*), An adaptive finite element method for the Stokes equations including control of the iteration error, in *ENUMATH'97* (H. G. Bock *et al.*, eds), World Scientific, Singapore, pp. 609–620.

R. Becker (1998*b*), Weighted error estimators for finite element approximations of the incompressible Navier–Stokes equations, Technical Report RR-3458, INRIA Sophia-Antipolis.

R. Becker (2001) Adaptive finite elements for optimal control problems, Habilitation thesis, Universität Heidelberg.

R. Becker (200x), 'Mesh adaptation for stationary flow control', *J. Math. Fluid Mech.* To appear.

R. Becker and M. Braack (2000), 'Multigrid techniques for finite elements on locally refined meshes', *Numer. Linear Algebra Appl.* **7**, 363–379.

R. Becker and R. Rannacher (1996*a*), Weighted a posteriori error control in FE methods. Lecture at ENUMATH-95, Paris, September 18–22, 1995, Preprint 96-01, SFB 359, Universität Heidelberg. In *Proc. ENUMATH'97* (H. G. Bock *et al.*, eds), World Scientific, Singapore (1998), pp. 621–637.

R. Becker and R. Rannacher (1996*b*), 'A feed-back approach to error control in finite element methods: Basic analysis and examples', *East–West J. Numer. Math.* **4**, 237–264.

R. Becker, C. Johnson and R. Rannacher (1995), 'Adaptive error control for multigrid finite element methods', *Computing* **55**, 271–288.

R. Becker, M. Braack, R. Rannacher and C. Waguet (1999), 'Fast and reliable solution of the Navier–Stokes equations including chemistry', *Comput. Visual. Sci.* **2**, 107–122.

R. Becker, M. Braack and R. Rannacher (2000*a*), Adaptive finite element methods for flow problems, in *FoCM'99* (A. Iserles, ed.), Cambridge University Press.

R. Becker, H. Kapp and R. Rannacher (2000*b*), Adaptive finite element methods for optimization problems, in *Numerical Analysis 1999* (D. F. Griffiths and G. A. Watson, eds), Chapman and Hall/CRC, London, pp. 21–42.

R. Becker, H. Kapp and R. Rannacher (2000*c*), 'Adaptive finite element methods for optimal control of partial differential equations: Basic concepts', *SIAM J. Control Optimiz.* **39**, 113–132.

R. Becker, G. Kanschat and F.-T. Suttmeier (2001*a*) DEAL: Differential Equations Analysis Library. Available from:
www.mathematik.uni-dortmund.de/user/lsx/suttmeier/deal/deal.html

R. Becker, V. Heuveline and R. Rannacher (2001*b*) An optimal control approach to adaptivity in computational fluid mechanics (or the power of elementary calculus), in *Proc. ICFD 2001, March 26–29, 2001, Oxford, UK*, Oxford University Press, to appear.

C. Bernardi, O. Bonnon, C. Langouët and B. Métivet (1995), Residual error indicators for linear problems: Extension to the Navier–Stokes equations, in *9th Int. Conf. Finite Elements in Fluids*.

H. Blum and F.-T. Suttmeier (1999), 'An adaptive finite element discretisation for a simplified Signorini problem', *Calcolo* **37**, 65–77.

H. Blum and F.-T. Suttmeier (2000), 'Weighted error estimates for finite element solutions of variational inequalities', *Computing* **65** 119–134.

K. Böttcher and R. Rannacher (1996), Adaptive error control in solving ordinary differential equations by the discontinuous Galerkin method, Technical Report Preprint 96-53, SFB 359, Universität Heidelberg.

M. Braack (1998), 'An adaptive finite element method for reactive flow problems', Doktorarbeit, Institut für Angewandte Mathematik, Universität Heidelberg.

M. Braack and R. Becker (2001) GASCOIGNE: Simulation tool for flows including chemical reactions. Available from: `http://gascoigne.uni-hd.de/`

M. Braack and R. Rannacher (1999), Adaptive finite element methods for low-Mach-number flows with chemical reactions, in 30*th Computational Fluid Dynamics* (H. Deconinck, ed.), Vol. 1999-03 of *Lecture Series*, von Karman Institute for Fluid Dynamics.

S. Brenner and R. L. Scott (1994), *The Mathematical Theory of Finite Element Methods*, Springer, Berlin/Heidelberg/New York.

G. Caloz and J. Rappaz (1997), Numerical analysis for nonlinear and bifurcation problems, in *Handbook of Numerical Analysis*, Vol. 5, *Techniques of Scientific Computing*, Part 2 (P. G. Ciarlet and J. L. Lions, eds), North-Holland, Amsterdam.

G. F. Carey and J. T. Oden (1984), *Finite Elements, Computational Aspects*, Vol. III, Prentice-Hall.

C. Carstensen and R. Verfürth (1999), 'Edge residuals dominate a posteriori error estimators for low-order finite element methods', *SIAM J. Numer. Anal.* **36**, 1571–1587.

P. G. Ciarlet (1978), *Finite Element Methods for Elliptic Problems*, North-Holland, Amsterdam.

W. Dörfler and M. Rumpf (1998), 'An adaptive strategy for elliptic problems including a posteriori controlled boundary approximation', *Math. Comp.* **224**, 1361–1382.

K. Eriksson and C. Johnson (1988), 'An adaptive finite element method for linear elliptic problems', *Math. Comp.* **50**, 361–383.

K. Eriksson and C. Johnson (1991), 'Adaptive finite element methods for parabolic problems, I: A linear model problem', *SIAM J. Numer. Anal.* **28**, 43–77.

K. Eriksson and C. Johnson (1993), 'Adaptive streamline diffusion finite element methods for stationary convection-diffusion problems', *Math. Comp.* **60**, 167–188.

K. Eriksson and C. Johnson (1995*a*), 'Adaptive finite element methods for parabolic problems, II: Optimal error estimates in $l_\infty l_2$ and $l_\infty l_\infty$', *SIAM J. Numer. Anal.* **32**, 706–740.

K. Eriksson and C. Johnson (1995*b*), 'Adaptive finite element methods for parabolic problems, IV: Nonlinear problems', *SIAM J. Numer. Anal.* **32**, 1729–1749.

K. Eriksson and C. Johnson (1995*c*), 'Adaptive finite element methods for parabolic problems, V: Long-time integration', *SIAM J. Numer. Anal.* **32**, 1750–1763.

K. Eriksson, D. Estep, P. Hansbo and C. Johnson (1995), Introduction to adaptive methods for differential equations, in *Acta Numerica* (A. Iserles, ed.), Vol. 4, Cambridge University Press, pp. 105–158.

K. Eriksson, C. Johnson and S. Larsson (1998), 'Adaptive finite element methods for parabolic problems, VI: Analytic semigroups', *SIAM J. Numer. Anal.* **35**, 1315–1325.

D. Estep (1995), 'A posteriori error bounds and global error control for approximation of ordinary differential equations', *SIAM J. Numer. Anal.* **32**, 1–48.

D. Estep and D. French (1994), 'Global error control for the continuous Galerkin finite element method for ordinary differential equations', *Modél. Math. Anal. Numér.* **28**, 815–852.

D. Estep and S. Larsson (1993), 'The discontinuous Galerkin method for semilinear parabolic problems', *Modél. Math. Anal. Numér.* **27**, 611–643.

C. Führer (1996), 'Error control in finite element methods for hyperbolic problems', Doktorarbeit, Institut für Angewandte Mathematik, Universität Heidelberg.

C. Führer and G. Kanschat (1997), 'A posteriori error control in radiative transfer', *Computing* **58**, 317–334.

C. Führer and R. Rannacher (1997), 'An adaptive streamline-diffusion finite element method for hyperbolic conservation laws', *East–West J. Numer. Math.* **5**, 145–162.

M. Giles, M. Larsson, M. Levenstam and E. Süli (1997), Adaptive error control for finite element approximations of the lift and drag coefficients in viscous flow, Technical Report NA-76/06, Oxford University Computing Laboratory.

P. Hansbo (200x), Three lectures on error estimation and adaptivity, in *Adaptive Finite Elements in Linear and Nonlinear Solid and Structural Mechanics* (E. Stein, ed.), Vol. 416 of *CISM Courses and Lectures*, Springer. To appear.

P. Hansbo and C. Johnson (1991), 'Adaptive streamline diffusion finite element methods for compressible flow using conservative variables', *Comput. Meth. Appl. Mech. Eng.* **87**, 267–280.

R. Hartmann (1998), 'A posteriori Fehlerschätzung und adaptive Schrittweiten- und Ortsgittersteuerung bei Galerkin-Verfahren für die Wärmeleitungsgleichung', Diplomarbeit, Institut für Angewandte Mathematik, Universität Heidelberg.

R. Hartmann (2000), A comparison of adaptive SDFEM and DG(r) method for linear transport problems, Technical report, Institut für Angewandte Mathematik, Universität Heidelberg.

R. Hartmann (200x), Adaptive FE-methods for conservation equations, in *8th International Conference on Hyperbolic Problems: Theory, Numerics, Applications* (HYP2000) (G. Warnecke, ed.) To appear.

F.-K. Hebeker and R. Rannacher (1999), 'An adaptive finite element method for unsteady convection-dominated flows with stiff source terms', *SIAM J. Sci. Comput.* **21**, 799–818.

V. Heuveline and R. Rannacher (2001*a*) A posteriori error control for finite element approximations of elliptic eigenvalue problems, Preprint 2001-08 (SFB 359), Universität Heidelberg. To appear in *Adv. Comput. Math.*

V. Heuveline and R. Rannacher (2001*b*) Adaptive finite element discretization of eigenvalue problems in hydrodynamic stability theory, Preprint, Institut für Angewandte Mathematik, Universität Heidelberg.

P. Houston, R. Rannacher and E. Süli (2000), 'A posteriori error analysis for stabilized finite element approximation of transport problems', *Comput. Meth. Appl. Mech. Eng.* **190**, 1483–1508.

T. J. R. Hughes and A. N. Brooks (1982), 'Streamline upwind/Petrov Galerkin formulations for convection dominated flows with particular emphasis on the incompressible Navier–Stokes equation', *Comput. Meth. Appl. Mech. Eng.* **32**, 199–259.

T. J. R. Hughes, L. P. Franca and M. Balestra (1986), 'A new finite element formulation for computational fluid dynamics, V: Circumvent the Babuška–Brezzi condition: A stable Petrov–Galerkin formulation for the Stokes problem accommodating equal order interpolation', *Comput. Meth. Appl. Mech. Eng.* **59**, 89–99.

T. J. R. Hughes, G. R. Feijoo, L. Mazzei and J.-B. Quincy (1998), 'The variational multiscale method: a paradigm for computational mechanics', *Comput. Meth. Appl. Mech. Eng.* **166**, 3–24.

C. Johnson (1987), *Numerical Solution of Partial Differential Equations by the Finite Element Method*, Cambridge University Press, Cambridge.

C. Johnson (1990), 'Adaptive finite element methods for diffusion and convection problems', *Comput. Meth. Appl. Mech. Eng.* **82**, 301–322.

C. Johnson (1992), 'Adaptive finite element methods for the obstacle problem', *Math. Models Meth. Appl. Sci.* **2**, 483–487.

C. Johnson (1993), 'Discontinuous Galerkin finite element methods for second order hyperbolic problems', *Comput. Meth. Appl. Mech. Eng.* **107**, 117–129.

C. Johnson (1994), A new paradigm for adaptive finite element methods, in *Proc. MAFELAP 93* (J. Whiteman, ed.), Wiley, pp. 105–120.

C. Johnson and P. Hansbo (1992a), Adaptive finite element methods for small strain elasto-plasticity, in *Finite Inelastic Deformations: Theory and Applications* (D. Besdo and E. Stein, eds), Springer, Berlin, pp. 273–288.

C. Johnson and P. Hansbo (1992b), 'Adaptive finite element methods in computational mechanics', *Comput. Meth. Appl. Mech. Eng.* **101**, 143–181.

C. Johnson and R. Rannacher (1994), On error control in CFD, in *Int. Workshop Numerical Methods for the Navier–Stokes Equations* (F.-K. Hebeker *et al.*, eds), Vol. 47 of *Notes Numer. Fluid Mech*, Vieweg, Braunschweig, pp. 133–144.

C. Johnson and A. Szepessy (1995), 'Adaptive finite element methods for conservation laws based on a posteriori error estimates', *Comm. Pure Appl. Math.* **48**, 199–234.

C. Johnson, R. Rannacher and M. Boman (1995), 'Numerics and hydrodynamic stability: Towards error control in CFD', *SIAM J. Numer. Anal.* **32**, 1058–1079.

G. Kanschat (1996), 'Parallel and adaptive Galerkin methods for radiative transfer problems', Doktorarbeit, Institut für Angewandte Mathematik, Universität Heidelberg.

G. Kanschat (1998), 'A robust finite element discretization for radiative transfer problems with scattering', *East–West J. Numer. Math.* **6**, 265–272.

G. Kanschat (2000), Solution of multi-dimensional radiative transfer problems on parallel computers, in *Parallel Solution of Partial Differential Equations* (P. Bjørstad and M. Luskin, eds), Vol. 120 of *IMA Volumes in Mathematics and its Applications*, Springer, New York, pp. 85–96.

H. Kapp (2000), 'Adaptive finite element methods for optimization in partial differential equations', Doktorarbeit, Institut für Angewandte Mathematik, Universität Heidelberg.

P. Ladeveze and D. Leguillon (1983), 'Error estimate procedure in the finite element method and applications', *SIAM J. Numer. Anal.* **20**, 485–509.

J. Lang (1999), 'Adaptive multilevel solution of nonlinear parabolic PDE systems: Theory, algorithms, and applications', Habilitationsschrift, Konrad-Zuse-Zentrum für Informationstechnik, Berlin.

M. G. Larson (2000), 'A posteriori and a priori error estimates for finite element approximations of selfadjoint eigenvalue problems', *SIAM J. Numer. Anal.* **38** 608–625.

M. G. Larson and A. J. Niklasson (1999), 'Adaptive multilevel finite element approximations of semilinear elliptic boundary value problems', *Numer. Math.* **84**, 249–274.

L. Machiels, A. T. Patera and J. Peraire (1998), Output bound approximation for partial differential equations: Application to the incompressible Navier–Stokes equations, in *Industrial and Environmental Applications of Direct and Large Eddy Numerical Simulation* (S. Biringen, H. Örs, A. Tezel and J. H. Ferziger, eds), Springer, Berlin/Heidelberg/New York, pp. 93–108.

L. Machiels, J. Peraire and A. T. Patera (1999), A posteriori finite element output bounds for the incompressible Navier–Stokes equations: Application to a natural convection problem, Technical Report 99-4, MIT FML.

J. Medina, M. Picasso and J. Rappaz (1996), 'Error estimates and adaptive finite elements for nonlinear diffusion–convection problems', *Math. Models Meth. Appl. Sci.* **5**, 689–712.

C. Nystedt (1995), A priori and a posteriori error estimates and adaptive finite element methods for a model eigenvalue problem, Technical Report Preprint NO 1995-05, Department of Mathematics, Chalmers University of Technology.

J. T. Oden and S. Prudhomme (1999), 'On goal-oriented error estimation for elliptic problems: Application to the control of pointwise errors', *Comput. Meth. Appl. Mech. Eng.* **176**, 313–331.

J. T. Oden, W. Wu and M. Ainsworth (1993), 'An a posteriori error estimate for finite element approximations of the Navier–Stokes equations', *Comput. Meth. Appl. Mech. Eng.* **111**, 185–202.

M. Paraschivoiu and A. T. Patera (1998), 'Hierarchical duality approach to bounds for the outputs of partial differential equations', *Comput. Meth. Appl. Mech. Eng.* **158**, 389–407.

J. Peraire and A. T. Patera (1998), Bounds for linear-functional outputs of coercive partial differential equations: Local indicators and adaptive refinement, in *Advances in Adaptive Computational Methods in Mechanics* (P. Ladeveze and J. T. Oden, eds), Elsevier, Amsterdam, pp. 199–215.

P. Le Quere and H. Paillere (2000), Modelling and simulation of natural convection flows with large temperature differences: A benchmark problem for low Mach number solvers. Available from:
    http://m17.limsi.fr/Individu/plq/nonbou/bench_english.html

R. Rannacher (1998a), Error control in finite element computations, in *Proc. Sum-*

mer School Error Control and Adaptivity in Scientific Computing (H. Bulgak and C. Zenger, eds), Kluwer Academic Publishers, pp. 247–278.

R. Rannacher (1998*b*), 'A posteriori error estimation in least-squares stabilized finite element schemes', *Comput. Meth. Appl. Mech. Eng.* **166**, 99–114.

R. Rannacher (200x), Duality techniques for error estimation and mesh adaptation in finite element methods, in *Adaptive Finite Elements in Linear and Nonlinear Solid and Structural Mechanics* (E. Stein, ed.), Vol. 416 of *CISM Courses and Lectures*, Springer. To appear.

R. Rannacher and L. R. Scott (1982), 'Some optimal error estimates for piecewise linear finite element approximations', *Math. Comp.* **38**, 437–445.

R. Rannacher and F.-T. Suttmeier (1997), 'A feed-back approach to error control in finite element methods: Application to linear elasticity', *Comput. Mechanics* **19**, 434–446.

R. Rannacher and F.-T. Suttmeier (1998), 'A posteriori error control in finite element methods via duality techniques: Application to perfect plasticity', *Comput. Mechanics* **21**, 123–133.

R. Rannacher and F.-T. Suttmeier (1999), 'A posteriori error estimation and mesh adaptation for finite element models in elasto-plasticity', *Comput. Meth. Appl. Mech. Eng.* **176**, 333–361.

R. Rannacher and F.-T. Suttmeier (200x), Error estimation and adaptive mesh design for FE models in elasto-plasticity, in *Error-Controlled Adaptive FEMs in Solid Mechanics* (E. Stein, ed.), Wiley. To appear.

M. Schäfer and S. Turek (1996), Benchmark computations of laminar flow around a cylinder (with support by F. Durst, E. Krause and R. Rannacher), in *Flow Simulation with High-Performance Computers, II: DFG Priority Research Program Results 1993–1995* (E. H. Hirschel, ed.), Vol. 52 of *Notes Numer. Fluid Mech.*, Vieweg, Wiesbaden, pp. 547–566.

J. Segatz, R. Rannacher, J. Wichmann, C. Orlemann, T. Dreier and J. Wolfrum (1996), 'Detailed numerical simulation in flow reactors: A new approach in measuring absolute rate constants', *J. Phys. Chem.* **100**, 9323–9333.

E. Stein and S. Ohnimus (1997), 'Coupled model- and solution-adaptivity in the finite-element method', *Comput. Meth. Appl. Mech. Eng.* **150**, 327–350.

F.-T. Suttmeier (1996), 'Adaptive finite element approximation of problems in elasto-plasticity theory', Doktorarbeit, Institut für Angewandte Mathematik, Universität Heidelberg.

F.-T. Suttmeier (1998), 'An adaptive displacement/pressure finite element scheme for treating incompressibility effects in elasto-plastic materials'. Preprint 1998-33, SFB 359, Universität Heidelberg. To appear in *Numer. Meth. Part. Diff. Equ.*

F.-T. Suttmeier (2001) General approach for a posteriori error estimates for finite element solutions of variational inequalities, *Comp. Mech.* **27**, 317–323.

R. Verfürth (1993), A posteriori error estimates for nonlinear problems, in *Numerical Methods for the Navier–Stokes Equations* (F.-K. Hebeker *et al.*, eds), Vol. 47 of *Notes Numer. Fluid Mech.*, Vieweg, Braunschweig, pp. 288–297.

R. Verfürth (1994), 'A posteriori error estimates for nonlinear problems: Finite element discretization of elliptic equations', *Math. Comp.* **62**, 445–475.

R. Verfürth (1996), *A Review of A Posteriori Error Estimation and Adaptive Mesh-Refinement Techniques*, Wiley/Teubner, New York/Stuttgart.

R. Verfürth (2000), 'A posteriori error estimation techniques for nonlinear elliptic and parabolic pde's', *Rev. Eur. Elem. Finis* **9**, 377–402.

B. Vexler (2000), 'A posteriori Fehlerschätzung und Gitteradaption bei Finite-Elemente-Approximationen nichtlinearer elliptischer Differentialgleichungen', Diplomarbeit, Institut für Angewandte Mathematik, Universität Heidelberg.

C. Waguet (2000), 'Adaptive finite element computation of chemical flow reactors', Doktorarbeit, Institut für Angewandte Mathematik, Universität Heidelberg.

R. Wehrse, E. Meinkoehn and G. Kanschat (1999), Recent work in Heidelberg on the solution of the radiative transfer equation, in *Forum du GRETA, Transfert de Rayonnement en Astrophysique* (P. Stee, ed.), CNRS-INSU-ASPS, pp. 82–99.

O. C. Zienkiewicz and J. Z. Zhu (1987), 'A simple error estimator and adaptive procedure for practical engineering analysis', *Int. J. Numer. Meth. Eng.* **24**, 337–357.

S. Ziukas and N.-E. Wiberg (1998), Adaptive procedure with superconvergent patch recovery for linear parabolic problems, in *Finite Element Methods: Superconvergence, Post-Processing, and A Posteriori Estimates* (M. Krizek *et al.*, eds), Vol. 196 of *Inc. Lect. Notes Pure Appl. Math.*, Marcel Dekker, pp. 303–314.

Acta Numerica (2001), pp. 103–214

Mathematical modelling of linearly elastic shells

Philippe G. Ciarlet

Laboratoire d'Analyse Numérique,

Université Pierre et Marie Curie,

4, Place Jussieu, 75005 Paris, France

E-mail: `pgc@ann.jussieu.fr`

The objective of this article is to lay down the proper *mathematical foundations of the two-dimensional theory of linearly elastic shells*. To this end, it provides, without any recourse to any *a priori* assumptions of a geometrical or mechanical nature, a *mathematical justification* of *two-dimensional linear shell theories*, by means of *asymptotic methods*, with the thickness as the 'small' parameter.

A major virtue of this approach is that it naturally leads to precise *mathematical definitions* of linearly elastic 'membrane' and 'flexural' shells. Another noteworthy feature is that it highlights in particular the role played by *two fundamental tensors*, each associated with a displacement field of the middle surface, the *linearized change of metric* and *linearized change of curvature tensors*.

More specifically, under fundamentally distinct sets of assumptions bearing on the *geometry of the middle surface*, on the *boundary conditions*, and on the *order of magnitude of the applied forces*, it is shown that the three-dimensional displacements, once properly *scaled*, *converge* (in H^1, or in L^2, or in *ad hoc* completions) as the thickness approaches zero towards a 'two-dimensional' limit that satisfies either the *linear two-dimensional equations of a 'membrane' shell* (themselves divided into two subclasses) or the *linear two-dimensional equations of a 'flexural' shell*. Note that this *asymptotic analysis* automatically provides in each case the 'limit' *two-dimensional equations*, together with the *function space* over which they are well-posed.

The linear two-dimensional shell equations that are most commonly used in numerical simulations, namely *Koiter's equations, Naghdi's equations*, and *'shallow' shell equations*, are then carefully described, mathematically analysed, and likewise justified by means of asymptotic analyses.

The *existence* and *uniqueness* of solutions to each one of these linear two-dimensional shell equations are also established by means of crucial *inequalities of Korn's type on surfaces*, which are proved in detail at the beginning of the article.

This article serves as a mathematical basis for the numerically oriented companion article by Dominique Chapelle, also in this issue of *Acta Numerica*.

CONTENTS

1. Ubiquity of shells

A *shell* is a three-dimensional elastic body that is geometrically character-ized by its middle surface and its 'small' thickness.

The *middle surface* S is a compact surface in $\mathbb{R}^3$ not contained in a plane (otherwise the shell is a plate) and it may or may not have a 'boundary' (for instance, the middle surface of a sail has a boundary, while that of a basketball has no boundary).

At each point $s \in S$, let $\boldsymbol{a}(s)$ denote a unit vector normal to S. Then the *reference configuration* of the shell, *i.e.*, the subset of $\mathbb{R}^3$ that it occupies 'before forces are applied to it', is a set of the form $\{(s + \zeta\boldsymbol{a}(s)) \in \mathbb{R}^3 : s \in S, |\zeta| \leq e(s)\}$, where the function $e : S \to \mathbb{R}$ is sufficiently smooth and satisfies $0 < e(s) \leq \varepsilon$ for all $s \in S$ and $\varepsilon > 0$ is thought of as being 'small' compared to some 'characteristic' length of S (its diameter for instance). If $e(s) = \varepsilon$ for all $s \in S$, the shell is said to have a *constant thickness* 2ε. If e is not a constant function, the shell is said to have a *variable thickness*.

Note that, since ε will essentially be used as a *dimensionless parameter* in the rest of this article, 2ε should thus be interpreted as the *ratio* between the actual thickness and a characteristic dimension of S, rather than as the thickness itself.

Shells and their assemblages constitute, or are found in, a wide variety of structures of considerable interest in contemporary engineering such as the blades of a rotor, an inner tube, a cooling tower, cylindrical tanks, balls used in various games, the sails and the hull of a sailing boat, a high-altitude scientific balloon (Figures 1.1 to 1.7); the doors, bumpers (fenders), bonnet (hood), windscreen (windshield), found in a car body; the wings, the tail, found in an aircraft; dams; parachutes.

Incidentally, these examples illustrate that actual shells generally have a variable thickness. For the sake of simplicity, we shall, however, only consider shells of constant thickness in this article, keeping in mind that this is not a serious restriction, as the effect of considering a variable thickness usually requires identical analyses, albeit involving substantially lengthier expressions at times.

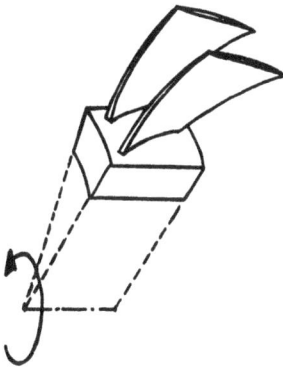

Fig. 1.1. A rotor and its blades provide an example of an elastic multi-structure, composed of a three-dimensional substructure (the rotor) and 'two-dimensional' substructures (the blades). Blades are often modelled as nonlinearly elastic 'shallow' shells

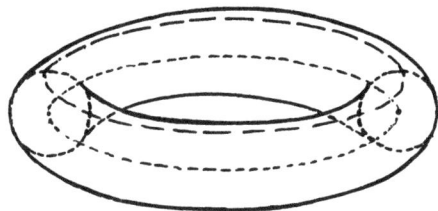

Fig. 1.2. An inner tube inside a tyre provides an example of a shell whose middle surface (a torus) has no boundary

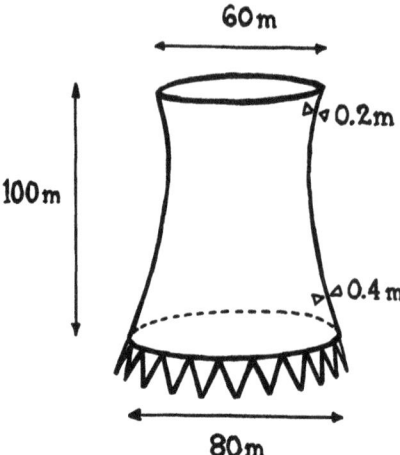

Fig. 1.3. A cooling tower in a utility plant: the middle surface is
approximately a ruled hyperboloid of revolution; the height is of
the order of 100 m, while the thickness varies from about 0.2 m at
the top to about 0.4 m at the bottom, thus providing an instance
of a ratio 2ε of approximately 1/500. Together with its supporting
rods, a cooling tower constitutes another elastic multi-structure,
composed of a 'two-dimensional' substructure (the shell) and 'one-
dimensional' substructures (the rods). Although an instance of
a *generalized membrane* shell (Section 9), such a shell is advan-
tageously modelled by Koiter's equations (Section 11)

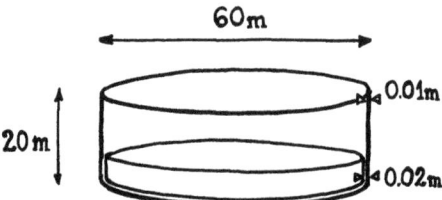

Fig. 1.4. A cylindrical tank for storing fuel in an oil refinery is
another elastic multi-structure, composed of two 'two-dimensional'
substructures: a cylindrical shell and a circular plate. In contemporary engineering, such a tank typically has a diameter of about
60 m, a height of 20 m, and a thickness varying from 0.04 m at the
top to 0.02 m at the bottom, thus providing an instance where the
ratio 2ε is approximately 1/5000

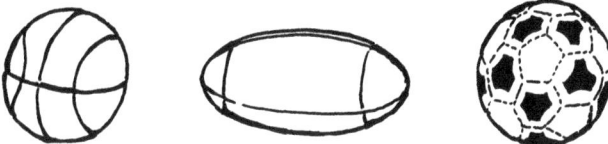

Fig. 1.5. Like an inner tube, balls used in various games provide examples of shells whose middle surface (a sphere or an ellipsoid-like surface) has no boundary. Another noteworthy feature, this time of a mechanical nature, of such shells is that they offer no resistance to crumpling when they are deflated. This observation alone suggests that they cannot be appropriately modelled by linear equations

Fig. 1.6. The sails and the hull of a sailing boat provide two strikingly different instances of shells. Like a balloon, a sail offers no resistance unless it is already under tension (think of a spinnaker); thus it must also be modelled by nonlinear equations. By contrast, linear equations should suffice for the modelling of the hull, because it is not expected to undergo large displacements. But, even within the linear realm, the mathematical modelling of such a shell is an extremely challenging problem, for such a shell is usually made of 'composite', 'multi-layered' elastic materials

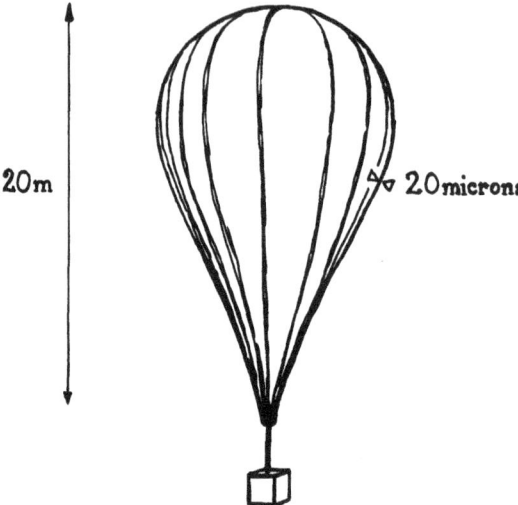

Fig. 1.7. A high-altitude scientific balloon provides a fascinating example. It is made by sealing together long, tapered, and originally flat sheets of polyethylene. The resulting structure is an incredibly thin shell, with an average thickness of about 20 microns and a height of about 20 m. The corresponding ratio 2ε is thus of the order of 10^{-6}, probably a world record! (This spectacular example was kindly brought to the author's attention by Frank Baginski, The George Washington University, Washington, DC.)

2. Why two-dimensional shell theories?

If any one of the structures described in Section 1 is viewed as a *three-dimensional elastic body*, the situation is on firm ground as regards its *mathematical modelling* (see, *e.g.*, Ciarlet (1988)). However, the situation is far from being idyllic as regards its *mathematical analysis*, at least if it is viewed as a *nonlinearly* elastic body. After the fundamental ideas set forth by Ball (1977) and his landmark existence result, there indeed remain various unresolved, and often exceedingly challenging, mathematical problems in nonlinear three-dimensional elasticity.

The *numerical analysis*, that is, the conception and mathematical analysis of convergent approximation schemes, most often finite element methods, is likewise well developed in three-dimensional elasticity, especially in the *linear* case (see in particular Ciarlet (1978, 1991), Glowinski (1984), Hughes (1987), Robert and Thomas (1991), Brezzi and Fortin (1991), Brenner and Scott (1994), Bathe (1996)), but also in the *nonlinear* case (see Le Tallec (1994) for an overview). There is nevertheless a strong proviso: three-dimensional numerical schemes almost invariably fail when they are applied to elastic structures that have a 'small' thickness, such as plates, shells, rods, and their assemblages.

The *'small' thickness* of a shell (or of a plate for that matter) makes it natural to 'replace' the genuine *three-dimensional model* by a 'simpler' *two-dimensional model*, that is, one that is posed over the middle surface of the shell. First, such a 'lower-dimensional' theory is of a *simpler mathematical structure*, which in turn generates a richer variety of results. Thus, while the 'global analysis', that is, the theories of existence, regularity, bifurcation, eversion phenomena, *etc.*, are still partly in their infancy in nonlinear *three-dimensional* elasticity (see in particular Marsden and Hughes (1983) and Ciarlet (1988)), such theories are by now on much firmer mathematical ground for the *two-dimensional* equations of nonlinearly elastic shells (see in particular Antman (1995) and Ciarlet (2000)).

In fact, not only is this replacement natural from a *theoretical* viewpoint, but it becomes a necessity when *numerical methods* must be devised for computing approximate displacements and stresses: any reasonably accurate three-dimensional discretization necessarily involves an astronomical number of unknowns, which renders it prohibitively expensive and makes its implementation extremely delicate, if not utterly impossible.

By contrast, the situation is on fairly safe ground, at least on the theoretical side, as regards the application of finite element methods to two-dimensional linear shell models: see in this respect Bernadou (1994) and the 'companion article' by Dominique Chapelle in this issue of *Acta Numerica*.

The above reasons clearly show why two-dimensional shell models are by and large preferred. Accordingly, three major questions naturally arise.

(i) How do we derive two-dimensional shell models in a systematic and rational manner from three-dimensional elasticity?

(ii) Has the mathematical analysis (existence, uniqueness, regularity, buckling, *etc.*, of solutions) of any known two-dimensional shell model reached a satisfactory stage?

(iii) In a given physical situation, how do we choose between the various 'available' two-dimensional shell models so that the chosen one be an 'as good as possible' approximation of the three-dimensional model it is supposed to 'replace'?

This last question is of paramount *practical* importance: it makes no sense to devise sophisticated numerical methods for accurately approximating the solution of the 'wrong' model!

The purpose of this article is to show how well-known, and sometimes not so well-known, two-dimensional linear shell equations can be fully justified by an asymptotic analysis of the three-dimensional equations, with the thickness as the 'small' parameter. It also provides a careful description of the physical situations where each kind of such equations should be safely employed.

This article thus only considers *linear* two-dimensional shell theories. A detailed justification from the same 'asymptotic' viewpoint, and a thorough mathematical analysis, of *nonlinear* two-dimensional shell theories are found in Ciarlet (2000).

Only recent references closely related to the 'asymptotic' approach followed here are listed in this article. The readers interested in an overview of the literature on shell theory may consult the reasonably complete bibliography provided, together with various historical commentaries, in Ciarlet (2000).

3. The three-dimensional Korn inequality in curvilinear coordinates

Although Sections 3 to 6 have a prelimininary character, they are essential: they provide an analysis of *Korn's inequalities in curvilinear coordinates*, whether in a three-dimensional domain or on a surface, which pervade most of the mathematical analysis of linearly elastic shells.

It is well known that the *three-dimensional Korn inequality* plays a fundamental role in establishing the *existence* and *uniqueness* of a solution in linearized three-dimensional elasticity in *Cartesian coordinates*. In essence, this inequality states that the $\mathbf{L}^2$-norm of the *linearized change of metric tensor* associated with displacement fields vanishing along a given portion, with area > 0, of the boundary of a domain in $\mathbb{R}^3$, is equivalent to the $\mathbf{H}^1$-norm of these fields, represented by means of their Cartesian components.

The objective of this section is to show that the three-dimensional Korn inequality can in fact be directly established in *curvilinear coordinates*; *cf.* Theorem 3.4.

A *domain* Ω in $\mathbb{R}^n$ is an open, bounded, connected subset of $\mathbb{R}^n$ with a Lipschitz-continuous boundary $\Gamma = \partial\Omega$, the set Ω being locally on one side of Γ. As Γ is Lipschitz-continuous, an *area element* $d\Gamma$ can be defined along Γ, and a *unit outer normal vector* $\boldsymbol{\nu} = (\nu_i)_{i=1}^n$ ('unit' meaning that its Euclidean norm is one) exists $d\Gamma$-almost everywhere along Γ.

Boldface letters denote vector-valued or matrix-valued functions and their associated function spaces. The norm in $L^2(\Omega)$ or $\mathbf{L}^2(\Omega)$ is denoted $|\cdot|_{0,\Omega}$ and the norm in the Sobolev spaces $H^m(\Omega)$ or $\mathbf{H}^m(\Omega), m \geq 1$, is denoted $\|\cdot\|_{m,\Omega}$. We also consider the Sobolev space

$$H^{-1}(\Omega) := \text{dual of space } H_0^1(\Omega).$$

It is clear that

$$v \in L^2(\Omega) \Rightarrow v \in H^{-1}(\Omega) \text{ and } \partial_i v \in H^{-1}(\Omega), 1 \leq i \leq n,$$

since (the duality between the spaces $\mathcal{D}(\Omega)$ and $\mathcal{D}'(\Omega)$ is denoted by $\langle \cdot, \cdot \rangle$)

$$|\langle v, \varphi \rangle| = \left| \int_\Omega v\varphi \, dx \right| \leq |v|_{0,\Omega} \|\varphi\|_{1,\Omega},$$

$$|\langle \partial_i v, \varphi \rangle| = |-\langle v, \partial_i \varphi \rangle| = \left| - \int_\Omega v \partial_i \varphi \, dx \right| \leq |v|_{0,\Omega} \|\varphi\|_{1,\Omega}$$

for all $\varphi \in \mathcal{D}(\Omega)$. It is *remarkable*, but also *remarkably difficult to prove*, that the converse implication holds.

Theorem 3.1: Lemma of J. L. Lions.   Let Ω be a domain in $\mathbb{R}^n$ and let v be a distribution on Ω. Then

$$\{v \in H^{-1}(\Omega) \text{ and } \partial_i v \in H^{-1}(\Omega), 1 \leq i \leq n\} \Rightarrow v \in L^2(\Omega). \qquad \square$$

This implication was first proved by J. L. Lions, as stated in Magenes and Stampacchia (1958, p. 320, Note 27). Its first published proof for domains with smooth boundaries appeared in Duvaut and Lions (1972, p. 111); another proof was also given by Tartar (1978). Extensions to 'genuine' domains, that is, with Lipschitz-continuous boundaries, were then given by Bolley and Camus (1976), Geymonat and Suquet (1986), Borchers and Sohr (1990), and Amrouche and Girault (1994).

From now on, Latin indices or exponents take their values in the set $\{1, 2, 3\}$ (except if they are used for indexing sequences) and the summation convention is used. The Euclidean inner product and the vector product of two vectors $\boldsymbol{u}, \boldsymbol{v} \in \mathbb{R}^3$ are denoted by $\boldsymbol{u} \cdot \boldsymbol{v}$ and $\boldsymbol{u} \wedge \boldsymbol{v}$; the Euclidean norm of $\boldsymbol{u} \in \mathbb{R}^3$ is denoted by $|\boldsymbol{u}|$.

Let Ω be a domain in $\mathbb{R}^3$, let $x = (x_i)$ denote a generic point in $\overline{\Omega}$, let $\partial_i = \partial/\partial x_i$, and let $\boldsymbol{\Theta} \in \mathcal{C}^2(\overline{\Omega}; \mathbb{R}^3)$ be a $\mathcal{C}^1$-diffeomorphism such that the three vectors $\boldsymbol{g}_i(x) := \partial_i \boldsymbol{\Theta}(x)$ are linearly independent at all points $x \in \overline{\Omega}$. The three vectors $\boldsymbol{g}_i(x)$ form the *covariant basis* at the point $\boldsymbol{\Theta}(x)$, while the three vectors $\boldsymbol{g}^i(x)$ defined by the relations $\boldsymbol{g}^i(x) \cdot \boldsymbol{g}_j(x) = \delta^i_j$ form the *contravariant basis* at the same point (δ^i_j designates the Kronecker symbol).

In particular, the mapping $\boldsymbol{\Theta} : \overline{\Omega} \to \mathbb{R}^3$ is injective, so that any point $\widehat{x} \in \boldsymbol{\Theta}(\overline{\Omega})$ is the image of a well-defined point $x \in \overline{\Omega}$. The three coordinates x_i of x then constitute the *curvilinear coordinates* of $\widehat{x}$.

Let $g_{ij} := \boldsymbol{g}_i \cdot \boldsymbol{g}_j$ and $g^{ij} := \boldsymbol{g}^i \cdot \boldsymbol{g}^j$ denote the covariant and contravariant components of the *metric tensor* of the set $\{\widehat{\Omega}\}^-$, where $\widehat{\Omega} := \boldsymbol{\Theta}(\Omega)$, let $g := \det(g_{ij})$, so that $\sqrt{g}\, dx$ denote the volume element in $\widehat{\Omega}$, and let $\Gamma^p_{ij} := \boldsymbol{g}^p \cdot \partial_i \boldsymbol{g}_j$ denote the *Christoffel symbols* (whenever no confusion should arise, the explicit dependence on $x \in \overline{\Omega}$ is henceforth omitted). The Christoffel symbols are used for computing the *first-order covariant derivatives*

$$v_{i\|j} := \partial_j v_i - \Gamma^p_{ij} v_p$$

of a vector field $v_i g^i$ defined over the set $\overline{\Omega}$ (for details about these classical notions, see, *e.g.*, Ciarlet (2000, Section 1.2)).

Consider a homogeneous, isotropic, elastic body whose reference configuration is the set $\{\widehat{\Omega}\}^-$ and assume furthermore that $\{\widehat{\Omega}\}^-$ is a *natural state*. When the equations of three-dimensional elasticity are stated 'in curvilinear coordinates', that is, in terms of the coordinates of the set $\overline{\Omega} = \Theta^{-1}(\{\widehat{\Omega}\}^-)$, the unknowns are the three covariant components $u_i : \overline{\Omega} \to \mathbb{R}$ of the displacement field $u_i g^i : \overline{\Omega} \to \mathbb{R}^3$ of the set $\{\widehat{\Omega}\}^-$. This means that, for each $x \in \overline{\Omega}$, $u_i(x) g^i(x)$ is the displacement of the point $\Theta(x) \in \Theta(\overline{\Omega}) = \{\widehat{\Omega}\}^-$.

In particular, the variational equations of linearized three-dimensional elasticity in curvilinear coordinates take the following form (see, *e.g.*, Ciarlet (2000, Theorem 1.3-1)). The field $u := (u_i)$ satisfies

$$u \in \mathbf{V}(\Omega) := \{v = (v_i) \in \mathbf{H}^1(\Omega) : v = 0 \text{ on } \Gamma_0\},$$

$$\int_\Omega A^{ijkl} e_{k\|l}(u) e_{i\|j}(v) \sqrt{g}\, dx = \int_\Omega f^i v_i \sqrt{g}\, dx$$

for all $v = (v_i) \in \mathbf{V}(\Omega)$, where Γ_0 is a given subset of the boundary Γ of Ω with area $\Gamma_0 > 0$, the contravariant components of the *three-dimensional elasticity tensor* of the body are denoted by

$$A^{ijkl} := \lambda g^{ij} g^{kl} + \mu(g^{ik} g^{jl} + g^{il} g^{jk}),$$

the *Lamé constants* of the constituent elastic material are denoted by λ and μ, the covariant components of the *linearized change of metric tensor* associated with an arbitrary displacement field $v_i g^i$ of the set $\{\widehat{\Omega}\}^-$ are denoted by

$$e_{i\|j}(v) := \frac{1}{2}(\partial_j v_i + \partial_i v_j) - \Gamma_{ij}^p v_p,$$

and the given functions $f^i \in L^2(\Omega)$ are the covariant components of the *applied body force* (we could as well consider surface forces acting on $\Gamma - \Gamma_0$). The functions $e_{i\|j}(v)$ are also called the *linearized strains in curvilinear coordinates*.

The interpretation of the functions $e_{i\|j}(v)$ is simple, yet crucial. Given an arbitrary displacement field $v_i g^i$ of the set $\Theta(\overline{\Omega})$ with sufficiently smooth covariant components $v_i : \overline{\Omega} \to \mathbb{R}$, let

$$g_{ij}(v) := \partial_i(\Theta + v_k g^k) \cdot \partial_j(\Theta + v_l g^l)$$

denote the covariant components of the *metric tensor* of the 'deformed' set $(\Theta + v_i g^i)(\overline{\Omega})$ associated with this displacement field. Then

$$e_{i\|j}(v) = \frac{1}{2}[g_{ij}(v) - g_{ij}]^{\text{lin}},$$

where $[\cdots]^{\text{lin}}$ denotes the linear part with respect to $v = (v_i)$ in the expression $[\cdots]$ (for a proof, see Ciarlet (2000, Theorem 1.5-1)). The components

$e_{i\|j}(\boldsymbol{v})$ are thus aptly called those of the 'linearized', 'change of metric', tensor associated with the displacement field $v_i \boldsymbol{g}^i$ of the set $\boldsymbol{\Theta}(\overline{\Omega})$.

The boundary condition $\boldsymbol{u} = \boldsymbol{0}$ on Γ_0, or the equivalent relation $u_i \boldsymbol{g}^i = \boldsymbol{0}$ on Γ_0, constitutes a (homogeneous) boundary condition of place. It states that the displacement field vanishes on the portion $\boldsymbol{\Theta}(\Gamma_0)$ of the boundary of the reference configuration $\boldsymbol{\Theta}(\overline{\Omega}) = \{\widehat{\Omega}\}^-$.

Naturally, the usual equations of linearized three-dimensional elasticity in Cartesian coordinates are recovered by letting $\boldsymbol{\Theta} = \boldsymbol{id}$, in which case $g^{ij} = \delta^{ij}$, $\Gamma^p_{ij} = 0$, and $g = 1$.

Since there exists a constant $C_e = C_e(\Omega, \boldsymbol{\Theta}, \mu)$ such that

$$\sum_{i,j} |t_{ij}|^2 \le C_e A^{ijkl}(x) t_{kl} t_{ij}$$

for all $x \in \overline{\Omega}$ and all symmetric matrices (t_{ij}) (see, e.g., Ciarlet (2000, Theorem 1.8-1)), establishing the existence and uniqueness of a solution to the above variational problem thus amounts to establishing the existence of a constant C such that

$$\|\boldsymbol{v}\|_{1,\Omega} \le C \left\{ \sum_{i,j} |e_{i\|j}(\boldsymbol{v})|^2_{0,\Omega} \right\}^{1/2}$$

for all $\boldsymbol{v} \in \mathbf{V}(\Omega)$ (all the other assumptions of the Lax–Milgram lemma are clearly satisfied). Our objective consists in proving that such a three-dimensional Korn inequality in curvilinear coordinates indeed holds (Theorem 3.4). Here, we follow Ciarlet (1993, 2000).

Such a Korn inequality is obtained in *three stages* (Theorems 3.2 to 3.4), the first one consisting in establishing, as a consequence of the Lemma of J. L. Lions (Theorem 3.1), a *Korn inequality valid for all vector fields* $\boldsymbol{v} = (v_i) \in \mathbf{H}^1(\Omega)$, i.e., that need not satisfy any boundary condition on Γ.

As its Cartesian special case, this inequality is truly remarkable, as only *six* different combinations of first-order partial derivatives, that is, $\frac{1}{2}(\partial_j v_i + \partial_i v_j)$, occur on its right-hand side, while all *nine* partial derivatives $\partial_j v_i$ occur on its left-hand side! A similarly striking observation applies to part (ii) of the proof of Theorem 3.2.

Theorem 3.2: Korn's inequality 'without boundary conditions' in curvilinear coordinates. Let Ω be a domain in $\mathbb{R}^3$ and let $\boldsymbol{\Theta} \in \mathcal{C}^2(\overline{\Omega}; \mathbb{R}^3)$ be a $\mathcal{C}^1$-diffeomorphism of $\overline{\Omega}$ onto $\{\widehat{\Omega}\}^- = \boldsymbol{\Theta}(\overline{\Omega})$ such that the three vectors $\boldsymbol{g}_i = \partial_i \boldsymbol{\Theta}$ are linearly independent at all points of $\overline{\Omega}$. Given $\boldsymbol{v} = (v_i) \in \mathbf{H}^1(\Omega)$, let

$$e_{i\|j}(\boldsymbol{v}) := \left\{ \frac{1}{2}(\partial_j v_i + \partial_i v_j) - \Gamma^p_{ij} v_p \right\} \in L^2(\Omega)$$

denote the covariant components of the linearized change of metric tensor associated with the displacement field $v_i g^i$ of the set $\Theta(\overline{\Omega})$. Then there exists a constant $C_0 = C_0(\Omega, \Theta)$ such that

$$\|v\|_{1,\Omega} \le C_0 \left\{ \sum_i |v_i|_{0,\Omega}^2 + \sum_{i,j} |e_{i\|j}(v)|_{0,\Omega}^2 \right\}^{1/2} \quad \text{for all } v \in \mathbf{H}^1(\Omega).$$

Proof. The proof given here is essentially an extension of that given in Duvaut and Lions (1972, p. 110) for proving Korn's inequality without boundary conditions in Cartesian coordinates.

(i) Define the space

$$\mathbf{W}(\Omega) := \left\{ v = (v_i) \in \mathbf{L}^2(\Omega) : e_{i\|j}(v) \in L^2(\Omega) \right\}.$$

Then, $\mathbf{W}(\Omega)$ is a Hilbert space when equipped with the norm $\| \cdot \|_{\mathbf{W}(\Omega)}$ defined by

$$\|v\|_{\mathbf{W}(\Omega)} := \left\{ \sum_i |v_i|_{0,\Omega}^2 + \sum_{i,j} |e_{i\|j}(v)|_{0,\Omega}^2 \right\}^{1/2}.$$

Note that the relations '$e_{i\|j}(v) \in L^2(\Omega)$' are understood in the sense of distributions. They mean that there exist functions $e_{i\|j}(v)$ in $L^2(\Omega)$ such that

$$\int_\Omega e_{i\|j}(v)\varphi \, dx = - \int_\Omega \left\{ \frac{1}{2}(v_i \partial_j \varphi + v_j \partial_i \varphi) + \Gamma_{ij}^p v_p \varphi \right\} dx \text{ for all } \varphi \in \mathcal{D}(\Omega).$$

Consider a Cauchy sequence $(v^k)_{k=1}^\infty$ with elements $v^k = (v_i^k) \in \mathbf{W}(\Omega)$. By definition of the norm $\| \cdot \|_{\mathbf{W}(\Omega)}$, there exist functions $v_i \in L^2(\Omega)$ and $e_{i\|j} \in L^2(\Omega)$ such that

$$v_i^k \to v_i \text{ in } L^2(\Omega) \text{ and } e_{i\|j}(v^k) \to e_{i\|j} \text{ in } L^2(\Omega) \text{ as } k \to \infty,$$

since the space $L^2(\Omega)$ is complete. Given a function $\varphi \in \mathcal{D}(\Omega)$, letting $k \to \infty$ in the relations

$$\int_\Omega e_{i\|j}(v^k)\varphi \, dx = - \int_\Omega \left\{ \frac{1}{2}(v_i^k \partial_j \varphi + v_j^k \partial_i \varphi) + \Gamma_{ij}^p v_p^k \varphi \right\} dx, \ k \ge 1,$$

shows that $e_{i\|j} = e_{i\|j}(v)$.

(ii) The spaces $\mathbf{W}(\Omega)$ and $\mathbf{H}^1(\Omega)$ coincide.

Clearly, $\mathbf{H}^1(\Omega) \subset \mathbf{W}(\Omega)$. To prove the other inclusion, let $v = (v_i) \in \mathbf{W}(\Omega)$. Then

$$e_{ij}(v) := \frac{1}{2}(\partial_j v_i + \partial_i v_j) = \left\{ e_{i\|j}(v) + \Gamma_{ij}^p v_p \right\} \in L^2(\Omega),$$

since $e_{i\|j}(\boldsymbol{v}) \in L^2(\Omega), \Gamma^p_{ij} \in \mathcal{C}^0(\overline{\Omega})$, and $v_p \in L^2(\Omega)$. We thus have

$$\partial_k v_i \in H^{-1}(\Omega),$$
$$\partial_j(\partial_k v_i) = \{\partial_j e_{ik}(\boldsymbol{v}) + \partial_k e_{ij}(\boldsymbol{v}) - \partial_i e_{jk}(\boldsymbol{v})\} \in H^{-1}(\Omega),$$

since $w \in L^2(\Omega)$ implies $\partial_k w \in H^{-1}(\Omega)$. Hence $\partial_k v_i \in L^2(\Omega)$ by the Lemma of J. L. Lions (Theorem 3.1) and thus $\boldsymbol{v} \in \mathbf{H}^1(\Omega)$.

(iii) Korn's inequality without boundary conditions.

The identity mapping ι from the space $\mathbf{H}^1(\Omega)$ equipped with $\|\cdot\|_{1,\Omega}$ into the space $\mathbf{W}(\Omega)$ equipped with $\|\cdot\|_{\mathbf{W}(\Omega)}$ is injective, continuous (there clearly exists a constant c such that $\|\boldsymbol{v}\|_{\mathbf{W}(\Omega)} \leq c\|\boldsymbol{v}\|_{1,\Omega}$ for all $\boldsymbol{v} \in \mathbf{H}^1(\Omega)$), and surjective by (ii). Since both spaces are complete (*cf.* (i)), the closed graph theorem then shows that the inverse mapping ι^{-1} is also continuous. This continuity is exactly what Korn's inequality without boundary conditions states. □

Our next objective is to 'get rid' of the norms $|v_i|_{0,\Omega}$ on the right-hand side of the Korn inequality established in Theorem 3.2 when the fields $\boldsymbol{v} = (v_i) \in \mathbf{H}^1(\Omega)$ are subjected to the boundary condition $\boldsymbol{v} = \boldsymbol{0}$ on $\Gamma_0 \subset \Gamma$ and area $\Gamma_0 > 0$. As a preliminary, we establish the weaker property that the *seminorm* $\boldsymbol{v} \to \{\sum_{i,j} |e_{i\|j}(\boldsymbol{v})|^2_{0,\Omega}\}^{1/2}$ becomes a *norm* for such fields, by generalizing to curvilinear coordinates the well-known *linearized rigid displacement lemma in Cartesian coordinates*. 'Linearized' reminds us that if $e_{i\|j}(\boldsymbol{v}) = 0$ in Ω, that is, if only the *linearized* part of the change of metric tensor vanishes, the corresponding displacement field $v_i \boldsymbol{g}^i$ is likewise only the *linearized approximation* to a genuine rigid displacement.

Part (a) in the next theorem is a linearized rigid displacement lemma *without boundary conditions*, while part (b) is a linearized rigid displacement lemma *with boundary conditions*.

Theorem 3.3: Linearized rigid displacement lemma in curvilinear coordinates. Let the assumptions be as in Theorem 3.2.

(a) Let $\boldsymbol{v} = (v_i) \in \mathbf{H}^1(\Omega)$ be such that

$$e_{i\|j}(\boldsymbol{v}) = 0 \text{ in } \Omega.$$

Then the vector field $v_i \boldsymbol{g}^i : \overline{\Omega} \to \mathbb{R}^3$ is a 'linearized rigid displacement' of the set $\boldsymbol{\Theta}(\overline{\Omega})$, in the sense that there exist two vectors $\widehat{\boldsymbol{c}}, \widehat{\boldsymbol{d}} \in \mathbb{R}^3$ such that

$$v_i(x)\boldsymbol{g}^i(x) = \widehat{\boldsymbol{c}} + \widehat{\boldsymbol{d}} \wedge \boldsymbol{\Theta}(x) \text{ for all } x \in \overline{\Omega}.$$

(b) Let Γ_0 be a $d\Gamma$-measurable subset of $\Gamma = \partial\Omega$ that satisfies

$$\text{area } \Gamma_0 > 0.$$

Then

$$v = (v_i) \in \mathbf{H}^1(\Omega), v = 0 \text{ on } \Gamma_0, \\ e_{i\|j}(v) = 0 \text{ in } \Omega \Bigg\} \Rightarrow v = 0 \text{ in } \Omega.$$

Proof. Let $\widehat{e}_i = \widehat{e}^i$ denote the basis vectors of the Cartesian frame. It is verified that the following relations hold:

$$\widehat{e}_{ij}(\widehat{v})(\widehat{x}) = \big(e_{k\|l}(v)[g^k]_i[g^l]_j\big)(x) \text{ for all } \widehat{x} = (\widehat{x}_i) := \Theta(x), x \in \Omega,$$

where $\widehat{e}_{ij}(\widehat{v}) := \frac{1}{2}(\widehat{\partial}_j \widehat{v}_i + \widehat{\partial}_i \widehat{v}_j), \widehat{\partial}_i := \partial/\partial \widehat{x}_i, [g^k]_i := g^k \cdot \widehat{e}_i$ denote the ith Cartesian component of the vector g^k, and the vector fields $\widehat{v} = (\widehat{v}_i) \in \mathbf{H}^1(\widehat{\Omega})$ and $v = (v_i) \in \mathbf{H}^1(\Omega)$ are related by

$$\widehat{v}_i(\widehat{x})\widehat{e}^i = v_i(x)g^i(x) \text{ for all } \widehat{x} = \Theta(x), x \in \Omega.$$

Hence

$$e_{i\|j}(v) = 0 \text{ in } \Omega \Rightarrow \widehat{e}_{ij}(\widehat{v}) = 0 \text{ in } \widehat{\Omega},$$

and the identity (the same as in the proof of Theorem 3.2)

$$\widehat{\partial}_j(\widehat{\partial}_k \widehat{v}_i) = \widehat{\partial}_j \widehat{e}_{ik}(\widehat{v}) + \widehat{\partial}_k \widehat{e}_{ij}(\widehat{v}) - \widehat{\partial}_i \widehat{e}_{jk}(\widehat{v}) \text{ in } \mathcal{D}'(\widehat{\Omega})$$

further shows that

$$\widehat{e}_{ij}(\widehat{v}) = 0 \text{ in } \widehat{\Omega} \Rightarrow \widehat{\partial}_j(\widehat{\partial}_k \widehat{v}_i) = 0 \text{ in } \mathcal{D}'(\widehat{\Omega}).$$

By a classical result from distribution theory (Schwartz 1966, p. 60), each function $\widehat{v}_i$ is therefore a polynomial of degree ≤ 1 (the set $\widehat{\Omega}$ is connected). In other words, there exist constants $\widehat{c}_i$ and $\widehat{d}_{ij}$ such that

$$\widehat{v}_i(\widehat{x}) = \widehat{c}_i + \widehat{d}_{ij}\widehat{x}_j \text{ for all } \widehat{x} = (\widehat{x}_i) \in \widehat{\Omega}.$$

But $\widehat{e}_{ij}(\widehat{v}) = 0$ also implies that $\widehat{d}_{ij} = -\widehat{d}_{ji}$; hence there exist two vectors $\widehat{c}, \widehat{d} \in \mathbb{R}^3$ such that

$$\widehat{v}_i(\widehat{x})\widehat{e}^i = \widehat{c} + \widehat{d} \wedge \widehat{O}\widehat{x} \text{ for all } \widehat{x} \in \widehat{\Omega},$$

and hence such that

$$v_i(x)g^i(x) = \widehat{c} + \widehat{d} \wedge \Theta(x) \text{ for all } x \in \Omega.$$

Since the set where such a vector field $\widehat{v}_i \widehat{e}^i$ vanishes is always of zero area unless $\widehat{c} = \widehat{d} = 0$ (as is easily proved), it follows that $\widehat{v} = 0$ in $\widehat{\Omega}$, and hence that $v = 0$ in Ω, when area $\Gamma_0 > 0$. □

We are now in a position to prove a fundamental inequality in curvilinear coordinates.

Theorem 3.4: Three-dimensional Korn's inequality in curvilinear coordinates. Let the assumptions be as in Theorem 3.2, let Γ_0 be a $d\Gamma$-measurable subset of $\Gamma = \partial\Omega$ that satisfies

$$\text{area } \Gamma_0 > 0,$$

and let the space $\mathbf{V}(\Omega)$ be defined by

$$\mathbf{V}(\Omega) := \left\{ v = (v_i) \in \mathbf{H}^1(\Omega) : v = 0 \text{ on } \Gamma_0 \right\}.$$

Then there exists a constant $C = C(\Omega, \Gamma_0, \Theta)$ such that

$$\|v\|_{1,\Omega} \leq C \left\{ \sum_{i,j} |e_{i\|j}(v)|^2_{0,\Omega} \right\}^{1/2} \quad \text{for all } v \in \mathbf{V}(\Omega).$$

Proof. Given $v = (v_i) \in \mathbf{H}^1(\Omega)$, let

$$|v|_{\mathbf{W}(\Omega)} := \left\{ \sum_{i,j} |e_{i\|j}(v)|^2_{0,\Omega} \right\}^{1/2}.$$

If the stated inequality is false, then there exists a sequence $(v^k)_{k=1}^{\infty}$ of elements $v^k \in \mathbf{V}(\Omega)$ such that

$$\|v^k\|_{1,\Omega} = 1 \text{ for all } k \text{ and } \lim_{k\to\infty} |v^k|_{\mathbf{W}(\Omega)} = 0.$$

Since the sequence $(v^k)_{k=1}^{\infty}$ is bounded in $\mathbf{H}^1(\Omega)$, there exists a subsequence $(v^l)_{l=1}^{\infty}$ that converges in $\mathbf{L}^2(\Omega)$ by the Rellich–Kondrašov theorem; furthermore, since $\lim_{l\to\infty} |v^l|_{\mathbf{W}(\Omega)} = 0$, each sequence $(e_{i\|j}(v^l))_{l=1}^{\infty}$ also converges in $L^2(\Omega)$ (to 0, but this information is not used at this stage). The subsequence $(v^l)_{l=1}^{\infty}$ is thus a Cauchy sequence with respect to the norm

$$v = (v_i) \to \left\{ \sum_{i} |v_i|^2_{0,\Omega} + \sum_{i,j} |e_{i\|j}(v)|^2_{0,\Omega} \right\}^{1/2},$$

and hence with respect to the norm $\| \cdot \|_{1,\Omega}$ by Korn's inequality without boundary conditions (Theorem 3.2).

The space $\mathbf{V}(\Omega)$ is complete, being a closed subspace of $\mathbf{H}^1(\Omega)$; thus there exists $v \in \mathbf{V}(\Omega)$ such that

$$v^l \to v \text{ in } \mathbf{H}^1(\Omega),$$

and the limit v satisfies $|e_{i\|j}(v)|_{0,\Omega} = \lim_{l\to\infty} |e_{i\|j}(v^l)|_{0,\Omega} = 0$; hence $v = 0$ by Theorem 3.3. But this contradicts the relations $\|v^l\|_{1,\Omega} = 1$ for all $l \geq 1$, and the proof is complete. $\qquad\square$

Letting $\Theta = id$ shows that Theorems 3.2, 3.3, and 3.4 contain as special cases the Korn inequalities and the linearized rigid displacement lemma in Cartesian coordinates (see, *e.g.*, Duvaut and Lions (1972)).

4. Inequality of Korn's type on a general surface

The theory of linearly elastic shells leads to 'two-dimensional' models, *i.e.*, that are defined in terms of *curvilinear coordinates* of the middle surface of the shell. The objective of Sections 4 to 6 is to show that *inequalities of Korn's type on a surface* can be established in terms of its curvilinear coordinates. As we shall see, such inequalities play a fundamental role in establishing the *existence* and *uniqueness* of solutions to such two-dimensional shell equations as the *Koiter, flexural,* and *membrane* ones. They also play a crucial role in the asymptotic analysis of the three-dimensional equations that justifies such two-dimensional models.

While a three-dimensional domain in $\mathbb{R}^3$ is unambiguously defined by a *single* tensor field, the metric tensor field (up to rigid deformations, of course) of a surface instead requires *two* tensor fields for its definition: the *metric tensor* field again and in addition the *curvature tensor* field, also called the *first* and *second fundamental forms* of the surface.

An inequality of Korn's type on a *general* surface can then be established. In essence, it states that, for a general surface S, the $\mathbf{L}^2$-norm of the *linearized change of metric tensor*, plus the $\mathbf{L}^2$-norm of the *linearized change of curvature tensor* (associated with displacement fields of S vanishing together with the normal derivative of their normal component along a given portion, with length > 0, of the 'boundary' of S) is equivalent to the $(H^1 \times H^1 \times H^2)$-norm of these fields, expressed here in curvilinear coordinates (both tangential components of the displacement fields are in H^1 and their normal components are in H^2); *cf.* Theorem 4.4.

To begin with, we briefly recall some basic results on the differential geometry of surfaces in $\mathbb{R}^3$; for references, see, *e.g.*, Stoker (1969), Klingenberg (1973), do Carmo (1976), Berger and Gostiaux (1992), Sanchez-Hubert and Sanchez-Palencia (1997), or Ciarlet (2000, Sections 2.1 to 2.5). Latin indices or components vary as before in the set $\{1, 2, 3\}$; in addition, Greek indices (except ν in ∂_ν) or exponents (except ε) vary in the set $\{1, 2\}$, and the summation convention now applies to both kinds of indices and exponents.

Let ω be a two-dimensional domain with boundary γ, let $y = (y_\alpha)$ denote a generic point in $\overline{\omega}$, let $\partial_\alpha = \partial/\partial y_\alpha$ and $\partial_{\alpha\beta} = \partial^2/\partial y_\alpha \partial y_\beta$, and let an injective mapping $\boldsymbol{\theta} \in \mathcal{C}^3(\overline{\omega}; \mathbb{R}^3)$ be given such that the two vectors $\boldsymbol{a}_\alpha(y) := \partial_\alpha \boldsymbol{\theta}(y)$ are linearly independent at all points $y \in \overline{\omega}$. They then form the *covariant basis of the tangent plane* to the surface $S := \boldsymbol{\theta}(\overline{\omega})$ at the point $\boldsymbol{\theta}(y)$, while the two vectors $\boldsymbol{a}^\alpha(y)$ of the tangent plane defined by the relations $\boldsymbol{a}^\alpha(y) \cdot \boldsymbol{a}_\beta(y) = \delta_\beta^\alpha$ form the *contravariant basis of the tangent plane at* $\boldsymbol{\theta}(y)$ (δ^α_β designates the Kronecker symbol). Let

$$\boldsymbol{a}^3(y) := \frac{\boldsymbol{a}_1(y) \wedge \boldsymbol{a}_2(y)}{|\boldsymbol{a}_1(y) \wedge \boldsymbol{a}_2(y)|};$$

then the vectors $\boldsymbol{a}^i(y)$ form the contravariant basis at the point $\boldsymbol{\theta}(y) \in S$.

The mapping $\boldsymbol{\theta} : \overline{\omega} \to \mathbb{R}^3$ being in particular injective, any point $\widehat{y}$ of the surface $S = \boldsymbol{\theta}(\overline{\omega})$ is the image of a well-defined point y in the set $\overline{\omega}$. The two coordinates y_α of y then constitute the *curvilinear coordinates* of $\widehat{y}$.

The *metric tensor*, or *first fundamental form*, of the surface S is defined by its *covariant components*

$$a_{\alpha\beta} := \boldsymbol{a}_\alpha \cdot \boldsymbol{a}_\beta = a_{\beta\alpha},$$

or by its *contravariant components*

$$a^{\alpha\beta} := \boldsymbol{a}^\alpha \cdot \boldsymbol{a}^\beta = a^{\beta\alpha}$$

(we omit the explicit dependence on $y \in \overline{\omega}$ when no confusion should arise). Note that the determinant

$$a := \det(a_{\alpha\beta})$$

is everywhere > 0 in $\overline{\omega}$ since the symmetric matrix $(a_{\alpha\beta})$ is positive definite in $\overline{\omega}$. The area element along S is $\sqrt{a}\, dy$.

The *curvature tensor*, or *second fundamental form*, of S is defined by its *covariant components*

$$b_{\alpha\beta} := \boldsymbol{a}_3 \cdot \partial_\alpha \boldsymbol{a}_\beta = -\partial_\alpha \boldsymbol{a}_3 \cdot \boldsymbol{a}_\beta = b_{\beta\alpha},$$

or by its *mixed components*

$$b^\beta_\alpha := a^{\beta\sigma} b_{\sigma\alpha}.$$

The *Christoffel symbols*

$$\Gamma^\sigma_{\alpha\beta} := \boldsymbol{a}^\sigma \cdot \partial_\alpha \boldsymbol{a}_\beta = \Gamma^\sigma_{\beta\alpha}$$

are used for computing the functions

$$\eta_{\beta|\alpha} := \partial_\alpha \eta_\beta - \Gamma^\sigma_{\alpha\beta} \eta_\sigma \quad \text{and} \quad \eta_{3|\alpha\beta} := \partial_{\alpha\beta} \eta_3 - \Gamma^\sigma_{\alpha\beta} \partial_\sigma \eta_3,$$

which are instances of first-order and second-order covariant derivatives of a vector field $\eta_i \boldsymbol{a}^i$ defined over the surface S, or for computing the functions

$$b^\tau_{\beta|\alpha} := \partial_\alpha b^\tau_\beta + \Gamma^\tau_{\alpha\sigma} b^\sigma_\beta + \Gamma^\sigma_{\alpha\beta} b^\tau_\sigma,$$

which are instances of first-order covariant derivatives of the curvature tensor of S, defined here by means of its mixed components.

The two-dimensional Koiter equations for a linearly elastic shell, which have been proposed by Koiter (1970), take the following form. The *unknowns* are the covariant components $\zeta^\varepsilon_{i,K} : \overline{\omega} \to \mathbb{R}$ of the displacement field

$\zeta_{i,K}^{\varepsilon} a^i : \overline{\omega} \to \mathbb{R}^3$ of the middle surface $S = \boldsymbol{\theta}(\overline{\omega})$ of the shell; $\boldsymbol{\zeta}_K^{\varepsilon} := (\zeta_{i,K}^{\varepsilon})$ satisfies

$$\boldsymbol{\zeta}_K^{\varepsilon} \in \mathbf{V}_K(\omega) := \{\boldsymbol{\eta} = (\eta_i) \in H^1(\omega) \times H^1(\omega) \times H^2(\omega) :$$
$$\eta_i = \partial_\nu \eta_3 = 0 \text{ on } \gamma_0\},$$

$$\int_\omega \left\{ \varepsilon a^{\alpha\beta\sigma\tau,\varepsilon} \gamma_{\sigma\tau}(\boldsymbol{\zeta}) \gamma_{\alpha\beta}(\boldsymbol{\eta}) + \frac{\varepsilon^3}{3} a^{\alpha\beta\sigma\tau,\varepsilon} \rho_{\sigma\tau}(\boldsymbol{\zeta}) \rho_{\alpha\beta}(\boldsymbol{\eta}) \right\} \sqrt{a}\, dy$$
$$= \int_\omega p^{i,\varepsilon} \eta_i \sqrt{a}\, dy$$

(∂_ν denoting the outer normal derivative operator along γ) for all $\boldsymbol{\eta} = (\eta_i) \in \mathbf{V}_K(\omega)$; γ_0 is a subset of γ with length $\gamma_0 > 0$; $2\varepsilon > 0$ is the thickness of the shell;

$$a^{\alpha\beta\sigma\tau,\varepsilon} := \frac{4\lambda^\varepsilon \mu^\varepsilon}{\lambda^\varepsilon + 2\mu^\varepsilon} a^{\alpha\beta} a^{\sigma\tau} + 2\mu^\varepsilon (a^{\alpha\sigma} a^{\beta\tau} + a^{\alpha\tau} a^{\beta\sigma})$$

denote the contravariant components of the two-dimensional elasticity tensor of the shell, λ^ε and μ^ε being the Lamé constants of the elastic material constituting the shell; the given functions $p^{i,\varepsilon} \in L^2(\omega)$ account for the applied forces. Finally, $\gamma_{\alpha\beta}(\boldsymbol{\eta})$ and $\rho_{\alpha\beta}(\boldsymbol{\eta})$ denote the covariant components of the linearized change of metric and linearized change of curvature tensors of S:

$$\gamma_{\alpha\beta}(\boldsymbol{\eta}) := \frac{1}{2}(\eta_{\alpha|\beta} + \eta_{\beta|\alpha}) - b_{\alpha\beta} \eta_3$$
$$= \frac{1}{2}(\partial_\beta \eta_\alpha + \partial_\alpha \eta_\beta) - \Gamma_{\alpha\beta}^\sigma \eta_\sigma - b_{\alpha\beta} \eta_3,$$

$$\rho_{\alpha\beta}(\boldsymbol{\eta}) := \eta_{3|\alpha\beta} - b_\alpha^\sigma b_{\sigma\beta} \eta_3 + b_\alpha^\sigma \eta_{\sigma|\beta} + b_\beta^\tau \eta_{\tau|\alpha} + b_\beta^\tau|_\alpha \eta_\tau$$
$$= \partial_{\alpha\beta} \eta_3 - \Gamma_{\alpha\beta}^\sigma \partial_\sigma \eta_3 - b_\alpha^\sigma b_{\sigma\beta} \eta_3$$
$$+ b_\alpha^\sigma (\partial_\beta \eta_\sigma - \Gamma_{\beta\sigma}^\tau \eta_\tau) + b_\beta^\tau (\partial_\alpha \eta_\tau - \Gamma_{\alpha\tau}^\sigma \eta_\sigma)$$
$$+ (\partial_\alpha b_\beta^\tau + \Gamma_{\alpha\sigma}^\tau b_\beta^\sigma - \Gamma_{\alpha\beta}^\sigma b_\sigma^\tau) \eta_\tau.$$

These functions play a fundamental role in linearized shell theory. As we shall see, they systematically appear in the linear two-dimensional shell equations later justified in this article!

Their interpretation, which is thus crucial for the understanding of these equations, is the following. Given an arbitrary displacement field $\eta_i a^i$ of the surface $S = \boldsymbol{\theta}(\overline{\omega})$ with sufficiently smooth covariant components $\eta_i : \overline{\omega} \to \mathbb{R}$, let

$$\boldsymbol{a}_\alpha(\boldsymbol{\eta}) := \partial_\alpha(\boldsymbol{\theta} + \eta_i a^i) \text{ and } \boldsymbol{a}_3(\boldsymbol{\eta}) := \frac{\boldsymbol{a}_1(\boldsymbol{\eta}) \wedge \boldsymbol{a}_2(\boldsymbol{\eta})}{|\boldsymbol{a}_1(\boldsymbol{\eta}) \wedge \boldsymbol{a}_2(\boldsymbol{\eta})|}$$

denote the vectors of the covariant bases attached to the 'deformed' surface $(\boldsymbol{\theta} + \eta_i a^i)(\overline{\omega})$ associated with this displacement field. Since the vectors

$a_\alpha = \partial_\alpha \theta$ are linearly independent in $\bar{\omega}$ by assumption, so are the vectors $a_\alpha(\eta)$ provided the fields $\eta = (\eta_i)$ are sufficiently small (*e.g.*, with respect to the norm of the space $\mathcal{C}^1(\bar{\omega}; \mathbb{R}^3)$); hence the vector $a_3(\eta)$ is well defined for such fields. The following interpretation is thus legitimate, because it only pertains to the linearized theory 'around $\eta = 0$'.

Let

$$a_{\alpha\beta}(\eta) := a_\alpha(\eta) \cdot a_\beta(\eta) \quad \text{and} \quad b_{\alpha\beta}(\eta) := a_3(\eta) \cdot \partial_\alpha a_\beta(\eta)$$

denote the covariant components of the metric and curvature tensors of the deformed surface $(\theta + \eta_i a^i)(\bar{\omega})$. Then

$$\gamma_{\alpha\beta}(\eta) = \frac{1}{2}[a_{\alpha\beta}(\eta) - a_{\alpha\beta}]^{\text{lin}},$$

$$\rho_{\alpha\beta}(\eta) = [b_{\alpha\beta}(\eta) - b_{\alpha\beta}]^{\text{lin}},$$

where $[\cdots]^{\text{lin}}$ denotes the linear part with respect to $\eta = (\eta_i)$ in the expression $[\cdots]$ (for a proof, see Ciarlet (2000, Theorems 2.4-1 and 2.5-1)). The components $\gamma_{\alpha\beta}(\eta)$ and $\rho_{\alpha\beta}(\eta)$ are thus aptly called those of the 'linearized', 'change of metric' and 'change of curvature' tensors associated with the displacement field $\eta_i a^i$ of the surface S.

Koiter's equations are of paramount importance in engineering practice, as they are very often used in numerical simulations of shell structures. They are further studied, and in particular fully justified, in Section 11.

As is easily seen (see *e.g.* Bernadou, Ciarlet and Miara (1994, Lemma 2.1), or Ciarlet (2000, Theorem 3.3-2)), there exists a constant $c_e = c_e(\omega, \theta, \mu^\varepsilon)$ such that

$$\sum_{\alpha,\beta} |t_{\alpha\beta}|^2 \leq c_e a^{\alpha\beta\sigma\tau,\varepsilon}(y) t_{\sigma\tau} t_{\alpha\beta}$$

for all $y \in \bar{\omega}$ and all symmetric matrices $(t_{\alpha\beta})$ and there exists a constant a_0 such that $a(y) \geq a_0 > 0$ for all $y \in \bar{\omega}$. Establishing the existence and uniqueness of a solution to this variational problem by the Lax–Milgram lemma thus amounts to establishing the existence of a constant c such that

$$\left\{ \sum_\alpha \|\eta_\alpha\|_{1,\omega}^2 + \|\eta_3\|_{2,\omega}^2 \right\}^{1/2} \leq c \left\{ \sum_{\alpha,\beta} |\gamma_{\alpha\beta}(\eta)|_{0,\omega}^2 + \sum_{\alpha,\beta} |\rho_{\alpha\beta}(\eta)|_{0,\omega}^2 \right\}^{1/2}$$

for all $\eta \in \mathbf{V}_K(\omega)$.

The objective of this section consists in showing that such an inequality of Korn's type indeed holds for a general surface (Theorem 4.4).

As is readily checked, the same inequality of Korn's type on a surface also provides an existence and uniqueness theorem for the two-dimensional equations of a linearly elastic 'flexural' shell. These equations, which will be fully justified in Section 10 through an asymptotic analysis of the three-dimensional solutions under the assumption that the space

$$\mathbf{V}_F(\omega) := \{\boldsymbol{\eta} \in \mathbf{V}_K(\omega) : \gamma_{\alpha\beta}(\boldsymbol{\eta}) = 0 \text{ in } \omega\}$$

does not reduce to $\{\mathbf{0}\}$, consist in finding the solution $\boldsymbol{\zeta}^\varepsilon = (\zeta_i^\varepsilon)$ of the following variational problem:

$$\boldsymbol{\zeta}^\varepsilon \in \mathbf{V}_F(\omega),$$

$$\frac{\varepsilon^3}{3} \int_\omega a^{\alpha\beta\sigma\tau,\varepsilon} \rho_{\sigma\tau}(\boldsymbol{\zeta}^\varepsilon) \rho_{\alpha\beta}(\boldsymbol{\eta}) \sqrt{a}\, dy = \int_\omega p^{i,\varepsilon} \eta_i \sqrt{a}\, dy$$

for all $\boldsymbol{\eta} = (\eta_i) \in \mathbf{V}_F(\omega)$.

In Section 3, we established 'three-dimensional' Korn inequalities, first without (Theorem 3.2), then with (Theorem 3.4), boundary conditions (the second one depending on a three-dimensional linearized rigid displacement lemma; cf. Theorem 3.3). Both inequalities involved the covariant components $e_{i\|j}(\boldsymbol{v})$ of the three-dimensional linearized change of metric tensor.

But while only *one* tensor, the metric tensor, is attached to a three-dimensional domain in $\mathbb{R}^3$, *two* tensors, the metric and curvature tensors, are attached to a surface in $\mathbb{R}^3$. It is thus natural to likewise establish inequalities of Korn's type *on a surface*, first *without* (Theorem 4.1), then *with* (Theorem 4.4), boundary conditions (the second one again depending on a linearized rigid displacement lemma, this time on a surface; cf. Theorem 4.3), such inequalities now involving the covariant components $\gamma_{\alpha\beta}(\boldsymbol{\eta})$ and $\rho_{\alpha\beta}(\boldsymbol{\eta})$ of both its linearized change of metric tensor and linearized change of curvature tensor.

We shall establish that these inequalities are valid for a 'general' surface $S = \boldsymbol{\theta}(\overline{\omega})$, that is, corresponding to a general mapping $\boldsymbol{\theta}$ (except that $\boldsymbol{\theta}$ should be sufficiently smooth; 'less smooth' mappings $\boldsymbol{\theta}$ are considered in Section 5). In other words, no restriction is imposed on the 'geometry' of S (in contrast, such a restriction holds for the inequality of Korn's type that will be established in Section 6).

The linearized rigid displacement lemma (Theorem 4.3) and the inequality of Korn's type on a general surface (Theorem 4.4) were first established by Bernadou and Ciarlet (1976). A simpler presentation, which we follow here, was then proposed by Ciarlet and Miara (1992b) (see also Bernadou, Ciarlet and Miara (1994)). Its first stage consists in establishing an inequality of Korn's type 'without boundary conditions', again as a consequence of the Lemma of J. L. Lions (as in dimension three; cf. Theorem 3.2).

Theorem 4.1: Inequality of Korn's type 'without boundary conditions' on a general surface. Let ω be a domain in $\mathbb{R}^2$ and let $\boldsymbol{\theta} \in C^3(\overline{\omega}; \mathbb{R}^3)$ be an injective mapping such that the two vectors $\boldsymbol{a}_\alpha = \partial_\alpha \boldsymbol{\theta}$ are linearly independent at all points of $\overline{\omega}$. Given $\boldsymbol{\eta} = (\eta_i) \in H^1(\omega) \times$

$H^1(\omega) \times H^2(\omega)$, let

$$\gamma_{\alpha\beta}(\boldsymbol{\eta}) := \left\{\frac{1}{2}(\partial_\beta\eta_\alpha + \partial_\alpha\eta_\beta) - \Gamma_{\alpha\beta}^\sigma\eta_\sigma - b_{\alpha\beta}\eta_3\right\} \in L^2(\omega),$$

$$\rho_{\alpha\beta}(\boldsymbol{\eta}) := \{\partial_{\alpha\beta}\eta_3 - \Gamma_{\alpha\beta}^\sigma\partial_\sigma\eta_3 - b_\alpha^\sigma b_{\sigma\beta}\eta_3$$

$$+ b_\alpha^\sigma(\partial_\beta\eta_\sigma - \Gamma_{\beta\sigma}^\tau\eta_\tau) + b_\beta^\tau(\partial_\alpha\eta_\tau - \Gamma_{\alpha\tau}^\sigma\eta_\sigma)$$

$$+ (\partial_\alpha b_\beta^\tau + \Gamma_{\alpha\sigma}^\tau b_\beta^\sigma - \Gamma_{\alpha\beta}^\sigma b_\sigma^\tau)\eta_\tau\} \in L^2(\omega)$$

denote the covariant components of the linearized change of metric and linearized change of curvature tensors associated with the displacement field $\eta_i \boldsymbol{a}^i$ of the surface $\boldsymbol{\theta}(\overline{\omega})$. Then there exists a constant $c_0 = c_0(\omega, \boldsymbol{\theta})$ such that

$$\left\{\sum_\alpha \|\eta_\alpha\|_{1,\omega}^2 + \|\eta_3\|_{2,\omega}^2\right\}^{1/2}$$

$$\leq c_0\left\{\sum_\alpha |\eta_\alpha|_{0,\omega}^2 + \|\eta_3\|_{1,\omega}^2 + \sum_{\alpha,\beta}|\gamma_{\alpha\beta}(\boldsymbol{\eta})|_{0,\omega}^2 + \sum_{\alpha,\beta}|\rho_{\alpha\beta}(\boldsymbol{\eta})|_{0,\omega}^2\right\}^{1/2}$$

for all $\boldsymbol{\eta} = (\eta_i) \in H^1(\omega) \times H^1(\omega) \times H^2(\omega)$.

Proof. **(i)** Define the space

$$\mathbf{W}_K(\omega) = \{\boldsymbol{\eta} = (\eta_i) \in L^2(\omega) \times L^2(\omega) \times H^1(\omega) :$$

$$\gamma_{\alpha\beta}(\boldsymbol{\eta}) \in L^2(\omega), \rho_{\alpha\beta}(\boldsymbol{\eta}) \in L^2(\omega)\}.$$

Then $\mathbf{W}_K(\omega)$ is a Hilbert space when equipped with the norm $\|\cdot\|_\omega^K$ defined by

$$\|\boldsymbol{\eta}\|_\omega^K := \left\{\sum_\alpha |\eta_\alpha|_{0,\omega}^2 + \|\eta_3\|_{1,\omega}^2 + \sum_{\alpha,\beta}|\gamma_{\alpha\beta}(\boldsymbol{\eta})|_{0,\omega}^2 + \sum_{\alpha,\beta}|\rho_{\alpha\beta}(\boldsymbol{\eta})|_{0,\omega}^2\right\}^{1/2}.$$

The relations '$\gamma_{\alpha\beta}(\boldsymbol{\eta}) \in L^2(\omega)$' and '$\rho_{\alpha\beta}(\boldsymbol{\eta}) \in L^2(\omega)$' appearing in the definition of the space $\mathbf{W}_K(\omega)$ are to be understood in the sense of distributions. They mean that $\boldsymbol{\eta} = (\eta_i) \in L^2(\omega) \times L^2(\omega) \times H^1(\omega)$ belongs to $\mathbf{W}_K(\omega)$ if there exist functions in $L^2(\omega)$, denoted by $\gamma_{\alpha\beta}(\boldsymbol{\eta})$ and $\rho_{\alpha\beta}(\boldsymbol{\eta})$, such that, for all $\varphi \in \mathcal{D}(\omega)$,

$$\int_\omega \gamma_{\alpha\beta}(\boldsymbol{\eta})\varphi \, d\omega = -\int_\omega \left\{\frac{1}{2}(\eta_\beta\partial_\alpha\varphi + \eta_\alpha\partial_\beta\varphi) + \Gamma_{\alpha\beta}^\sigma\eta_\sigma\varphi + b_{\alpha\beta}\eta_3\varphi\right\} d\omega,$$

$$\int_\omega \rho_{\alpha\beta}(\boldsymbol{\eta})\varphi \, d\omega = -\int_\omega \{\partial_\alpha\eta_3\partial_\beta\varphi + \Gamma_{\alpha\beta}^\sigma\partial_\sigma\eta_3\varphi + b_\alpha^\sigma b_{\sigma\beta}\eta_3\varphi$$

$$+ b_\alpha^\sigma(\eta_\sigma\partial_\beta\varphi + \Gamma_{\beta\sigma}^\tau\eta_\tau\varphi) + b_\beta^\tau(\eta_\tau\partial_\alpha\varphi + \Gamma_{\alpha\tau}^\sigma\eta_\sigma\varphi)$$

$$+ (\partial_\alpha b_\beta^\tau + \Gamma_{\alpha\sigma}^\tau b_\beta^\sigma - \Gamma_{\alpha\beta}^\sigma b_\sigma^\tau)\eta_\tau\varphi\} d\omega.$$

Consider a Cauchy sequence $(\boldsymbol{\eta}^k)_{k=1}^{\infty}$ with elements $\boldsymbol{\eta}^k = (\eta_i^k) \in \mathbf{W}_K(\omega)$. The definition of the norm $\| \cdot \|_{\omega}^K$ shows that there exist $\eta_{\alpha} \in L^2(\omega), \eta_3 \in H^1(\omega), \gamma_{\alpha\beta} \in L^2(\omega)$, and $\rho_{\alpha\beta} \in L^2(\omega)$ such that

$$\eta_{\alpha}^k \to \eta_{\alpha} \text{ in } L^2(\omega), \qquad \eta_3^k \to \eta_3 \text{ in } H^1(\omega),$$
$$\gamma_{\alpha\beta}(\boldsymbol{\eta}^k) \to \gamma_{\alpha\beta} \text{ in } L^2(\omega), \qquad \rho_{\alpha\beta}(\boldsymbol{\eta}^k) \to \rho_{\alpha\beta} \text{ in } L^2(\omega)$$

as $k \to \infty$. Given a function $\varphi \in \mathcal{D}(\omega)$, letting $k \to \infty$ in the relations $\int_{\omega} \gamma_{\alpha\beta}(\boldsymbol{\eta}^k)\varphi \, d\omega = \cdots$ and $\int_{\omega} \rho_{\alpha\beta}(\boldsymbol{\eta}^k)\varphi \, d\omega = \cdots$ then shows that $\gamma_{\alpha\beta} = \gamma_{\alpha\beta}(\boldsymbol{\eta})$ and $\rho_{\alpha\beta} = \rho_{\alpha\beta}(\boldsymbol{\eta})$.

(ii) The spaces $\mathbf{W}_K(\omega)$ and $H^1(\omega) \times H^1(\omega) \times H^2(\omega)$ coincide.

Clearly, $H^1(\omega) \times H^1(\omega) \times H^2(\omega) \subset \mathbf{W}_K(\omega)$. To prove the other inclusion, let $\boldsymbol{\eta} = (\eta_i) \in \mathbf{W}_K(\omega)$. The relations

$$e_{\alpha\beta}(\boldsymbol{\eta}) := \frac{1}{2}(\partial_{\alpha}\eta_{\beta} + \partial_{\beta}\eta_{\alpha}) = \gamma_{\alpha\beta}(\boldsymbol{\eta}) + \Gamma_{\alpha\beta}^{\sigma}\eta_{\sigma} + b_{\alpha\beta}\eta_3$$

then imply that $e_{\alpha\beta}(\boldsymbol{\eta}) \in L^2(\omega)$ since the functions $\Gamma_{\alpha\beta}^{\sigma}$ and $b_{\alpha\beta}$ are continuous on $\bar{\omega}$ (in fact, even continuously differentiable; recall that we assume $\boldsymbol{\theta} \in C^3(\bar{\omega}; \mathbb{R}^3)$). Therefore

$$\partial_{\sigma}\eta_{\alpha} \in H^{-1}(\omega),$$
$$\partial_{\beta}(\partial_{\sigma}\eta_{\alpha}) = \{\partial_{\beta}e_{\alpha\sigma}(\boldsymbol{\eta}) + \partial_{\sigma}e_{\alpha\beta}(\boldsymbol{\eta}) - \partial_{\alpha}e_{\beta\sigma}(\boldsymbol{\eta})\} \in H^{-1}(\omega),$$

since $\theta \in L^2(\omega)$ implies $\partial_{\sigma}\theta \in H^{-1}(\omega)$. Hence $\partial_{\sigma}\eta_{\sigma} \in L^2(\omega)$ by the Lemma of J. L. Lions (Theorem 3.1), and thus $\eta_{\alpha} \in H^1(\omega)$.

The definition of the functions $\rho_{\alpha\beta}(\boldsymbol{\eta})$, the continuity over $\bar{\omega}$ of the functions $\Gamma_{\alpha\beta}^{\sigma}, b_{\sigma\beta}, b_{\alpha}^{\sigma}$, and $\partial_{\alpha}b_{\beta}^{\tau}$, and the relations $\rho_{\alpha\beta}(\boldsymbol{\eta}) \in L^2(\omega)$ then imply that $\partial_{\alpha\beta}\eta_3 \in L^2(\omega)$, and hence that $\eta_3 \in H^2(\omega)$.

(iii) Inequality of Korn's type without boundary conditions.

The identity mapping ι from the space $H^1(\omega) \times H^1(\omega) \times H^2(\omega)$ equipped with its product norm $\boldsymbol{\eta} = (\eta_i) \to \{\sum_{\alpha} \|\eta_{\alpha}\|_{1,\omega}^2 + \|\eta_3\|_{2,\omega}^2\}^{1/2}$ into the space $\mathbf{W}_K(\omega)$ equipped with $\| \cdot \|_{\omega}^K$ is injective, continuous, and surjective by (ii). Since both spaces are complete (cf. (i)), the closed graph theorem then shows that the inverse mapping ι^{-1} is also continuous or, equivalently, that the inequality of Korn's type without boundary conditions holds. □

In order to establish an inequality of Korn's type 'with boundary conditions', we have to identify classes of boundary conditions to be imposed on the fields $\boldsymbol{\eta} = (\eta_i) \in H^1(\omega) \times H^1(\omega) \times H^2(\omega)$ in order that we can 'get rid' of the norms $|\eta_{\alpha}|_{0,\omega}$ and $\|\eta_3\|_{1,\omega}$ on the right-hand side of the above inequality, that is, situations where the seminorm

$$\boldsymbol{\eta} = (\eta_i) \to \left\{ \sum_{\alpha,\beta} |\gamma_{\alpha\beta}(\boldsymbol{\eta})|^2_{0,\omega} + \sum_{\alpha,\beta} |\rho_{\alpha\beta}(\boldsymbol{\eta})|^2_{0,\omega} \right\}^{1/2}$$

becomes a *norm*, which should in addition be *equivalent* to the product norm.

To this end, we begin by establishing (as in dimension three; *cf.* Theorem 3.3) a linearized rigid displacement lemma (Theorem 4.3), which provides in particular one instance of boundary conditions implying that this seminorm becomes a norm; as stated here, this lemma is due to Bernadou and Ciarlet (1976, Theorems 5.1-1 and 5.2-1).

The elegant proof of this lemma given here is based on an idea of Chapelle (1994). It relies on the preliminary observation that a vector field $\eta_i \boldsymbol{a}^i$ on a surface may be 'canonically' extended to a three-dimensional vector field $v_i \boldsymbol{g}^i$, in such a way that all the components $e_{i\|j}(\boldsymbol{v})$ of the associated three-dimensional linearized change of metric tensor have remarkable expressions in terms of the components $\gamma_{\alpha\beta}(\boldsymbol{\eta})$ and $\rho_{\alpha\beta}(\boldsymbol{\eta})$ of the linearized change of metric and linearized change of curvature tensors of the surface field.

Theorem 4.2: 'Canonical' three-dimensional extension of a surface vector field. Let the assumptions on the mapping $\boldsymbol{\theta} : \overline{\omega} \to \mathbb{R}^3$ be as in Theorem 4.1 and let

$$\boldsymbol{a}_3(y) = \frac{\boldsymbol{a}_1(y) \wedge \boldsymbol{a}_2(y)}{|\boldsymbol{a}_1(y) \wedge \boldsymbol{a}_2(y)|}.$$

There exists $\varepsilon_0 > 0$ such that the mapping $\boldsymbol{\Theta} : \overline{\omega} \times [-\varepsilon_0, \varepsilon_0] \to \mathbb{R}^3$ defined by

$$\boldsymbol{\Theta}(y, x_3) := \boldsymbol{\theta}(y) + x_3 \boldsymbol{a}_3(y) \text{ for all } (y, x_3) \in \overline{\omega} \times [-\varepsilon_0, \varepsilon_0]$$

is a C^1-diffeomorphism. With any vector field $\eta_i \boldsymbol{a}^i : \overline{\omega} \to \mathbb{R}^3$ with covariant components η_α in $H^1(\omega)$ and $\eta_3 \in H^2(\omega)$, let there be associated the vector field $v_i \boldsymbol{g}^i : \overline{\Omega} \to \mathbb{R}^3$ defined by

$$v_i(y, x_3) \boldsymbol{g}^i(y, x_3) = \eta_i(y) \boldsymbol{a}^i(y) + x_3 \mathcal{X}_\alpha(y) \boldsymbol{a}^\alpha(y)$$

for all $(y, x_3) \in \overline{\Omega}$, where $\Omega := \omega \times] - \varepsilon_0, \varepsilon_0[$, the vectors $\boldsymbol{g}^i$ form the contravariant basis associated with the mapping $\boldsymbol{\Theta}$ (Section 3), and

$$\mathcal{X}_\alpha := -(\partial_\alpha \eta_3 + b_\alpha^\sigma \eta_\sigma).$$

Then the covariant components v_i of the vector field $v_i \boldsymbol{g}^i$ are in $H^1(\Omega)$ and the covariant components $e_{i\|j}(\boldsymbol{v}) \in L^2(\Omega)$ of the associated linearized change of metric tensor are given by

$$e_{\alpha\|\beta}(\boldsymbol{v}) = \gamma_{\alpha\beta}(\boldsymbol{\eta}) - x_3 \rho_{\alpha\beta}(\boldsymbol{\eta})$$

$$+ \frac{x_3^2}{2} \{ b_\alpha^\sigma \rho_{\beta\sigma}(\boldsymbol{\eta}) + b_\beta^\tau \rho_{\alpha\tau}(\boldsymbol{\eta}) - 2 b_\alpha^\sigma b_\beta^\tau \gamma_{\sigma\tau}(\boldsymbol{\eta}) \},$$

$$e_{i\|3}(\boldsymbol{v}) = 0.$$

Proof. (i) *Preliminaries.* The mapping $\Theta : \bar{\omega} \times [-\varepsilon, \varepsilon] \to \mathbb{R}^3$ defined above is a C^1-diffeomorphism if $\varepsilon > 0$ is sufficiently small; *cf.* Ciarlet (2000, Theorem 3.1-1). Since

$$\partial_\alpha a_3 = -b_\alpha^\sigma a_\sigma$$

by Weingarten's formula, the vectors of the covariant basis associated with the mapping $\Theta = \theta + x_3 a_3$ are given by

$$g_\alpha = a_\alpha - x_3 b_\alpha^\sigma a_\sigma \text{ and } g_3 = a_3.$$

(ii) Given functions $\eta_\alpha, \mathcal{X}_\alpha \in H^1(\omega)$ and $\eta_3 \in H^2(\omega)$, let the vector field $v_i g^i : \bar{\Omega} \to \mathbb{R}^3$ be defined by

$$v_i g^i = \eta_i a^i + x_3 \mathcal{X}_\alpha a^\alpha$$

(in other words, we momentarily ignore the specific forms of the functions $\mathcal{X}_\alpha$ indicated in the theorem). Then the functions v_i are in $H^1(\Omega)$ and the covariant components $e_{i\|j}(v)$ of the linearized change of metric tensor associated with the field $v_i g^i$ are given by

$$e_{\alpha\|\beta}(v) = \frac{1}{2}\left\{(\eta_{\alpha|\beta} + \eta_{\beta|\alpha}) - b_{\alpha\beta}\eta_3\right\}$$

$$+x_3\left\{\frac{1}{2}(\mathcal{X}_{\alpha|\beta} + \mathcal{X}_{\beta|\alpha}) - \frac{1}{2}b_\alpha^\sigma(\eta_{\sigma|\beta} - b_{\beta\sigma}\eta_3) - \frac{1}{2}b_\beta^\tau(\eta_{\tau|\alpha} - b_{\alpha\tau}\eta_3)\right\}$$

$$+\frac{x_3^2}{2}\left\{-b_\alpha^\sigma\mathcal{X}_{\sigma|\beta} - b_\beta^\tau\mathcal{X}_{\tau|\alpha}\right\},$$

$$e_{\alpha\|3}(v) = \frac{1}{2}(\mathcal{X}_\alpha + \partial_\alpha\eta_3 + b_\alpha^\sigma\eta_\sigma),$$

$$e_{3\|3}(v) = 0.$$

The assumed regularities of the functions η_i and $\mathcal{X}_\alpha$ imply that

$$v_i = (v_j g^j) \cdot g_i = (\eta_i a^i + x_3 \mathcal{X}_\alpha a^\alpha) \cdot g_i \in H^1(\Omega),$$

since $g_i \in C^1(\bar{\Omega}; \mathbb{R}^3)$. The stated expressions for the functions $e_{i\|j}(v)$ are obtained by simple computations, based on the relations

$$e_{i\|j}(v) = \frac{1}{2}(v_{i\|j} + v_{j\|i}) \text{ and } v_{i\|j} = \{\partial_j(v_k g^k)\} \cdot g_i$$

(the vectors g_i having been computed in (i)).

(iii) When $\mathcal{X}_\alpha = -(\partial_\alpha\eta_3 + b_\alpha^\sigma\eta_\sigma)$, the functions $e_{i\|j}(v)$ found in (ii) take the expressions stated in the theorem.

We first note that $\mathcal{X}_\alpha \in H^1(\omega)$ (since $b_\alpha^\sigma \in C^1(\bar{\omega})$) and that $e_{\alpha\|3}(v) = 0$ when $\mathcal{X}_\alpha = -(\partial_\alpha\eta_3 + b_\alpha^\sigma\eta_\sigma)$. It thus remains to find the explicit forms of the functions $e_{\alpha\|\beta}(v)$. Replacing the functions $\mathcal{X}_\alpha$ by their expressions and using the symmetry relations $b_\alpha^\sigma|_\beta = b_\beta^\sigma|_\alpha$, we find that

$$\frac{1}{2}(\mathcal{X}_{\alpha|\beta} + \mathcal{X}_{\beta|\alpha}) - \frac{1}{2}b_\alpha^\sigma(\eta_{\sigma|\beta} - b_{\beta\sigma}\eta_3) - \frac{1}{2}b_\beta^\tau(\eta_{\tau|\alpha} - b_{\alpha\tau}\eta_3)$$
$$= -\eta_{3|\alpha\beta} - \beta_\alpha^\sigma\eta_{\sigma|\beta} - b_\beta^\tau\eta_{\tau|\alpha} - b_\beta^\tau|_\alpha\eta_\tau + b_\alpha^\sigma b_{\sigma\beta}\eta_3,$$

that is, the factor of x_3 in $e_{\alpha\|\beta}(v)$ is precisely equal to $-\rho_{\alpha\beta}(\eta)$. Finally,

$$-b_\alpha^\sigma\mathcal{X}_{\sigma|\beta} - b_\beta^\tau\mathcal{X}_{\tau|\alpha}$$
$$= b_\alpha^\sigma(\eta_{3|\beta\sigma} + b_{\sigma|\beta}^\tau\eta_\tau + b_\alpha^\tau\eta_{\tau|\beta}) + b_\beta^\tau(\eta_{3|\alpha\tau} + b_\tau^\sigma|_\alpha\eta_\sigma + b_\tau^\sigma\eta_{\sigma|\alpha})$$
$$= b_\alpha^\sigma(\rho_{\beta\sigma}(\eta) - b_\beta^\tau\eta_{\tau|\sigma} + b_\beta^\tau b_{\tau\sigma}\eta_3) + b_\beta^\tau(\rho_{\alpha\tau}(\eta) - b_\alpha^\sigma\eta_{\sigma|\tau} + b_\alpha^\sigma b_{\sigma\tau}\eta_3)$$
$$= b_\alpha^\sigma\rho_{\beta\sigma}(\eta) + b_\beta^\tau\rho_{\alpha\tau}(\eta) - 2b_\alpha^\sigma b_\beta^\tau\gamma_{\sigma\tau}(\eta),$$

that is, the factor of $x_3^2/2$ in $e_{\alpha\|\beta}(v)$ is precisely the combination of functions $\gamma_{\alpha\beta}(\eta)$ and $\rho_{\alpha\beta}(\eta)$ stated in the theorem. □

We now establish a linearized rigid displacement lemma on a general surface. 'Linearized' reminds us that only the *linearized* parts of the change of metric and change of curvature tensors are required to vanish. Thanks to Theorem 4.2 this lemma becomes a direct corollary to the 'three-dimensional' linearized rigid displacement lemma (Theorem 3.3), to which it may be profitably compared.

Part (a) of the next theorem is a linearized rigid displacement lemma *without boundary conditions*, while part (b) is a linearized rigid displacement lemma *with boundary conditions*.

Theorem 4.3: Linearized rigid displacement lemma on a general surface. Let the assumptions on the mapping $\boldsymbol{\theta} : \overline{\omega} \to \mathbb{R}^3$ be as in Theorem 4.1.

(a) Let $\boldsymbol{\eta} = (\eta_i) \in H^1(\omega) \times H^1(\omega) \times H^2(\omega)$ be such that

$$\gamma_{\alpha\beta}(\boldsymbol{\eta}) = \rho_{\alpha\beta}(\boldsymbol{\eta}) = 0 \text{ in } \omega.$$

Then the vector field $\eta_i\boldsymbol{a}^i : \overline{\omega} \to \mathbb{R}^3$ is a 'linearized rigid displacement' of the surface $S = \boldsymbol{\theta}(\overline{\omega})$, in the sense that there exist two vectors $\widehat{\boldsymbol{c}}, \widehat{\boldsymbol{d}} \in \mathbb{R}^3$ such that

$$\eta_i(y)\boldsymbol{a}^i(y) = \widehat{\boldsymbol{c}} + \widehat{\boldsymbol{d}} \wedge \widehat{\boldsymbol{\theta}}(y) \text{ for all } y \in \overline{\omega}.$$

(b) Let γ_0 be a $d\gamma$-measurable subset of $\gamma = \partial\omega$ that satisfies

$$\text{length } \gamma_0 > 0.$$

Then

$$\left.\begin{array}{r} \boldsymbol{\eta} = (\eta_i) \in H^1(\omega) \times H^1(\omega) \times H^2(\omega), \\ \eta_i = \partial_\nu\eta_3 = 0 \text{ on } \gamma_0, \\ \gamma_{\alpha\beta}(\boldsymbol{\eta}) = \rho_{\alpha\beta}(\boldsymbol{\eta}) = 0 \text{ in } \omega \end{array}\right\} \Rightarrow \boldsymbol{\eta} = \boldsymbol{0} \text{ in } \omega.$$

Proof. Let the set $\Omega = \omega \times] - \varepsilon_0, \varepsilon_0[$ and the field $\boldsymbol{v} = (v_i) \in \mathbf{H}^1(\Omega)$ be defined as in Theorem 4.2. By this theorem,

$$\gamma_{\alpha\beta}(\boldsymbol{\eta}) = \rho_{\alpha\beta}(\boldsymbol{\eta}) = 0 \text{ in } \omega \Rightarrow e_{i\|j}(\boldsymbol{v}) = 0 \text{ in } \Omega,$$

and thus, by Theorem 3.3(a), there exist two vectors $\widehat{\boldsymbol{c}}, \widehat{\boldsymbol{d}} \in \mathbb{R}^3$ such that

$$v_i(y, x_3) \boldsymbol{g}^i(y, x_3) = \widehat{\boldsymbol{c}} + \widehat{\boldsymbol{d}} \wedge \{\boldsymbol{\theta}(y) + x_3 \boldsymbol{a}_3(y)\} \text{ for all } (y, x_3) \in \overline{\Omega}.$$

Hence

$$\eta_i(y) \boldsymbol{a}^i(y) = v_i(y, x_3) \boldsymbol{g}^i(y, x_3)|_{x_3=0} = \widehat{\boldsymbol{c}} + \widehat{\boldsymbol{d}} \wedge \boldsymbol{\theta}(y) \text{ for all } y \in \overline{\omega},$$

and part (a) is established.

If in addition $\eta_i = \partial_\nu \eta_3 = 0$ on γ_0, then $\mathcal{X}_\alpha = -(\partial_\alpha \eta_3 + b_\alpha^\sigma \eta_\sigma) = 0$ on γ_0, since $\eta_3 = \partial_\nu \eta_3 = 0$ on γ_0 implies $\partial_\alpha \eta_3 = 0$ on γ_0; consequently,

$$v_i = (v_j \boldsymbol{g}^j) \cdot \boldsymbol{g}_i = (\eta_j \boldsymbol{a}^j + x_3 \mathcal{X}_\alpha \boldsymbol{a}^\alpha) \cdot \boldsymbol{g}_i = 0 \text{ on } \Gamma_0 := \gamma_0 \times [-\varepsilon_0, \varepsilon_0].$$

Since area $\Gamma_0 > 0$, Theorem 3.3(b) implies that $\boldsymbol{v} = \boldsymbol{0}$ in $\overline{\Omega}$, hence that $\boldsymbol{\eta} = \boldsymbol{0}$ on $\overline{\omega}$. □

We are now in a position to prove an inequality that plays a fundamental role in the analysis of linearly elastic shells, in particular in establishing the existence and uniqueness of the solution to the two-dimensional shell equations of W. T. Koiter and of the solution to the two-dimensional equations of a 'flexural' shell, as already observed at the beginning of this section. This inequality is due to Bernadou and Ciarlet (1976); see also Bernadou, Ciarlet and Miara (1994).

Theorem 4.4: Inequality of Korn's type on a general surface. Let the assumptions on the mapping $\boldsymbol{\theta} : \overline{\omega} \to \mathbb{R}^3$ be as in Theorem 4.1, let γ_0 be a $d\gamma$-measurable subset of $\gamma = \partial\omega$ that satisfies

$$\text{length } \gamma_0 > 0,$$

and let the space $\mathbf{V}_K(\omega)$ be defined by

$$\mathbf{V}_K(\omega) := \{\boldsymbol{\eta} = (\eta_i) \in H^1(\omega) \times H^1(\omega) \times H^2(\omega) : \eta_i = \partial_\nu \eta_3 = 0 \text{ on } \gamma_0\}.$$

Then there exists a constant $c = c(\omega, \gamma_0, \boldsymbol{\theta})$ such that

$$\left\{\sum_\alpha \|\eta_\alpha\|_{1,\omega}^2 + \|\eta_3\|_{2,\omega}^2\right\}^{1/2} \leq c \left\{\sum_{\alpha,\beta} |\gamma_{\alpha\beta}(\boldsymbol{\eta})|_{0,\omega}^2 + \sum_{\alpha,\beta} |\rho_{\alpha\beta}(\boldsymbol{\eta})|_{0,\omega}^2\right\}^{1/2}$$

for all $\boldsymbol{\eta} \in \mathbf{V}_K(\omega)$.

Proof. Let

$$\|\boldsymbol{\eta}\|_{H^1(\omega) \times H^1(\omega) \times H^2(\omega)} := \left\{\sum_\alpha \|\eta_\alpha\|_{1,\omega}^2 + \|\eta_3\|_{2,\omega}^2\right\}^{1/2},$$

and let

$$|\eta|_\omega^K := \left\{ \sum_{\alpha,\beta} |\gamma_{\alpha\beta}(\eta)|_{0,\omega}^2 + \sum_{\alpha,\beta} |\rho_{\alpha\beta}(\eta)|_{0,\omega}^2 \right\}^{1/2}.$$

If the stated inequality is false, there exists a sequence $(\eta^k)_{k=1}^\infty$ of functions $\eta^k \in \mathbf{V}_K(\omega)$ such that

$$\|\eta^k\|_{H^1(\omega) \times H^1(\omega) \times H^2(\omega)} = 1 \text{ for all } k \text{ and } \lim_{k\to\infty} |\eta^k|_\omega^K = 0.$$

Since the sequence $(\eta^k)_{k=1}^\infty$ is bounded in $H^1(\omega) \times H^1(\omega) \times H^2(\omega)$, there exists a subsequence $(\eta^l)_{l=1}^\infty$ that converges in $L^2(\omega) \times L^2(\omega) \times H^1(\omega)$ by the Rellich–Kondrašov theorem; furthermore, since $\lim_{l\to\infty} |\eta^l|_\omega^K = 0$, each sequence $(\gamma_{\alpha\beta}(\eta^l))_{l=1}^\infty$ and $(\rho_{\alpha\beta}(\eta^l))_{l=1}^\infty$ also converges in $L^2(\omega)$ (to 0, but this information is not used at this stage). The subsequence $(\eta^l)_{l=1}^\infty$ is thus a Cauchy sequence with respect to the norm

$$\eta = (\eta_i) \to \left\{ \sum_\alpha |\eta_\alpha|_{0,\omega}^2 + \|\eta_3\|_{1,\omega}^2 + \sum_{\alpha,\beta} |\gamma_{\alpha\beta}(\eta)|_{0,\omega}^2 + \sum_{\alpha,\beta} |\rho_{\alpha\beta}(\eta)|_{0,\omega}^2 \right\}^{1/2},$$

and hence with respect to the norm $\| \cdot \|_{H^1(\omega) \times H^1(\omega) \times H^2(\omega)}$ by Korn's inequality without boundary conditions (Theorem 4.1).

The space $\mathbf{V}_K(\omega)$ being complete as a closed subspace of $H^1(\omega) \times H^1(\omega) \times H^2(\omega)$, there exists $\eta \in \mathbf{V}_K(\omega)$ such that

$$\eta^l \to \eta \text{ in } H^1(\omega) \times H^1(\omega) \times H^2(\omega),$$

and the limit η satisfies

$$|\gamma_{\alpha\beta}(\eta)|_{0,\omega} = \lim_{l\to\infty} |\gamma_{\alpha\beta}(\eta^l)|_{0,\omega} = 0,$$

$$|\rho_{\alpha\beta}(\eta)|_{0,\omega} = \lim_{l\to\infty} |\rho_{\alpha\beta}(\eta^l)|_{0,\omega} = 0.$$

Hence $\eta = \mathbf{0}$ by Theorem 4.3.

But this contradicts the relations $\|\eta^l\|_{H^1(\omega) \times H^1(\omega) \times H^2(\omega)} = 1$ for all $l \geq 1$, and the proof is complete. □

It has recently been shown by Ciarlet and Mardare (2000, 200x) that the canonical three-dimensional extension of a surface vector field used in Theorem 4.2 can be further put to use, to the extent that it provides a new proof of the inequality of Korn's type on a general surface itself (Theorem 4.4), directly as a corollary to the three-dimensional Korn inequality in curvilinear coordinates (Theorem 3.4).

For another, 'intrinsic', approach to inequalities of Korn's type on surfaces, see Delfour (200x).

5. Inequality of Korn's type on a surface with little regularity

As shown by Blouza and Le Dret (1999), the *regularity* assumptions made in the previous section on the mapping $\boldsymbol{\theta}$ and on the field $\boldsymbol{\eta} = (\eta_i)$, in both the linearized rigid displacement lemma and the inequality of Korn's type (Theorems 4.3 and 4.4), can be substantially weakened.

This improvement relies on the observation that the covariant components of the linearized change of metric and change of curvature tensors, that is,

$$\gamma_{\alpha\beta}(\boldsymbol{\eta}) = \frac{1}{2}(\partial_\beta \eta_\alpha + \partial_\alpha \eta_\beta) - \Gamma^\sigma_{\alpha\beta}\eta_\sigma - b_{\alpha\beta}\eta_3$$

and

$$
\begin{aligned}
\rho_{\alpha\beta}(\boldsymbol{\eta}) = {} & \partial_{\alpha\beta}\eta_3 - \Gamma^\sigma_{\alpha\beta}\partial_\sigma\eta_3 - b^\sigma_\alpha b_{\sigma\beta}\eta_3 \\
& + b^\sigma_\alpha(\partial_\beta\eta_\sigma - \Gamma^\tau_{\beta\sigma}\eta_\tau) + b^\tau_\beta(\partial_\alpha\eta_\tau - \Gamma^\sigma_{\alpha\tau}\eta_\sigma) \\
& + (\partial_\alpha b^\tau_\beta + \Gamma^\tau_{\alpha\sigma}b^\sigma_\beta - \Gamma^\sigma_{\alpha\beta}b^\tau_\sigma)\eta_\tau,
\end{aligned}
$$

can be also written as

$$\gamma_{\alpha\beta}(\boldsymbol{\eta}) = \frac{1}{2}(\partial_\beta\tilde{\boldsymbol{\eta}}\cdot\boldsymbol{a}_\alpha + \partial_\alpha\tilde{\boldsymbol{\eta}}\cdot\boldsymbol{a}_\beta) =: \tilde{\gamma}_{\alpha\beta}(\tilde{\boldsymbol{\eta}})$$

and

$$\rho_{\alpha\beta}(\boldsymbol{\eta}) = (\partial_{\alpha\beta}\tilde{\boldsymbol{\eta}} - \Gamma^\sigma_{\alpha\beta}\partial_\sigma\tilde{\boldsymbol{\eta}})\cdot\boldsymbol{a}_3 =: \tilde{\rho}_{\alpha\beta}(\tilde{\boldsymbol{\eta}}),$$

in terms of the field

$$\tilde{\boldsymbol{\eta}} := \eta_i\boldsymbol{a}^i.$$

The interest of the new expressions $\tilde{\gamma}_{\alpha\beta}(\tilde{\boldsymbol{\eta}})$ and $\tilde{\rho}_{\alpha\beta}(\tilde{\boldsymbol{\eta}})$ is that they still define *bona fide* distributions under significantly weaker smoothness assumptions than those made so far, that is, $\boldsymbol{\theta} \in \mathcal{C}^3(\overline{w};\mathbb{R}^3)$ and $\boldsymbol{\eta} = (\eta_i) \in H^1(w) \times H^1(w) \times H^2(w)$. More specifically, it is easily verified that $\tilde{\gamma}_{\alpha\beta}(\tilde{\boldsymbol{\eta}}) \in L^2(w)$ and $\tilde{\rho}_{\alpha\beta}(\tilde{\boldsymbol{\eta}}) \in H^{-1}(w)$ if $\boldsymbol{\theta} \in W^{2,\infty}(w;\mathbb{R}^3)$ and $\tilde{\boldsymbol{\eta}} \in \mathbf{H}^1(w)$.

Note that, to avoid any confusion, we intentionally employ the new notation $\tilde{\gamma}_{\alpha\beta}(\tilde{\boldsymbol{\eta}})$ and $\tilde{\rho}_{\alpha\beta}(\tilde{\boldsymbol{\eta}})$.

Using this observation, Blouza and Le Dret (1999, Theorem 6) first establish the following extension of Theorem 4.3.

Theorem 5.1: Linearized rigid displacement lemma on a surface with little regularity. Let w be a domain in $\mathbb{R}^2$ and let $\boldsymbol{\theta} \in W^{2,\infty}(w;\mathbb{R}^3)$ be an injective mapping such that the two vectors $\boldsymbol{a}_\alpha = \partial_\alpha\boldsymbol{\theta}$ are linearly independent at all points of $\overline{w}$.

Given $\tilde{\boldsymbol{\eta}} \in \mathbf{H}^1(\omega)$, let the distributions $\tilde{\gamma}_{\alpha\beta}(\tilde{\boldsymbol{\eta}}) \in L^2(\omega)$ and $\tilde{\rho}_{\alpha\beta}(\tilde{\boldsymbol{\eta}}) \in H^{-1}(\omega)$ be defined by

$$\tilde{\gamma}_{\alpha\beta}(\tilde{\boldsymbol{\eta}}) := \frac{1}{2}(\partial_\beta\tilde{\boldsymbol{\eta}} \cdot \boldsymbol{a}_\alpha + \partial_\alpha\tilde{\boldsymbol{\eta}} \cdot \boldsymbol{a}_\beta),$$

$$\tilde{\rho}_{\alpha\beta}(\tilde{\boldsymbol{\eta}}) := (\partial_{\alpha\beta}\tilde{\boldsymbol{\eta}} - \Gamma^\sigma_{\alpha\beta}\partial_\sigma\tilde{\boldsymbol{\eta}}) \cdot \boldsymbol{a}_3.$$

Let $\tilde{\boldsymbol{\eta}} \in \mathbf{H}^1(\omega)$ be such that

$$\tilde{\gamma}_{\alpha\beta}(\tilde{\boldsymbol{\eta}}) = \tilde{\rho}_{\alpha\beta}(\tilde{\boldsymbol{\eta}}) = 0 \text{ in } \omega.$$

Then $\tilde{\boldsymbol{\eta}}$ is a 'linearized rigid displacement' of the surface $S = \boldsymbol{\theta}(\overline{\omega})$, in the sense that there exist two vectors $\widehat{\boldsymbol{c}}, \widehat{\boldsymbol{d}} \in \mathbb{R}^3$ such that

$$\tilde{\boldsymbol{\eta}}(y) = \widehat{\boldsymbol{c}} + \widehat{\boldsymbol{d}} \wedge \boldsymbol{\theta}(y) \text{ for all } y \in \overline{\omega}. \qquad \square$$

Blouza and Le Dret (1999, Lemma 11) then proceed to establish the following variant of Theorem 4.4, which, for convenience, is stated here with boundary conditions corresponding to a shell that is simply supported along its entire boundary, that is, $\tilde{\boldsymbol{\eta}} = \mathbf{0}$ on γ.

Boundary conditions of *clamping*, as considered in Theorem 4.4, can also be handled via the present approach, provided they are first re-interpreted so as to make sense for vector fields $\tilde{\boldsymbol{\eta}}$ that only satisfy $\tilde{\boldsymbol{\eta}} \in \mathbf{H}^1(\omega)$ and $\partial_{\alpha\beta}\tilde{\boldsymbol{\eta}} \cdot \boldsymbol{a}_3 \in L^2(\omega)$; cf. Blouza and Le Dret (1999, Section 6).

Theorem 5.2: Inequality of Korn's type on a surface with little regularity.   Let the assumptions on the mapping $\boldsymbol{\theta}$ be as in Theorem 5.1 and let the space $\tilde{\mathbf{V}}^s_K(\omega)$ be defined by

$$\tilde{\mathbf{V}}^s_K(\omega) = \{\tilde{\boldsymbol{\eta}} \in \mathbf{H}^1_0(\omega) : \partial_{\alpha\beta}\tilde{\boldsymbol{\eta}} \cdot \boldsymbol{a}_3 \in L^2(\omega)\}.$$

Then there exists a constant c such that

$$\left\{\|\tilde{\boldsymbol{\eta}}\|^2_{1,\omega} + \sum_{\alpha,\beta} |\partial_{\alpha\beta}\tilde{\boldsymbol{\eta}} \cdot \boldsymbol{a}_3|^2_{0,\omega}\right\}^{1/2}$$

$$\leq c\left\{\sum_{\alpha,\beta} |\tilde{\gamma}_{\alpha\beta}(\tilde{\boldsymbol{\eta}})|^2_{0,\omega} + \sum_{\alpha,\beta} |\tilde{\rho}_{\alpha\beta}(\tilde{\boldsymbol{\eta}})|^2_{0,\omega}\right\}^{1/2} \text{ for all } \tilde{\boldsymbol{\eta}} \in \tilde{\mathbf{V}}^s_K(\omega),$$

where the distributions $\tilde{\gamma}_{\alpha\beta}(\tilde{\boldsymbol{\eta}})$ and $\tilde{\rho}_{\alpha\beta}(\tilde{\boldsymbol{\eta}})$ are defined as in Theorem 5.1 (note that $\tilde{\rho}_{\alpha\beta}(\tilde{\boldsymbol{\eta}}) \in L^2(\omega)$ if $\tilde{\boldsymbol{\eta}} \in \tilde{\mathbf{V}}^s_K(\omega)$). $\qquad \square$

This theorem establishes as a corollary the existence and uniqueness of the solution to the two-dimensional Koiter equations for a simply supported shell whose middle surface has little regularity, once these equations are re-written in terms of the expressions $\tilde{\gamma}_{\alpha\beta}(\tilde{\boldsymbol{\eta}})$ and $\tilde{\rho}_{\alpha\beta}(\tilde{\boldsymbol{\eta}})$; cf. Section 11.3.

6. Inequality of Korn's type on an elliptic surface

The two-dimensional equations of a linearly elastic 'membrane' shell take the following form. The unknowns are the covariant components $\zeta_i^\varepsilon : \overline{\omega} \to \mathbb{R}$ of the displacement $\zeta_i^\varepsilon a^i : \overline{\omega} \to \mathbb{R}^3$ of the middle surface $S = \boldsymbol{\theta}(\overline{\omega})$ of the shell, and $\boldsymbol{\zeta}^\varepsilon := (\zeta_i^\varepsilon)$ satisfies

$$\boldsymbol{\zeta}^\varepsilon \in \mathbf{V}_M(\omega) := H_0^1(\omega) \times H_0^1(\omega) \times L^2(\omega),$$

$$\int_\omega \varepsilon a^{\alpha\beta\sigma\tau,\varepsilon} \gamma_{\sigma\tau}(\boldsymbol{\zeta}^\varepsilon) \gamma_{\alpha\beta}(\boldsymbol{\eta}) \sqrt{a} \, dy = \int_\omega p^{i,\varepsilon} \eta_i \sqrt{a} \, dy,$$

for all $\boldsymbol{\eta} = (\eta_i) \in \mathbf{V}_M(\omega)$, where $2\varepsilon > 0$ is the thickness of the shell,

$$a^{\alpha\beta\sigma\tau,\varepsilon} := \frac{4\lambda^\varepsilon \mu^\varepsilon}{\lambda^\varepsilon + 2\mu^\varepsilon} a^{\alpha\beta} a^{\sigma\tau} + 2\mu^\varepsilon (a^{\alpha\sigma} a^{\beta\tau} + a^{\alpha\tau} a^{\beta\sigma})$$

denote (as in Section 4) the contravariant components of the two-dimensional elasticity tensor of the shell,

$$\gamma_{\alpha\beta}(\boldsymbol{\eta}) := \frac{1}{2}(\partial_\beta \eta_\alpha + \partial_\alpha \eta_\beta) - \Gamma_{\alpha\beta}^\sigma \eta_\sigma - b_{\alpha\beta} \eta_3$$

denote (again as in Section 4) the covariant components of the linearized change of metric tensor of S, and the given functions $p^{i,\varepsilon} \in L^2(\omega)$ account for the applied forces.

These equations will be further studied in Section 8, where it will be shown in particular that they can be fully justified through an asymptotic analysis of the three-dimensional solutions.

As already noted in Section 4, there exist constants c_e and a_0 such that

$$\sum_{\alpha,\beta} |t_{\alpha\beta}|^2 \le c_e a^{\alpha\beta\sigma\tau,\varepsilon}(y) t_{\sigma\tau} t_{\alpha\beta}$$

for all $y \in \overline{\omega}$ and all symmetric matrices $(t_{\alpha\beta})$ and such that $a(y) \ge a_0 > 0$ for all $y \in \overline{\omega}$. Establishing the existence and uniqueness of a solution to the above variational problem by the Lax–Milgram lemma thus amounts to proving the existence of a constant c_M such that

$$\left\{ \sum_\alpha \|\eta_\alpha\|_{1,\omega}^2 + |\eta_3|_{0,\omega}^2 \right\}^{1/2} \le c_M \left\{ \sum_{\alpha,\beta} |\gamma_{\alpha\beta}(\boldsymbol{\eta})|_{0,\omega}^2 \right\}^{1/2}$$

for all $\boldsymbol{\eta} = (\eta_i) \in \mathbf{V}_M(\omega)$.

The objective of this section, based on Ciarlet and Lods (1996a) and Ciarlet and Sanchez-Palencia (1996), is to find sufficient conditions, essentially bearing on the 'geometry' of the surface S, guaranteeing that such an inequality of Korn's type holds. It is also worth noticing that the justification alluded to above of these two-dimensional 'membrane' shell equations from three-dimensional elasticity is performed under precisely the *same* assumptions on the geometry of S, as we shall see in Section 8.

We follow the usual pattern, that is, we begin by proving an inequality of Korn's type *without boundary condition*, which remarkably holds for 'arbitrary' geometries, although it only involves the linearized change of metric tensor (compare with Theorem 4.1).

Theorem 6.1: Second inequality of Korn's type without boundary conditions on a general surface. Let ω be a domain in $\mathbb{R}^2$ and let $\boldsymbol{\theta} \in C^2(\overline{\omega}; \mathbb{R}^3)$ be an injective mapping such that the two vectors $\boldsymbol{a}_\alpha = \partial_\alpha \boldsymbol{\theta}$ are linearly independent at all points of $\overline{\omega}$. Given $\boldsymbol{\eta} = (\eta_i) \in H^1(\omega) \times H^1(\omega) \times L^2(\omega)$, let

$$\gamma_{\alpha\beta}(\boldsymbol{\eta}) = \left\{ \frac{1}{2}(\partial_\beta \eta_\alpha + \partial_\alpha \eta_\beta) - \Gamma^\sigma_{\alpha\beta}\eta_\sigma - b_{\alpha\beta}\eta_3 \right\} \in L^2(\omega)$$

denote the covariant components of the linearized change of metric tensor associated with the displacement field $\eta_i \boldsymbol{a}^i$ of the surface $S = \boldsymbol{\theta}(\overline{\omega})$. Then there exists a constant $c_0 = c_0(\omega, \boldsymbol{\theta})$ such that

$$\left\{ \sum_\alpha \|\eta_\alpha\|^2_{1,\omega} + |\eta_3|^2_{0,\omega} \right\}^{1/2} \leq c_0 \left\{ \sum_i |\eta_i|^2_{0,\omega} + \sum_{\alpha,\beta} |\gamma_{\alpha\beta}(\boldsymbol{\eta})|^2_{0,\omega} \right\}^{1/2}$$

for all $\boldsymbol{\eta} = (\eta_i) \in H^1(\omega) \times H^1(\omega) \times L^2(\omega)$.

Proof. The proof is analogous to that of Theorem 4.1 and, for this reason, is only sketched. It relies on the following steps. First, the space

$$\mathbf{W}_M(\omega) := \left\{ \boldsymbol{\eta} = (\eta_i) \in \mathbf{L}^2(\omega) : \gamma_{\alpha\beta}(\boldsymbol{\eta}) \in L^2(\omega) \right\}$$

becomes a Hilbert space when it is equipped with the norm $\|\cdot\|^M_\omega$ defined by

$$\|\boldsymbol{\eta}\|^M_\omega := \left\{ \sum_i |\eta_i|^2_{0,\omega} + \sum_{\alpha,\beta} |\gamma_{\alpha\beta}(\boldsymbol{\eta})|^2_{0,\omega} \right\}^{1/2}.$$

Next, the two spaces $\mathbf{W}_M(\omega)$ and $H^1(\omega) \times H^1(\omega) \times L^2(\omega)$ coincide, thanks again to the identities

$$\partial_{\alpha\beta}\eta_\sigma = \partial_\alpha e_{\beta\sigma}(\boldsymbol{\eta}) + \partial_\beta e_{\alpha\sigma}(\boldsymbol{\eta}) - \partial_\sigma e_{\alpha\beta}(\boldsymbol{\eta})$$

and to the Lemma of J. L. Lions (Theorem 3.1).

Finally, the closed graph theorem shows that the identity mapping from the space $H^1(\omega) \times H^1(\omega) \times L^2(\omega)$ equipped with the product norm

$$\boldsymbol{\eta} = (\eta_i) \rightarrow \left\{ \sum_\alpha \|\eta_\alpha\|^2_{1,\omega} + |\eta_3|^2_{0,\omega} \right\}^{1/2}$$

onto the space $\mathbf{W}_M(\omega)$ equipped with the norm $\|\cdot\|_\omega^M$ has a continuous inverse. Hence the stated inequality holds.   □

The next step consists in identifying sufficient conditions allowing the 'elimination' of the norms $|\eta_i|_{0,\omega}$ on the right-hand side of the above inequality of Korn's type. Whether it be for the three-dimensional Korn inequality in curvilinear coordinates (Theorem 3.4) or for the inequality of Korn's type on a general surface (Theorem 4.4), the corresponding eliminations simply resulted from imposing *ad hoc* boundary conditions on the displacement fields, in such a way that a linearized rigid displacement lemma with boundary conditions holds (see Theorems 3.3(b) and 4.3(b)).

In other words, we are facing the problem of finding boundary conditions such that the seminorm

$$\boldsymbol{\eta} = (\eta_i) \rightarrow \left\{ \sum_{\alpha,\beta} |\gamma_{\alpha\beta}(\boldsymbol{\eta})|_{0,\omega}^2 \right\}^{1/2}$$

becomes a *norm* for the displacement fields $\eta_i \boldsymbol{a}^i$ that satisfy them. Since η_3 is only in $L^2(\omega)$ and η_α is in $H^1(\omega)$, the only possibility consists in trying boundary conditions of the form

$$\eta_\alpha = 0 \text{ on } \gamma_0 \subset \gamma, \text{ with area } \gamma_0 > 0.$$

It then turns out that such a linearized rigid displacement lemma does hold, but only for *special geometries of the surface S* and *special subsets* $\boldsymbol{\theta}(\gamma_0)$ of *the boundary of S*. In this direction, we refer to Sanchez-Palencia (1993), Sanchez-Hubert and Sanchez-Palencia (1997), Lods and Mardare (1998*a*), Mardare (1998*c*), and Şlicaru (1998), who have identified various situations of interest where this lemma holds.

But even though such a linearized rigid displacement lemma *often* holds, it *very seldom* implies that the norm

$$\boldsymbol{\eta} \rightarrow \left\{ \sum_{\alpha,\beta} |\gamma_{\alpha\beta}(\boldsymbol{\eta})|_{0,\omega}^2 \right\}^{1/2}$$

is equivalent to the norm

$$\boldsymbol{\eta} \rightarrow \left\{ \sum_\alpha \|\eta_\alpha\|_{1,\omega}^2 + |\eta_3|_{0,\omega}^2 \right\}^{1/2}.$$

More precisely, we shall prove (in Theorem 6.3) that, under *ad hoc* regularity assumptions on the mapping $\boldsymbol{\theta}$ and on the boundary γ, these two norms are equivalent if $\gamma = \gamma_0$ and the surface S is *elliptic* according to the definition given below. Conversely, Şlicaru (1998) has shown the remarkable result that, even under the 'minimal' regularity assumptions '$\boldsymbol{\theta} \in \mathcal{C}^2(\overline{\omega}; \mathbb{R}^3)$

and γ Lipschitz-continuous', the same sufficient conditions are also necessary for the equivalence of the norms, which thus very seldom occurs indeed!

We now prove the stated 'linearized rigid displacement lemma', directly under the assumptions ($\gamma_0 = \gamma$ and S elliptic) that will eventually lead to the equivalence of norms. We begin with a definition.

Let a surface $S = \boldsymbol{\theta}(\bar{\omega})$ be given, where $\boldsymbol{\theta} \in C^2(\bar{\omega}; \mathbb{R}^3)$ is an injective mapping such that the two vectors $\boldsymbol{a}_\alpha$ are linearly independent at all points of $\bar{\omega}$. Then S is *elliptic* if the symmetric matrix $(b_{\alpha\beta}(y))$ formed by the covariant components of the curvature tensor of S is positive, *or* negative, definite at all points $y \in \bar{\omega}$, or equivalently, if there exists a constant c such that

$$c > 0 \text{ and } |b_{\alpha\beta}(y)\xi^\alpha\xi^\beta| \geq c\sum_\alpha |\xi^\alpha|^2,$$

for all $y \in \bar{\omega}$ and all $(\xi^\alpha) \in \mathbb{R}^2$. Geometrically, this means that the *Gaussian curvature* of the surface S is everywhere > 0, or equivalently, that the two principal radii of curvature are of the same sign at each point of S (for details about these classical notions, see, e.g., Ciarlet (2000, Section 2.2)). A portion of an *ellipsoid* provides an instance of elliptic surface.

In the next proof of the theorem, analytic functions of two real variables in an open subset of $\mathbb{R}^2$ are considered; we refer to Dieudonné (1968) for a particularly elegant treatment of analytic functions of any finite number of real or complex variables.

Theorem 6.2: Linearized rigid displacement lemma on an elliptic surface. Let there be given a domain ω in $\mathbb{R}^2$ and an injective mapping $\boldsymbol{\theta} \in C^{2,1}(\bar{\omega}; \mathbb{R}^3)$ such that the two vectors $\boldsymbol{a}_\alpha = \partial_\alpha\boldsymbol{\theta}$ are linearly independent at all points of $\bar{\omega}$ and such that the surface $S = \boldsymbol{\theta}(\bar{\omega})$ is elliptic. Then

$$\left.\begin{array}{l} \boldsymbol{\eta} = (\eta_i) \in H_0^1(\omega) \times H_0^1(\omega) \times L^2(\omega), \\ \gamma_{\alpha\beta}(\boldsymbol{\eta}) = 0 \text{ in } \omega \end{array}\right\} \Rightarrow \boldsymbol{\eta} = \mathbf{0} \text{ in } \omega.$$

Proof. We give the proof under the additional assumptions that the boundary γ is of class C^3 and that the components of the mapping $\boldsymbol{\theta}$ are restrictions to $\bar{\omega}$ of analytic functions in an open set $\omega' \subset \mathbb{R}^2$ containing $\bar{\omega}$. We refer to Lods and Mardare (1998a) for a proof (then more 'technical') under the more general assumptions stated in the theorem. An earlier version of this lemma is due to Vekua (1962), who proved it under the assumptions that γ is of class C^3 and $\boldsymbol{\theta} \in W^{3,p}(\omega; \mathbb{R}^3)$ for some $p > 1$, using the theory of 'generalized analytic functions'.

(i) We first note that establishing this implication is equivalent to proving a *uniqueness theorem*, that is, $\boldsymbol{\eta} = (\eta_i) = \mathbf{0}$ is the only solution in the space $H^1(\omega) \times H^1(\omega) \times L^2(\omega)$ of the linear system formed by the three partial differential equations $\gamma_{\alpha\beta}(\boldsymbol{\eta}) = 0$ in ω together with the two boundary conditions (understood in the sense of traces) $\eta_\alpha = 0$ on γ, or, 'in full',

$$\partial_1 \eta_1 - \Gamma^\sigma_{11}\eta_\sigma - b_{11}\eta_3 = 0 \text{ in } \omega,$$

$$\frac{1}{2}\partial_2\eta_1 + \frac{1}{2}\partial_1\eta_2 - \Gamma^\sigma_{12}\eta_\sigma - b_{12}\eta_3 = 0 \text{ in } \omega,$$

$$\partial_2\eta_2 - \Gamma^\sigma_{22}\eta_\sigma - b_{22}\eta_3 = 0 \text{ in } \omega,$$

$$\eta_1 = 0 \text{ on } \gamma,$$

$$\eta_2 = 0 \text{ on } \gamma.$$

(ii) Any solution $\boldsymbol{\eta} = (\eta_i) \in H^1_0(\omega) \times H^1_0(\omega) \times L^2(\omega)$ of the system

$$\gamma_{\alpha\beta}(\boldsymbol{\eta}) = 0 \text{ in } \omega \text{ and } \eta_\alpha = 0 \text{ on } \gamma$$

is in the space $C^1(\overline{\omega}) \times C^1(\overline{\omega}) \times C^0(\overline{\omega})$.

This regularity result relies on a crucial observation made by Geymonat and Sanchez-Palencia (1991). The partial differential equations $\gamma_{\alpha\beta}(\boldsymbol{\eta}) = 0$ in ω constitute a first-order system that is 'uniformly elliptic' in the sense of Agmon, Douglis and Nirenberg (1964). This means that there exists a constant $A > 0$ such that (here and subsequently, we use the notation of Agmon, Douglis and Nirenberg (1964))

$$A^{-1}\sum_\alpha |\xi_\alpha|^2 \le |L(y,\boldsymbol{\xi})| \le A\sum_\alpha |\xi_\alpha|^2,$$

for all $y \in \overline{\omega}$ and $\boldsymbol{\xi} = (\xi_\alpha) \in \mathbb{R}^2$, where

$$L(y,\boldsymbol{\xi}) := \det \begin{pmatrix} \xi_1 & 0 & -b_{11}(y) \\ \frac{1}{2}\xi_2 & \frac{1}{2}\xi_1 & -b_{12}(y) \\ 0 & \xi_2 & -b_{22}(y) \end{pmatrix}.$$

The way the above matrix of order three is constructed from the equations $\gamma_{\alpha\beta}(\boldsymbol{\eta}) = 0$ should be clear; suffice it to specify that only the coefficients of the partial derivatives of the highest order for each unknown (one for η_α and zero for η_3) are taken into account. The uniform ellipticity of the system formed by the partial differential equations $\gamma_{\alpha\beta}(\boldsymbol{\eta}) = 0$ in ω thus holds, since

$$L(y,\boldsymbol{\xi}) = -\frac{1}{2}(\xi_2 - \xi_1)\begin{pmatrix} b_{11}(y) & b_{12}(y) \\ b_{21}(y) & b_{22}(y) \end{pmatrix}\begin{pmatrix} \xi_2 \\ -\xi_1 \end{pmatrix}$$

in the present case, and since the symmetric matrix $(b_{\alpha\beta}(y))$ is either positive, or negative, definite at all points $y \in \overline{\omega}$ by the assumed *ellipticity* of the surface S.

In addition, the 'supplementary condition on L' (which needs to be verified only in two dimensions, as here) is also satisfied. The degree m of the polynomial L with respect to ξ_1 and ξ_2 being two, the polynomial

$$\tau \in \mathbb{C} \to L(y, \boldsymbol{\xi} + \tau\boldsymbol{\eta}) \in \mathbb{C}$$

has exactly $\frac{m}{2} = 1$ root τ^+ with $\operatorname{Im}\tau^+ > 0$, for all $y \in \bar{\omega}$ and all linearly independent vectors $\boldsymbol{\xi} = (\xi_\alpha)$ and $\boldsymbol{\eta} = (\eta_\alpha)$ in $\mathbb{R}^2$.

Finally, when $\frac{m}{2}$, i.e., *one* of the two boundary conditions $\eta_\alpha = 0$ on γ is appended to the equations $\gamma_{\alpha\beta}(\boldsymbol{\eta}) = 0$ in ω, the 'complementary boundary condition' is also satisfied. Thus the polynomial $\tau \in \mathbb{C} \to (\tau - \tau^+)$ divides the polynomials $\tau \to c(\xi_1 + \tau\eta_1)$ and $\tau \to c(\xi_2 + \tau\eta_2)$ only if the constant c vanishes.

It then follows from Agmon, Douglis and Nirenberg (1964, Theorem 10.5) that, if γ is of class $\mathcal{C}^3$ and the coefficients of the uniformly elliptic system $\gamma_{\alpha\beta}(\boldsymbol{\eta}) = 0$ are in $\mathcal{C}^2(\bar{\omega})$, any solution $\boldsymbol{\eta} \in H^1(\omega) \times H^1(\omega) \times L^2(\omega)$ of $\gamma_{\alpha\beta}(\boldsymbol{\eta}) = 0$ in ω together with, for instance, $\eta_1 = 0$ on γ is in the space $H^3(\omega) \times H^3(\omega) \times H^2(\omega)$. The assertion then follows from the continuous embeddings $H^m(\omega) \hookrightarrow \mathcal{C}^{m-2}(\bar{\omega}), m = 2, 3$.

(iii) 'Local' uniqueness of the solution of the system

$$\gamma_{\alpha\beta}(\boldsymbol{\eta}) = 0 \text{ in } \omega \text{ and } \eta_\alpha = 0 \text{ on } \gamma.$$

The assumed ellipticity of the surface S shows that there exists a constant $c > 0$ such that $|b_{11}(y)| \geq c$ for all $y \in \bar{\omega}$. Hence the unknown η_3 may be eliminated, for instance by means of the equation $\gamma_{11}(\boldsymbol{\eta}) = 0$. This elimination shows that

$$\eta_3 = \frac{1}{b_{11}}(\partial_1\eta_1 - \Gamma^\sigma_{11}\eta_\sigma)$$

and that η_1 and η_2 are solutions of the *reduced system*

$$-2\frac{b_{12}}{b_{11}}\partial_1\eta_1 + \partial_2\eta_1 + \partial_1\eta_2 - 2\left(\Gamma^\sigma_{12} - \frac{b_{12}}{b_{11}}\Gamma^\sigma_{11}\right)\eta_\sigma = 0 \text{ in } \omega,$$

$$-\frac{b_{22}}{b_{11}}\partial_1\eta_1 + \partial_2\eta_2 - \left(\Gamma^\sigma_{22} - \frac{b_{22}}{b_{11}}\Gamma^\sigma_{11}\right)\eta_\sigma = 0 \text{ in } \omega,$$

$$\eta_1 = 0 \text{ on } \gamma,$$

$$\eta_2 = 0 \text{ on } \gamma.$$

Since the coefficients of this reduced system are analytic in ω', since the boundary γ is of class $\mathcal{C}^3$ and is not a characteristic curve for this system, as is easily verified by again using the assumed *ellipticity* of the surface S, Holmgren's Uniqueness Theorem (see, e.g., Courant and Hilbert (1962, p. 238) or Bers, John and Schechter (1964, p. 47)) shows that 'locally', i.e., in a sufficiently small neighbourhood $\tilde{\omega} \subset \omega'$ of any point $\tilde{y}$ of γ,

$\eta_1 = \eta_2 = 0$ is the unique solution in $\mathcal{C}^1(\widetilde{\omega})$. Recalling that any solution $\boldsymbol{\eta} = (\eta_i)$ of the 'full' system is such that $\eta_\alpha \in \mathcal{C}^1(\overline{\omega})$ by (ii), we have thus reached the following conclusion. Given any point $\widetilde{y} \in \gamma$, there exists a neighbourhood $\widetilde{\omega} \subset \omega$ of $\widetilde{y}$ such that $\boldsymbol{\eta} = \mathbf{0}$ is the only solution $\boldsymbol{\eta} = (\eta_i) \in H^1(\omega) \times H^1(\omega) \times L^2(\omega)$ in $\widetilde{\omega} \cap \overline{\omega}$ of the 'full' system

$$\gamma_{\alpha\beta}(\boldsymbol{\eta}) = 0 \text{ in } \omega \quad \text{and} \quad \eta_\alpha = 0 \text{ on } \gamma.$$

(iv) 'Global' uniqueness of the solution of the system

$$\gamma_{\alpha\beta}(\boldsymbol{\eta}) = 0 \text{ in } \omega \quad \text{and} \quad \eta_\alpha = 0 \text{ on } \gamma.$$

By a theorem of Morrey and Nirenberg (1957), any solution of a uniformly elliptic system whose coefficients are analytic in ω is analytic in ω. Since $\boldsymbol{\eta} = \mathbf{0}$ is an analytic solution, the Analytic Continuation Theorem for analytic functions of several variables (see, *e.g.*, Dieudonné (1968, Theorem 9.4.2)) thus shows that $\boldsymbol{\eta} = \mathbf{0}$ is the only solution. $\qquad\square$

We are now in a position to prove the main result of this section, due to Ciarlet and Lods (1996*a*) and Ciarlet and Sanchez-Palencia (1996), who provided two different proofs. Special mention must also be made of the early existence and uniqueness theorem of Destuynder (1985, Theorems 6.1 and 6.5), obtained under the additional assumptions that the elliptic surface S can be covered by a single system of lines of curvature and that the $\mathcal{C}^0(\overline{\omega})$-norms of the Christoffel symbols of S are sufficiently small.

It is indeed remarkable that, if the surface S is elliptic and the tangential components of the admissible displacement fields of S vanish over the entire boundary of S, the L^2-norm of the linearized change of metric tensor alone is 'already' equivalent to the $(H^1 \times H^1 \times L^2)$-norm of these fields (compare with Theorem 4.4; note, however, that the H^2-norm of the normal components that appears there in the inequality of Korn's type on a general surface is now replaced by the L^2-norm).

Theorem 6.3: Inequality of Korn's type on an elliptic surface. Let the assumptions be as in Theorem 6.2. Then there exists a constant $c_M = c_M(\omega, \boldsymbol{\theta})$ such that

$$\left\{ \sum_\alpha \|\eta_\alpha\|_{1,\omega}^2 + |\eta_3|_{0,\omega}^2 \right\}^{1/2} \leq c_M \left\{ \sum_{\alpha,\beta} |\gamma_{\alpha\beta}(\boldsymbol{\eta})|_{0,\omega}^2 \right\}^{1/2}$$

for all $\boldsymbol{\eta} \in \mathbf{V}_M(\omega) := H_0^1(\omega) \times H_0^1(\omega) \times L^2(\omega)$.

Proof. (i) By the second inequality of Korn's type without boundary conditions on a general surface (Theorem 6.1), there exists a constant c_0 such that

$$\|\boldsymbol{\eta}\|_{H^1(\omega)\times H^1(\omega)\times L^2(\omega)} := \left\{\sum_\alpha \|\eta_\alpha\|^2_{1,\omega} + |\eta_3|^2_{0,\omega}\right\}^{1/2}$$

$$\leq c_0\left\{\sum_i |\eta_i|^2_{0,\omega} + \sum_{\alpha,\beta} |\gamma_{\alpha\beta}(\boldsymbol{\eta})|^2_{0,\omega}\right\}^{1/2}$$

for all $\boldsymbol{\eta} \in \mathbf{V}_M(\omega)$, since $\mathbf{V}_M(\omega) \subset H^1(\omega) \times H^1(\omega) \times L^2(\omega)$. Hence it suffices to show that there exists a constant c such that

$$\left\{\sum_i |\eta_i|^2_{0,\omega}\right\}^{1/2} \leq c\left\{\sum_{\alpha,\beta} |\gamma_{\alpha\beta}(\boldsymbol{\eta})|^2_{0,\omega}\right\}^{1/2} \quad \text{for all } \boldsymbol{\eta} \in \mathbf{V}_M(\omega).$$

(ii) If the last inequality is false, there exists a sequence $(\boldsymbol{\eta}^k)_{k=1}^\infty$ of functions $\boldsymbol{\eta}^k = (\eta_i^k) \in \mathbf{V}_M(\omega)$ such that

$$\left\{\sum_i |\eta_i^k|^2_{0,\omega}\right\}^{1/2} = 1 \text{ for all } k \quad \text{and} \quad \lim_{k\to\infty}\left\{\sum_{\alpha,\beta} |\gamma_{\alpha\beta}(\boldsymbol{\eta}^k)|^2_{0,\omega}\right\}^{1/2} = 0.$$

In particular, then, the sequence $(\boldsymbol{\eta}^k)_{k=1}^\infty$ is bounded with respect to the norm $\|\cdot\|_{H^1(\omega)\times H^1(\omega)\times L^2(\omega)}$, thanks again to the second inequality of Korn's type of Theorem 6.1. Since any bounded sequence in a Hilbert space contains a weakly convergent sequence, there exists a subsequence $(\boldsymbol{\eta}^l)_{l=1}^\infty$ and an element $\boldsymbol{\eta} = (\eta_i) \in \mathbf{V}_M(\omega)$ such that

$$\eta_\alpha^l \rightharpoonup \eta_\alpha \text{ in } H^1(\omega) \text{ and } \eta_\alpha^l \to \eta_\alpha \text{ in } L^2(\omega),$$
$$\eta_3^l \rightharpoonup \eta_3 \text{ in } L^2(\omega),$$

where $\rightharpoonup$ and $\to$ denote weak and strong convergence (the compact embedding $H^1(\omega) \Subset L^2(\omega)$ is also used here).

(iii) Naturally, the difficulty rests with the subsequence $(\eta_3^l)_{l=1}^\infty$, which converges only *weakly* in $L^2(\omega)$. Our recourse for showing that it in fact *strongly* converges in $L^2(\omega)$ will be (*cf.* (iv)) the assumed *ellipticity of the surface S*; but first, we prove that $\boldsymbol{\eta} = (\eta_i) = \mathbf{0}$. To this end, we simply note that

$$\eta_\alpha^l \rightharpoonup \eta_\alpha \text{ in } H^1(\omega) \text{ and } \eta_3^l \rightharpoonup \eta_3 \text{ in } L^2(\omega) \Rightarrow \gamma_{\alpha\beta}(\boldsymbol{\eta}^l) \rightharpoonup \gamma_{\alpha\beta}(\boldsymbol{\eta}) \text{ in } L^2(\omega),$$

on the one hand; since

$$\gamma_{\alpha\beta}(\boldsymbol{\eta}^l) \to 0 \text{ in } L^2(\omega),$$

on the other, we conclude that $\gamma_{\alpha\beta}(\boldsymbol{\eta}) = 0$. Hence $\boldsymbol{\eta} = \mathbf{0}$ by Theorem 6.2.

(iv) We next show that $\eta_3^l \to 0$ in $L^2(\omega)$. The strong convergence

$$\gamma_{\alpha\beta}(\boldsymbol{\eta}^l) \to 0 \text{ in } L^2(\omega) \text{ and } \eta_\alpha^l \to 0 \text{ in } L^2(\omega)$$

combined with the definition of the functions $\gamma_{\alpha\beta}(\boldsymbol{\eta})$ implies the following strong convergence:

$$\partial_1\eta_1^l - b_{11}\eta_3^l = \{\gamma_{11}(\boldsymbol{\eta}^l) + \Gamma_{11}^\sigma\eta_\sigma^l\} \quad \to 0 \text{ in } L^2(\omega),$$
$$\partial_2\eta_1^l + \partial_1\eta_2^l - 2b_{12}\eta_3^l = \{2\gamma_{12}(\boldsymbol{\eta}^l) + 2\Gamma_{12}^\sigma\eta_\sigma^l\} \to 0 \text{ in } L^2(\omega),$$
$$\partial_2\eta_2^l - b_{22}\eta_3^l = \{\gamma_{22}(\boldsymbol{\eta}^l) + \Gamma_{22}^\sigma\eta_\sigma^l\} \quad \to 0 \text{ in } L^2(\omega).$$

As the function $b_{11} \in C^0(\overline{\omega})$ does not vanish in $\overline{\omega}$ by the assumed *ellipticity* of the surface S, we can eliminate η_3^l between the first and second, and between the first and third, relations; this elimination yields

$$\left\{\partial_2\eta_1^l + \partial_1\eta_2^l - 2\frac{b_{12}}{b_{11}}\partial_1\eta_1^l\right\} \to 0 \text{ in } L^2(\omega),$$

$$\left\{\partial_2\eta_2^l - \frac{b_{22}}{b_{11}}\partial_1\eta_1^l\right\} \to 0 \text{ in } L^2(\omega).$$

Multiplying the first relation by $\partial_2\eta_1^l$ and the second by $\partial_1\eta_1^l$, then integrating over ω, we get

$$\int_\omega \left\{(\partial_2\eta_1^l)^2 + \partial_2\eta_1^l\partial_1\eta_2^l - 2\frac{b_{12}}{b_{11}}\partial_1\eta_1^l\partial_2\eta_1^l\right\} dy \to 0,$$

$$\int_\omega \left\{\partial_1\eta_1^l\partial_2\eta_2^l - \frac{b_{22}}{b_{11}}(\partial_1\eta_1^l)^2\right\} dy \to 0,$$

since each sequence $(\partial_\alpha\eta_1^l)_{l=1}^\infty$ is bounded in $L^2(\omega)$ (each sequence even weakly converges to 0 in $L^2(\omega)$). Subtracting the last two relations and using the relation $\int_\omega \partial_2\eta_1^l\partial_1\eta_2^l\, dy = \int_\omega \partial_1\eta_1^l\partial_2\eta_2^l\, dy$, we thus obtain

$$\int_\omega \left\{\left(\partial_2\eta_1^l - \frac{b_{12}}{b_{11}}\partial_1\eta_1^l\right)^2 + \frac{1}{(b_{11})^2}(b_{11}b_{22} - (b_{12})^2)(\partial_1\eta_1^l)^2\right\} dy \to 0,$$

and consequently

$$\partial_1\eta_1^l \to 0 \text{ in } L^2(\omega),$$

since $b_{11}b_{22} - (b_{12})^2 = \det(b_{\alpha\beta}) \in C^0(\overline{\omega})$ does not vanish in $\overline{\omega}$ by the assumed ellipticity of S. Hence

$$\eta_3^l = \left\{\frac{1}{b_{11}}\partial_1\eta_1^l - \frac{1}{b_{11}}(\partial_1\eta_1^l - b_{11}\eta_3^l)\right\} \to 0 \text{ in } L^2(\omega).$$

(v) The relations $\eta_i^l \to 0$ in $L^2(\omega)$ established in parts (ii)–(iv) thus contradict the relations $\{\sum_i |\eta_i^l|_{0,\omega}^2\}^{1/2} = 1$ for all l, and the proof is complete.
∎

When the surface S is elliptic, the covariant components η_α of the displacement field vanish over the entire boundary γ, and the assumptions on

ω and $\boldsymbol{\theta}$ are as in Theorem 6.2, the two-dimensional equations of a 'membrane' shell (described at the beginning of this section) thus have exactly one solution.

7. Preliminaries to the asymptotic analysis of linearly elastic shells

The purpose of this section is to gather the fundamental preliminaries needed in Sections 8 to 10 for carrying out the asymptotic analysis of all kinds of linearly elastic shells. After *ad hoc* 'scalings' of the unknowns (the covariant components of the three-dimensional displacement field) and *ad hoc* 'asymptotic' assumptions on the data (the Lamé constants and applied force densities) have been made, the problem of a linearly elastic clamped shell with thickness $2\varepsilon > 0$ is transformed into a *scaled problem*, defined over a domain that is *independent of ε*.

Recall that ε is *not* subjected to the rule governing Greek exponents.

7.1. *The three-dimensional equations*

As in Section 4, let ω be a domain in $\mathbb{R}^2$ with boundary γ, let $y = (y_\alpha)$ denote a generic point in the set $\bar{\omega}$, and let $\partial_\alpha := \partial/\partial y_\alpha$. Let $\boldsymbol{\theta} \in C^2(\bar{\omega}; \mathbb{R}^3)$ be an *injective* mapping such that the two vectors $\boldsymbol{a}_\alpha(y) := \partial_\alpha \boldsymbol{\theta}(y)$ are *linearly independent at all points* $y \in \bar{\omega}$. These two vectors form the *covariant basis* of the tangent plane to the surface $S := \boldsymbol{\theta}(\bar{\omega})$ at the point $\boldsymbol{\theta}(y)$ and the two vectors $\boldsymbol{a}^\alpha(y)$ of the tangent plane at $\boldsymbol{\theta}(y)$ defined by the relations $\boldsymbol{a}^\alpha(y) \cdot \boldsymbol{a}_\beta(y) = \delta^\alpha_\beta$ form its *contravariant basis*. Also, let

$$\boldsymbol{a}_3(y) = \boldsymbol{a}^3(y) := \frac{\boldsymbol{a}_1(y) \wedge \boldsymbol{a}_2(y)}{|\boldsymbol{a}_1(y) \wedge \boldsymbol{a}_2(y)|}.$$

Then $|\boldsymbol{a}_3(y)| = 1$, the vector $\boldsymbol{a}_3(y)$ is normal to S at the point $\boldsymbol{\theta}(y)$, and the three vectors $\boldsymbol{a}^i(y)$ form the *contravariant basis at* $\boldsymbol{\theta}(y)$. Recall that (y_1, y_2) constitutes a system of curvilinear coordinates for describing the surface S.

Let γ_0 denote a $d\gamma$-measurable subset of the boundary γ of ω satisfying

$$\text{length } \gamma_0 > 0.$$

For each $\varepsilon > 0$, we define the sets

$$\Omega^\varepsilon := \omega \times]-\varepsilon, \varepsilon[\text{ and } \Gamma_0^\varepsilon := \gamma_0 \times [-\varepsilon, \varepsilon].$$

Let $x^\varepsilon = (x_i^\varepsilon)$ denote a generic point in the set $\bar{\Omega}^\varepsilon$ and let $\partial_i^\varepsilon := \partial/\partial x_i^\varepsilon$; hence $x_\alpha^\varepsilon = y_\alpha$ and $\partial_\alpha^\varepsilon = \partial_\alpha$.

Consider an *elastic shell* with middle surface $S = \boldsymbol{\theta}(\bar{\omega})$ and (constant) thickness $2\varepsilon > 0$, that is, an elastic body whose reference configuration consists of all points within a distance $\leq \varepsilon$ from S. In other words, the reference configuration of the shell is the image $\Theta(\overline{\Omega}^\varepsilon) \subset \mathbb{R}^3$ of the set $\overline{\Omega}^\varepsilon \subset \mathbb{R}^3$ through the mapping $\Theta : \overline{\Omega}^\varepsilon \to \mathbb{R}^3$ given by

$$\Theta(x^\varepsilon) := \boldsymbol{\theta}(y) + x_3^\varepsilon \boldsymbol{a}_3(y) \text{ for all } x^\varepsilon = (y, x_3^\varepsilon) = (y_1, y_2, x_3^\varepsilon) \in \overline{\Omega}^\varepsilon.$$

It can then be shown (Ciarlet 2000, Theorem 3.1-1) that the mapping $\Theta : \overline{\Omega}^\varepsilon \to \mathbb{R}^3$ is injective for sufficiently small $\varepsilon > 0$. In other words, if $\varepsilon > 0$ is sufficiently small, $(y_1, y_2, x_3^\varepsilon)$ constitutes a *bona fide* system of curvilinear coordinates (Section 3) for describing the reference configuration $\Theta(\overline{\Omega}^\varepsilon)$ of the shell and the physical problem is meaningful since the set $\Theta(\overline{\Omega}^\varepsilon)$ 'does not interpenetrate itself'. These curvilinear coordinates are called the *'natural' curvilinear coordinates of the shell* and the curvilinear coordinate $x_3^\varepsilon \in [-\varepsilon, \varepsilon]$ is called the *transverse variable*. We shall also use the notation x_i^ε for the 'natural' curvilinear coordinates of the shell, that is, we shall let $x_\alpha^\varepsilon = y_\alpha$, so that a generic point in the set $\overline{\Omega}^\varepsilon$ may be written as $x^\varepsilon = (x_i^\varepsilon)$.

It can likewise be shown (see Ciarlet (2000, Theorem 3.1-1)) that, again for sufficiently small $\varepsilon > 0$, the three vectors $\boldsymbol{g}_i^\varepsilon(x^\varepsilon) := \partial_i^\varepsilon \Theta(x^\varepsilon)$ form the *covariant basis* at each point $\Theta(x^\varepsilon)$, $x^\varepsilon \in \overline{\Omega}^\varepsilon$, of the reference configuration, while the three vectors $\boldsymbol{g}^{i,\varepsilon}(x^\varepsilon)$ defined by $\boldsymbol{g}^{i,\varepsilon}(x^\varepsilon) \cdot \boldsymbol{g}_j^\varepsilon(x^\varepsilon) = \delta_j^i$ form the *contravariant basis* at $\Theta(x^\varepsilon)$.

As in Section 3, we define the covariant and contravariant components g_{ij}^ε and $g^{ij,\varepsilon}$ of the *metric tensor* of the set $\Theta(\overline{\Omega}^\varepsilon)$ and the *Christoffel symbols* $\Gamma_{ij}^{p,\varepsilon}$ by letting

$$g_{ij}^\varepsilon := \boldsymbol{g}_i^\varepsilon \cdot \boldsymbol{g}_j^\varepsilon, \quad g^{ij,\varepsilon} := \boldsymbol{g}^{i,\varepsilon} \cdot \boldsymbol{g}^{j,\varepsilon}, \text{ and } \Gamma_{ij}^{p,\varepsilon} := \boldsymbol{g}^{p,\varepsilon} \cdot \partial_i^\varepsilon \boldsymbol{g}_j^\varepsilon$$

(we omit the explicit dependence on x^ε).

The *volume element* in the set $\Theta(\overline{\Omega}^\varepsilon)$ is $\sqrt{g^\varepsilon}\, dx^\varepsilon$, where $g^\varepsilon := \det(g_{ij}^\varepsilon)$.

We assume that the material constituting the shell is homogeneous and isotropic and that $\Theta(\overline{\Omega}^\varepsilon)$ is a natural state, so that the material is characterized by its two Lamé constants $\lambda^\varepsilon > 0$ and $\mu^\varepsilon > 0$ (Ciarlet 1988, Section 6.2). The unknown of the problem is the vector field $\boldsymbol{u}^\varepsilon = (u_i^\varepsilon) : \overline{\Omega}^\varepsilon \to \mathbb{R}^3$, where the three functions $u_i^\varepsilon : \overline{\Omega}^\varepsilon \to \mathbb{R}$ are the covariant components (with respect to the contravariant bases $\{\boldsymbol{g}^{i,\varepsilon}\}$) of the displacement field $u_i^\varepsilon \boldsymbol{g}^{i,\varepsilon} : \overline{\Omega}^\varepsilon \to \mathbb{R}^3$ experienced by the shell under the influence of *applied forces*. This means that $u_i^\varepsilon(x^\varepsilon)\boldsymbol{g}^{i,\varepsilon}(x^\varepsilon)$ is the displacement of the point $\Theta(x^\varepsilon)$; see Figure 7.1.

Finally, we assume that the shell is subjected to a boundary condition of place $\boldsymbol{u}^\varepsilon = \boldsymbol{0}$ on Γ_0^ε, that is, that the displacement field $u_i^\varepsilon \boldsymbol{g}^{i,\varepsilon}$ vanishes along the portion $\Theta(\Gamma_0^\varepsilon)$ of its *lateral face* $\Theta(\gamma \times [-\varepsilon, \varepsilon])$.

In linearized elasticity, the unknown $\boldsymbol{u}^\varepsilon = (u_i^\varepsilon)$ then satisfies the following three-dimensional variational problem $\mathcal{P}(\Omega^\varepsilon)$ for a linearly elastic shell in curvilinear coordinates, that is, stated in terms of the 'natural' curvilinear

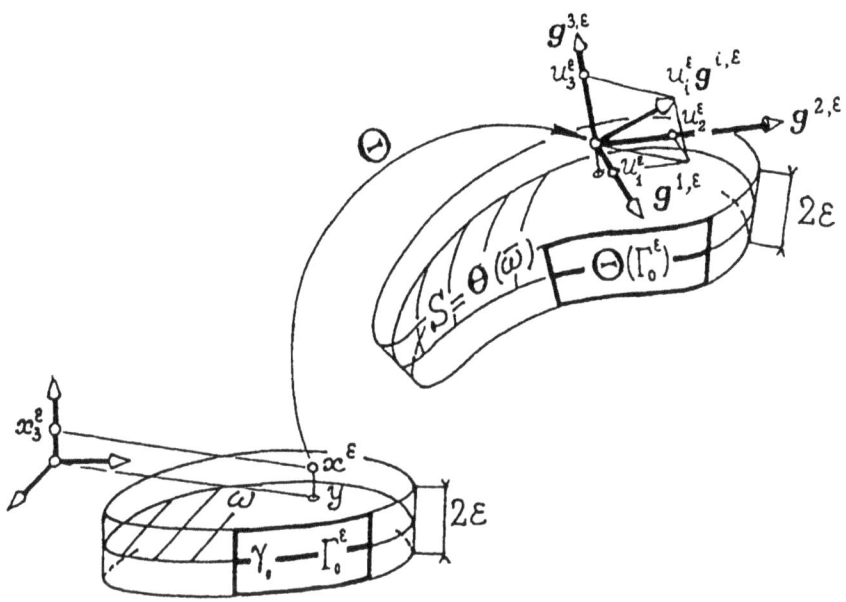

Fig. 7.1. A three-dimensional shell problem. Let $\Omega^\varepsilon = \omega \times]-\varepsilon, \varepsilon[$. The set $\Theta(\overline{\Omega}^\varepsilon)$, where $\Theta(y, x_3^\varepsilon) = \theta(y) + x_3^\varepsilon a_3(y)$ for all $x^\varepsilon = (y, x_3^\varepsilon) \in \overline{\Omega}^\varepsilon$, is the reference configuration of a shell, with thickness 2ε and middle surface $S = \theta(\overline{\omega})$. The unknowns of the problem are the three covariant components $u_i^\varepsilon : \overline{\Omega}^\varepsilon \to \mathbb{R}$ of the displacement field $u_i^\varepsilon g^{i,\varepsilon} : \overline{\Omega}^\varepsilon \to \mathbb{R}^3$ of the points of $\Theta(\overline{\Omega}^\varepsilon)$. This means that, for each $x^\varepsilon \in \overline{\Omega}^\varepsilon$, $u_i^\varepsilon(x^\varepsilon) g^{i,\varepsilon}(x^\varepsilon)$ is the displacement of the point $\Theta(x^\varepsilon) \in \Theta(\overline{\Omega}^\varepsilon)$

coordinates x_i^ε of the shell,

$$\boldsymbol{u}^\varepsilon \in \mathbf{V}(\Omega^\varepsilon) := \{\boldsymbol{v}^\varepsilon = (v_i^\varepsilon) \in \mathbf{H}^1(\Omega^\varepsilon) : \boldsymbol{v}^\varepsilon = \mathbf{0} \text{ on } \Gamma_0^\varepsilon\},$$

$$\int_{\Omega^\varepsilon} A^{ijkl,\varepsilon} e_{k\|l}^\varepsilon(\boldsymbol{u}^\varepsilon) e_{i\|j}^\varepsilon(\boldsymbol{v}^\varepsilon) \sqrt{g^\varepsilon} \, dx^\varepsilon = \int_{\Omega^\varepsilon} f^{i,\varepsilon} v_i^\varepsilon \sqrt{g^\varepsilon} \, dx^\varepsilon$$

for all $\boldsymbol{v}^\varepsilon \in \mathbf{V}(\Omega^\varepsilon)$, where

$$A^{ijkl,\varepsilon} := \lambda^\varepsilon g^{ij,\varepsilon} g^{kl,\varepsilon} + \mu^\varepsilon (g^{ik,\varepsilon} g^{jl,\varepsilon} + g^{il,\varepsilon} g^{jk,\varepsilon})$$

designate the contravariant components of the *three-dimensional elasticity tensor* of the shell, and

$$e_{i\|j}^\varepsilon(\boldsymbol{v}^\varepsilon) := \frac{1}{2} \left(\partial_j^\varepsilon v_i^\varepsilon + \partial_i^\varepsilon v_j^\varepsilon \right) - \Gamma_{ij}^{p,\varepsilon} v_p^\varepsilon$$

are the *linearized strains in curvilinear coordinates* associated with an arbitrary displacement field $v_i^\varepsilon g^{i,\varepsilon}$ of the set $\Theta(\overline{\Omega}^\varepsilon)$; $f^{i,\varepsilon} \in L^2(\Omega^\varepsilon)$ denote the contravariant components of the *applied body force density*, applied to the interior $\Theta(\Omega^\varepsilon)$ of the shell, and $d\Gamma^\varepsilon$ designates the area element along $\partial\Omega^\varepsilon$. See Ciarlet (2000, Chapters 1 and 3) for details.

Surface forces over the 'upper' and 'lower' faces $\Theta(\omega \times \{\varepsilon\})$ and $\Theta(\omega \times \{-\varepsilon\})$ of the shell could as well be considered without much further ado, other than 'technical': their consideration simply results in extra terms on the right-hand sides of the two-dimensional equations that will eventually be obtained (see Ciarlet (2000) for details). By contrast, we *assume* that there are *no* surface forces applied to the portion $\Theta((\gamma - \gamma_0) \times [-\varepsilon, \varepsilon])$ of the lateral face of the shell, as their consideration would substantially modify the subsequent analyses.

The above three-dimensional equations of a linearly elastic shell have exactly one solution. This existence and uniqueness result relies on the three-dimensional Korn inequality in curvilinear coordinates (Theorem 3.4), combined with the uniform positive definiteness of the three-dimensional elasticity tensor, already mentioned in Section 4.

Our basic objective consists in showing that, if $\varepsilon > 0$ is small enough and the data are of appropriate orders with respect to ε, the above three-dimensional problems are 'asymptotically equivalent' to a *two-dimensional problem* posed over the middle surface of the shell. This means that the new unknown should be $\boldsymbol{\zeta}^\varepsilon = (\zeta_i^\varepsilon)$, where ζ_i^ε are the covariant components (*i.e.*, over the covariant bases $\{\boldsymbol{a}^i\}$) of the displacement field $\zeta_i^\varepsilon \boldsymbol{a}^i : \overline{\omega} \to \mathbb{R}^3$ of the points of the middle surface $S = \boldsymbol{\theta}(\overline{\omega})$. In other words, $\zeta_i^\varepsilon(y)\boldsymbol{a}^i(y)$ is the displacement of the point $\boldsymbol{\theta}(y) \in S$; see Figure 7.2.

7.2. *The three-dimensional equations over a fixed domain; the fundamental scalings and assumptions on the data*

We now describe the basic preliminaries of the asymptotic analysis of a linearly elastic shell, as set forth by Sanchez-Palencia (1990, 1992) in a slightly different, but in fact equivalent, framework of a 'multi-scale' asymptotic analysis, then by Miara and Sanchez-Palencia (1996), Ciarlet and Lods (1996*b*, 1996*d*), and Ciarlet, Lods and Miara (1996).

'Asymptotic analysis' means that the objective is to study the behaviour of the displacement field $u_i^\varepsilon \boldsymbol{g}^{i,\varepsilon} : \overline{\Omega}^\varepsilon \to \mathbb{R}^3$ as $\varepsilon \to 0$, an endeavour that will be achieved by studying the behaviour as $\varepsilon \to 0$ of the covariant components $u_i^\varepsilon : \overline{\Omega}^\varepsilon \to \mathbb{R}$ of the displacement field, that is, the behaviour of the unknown $\boldsymbol{u}^\varepsilon = (u_i^\varepsilon) : \overline{\Omega}^\varepsilon \to \mathbb{R}^3$ of the three-dimensional variational problem $\mathcal{P}(\Omega^\varepsilon)$ described above.

Since these fields are defined on sets $\overline{\Omega}^\varepsilon$ that themselves vary with ε, our first task naturally consists in transforming the three-dimensional problems $\mathcal{P}(\Omega^\varepsilon)$ into problems posed over a set that does not depend on ε. The underlying principle is thus identical to that followed for *plates*, albeit with differences in the way it is put to use (*cf.* Ciarlet (1997, Section 1.3)).

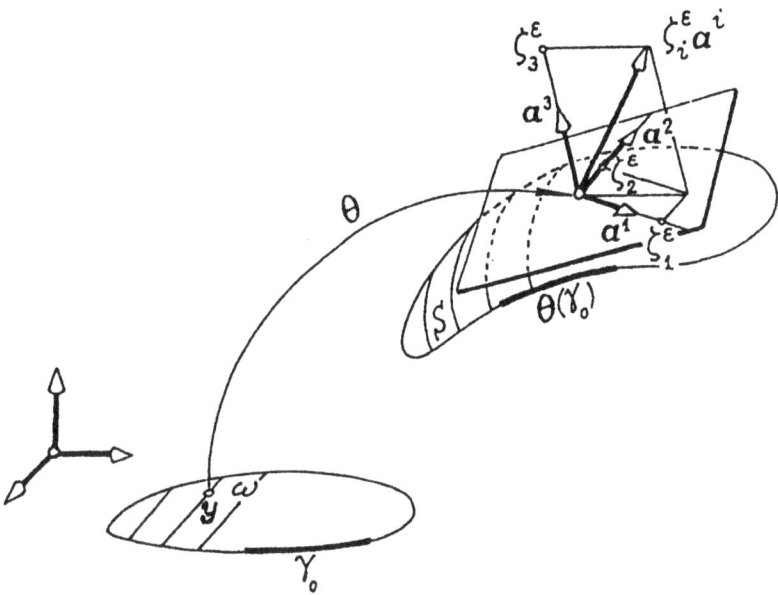

Fig. 7.2. A two-dimensional shell problem. The unknowns are the three covariant components $\zeta_i^\varepsilon : \overline{\omega} \to \mathbb{R}$ of the displacement field $\zeta_i^\varepsilon a^i : \overline{\omega} \to \mathbb{R}^3$ of the points of the middle surface $S = \boldsymbol{\theta}(\overline{\omega})$. This means that, for each $y \in \overline{\omega}$, $\zeta_i^\varepsilon(y)a^i(y)$ is the displacement of the point $\boldsymbol{\theta}(y) \in S$

Let

$$\Omega := \omega \times] - 1, 1[\text{ and } \Gamma_0 := \gamma_0 \times [-1, 1].$$

Let $x = (x_1, x_2, x_3)$ denote a generic point in the set $\overline{\Omega}$ and let $\partial_i := \partial/\partial x_i$; hence $x_\alpha = y_\alpha$, since a generic point in the set $\overline{\omega}$ is denoted by $y = (y_1, y_2)$. The coordinate $x_3 \in [-1, 1]$ will also be called *transverse variable*, like $x_3^\varepsilon \in [-\varepsilon, \varepsilon]$). With each point $x = (x_i) \in \overline{\Omega}$, we associate the point $x^\varepsilon = (x_i^\varepsilon) \in \overline{\Omega}^\varepsilon$ through the bijection (see Figure 7.3)

$$\pi^\varepsilon : x = (x_1, x_2, x_3) \in \overline{\Omega} \longrightarrow x^\varepsilon = (x_i^\varepsilon) = (x_1, x_2, \varepsilon x_3) \in \overline{\Omega}^\varepsilon.$$

Note in passing that we therefore have $x_\alpha^\varepsilon = x_\alpha = y_\alpha$, $\partial_\alpha^\varepsilon = \partial_\alpha$ and $\partial_3^\varepsilon = \frac{1}{\varepsilon}\partial_3$.

In order to carry out our asymptotic treatment of the solutions $\boldsymbol{u}^\varepsilon = (u_i^\varepsilon)$ of problems $\mathcal{P}(\Omega^\varepsilon)$ by considering ε as a small parameter, we must:

(i) specify the way the unknown $\boldsymbol{u}^\varepsilon = (u_i^\varepsilon)$ and more generally the vector fields $\boldsymbol{v} = (v_i^\varepsilon)$ appearing in the formulation of problems $\mathcal{P}(\Omega^\varepsilon)$ are mapped into vector fields over the set $\overline{\Omega}$;

(ii) control the way the Lamé constants and the applied body forces depend on the parameter ε.

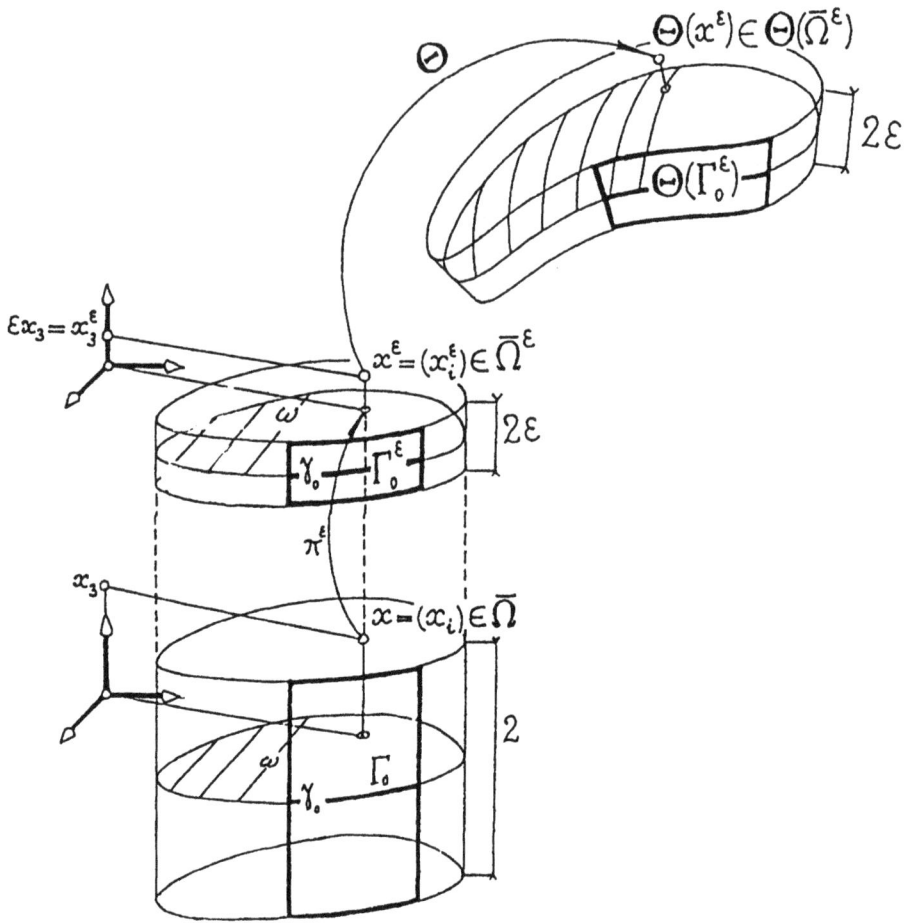

Fig. 7.3. Transformation of the three-dimensional shell problem into a 'scaled' problem, posed over the fixed domain $\Omega = \omega \times] - 1, 1[$

With the unknown $\boldsymbol{u}^\varepsilon = (u_i^\varepsilon) : \overline{\Omega}^\varepsilon \to \mathbb{R}^3$ and the vector fields $\boldsymbol{v}^\varepsilon = (v_i^\varepsilon) : \overline{\Omega}^\varepsilon \to \mathbb{R}^3$ appearing in the three-dimensional variational problem $\mathcal{P}(\Omega^\varepsilon)$, we associate the *scaled unknown* $\boldsymbol{u}(\varepsilon) = (u_i(\varepsilon)) : \overline{\Omega} \to \mathbb{R}^3$ and the *scaled vector fields* $\boldsymbol{v} = (v_i) : \overline{\Omega} \to \mathbb{R}^3$ defined by the *scalings*

$$u_i(\varepsilon)(x) := u_i^\varepsilon(x^\varepsilon) \text{ and } v_i(x) := v_i^\varepsilon(x^\varepsilon) \text{ for all } x^\varepsilon = \pi^\varepsilon x \in \overline{\Omega}^\varepsilon.$$

The three components $u_i(\varepsilon)$ of the scaled unknown $\boldsymbol{u}(\varepsilon)$ are called the *scaled displacements*.

We next make the following assumptions on the data, that is, on the Lamé constants and on the applied body forces. There exist constants $\lambda > 0$ and $\mu > 0$ independent of ε and there exist functions $f^i \in L^2(\Omega)$ independent of

ε such that, for all $\varepsilon > 0$,

$$\lambda^\varepsilon = \lambda \quad \text{and} \quad \mu^\varepsilon = \mu,$$
$$f^{i,\varepsilon}(x^\varepsilon) = \varepsilon^a f^i(x) \quad \text{for all} \quad x^\varepsilon = \pi^\varepsilon x \in \Omega^\varepsilon,$$

where the exponent a is for the time being left unspecified (needless to say, a is *not* subjected to the usual rule governing Latin exponents!)

Since the problem is linear, we assume without loss of generality that the scaled unknown $\boldsymbol{u}(\varepsilon)$ is 'of order 0 with respect to ε'. This means that the limit of $\boldsymbol{u}(\varepsilon)$ as ε approaches zero (assuming that such a limit exists, in an *ad hoc* function space) is *a priori* assumed to be of order 0, when the applied forces are of the right orders.

We have assumed that the Lamé constants are independent of ε. However, this assumption is merely a special case among a more general class of assumptions, which permit in particular the Lamé constants to vary with ε as $\varepsilon \to 0$ if one so wishes. More precisely, a multiplication of both Lamé constants by a factor $\varepsilon^t, t \in \mathbb{R}$, is always possible, as we shall see after Theorem 7.1. The choice $t = 0$ is merely made here for simplicity.

For sufficiently small $\varepsilon > 0$ (so that the mapping $\boldsymbol{\Theta}$ that defines the reference configuration of the shell is injective), a simple computation then produces the equations that the scaled unknown $\boldsymbol{u}(\varepsilon)$ satisfies over the set Ω, thus over a domain that is *independent of ε* (the Christoffel symbols $\Gamma^{3,\varepsilon}_{\alpha 3}$ and $\Gamma^{p,\varepsilon}_{33}$ vanish in Ω^ε for the special class of mappings $\boldsymbol{\Theta}$ considered here; consequently, the functions $\Gamma^3_{\alpha 3}(\varepsilon)$ and $\Gamma^p_{33}(\varepsilon)$ defined below likewise vanish in Ω).

Theorem 7.1: The three-dimensional shell problem over the fixed domain $\Omega = \omega \times] - 1, 1[$. With the functions $\Gamma^{p,\varepsilon}_{ij}, g^\varepsilon, A^{ijkl,\varepsilon} : \overline{\Omega}^\varepsilon \to \mathbb{R}$ appearing in the equations of problem $\mathcal{P}(\Omega^\varepsilon)$, we associate the 'scaled' functions $\Gamma^p_{ij}(\varepsilon), g(\varepsilon), A^{ijkl}(\varepsilon) : \overline{\Omega} \to \mathbb{R}$ defined by

$$\Gamma^p_{ij}(\varepsilon)(x) := \Gamma^{p,\varepsilon}_{ij}(x^\varepsilon), \quad g(\varepsilon)(x) := g^\varepsilon(x^\varepsilon), \quad A^{ijkl}(\varepsilon)(x) := A^{ijkl,\varepsilon}(x^\varepsilon)$$

for all $x^\varepsilon = \pi^\varepsilon x \in \overline{\Omega}^\varepsilon$.

With any vector field $\boldsymbol{v} = (v_i) \in \mathbf{H}^1(\Omega)$, we associate the 'scaled linearized strains' $e_{i\|j}(\varepsilon; \boldsymbol{v}) = e_{j\|i}(\varepsilon; \boldsymbol{v}) \in L^2(\Omega)$ defined by

$$e_{\alpha\|\beta}(\varepsilon; \boldsymbol{v}) := \frac{1}{2}(\partial_\beta v_\alpha + \partial_\alpha v_\beta) - \Gamma^p_{\alpha\beta}(\varepsilon)v_p,$$

$$e_{\alpha\|3}(\varepsilon; \boldsymbol{v}) := \frac{1}{2}\left(\frac{1}{\varepsilon}\partial_3 v_\alpha + \partial_\alpha v_3\right) - \Gamma^\sigma_{\alpha 3}(\varepsilon)v_\sigma,$$

$$e_{3\|3}(\varepsilon; \boldsymbol{v}) := \frac{1}{\varepsilon}\partial_3 v_3.$$

Let the assumptions on the data be as above. Then, for sufficiently small ε, the scaled unknown $\boldsymbol{u}(\varepsilon)$ satisfies the following scaled three-dimensional variational problem $\mathcal{P}(\varepsilon; \Omega)$ of a linearly elastic shell:

$$\boldsymbol{u}(\varepsilon) \in \mathbf{V}(\Omega) := \{\boldsymbol{v} = (v_i) \in \mathbf{H}^1(\Omega) : \boldsymbol{v} = \mathbf{0} \text{ on } \Gamma_0\},$$

$$\int_\Omega A^{ijkl}(\varepsilon) e_{k\|l}(\varepsilon; \boldsymbol{u}(\varepsilon)) e_{i\|j}(\varepsilon; \boldsymbol{v}) \sqrt{g(\varepsilon)} \, dx = \varepsilon^a \int_\Omega f^i v_i \sqrt{g(\varepsilon)} \, dx$$

for all $\boldsymbol{v} \in \mathbf{V}(\Omega)$.   □

The functions $A^{ijkl}(\varepsilon)$ are called the *contravariant components* of the scaled three-dimensional elasticity tensor of the shell. The functions $e_{i\|j}(\varepsilon; \boldsymbol{v})$ are called *'scaled' linearized strains* because they satisfy

$$e_{i\|j}(\varepsilon; \boldsymbol{v})(x) = e_{i\|j}^\varepsilon(\boldsymbol{v}^\varepsilon)(x^\varepsilon) \quad \text{for all } x^\varepsilon = \pi^\varepsilon x \in \overline{\Omega}^\varepsilon.$$

Note that the scaled strains $e_{i\|3}(\varepsilon; \boldsymbol{v})$ are *not* defined for $\varepsilon = 0$. Hence problems $\mathcal{P}(\varepsilon; \Omega)$ provide instances of singular perturbation problems in variational form, as considered and extensively studied by Lions (1973).

Note also that *exactly the same* scaled three-dimensional problem $\mathcal{P}(\varepsilon; \Omega)$ is evidently obtained if the scaled unknown is defined as before, but the following more general assumptions on the data are made:

$$\lambda^\varepsilon = \varepsilon^t \lambda \quad \text{and} \quad \mu^\varepsilon = \varepsilon^t \mu,$$
$$f^{i,\varepsilon}(x^\varepsilon) = \varepsilon^{a+t} f^i(x) \quad \text{for all } x^\varepsilon = \pi^\varepsilon x \in \Omega^\varepsilon,$$

where the constants $\lambda > 0$ and $\mu > 0$ and the functions $f^i \in L^2(\Omega)$ are as before independent of ε, but t is an *arbitrary real number*.

Our main objective now consists in establishing the *convergence* of the scaled unknown $\boldsymbol{u}(\varepsilon)$ in *ad hoc* function spaces as ε approaches zero. We shall see in this respect that there are essentially two distinct possible types of limit behaviour of $\boldsymbol{u}(\varepsilon)$, corresponding either to linearly elastic *'membrane'* shells (Sections 8 and 9) or to linearly elastic *'flexural'* shells (Section 10).

8. 'Elliptic membrane' shells

As we shall see, the classification of *linearly elastic shells* critically hinges on whether there exist nonzero displacement fields $\eta_i \boldsymbol{a}^i : \overline{\omega} \to \mathbb{R}$ of the middle surface $S = \boldsymbol{\theta}(\overline{\omega})$ that are both *linearized inextensional* ones, i.e., that satisfy $\gamma_{\alpha\beta}(\boldsymbol{\eta}) = 0$ in ω, and *admissible*, i.e., that satisfy the boundary conditions $\eta_i = \partial_\nu \eta_3 = 0$ on γ_0.

More specifically, define the space

$$\mathbf{V}_F(\omega) := \{\boldsymbol{\eta} = (\eta_i) \in H^1(\omega) \times H^1(\omega) \times H^2(\omega) : \eta_i = \partial_\nu \eta_3 = 0 \text{ on } \gamma_0,$$
$$\gamma_{\alpha\beta}(\boldsymbol{\eta}) = 0 \text{ in } \omega\}.$$

Then a shell is called either a *linearly elastic 'membrane' shell* if $\mathbf{V}_F(\omega) = \{\mathbf{0}\}$, that is, if $\mathbf{V}_F(\omega)$ contains only $\eta = \mathbf{0}$, or a *linearly elastic 'flexural' shell* if $\mathbf{V}_F(\omega) \neq \{\mathbf{0}\}$, that is, if $\mathbf{V}_F(\omega)$ contains *nonzero* elements.

A first instance where $\mathbf{V}_F(\omega) = \{\mathbf{0}\}$ is provided by a *linearly elastic 'elliptic membrane' shell*, that is, one whose middle surface $S = \boldsymbol{\theta}(\overline{\omega})$ is *elliptic* (equivalently, the Gaussian curvature of S is everywhere > 0) and which is subjected to a boundary condition of place along its *entire* lateral face: that $\mathbf{V}_F(\omega) = \{\mathbf{0}\}$ simply follows from the linearized rigid displacement lemma on an elliptic surface (Theorem 6.2).

The other instances where $\mathbf{V}_F(\omega) = \{\mathbf{0}\}$ constitute the *linearly elastic 'generalized membrane' shells*, which will be studied in the next section.

The purpose of this section is to identify and to mathematically justify the *two-dimensional equations of a linearly elastic elliptic membrane shell*, by showing how the convergence of the three-dimensional displacements can be established in *ad hoc* function spaces as the thickness of such a shell approaches zero.

8.1. Definition and example

Let ω be a domain in $\mathbb{R}^2$ with boundary γ and let $\boldsymbol{\theta} \in \mathcal{C}^2(\overline{\omega}; \mathbb{R}^3)$ be an injective mapping such that the two vectors $\partial_\alpha \boldsymbol{\theta}(y)$ are linearly independent at all points $y \in \overline{\omega}$. A shell with middle surface $S = \boldsymbol{\theta}(\overline{\omega})$ is called a *linearly elastic elliptic membrane shell* if the following two conditions are simultaneously satisfied.

(i) The shell is subjected to a (homogeneous) *boundary condition of place* along its *entire* lateral face $\Theta(\gamma \times [-\varepsilon, \varepsilon])$, that is, the displacement field vanishes there; equivalently,

$$\gamma_0 = \gamma.$$

(ii) The middle surface S is elliptic, in the sense that there exists a constant c such that

$$\sum_\alpha |\xi^\alpha|^2 \leq c|b_{\alpha\beta}(y)\xi^\alpha\xi^\beta| \text{ for all } y \in \overline{\omega} \text{ and all } (\xi^\alpha) \in \mathbb{R}^2,$$

where the functions $b_{\alpha\beta} : \overline{\omega} \to \mathbb{R}$ are the covariant components of the curvature tensor of S (this definition was given in Section 6). This assumption means that the Gaussian curvature of S is everywhere > 0; equivalently, the two principal radii of curvature are either both > 0 at all points of S, or both < 0 at all points of S (see, *e.g.*, Ciarlet (2000, Section 2.2) for a detailed exposition of these notions).

A shell whose middle surface $S = \boldsymbol{\theta}(\overline{\omega})$ is a portion of an ellipsoid, and which is subjected to a boundary condition of place, that is, of vanishing displacement field along its entire lateral face $\Theta(\gamma \times [-\varepsilon, \varepsilon])$ (solid black in

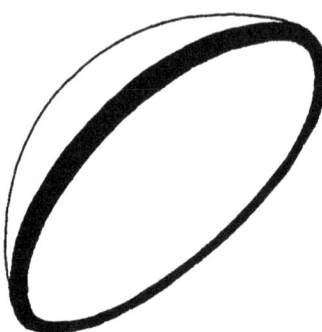

Fig. 8.1. A linearly elastic 'elliptic membrane' shell

the figure), provides an instance of a linearly elastic *elliptic membrane* shell; see Figure 8.1.

The definition of a linearly elastic elliptic membrane shell thus depends only on the subset of the lateral face where the shell is subjected to a boundary condition of place (via the set γ) and on the geometry of its middle surface.

If assumptions (i) and (ii) are satisfied and $\boldsymbol{\theta} \in \mathcal{C}^{2,1}(\overline{\omega}; \mathbb{R}^3)$, the linearized rigid displacement lemma on an elliptic surface (Theorem 6.2) shows that

$$\{\boldsymbol{\eta} = (\eta_i) \in H_0^1(\omega) \times H_0^1(\omega) \times L^2(\omega) : \gamma_{\alpha\beta}(\boldsymbol{\eta}) = 0 \text{ in } \omega\} = \{\boldsymbol{0}\}.$$

Hence linearly elastic elliptic membrane shells indeed provide a first instance where the space

$$\mathbf{V}_F(\omega) := \{\boldsymbol{\eta} = (\eta_i) \in H^1(\omega) \times H^1(\omega) \times H^2(\omega) :$$
$$\eta_i = \partial_\nu \eta_3 = 0 \text{ on } \gamma_0, \gamma_{\alpha\beta}(\boldsymbol{\eta}) = 0 \text{ in } \omega\}$$

a fortiori reduces to $\{\boldsymbol{0}\}$. We recall that ∂_ν denotes the outer normal derivative operator along γ; the subscript 'F' reminds us that this space will be central to the study of *linearly elastic 'flexural' shells* in Section 10.

8.2. *Convergence of the scaled displacements as the thickness approaches zero*

We now establish the main results of this section. Consider a family of linearly elastic elliptic membrane shells with thickness $2\varepsilon > 0$ and with each having the same middle surface $S = \boldsymbol{\theta}(\overline{\omega})$, the assumption on the data being as in Theorem 8.1 below.

Then the solutions $\boldsymbol{u}(\varepsilon)$ of the associated scaled three-dimensional problems $\mathcal{P}(\varepsilon; \Omega)$ (Theorem 7.1) converge in $H^1(\Omega) \times H^1(\Omega) \times L^2(\Omega)$ as $\varepsilon \to 0$

toward a limit u and this limit, which is independent of the transverse variable x_3, can be identified with the solution $\bar{u}$ of a two-dimensional variational problem $\mathcal{P}_M(\omega)$ posed over the set ω.

The functions $\gamma_{\alpha\beta}(\eta)$ appearing in the next theorem represent the covariant components of the *linearized change of metric tensor* associated with a displacement field $\eta_i a^i$ of the middle surface S.

Note that the assumption on the applied body forces made in the next theorem corresponds to letting $a = 0$ in Theorem 7.1. That $a = 0$ is indeed the 'correct' exponent in this case can be justified in two different ways:

It is easily checked that this choice is the only one that let the applied body forces enter (via the functions p^i) the right-hand sides of the variational equations in the 'limit' variational problem $\mathcal{P}_M(\omega)$ satisfied by $\bar{u}$.

Otherwise, the number a can be considered *a priori* as an unknown. Then a *formal* asymptotic analysis of the scaled unknown $u(\varepsilon)$ shows that, for a family of linearly elastic membrane shells (thus of the type considered here), the exponent a must be set equal to 0, again in order that the applied body forces contribute to the 'limit' variational problem; *cf.* Miara and Sanchez-Palencia (1996).

The following result is due to Ciarlet and Lods (1996b, Theorem 5.1); a complete proof is also given in Ciarlet (2000, Theorem 4.4-1).

Theorem 8.1: Convergence of the scaled displacements.   Assume that $\boldsymbol{\theta} \in \mathcal{C}^3(\overline{\omega}; \mathbb{R}^3)$. Consider a family of linearly elastic elliptic membrane shells with thickness 2ε approaching zero and with each having the same elliptic middle surface $S = \boldsymbol{\theta}(\overline{\omega})$, and assume that there exist constants $\lambda > 0$ and $\mu > 0$ and functions $f^i \in L^2(\Omega)$ independent of ε such that

$$\lambda^\varepsilon = \lambda \quad \text{and} \quad \mu^\varepsilon = \mu,$$
$$f^{i,\varepsilon}(x^\varepsilon) = f^i(x) \quad \text{for all} \quad x^\varepsilon = \pi^\varepsilon x \in \Omega^\varepsilon.$$

(the notation is that of Section 7). Let $u(\varepsilon)$ denote for sufficiently small $\varepsilon > 0$ the solution of the associated scaled three-dimensional problems $\mathcal{P}(\varepsilon; \Omega)$ (Theorem 7.1). Then there exist functions $u_\alpha \in H^1(\Omega)$ satisfying $u_\alpha = 0$ on $\gamma \times [-1, 1]$ and a function $u_3 \in L^2(\Omega)$ such that

$$u_\alpha(\varepsilon) \to u_\alpha \text{ in } H^1(\Omega) \text{ and } u_3(\varepsilon) \to u_3 \text{ in } L^2(\Omega) \text{ as } \varepsilon \to 0,$$
$$u := (u_i) \text{ is independent of the transverse variable } x_3.$$

Furthermore, the average $\bar{u} := \frac{1}{2} \int_{-1}^{1} u \, dx_3$ satisfies the following two-dimensional variational problem $\mathcal{P}_M(\omega)$:

$$\bar{u} = (\bar{u}_i) \in \mathbf{V}_M(\omega) := H_0^1(\omega) \times H_0^1(\omega) \times L^2(\omega),$$
$$\int_\omega a^{\alpha\beta\sigma\tau} \gamma_{\sigma\tau}(\bar{u}) \gamma_{\alpha\beta}(\eta) \sqrt{a} \, dy = \int_\omega p^i \eta_i \sqrt{a} \, dy$$

for all $\boldsymbol{\eta} = (\eta_i) \in \mathbf{V}_M(\omega)$. Here

$$\gamma_{\alpha\beta}(\boldsymbol{\eta}) := \frac{1}{2}(\partial_\beta \eta_\alpha + \partial_\alpha \eta_\beta) - \Gamma_{\alpha\beta}^\sigma \eta_\sigma - b_{\alpha\beta}\eta_3,$$

$$a^{\alpha\beta\sigma\tau} := \frac{4\lambda\mu}{\lambda + 2\mu} a^{\alpha\beta} a^{\sigma\tau} + 2\mu(a^{\alpha\sigma}a^{\beta\tau} + a^{\alpha\tau}a^{\beta\sigma}),$$

$$p^i := \int_{-1}^{1} f^i \, dx_3.$$

Sketch of proof. **(i)** The proof rests on a crucial three-dimensional inequality of Korn's type for a family of linearly elastic elliptic membrane shells with each having the *same* elliptic middle surface $S = \boldsymbol{\theta}(\bar\omega)$. For such a family, there exists a constant C such that, for sufficiently small $\varepsilon > 0$,

$$\left\{ \sum_\alpha \|v_\alpha\|_{1,\Omega}^2 + |v_3|_{0,\Omega}^2 \right\}^{1/2} \le C \left\{ \sum_{i,j} |e_{i\|j}(\varepsilon; \boldsymbol{v})|_{0,\Omega}^2 \right\}^{1/2}$$

for all $\boldsymbol{v} = (v_i) \in \mathbf{V}(\Omega)$, where

$$\mathbf{V}(\Omega) = \{\boldsymbol{v} = (v_i) \in \mathbf{H}^1(\Omega) : \boldsymbol{v} = \mathbf{0} \text{ on } \gamma \times [-1, 1]\},$$

and the functions $e_{i\|j}(\varepsilon; \boldsymbol{v})$ are the scaled linearized strains appearing in Theorem 7.1. Note that the proof of this inequality relies in a critical way on the inequality of Korn's type on an elliptic surface (Theorem 6.3).

(ii) The special form of the mapping $\boldsymbol{\Theta}$ that defines the reference configurations of the shells (Section 7) implies that there exists a constant C_e such that, for sufficiently small $\varepsilon > 0$,

$$\sum_{i,j} |t_{ij}|^2 \le C_e A^{ijkl}(\varepsilon)(x) t_{kl} t_{ij}$$

for all $x \in \bar\Omega$ and all symmetric matrices (t_{ij}).

Letting $\boldsymbol{v} = \boldsymbol{u}(\varepsilon)$ in the variational equations of problem $\mathcal{P}(\varepsilon; \Omega)$ (Theorem 7.1) and combining the three-dimensional inequality of Korn's type of (i) with the above inequality then yields a chain of inequalities implying that the norms $\|u_\alpha(\varepsilon)\|_{1,\Omega}$, $|u_3(\varepsilon)|_{0,\Omega}$, and $|e_{i\|j}(\varepsilon; \boldsymbol{u}(\varepsilon))|_{0,\Omega}$ are bounded independently of ε.

Thus there exists a subsequence, still denoted by $(\boldsymbol{u}(\varepsilon))_{\varepsilon>0}$ for notational convenience, such that

$$u_\alpha(\varepsilon) \rightharpoonup u_\alpha \text{ in } H^1(\Omega), \qquad u_\alpha(\varepsilon) \to u_\alpha \text{ in } L^2(\Omega),$$
$$u_3(\varepsilon) \rightharpoonup u_3 \text{ in } L^2(\Omega), \quad e_{i\|j}(\varepsilon; \boldsymbol{u}(\varepsilon)) \rightharpoonup e_{i\|j} \text{ in } L^2(\Omega)$$

(strong and weak convergence being respectively denoted by $\to$ and $\rightharpoonup$).

(iii) The above convergence, combined with the asymptotic behaviour of the functions $\Gamma_{ij}^p(\varepsilon)$, $A^{ijkl}(\varepsilon)$, and $g(\varepsilon)$, then implies that the functions u_i and $e_{i\|j}$ are independent of x_3 and that they are related by

$$e_{\alpha\|\beta} = \frac{1}{2}(\partial_\alpha u_\beta + \partial_\beta u_\alpha) - \Gamma_{\alpha\beta}^\sigma u_\sigma - b_{\alpha\beta} u_3,$$

$$e_{\alpha\|3} = 0, \quad e_{3\|3} = -\frac{\lambda}{\lambda + 2\mu} a^{\alpha\beta} e_{\alpha\|\beta}.$$

(iv) In the variational equations of problem $\mathcal{P}(\varepsilon; \Omega)$, keep a function $v \in \mathbf{V}(\Omega)$ fixed and let ε approach zero. Then the asymptotic behaviour of the functions $A^{ijkl}(\varepsilon)$ and $g(\varepsilon)$, combined with the relations found in (iii), shows that the average $\bar{\boldsymbol{u}} = \frac{1}{2}\int_{-1}^{1} \boldsymbol{u}\, dx_3 \in \mathbf{V}_M(\omega)$ indeed satisfies the variational equations of the two-dimensional problem $\mathcal{P}_M(\omega)$ stated in the theorem.

The solution to $\mathcal{P}_M(\omega)$ being *unique*, the convergence established in (ii) for a subsequence thus holds for the *whole* family $(\boldsymbol{u}(\varepsilon))_{\varepsilon>0}$.

(v) Again letting $v = \boldsymbol{u}(\varepsilon)$ in the variational equations of $\mathcal{P}(\varepsilon; \Omega)$ and using the results obtained in (ii)–(iv), we obtain the following *strong convergence*:

$$e_{i\|j}(\varepsilon; \boldsymbol{u}(\varepsilon)) \to e_{i\|j} \text{ in } L^2(\Omega),$$

$$\frac{1}{2}\int_{-1}^{1} \boldsymbol{u}(\varepsilon)\, dx_3 \to \frac{1}{2}\int_{-1}^{1} \boldsymbol{u}\, dx_3 \text{ in } H^1(\omega) \times H^1(\omega) \times L^2(\omega),$$

$$u_3(\varepsilon) \to u_3 \text{ in } L^2(\omega).$$

(vi) The strong convergence

$$u_\alpha(\varepsilon) \to u_\alpha \text{ in } H^1(\Omega),$$

is then obtained as a consequence of the classical three-dimensional Korn inequality in Cartesian coordinates, combined with another use of the Lemma of J. L. Lions (Theorem 3.1).   ☐

8.3. The two-dimensional equations of a linearly elastic 'elliptic membrane' shell

The next theorem recapitulates the definition and assembles the main properties of the 'limit' two-dimensional variational problem $\mathcal{P}_M(\omega)$ found at the outcome of the asymptotic analysis carried out in Theorem 8.1. Note that $\mathcal{P}_M(\omega)$ is an *atypical variational problem*, in that one of the unknowns, namely, the third component ζ_3 of the vector field $\boldsymbol{\zeta} = (\zeta_i)$, 'only' lies in the space $L^2(\omega)$.

The existence and uniqueness theory, which is quickly reviewed in this theorem, is expounded in detail in Section 6, where *ad hoc* references are also provided.

Theorem 8.2: Existence, uniqueness, and regularity of solutions; formulation as a boundary value problem. Let ω be a domain in $\mathbb{R}^2$ and let $\boldsymbol{\theta} \in C^{2,1}(\overline{\omega}; \mathbb{R}^3)$ be an injective mapping such that the two vectors $\boldsymbol{a}_\alpha = \partial_\alpha \boldsymbol{\theta}$ are linearly independent at all points of $\overline{\omega}$ and such that the surface $S = \boldsymbol{\theta}(\overline{\omega})$ is elliptic.

(a) The associated two-dimensional variational problem $\mathcal{P}_M(\omega)$ found in Theorem 8.1 is as follows. Given functions $p^i \in L^2(\omega)$, find $\boldsymbol{\zeta} = (\zeta_i)$ satisfying

$$\boldsymbol{\zeta} \in \mathbf{V}_M(\omega) := H_0^1(\omega) \times H_0^1(\omega) \times L^2(\omega),$$

$$\int_\omega a^{\alpha\beta\sigma\tau} \gamma_{\sigma\tau}(\boldsymbol{\zeta}) \gamma_{\alpha\beta}(\boldsymbol{\eta}) \sqrt{a} \, dy = \int_\omega p^i \eta_i \sqrt{a} \, dy$$

for all $\boldsymbol{\eta} = (\eta_i) \in \mathbf{V}_M(\omega)$, where

$$\gamma_{\alpha\beta}(\boldsymbol{\eta}) := \frac{1}{2}(\partial_\beta \eta_\alpha + \partial_\alpha \eta_\beta) - \Gamma_{\alpha\beta}^\sigma \eta_\sigma - b_{\alpha\beta}\eta_3,$$

$$a^{\alpha\beta\sigma\tau} := \frac{4\lambda\mu}{\lambda + 2\mu} a^{\alpha\beta} a^{\sigma\tau} + 2\mu(a^{\alpha\sigma} a^{\beta\tau} + a^{\alpha\tau} a^{\beta\sigma}).$$

This problem has exactly one solution, which is also the unique solution of the minimization problem:
Find $\boldsymbol{\zeta}$ such that

$$\boldsymbol{\zeta} \in \mathbf{V}_M(\omega) \quad \text{and} \quad j_M(\boldsymbol{\zeta}) = \inf_{\boldsymbol{\eta} \in \mathbf{V}_M(\omega)} j_M(\boldsymbol{\eta}), \quad \text{where}$$

$$j_M(\boldsymbol{\eta}) := \frac{1}{2} \int_\omega a^{\alpha\beta\sigma\tau} \gamma_{\sigma\tau}(\boldsymbol{\eta}) \gamma_{\alpha\beta}(\boldsymbol{\eta}) \sqrt{a} \, dy - \int_\omega p^i \eta_i \sqrt{a} \, dy.$$

(b) If the solution $\boldsymbol{\zeta} = (\zeta_i)$ of $\mathcal{P}_M(\omega)$ is sufficiently smooth, it also satisfies the boundary value problem

$$-n^{\alpha\beta}|_\beta = p^\alpha \text{ in } \omega,$$
$$-b_{\alpha\beta}n^{\alpha\beta} = p^3 \text{ in } \omega,$$
$$\zeta_\alpha = 0 \text{ on } \gamma,$$

where

$$n^{\alpha\beta} := a^{\alpha\beta\sigma\tau} \gamma_{\sigma\tau}(\boldsymbol{\zeta}) \quad \text{and} \quad n^{\alpha\beta}|_\sigma := \partial_\sigma n^{\alpha\beta} + \Gamma_{\sigma\tau}^\alpha n^{\beta\tau} + \Gamma_{\sigma\tau}^\beta n^{\alpha\tau}.$$

(c) Assume that there exist an integer $m \geq 0$ and a real number $q > 1$ such that γ is of class C^{m+3}, $\boldsymbol{\theta} \in C^{m+3}(\overline{\omega}; \mathbb{R}^3)$, $p^\alpha \in W^{m,q}(\omega)$, and $p^3 \in W^{m+1,q}(\omega)$. Then

$$\boldsymbol{\zeta} = (\zeta_i) \in W^{m+2,q}(\omega) \times W^{m+2,q}(\omega) \times W^{m+1,q}(\omega).$$

Proof. The existence and uniqueness of a solution to the variational problem $\mathcal{P}_M(\omega)$, or to its equivalent minimization problem, is a consequence of the inequality of Korn's type on an elliptic surface (Theorem 6.3), of the existence of constants c_e and a_0 such that

$$\sum_{\alpha,\beta} |t_{\alpha\beta}|^2 \leq c_e a^{\alpha\beta\sigma\tau}(y) t_{\sigma\tau} t_{\alpha\beta}$$

for all $y \in \bar{\omega}$ and all symmetric matrices $(t_{\alpha\beta})$ (Ciarlet 2000, Theorem 3.3-2) and $a(y) \geq a_0 > 0$ for all $y \in \bar{\omega}$, and of the Lax–Milgram lemma.

In view of finding the associated boundary value problem stated in part (b), we first note that

$$\partial_\alpha \sqrt{a} = \sqrt{a}\, \Gamma^\sigma_{\sigma\alpha},$$

as is easily verified.

Using Green's formula in Sobolev space and assuming that the functions $n^{\alpha\beta} = n^{\beta\alpha}$ are in $H^1(\omega)$, we next obtain

$$\int_\omega a^{\alpha\beta\sigma\tau} \gamma_{\sigma\tau}(\zeta) \gamma_{\alpha\beta}(\eta) \sqrt{a}\, dy = \int_\omega n^{\alpha\beta} \gamma_{\alpha\beta}(\eta) \sqrt{a}\, dy$$

$$= \int_\omega \sqrt{a}\, n^{\alpha\beta} \left(\frac{1}{2}(\partial_\beta \eta_\alpha + \partial_\alpha \eta_\beta) - \Gamma^\sigma_{\alpha\beta} \eta_\sigma - b_{\alpha\beta} \eta_3 \right) dy$$

$$= \int_\omega \sqrt{a}\, n^{\alpha\beta} \partial_\beta \eta_\alpha\, dy - \int_\omega \sqrt{a}\, n^{\alpha\beta} \Gamma^\sigma_{\alpha\beta} \eta_\sigma\, dy - \int_\omega \sqrt{a}\, n^{\alpha\beta} b_{\alpha\beta} \eta_3\, dy$$

$$= -\int_\omega \partial_\beta (\sqrt{a}\, n^{\alpha\beta}) \eta_\alpha\, dy - \int_\omega \sqrt{a}\, n^{\alpha\beta} \Gamma^\sigma_{\alpha\beta} \eta_\sigma\, dy - \int_\omega \sqrt{a}\, n^{\alpha\beta} b_{\alpha\beta} \eta_3\, dy$$

$$= -\int_\omega \sqrt{a} (\partial_\beta n^{\alpha\beta} + \Gamma^\alpha_{\tau\beta} n^{\tau\beta} + \Gamma^\beta_{\beta\tau} n^{\alpha\tau}) \eta_\alpha\, dy - \int_\omega \sqrt{a}\, n^{\alpha\beta} b_{\alpha\beta} \eta_3\, dy$$

$$= -\int_\omega \sqrt{a} \{ (n^{\alpha\beta}|_\beta) \eta_\alpha + b_{\alpha\beta} n^{\alpha\beta} \eta_3 \}\, dy$$

for all $\eta = (\eta_i) \in \mathbf{V}_M(\omega)$. Hence the variational equations imply that

$$\int_\omega \sqrt{a} \{ (n^{\alpha\beta}|_\beta + p^\alpha) \eta_\alpha + (b_{\alpha\beta} n^{\alpha\beta} + p^3) \eta_3 \}\, dy = 0$$

for all $(\eta_i) \in \mathbf{V}_M(\omega)$, and thus $n^{\alpha\beta}|_\beta = p^\alpha$ and $b_{\alpha\beta} n^{\alpha\beta} = p^3$ in ω.

The regularity result of part (c) is due to Genevey (1996).   ☐

Note that the functions $n^{\alpha\beta}|_\sigma$ are instances of *first-order covariant derivatives of a tensor field*, defined here by means of its contravariant components $n^{\alpha\beta}$.

In order to get physically meaningful formulas, it remains to 'de-scale' the unknowns ζ_i that satisfy the limit 'scaled' problem $\mathcal{P}_M(\omega)$ found in Theorem 8.2. In view of the scalings $u_i(\varepsilon)(x) = u_i^\varepsilon(x^\varepsilon)$ for all $x^\varepsilon = \pi^\varepsilon x \in \overline{\Omega^\varepsilon}$ made on the covariant components of the displacement field (Section 7), we are led to defining for each $\varepsilon > 0$ the covariant components $\zeta_i^\varepsilon : \bar{\omega} \to \mathbb{R}$ of the 'limit displacement field' $\zeta_i^\varepsilon a^i : \bar{\omega} \to \mathbb{R}^3$ of the middle surface S of the shell by letting

$$\zeta_i^\varepsilon := \zeta_i$$

(the vectors a^i forming the contravariant basis at each point of S).

Naturally, the field (ζ_i^ε) and the field $\zeta_i^\varepsilon a^i$ must be carefully distinguished! The former is essentially a convenient mathematical 'intermediary', while only the latter has physical significance.

Recall that $f^{i,\varepsilon} \in L^2(\Omega^\varepsilon)$ represent the contravariant components of the applied body forces *actually* acting on the shell and that λ^ε and μ^ε denote the *actual* Lamé constants of its constituent material. We then have the following immediate corollary to Theorems 8.1 and 8.2. Naturally, the existence, uniqueness and regularity results of Theorem 8.2 apply *verbatim* to the solution of the 'de-scaled' problem $\mathcal{P}_M^\varepsilon(\omega)$ found in the next theorem (for this reason, they are not reproduced here).

Theorem 8.3: The two-dimensional equations of a linearly elastic 'elliptic membrane' shell. Let the assumptions on the data be as in Theorem 8.1. Then the vector field $\boldsymbol{\zeta}^\varepsilon := (\zeta_i^\varepsilon)$ formed by the covariant components of the limit displacement field $\zeta_i^\varepsilon a^i$ of the middle surface S satisfies the following two-dimensional variational problem $\mathcal{P}_M^\varepsilon(\omega)$ of a linearly elastic elliptic membrane shell:

$$\boldsymbol{\zeta}^\varepsilon \in \mathbf{V}_M(\omega) := H_0^1(\omega) \times H_0^1(\omega) \times L^2(\omega),$$

$$\varepsilon \int_\omega a^{\alpha\beta\sigma\tau,\varepsilon} \gamma_{\sigma\tau}(\boldsymbol{\zeta}^\varepsilon) \gamma_{\alpha\beta}(\boldsymbol{\eta}) \sqrt{a}\, dy = \int_\omega p^{i,\varepsilon} \eta_i \sqrt{a}\, dy$$

for all $\boldsymbol{\eta} = (\eta_i) \in \mathbf{V}_M(\omega)$, where

$$\gamma_{\alpha\beta}(\boldsymbol{\eta}) := \frac{1}{2}(\partial_\beta \eta_\alpha + \partial_\alpha \eta_\beta) - \Gamma_{\alpha\beta}^\sigma \eta_\sigma - b_{\alpha\beta} \eta_3,$$

$$a^{\alpha\beta\sigma\tau,\varepsilon} := \frac{4\lambda^\varepsilon \mu^\varepsilon}{\lambda^\varepsilon + 2\mu^\varepsilon} a^{\alpha\beta} a^{\sigma\tau} + 2\mu^\varepsilon (a^{\alpha\sigma} a^{\beta\tau} + a^{\alpha\tau} a^{\beta\sigma}),$$

$$p^{i,\varepsilon} := \int_{-\varepsilon}^\varepsilon f^{i,\varepsilon}\, dx_3^\varepsilon.$$

Equivalently, the field $\boldsymbol{\zeta}^\varepsilon$ satisfies the minimization problem

$$\boldsymbol{\zeta}^\varepsilon \in \mathbf{V}_M(\omega) \quad \text{and} \quad j_M^\varepsilon(\boldsymbol{\zeta}^\varepsilon) = \inf_{\boldsymbol{\eta} \in \mathbf{V}_M(\omega)} j_M^\varepsilon(\boldsymbol{\eta}), \quad \text{where}$$

$$j_M^\varepsilon(\boldsymbol{\eta}) := \frac{\varepsilon}{2} \int_\omega a^{\alpha\beta\sigma\tau,\varepsilon} \gamma_{\sigma\tau}(\boldsymbol{\eta}) \gamma_{\alpha\beta}(\boldsymbol{\eta}) \sqrt{a}\, dy - \int_\omega p^{i,\varepsilon} \eta_i \sqrt{a}\, dy.$$

If the field $\boldsymbol{\zeta}^\varepsilon = (\zeta_i^\varepsilon)$ is sufficiently smooth, it also satisfies the following boundary value problem:

$$-n^{\alpha\beta,\varepsilon}|_{\beta} = p^{\alpha,\varepsilon} \text{ in } \omega,$$

$$-b_{\alpha\beta}n^{\alpha\beta,\varepsilon} = p^{3,\varepsilon} \text{ in } \omega,$$

$$\zeta_{\alpha}^{\varepsilon} = 0 \text{ on } \gamma,$$

where

$$n^{\alpha\beta,\varepsilon} := \varepsilon a^{\alpha\beta\sigma\tau,\varepsilon}\gamma_{\sigma\tau}(\boldsymbol{\zeta}^{\varepsilon}),$$

$$n^{\alpha\beta,\varepsilon}|_{\sigma} := \partial_{\sigma}n^{\alpha\beta,\varepsilon} + \Gamma^{\alpha}_{\tau\sigma}n^{\tau\beta,\varepsilon} + \Gamma^{\beta}_{\sigma\tau}n^{\alpha\tau,\varepsilon}. \qquad \square$$

Each one of the three formulations found in Theorem 8.3 constitutes one version of the two-dimensional equations of a linearly elastic elliptic membrane shell. The functions $\gamma_{\alpha\beta}(\boldsymbol{\eta})$ are the *covariant components* of the *linearized change of metric tensor* associated with a displacement field $\eta_i \boldsymbol{a}^i$ of the middle surface S, the functions $a^{\alpha\beta\sigma\tau,\varepsilon}$ are the *contravariant components* of the *two-dimensional elasticity tensor of the shell*, and the functions $n^{\alpha\beta,\varepsilon}$ are the contravariant components of the *stress resultant tensor* field.

The functional $j_M^{\varepsilon} : \mathbf{V}_M(\omega) \rightarrow \mathbb{R}$ is the *two-dimensional energy*, and the functional

$$\boldsymbol{\eta} \in \mathbf{V}_M(\omega) \rightarrow \frac{\varepsilon}{2}\int_{\omega}a^{\alpha\beta\sigma\tau,\varepsilon}\gamma_{\sigma\tau}(\boldsymbol{\eta})\gamma_{\alpha\beta}(\boldsymbol{\eta})\sqrt{a}\,dy$$

is the *two-dimensional strain energy* of a *linearly elastic elliptic membrane shell*.

Finally, the equations $-n^{\alpha\beta,\varepsilon}|_{\beta} = p^{\alpha,\varepsilon}$ and $-b_{\alpha\beta}n^{\alpha\beta,\varepsilon} = p^{3,\varepsilon}$ in ω constitute the *two-dimensional equations of equilibrium*, and the relations $n^{\alpha\beta,\varepsilon} = \varepsilon a^{\alpha\beta\sigma\tau,\varepsilon}\gamma_{\sigma\tau}(\boldsymbol{\zeta}^{\varepsilon})$ constitute the *two-dimensional constitutive equation* of a *linearly elastic elliptic membrane shell*.

Under the essential assumptions that $\gamma_0 = \gamma$ and that the surface S is elliptic, we have therefore justified by a convergence result (Theorem 8.1) two-dimensional equations that are called those of a linearly elastic 'membrane' shell in the literature (which, however, usually ignores the distinction between 'elliptic' and 'generalized' membrane shells); see, *e.g.*, Koiter (1966, equations (9.14) and (9.15)), Green and Zerna (1968, Section 11.1), Dikmen (1982, equations (7.10)), or Niordson (1985, equation (10.3)).

In so doing, we have also justified the formal asymptotic approach of Sanchez-Palencia (1990) (see also Miara and Sanchez-Palencia (1996), Caillerie and Sanchez-Palencia (1995*b*), Faou (2000*a*, 2000*b*)) when 'bending is well-inhibited', according to the terminology of E. Sanchez-Palencia.

Note that the above convergence analysis also substantiates an important observation. In an *elliptic membrane shell*, body forces of order $O(1)$ with respect to ε produce a limit displacement field that is also $O(1)$. By contrast, body forces must be of order $O(\varepsilon^2)$ in order to produce an $O(1)$ limit displacement field in a *flexural shell*. See Section 10.

After the original work of Ciarlet and Lods (1996b) described in this section, the asymptotic analysis of linearly elastic membrane shells underwent several refinements and generalizations.

First, Genevey (1999) has shown that the convergence result of Theorem 8.1 can also be obtained by resorting to Γ-convergence theory.

Using the techniques of Lions (1973), Mardare (1998a) was able to compute a *corrector*, so as to obtain in this fashion the following remarkable error estimate. In addition to the hypotheses made in Theorem 8.1, assume that the boundary of the domain w is of class $\mathcal{C}^2$, that $\partial_\alpha f^\alpha \in L^2(\Omega)$, and that

$$\zeta = \frac{1}{2} \int_{-1}^{1} u \, \mathrm{d}x_3 \in \mathbf{H}^2(w) \cap \mathbf{V}_M(w).$$

Then there exists a constant $C = C(w, \boldsymbol{\theta}, f^i, \boldsymbol{\zeta})$ independent of ε such that

$$\|u(\varepsilon) - u\|_{H^1(\Omega) \times H^1(\Omega) \times L^2(\Omega)} \le C\varepsilon^{1/6},$$

and moreover, the exponent $1/6$ is the best possible.

Other useful extensions include the justification by an asymptotic analysis of linearly elastic membrane shells with variable thickness (Busse 1998), the convergence of the stresses and the explicit forms of the limit stresses (Collard and Miara 1999), an asymptotic analysis of the associated time-dependent problem (Xiao Li-Ming 1998), and the extension of the present analysis to shells whose middle surface is elliptic but has 'no boundary', such as an entire ellipsoid (Ramos (1995) and Şlicaru (1997)).

The variational formulation of the limit two-dimensional problem of a linearly elastic elliptic membrane shell (Theorem 8.3) possesses the unusual feature that its third unknown ζ_3 'only' belongs to the space $L^2(w)$. This explains why the averaged three-dimensional boundary condition

$$\overline{u}_3^\varepsilon := \frac{1}{2\varepsilon} \int_{-\varepsilon}^{\varepsilon} u_3^\varepsilon \, \mathrm{d}x_3^\varepsilon = 0 \text{ on } \gamma$$

is 'lost' as $\varepsilon \to 0$, since $\zeta_3^\varepsilon = 0$ on γ does not make sense. As expected, this loss is compensated by the appearance of a *boundary layer* in the unknown ζ_3^ε.

Again because the third unknown ζ_3^ε is only in $L^2(w)$, the linear operator associated with the variational problem of a linearly elastic elliptic membrane shell is *not* compact and thus the analysis of the corresponding eigenvalue problem requires special care; see Sanchez-Hubert and Sanchez-Palencia (1997, Chapter 10).

9. 'Generalized membrane' shells

A shell with middle surface $S = \boldsymbol{\theta}(\bar{\omega})$, subjected to a boundary condition of place along a portion of its lateral face with $\boldsymbol{\theta}(\gamma_0)$, where $\gamma_0 \subset \gamma$, as its middle curve, is a *linearly elastic 'generalized membrane' shell* if it is *not* an elliptic membrane shell according to the definition given in Section 8, yet if its associated space

$$\mathbf{V}_F(\omega) = \{\boldsymbol{\eta} = (\eta_i) \in H^1(\omega) \times H^1(\omega) \times H^2(\omega) :$$
$$\eta_i = \partial_\nu \eta_3 = 0 \text{ on } \gamma_0, \gamma_{\alpha\beta}(\boldsymbol{\eta}) = 0 \text{ in } \omega\}$$

still reduces to $\{\mathbf{0}\}$. As shown in Section 9.1, examples of generalized membrane shells abound.

The purpose of this section is to identify and to mathematically justify the *two-dimensional equations of a linearly elastic generalized membrane shell*, by establishing the convergence of the three-dimensional displacements in *ad hoc* function spaces as the thickness of such a shell approaches zero.

9.1. Definition and examples

Let ω be a domain in $\mathbb{R}^2$ with boundary γ and let $\boldsymbol{\theta} \in \mathcal{C}^2(\bar{\omega}; \mathbb{R}^3)$ be an injective mapping such that the two vectors $\partial_\alpha \boldsymbol{\theta}(y)$ are linearly independent at all points $y \in \bar{\omega}$. A shell with middle surface $S = \boldsymbol{\theta}(\bar{\omega})$ is called a *linearly elastic generalized membrane shell* if the following three conditions are simultaneously satisfied.

(i) The shell is subjected to a (homogeneous) *boundary condition of place* (*i.e.*, of vanishing displacement field) along a portion of its lateral face with $\boldsymbol{\theta}(\gamma_0)$ as its middle curve, where the subset $\gamma_0 \subset \gamma$ satisfies

$$\text{length } \gamma_0 > 0.$$

(ii) Define the space

$$\mathbf{V}_F(\omega) := \{\boldsymbol{\eta} = (\eta_i) \in H^1(\omega) \times H^1(\omega) \times H^2(\omega) :$$
$$\eta_i = \partial_\nu \eta_3 = 0 \text{ on } \gamma_0, \gamma_{\alpha\beta}(\boldsymbol{\eta}) = 0 \text{ in } \omega\}$$

(∂_ν denoting the outer normal derivative operator along γ). Then the space $\mathbf{V}_F(\omega)$ contains only $\boldsymbol{\eta} = \mathbf{0}$.

(iii) The shell is not an elliptic membrane shell. We recall that a linearly elastic shell is an 'elliptic membrane' shell if $\gamma_0 = \gamma$ and S is elliptic (Section 8.1) and that an elliptic membrane shell also provides an instance where the space $\mathbf{V}_F(\omega)$, which in this case is the space $H_0^1(\omega) \times H_0^1(\omega) \times H_0^2(\omega)$, reduces to $\{\mathbf{0}\}$ (at least if $\boldsymbol{\theta} \in \mathcal{C}^{2,1}(\omega; \mathbb{R}^3)$; *cf.* Theorem 6.2). Generalized membrane shells thus exhaust all the remaining cases of linearly elastic membrane shells, *i.e.*, those for which $\mathbf{V}_F(\omega) = \{\mathbf{0}\}$.

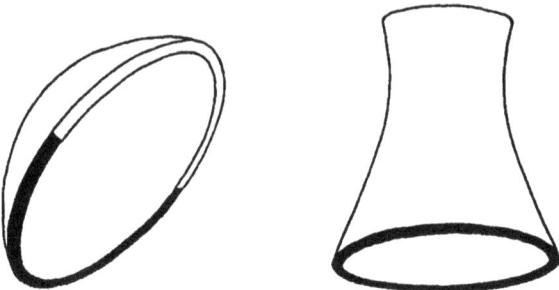

Fig. 9.1. Two examples of linearly elastic 'generalized membrane' shells

The definition of a linearly elastic 'generalized membrane' shell thus depends only on the subset of the lateral face where the shell is subjected to a boundary condition of place (via the set γ_0) and on the geometry of the middle surface of the shell.

As shown by Vekua (1962) under the assumptions that $\boldsymbol{\theta} \in W^{3,p}(\omega; \mathbb{R}^3)$ for some $p > 2$ and that γ is of class $\mathcal{C}^3$, then by Lods and Mardare (1998a) under the assumption that $\boldsymbol{\theta} \in \mathcal{C}^{2,1}(\overline{\omega}; \mathbb{R}^3)$ and that γ is Lipschitz-continuous, a shell whose middle surface $S = \boldsymbol{\theta}(\overline{\omega})$ is a portion of an ellipsoid and which is subjected to a boundary condition of place along a portion (solid black in the figure) of its lateral face whose middle curve $\boldsymbol{\theta}(\gamma_0)$ is such that $0 < \text{length } \gamma_0 < \text{length } \gamma$, provides an instance of a linearly elastic generalized membrane shell; see Figure 9.1. A comparison with Figure 8.1 illustrates the crucial role played by the set $\boldsymbol{\theta}(\gamma_0)$ in determining the type of shell!

As shown by Mardare (1998c) under the assumption that $\boldsymbol{\theta} \in \mathcal{C}^{2,1}(\overline{\omega}; \mathbb{R}^3)$ (see also Vekua (1962) and Sanchez-Hubert and Sanchez-Palencia (1997, Chapter 7, Section 2)), a shell whose middle surface $S = \boldsymbol{\theta}(\overline{\omega})$ is a portion of a hyperboloid of revolution and which is subjected to a boundary condition of place along its entire 'lower' lateral face provides another instance of a linearly elastic generalized membrane shell; see Figure 9.1.

A shell whose middle surface $S = \boldsymbol{\theta}(\overline{\omega})$ is a portion of a cone or a cylinder and which is subjected to a boundary condition of place along a portion (solid black in the figure) of its lateral face with $\boldsymbol{\theta}(\gamma_0)$ as its middle curve is a linearly elastic generalized membrane shell if $\boldsymbol{\theta}(\gamma_0)$ intersects all the generatrices of S; see Figure 9.2.

As for elliptic membrane shells (Section 8), the formal asymptotic analysis of Miara and Sanchez-Palencia (1996) again suggests making the following assumptions on the data for a family of generalized membrane shells. We require that the Lamé constants and the applied body densities appearing

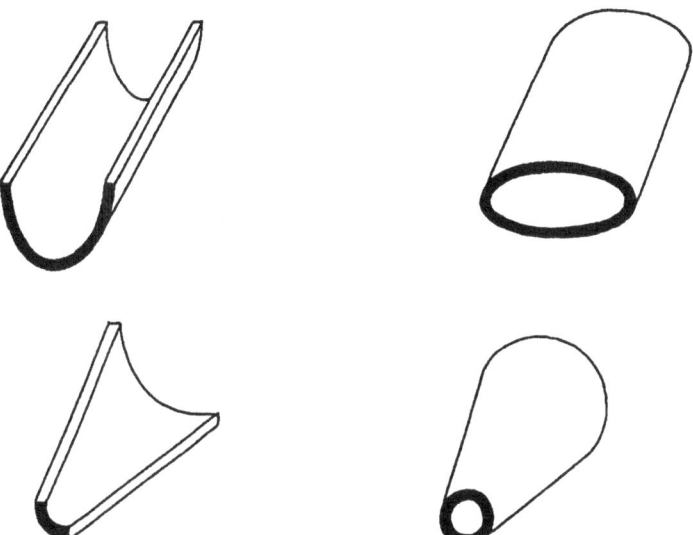

Fig. 9.2. Other examples of linearly elastic 'generalized membrane' shells

in the three-dimensional problems $\mathcal{P}(\Omega^\varepsilon)$ (Section 7) satisfy

$$\lambda^\varepsilon = \lambda \quad \text{and} \quad \mu^\varepsilon = \mu,$$
$$f^{i,\varepsilon}(x^\varepsilon) = f^i(x) \quad \text{for all} \quad x^\varepsilon = \pi^\varepsilon x \in \Omega^\varepsilon,$$

where the constants $\lambda > 0$ and $\mu > 0$ and the functions $f^i \in L^2(\Omega)$ are independent of ε. In other words, the exponent a in Theorem 7.1 again vanishes.

It turns out, however, that in order to carry out the asymptotic analysis of generalized membrane shells, we have to make a specific, and rather stringent, *assumption on the applied forces*, which supersedes in fact the above one, in such a way that the linear form appearing in the variational problem $\mathcal{P}(\varepsilon; \Omega)$ of Theorem 7.1 becomes continuous with respect to an *ad hoc* norm, and uniformly so with respect to ε. We now describe this assumption, particular to generalized membrane shells.

9.2. 'Admissible' applied forces

Consider a *family of linearly elastic generalized membrane shells*, with thickness 2ε, with each having the same middle surface $S = \boldsymbol{\theta}(\overline{\omega})$, and with each subjected to a boundary condition of place along a portion of its lateral face having the same set $\boldsymbol{\theta}(\gamma_0)$ as its middle curve, and let the assumptions on the data be as in Section 9.1.

Let

$$\mathbf{V}(\Omega) := \{\boldsymbol{v} = (v_i) \in \mathbf{H}^1(\Omega) : \boldsymbol{v} = \mathbf{0} \text{ on } \Gamma_0\},$$

and, for each $\varepsilon > 0$, let the linear form $L(\varepsilon) : \mathbf{V}(\Omega) \to \mathbb{R}$ be defined by

$$L(\varepsilon)(\boldsymbol{v}) := \int_\Omega f^i v_i \sqrt{g(\varepsilon)}\, \mathrm{d}x \quad \text{for all}\ \boldsymbol{v} \in \mathbf{V}(\Omega).$$

In other words, $L(\varepsilon)(\boldsymbol{v})$ is the right-hand side in problem $\mathcal{P}(\varepsilon; \Omega)$ (Theorem 7.1), which takes into account the applied body forces through the functions $f^i \in L^2(\Omega)$. Then each linear form $L(\varepsilon) : \mathbf{V}(\Omega) \to \mathbb{R}$ is clearly continuous with respect to the norm $\|\cdot\|_{1,\Omega}$ and uniformly so with respect to sufficiently small $\varepsilon > 0$.

It so happens, however, that an essentially stronger property is needed. The linear forms $L(\varepsilon)$ should also be continuous, and uniformly so, with respect to sufficiently small $\varepsilon > 0$, and with respect to the norm (itself dependent on ε)

$$\boldsymbol{v} \to \left\{ \sum_{i,j} |e_{i\|j}(\varepsilon; \boldsymbol{v})|^2_{0,\Omega} \right\}^{1/2}.$$

In order to fulfil this requirement in a concrete manner, we set the following definition. Applied forces acting on a family of linearly elastic generalized membrane shells are said to be 'admissible' if there exist functions $F^{ij}(\varepsilon) = F^{ji}(\varepsilon) \in L^2(\Omega)$ and functions $F^{ij} = F^{ji} \in L^2(\Omega)$ such that

$$L(\varepsilon)(\boldsymbol{v}) = \int_\Omega F^{ij}(\varepsilon) e_{i\|j}(\varepsilon; \boldsymbol{v}) \sqrt{g(\varepsilon)}\, \mathrm{d}x$$

for all $0 < \varepsilon \leq \varepsilon_0$ and for all $\boldsymbol{v} \in \mathbf{V}(\Omega)$, and

$$F^{ij}(\varepsilon) \to F^{ij} \text{ in } L^2(\Omega) \text{ as } \varepsilon \to 0.$$

If the applied forces are admissible, there thus exists a constant κ_0 such that

$$|L(\varepsilon)(\boldsymbol{v})| \leq \kappa_0 \left\{ \sum_{i,j} |e_{i\|j}(\varepsilon; \boldsymbol{v})|^2_{0,\Omega} \right\}^{1/2}$$

for all $0 < \varepsilon \leq \varepsilon_0$ and for all $\boldsymbol{v} \in \mathbf{V}(\Omega)$, as was required.

This inequality will be put to an essential use in Theorem 9.1 for finding the a priori bounds that the family of scaled unknowns satisfies.

The convergence $F^{ij}(\varepsilon) \to F^{ij}$ in $L^2(\Omega)$ serves a further purpose, that of defining the right-hand sides appearing in the 'limit' two-dimensional problems (see again Theorem 9.1).

Naturally, admissible forces have to be identified for each instance of generalized membrane shells; see in this respect the references given in Section 9.4.

9.3. Convergence of the scaled displacements as the thickness approaches zero

A generalized membrane shell is 'of the first kind' if the space

$$\mathbf{V}_0(\omega) = \{\boldsymbol{\eta} = (\eta_i) \in \mathbf{H}^1(\omega) : \boldsymbol{\eta} = \mathbf{0} \text{ on } \gamma_0, \gamma_{\alpha\beta}(\boldsymbol{\eta}) = 0 \text{ in } \omega\},$$

which is larger than the space $\mathbf{V}_F(\omega)$, 'already' reduces to $\{\mathbf{0}\}$. Equivalently, the seminorm $|\cdot|_\omega^M$ defined by

$$|\boldsymbol{\eta}|_\omega^M = \left\{\sum_{\alpha,\beta} |\gamma_{\alpha\beta}(\boldsymbol{\eta})|_{0,\omega}^2\right\}^{1/2}$$

is 'already' a norm over the space

$$\mathbf{V}(\omega) = \{\boldsymbol{\eta} = (\eta_i) \in \mathbf{H}^1(\omega) : \boldsymbol{\eta} = \mathbf{0} \text{ on } \gamma_0\}.$$

As *all* the known examples of generalized membrane shells satisfy this assumption, we shall not consider here the generalized membrane shells 'of the second kind', *i.e.*, those for which $\mathbf{V}_F(\omega)$ contains only $\boldsymbol{\eta} = \mathbf{0}$, but $\mathbf{V}_0(\omega)$ contains *nonzero* elements. Such shells are analysed in Ciarlet and Lods (1996d, Theorem 5.1); see also Ciarlet (2000, Theorem 5.6-2).

We now establish the main results of this section. Consider a *family of linearly elastic generalized membrane shells of the first kind*, with thickness $2\varepsilon > 0$, with each having the *same* middle surface $S = \boldsymbol{\theta}(\overline{\omega})$, and with each subjected to a boundary condition of place along a portion of its lateral face having the *same* set $\boldsymbol{\theta}(\gamma_0)$ as its middle curve, the applied forces being admissible. Then the averages

$$\overline{\boldsymbol{u}(\varepsilon)} = \frac{1}{2} \int_{-1}^{1} \boldsymbol{u}(\varepsilon) \, dx_3$$

of the scaled unknowns converge in an 'abstract' completion $\mathbf{V}_M^\sharp(\omega)$ as $\varepsilon \to 0$ and their limit satisfies an 'abstract' variational problem posed over the same space $\mathbf{V}_M^\sharp(\omega)$.

The functions $\gamma_{\alpha\beta}(\boldsymbol{\eta})$ appearing in the next theorem represent the covariant components of the *linearized change of metric tensor* associated with a displacement field $\eta_i \boldsymbol{a}^i$ of the surface S. Hence the bilinear form B_M defined below coincides with that found in the scaled variational problem of a linearly elastic elliptic membrane shell (Theorem 8.2).

The following result is due to Ciarlet and Lods (1996d, Theorem 5.1); a complete proof is also given in Ciarlet (2000, Theorem 5.6-1). In these references, it is also shown how the convergence of the scaled unknowns $\boldsymbol{u}(\varepsilon)$ themselves can be also established, in an *ad hoc* completion.

Theorem 9.1: Convergence of the scaled displacement. Assume that $\boldsymbol{\theta} \in C^3(\overline{\omega}; \mathbb{R}^3)$. Consider a family of linearly elastic generalized membrane shells of the first kind, with thickness 2ε approaching zero, with each having the same middle surface $S = \boldsymbol{\theta}(\bar{\omega})$, and with each subjected to a boundary condition of place along a portion of its lateral face having the same set $\boldsymbol{\theta}(\gamma_0)$ as its middle curve. Assume that there exist constants $\lambda > 0$ and $\mu > 0$ such that

$$\lambda^\varepsilon = \lambda \text{ and } \mu^\varepsilon = \mu.$$

Finally, assume that the applied forces are admissible (Section 9.2). For sufficiently small $\varepsilon > 0$, let $\boldsymbol{u}(\varepsilon)$ denote the solution of the associated scaled three-dimensional problems $\mathcal{P}(\varepsilon; \Omega)$ (Theorem 7.1).

Define the space

$$\mathbf{V}_M^\sharp(\omega) := \text{completion of } \mathbf{V}(\omega) \text{ with respect to } | \cdot |_\omega^M.$$

Then there exists $\boldsymbol{\zeta} \in \mathbf{V}_M^\sharp(\omega)$ such that

$$\overline{\boldsymbol{u}(\varepsilon)} := \frac{1}{2} \int_{-1}^1 \boldsymbol{u}(\varepsilon) \, dx_3 \to \boldsymbol{\zeta} \quad \text{in } \mathbf{V}_M^\sharp(\omega) \quad \text{as } \varepsilon \to 0.$$

Let

$$a^{\alpha\beta\sigma\tau} := \frac{4\lambda\mu}{\lambda + 2\mu} a^{\alpha\beta} a^{\sigma\tau} + 2\mu(a^{\alpha\sigma} a^{\beta\tau} + a^{\alpha\tau} a^{\beta\sigma}),$$

$$\gamma_{\alpha\beta}(\boldsymbol{\eta}) := \frac{1}{2}(\partial_\beta \eta_\alpha + \partial_\alpha \eta_\beta) - \Gamma_{\alpha\beta}^\sigma \eta_\sigma - b_{\alpha\beta}\eta_3,$$

$$B_M(\boldsymbol{\zeta}, \boldsymbol{\eta}) := \int_\omega a^{\alpha\beta\sigma\tau} \gamma_{\sigma\tau}(\boldsymbol{\zeta}) \gamma_{\alpha\beta}(\boldsymbol{\eta}) \sqrt{a} \, dy \text{ for } \boldsymbol{\zeta}, \boldsymbol{\eta} \in \mathbf{V}(\omega),$$

$$L_M(\boldsymbol{\eta}) := \int_\omega \varphi^{\alpha\beta} \gamma_{\alpha\beta}(\boldsymbol{\eta}) \sqrt{a} \, dy \text{ for } \boldsymbol{\eta} \in \mathbf{V}(\omega),$$

$$\varphi^{\alpha\beta} := \int_{-1}^1 \left\{ F^{\alpha\beta} - \frac{\lambda}{\lambda + 2\mu} a^{\alpha\beta} F^{33} \right\} dx_3 \in L^2(\omega),$$

where the functions $F^{ij} \in L^2(\Omega)$ are those used in the definition of admissible forces, and let $B_M^\sharp$ and $L_M^\sharp$ denote the unique continuous extensions from $\mathbf{V}(\omega)$ to $\mathbf{V}_M^\sharp(\omega)$ of the bilinear form B_M and linear form L_M. Then the limit $\boldsymbol{\zeta}$ satisfies the following two-dimensional variational problem $\mathcal{P}_M^\sharp(\omega)$:

$$\boldsymbol{\zeta} \in \mathbf{V}_M^\sharp(\omega) \text{ and } B_M^\sharp(\boldsymbol{\zeta}, \boldsymbol{\eta}) = L_M^\sharp(\boldsymbol{\eta}) \text{ for all } \boldsymbol{\eta} \in \mathbf{V}_M^\sharp(\omega).$$

Sketch of proof. (i) The proof rests on a crucial three-dimensional inequality of Korn's type for a family of linearly elastic shells, with each having the *same* middle surface $S = \boldsymbol{\theta}(\bar{\omega})$, and with each subjected to a boundary condition of place along a portion of its lateral face having the *same* set $\boldsymbol{\theta}(\gamma_0)$ as its middle curve. For such a family, there exists a constant C such that, for sufficiently small $\varepsilon > 0$,

$$\|v\|_{1,\Omega} \leq \frac{C}{\varepsilon}\left\{\sum_{i,j}|e_{i\|j}(\varepsilon;v)|^2_{0,\Omega}\right\}^{1/2}$$

for all $v \in \mathbf{V}(\Omega)$, where

$$\mathbf{V}(\Omega) = \{v \in \mathbf{H}^1(\Omega) : v = \mathbf{0} \text{ on } \gamma_0 \times [-1,1]\},$$

and the functions $e_{i\|j}(\varepsilon;v)$ are the scaled linearized strains appearing in Theorem 7.1. The proof of this inequality relies in a critical way on the linearized rigid displacement lemma on a general surface (Theorem 4.3).

(ii) Given a function $v \in L^2(\Omega)$, let

$$\bar{v} := \frac{1}{2}\int_{-1}^1 v\,dx_3 \in L^2(\omega)$$

denote its *average* with respect to the transverse variable x_3; the same notation is used for vector-valued functions. Letting $v = u(\varepsilon)$ in the variational equations of problem $\mathcal{P}(\varepsilon;\Omega)$ (Theorem 7.1) and using the three-dimensional inequality of Korn's type of (i) then yields a chain of inequalities showing that the norms $|\partial_3 u(\varepsilon)|_{0,\Omega}$, $|\overline{u(\varepsilon)}|^M_\omega$, $|e_{i\|j}(\varepsilon;u(\varepsilon)|_{0,\Omega}$, and $\|\varepsilon u(\varepsilon)\|_{1,\Omega}$ are bounded independently of ε. Note that the assumption that the applied forces are 'admissible' is crucial here.

Thus there exists a subsequence, still denoted by $(u(\varepsilon))_{\varepsilon>0}$ for notational convenience, such that $u(\varepsilon) \rightharpoonup u$ in the completion of the space $\mathbf{V}(\Omega)$ with respect to the norm $|\cdot|^M_\Omega$ defined by $|v|^M_\Omega := \{|\partial_3 v|^2_{0,\Omega} + (|\bar{v}|^M_\omega)^2\}^{1/2}$, and such that

$$e_{i\|j}(\varepsilon;u(\varepsilon)) \rightharpoonup e_{i\|j} \text{ in } L^2(\Omega), \qquad \varepsilon u(\varepsilon) \rightharpoonup u^{-1} \text{ in } \mathbf{H}^1(\Omega),$$

$$\partial_3 u_3(\varepsilon) = \varepsilon e_{33}(\varepsilon;u(\varepsilon)) \to 0 \text{ in } L^2(\Omega), \qquad \overline{u(\varepsilon)} \rightharpoonup \zeta \text{ in } \mathbf{V}^\sharp_M(\omega).$$

(iii) The above convergence, combined with the asymptotic behaviour of the functions $\Gamma^p_{ij}(\varepsilon)$, $A^{ijkl}(\varepsilon)$, and $g(\varepsilon)$, then implies that

$$e_{\alpha\|3} = \frac{1}{2\mu}a_{\alpha\beta}F^{\beta3},$$

$$e_{3\|3} := -\frac{\lambda}{\lambda+2\mu}a^{\alpha\beta}e_{\alpha\|\beta} + \frac{F^{33}}{\lambda+2\mu},$$

$$\gamma_{\alpha\beta}(\overline{u(\varepsilon)}) \rightharpoonup \overline{e_{\alpha\|\beta}} \text{ in } L^2(\omega),$$

$$\varepsilon u(\varepsilon) \rightharpoonup 0 \text{ in } \mathbf{H}^1(\Omega),$$

$$\partial_3 u_\alpha(\varepsilon) \rightharpoonup 0 \text{ in } L^2(\omega),$$

$e_{\alpha\|\beta}$ is independent of the transverse variable x_3, where the functions $F^{ij} \in L^2(\Omega)$ are those appearing in the definition of 'admissible' forces.

(iv) In the variational equations of problem $\mathcal{P}(\varepsilon; \Omega)$, let $v \in V(\Omega)$ be independent of the transverse variable x_3. Keep such a function v *fixed* and let ε approach zero. Then the asymptotic behaviour of the functions $A^{ijkl}(\varepsilon)$ and $g(\varepsilon)$ combined with the relations found in (ii) and (iii) together show that the limits $e_{\alpha\|\beta}$ found in part (ii) satisfy

$$\int_\omega a^{\alpha\beta\sigma\tau} \overline{e_{\sigma\|\tau}} \gamma_{\alpha\beta}(\eta) \sqrt{a}\, dy = \int_\omega \varphi^{\alpha\beta} \gamma_{\alpha\beta}(\eta) \sqrt{a}\, dy \text{ for all } \eta \in V(\omega),$$

where

$$\varphi^{\alpha\beta} := \int_{-1}^1 \left\{ F^{\alpha\beta} - \frac{\lambda}{\lambda + 2\mu} a^{\alpha\beta} F^{33} \right\} dx_3 \in L^2(\omega).$$

(v) Again letting $v = u(\varepsilon)$ in the variational equations of $\mathcal{P}(\varepsilon; \Omega)$ and using the results obtained in (ii)–(iv), we obtain the following strong convergence:

$$e_{i\|j}(\varepsilon; u(\varepsilon)) \to e_{i\|j} \text{ in } L^2(\Omega),$$

$$\varepsilon u(\varepsilon) \to 0 \text{ in } \mathbf{H}^1(\Omega),$$

$$\gamma_{\alpha\beta}(\overline{u(\varepsilon)}) \to \overline{e_{\alpha\|\beta}} \text{ in } L^2(\omega),$$

$$\overline{u(\varepsilon)} \to \zeta \text{ in } \mathbf{V}_M^\sharp(\omega).$$

(vi) The convergence $\gamma_{\alpha\beta}(\overline{u(\varepsilon)}) \to \overline{e_{\alpha\|\beta}}$ in $L^2(\omega)$ implies that the limit $\zeta \in \mathbf{V}_M^\sharp(\omega)$ found in (v) satisfies the equations

$$B_M^\sharp(\zeta, \eta) = L_M^\sharp(\eta) \text{ for all } \eta \in \mathbf{V}_M^\sharp(\omega),$$

which have a unique solution. Consequently, the convergence

$$\overline{u(\varepsilon)} \to \zeta \text{ in } \mathbf{V}_M^\sharp(\omega)$$

established in (v) holds for the *whole* family $(\overline{u(\varepsilon)})_{\varepsilon > 0}$. □

9.4. The two-dimensional equations of a linearly elastic 'generalized membrane' shell

Again, we only consider generalized membrane shells of the first kind. The next theorem recapitulates the definition and assembles the main properties of the 'limit' two-dimensional problem found at the outcome of the asymptotic analysis carried out in Theorem 9.1.

Theorem 9.2: Existence and uniqueness of solutions. Let ω be a domain in $\mathbb{R}^2$, let γ_0 be a subset of the boundary of ω with length $\gamma_0 > 0$, and let $\boldsymbol{\theta} \in C^3(\overline{\omega}; \mathbb{R}^3)$ be an injective mapping such that the two vectors

$a_1 = \partial_1 \boldsymbol{\theta}, a_2 = \partial_2 \boldsymbol{\theta}$ are linearly independent at all points of $\bar{\omega}$. Assume that $\mathbf{V}_0(\omega) = \{\mathbf{0}\}$, where

$$\mathbf{V}_0(\omega) := \{\boldsymbol{\eta} = (\eta_i) \in \mathbf{H}^1(\omega) : \boldsymbol{\eta} = \mathbf{0} \text{ on } \gamma_0, \gamma_{\alpha\beta}(\boldsymbol{\eta}) = 0 \text{ in } \omega\},$$

$$\gamma_{\alpha\beta}(\boldsymbol{\eta}) := \frac{1}{2}(\partial_\beta \eta_\alpha + \partial_\alpha \eta_\beta) - \Gamma_{\alpha\beta}^\sigma \eta_\sigma - b_{\alpha\beta}\eta_3,$$

and define the spaces

$$\mathbf{V}(\omega) := \{\boldsymbol{\eta} = (\eta_i) \in \mathbf{H}^1(\omega) : \boldsymbol{\eta} = \mathbf{0} \text{ on } \gamma_0\},$$

$$\mathbf{V}_M^\sharp(\omega) := \text{completion of } \mathbf{V}(\omega) \text{ with respect to } |\cdot|_\omega^M, \text{ where}$$

$$|\boldsymbol{\eta}|_\omega^M := \left\{ \sum_{\alpha,\beta} |\gamma_{\alpha\beta}(\boldsymbol{\eta})|_{0,\omega}^2 \right\}^{1/2}.$$

Let

$$a^{\alpha\beta\sigma\tau} := \frac{4\lambda\mu}{\lambda+2\mu} a^{\alpha\beta} a^{\sigma\tau} + 2\mu(a^{\alpha\sigma} a^{\beta\tau} + a^{\alpha\tau} a^{\beta\sigma}),$$

$$B_M(\boldsymbol{\zeta}, \boldsymbol{\eta}) := \int_\omega a^{\alpha\beta\sigma\tau} \gamma_{\sigma\tau}(\boldsymbol{\zeta}) \gamma_{\alpha\beta}(\boldsymbol{\eta}) \sqrt{a} \, dy \text{ for } \boldsymbol{\zeta}, \boldsymbol{\eta} \in \mathbf{V}(\omega),$$

$$L(\boldsymbol{\eta}) := \int_\omega \varphi^{\alpha\beta} \gamma_{\alpha\beta}(\boldsymbol{\eta}) \sqrt{a} \, dy \text{ for } \boldsymbol{\eta} \in \mathbf{V}(\omega),$$

where the functions $\varphi^{\alpha\beta} \in L^2(\omega)$ are given, and let $B_M^\sharp$ and $L_M^\sharp$ denote the unique continuous extensions from $\mathbf{V}(\omega)$ to $\mathbf{V}_M^\sharp(\omega)$ of the bilinear form B_M and linear form L.

Then there is exactly one solution to the associated two-dimensional variational problem $\mathcal{P}_M^\sharp(\omega)$ of Theorem 9.1:
Find $\boldsymbol{\zeta}$ such that

$$\boldsymbol{\zeta} \in \mathbf{V}_M^\sharp(\omega) \text{ and } B_M^\sharp(\boldsymbol{\zeta}, \boldsymbol{\eta}) = L_M^\sharp(\boldsymbol{\eta})$$

for all $\boldsymbol{\eta} \in \mathbf{V}_M^\sharp(\omega)$.

Proof. The assumption $\mathbf{V}_0(\omega) = \{\mathbf{0}\}$ means that the seminorm $|\cdot|_\omega^M$ is a norm over the space $\mathbf{V}(\omega)$. The linear form $L : \mathbf{V}(\omega) \to \mathbb{R}$ and the bilinear form $B_M : \mathbf{V}(\omega) \times \mathbf{V}(\omega) \to \mathbb{R}$ are clearly continuous with respect to this norm. Besides,

$$B_M(\boldsymbol{\eta}, \boldsymbol{\eta}) \geq c_e^{-1} \sqrt{a_0}(|\boldsymbol{\eta}|_\omega^M)^2 \text{ for all } \boldsymbol{\eta} \in \mathbf{V}(\omega),$$

since there exist constants c_e and a_0 such that

$$\sum_{\alpha,\beta} |t_{\alpha\beta}|^2 \leq c_e a^{\alpha\beta\sigma\tau}(y) t_{\sigma\tau} t_{\alpha\beta}$$

for all $y \in \bar{\omega}$ and all symmetric matrices $(t_{\alpha\beta})$ and $a(y) \geq a_0 > 0$ for all $y \in \bar{\omega}$. These properties remain valid on the space $\mathbf{V}_M^{\sharp}(\omega)$ since $\mathbf{V}(\omega)$ is by construction dense in $\mathbf{V}_M^{\sharp}(\omega)$, again with respect to $|\cdot|_\omega^M$. The conclusion thus follows from the Lax–Milgram lemma. $\qquad\qquad\qquad\qquad\qquad\square$

In order to get physically meaningful formulas, it remains to 'de-scale' the unknown $\boldsymbol{\zeta}$ that satisfies the limit 'scaled' problem $\mathcal{P}_M^{\sharp}(\omega)$ found in Theorem 9.1. In view of the scaling $\boldsymbol{u}(\varepsilon)(x) = \boldsymbol{u}^\varepsilon(x^\varepsilon)$ for all $x^\varepsilon = \pi^\varepsilon x \in \overline{\Omega}^\varepsilon$ made on the displacement field (Section 7), we are naturally led to defining for each $\varepsilon > 0$ the 'limit' vector field $\boldsymbol{\zeta}^\varepsilon$ by letting

$$\boldsymbol{\zeta}^\varepsilon := \boldsymbol{\zeta}.$$

Recall that λ^ε and μ^ε denote for each $\varepsilon > 0$ the *actual* Lamé constants of the elastic material constituting the shell. We then have the following immediate corollary to Theorems 9.1 and 9.2; naturally, the existence and uniqueness results of Theorem 9.2 apply *verbatim* to the de-scaled problem $\mathcal{P}_M^{\sharp\varepsilon}(\omega)$ (for this reason, they are not reproduced here).

Theorem 9.3: The two-dimensional equations of a linearly elastic 'generalized membrane' shell. Let the assumptions and definitions not repeated here be as in Theorems 9.1 and 9.2. Let

$$a^{\alpha\beta\sigma\tau,\varepsilon} := \frac{4\lambda^\varepsilon \mu^\varepsilon}{\lambda^\varepsilon + 2\mu^\varepsilon} a^{\alpha\beta} a^{\sigma\tau} + 2\mu^\varepsilon (a^{\alpha\sigma} a^{\beta\tau} + a^{\alpha\tau} a^{\beta\sigma}),$$

$$B_M^\varepsilon(\boldsymbol{\zeta}, \boldsymbol{\eta}) := \varepsilon \int_\omega a^{\alpha\beta\sigma\tau,\varepsilon} \gamma_{\sigma\tau}(\boldsymbol{\zeta}) \gamma_{\alpha\beta}(\boldsymbol{\eta}) \sqrt{a}\, dy \text{ for } \boldsymbol{\zeta}, \boldsymbol{\eta} \in \mathbf{V}(\omega),$$

$$L_M^\varepsilon(\boldsymbol{\eta}) := \int_\omega \varphi^{\alpha\beta,\varepsilon} \gamma_{\alpha\beta}(\boldsymbol{\eta}) \sqrt{a}\, dy \text{ for } \boldsymbol{\eta} \in \mathbf{V}(\omega),$$

$$\varphi^{\alpha\beta,\varepsilon} := \varepsilon \varphi^{\alpha\beta},$$

and let $B_M^{\sharp\varepsilon}$ and $L_M^{\sharp\varepsilon}$ denote the unique continuous extensions from $\mathbf{V}(\omega)$ to $\mathbf{V}_M^{\sharp}(\omega)$ of the bilinear form B_M^ε and linear form L_M^ε. Then the limit vector field $\boldsymbol{\zeta}^\varepsilon$ satisfies the following two-dimensional variational problem $\mathcal{P}_M^{\sharp\varepsilon}(\omega)$ of a linearly elastic generalized membrane shell:

$$\boldsymbol{\zeta}^\varepsilon \in \mathbf{V}_M^{\sharp}(\omega) \text{ and } B_M^{\sharp\varepsilon}(\boldsymbol{\zeta}^\varepsilon, \boldsymbol{\eta}) = L_M^{\sharp\varepsilon}(\boldsymbol{\eta}) \text{ for all } \boldsymbol{\eta} \in \mathbf{V}_M^{\sharp}(\omega).$$

Equivalently, the field $\boldsymbol{\zeta}^\varepsilon$ satisfies the following minimization problem

$$\boldsymbol{\zeta}^\varepsilon \in \mathbf{V}_M^{\sharp}(\omega) \text{ and } j_M^{\sharp\varepsilon}(\boldsymbol{\zeta}^\varepsilon) = \inf_{\boldsymbol{\eta} \in \mathbf{V}_M^{\sharp}(\omega)} j_M^{\sharp\varepsilon}(\boldsymbol{\eta}), \text{ where}$$

$$j_M^{\sharp\varepsilon}(\boldsymbol{\eta}) := \frac{1}{2} B_M^{\sharp\varepsilon}(\boldsymbol{\eta}, \boldsymbol{\eta}) - L_M^{\sharp\varepsilon}(\boldsymbol{\eta}). \qquad\qquad\qquad\square$$

Each one of the two formulations found in Theorem 9.3 constitutes the two-dimensional equations of a linearly elastic generalized membrane shell. The functional $j_M^{\sharp\varepsilon} : \mathbf{V}_M^{\sharp}(\omega) \to \mathbb{R}$ is the *two-dimensional energy* and the functional

$$\boldsymbol{\eta} \in \mathbf{V}_M^{\sharp}(\omega) \to \frac{1}{2} B_M^{\sharp\varepsilon}(\boldsymbol{\eta}, \boldsymbol{\eta})$$

is the *two-dimensional strain energy* of a *linearly elastic generalized membrane shell*. The functions $a^{\alpha\beta\sigma\tau,\varepsilon}$ are the *contravariant components* of the *two-dimensional elasticity tensor of the shell*, already encountered in the two-dimensional equations of a linearly elastic elliptic membrane shell (Theorem 8.3).

The bilinear form $B_M^{\sharp\varepsilon}$ found in the variational equations of a linearly elastic *generalized membrane* shell is an extension of the bilinear form B_M^{ε} already found in the variational equations of a linearly elastic *elliptic membrane* shell (Theorem 8.3). Recall that both kinds constitute together the *linearly elastic membrane shells*.

Under the essential assumptions that the space $\mathbf{V}_F(\omega)$ reduces to $\{0\}$ and that the forces are admissible, we have therefore justified by a convergence result (Theorem 9.1) the two-dimensional equations of a linearly elastic generalized membrane shell. In so doing, we have also justified the formal asymptotic approach of Caillerie and Sanchez-Palencia (1995*b*) when 'bending is badly inhibited', according to the terminology of E. Sanchez-Palencia.

The asymptotic analysis of Ciarlet and Lods (1996*d*) described in this section has been extended by Şlicaru (1998) to linearly elastic shells whose middle surface 'has no boundary', such as a torus.

Among linearly elastic shells, generalized membrane shells possess distinctive characteristics that set them apart.

While *forces* applied to a family of elliptic membrane or flexural shells are not subjected to any restriction (see Sections 8 and 10), body forces applied to a family of generalized membrane shells can no longer be accounted for by an arbitrary linear form of the form

$$\boldsymbol{v}^\varepsilon = (v_i^\varepsilon) \to \int_{\Omega^\varepsilon} f^{i,\varepsilon} v_i^\varepsilon \sqrt{g^\varepsilon} \, \mathrm{d}x^\varepsilon,$$

that is, with *arbitrary* contravariant components $f^{i,\varepsilon} \in L^2(\Omega^\varepsilon)$. They must be *admissible for the three-dimensional equations*, in order that the associated scaled linear forms be in particular continuous with respect to the norm

$$\boldsymbol{v} \to \left\{ \sum_{i,j} |e_{i\|j}(\varepsilon; \boldsymbol{v})|_{0,\Omega}^2 \right\}^{1/2}$$

and uniformly so with respect to $\varepsilon > 0$ (Section 9.2).

The linear form found in the variational equations of the limit two-dimensional problem for such a shell is likewise subjected to a restriction. On the dense subspace $\mathbf{V}(\omega)$ of the space $\mathbf{V}_M^\sharp(\omega)$, it must be of the form

$$\boldsymbol{\eta} \to \int_\omega \varphi^{\alpha\beta} \gamma_{\alpha\beta}(\boldsymbol{\eta}) \sqrt{a}\, dy$$

(Theorem 9.1). In other words, the applied forces must also be *admissible for the two-dimensional equations*, in such a way that the linear form appearing therein must be an element of the *dual space* of $\mathbf{V}_M^\sharp(\omega)$.

As this dual space may be quite 'small', the limit variational problem, which otherwise satisfies all the assumptions of the Lax–Milgram lemma (Theorem 9.2), possesses the unusual feature that its solution may no longer exist if the data undergo arbitrarily small, yet arbitrarily smooth, perturbations! Another unusual feature of this problem is that the space $\mathbf{V}_M^\sharp(\omega)$ in which its solution is sought may not necessarily be a space of distributions!

Such variational problems fall in the category of 'sensitive problems' introduced by Lions and Sanchez-Palencia (1994). Since then, such problems have been extensively studied. See, in particular, Lions and Sanchez-Palencia (1996, 1997a,b, 1998, 2000), Pitkäranta and Sanchez-Palencia (1997), Sanchez-Palencia (1999, 2000), Leguillon, Sanchez-Hubert and Sanchez-Palencia (1999), Delfour (1999).

Examples of linearly elastic generalized membrane shells are numerous and, in this respect, those given in Section 9.1 constitute only a small sample. In each case, however, the proof that the space $\mathbf{V}_F(\omega)$ reduces to $\{\mathbf{0}\}$, the identification of the corresponding space $\mathbf{V}_M^\sharp(\omega)$, and the identification of 'admissible' applied forces usually require delicate analyses. In this respect, see notably Sanchez-Hubert and Sanchez-Palencia (1997, Chapter 7, Sections 2 and 4), Lions and Sanchez-Palencia (1997b, 1998), Karamian (1998b), Lods and Mardare (1998a), Mardare (1998c), Gérard and Sanchez-Palencia (2000).

The occurrence of boundary layers in generalized membrane shells is studied in Karamian, Sanchez-Hubert and Sanchez-Palencia (2000).

10. 'Flexural' shells

A shell with middle surface $S = \boldsymbol{\theta}(\overline{\omega})$, subjected to a boundary condition of place along a portion of its lateral face with $\boldsymbol{\theta}(\gamma_0)$, where $\gamma_0 \subset \gamma$, as its middle curve, is called a *linearly elastic 'flexural' shell* if its associated space

$$\mathbf{V}_F(\omega) = \big\{ \boldsymbol{\eta} = (\eta_i) \in H^1(\omega) \times H^1(\omega) \times H^2(\omega) :$$
$$\eta_i = \partial_\nu \eta_3 = 0 \text{ on } \gamma_0, \gamma_{\alpha\beta}(\boldsymbol{\eta}) = 0 \text{ in } \omega \big\}$$

contains *nonzero* functions.

The purpose of this section is to identify and to mathematically justify the *two-dimensional equations of a linearly elastic flexural shell*, by showing how the convergence of the three-dimensional displacements can be established in *ad hoc* function spaces as the thickness of such a shell approaches zero.

10.1. Definition and examples

Let ω be a domain in $\mathbb{R}^2$ with boundary γ and let $\boldsymbol{\theta} \in \mathcal{C}^2(\overline{\omega}; \mathbb{R}^3)$ be an injective mapping such that the two vectors $\partial_1 \boldsymbol{\theta}(y), \partial_2 \boldsymbol{\theta}(y)$ are linearly independent at every point $y \in \overline{\omega}$. A shell with middle surface $S = \boldsymbol{\theta}(\overline{\omega})$ is called a *linearly elastic 'flexural' shell* if the following two conditions are simultaneously satisfied.

(i) The shell is subjected to a (homogeneous) *boundary condition of place* along a portion of its lateral face with $\boldsymbol{\theta}(\gamma_0)$ as its middle curve (*i.e.*, the displacement vanishes on this portion), where the subset $\gamma_0 \subset \gamma$ satisfies

$$\text{length } \gamma_0 > 0.$$

(ii) Define the space

$$\mathbf{V}_F(\omega) := \{ \boldsymbol{\eta} = (\eta_i) \in H^1(\omega) \times H^1(\omega) \times H^2(\omega) : \\ \eta_i = \partial_\nu \eta_3 = 0 \text{ on } \gamma_0, \gamma_{\alpha\beta}(\boldsymbol{\eta}) = 0 \text{ in } \omega \}$$

(∂_ν denoting the outer normal derivative operator along γ). Then the space $\mathbf{V}_F(\omega)$ contains nonzero functions; equivalently,

$$\mathbf{V}_F(\omega) \neq \{\mathbf{0}\}.$$

We recall that the functions

$$\gamma_{\alpha\beta}(\boldsymbol{\eta}) = \frac{1}{2}(\partial_\beta \eta_\alpha + \partial_\alpha \eta_\beta) - \Gamma^\sigma_{\alpha\beta}\eta_\sigma - b_{\alpha\beta}\eta_3$$

denote the covariant components of the *linearized change of metric tensor* associated with a displacement field $\eta_i \mathbf{a}^i$ of the surface S.

In other words, there exist nonzero admissible linearized inextensional displacements $\eta_i \mathbf{a}^i$ of the middle surface S. 'Admissible' means that they satisfy *two-dimensional boundary conditions of clamping* along the curve $\boldsymbol{\theta}(\gamma_0)$, expressed here by means of the boundary conditions $\eta_i = \partial_\nu \eta_3$ on γ_0 on the associated field $\boldsymbol{\eta} = (\eta_i)$ (these boundary conditions will be interpreted later). 'Linearized inextensional' indicates that the functions $\gamma_{\alpha\beta}(\boldsymbol{\eta})$ are the *linearizations* with respect to $\boldsymbol{\eta} = (\eta_i)$ of the covariant components of the *exact* change of metric tensor associated with a displacement field $\eta_i \mathbf{a}^i$ of the surface S; cf. Section 4.

A shell whose middle surface $S = \boldsymbol{\theta}(\overline{\omega})$ is a portion of a cylinder and which is subjected to a boundary condition of place (*i.e.*, of vanishing displacement field) along a portion (solid black in the figure) of its lateral face whose

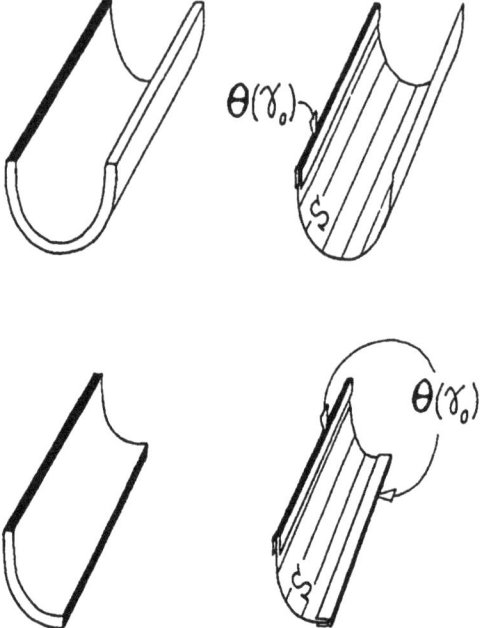

Fig. 10.1. Linearly elastic 'flexural' shells

middle curve $\boldsymbol{\theta}(\gamma_0)$ is contained in one or two generatrices of S provides
an instance of a linearly elastic *flexural* shell, that is, one for which the
associated space $\mathbf{V}_F(\omega)$ contains nonzero functions $\boldsymbol{\eta}$; see Figure 10.1. The
two-dimensional boundary conditions of clamping $\eta_i = \partial_\nu \eta_3 = 0$ on γ_0 that
will eventually be inherited by the limit two-dimensional equations are so
named because they mean that the points of, and the tangent spaces to, the
deformed and undeformed middle surfaces coincide along the set $\boldsymbol{\theta}(\gamma_0)$, as
suggested in the 'two-dimensional' figures.

A shell whose middle surface $S = \boldsymbol{\theta}(\overline{\omega})$ is a portion of a cone excluding
its vertex and which is subjected to a boundary condition of place along
a portion (solid black in the figure) of its lateral face whose middle curve
$\boldsymbol{\theta}(\gamma_0)$ is contained in one generatrix of S provides another example of a
linearly elastic *flexural* shell, since again $\mathbf{V}_F(\omega) \neq \{\mathbf{0}\}$ in this case. See
Figure 10.2, where the two-dimensional boundary conditions of clamping
inherited by the limit two-dimensional equations are again suggested in the
'two-dimensional' figure.

Incidentally, a comparison with the cylindrical and conical shells shown in
Figure 9.2 illustrates the crucial role played by the set $\boldsymbol{\theta}(\gamma_0)$ in determining
the type of shell!

A plate, subjected to a boundary condition of place along any portion
(solid black in the figure) of its lateral face whose middle line γ_0 satisfies

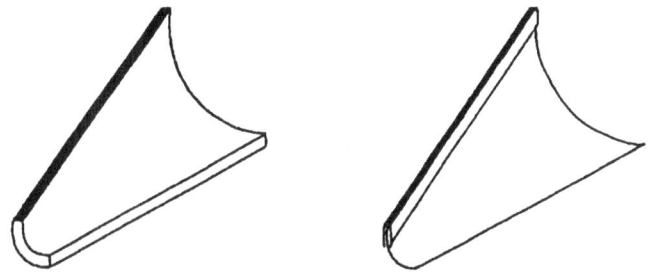

Fig. 10.2. Another example of a linearly elastic 'flexural' shell

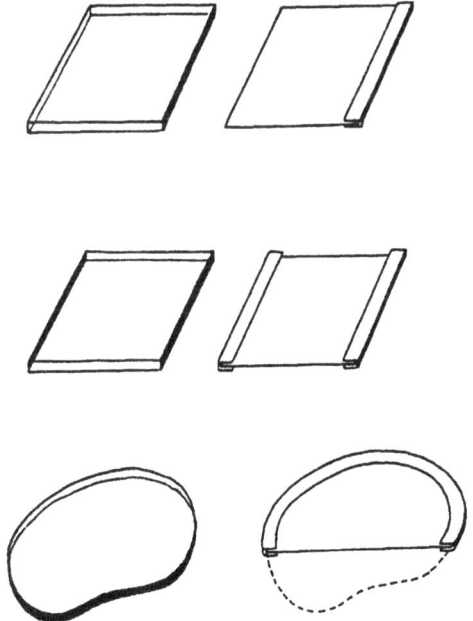

Fig. 10.3. Another example of a linearly elastic 'flexural' shell: a plate

length $\gamma_0 > 0$, provides an instance of a linearly elastic *flexural* shell since

$$\mathbf{V}_F(\omega) \supset \{\boldsymbol{\eta} = (0,0,\eta_3) : \eta_3 \in H_0^2(\omega)\} \neq \{\mathbf{0}\}$$

in each case. See Figure 10.3, where the two-dimensional boundary conditions of clamping inherited by the limit two-dimensional equations are again suggested in the 'two-dimensional' figures.

The definition of a linearly elastic flexural shell thus depends only on the subset of the lateral face where the shell is subjected to a boundary condition of place (via the set γ_0) and on the geometry of the middle surface of the shell.

10.2. Convergence of the scaled displacements as the thickness approaches zero

We now establish the main results of this section. Consider a family of linearly elastic flexural shells with thickness $2\varepsilon > 0$, with each having the same middle surface $S = \boldsymbol{\theta}(\omega)$, and with each subjected to a boundary condition of place along a portion of its lateral face having the same set $\boldsymbol{\theta}(\gamma_0)$ as its middle curve, the assumptions on the data being as in Theorem 10.1 below.

Then the solutions $\boldsymbol{u}(\varepsilon)$ of the associated scaled three-dimensional problems $\mathcal{P}(\varepsilon; \Omega)$ (Theorem 7.1) converge in $\mathbf{H}^1(\Omega)$ as $\varepsilon \to 0$ toward a limit $\boldsymbol{u}$ and this limit, which is independent of the transverse variable x_3, can be identified with the solution $\bar{\boldsymbol{u}}$ of a two-dimensional variational problem $\mathcal{P}_F(\omega)$ posed over the set ω.

The functions $\gamma_{\alpha\beta}(\boldsymbol{\eta})$ and $\rho_{\alpha\beta}(\boldsymbol{\eta})$ appearing in the next theorem respectively represent the covariant components of the *linearized change of metric* and *linearized change of curvature tensors* associated with a displacement field $\eta_i \boldsymbol{a}^i$ of the middle surface S.

Note that the assumption on the applied body forces made in the next theorem corresponds to letting $a = 2$ in Theorem 7.1. That $a = 2$ is indeed the 'correct' exponent in this case can be justified in two different ways.

It is easily checked that this choice is the only one that lets the applied body forces enter (via the functions p^i) the right-hand sides of the variational equations in the 'limit' variational problem $\mathcal{P}_F(\omega)$ satisfied by $\bar{\boldsymbol{u}}$.

Otherwise, the number a can be considered *a priori* as an unknown. Then a *formal* (but careful!) asymptotic analysis of the scaled unknown $\boldsymbol{u}(\varepsilon)$ shows that, for a family of linearly elastic flexural shells, the exponent a must be set equal to 2, again in order that the applied body forces contribute to the 'limit' variational problem; cf. Miara and Sanchez-Palencia (1996).

The next result is due to Ciarlet, Lods and Miara (1996, Theorem 5.1); a complete proof is also given in Ciarlet (2000, Theorem 6.2-1).

Theorem 10.1: Convergence of the scaled displacements.   Assume that $\boldsymbol{\theta} \in \mathcal{C}^3(\overline{\omega}; \mathbb{R}^3)$. Consider a family of linearly elastic flexural shells with thickness 2ε approaching zero, with each having the same middle surface $S = \boldsymbol{\theta}(\overline{\omega})$, and with each subjected to a boundary condition of place along a portion of its lateral face having the same set $\boldsymbol{\theta}(\gamma_0)$ as its middle curve. Further, assume that there exist constants $\lambda > 0$ and $\mu > 0$ and functions

$f^i \in L^2(\Omega)$ independent of ε such that

$$\lambda^\varepsilon = \lambda \quad \text{and} \quad \mu^\varepsilon = \mu,$$
$$f^{i,\varepsilon}(x^\varepsilon) = \varepsilon^2 f^i(x) \quad \text{for all} \quad x^\varepsilon = \pi^\varepsilon x \in \Omega^\varepsilon$$

(the notation is that of Section 7).

Let $\boldsymbol{u}(\varepsilon)$ denote, for sufficiently small $\varepsilon > 0$, the solution of the associated scaled three-dimensional problem $\mathcal{P}(\varepsilon; \Omega)$ (Theorem 7.1). Then there exists $\boldsymbol{u} \in \mathbf{H}^1(\Omega)$ satisfying $\boldsymbol{u} = \boldsymbol{0}$ on $\Gamma_0 = \gamma_0 \times [-1, 1]$ such that

$$\boldsymbol{u}(\varepsilon) \to \boldsymbol{u} \text{ in } \mathbf{H}^1(\Omega) \text{ as } \varepsilon \to 0,$$

where $\boldsymbol{u} = (u_i)$ is independent of the transverse variable x_3.

Furthermore, the average $\overline{\boldsymbol{u}} := \frac{1}{2} \int_{-1}^{1} \boldsymbol{u} \, dx_3$ satisfies the following two-dimensional variational problem $\mathcal{P}_F(\omega)$:

$$\overline{\boldsymbol{u}} = (\overline{u}_i) \in \mathbf{V}_F(\omega) := \{\boldsymbol{\eta} = (\eta_i) \in H^1(\omega) \times H^1(\omega) \times H^2(\omega) :$$
$$\eta_i = \partial_\nu \eta_3 = 0 \text{ on } \gamma_0, \gamma_{\alpha\beta}(\boldsymbol{\eta}) = 0 \text{ in } \omega\},$$

$$\frac{1}{3} \int_\omega a^{\alpha\beta\sigma\tau} \rho_{\sigma\tau}(\overline{\boldsymbol{u}}) \rho_{\alpha\beta}(\boldsymbol{\eta}) \sqrt{a} \, dy = \int_\omega p^i \eta_i \sqrt{a} \, dy$$

for all $\boldsymbol{\eta} = (\eta_i) \in \mathbf{V}_F(\omega)$. Here

$$\gamma_{\alpha\beta}(\boldsymbol{\eta}) := \frac{1}{2}(\partial_\beta \eta_\alpha + \partial_\alpha \eta_\beta) - \Gamma^\sigma_{\alpha\beta}\eta_\sigma - b_{\alpha\beta}\eta_3,$$

$$\rho_{\alpha\beta}(\boldsymbol{\eta}) := \partial_{\alpha\beta}\eta_3 - \Gamma^\sigma_{\alpha\beta}\partial_\sigma\eta_3 - b^\sigma_\alpha b_{\sigma\beta}\eta_3 + b^\sigma_\alpha(\partial_\beta\eta_\sigma - \Gamma^\tau_{\beta\sigma}\eta_\tau)$$
$$+ b^\tau_\beta(\partial_\alpha\eta_\tau - \Gamma^\sigma_{\alpha\tau}\eta_\sigma) + (\partial_\alpha b^\tau_\beta + \Gamma^\tau_{\alpha\sigma}b^\sigma_\beta - \Gamma^\sigma_{\alpha\beta}b^\tau_\sigma)\eta_\tau,$$

$$a^{\alpha\beta\sigma\tau} := \frac{4\lambda\mu}{\lambda + 2\mu} a^{\alpha\beta}a^{\sigma\tau} + 2\mu(a^{\alpha\sigma}a^{\beta\tau} + a^{\alpha\tau}a^{\beta\sigma}),$$

$$p^i := \int_{-1}^{1} f^i \, dx_3.$$

Sketch of proof. **(i)** The proof rests on the same crucial three-dimensional inequality of Korn's type that was already needed for the asymptotic analysis of 'generalized membrane' shells (Theorem 9.1). For a family of linearly elastic shells, with each having the same middle surface $S = \boldsymbol{\theta}(\overline{\omega})$, and with each subjected to a boundary condition of place along a portion of its lateral face having the same set $\boldsymbol{\theta}(\gamma_0)$ as its middle curve, there exists a constant C such that, for sufficiently small $\varepsilon > 0$,

$$\|\boldsymbol{v}\|_{1,\Omega} \leq \frac{C}{\varepsilon} \left\{ \sum_{i,j} |e_{i\|j}(\varepsilon; \boldsymbol{v})|_{0,\Omega}^2 \right\}^{1/2}$$

for all $\boldsymbol{v} \in \mathbf{V}(\Omega)$, where

$$\mathbf{V}(\Omega) = \{\boldsymbol{v} \in \mathbf{H}^1(\Omega) : \boldsymbol{v} = \boldsymbol{0} \text{ on } \gamma_0 \times [-1, 1]\},$$

and the functions $e_{i\|j}(\varepsilon; v)$ are the scaled linearized strains appearing in Theorem 7.1.

(ii) Letting $v = u(\varepsilon)$ in the variational equations of problem $\mathcal{P}(\varepsilon; \Omega)$ (Theorem 7.1) and using the three-dimensional inequality of Korn's type used in (i), we obtain a chain of inequalities showing that the norms $\|u(\varepsilon)\|_{1,\Omega}$ and $|\frac{1}{\varepsilon}e_{i\|j}(\varepsilon; u(\varepsilon))|_{0,\Omega}$ are bounded independently of ε.

Thus there exists a subsequence, still denoted by $(u(\varepsilon))_{\varepsilon>0}$ for notational convenience, such that

$$u(\varepsilon) \rightharpoonup u \text{ in } \mathbf{H}^1(\Omega), \text{ and thus } u(\varepsilon) \to u \text{ in } \mathbf{L}^2(\Omega),$$

$$\frac{1}{\varepsilon}e_{i\|j}(\varepsilon; u(\varepsilon)) \rightharpoonup e_{i\|j}^1 \text{ in } L^2(\Omega).$$

(iii) The above convergence, combined with the asymptotic behaviour of the functions $\Gamma_{ij}^p(\varepsilon)$, $A^{ijkl}(\varepsilon)$, and $g(\varepsilon)$, then implies that the vector field u is independent of x_3. Further, the average $\bar{u} = \frac{1}{2}\int_{-1}^{1} u \, dx_3$ belongs to the space $\mathbf{V}_F(\omega)$, and the field u and the functions $e_{i\|j}^1$ are related by

$$-\partial_3 e_{\alpha\|\beta}^1 = \rho_{\alpha\beta}(u),$$

$$e_{\alpha\|3}^1 = 0, \quad e_{3\|3}^1 = -\frac{\lambda}{\lambda+2\mu}a^{\alpha\beta}e_{\alpha\|\beta}^1.$$

(iv) In the variational equations of problem $\mathcal{P}(\varepsilon; \Omega)$, let $v = (v_i(\varepsilon))$, where the functions $v_i(\varepsilon)$ are of the form

$$v_\alpha(\varepsilon) = \eta_\alpha - \varepsilon x_3(\partial_\alpha \eta_3 + 2b_\alpha^\sigma \eta_\sigma) \text{ and } v_3(\varepsilon) = \eta_3$$

for some *fixed* $\eta = (\eta_i)$ in the space $\mathbf{V}_F(\omega)$, and let ε approach zero. Then the asymptotic behaviour of the functions $A^{ijkl}(\varepsilon)$ and $g(\varepsilon)$, combined with the relations found in (iii), shows that the average $\bar{u} \in \mathbf{V}_F(\omega)$ indeed satisfies the variational equations of the two-dimensional problem $\mathcal{P}_F(\omega)$ stated in the statement of the theorem.

The solution to $\mathcal{P}_F(\omega)$ being *unique*, the convergence $u(\varepsilon) \rightharpoonup u$ in $\mathbf{H}^1(\Omega)$ and $u(\varepsilon) \to u$ in $\mathbf{L}^2(\Omega)$ established in (ii) for a subsequence thus holds for the *whole* family $(u(\varepsilon))_{\varepsilon>0}$.

(v) Again letting $v = u(\varepsilon)$ in the variational equations of $\mathcal{P}(\varepsilon; \Omega)$ and using the results obtained in (ii)–(iv), we obtain the strong convergence

$$\frac{1}{\varepsilon}e_{i\|j}(\varepsilon; u(\varepsilon)) \to e_{i\|j}^1 \text{ in } L^2(\Omega),$$

which in turn implies that

$$u(\varepsilon) \to u \text{ in } \mathbf{H}^1(\Omega),$$

as was to be proved.   □

10.3. The two-dimensional equations of a linearly elastic 'flexural' shell

The next theorem recapitulates the definition and assembles the main features of the 'limit' two-dimensional variational problem $\mathcal{P}_F(\omega)$ found at the outcome of the asymptotic analysis carried out in Theorem 10.1.

Theorem 10.2: Existence and uniqueness of solutions. Let ω be a domain in $\mathbb{R}^2$, let γ_0 be a subset of the boundary of ω with length $\gamma_0 > 0$, and let $\boldsymbol{\theta} \in C^3(\overline{\omega}; \mathbb{R}^3)$ be an injective mapping such that the two vectors $\boldsymbol{a}_\alpha = \partial_\alpha \boldsymbol{\theta}$ are linearly independent at all points of $\overline{\omega}$ and such that

$$\mathbf{V}_F(\omega) := \{\boldsymbol{\eta} = (\eta_i) \in H^1(\omega) \times H^1(\omega) \times H^2(\omega) : $$
$$\eta_i = \partial_\nu \eta_3 = 0 \text{ on } \gamma_0, \gamma_{\alpha\beta}(\boldsymbol{\eta}) = 0 \text{ in } \omega\} \neq \{\mathbf{0}\},$$

where

$$\gamma_{\alpha\beta}(\boldsymbol{\eta}) := \frac{1}{2}(\partial_\beta \eta_\alpha + \partial_\alpha \eta_\beta) - \Gamma^\sigma_{\alpha\beta}\eta_\sigma - b_{\alpha\beta}\eta_3.$$

The associated two-dimensional variational problem $\mathcal{P}_F(\omega)$ found in Theorem 10.1 is as follows. Given $p^i \in L^2(\omega)$, find $\boldsymbol{\zeta} = (\zeta_i)$ satisfying

$$\boldsymbol{\zeta} \in \mathbf{V}_F(\omega),$$

$$\frac{1}{3}\int_\omega a^{\alpha\beta\sigma\tau}\rho_{\sigma\tau}(\boldsymbol{\zeta})\rho_{\alpha\beta}(\boldsymbol{\eta})\sqrt{a}\,\mathrm{d}y = \int_\omega p^i \eta_i \sqrt{a}\,\mathrm{d}y$$

for all $\boldsymbol{\eta} = (\eta_i) \in \mathbf{V}_F(\omega)$, where

$$\rho_{\alpha\beta}(\boldsymbol{\eta}) := \partial_{\alpha\beta}\eta_3 - \Gamma^\sigma_{\alpha\beta}\partial_\sigma\eta_3 - b^\sigma_\alpha b_{\sigma\beta}\eta_3 + b^\sigma_\alpha(\partial_\beta\eta_\sigma - \Gamma^\tau_{\beta\sigma}\eta_\tau)$$
$$+ b^\tau_\beta(\partial_\alpha\eta_\tau - \Gamma^\sigma_{\alpha\tau}\eta_\sigma) + (\partial_\alpha b^\tau_\beta + \Gamma^\tau_{\alpha\sigma}b^\sigma_\beta - \Gamma^\sigma_{\alpha\beta}b^\tau_\sigma)\eta_\tau,$$

$$a^{\alpha\beta\sigma\tau} := \frac{4\lambda\mu}{\lambda + 2\mu}a^{\alpha\beta}a^{\sigma\tau} + 2\mu(a^{\alpha\sigma}a^{\beta\tau} + a^{\alpha\tau}a^{\beta\sigma}).$$

This problem has exactly one solution, which is also the unique solution of the minimization problem:
Find $\boldsymbol{\zeta}$ such that

$$\boldsymbol{\zeta} \in \mathbf{V}_F(\omega) \text{ and } j_F(\boldsymbol{\zeta}) = \inf_{\boldsymbol{\eta} \in \mathbf{V}_F(\omega)} j_F(\boldsymbol{\eta}), \text{ where}$$

$$j_F(\boldsymbol{\eta}) := \frac{1}{6}\int_\omega a^{\alpha\beta\sigma\tau}\rho_{\sigma\tau}(\boldsymbol{\eta})\rho_{\alpha\beta}(\boldsymbol{\eta})\sqrt{a}\,\mathrm{d}y - \int_\omega p^i \eta_i \sqrt{a}\,\mathrm{d}y.$$

Proof. The existence and uniqueness of a solution to the variational problem $\mathcal{P}_F(\omega)$, or to its equivalent minimization problem, is a consequence of the inequality of Korn's type on a general surface (Theorem 4.4), of the existence of constants c_e and a_0 such that

$$\sum_{\alpha,\beta}|t_{\alpha\beta}|^2 \leq c_e a^{\alpha\beta\sigma\tau}(y)t_{\sigma\tau}t_{\alpha\beta}$$

for all $y \in \bar{\omega}$ and all symmetric matrices $(t_{\alpha\beta})$ and $a(y) \geq a_0 > 0$ for all $y \in \bar{\omega}$, and of the Lax–Milgram lemma.   □

The minimization problem encountered in Theorem 10.2 (or that in Theorem 10.3 below in its 'de-scaled' formulation) provides an interesting example of a minimization problem with 'equality constraints', namely the relations

$$\gamma_{\alpha\beta}(\boldsymbol{\eta}) = 0 \text{ in } \omega$$

to be satisfied by the elements $\boldsymbol{\eta}$ of the space $\mathbf{V}_F(\omega)$ over which the functional is to be minimized.

In order to get physically meaningful formulas, we must 'de-scale' the unknowns ζ_i that satisfy the limit 'scaled' problem $\mathcal{P}_F(\omega)$ found in Theorem 10.2. In view of the scalings $u_i(\varepsilon)(x) = u_i^\varepsilon(x^\varepsilon)$ for all $x^\varepsilon = \pi^\varepsilon x \in \overline{\Omega}^\varepsilon$ made on the covariant components of the displacement field (Section 7), we are naturally led to defining, for each $\varepsilon > 0$, the covariant components $\zeta_i^\varepsilon : \bar{\omega} \to \mathbb{R}$ of the 'limit displacement field' $\zeta_i^\varepsilon \boldsymbol{a}^i : \bar{\omega} \to \mathbb{R}^3$ of the middle surface S of the shell by letting

$$\zeta_i^\varepsilon := \zeta_i$$

(the vectors $\boldsymbol{a}^i$ forming the contravariant basis at each point of S).

Like those found in the analysis of linearly elastic elliptic membrane shells (Section 8), the fields $\boldsymbol{\zeta}^\varepsilon := (\zeta_i^\varepsilon)$ and $\zeta_i^\varepsilon \boldsymbol{a}^i$ must be carefully distinguished! The former is essentially a convenient mathematical 'intermediary', but only the latter has physical significance.

Recall that $f^{i,\varepsilon} \in L^2(\Omega^\varepsilon)$ represent the contravariant components of the applied body forces *actually* acting on the shell and that λ^ε and μ^ε denote the *actual* Lamé constants of its constituent material. We then have the following immediate corollary to Theorems 10.1 and 10.2; naturally, the existence and uniqueness results of Theorem 10.2 apply *verbatim* to the solution of the 'de-scaled' problem $\mathcal{P}_F^\varepsilon(\omega)$ found in the next theorem (for this reason, they are not reproduced here).

Theorem 10.3: The two-dimensional equations of a linearly elastic 'flexural' shell. Let the assumptions on the data and the definitions of the functions $\gamma_{\alpha\beta}(\boldsymbol{\eta})$ and $\rho_{\alpha\beta}(\boldsymbol{\eta})$ be as in Theorem 10.2. Then the vector field $\boldsymbol{\zeta}^\varepsilon := (\zeta_i^\varepsilon)$ formed by the covariant components of the limit displacement field $\zeta_i^\varepsilon \boldsymbol{a}^i$ of the middle surface S satisfies the following two-dimensional variational problem $\mathcal{P}_F^\varepsilon(\omega)$ of a linearly elastic flexural shell:

$$\boldsymbol{\zeta}^\varepsilon \in \mathbf{V}_F(\omega) := \{\boldsymbol{\eta} = (\eta_i) \in H^1(\omega) \times H^1(\omega) \times H^2(\omega) :$$
$$\eta_i = \partial_\nu \eta_3 = 0 \text{ on } \gamma_0, \gamma_{\alpha\beta}(\boldsymbol{\eta}) = 0 \text{ in } \omega\},$$

$$\frac{\varepsilon^3}{3} \int_\omega a^{\alpha\beta\sigma\tau,\varepsilon} \rho_{\sigma\tau}(\boldsymbol{\zeta}^\varepsilon) \rho_{\alpha\beta}(\boldsymbol{\eta}) \sqrt{a} \, dy = \int_\omega p^{i,\varepsilon} \eta_i \sqrt{a} \, dy$$

for all $\boldsymbol{\eta} = (\eta_i) \in \mathbf{V}_F(\omega)$, where

$$a^{\alpha\beta\sigma\tau,\varepsilon} := \frac{4\lambda^\varepsilon \mu^\varepsilon}{\lambda^\varepsilon + 2\mu^\varepsilon} a^{\alpha\beta} a^{\sigma\tau} + 2\mu^\varepsilon (a^{\alpha\sigma} a^{\beta\tau} + a^{\alpha\tau} a^{\beta\sigma}),$$

$$p^{i,\varepsilon} := \int_{-\varepsilon}^{\varepsilon} f^{i,\varepsilon} \, dx_3^\varepsilon.$$

Equivalently, the field $\boldsymbol{\zeta}^\varepsilon = (\zeta_i^\varepsilon)$ satisfies the minimization problem

$$\boldsymbol{\zeta}^\varepsilon \in \mathbf{V}_F(\omega) \quad \text{and} \quad j_F^\varepsilon(\boldsymbol{\zeta}^\varepsilon) = \inf_{\boldsymbol{\eta} \in \mathbf{V}_F(\omega)} j_F^\varepsilon(\boldsymbol{\eta}), \quad \text{where}$$

$$j_F^\varepsilon(\boldsymbol{\eta}) := \frac{\varepsilon^3}{6} \int_\omega a^{\alpha\beta\sigma\tau,\varepsilon} \rho_{\sigma\tau}(\boldsymbol{\eta}) \rho_{\alpha\beta}(\boldsymbol{\eta}) \sqrt{a} \, dy - \int_\omega p^{i,\varepsilon} \eta_i \sqrt{a} \, dy. \qquad \square$$

Each one of the two formulations found in Theorem 10.3 constitutes the two-dimensional equations of a linearly elastic flexural shell.

We recall that the condition $\mathbf{V}_F(\omega) \neq \{\mathbf{0}\}$, which is the basis of the definition of a linearly elastic flexural shell, means that there exist nonzero 'admissible linearized inextensional displacements' of the middle surface, since the functions $\gamma_{\alpha\beta}(\boldsymbol{\eta})$ used in the definition of $\mathbf{V}_F(\omega)$ are the covariant components of the *linearized change of metric tensor* associated with a displacement field $\eta_i \boldsymbol{a}^i$ of the middle surface S; 'admissible' means that the fields $\boldsymbol{\eta} = (\eta_i) \in \mathbf{V}_F(\omega)$ must also satisfy the boundary conditions $\eta_i = \partial_\nu \eta_3 = 0$ on γ_0.

In order to interpret these boundary conditions, let $\eta_i \boldsymbol{a}^i$ be a displacement field of the middle surface $S = \boldsymbol{\theta}(\overline{\omega})$ with smooth enough, but otherwise arbitrary, covariant components $\eta_i : \overline{\omega} \to \mathbb{R}$. The tangent plane at an arbitrary point $\boldsymbol{\theta}(y) + \eta_i(y) \boldsymbol{a}^i(y), y \in \overline{\omega}$, of the deformed surface $(\boldsymbol{\theta} + \eta_i \boldsymbol{a}^i)(\overline{\omega})$ is thus spanned by the vectors

$$\partial_\alpha(\boldsymbol{\theta} + \eta_i \boldsymbol{a}^i)(y) = \boldsymbol{a}_\alpha(y) + \partial_\alpha \eta_i(y) \boldsymbol{a}^i(y) + \eta_i(y) \partial_\alpha \boldsymbol{a}^i(y),$$

if these are linearly independent. Since

$$\eta_i = \partial_\nu \eta_3 = 0 \text{ on } \gamma_0 \Rightarrow \eta_i = \partial_\alpha \eta_3 = 0 \text{ on } \gamma_0,$$

it follows that

$$\boldsymbol{\theta}(y) + \eta_i(y) \boldsymbol{a}^i(y) = \boldsymbol{\theta}(y) \text{ for all } y \in \gamma_0,$$
$$\partial_\alpha(\boldsymbol{\theta} + \eta_i \boldsymbol{a}^i)(y) = \boldsymbol{a}_\alpha(y) + \partial_\alpha \eta_\beta(y) \boldsymbol{a}^\beta(y) \text{ for all } y \in \gamma_0.$$

These relations thus show that the points of the deformed and undeformed middle surfaces, and their tangent spaces at those points where the vectors $\partial_\alpha(\boldsymbol{\theta} + \eta_i \boldsymbol{a}^i)$ are linearly independent, coincide along the set $\boldsymbol{\theta}(\gamma_0)$. Such 'two-dimensional boundary conditions of clamping' are suggested in Figures 10.1 to 10.3.

The functions $\rho_{\alpha\beta}(\boldsymbol{\eta})$ are the *covariant components* of the *linearized change of curvature tensor* associated with a displacement field $\eta_i \boldsymbol{a}^i$ of the middle surface S and the functions $a^{\alpha\beta\sigma\tau,\varepsilon}$ are the *contravariant components* of the *two-dimensional elasticity tensor of the shell*, already encountered in the two-dimensional equations of linearly elastic elliptic membrane and generalized membrane shells (Theorems 8.3 and 9.3).

Finally, the functional $j_F^\varepsilon : \mathbf{V}_F(\omega) \to \mathbb{R}$ is the *two-dimensional energy* and the functional

$$\boldsymbol{\eta} \in \mathbf{V}_F(\omega) \to \frac{\varepsilon^3}{6} \int_\omega a^{\alpha\beta\sigma\tau,\varepsilon} \rho_{\sigma\tau}(\boldsymbol{\eta}) \rho_{\alpha\beta}(\boldsymbol{\eta}) \sqrt{a}\,dy$$

is the *two-dimensional strain energy* of a *linearly elastic flexural shell*.

Under the essential assumptions that the space $\mathbf{V}_F(\omega)$ contains nonzero elements, we have therefore justified by a convergence result (Theorem 10.1) the two-dimensional equations of a linearly elastic flexural shell. In so doing, we have justified the formal asymptotic approach of Sanchez-Palencia (1990) (see also Miara and Sanchez-Palencia (1996) and Caillerie and Sanchez-Palencia (1995b)) when 'bending is not inhibited', according to the terminology of E. Sanchez-Palencia.

Due credit should be given in this respect to Sanchez-Palencia (1989a) for recognizing the central role played by the space $\mathbf{V}_F(\omega)$ in the classification of linearly elastic shells.

The above convergence analysis also substantiate an important observation. In a *flexural shell*, body forces of order $O(\varepsilon^2)$ produce an $O(1)$ limit displacement field. By contrast, body forces of order $O(1)$ are required to also produce an $O(1)$ limit displacement field in an *elliptic membrane shell*; *cf.* Section 8.

Membrane and flexural shells thus exhibit strikingly different limit behaviour!

After the original work of Ciarlet, Lods and Miara (1996) described in this section, the asymptotic analysis of linearly elastic flexural shells underwent several refinements and generalizations, which include another proof of Theorem 10.1 by means of Γ-convergence theory (Genevey 1999), an asymptotic analysis of linearly elastic flexural shells with variable thickness (Busse 1998) or made with a nonhomogeneous and anisotropic material (Giroud 1998), the convergence of the stresses and the explicit forms of the limit stresses (Collard and Miara 1999), and an asymptotic analysis of the associated eigenvalue problem (Kesavan and Sabu 2000) and time-dependent problem (Xiao Li-Ming 200x a).

11. Koiter's equations

Founding his approach on *a priori* assumptions of a *geometrical* and *mechanical* nature about the three-dimensional displacements and stresses when the thickness is 'small', W. T. Koiter proposed in the sixties a two-dimensional shell model that has quickly acquired widespread popularity within the computational mechanics community.

After briefly describing the genesis of these equations, which were en-countered in Section 4, we review in this section their main mathematical properties, such as the existence, uniqueness, and regularity of their solu-tion, or their formulation as a boundary value problem. We also show how they can be extended to shells whose middle surface has little regularity and we describe the closely related Budiansky–Sanders equations.

It is remarkable that Koiter's equations can be fully justified for all types of shells, even though it is clear that these equations cannot be recovered as the outcome of an asymptotic analysis of the three-dimensional equations, since Sections 8 to 10 have exhausted all such possible outcomes!

More specifically, we also show in this section that, for each category of linearly elastic shells (elliptic membrane, generalized membrane, or flexural), the solution of Koiter's equation and the average through the thickness of the three-dimensional solution have the same asymptotic behaviour in *ad hoc* function spaces as $\varepsilon \to 0$.

So, even though Koiter's linear model is not a limit model, it is in this sense the 'best' two-dimensional one for linearly elastic shells!

11.1. *Genesis; existence, uniqueness, and regularity of solutions; formulation as a boundary value problem*

Let ω be a domain in $\mathbb{R}^2$ with boundary γ, let $\boldsymbol{\theta} \in \mathcal{C}^3(\overline{\omega}; \mathbb{R}^3)$ be an injective mapping such that the two vectors $\boldsymbol{a}_\alpha = \partial_\alpha \boldsymbol{\theta}$ are linearly independent at all points of $\overline{\omega}$, and let γ_0 be a portion of γ that satisfies length $\gamma_0 > 0$.

Consider as in the previous sections a linearly elastic shell with middle surface $S = \boldsymbol{\theta}(\overline{\omega})$ and thickness $2\varepsilon > 0$, that is, a linearly elastic body whose reference configuration is the set $\boldsymbol{\Theta}(\overline{\Omega}^\varepsilon)$, where

$$\Omega^\varepsilon := \omega \times] - \varepsilon, \varepsilon[,$$
$$\boldsymbol{\Theta}(y, x_3^\varepsilon) := \boldsymbol{\theta}(y) + x_3^\varepsilon \boldsymbol{a}_3(y) \text{ for all } (y, x_3^\varepsilon) \in \overline{\Omega}^\varepsilon.$$

The material constituting the shell is *homogeneous* and *isotropic* and the reference configuration is a *natural state*, so that the material is characterized by its two Lamé constants $\lambda^\varepsilon > 0$ and $\mu^\varepsilon > 0$. The shell is subjected to a *boundary condition of place* along the portion $\boldsymbol{\Theta}(\Gamma_0^\varepsilon)$ of its lateral face, where $\Gamma_0^\varepsilon := \gamma_0 \times [-\varepsilon, \varepsilon]$, that is, the three-dimensional displacement vanishes on $\boldsymbol{\Theta}(\Gamma_0^\varepsilon)$. Finally, the shell is subjected to *applied body forces* in its interior $\boldsymbol{\Theta}(\Omega^\varepsilon)$, their densities being given by their contravariant components $f^{i,\varepsilon} \in L^2(\Omega^\varepsilon)$.

In a seminal work, John (1965, 1971) showed that, if the thickness of such a shell is small enough, the state of stress is 'approximately planar' and the stresses parallel to the middle surface vary 'approximately linearly' across the thickness, at least 'away from the lateral face'. In Koiter's approach (Koiter 1960, 1966, 1970), these approximations are taken as an *a priori* assumption of a *mechanical* nature and combined with another *a priori* assumption of a *geometrical* nature, called the Kirchhoff–Love assumption: any point on a normal to the middle surface remains on the normal to the deformed middle surface after the deformation has taken place and the distance between such a point and the middle surface remains constant. In fact, this assumption is required to hold only 'to within the first order' in the linearized theory considered in this section.

Taking these two *a priori* assumptions into account, Koiter then shows that the displacement field across the thickness of the shell can be completely determined from the sole knowledge of the displacement field of the middle surface S, and he identifies the two-dimensional problem, that is, posed over the two-dimensional set $\bar{\omega}$, that this displacement field should satisfy. As in the two-dimensional theories encountered so far, the unknown is a vector field, now denoted by $\boldsymbol{\zeta}_K^\varepsilon = (\zeta_{i,K}^\varepsilon) : \bar{\omega} \to \mathbb{R}^3$, whose components $\zeta_{i,K}^\varepsilon : \bar{\omega} \to \mathbb{R}$ are the covariant components of the displacement field of the middle surface S. This means that $\zeta_{i,K}^\varepsilon(y)\boldsymbol{a}^i(y)$ is the displacement of the point $\boldsymbol{\theta}(y)$; see Figure 11.1.

In their linearized version, the equations found by Koiter consist in solving the following variational problem $\mathcal{P}_K^\varepsilon(\omega)$:

Find $\boldsymbol{\zeta}_K^\varepsilon = (\zeta_{K,i}^\varepsilon)$ such that

$$\boldsymbol{\zeta}_K^\varepsilon \in \mathbf{V}_K(\omega) := \{\boldsymbol{\eta} = (\eta_i) \in H^1(\omega) \times H^1(\omega) \times H^2(\omega) :$$
$$\eta_i = \partial_\nu \eta_3 = 0 \text{ on } \gamma_0\},$$

$$\int_\omega \left\{ \varepsilon a^{\alpha\beta\sigma\tau,\varepsilon} \gamma_{\sigma\tau}(\boldsymbol{\zeta}_K^\varepsilon)\gamma_{\alpha\beta}(\boldsymbol{\eta}) + \frac{\varepsilon^3}{3} a^{\alpha\beta\sigma\tau,\varepsilon} \rho_{\sigma\tau}(\boldsymbol{\zeta}_K^\varepsilon)\rho_{\alpha\beta}(\boldsymbol{\eta}) \right\} \sqrt{a}\, dy$$
$$= \int_\omega p^{i,\varepsilon}\eta_i \sqrt{a}\, dy$$

for all $\boldsymbol{\eta} = (\eta_i) \in \mathbf{V}_K(\omega)$, where

$$a^{\alpha\beta\sigma\tau,\varepsilon} := \frac{4\lambda^\varepsilon \mu^\varepsilon}{\lambda^\varepsilon + 2\mu^\varepsilon} a^{\alpha\beta} a^{\sigma\tau} + 2\mu^\varepsilon (a^{\alpha\sigma} a^{\beta\tau} + a^{\alpha\tau} a^{\beta\sigma}),$$

$$\gamma_{\alpha\beta}(\boldsymbol{\eta}) := \frac{1}{2}(\partial_\beta \eta_\alpha + \partial_\alpha \eta_\beta) - \Gamma_{\alpha\beta}^\sigma \eta_\sigma - b_{\alpha\beta}\eta_3,$$

$$\rho_{\alpha\beta}(\boldsymbol{\eta}) := \partial_{\alpha\beta}\eta_3 - \Gamma_{\alpha\beta}^\sigma \partial_\sigma \eta_3 - b_\alpha^\sigma b_{\sigma\beta}\eta_3$$
$$+ b_\alpha^\sigma(\partial_\beta \eta_\sigma - \Gamma_{\beta\sigma}^\tau \eta_\tau) + b_\beta^\tau(\partial_\alpha \eta_\tau - \Gamma_{\alpha\tau}^\sigma \eta_\sigma)$$
$$+ (\partial_\alpha b_\beta^\tau + \Gamma_{\alpha\sigma}^\tau b_\beta^\sigma - \Gamma_{\alpha\beta}^\sigma b_\sigma^\tau)\eta_\tau,$$

$$p^{i,\varepsilon} := \int_{-\varepsilon}^\varepsilon f^{i,\varepsilon}\, dx_3^\varepsilon$$

(the functions $a^{\alpha\beta}, b_{\alpha\beta}, b_\alpha^\sigma, \Gamma_{\alpha\beta}^\sigma$, and a defined as usual: see Section 4).

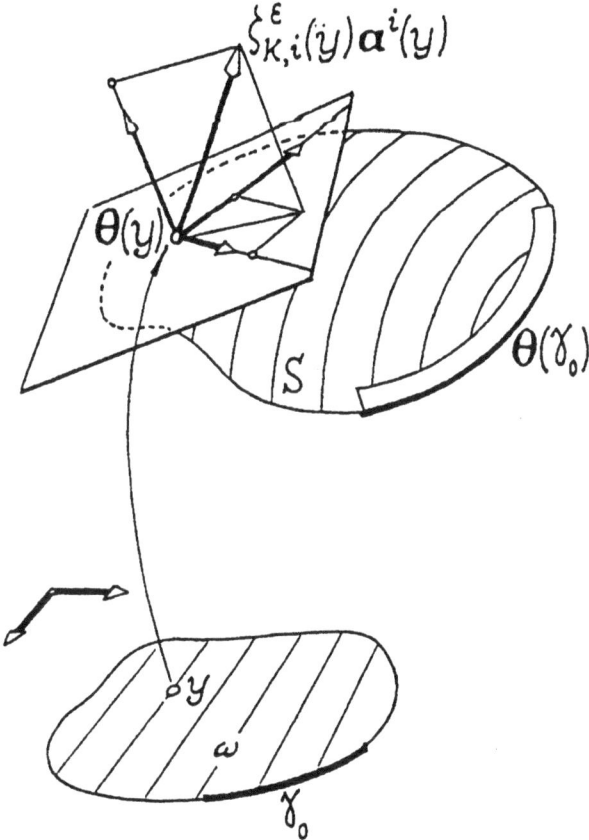

Fig. 11.1. The three unknowns in Koiter's equations are the covariant components $\zeta_{i,K}^{\varepsilon} : \overline{\omega} \to \mathbb{R}$ of the displacement field $\zeta_{i,K}^{\varepsilon} a^i : \overline{\omega} \to \mathbb{R}^3$ of the middle surface S; this means that, for each $y \in \overline{\omega}$, $\zeta_{i,K}^{\varepsilon}(y)a^i(y)$ is the displacement of the point $\theta(y) \in S$

The functions $\gamma_{\alpha\beta}(\boldsymbol{\eta})$ and $\rho_{\alpha\beta}(\boldsymbol{\eta})$ are the customary covariant components of the *linearized change of metric* and *linearized change of curvature tensors* associated with a displacement field $\eta_i a^i$ of the middle surface S and the functions $a^{\alpha\beta\sigma\tau,\varepsilon}$ are the customary contravariant components of the *two-dimensional elasticity tensor of the shell*.

Note that Destuynder (1985, 1990) has found an illuminating way of deriving the same linear Koiter equations from three-dimensional elasticity, which uses *a priori* assumptions only of a geometrical nature. Note also that the linearized Kirchhoff–Love assumption has been *a posteriori* justified for linearly elastic elliptic membrane shells by Lods and Mardare (1998*b*, 2000*b*).

The existence and uniqueness of a solution to problem $\mathcal{P}_K^\varepsilon(\omega)$, which essentially follow from the $\mathbf{V}_K(\omega)$-ellipticity of the bilinear form, was first established by Bernadou and Ciarlet (1976); a more natural proof was subsequently proposed by Ciarlet and Miara (1992b), then combined with the first one in Bernadou, Ciarlet and Miara (1994). The existence and uniqueness of the solution to the time-dependent Koiter equations have recently been established by Xiao Li-Ming (1999).

Theorem 11.1: Existence and uniqueness of solutions. Let ω be a domain in $\mathbb{R}^2$, let γ_0 be a subset of $\gamma = \partial\omega$ with length $\gamma_0 > 0$, and let $\boldsymbol{\theta} \in \mathcal{C}^3(\overline{\omega}; \mathbb{R}^3)$ be an injective mapping such that the two vectors $\mathbf{a}_\alpha = \partial_\alpha\boldsymbol{\theta}$ are linearly independent at all points of $\overline{\omega}$.

Then the variational problem $\mathcal{P}_K^\varepsilon(\omega)$ has exactly one solution, which is also the unique solution to the minimization problem:
Find $\boldsymbol{\zeta}_K^\varepsilon = (\zeta_{K,i}^\varepsilon)$ such that

$$\boldsymbol{\zeta}_K^\varepsilon \in \mathbf{V}_K(\omega) \text{ and } j_K^\varepsilon(\boldsymbol{\zeta}_K^\varepsilon) = \inf_{\boldsymbol{\eta} \in \mathbf{V}_K(\omega)} j_K^\varepsilon(\boldsymbol{\eta}), \text{ where}$$

$$j_K^\varepsilon(\boldsymbol{\eta}) := \frac{1}{2}\int_\omega \left\{ \varepsilon a^{\alpha\beta\sigma\tau,\varepsilon}\gamma_{\sigma\tau}(\boldsymbol{\eta})\gamma_{\alpha\beta}(\boldsymbol{\eta}) \right.$$
$$\left. + \frac{\varepsilon^3}{3}a^{\alpha\beta\sigma\tau,\varepsilon}\rho_{\sigma\tau}(\boldsymbol{\eta})\rho_{\alpha\beta}(\boldsymbol{\eta}) \right\}\sqrt{a}\,dy - \int_\omega p^{i,\varepsilon}\eta_i\sqrt{a}\,dy.$$

Proof. The assumptions $f^{i,\varepsilon} \in L^2(\Omega^\varepsilon)$ imply that $p^{i,\varepsilon} \in L^2(\omega)$. The existence and uniqueness of a solution to the variational problem $\mathcal{P}_K^\varepsilon(\omega)$, or to its equivalent minimization problem, are consequences of the inequality of Korn's type on a general surface (Theorem 4.4), of the existence of a constant c_e such that

$$\sum|t_{\alpha\beta}|^2 \le c_e a^{\alpha\beta\sigma\tau}(y)t_{\sigma\tau}t_{\alpha\beta}$$

for all $y \in \overline{\omega}$ and all symmetric matrices $(t_{\alpha\beta})$, of the existence of a_0 such that $a(y) \ge a_0 > 0$ for all $y \in \overline{\omega}$, and of the Lax–Milgram lemma. □

We next derive the boundary value problem that is (at least formally) equivalent to Koiter's variational problem $\mathcal{P}_K^\varepsilon(\omega)$. We also state a regularity result that provides instances where the *weak solution* (the solution to the variational problem) becomes a *classical solution* (a solution to the boundary value problem).

Theorem 11.2: Regularity of solutions; formulation as a boundary value problem. (a) Assume that the boundary γ of ω and the functions $p^{i,\varepsilon}$ are sufficiently smooth. Then, if the solution $\boldsymbol{\zeta}_K^\varepsilon = (\zeta_{K,i}^\varepsilon)$ to the variational problem $\mathcal{P}_K^\varepsilon(\omega)$ (Theorem 11.1) is sufficiently smooth, it is also a solution to the following boundary value problem:

$$m^{\alpha\beta,\varepsilon}|_{\alpha\beta} - b_\alpha^\sigma b_{\sigma\beta}m^{\alpha\beta,\varepsilon} - b_{\alpha\beta}n^{\alpha\beta,\varepsilon} = p^{3,\varepsilon} \text{ in } \omega,$$

$$-(n^{\alpha\beta,\varepsilon} + b_\sigma^\alpha m^{\sigma\beta,\varepsilon})|_\beta - b_\sigma^\alpha(m^{\sigma\beta,\varepsilon}|_\beta) = p^{\alpha,\varepsilon} \text{ in } \omega,$$

$$\zeta_{i,K}^\varepsilon = \partial_\nu \zeta_{3,K}^\varepsilon = 0 \text{ on } \gamma_0,$$

$$m^{\alpha\beta,\varepsilon}\nu_\alpha\nu_\beta = 0 \text{ on } \gamma_1,$$

$$(m^{\alpha\beta,\varepsilon}|_\alpha)\nu_\beta + \partial_\tau(m^{\alpha\beta,\varepsilon}\nu_\alpha\tau_\beta) = 0 \text{ on } \gamma_1,$$

$$(n^{\alpha\beta,\varepsilon} + 2b_\sigma^\alpha m^{\sigma\beta,\varepsilon})\nu_\beta = 0 \text{ on } \gamma_1,$$

where $\gamma_1 := \gamma - \gamma_0, (\nu_\alpha)$ is the unit outer normal vector along γ, $\tau_1 := -\nu_2, \tau_2 := \nu_1, \partial_\tau\theta := \tau_\alpha\partial_\alpha\theta$ denotes the tangential derivative of θ in the direction of the vector (τ_α),

$$n^{\alpha\beta,\varepsilon} := \varepsilon a^{\alpha\beta\sigma\tau,\varepsilon}\gamma_{\sigma\tau}(\zeta_K^\varepsilon), \quad m^{\alpha\beta,\varepsilon} := \frac{\varepsilon^3}{3}a^{\alpha\beta\sigma\tau,\varepsilon}\rho_{\sigma\tau}(\zeta_K^\varepsilon),$$

and finally, for an arbitrary tensor with twice differentiable covariant components $n^{\alpha\beta}$,

$$n^{\alpha\beta}|_\beta := \partial_\beta n^{\alpha\beta} + \Gamma_{\beta\sigma}^\alpha n^{\beta\sigma} + \Gamma_{\beta\sigma}^\beta n^{\alpha\sigma},$$

$$n^{\alpha\beta}|_{\alpha\beta} := \partial_\alpha(n^{\alpha\beta}|_\beta) + \Gamma_{\alpha\sigma}^\sigma(n^{\alpha\beta}|_\beta).$$

(b) Assume that $\gamma = \gamma_0$ and that, for some integer $m \geq 0$ and some real number $q > 1$, γ is of class C^{m+4}, $\theta \in C^{m+4}(\overline{\omega}; \mathbb{R}^3)$, $p^{\alpha,\varepsilon} \in W^{m+1,q}(\omega)$, and $p^{3,\varepsilon} \in W^{m,q}(\omega)$. Then

$$\zeta_K^\varepsilon = (\zeta_i^\varepsilon) \in W^{m+3,q}(\omega) \times W^{m+3,q}(\omega) \times W^{m+4,q}(\omega).$$

Proof. For brevity, we give the proof of (a) when $\gamma_0 = \gamma$, in which case

$$\mathbf{V}_K(\omega) = H_0^1(\omega) \times H_0^1(\omega) \times H_0^2(\omega),$$

and we omit the exponents 'ε' and the indices 'K' throughout the proof, that is, we let

$$\zeta := \zeta_K^\varepsilon, \quad n^{\alpha\beta} := a^{\alpha\beta\sigma\tau}\gamma_{\sigma\tau}(\zeta), \quad m^{\alpha\beta} := \frac{1}{3}a^{\alpha\beta\sigma\tau}\rho_{\sigma\tau}(\zeta), \quad p^i := p^{i,\varepsilon}.$$

Assume that the solution ζ is smooth in the sense that $n^{\alpha\beta} \in H^1(\omega)$ and $m^{\alpha\beta} \in H^2(\omega)$.

We have already seen in the proof of Theorem 8.2 that

$$\int_\omega a^{\alpha\beta\sigma\tau}\gamma_{\sigma\tau}(\zeta)\gamma_{\alpha\beta}(\eta)\sqrt{a}\,dy = -\int_\omega \sqrt{a}\{(n^{\alpha\beta}|_\beta)\eta_\alpha + b_{\alpha\beta}n^{\alpha\beta}\eta_3\}\,dy$$

for all $\eta = (\eta_i) \in H_0^1(\omega) \times H_0^1(\omega) \times L^2(\omega)$, hence *a fortiori* for all $\eta \in H_0^1(\omega) \times H_0^1(\omega) \times H_0^2(\omega)$. It thus remains to transform the other integral appearing on the left-hand side of the variational equations, that is,

$$\frac{1}{3}\int_\omega a^{\alpha\beta\sigma\tau}\rho_{\sigma\tau}(\zeta)\rho_{\alpha\beta}(\eta)\sqrt{a}\,dy = \int_\omega m^{\alpha\beta}\rho_{\alpha\beta}(\eta)\sqrt{a}\,dy$$

$$= \int_\omega \sqrt{a}\,m^{\alpha\beta}\partial_{\alpha\beta}\eta_3\,dy$$

$$+ \int_\omega \sqrt{a}\,m^{\alpha\beta}(2b_\alpha^\sigma\partial_\beta\eta_\sigma - \Gamma_{\alpha\beta}^\sigma\partial_\sigma\eta_3)\,dy$$

$$+ \int_\omega \sqrt{a}\,m^{\alpha\beta}(-2b_\beta^\tau\Gamma_{\alpha\tau}^\sigma\eta_\sigma + b_\beta^\sigma|_\alpha\eta_\sigma - b_\alpha^\sigma b_{\sigma\beta}\eta_3)\,dy,$$

where $\eta = (\eta_i) \in H_0^1(\omega)\times H_0^1(\omega)\times H_0^2(\omega)$. Using the symmetry $m^{\alpha\beta}=m^{\beta\alpha}$, the relation $\partial_\beta\sqrt{a} = \sqrt{a}\,\Gamma_{\beta\sigma}^\sigma$, and Green's formula in Sobolev space, we obtain

$$\int_\omega m^{\alpha\beta}\rho_{\alpha\beta}(\eta)\sqrt{a}\,dy = -\int_\omega \sqrt{a}(\partial_\beta m^{\alpha\beta} + \Gamma_{\beta\sigma}^\sigma m^{\alpha\beta} + \Gamma_{\sigma\beta}^\alpha m^{\sigma\beta})\partial_\alpha\eta_3\,dy$$

$$+ 2\int_\omega \sqrt{a}\,m^{\alpha\beta}b_\alpha^\sigma\partial_\beta\eta_\sigma\,dy$$

$$+ \int_\omega \sqrt{a}\,m^{\alpha\beta}(-2b_\beta^\tau\Gamma_{\alpha\tau}^\sigma\eta_\sigma + b_\beta^\sigma|_\alpha\eta_\sigma - b_\alpha^\sigma b_{\sigma\beta}\eta_3)\,dy.$$

The same Green's formula shows that

$$-\int_\omega \sqrt{a}(\partial_\beta m^{\alpha\beta} + \Gamma_{\beta\sigma}^\sigma m^{\alpha\beta} + \Gamma_{\sigma\beta}^\alpha m^{\sigma\beta})\partial_\alpha\eta_3\,dy$$

$$= -\int_\omega \sqrt{a}(m^{\alpha\beta}|_\beta)\partial_\alpha\eta_3\,dy = \int_\omega \partial_\alpha(\sqrt{a}\,m^{\alpha\beta}|_\beta)\eta_3\,dy$$

$$= \int_\omega \sqrt{a}(m^{\alpha\beta}|_{\alpha\beta})\eta_3\,dy,$$

$$2\int_\omega \sqrt{a}\,m^{\alpha\beta}b_\alpha^\sigma\partial_\beta\eta_\sigma\,dy = -2\int_\omega \sqrt{a}\{\partial_\beta(b_\alpha^\sigma m^{\alpha\beta}) + \Gamma_{\beta\tau}^\tau b_\alpha^\sigma m^{\alpha\beta}\}\eta_\sigma\,dy.$$

Consequently,

$$\int_\omega m^{\alpha\beta}\rho_{\alpha\beta}(\eta)\sqrt{a}\,dy = \int_\omega \sqrt{a}\{-2(b_\sigma^\alpha m^{\sigma\beta})|_\beta + (b_\beta^\alpha|_\sigma)m^{\sigma\beta}\}\eta_\alpha\,dy$$

$$+ \int_\omega \sqrt{a}\{m^{\alpha\beta}|_{\alpha\beta} - b_\alpha^\sigma b_{\sigma\beta}m^{\alpha\beta}\}\eta_3\,dy.$$

Using in this relation the easily verified formula

$$(b_\sigma^\alpha m^{\sigma\beta})|_\beta = (b_\beta^\alpha|_\sigma)m^{\sigma\beta} + b_\sigma^\alpha(m^{\sigma\beta}|_\beta)$$

and the symmetry $b^{\alpha}_{\beta}|_{\sigma} = b^{\alpha}_{\sigma}|_{\beta}$, we finally obtain

$$\int_{\omega} m^{\alpha\beta}\rho_{\alpha\beta}(\boldsymbol{\eta})\sqrt{a}\,dy = -\int_{\omega} \sqrt{a}\{(b^{\alpha}_{\sigma}m^{\sigma\beta})|_{\beta} + b^{\alpha}_{\sigma}(m^{\sigma\beta}|_{\beta})\}\eta_{\alpha}\,dy$$

$$-\int_{\omega} \sqrt{a}\{b^{\sigma}_{\alpha}b_{\sigma\beta}m^{\alpha\beta} - m^{\alpha\beta}|_{\alpha\beta}\}\eta_3\,dy.$$

Hence the variational equations

$$\int_{\omega} \left\{ a^{\alpha\beta\sigma\tau}\gamma_{\sigma\tau}(\boldsymbol{\zeta})\gamma_{\alpha\beta}(\boldsymbol{\eta}) + \frac{1}{3}a^{\alpha\beta\sigma\tau}\rho_{\sigma\tau}(\boldsymbol{\zeta})\rho_{\alpha\beta}(\boldsymbol{\eta}) - p^i\eta_i \right\}\sqrt{a}\,dy = 0$$

imply that

$$\int_{\omega} \sqrt{a}\{(n^{\alpha\beta} + b^{\alpha}_{\sigma}m^{\sigma\beta})|_{\beta} + b^{\alpha}_{\sigma}(m^{\sigma\beta}|_{\beta}) + p^{\alpha}\}\eta_{\alpha}\,dy$$

$$+\int_{\omega} \sqrt{a}\{b_{\alpha\beta}n^{\alpha\beta} + b^{\sigma}_{\alpha}b_{\sigma\beta}m^{\alpha\beta} - m^{\alpha\beta}|_{\alpha\beta} + p^3\}\eta_3\,dy = 0$$

for all $(\eta_i) \in H^1_0(\omega) \times H^1_0(\omega) \times H^2_0(\omega)$. The stated partial differential equations are thus satisfied in ω.

The regularity result of part (b) is due to Alexandrescu (1994). $\qquad\square$

Note that the functions $n^{\alpha\beta}|_{\beta}$ and $m^{\alpha\beta}|_{\alpha\beta}$ appearing in the boundary value problem are instances of first-order and second-order covariant derivatives of tensor fields, defined here by means of their contravariant components $n^{\alpha\beta}$ or $m^{\alpha\beta}$. The covariant derivatives $n^{\alpha\beta}|_{\beta}$ also occurred in the boundary value problem associated with a linearly elastic elliptic membrane shell (Theorem 8.2).

Each one of the three formulations found in Theorems 11.1 and 11.2 constitutes the two-dimensional Koiter equations for a linearly elastic shell. We recall that the functions $\gamma_{\alpha\beta}(\boldsymbol{\eta})$ and $\rho_{\alpha\beta}(\boldsymbol{\eta})$ are the covariant components of the linearized change of metric and change of curvature tensors associated with a displacement field $\eta_i a^i$ of the middle surface S, the functions $a^{\alpha\beta\sigma\tau,\varepsilon}$ are the contravariant components of the two-dimensional elasticity tensor of the shell. The functions $n^{\alpha\beta,\varepsilon}$ and $m^{\alpha\beta,\varepsilon}$ are the contravariant components of the *stress resultant* and *stress couple*, or *bending moment, tensor* fields.

As shown at the end of Section 10, the 'two-dimensional boundary conditions of clamping' $\zeta^{\varepsilon}_{K,i} = \partial_{\nu}\zeta^{\varepsilon}_{K,3} = 0$ on γ_0 state that the points of, and the tangent spaces to, the deformed and undeformed middle surfaces coincide along the set $\boldsymbol{\theta}(\gamma_0)$, as suggested in Figure 11.1.

The functional $j^{\varepsilon}_K : \mathbf{V}_K(\omega) \to \mathbb{R}$ in Theorem 11.1 is the two-dimensional Koiter energy of a linearly elastic shell. The associated Koiter strain energy,

$$\boldsymbol{\eta} \in \mathbf{V}_K(\omega) \to \frac{1}{2}\int_{\omega} \left\{ \varepsilon a^{\alpha\beta\sigma\tau,\varepsilon}\gamma_{\sigma\tau}(\boldsymbol{\eta})\gamma_{\alpha\beta}(\boldsymbol{\eta}) \right.$$

$$\left. + \frac{\varepsilon^3}{3}a^{\alpha\beta\sigma\tau,\varepsilon}\rho_{\sigma\tau}(\boldsymbol{\eta})\rho_{\alpha\beta}(\boldsymbol{\eta}) \right\}\sqrt{a}\,dy,$$

is thus the *sum* of the strain energies of a linearly elastic elliptic membrane shell (Section 8) and of a linearly elastic flexural shell (Section 10).

Finally, note that the partial differential equations in ω together with the boundary conditions on γ_1 found in Theorem 11.2 may be viewed as two-dimensional equations of equilibrium, while the equations relating the unknown ζ_K^ε and the functions $n^{\alpha\beta,\varepsilon}$ and $m^{\alpha\beta,\varepsilon}$ may be viewed as two-dimensional constitutive equations.

11.2. Justification of Koiter's equations for all types of shells

When it is viewed as a three-dimensional body, the linearly elastic shell described at the beginning of this section is modelled by the variational problem $\mathcal{P}(\Omega^\varepsilon)$ that constituted the point of departure of the asymptotic analyses of Sections 8 to 10. This problem, described in Section 7.1, consists in finding $\boldsymbol{u}^\varepsilon = (u_i^\varepsilon)$ such that

$$\boldsymbol{u}^\varepsilon \in \mathbf{V}(\Omega^\varepsilon) = \{\boldsymbol{v}^\varepsilon = (v_i^\varepsilon) \in \mathbf{H}^1(\Omega^\varepsilon) : \boldsymbol{v}^\varepsilon = \boldsymbol{0} \text{ on } \Gamma_0^\varepsilon\},$$

$$\int_{\Omega^\varepsilon} A^{ijkl,\varepsilon} e^\varepsilon_{k\|l}(\boldsymbol{u}^\varepsilon) e^\varepsilon_{i\|j}(\boldsymbol{v}^\varepsilon) \sqrt{g^\varepsilon}\, \mathrm{d}x^\varepsilon = \int_{\Omega^\varepsilon} f^{i,\varepsilon} v_i^\varepsilon \sqrt{g^\varepsilon}\, \mathrm{d}x^\varepsilon$$

for all $\boldsymbol{v}^\varepsilon \in \mathbf{V}(\Omega^\varepsilon)$, where

$$A^{ijkl,\varepsilon} := \lambda^\varepsilon g^{ij,\varepsilon} g^{kl,\varepsilon} + \mu^\varepsilon (g^{ik,\varepsilon} g^{jl,\varepsilon} + g^{il,\varepsilon} g^{jk,\varepsilon}),$$

$$e^\varepsilon_{i\|j}(\boldsymbol{v}^\varepsilon) := \frac{1}{2}(\partial_j^\varepsilon v_i^\varepsilon + \partial_i^\varepsilon v_j^\varepsilon) - \Gamma_{ij}^{p,\varepsilon}(\boldsymbol{v}^\varepsilon)$$

(all notation not redefined here is defined in Section 7.1).

The unknown functions u_i^ε in problem $\mathcal{P}(\Omega^\varepsilon)$ represent the covariant components of the displacement field $u_i^\varepsilon \boldsymbol{g}^{i,\varepsilon}$ of the points of the reference configuration $\boldsymbol{\Theta}(\overline{\Omega}^\varepsilon)$; see Figure 7.1.

Now consider a family of such linearly elastic shells, with each having the same middle surface $S = \boldsymbol{\theta}(\overline{\omega})$, and with each subjected to a boundary condition of place along a portion of its lateral face having the same set $\boldsymbol{\theta}(\gamma_0)$ as its middle curve. All the linearly elastic shells in such a family are thus either *elliptic membrane*, or *generalized membrane*, or *flexural*, according to the definitions given in Sections 8, 9, and 10. Assume that the assumptions on the data are in each case those that guarantee the convergence of the scaled displacements as the thickness approaches zero (Theorems 8.1, 9.1, and 10.1).

It is then remarkable that, in each case, the asymptotic behaviour as $\varepsilon \to 0$ of the average $\frac{1}{2\varepsilon} \int_{-\varepsilon}^{\varepsilon} \boldsymbol{u}^\varepsilon \, \mathrm{d}x_3^\varepsilon$ of the solution to the three-dimensional variational problem $\mathcal{P}(\Omega^\varepsilon)$ and of the solution ζ_K^ε to the two-dimensional Koiter equations formulated as the variational problem $\mathcal{P}_K^\varepsilon(\omega)$ (Theorem 11.1) are identical.

To see this, we proceed as in Ciarlet and Lods (1996c, Theorems 2.1 and 2.2) and (1996d, Theorems 6.1 and 6.2). We compare the convergence theorems established in Sections 8, 9, and 10 with former results of Destuynder (1985), Sanchez-Palencia (1989a, 1989b, 1992), and Caillerie and Sanchez-Palencia (1995a) (see also Caillerie (1996)) about the asymptotic behaviour of the solution of Koiter's equation as ε approaches zero.

The forthcoming analyses have been recently extended by Xiao Li-Ming (200xb), who likewise justified the time-dependent Koiter equations for elliptic membrane and flexural shells.

To begin with, we consider *elliptic membrane shells*, as defined in Section 8.1.

Theorem 11.3: Justification of Koiter's equations for 'elliptic membrane' shells. Assume that $\boldsymbol{\theta} \in \mathcal{C}^3(\overline{\omega}; \mathbb{R}^3)$. Consider a family of linearly elastic elliptic membrane shells, with thickness 2ε approaching zero and with each having the same elliptic middle surface $S = \boldsymbol{\theta}(\overline{\omega})$, and let the assumptions on the data be as in Theorem 8.1 (in particular, $\gamma_0 = \gamma$).

For each $\varepsilon > 0$ let

$$(u_i^\varepsilon) \in \mathbf{H}^1(\Omega^\varepsilon) \text{ and } \boldsymbol{\zeta}_K^\varepsilon = (\zeta_{i,K}^\varepsilon) \in H_0^1(\omega) \times H_0^1(\omega) \times H_0^2(\omega)$$

respectively denote the solutions to the three-dimensional and two-dimensional variational problems $\mathcal{P}(\Omega^\varepsilon)$ and $\mathcal{P}_K^\varepsilon(\omega)$. Also, let

$$\boldsymbol{\zeta} = (\zeta_i) \in H_0^1(\omega) \times H_0^1(\omega) \times L^2(\omega)$$

denote the solution to the two-dimensional 'scaled' variational problem $\mathcal{P}_M(\omega)$ (Theorem 8.2), a solution which is thus independent of ε. Then

$$\frac{1}{2\varepsilon} \int_{-\varepsilon}^\varepsilon u_\alpha^\varepsilon \, dx_3^\varepsilon \to \zeta_\alpha \text{ in } H^1(\omega) \text{ and } \frac{1}{2\varepsilon} \int_{-\varepsilon}^\varepsilon u_3^\varepsilon \, dx_3^\varepsilon \to \zeta_3 \text{ in } L^2(\omega),$$

$$\zeta_{K,\alpha}^\varepsilon \to \zeta_\alpha \text{ in } H^1(\omega) \text{ and } \zeta_{K,3}^\varepsilon \to \zeta_3 \text{ in } L^2(\omega).$$

Proof. Under the assumptions that there exist constants $\lambda > 0$ and $\mu > 0$ and functions $f^i \in L^2(\Omega)$ independent of ε such that

$$\lambda^\varepsilon = \lambda \quad \text{and} \quad \mu^\varepsilon = \mu,$$
$$f^{i,\varepsilon}(x^\varepsilon) = f^i(x) \quad \text{for all} \quad x^\varepsilon = \pi^\varepsilon x \in \Omega^\varepsilon$$

(these are the assumptions on the data for a family of linearly elastic elliptic membrane shells) and that $\boldsymbol{\theta} \in \mathcal{C}^3(\overline{\omega}; \mathbb{R}^3)$, then

$$\frac{1}{2\varepsilon} \int_{-\varepsilon}^\varepsilon u_\alpha^\varepsilon \, dx_3^\varepsilon = \frac{1}{2} \int_{-1}^1 u_\alpha(\varepsilon) \, dx_3 \to \zeta_\alpha \text{ in } H^1(\omega)$$

and

$$\frac{1}{2\varepsilon} \int_{-\varepsilon}^\varepsilon u_3^\varepsilon \, dx_3^\varepsilon = \frac{1}{2} \int_{-1}^1 u_3(\varepsilon) \, dx_3 \to \zeta_3 \text{ in } L^2(\omega)$$

as $\varepsilon \to 0$ are easy corollaries to the fundamental convergence result of Theorem 8.1.

The convergence $\zeta_K^\varepsilon \to \zeta$ in $H^1(\omega) \times H^1(\omega) \times L^2(\omega)$ was first established by Destuynder (1985, Theorem 7.1); it was also noted by Sanchez-Palencia (1989a, Theorem 4.1) (see also Caillerie and Sanchez-Palencia (1995a)), who observed that it is a consequence of general results in perturbation theory, as found for instance in Sanchez-Palencia (1980). We give here a simple and self-contained proof. Let

$$a^{\alpha\beta\sigma\tau} := \frac{2\lambda\mu}{\lambda + 2\mu} a^{\alpha\beta} a^{\sigma\tau} + 2\mu(a^{\alpha\sigma} a^{\beta\tau} + a^{\alpha\tau} a^{\beta\sigma}),$$

$$B_M(\zeta, \eta) := \int_\omega a^{\alpha\beta\sigma\tau} \gamma_{\sigma\tau}(\zeta) \gamma_{\alpha\beta}(\eta) \sqrt{a}\, dy,$$

$$B_F(\zeta, \eta) := \frac{1}{3} \int_\omega a^{\alpha\beta\sigma\tau} \rho_{\sigma\tau}(\zeta) \rho_{\alpha\beta}(\eta) \sqrt{a}\, dy,$$

$$L(\eta) := \int_\omega p^i \eta_i \sqrt{a}\, dy, \quad \text{where } p^i := \int_{-1}^1 f^i\, dx_3$$

$$\mathbf{V}_M(\omega) := H_0^1(\omega) \times H_0^1(\omega) \times L^2(\omega),$$

$$\|\eta\|_{\mathbf{V}_M(\omega)} := \left\{ \sum_\alpha \|\eta_\alpha\|_{1,\omega}^2 + |\eta_3|_{0,\omega}^2 \right\}^{1/2}.$$

By virtue of the assumptions on the applied forces, the solution ζ_K^ε of the two-dimensional Koiter equations also satisfies the scaled Koiter equations for an elliptic membrane shell, namely,

$$B_M(\zeta_K^\varepsilon, \eta) + \varepsilon^2 B_F(\zeta_K^\varepsilon, \eta) = L(\eta) \text{ for all } \eta \in \mathbf{V}_K(\omega).$$

Recall that there exists a constant $c_e > 0$ such that

$$\sum_{\alpha,\beta} |t_{\alpha\beta}|^2 \le c_e a^{\alpha\beta\sigma\tau}(y) t_{\sigma\tau} t_{\alpha\beta}$$

for all $y \in \bar{\omega}$ and all symmetric matrices $(t_{\alpha\beta})$. Hence letting $\eta = \zeta_K^\varepsilon$ in these scaled equations and using the inequality of Korn's type on an elliptic surface (Theorem 6.3) shows that the family $(\zeta_K^\varepsilon)_{\varepsilon>0}$ is bounded in $\mathbf{V}_M(\omega)$ and that the families $(\varepsilon\rho_{\alpha\beta}(\zeta_K^\varepsilon))_{\varepsilon>0}$ are bounded in $L^2(\omega)$.

Consequently, there exists a subsequence, still denoted by $(\zeta_K^\varepsilon)_{\varepsilon>0}$ for convenience, and there exist $\zeta^* \in \mathbf{V}_M(\omega)$ and $\rho_{\alpha\beta}^{-1} \in L^2(\omega)$ such that

$$\zeta_K^\varepsilon \rightharpoonup \zeta^* \text{ in } \mathbf{V}_M(\omega) \text{ and } \varepsilon\rho_{\alpha\beta}(\zeta_K^\varepsilon) \rightharpoonup \rho_{\alpha\beta}^{-1} \text{ in } L^2(\omega)$$

(as usual weak convergence is denoted by $\rightharpoonup$).

Fix $\eta \in \mathbf{V}_K(\omega)$ in the scaled Koiter equations and let $\varepsilon \to 0$; then the above weak convergence yields $B_M(\zeta^*, \eta) = L(\eta)$. Since the space $\mathbf{V}_K(\omega)$ is dense in $\mathbf{V}_M(\omega)$, it follows that $B_M(\zeta^*, \eta) = L(\eta)$ for all $\eta \in \mathbf{V}_M(\omega)$.

Hence

$$\zeta^* = \zeta,$$

where $\zeta \in \mathbf{V}_M(\omega)$ is the unique solution to problem $\mathcal{P}_M(\omega)$ (Theorem 8.2). Furthermore, the weak convergence

$$\zeta_K^\varepsilon \rightharpoonup \zeta \text{ in } \mathbf{V}_M(\omega)$$

holds for the whole family $(\zeta_K^\varepsilon)_{\varepsilon>0}$.

By the inequality of Korn's type on an elliptic surface, establishing the *strong* convergence $\zeta_K^\varepsilon \to \zeta$ in $\mathbf{V}_M(\omega)$ is equivalent to establishing the convergence

$$B_M(\zeta_K^\varepsilon - \zeta, \zeta_K^\varepsilon - \zeta) \to 0,$$

which itself easily follows by letting $\eta = \zeta_K^\varepsilon$ in the scaled Koiter equations, by noting that $B_M(\zeta_K^\varepsilon, \zeta_K^\varepsilon) \le L(\zeta_K^\varepsilon)$, and by using the weak convergence $\zeta_K^\varepsilon \rightharpoonup \zeta$ in $\mathbf{V}_M(\omega)$. $\qquad\square$

Note that the convergence results of Theorem 11.3 have been improved by Lods and Mardare (1998b, 2000b), who showed that

$$\left\| \frac{1}{2\varepsilon} \int_{-\varepsilon}^{\varepsilon} \mathbf{u}^\varepsilon \, dx_3^\varepsilon - \zeta_K^\varepsilon \right\|_{H^1(\omega) \times H^1(\omega) \times L^2(\omega)} = O(\varepsilon^{1/5}),$$

and by Mardare (1998b, Theorem 5.1), who showed that

$$\|\zeta_K^\varepsilon - \zeta\|_{H^1(\omega) \times H^1(\omega) \times L^2(\omega)} = O(\varepsilon^{1/5}).$$

Under the assumptions of Theorem 11.3, the function $\zeta_{3,K}^\varepsilon$ thus 'loses its boundary condition' as ε approaches zero. We have already remarked in Section 8.3 that, under the same assumptions, a similar 'loss of boundary condition' is shared by the average $\frac{1}{2\varepsilon} \int_{-\varepsilon}^{\varepsilon} u_3^\varepsilon \, dx_3^\varepsilon$ as ε approaches zero.

We next consider *generalized membrane shells*, as defined in Section 9.1.

In the same way that in Section 9.2 we required the applied forces to be 'admissible' in order to carry out (in Theorem 9.1) the asymptotic analysis of the three-dimensional solutions, we need to assume that the applied forces enter Koiter's equations in such a way that the corresponding (scaled) linear forms are continuous with respect to the norm $|\cdot|_\omega^M$ of the 'limit' space $\mathbf{V}_M^\sharp(\omega)$, and uniformly so with respect to ε.

More specifically, we set the following definition, after Ciarlet and Lods (1996d) (notice the analogy with that given in Section 9.2). Applied forces are admissible for the two-dimensional Koiter equations if there exist functions $\varphi^{\alpha\beta} = \varphi^{\beta\alpha} \in L^2(\omega)$ such that, for each $\varepsilon > 0$, the right-hand side in Koiter's equations can also be written as

$$\int_\omega p^{i,\varepsilon} \eta_i \sqrt{a} \, dy = \varepsilon \int_\omega \varphi^{\alpha\beta} \gamma_{\alpha\beta}(\eta) \sqrt{a} \, dy \text{ for all } \eta = (\eta_i) \in \mathbf{V}_K(\omega).$$

As in Section 9.3, we let

$$\mathbf{V}_M^\sharp(\omega) := \text{completion of } \mathbf{V}(\omega) \text{ with respect to } |\cdot|_\omega^M,$$

where

$$\mathbf{V}(\omega) := \{\boldsymbol{\eta} = (\eta_i) \in \mathbf{H}^1(\omega) : \boldsymbol{\eta} = \mathbf{0} \text{ on } \gamma_0\},$$

$$|\boldsymbol{\eta}|_\omega^M := \left\{ \sum_{\alpha,\beta} |\gamma_{\alpha\beta}(\boldsymbol{\eta})|_{0,\omega}^2 \right\}^{1/2}.$$

As in Section 9.3, we restrict ourselves to generalized membrane shells 'of the first kind', since we have already noted that there is no loss of generality in doing so.

Theorem 11.4: Justification of Koiter's equations for 'generalized membrane' shells. Assume that $\boldsymbol{\theta} \in \mathcal{C}^3(\overline{\omega}; \mathbb{R}^3)$. Consider a family of linearly elastic generalized membrane shells of the first kind, with thickness 2ε approaching zero, with each having the same middle surface $S = \boldsymbol{\theta}(\overline{\omega})$, with each subjected to a boundary condition of place along a portion of its lateral face having the same set $\boldsymbol{\theta}(\gamma_0)$ as its middle curve, and subjected to applied forces that are admissible for both the three-dimensional equations (Section 9.2) and the two-dimensional Koiter equations, the functions $\varphi^{\alpha\beta} \in L^2(\omega)$ coinciding in addition with those found in Theorem 9.1.

For each $\varepsilon > 0$, let

$$\boldsymbol{u}^\varepsilon \in \mathbf{H}^1(\Omega^\varepsilon) \quad \text{and} \quad \boldsymbol{\zeta}_K^\varepsilon \in H^1(\omega) \times H^1(\omega) \times H^2(\omega),$$

respectively, denote the solutions to the three-dimensional and two-dimensional variational problems $\mathcal{P}(\Omega^\varepsilon)$ and $\mathcal{P}_K^\varepsilon(\omega)$. Let

$$\boldsymbol{\zeta} \in \mathbf{V}_M^\sharp(\omega)$$

denote the solution to the two-dimensional 'scaled' variational problem $\mathcal{P}_M^\sharp(\omega)$ (Theorem 9.2), a solution which is thus independent of ε. Then

$$\frac{1}{2\varepsilon} \int_{-\varepsilon}^{\varepsilon} \boldsymbol{u}^\varepsilon \, \mathrm{d}x_3^\varepsilon \longrightarrow \boldsymbol{\zeta} \text{ in } \mathbf{V}_M^\sharp(\omega) \text{ as } \varepsilon \to 0,$$

$$\boldsymbol{\zeta}_K^\varepsilon \longrightarrow \boldsymbol{\zeta} \text{ in } \mathbf{V}_M^\sharp(\omega) \text{ as } \varepsilon \to 0.$$

Proof. Under the assumptions that $\boldsymbol{\theta} \in \mathcal{C}^3(\overline{\omega}; \mathbb{R}^3)$ and that the applied forces are admissible in the sense of Section 9.2, the convergence

$$\frac{1}{2\varepsilon} \int_{-\varepsilon}^{\varepsilon} \boldsymbol{u}^\varepsilon \, \mathrm{d}x_3^\varepsilon = \frac{1}{2} \int_{-1}^{1} \boldsymbol{u}(\varepsilon) \, \mathrm{d}x_3 \longrightarrow \boldsymbol{\zeta} \text{ in } \mathbf{V}_M^\sharp(\omega)$$

as $\varepsilon \to 0$ was already established in Theorem 9.1.

The rest of the proof is an elaboration of Caillerie and Sanchez-Palencia (1995a, Theorem 4.5), who established the *weak* convergence $\zeta_K^\varepsilon \rightharpoonup \zeta$ in $\mathbf{V}_M^\sharp(\omega)$ as $\varepsilon \to 0$.

Since the space $\mathbf{V}_K(\omega)$ is dense in the space $\mathbf{V}(\omega)$ with respect to the norm $\| \cdot \|_{1,\omega}$ and there exists c such that $|\boldsymbol{\eta}|_\omega^M \le c\|\boldsymbol{\eta}\|_{1,\omega}$ for all $\boldsymbol{\eta} \in \mathbf{V}(\omega)$, the space $\mathbf{V}_K(\omega)$ is dense in $\mathbf{V}(\omega)$ with respect to $|\cdot|_\omega^M$ and thus the space $\mathbf{V}_M^\sharp(\omega)$ is also the completion of $\mathbf{V}_K(\omega)$ with respect to $|\cdot|_\omega^M$.

Let $B_M^\sharp$ and $L_M^\sharp$ denote the unique continuous extensions from $\mathbf{V}(\omega)$ to $\mathbf{V}_M^\sharp(\omega)$ of the bilinear and linear forms B_M and L_M defined by

$$B_M(\boldsymbol{\zeta},\boldsymbol{\eta}) := \int_\omega a^{\alpha\beta\sigma\tau}\gamma_{\sigma\tau}(\boldsymbol{\zeta})\gamma_{\alpha\beta}(\boldsymbol{\eta})\sqrt{a}\,dy,$$

$$L_M(\boldsymbol{\eta}) := \int_\omega \varphi^{\alpha\beta}\gamma_{\alpha\beta}(\boldsymbol{\eta})\sqrt{a}\,dy.$$

Since the applied forces are admissible for the two-dimensional Koiter equations, their solution $\boldsymbol{\zeta}_K^\varepsilon$ satisfies the scaled Koiter equations for a generalized membrane shell, namely,

$$B_M(\boldsymbol{\zeta}_K^\varepsilon,\boldsymbol{\eta}) + \varepsilon^2 B_F(\boldsymbol{\zeta}_K^\varepsilon,\boldsymbol{\eta}) = L_M(\boldsymbol{\eta}) \text{ for all } \boldsymbol{\eta} \in \mathbf{V}_K(\omega),$$

where

$$B_F(\boldsymbol{\zeta},\boldsymbol{\eta}) := \frac{1}{3}\int_\omega a^{\alpha\beta\sigma\tau}\rho_{\sigma\tau}(\boldsymbol{\zeta})\rho_{\alpha\beta}(\boldsymbol{\eta})\sqrt{a}\,dy.$$

Setting $\boldsymbol{\eta} = \boldsymbol{\zeta}_K^\varepsilon$ in these scaled equations then shows that the family $(\boldsymbol{\zeta}_K^\varepsilon)_{\varepsilon>0}$ is bounded in the space $\mathbf{V}_M^\sharp(\omega)$ and that the families $(\varepsilon\rho_{\alpha\beta}(\boldsymbol{\zeta}_K^\varepsilon))_{\varepsilon>0}$ are bounded in $L^2(\omega)$.

Consequently, there exists a subsequence, still denoted by $(\boldsymbol{\zeta}_K^\varepsilon)_{\varepsilon>0}$ for convenience, and there exist $\boldsymbol{\zeta}^* \in \mathbf{V}_M^\sharp(\omega)$ and $\rho_{\alpha\beta}^{-1} \in L^2(\omega)$ such that

$$\boldsymbol{\zeta}_K^\varepsilon \rightharpoonup \boldsymbol{\zeta}^* \text{ in } \mathbf{V}_M^\sharp(\omega) \text{ and } \varepsilon\rho_{\alpha\beta}(\boldsymbol{\zeta}_K^\varepsilon) \rightharpoonup \rho_{\alpha\beta}^{-1} \text{ in } L^2(\omega).$$

Fix $\boldsymbol{\eta} \in \mathbf{V}_K(\omega)$ in the scaled Koiter equations and let $\varepsilon \to 0$; then the above weak convergence yields $B_M^\sharp(\boldsymbol{\zeta}^*,\boldsymbol{\eta}) = L_M(\boldsymbol{\eta})$. Since $\mathbf{V}_K(\omega)$ is dense in $\mathbf{V}_M^\sharp(\omega)$, it follows that $B_M^\sharp(\boldsymbol{\zeta}^*,\boldsymbol{\eta}) = L_M^\sharp(\boldsymbol{\eta})$ for all $\boldsymbol{\eta} \in \mathbf{V}_M^\sharp(\omega)$. Hence

$$\boldsymbol{\zeta}^* = \boldsymbol{\zeta},$$

where $\boldsymbol{\zeta} \in \mathbf{V}_M^\sharp(\omega)$ is the unique solution to the scaled problem $\mathcal{P}_M^\sharp(\omega)$ (Theorem 9.2). Furthermore, the *weak* convergence

$$\boldsymbol{\zeta}_K^\varepsilon \rightharpoonup \boldsymbol{\zeta} \text{ in } \mathbf{V}_M^\sharp(\omega)$$

then holds for the *whole* family $(\boldsymbol{\zeta}_K^\varepsilon)_{\varepsilon>0}$.

By definition of the norm $|\cdot|_\omega^M$ and of the bilinear form B_M and of its extension $B_M^\sharp$, establishing the *strong* convergence $\zeta_K^\varepsilon \to \zeta$ in $\mathbf{V}_M^\sharp(\omega)$ is equivalent to establishing the convergence

$$B_M^\sharp(\zeta_K^\varepsilon - \zeta, \zeta_K^\varepsilon - \zeta) \to 0,$$

which itself easily follows by letting $\eta = \zeta_K^\varepsilon$ in the scaled Koiter equations, by noting that $B_M(\zeta_K^\varepsilon, \zeta_K^\varepsilon) \le L_M(\zeta_K^\varepsilon)$, and by using the weak convergence $\zeta_K^\varepsilon \rightharpoonup \zeta$ in $\mathbf{V}_M^\sharp(\omega)$. □

Finally, we consider *flexural shells*, as defined in Section 10.1

Theorem 11.5: Justification of Koiter's equations for 'flexural' shells. Assume that $\boldsymbol{\theta} \in C^3(\overline{\omega}; \mathbb{R}^3)$. Consider a family of linearly elastic flexural shells, with thickness 2ε approaching zero, with each having the same middle surface $S = \boldsymbol{\theta}(\overline{\omega})$, and with each subjected to a boundary condition of place along a portion of its lateral face having the same set $\boldsymbol{\theta}(\gamma_0)$ as its middle curve, and let the assumptions on the data be as in Theorem 10.1.

For each $\varepsilon > 0$, let

$$(u_i^\varepsilon) \in \mathbf{H}^1(\Omega^\varepsilon) \text{ and } \zeta_K^\varepsilon = (\zeta_{i,K}^\varepsilon) \in H^1(\omega) \times H^1(\omega) \times H^2(\omega),$$

respectively, denote the solutions to the three-dimensional and two-dimensional variational problems $\mathcal{P}(\Omega^\varepsilon)$ and $\mathcal{P}_K^\varepsilon(\omega)$. Also, let

$$\zeta = (\zeta_i) \in H^1(\omega) \times H^1(\omega) \times H^2(\omega)$$

denote the solution to the two-dimensional scaled variational problem $\mathcal{P}_F(\omega)$ (Theorem 10.2), a solution which is thus independent of ε. Then

$$\frac{1}{2\varepsilon} \int_{-\varepsilon}^{\varepsilon} u_i^\varepsilon \, dx_3^\varepsilon \longrightarrow \zeta_i \text{ in } H^1(\omega),$$

$$\zeta_{K,\alpha}^\varepsilon \longrightarrow \zeta_\alpha \text{ in } H^1(\omega) \text{ and } \zeta_{K,3}^\varepsilon \longrightarrow \zeta_3 \text{ in } H^2(\omega).$$

Proof. Under the assumptions that there exist constants $\lambda > 0$ and $\mu > 0$ and functions $f^i \in L^2(\Omega)$ independent of ε such that

$$\lambda^\varepsilon = \lambda \quad \text{and} \quad \mu^\varepsilon = \mu,$$
$$f^{i,\varepsilon}(x^\varepsilon) = \varepsilon^2 f^i(x) \text{ for all } x^\varepsilon = \pi^\varepsilon x \in \Omega^\varepsilon$$

(these are the assumptions on the data for a family of linearly elastic flexural shells) and that $\boldsymbol{\theta} \in C^3(\overline{\omega}; \mathbb{R}^3)$, then

$$\frac{1}{2\varepsilon} \int_{-\varepsilon}^{\varepsilon} u_i^\varepsilon \, dx_3^\varepsilon = \frac{1}{2\varepsilon} \int_{-1}^{1} u_i(\varepsilon) \, dx_3^\varepsilon \longrightarrow \zeta_i \text{ in } \mathbf{H}^1(\omega)$$

as $\varepsilon \to 0$ are easy corollaries to the fundamental convergence result of Theorem 10.1.

The *weak* convergence $\boldsymbol{\zeta}_K^\varepsilon \rightharpoonup \boldsymbol{\zeta}$ in $H^1(\omega) \times H^1(\omega) \times H^2(\omega)$ was first established by Sanchez-Palencia (1989a, Theorem 2.1), as a consequence of general results in perturbation theory.

We directly establish here that, in fact, the *strong* convergence $\boldsymbol{\zeta}_K^\varepsilon \to \boldsymbol{\zeta}$ in $\mathbf{V}_K(\omega)$ holds. Let the bilinear forms B_M and B_F and the linear form L be defined as in the proof of Theorem 11.3; in addition, let

$$\mathbf{V}_F(\omega) := \{\boldsymbol{\eta} \in \mathbf{V}_K(\omega) : \gamma_{\alpha\beta}(\boldsymbol{\eta}) = 0 \text{ in } \omega\} \subset \mathbf{V}_K(\omega),$$

the space $\mathbf{V}_K(\omega)$ being equipped with the norm

$$\boldsymbol{\eta} = (\eta_i) \to \left\{ \sum_\alpha \|\eta_\alpha\|_{1,\omega}^2 + \|\eta_3\|_{2,\omega}^2 \right\}.$$

By virtue of the assumptions on the applied forces, the solution $\boldsymbol{\zeta}_K^\varepsilon$ also satisfies the scaled Koiter equations for a flexural shell (it is instructive to compare them with those for an elliptic membrane shell introduced in the proof of Theorem 11.3), namely,

$$\frac{1}{\varepsilon^2} B_M(\boldsymbol{\zeta}_K^\varepsilon, \boldsymbol{\eta}) + B_F(\boldsymbol{\zeta}_K^\varepsilon, \boldsymbol{\eta}) = L(\boldsymbol{\eta}) \text{ for all } \boldsymbol{\eta} \in \mathbf{V}_K(\omega).$$

Letting $\boldsymbol{\eta} = \boldsymbol{\zeta}_K^\varepsilon$ in these scaled equations and using the inequality of Korn's type on a general surface (Theorem 4.4) then show that the family $(\boldsymbol{\zeta}_K^\varepsilon)_{\varepsilon>0}$ is bounded in $\mathbf{V}_K(\omega)$ and that the families $\left(\frac{1}{\varepsilon}\gamma_{\alpha\beta}(\boldsymbol{\zeta}_K^\varepsilon)\right)_{\varepsilon>0}$ and $(\rho_{\alpha\beta}(\boldsymbol{\zeta}_K^\varepsilon))_{\varepsilon>0}$ are bounded in $L^2(\omega)$.

Consequently, there exists a subsequence, still denoted by $(\boldsymbol{\zeta}_K^\varepsilon)_{\varepsilon>0}$ for convenience, and there exists $\boldsymbol{\zeta}^* \in \mathbf{V}_K(\omega)$ such that

$$\boldsymbol{\zeta}_K^\varepsilon \rightharpoonup \boldsymbol{\zeta}^* \text{ in } \mathbf{V}_K(\omega) \text{ and } \gamma_{\alpha\beta}(\boldsymbol{\zeta}_K^\varepsilon) \to 0 \text{ in } L^2(\omega).$$

The weak convergence $\boldsymbol{\zeta}_K^\varepsilon \rightharpoonup \boldsymbol{\zeta}^*$ in $\mathbf{V}_K(\omega)$ implies the weak convergence $\gamma_{\alpha\beta}(\boldsymbol{\zeta}^\varepsilon) \rightharpoonup \gamma_{\alpha\beta}(\boldsymbol{\zeta}^*)$ in $L^2(\omega)$; hence $\gamma_{\alpha\beta}(\boldsymbol{\zeta}^*) = 0$ and thus $\boldsymbol{\zeta}^* \in \mathbf{V}_F(\omega)$. Fix $\boldsymbol{\eta} \in \mathbf{V}_F(\omega)$ in the scaled Koiter equations and let $\varepsilon \to 0$; then the weak convergence $\boldsymbol{\zeta}_K^\varepsilon \rightharpoonup \boldsymbol{\zeta}^*$ in $\mathbf{V}_K(\omega)$ yields $B_F(\boldsymbol{\zeta}^*, \boldsymbol{\eta}) = L(\boldsymbol{\eta})$. Hence

$$\boldsymbol{\zeta}^* = \boldsymbol{\zeta},$$

where $\boldsymbol{\zeta} \in \mathbf{V}_F(\omega)$ is the unique solution to the scaled problem $\mathcal{P}_F(\omega)$ (Theorem 10.2) and the *weak* convergence then holds for the *whole* family $(\boldsymbol{\zeta}^\varepsilon)_{\varepsilon>0}$.

By the inequality of Korn's type on a general surface combined with the strong convergence $\gamma_{\alpha\beta}(\boldsymbol{\zeta}_K^\varepsilon) \to 0$ in $L^2(\omega)$ and the relations $\gamma_{\alpha\beta}(\boldsymbol{\zeta}) = 0$, establishing the *strong* convergence $\boldsymbol{\zeta}_K^\varepsilon \to \boldsymbol{\zeta}$ in $\mathbf{V}_K(\omega)$ is equivalent to establishing the convergence

$$B_F(\boldsymbol{\zeta}_K^\varepsilon - \boldsymbol{\zeta}, \boldsymbol{\zeta}_K^\varepsilon - \boldsymbol{\zeta}) \to 0,$$

which itself easily follows by letting $\boldsymbol{\eta} = \boldsymbol{\zeta}_K^\varepsilon$ in the scaled Koiter equations, by noting that $B_F(\zeta_K^\varepsilon, \zeta_K^\varepsilon) \leq L(\zeta_K^\varepsilon)$, and by using the weak convergence $\zeta_K^\varepsilon \rightharpoonup \zeta$ in $\mathbf{V}_K(\omega)$. □

A major conclusion emerging from Theorems 11.3, 11.4, and 11.5 is that the two-dimensional linear Koiter equations are thus justified for all kinds of shells, since, in each case, the average across the thickness of the three-dimensional solution and the solution of Koiter's equations have the same principal part, namely, in each case the solution ζ to the corresponding two-dimensional scaled problem, as the thickness approaches zero.

By virtue of the de-scalings, which are in each case of the form $\zeta^\varepsilon = \zeta$ (see Sections 8.3, 9.3, and 10.3), the above asymptotic analyses also show that the solution ζ_K^ε of Koiter's equations is 'asymptotically as good' as the solution ζ^ε obtained by solving either the two-dimensional problem $\mathcal{P}_M^\varepsilon(\omega)$, or the two-dimensional problem $\mathcal{P}_M^{\sharp\varepsilon}(\omega)$, or the two-dimensional problem $\mathcal{P}_F^\varepsilon(\omega)$ (see Theorems 8.3, 9.3, and 10.3), according to which category the shell falls into.

Compared to these limit two-dimensional equations, Koiter's equations thus possess two outstanding advantages: not only does using Koiter's equations avoid a 'preliminary' knowledge of the category in which a given linearly elastic shell falls into, but it also avoids the mathematical or numerical difficulties inherent to each such category, briefly summarized below.

(a) If the shell is an *elliptic membrane*, no boundary condition can be imposed on the normal component ζ_3^ε of the displacement field since ζ_3^ε is 'only' in $L^2(\omega)$!

(b) If the shell is a *generalized membrane*, the solution ζ^ε belongs to an 'abstract' completion $\mathbf{V}_M^\sharp(\omega)$; the boundary conditions on ζ^ε may thus be quite 'exotic'!

(c) If the shell is *flexural*, the unknown ζ^ε is subjected to the constraints $\gamma_{\alpha\beta}(\zeta^\varepsilon) = 0$ in ω, which certainly hinder its numerical approximation!

It is to be strongly emphasized that these conclusions could not be reached by an asymptotic analysis of Koiter's equations alone, for they definitely rely on an asymptotic analysis of the three-dimensional equations, namely, the content of Sections 8, 9, and 10!

Note that engineers and experts in computational mechanics often base their classification of linearly elastic shells on the relative orders of magnitudes of the 'membrane' and 'flexural' strain energies, namely,

$$\frac{\varepsilon}{2} \int_\omega a^{\alpha\beta\sigma\tau,\varepsilon} \gamma_{\sigma\tau}(\zeta_K^\varepsilon) \gamma_{\alpha\beta}(\zeta_K^\varepsilon) \sqrt{a}\, dy$$

and

$$\frac{\varepsilon^3}{6} \int_\omega a^{\alpha\beta\sigma\tau,\varepsilon} \rho_{\sigma\tau}(\zeta_K^\varepsilon) \rho_{\alpha\beta}(\zeta_K^\varepsilon) \sqrt{a}\, dy,$$

found in Koiter's energy j^ε_K (Section 11.1) evaluated at a given solution $\boldsymbol{\zeta}^\varepsilon_K$, rather than on an asymptotic analysis of the three-dimensional solution as here. This approach, in which the applied forces may thus also dictate either 'membrane-dominated' or 'flexural-dominated' behaviour, has recently been given a mathematical basis by Blouza, Brezzi and Lovadina (1999).

Koiter's equations are often used for identifying and approximating boundary layers in shells; see Hakula and Pitkäranta (1995), Hakula (1997), Gerdes, Matache and Schwab (1998).

By contrast with 'boundary' layers, *'interior' layers*, that is, 'away from the lateral face', may appear inside shells with a hyperbolic middle surface. This challenging phenomenon seems again to be well modelled by Koiter's equations, as suggested by Sanchez-Palencia and Sanchez-Hubert (1998). See also Karamian (1998a), Leguillon, Sanchez-Hubert and Sanchez-Palencia (1999), Pitkäranta, Matache and Schwab (2000).

Koiter's equations may be adapted to the modelling of shells with periodically varying thickness, by means of a homogenization procedure; see Telega and Lewiński (1998a, 1998b), and Lewiński and Telega (2000). They may likewise be adapted to shells made of anisotropic and nonhomogeneous elastic materials, in which case additional terms in the strain energy couple the linearized change of metric and linearized change of curvature tensors; see Caillerie and Sanchez-Palencia (1995a), Figueiredo and Leal (1998).

11.3. Koiter's equations for shells whose middle surface has little regularity

In Section 5, we described how Blouza and Le Dret (1999) showed that the introduction of new expressions $\widetilde{\gamma}_{\alpha\beta}(\widetilde{\boldsymbol{\eta}})$ and $\widetilde{\rho}_{\alpha\beta}(\widetilde{\boldsymbol{\eta}})$ (reproduced below) for the functions $\gamma_{\alpha\beta}(\boldsymbol{\eta})$ and $\rho_{\alpha\beta}(\boldsymbol{\eta})$ allows us to consider more general situations, where the mapping $\boldsymbol{\theta}$ need only be in the space $W^{2,\infty}(\omega; \mathbb{R}^3)$. See also Blouza and Le Dret (2000) for further developments of this approach.

For a linearly elastic shell, simply supported along its entire boundary (boundary conditions of clamping along a portion of its boundary can be handled as well, provided they are first re-interpreted in an *ad hoc* manner), the associated 'Koiter's equations for shells whose middle surface has little regularity' accordingly take the following form. The unknown $\widetilde{\boldsymbol{\zeta}}^\varepsilon_K$, which is now the displacement field of the middle surface, satisfies the variational problem $\widetilde{\mathcal{P}}^\varepsilon_K(\omega)$:

$$\widetilde{\boldsymbol{\zeta}}^\varepsilon_K \in \widetilde{\mathbf{V}}^s_K(\omega) := \{\widetilde{\boldsymbol{\eta}} \in \mathbf{H}^1_0(\omega) : \partial_{\alpha\beta}\widetilde{\boldsymbol{\eta}} \cdot \boldsymbol{a}_3 \in L^2(\omega)\},$$

$$\int_\omega \left\{ \varepsilon a^{\alpha\beta\sigma\tau,\varepsilon}\widetilde{\gamma}_{\sigma\tau}(\widetilde{\boldsymbol{\zeta}}^\varepsilon_K)\widetilde{\gamma}_{\alpha\beta}(\widetilde{\boldsymbol{\eta}}) + \frac{\varepsilon^3}{3}a^{\alpha\beta\sigma\tau,\varepsilon}\widetilde{\rho}_{\sigma\tau}(\widetilde{\boldsymbol{\zeta}}^\varepsilon_K)\widetilde{\rho}_{\alpha\beta}(\widetilde{\boldsymbol{\eta}}) \right\}\sqrt{a}\,dy$$

$$= \int_\omega \widetilde{\boldsymbol{p}}^\varepsilon \cdot \widetilde{\boldsymbol{\eta}}\sqrt{a}\,dy$$

for all $\widetilde{\boldsymbol{\eta}} \in \widetilde{\mathbf{V}}_K^s(\omega)$, where

$$\widetilde{\gamma}_{\alpha\beta}(\widetilde{\boldsymbol{\eta}}) := \frac{1}{2}(\partial_\beta \widetilde{\boldsymbol{\eta}} \cdot \boldsymbol{a}_\alpha + \partial_\alpha \widetilde{\boldsymbol{\eta}} \cdot \boldsymbol{a}_\beta),$$

$$\widetilde{\rho}_{\alpha\beta}(\widetilde{\boldsymbol{\eta}}) := (\partial_{\alpha\beta}\widetilde{\boldsymbol{\eta}} - \Gamma_{\alpha\beta}^\sigma \partial_\sigma \widetilde{\boldsymbol{\eta}}) \cdot \boldsymbol{a}_3,$$

the given function $\widetilde{\boldsymbol{p}}^\varepsilon \in \mathbf{L}^2(\omega)$ accounts for the applied forces, and $a^{\alpha\beta\sigma\tau,\varepsilon}$ are the usual contravariant components of the two-dimensional elasticity tensor of the shell.

Recall that $\widetilde{\gamma}_{\alpha\beta}(\widetilde{\boldsymbol{\eta}}) = \gamma_{\alpha\beta}(\boldsymbol{\eta})$ and $\widetilde{\rho}_{\alpha\beta}(\widetilde{\boldsymbol{\eta}}) = \rho_{\alpha\beta}(\boldsymbol{\eta})$ if $\widetilde{\boldsymbol{\eta}} = \eta_i \boldsymbol{a}^i$ is such that $\boldsymbol{\eta} = (\eta_i) \in H^1(\omega) \times H^1(\omega) \times H^2(\omega)$.

A proof similar to that of Theorem 11.1, now based on the inequality of Korn's type on a surface with little regularity (Theorem 5.2), then produces the following result.

Theorem 11.6: Existence and uniqueness of solutions. Let there be given a domain ω in $\mathbb{R}^2$ and an injective mapping $\boldsymbol{\theta} \in W^{2,\infty}(\omega; \mathbb{R}^3)$ such that the two vectors $\boldsymbol{a}_\alpha = \partial_\alpha \boldsymbol{\theta}$ are linearly independent at all points of $\overline{\omega}$.

Then the associated 'Koiter's equations $\widetilde{\mathcal{P}}_K^\varepsilon(\omega)$ for a shell with little regularity' have exactly one solution, which is also the unique solution to the minimization problem:
Find $\widetilde{\boldsymbol{\zeta}}_K^\varepsilon$ such that

$$\widetilde{\boldsymbol{\zeta}}_K^\varepsilon \in \widetilde{\mathbf{V}}_K^s(\omega) \text{ and } \widetilde{j}_K^\varepsilon(\widetilde{\boldsymbol{\zeta}}_K^\varepsilon) = \inf_{\widetilde{\boldsymbol{\eta}} \in \widetilde{\mathbf{V}}_K^s(\omega)} \widetilde{j}_K^\varepsilon(\widetilde{\boldsymbol{\eta}}), \text{ where}$$

$$\widetilde{j}_K^\varepsilon(\widetilde{\boldsymbol{\eta}}) := \frac{1}{2} \int_\omega \left\{ \varepsilon a^{\alpha\beta\sigma\tau,\varepsilon} \widetilde{\gamma}_{\sigma\tau}(\widetilde{\boldsymbol{\eta}}) \widetilde{\gamma}_{\alpha\beta}(\widetilde{\boldsymbol{\eta}}) \right.$$

$$\left. + \frac{\varepsilon^3}{3} a^{\alpha\beta\sigma\tau,\varepsilon} \widetilde{\rho}_{\sigma\tau}(\widetilde{\boldsymbol{\eta}}) \widetilde{\rho}_{\alpha\beta}(\widetilde{\boldsymbol{\eta}}) \right\} \sqrt{a}\, \mathrm{d}y - \int_\omega \widetilde{\boldsymbol{p}}^\varepsilon \cdot \widetilde{\boldsymbol{\eta}} \sqrt{a}\, \mathrm{d}y. \qquad \square$$

It must be emphasized that, in this approach, the unknown $\widetilde{\boldsymbol{\zeta}}_K^\varepsilon$ and the fields $\widetilde{\boldsymbol{\eta}}$ are displacement fields of the middle surface, no longer recovered in general as $\widetilde{\boldsymbol{\zeta}}_K^\varepsilon = \zeta_{K,i}^\varepsilon \boldsymbol{a}^i$ or $\widetilde{\boldsymbol{\eta}} = \eta_i \boldsymbol{a}^i$ by means of their covariant components $\zeta_{K,i}^\varepsilon$ or η_i.

11.4. Budiansky–Sanders equations

Sanders (1959) and Koiter (1960) have proposed a linear shell theory akin to Koiter's, where the covariant components $\rho_{\alpha\beta}(\boldsymbol{\eta})$ of the linearized change of curvature tensor are replaced by the covariant components $\rho_{\alpha\beta}^{BS}(\boldsymbol{\eta})$ of the 'Budiansky–Sanders linearized change of curvature tensor', defined by

$$\rho_{\alpha\beta}^{BS}(\boldsymbol{\eta}) := \rho_{\alpha\beta}(\boldsymbol{\eta}) - \frac{1}{2}(b_\alpha^\sigma \gamma_{\sigma\beta}(\boldsymbol{\eta}) + b_\beta^\tau \gamma_{\tau\alpha}(\boldsymbol{\eta})).$$

The remaining terms in the equations are otherwise identical to those in

Koiter's equations. In other words, the Budiansky–Sanders equations take the following form, when they are stated as a variational problem $\mathcal{P}_{BS}^{\varepsilon}(\omega)$: Find $\boldsymbol{\zeta}^{\varepsilon} = (\zeta_i^{\varepsilon})$ such that

$$\boldsymbol{\zeta}^{\varepsilon} \in \mathbf{V}_K(\omega) = \{\boldsymbol{\eta} = (\eta_i) \in H^1(\omega) \times H^1(\omega) \times H^2(\omega) :$$
$$\eta_i = \partial_\nu \eta_3 = 0 \text{ on } \gamma_0\},$$

$$\int_\omega \left\{ \varepsilon a^{\alpha\beta\sigma\tau,\varepsilon} \gamma_{\sigma\tau}(\boldsymbol{\zeta}^{\varepsilon}) \gamma_{\alpha\beta}(\boldsymbol{\eta}) + \frac{\varepsilon^3}{3} a^{\alpha\beta\sigma\tau,\varepsilon} \rho_{\sigma\tau}^{BS}(\boldsymbol{\zeta}^{\varepsilon}) \rho_{\alpha\beta}^{BS}(\boldsymbol{\eta}) \right\} \sqrt{a}\, dy$$
$$= \int_\omega p^{i,\varepsilon} \eta_i \sqrt{a}\, dy$$

for all $\boldsymbol{\eta} = (\eta_i) \in \mathbf{V}_K(\omega)$.

The interest of using the modified functions $\rho_{\alpha\beta}^{BS}(\boldsymbol{\eta})$, rather than the 'genuine' functions $\rho_{\alpha\beta}(\boldsymbol{\eta})$, has been discussed at length in Budiansky and Sanders (1967) and, for this reason, the resulting theory has become known as the Budiansky–Sanders theory.

In addition, Destuynder (1985) has shown how this theory can be derived from three-dimensional linearized elasticity, again on the basis of two *a priori* assumptions, *both* of a geometrical nature, one of them being the linearized Kirchoff–Love assumption (Section 11.1).

Theorem 11.7: Existence and uniqueness of solutions.   Let the assumptions be as in Theorem 11.1. Then the associated Budiansky–Sanders equations $\mathcal{P}_{BS}^{\varepsilon}(\omega)$ have exactly one solution (which is also the unique solution to a minimization problem, the form of which should be clear).

Proof.   The definition of the functions $\rho_{\alpha\beta}^{BS}(\boldsymbol{\eta})$ and the equivalence

$$\gamma_{\alpha\beta}(\boldsymbol{\eta}) = \rho_{\alpha\beta}^{BS}(\boldsymbol{\eta}) = 0 \text{ in } \omega \Leftrightarrow \gamma_{\alpha\beta}(\boldsymbol{\eta}) = \rho_{\alpha\beta}(\boldsymbol{\eta}) = 0 \text{ in } \omega$$

together imply that the proof of the existence and uniqueness of the solution to Koiter's equations (Section 4 and Theorem 11.1) extends almost *verbatim* to the Budiansky–Sanders equations.   □

12. Naghdi's equations

While Koiter's equations belong to the family of Kirchhoff–Love theories, two-dimensional shell equations that rely on the notion of one-director Cosserat surfaces were proposed by P. M. Naghdi, again in the sixties. Since then, they have appealed as much as Koiter's equations to the computational mechanics community. In particular, they seem to be quite effective in the numerical simulation of shells with a 'moderately small' thickness; in this respect, see the companion article by Dominique Chapelle.

After describing the associated two-dimensional Naghdi equations for a linearly elastic shell, we briefly review in this section the existence and uniqueness theory for these equations.

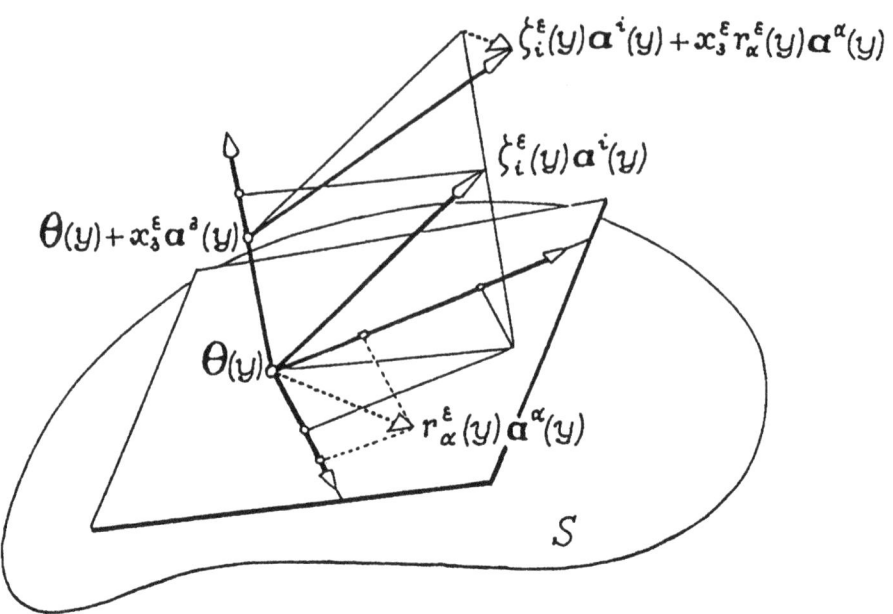

$$\zeta_i^\varepsilon(y)\boldsymbol{a}^i(y)+x_3^\varepsilon r_\alpha^\varepsilon(y)\boldsymbol{a}^\alpha(y)$$

$$\zeta_i^\varepsilon(y)\boldsymbol{a}^i(y)$$

$$\boldsymbol{\theta}(y)+x_3^\varepsilon\boldsymbol{a}^3(y)$$

$$\boldsymbol{\theta}(y)$$

$$r_\alpha^\varepsilon(y)\boldsymbol{a}^\alpha(y)$$

$$S$$

Fig. 12.1. The five unknowns in Naghdi's equations are the three co-variant components $\zeta_i^\varepsilon : \overline{\omega} \to \mathbb{R}$ of the displacement field of the middle surface S and the two covariant components $r_\alpha^\varepsilon : \overline{\omega} \to \mathbb{R}$ of the linear-ized rotation field of the unit normal vector along S; this means that, for each $y \in \overline{\omega}$, $\zeta_i^\varepsilon(y)\boldsymbol{a}^i(y) + x_3^\varepsilon r_\alpha^\varepsilon(y)\boldsymbol{a}^\alpha(y)$ is the displacement of the point $(\boldsymbol{\theta}(y) + x_3^\varepsilon\boldsymbol{a}^3(y))$ of the reference configuration of the shell

Consider as in Section 11.1 a shell with middle surface $S = \boldsymbol{\theta}(\overline{\omega})$ and thickness $2\varepsilon > 0$, constituted by a homogeneous and isotropic linear elastic material with Lamé constants $\lambda^\varepsilon > 0$ and $\mu^\varepsilon > 0$, and subjected to applied body forces with contravariant components $f^{i,\varepsilon} \in L^2(\Omega^\varepsilon)$.

In Naghdi's approach (Naghdi 1963, 1972), the *a priori* assumption of a *mechanical* nature about the stresses inside the shell is the same as in Koiter's approach (Section 11.1), but the *a priori* assumption of a *geomet-rical* nature is different. The points situated on a line normal to S remain on a line and the lengths are unmodified along this line after the deformation has taken place as in Koiter's approach, but this line need no longer remain normal to the deformed middle surface.

In the linearized version of this approach described here, there are *five* unknowns: the three covariant components $\zeta_i^\varepsilon : \overline{\omega} \to \mathbb{R}$ of the displacement field $\zeta_i^\varepsilon\boldsymbol{a}^i$ of the middle surface S and the two covariant components $r_\alpha^\varepsilon : \overline{\omega} \to \mathbb{R}$ of the linearized rotation field $r_\alpha^\varepsilon\boldsymbol{a}^\alpha$ of the unit normal vector along S. This means that the displacement of the point $(\boldsymbol{\theta}(y)+x_3^\varepsilon\boldsymbol{a}^3(y))$ is the vector

$(\zeta_i^\varepsilon(y)\boldsymbol{a}^i(y) + x_3^\varepsilon r_\alpha^\varepsilon(y)\boldsymbol{a}^\alpha(y))$; see Figure 12.1. The surface S thus becomes a Cosserat surface, in the sense that it is endowed with the field $r_\alpha^\varepsilon \boldsymbol{a}^\alpha$, then called a director field (it is easily seen that the rotation field of the unit normal should be indeed tangential in a linearized theory).

In their weak formulation, Naghdi's equations for a linearly elastic shell consist in solving the following variational problem $\mathcal{P}_N^\varepsilon(\omega)$:
Find $(\boldsymbol{\zeta}^\varepsilon, \boldsymbol{r}^\varepsilon) = ((\zeta_i^\varepsilon), (r_\alpha^\varepsilon))$ such that

$$(\boldsymbol{\zeta}^\varepsilon, \boldsymbol{r}^\varepsilon) \in \mathbf{V}_N(\omega) := \{(\boldsymbol{\eta}, \boldsymbol{s}) = ((\eta_i), (s_\alpha)) \in \mathbf{H}^1(\omega) : \eta_i = s_\alpha = 0 \text{ on } \gamma_0\},$$

$$\varepsilon \int_\omega \{a^{\alpha\beta\sigma\tau,\varepsilon}\gamma_{\sigma\tau}(\boldsymbol{\zeta}^\varepsilon)\gamma_{\alpha\beta}(\boldsymbol{\eta}) + c\mu^\varepsilon a^{\alpha\beta}\gamma_{\alpha3}(\boldsymbol{\zeta}^\varepsilon, \boldsymbol{r}^\varepsilon)\gamma_{\beta3}(\boldsymbol{\eta}, \boldsymbol{s})\}\sqrt{a}\,dy$$

$$+\frac{\varepsilon^3}{3}\int_\omega a^{\alpha\beta\sigma\tau,\varepsilon}\rho_{\sigma\tau}^N(\boldsymbol{\zeta}^\varepsilon, \boldsymbol{r}^\varepsilon)\rho_{\alpha\beta}^N(\boldsymbol{\eta}, \boldsymbol{s})\sqrt{a}\,dy = \int_\omega p^{i,\varepsilon}\eta_i\sqrt{a}\,dy$$

for all $(\boldsymbol{\eta}, \boldsymbol{s}) \in \mathbf{V}_N(\omega)$ (the notation $\mathbf{H}^1(\Omega)$ standing for the space $(H^1(\omega))^5$ in the definition of the space $\mathbf{V}_N(\omega)$), where

$$a^{\alpha\beta\sigma\tau,\varepsilon} := \frac{4\lambda^\varepsilon\mu^\varepsilon}{\lambda^\varepsilon + 2\mu^\varepsilon}a^{\alpha\beta}a^{\sigma\tau} + 2\mu^\varepsilon(a^{\alpha\sigma}a^{\beta\tau} + a^{\alpha\tau}a^{\beta\sigma}),$$

$$\gamma_{\alpha\beta}(\boldsymbol{\eta}) := \frac{1}{2}(\partial_\beta\eta_\alpha + \partial_\alpha\eta_\beta) - \Gamma_{\alpha\beta}^\sigma\eta_\sigma - b_{\alpha\beta}\eta_3,$$

$$\gamma_{\alpha3}(\boldsymbol{\eta}, \boldsymbol{s}) := \frac{1}{2}(\partial_\alpha\eta_3 + b_\alpha^\sigma\eta_\sigma + s_\alpha),$$

$$\rho_{\alpha\beta}^N(\boldsymbol{\eta}, \boldsymbol{s}) := -\frac{1}{2}(\partial_\beta s_\alpha + \partial_\alpha s_\beta) + \Gamma_{\alpha\beta}^\sigma s_\sigma - b_\alpha^\sigma b_{\sigma\beta}\eta_3$$

$$+ \frac{1}{2}b_\alpha^\sigma(\partial_\beta\eta_\sigma - \Gamma_{\beta\sigma}^\tau\eta_\tau) + \frac{1}{2}b_\beta^\tau(\partial_\alpha\eta_\tau - \Gamma_{\alpha\tau}^\sigma\eta_\sigma),$$

$$p^{i,\varepsilon} := \int_{-\varepsilon}^\varepsilon f^{i,\varepsilon}\,dx_3^\varepsilon$$

(the functions $a^{\alpha\beta}, b_{\alpha\beta}, b_\alpha^\sigma, \Gamma_{\alpha\beta}^\sigma$, and a defined as usual: see Section 4) and c is a strictly positive constant (what should be the 'best' constant seems to be an unresolved issue).

The functions $a^{\alpha\beta\sigma\tau,\varepsilon}$ are the contravariant components of the two-dimensional elasticity tensor of the shell and the functions $\gamma_{\alpha\beta}(\boldsymbol{\eta})$ are the covariant components of the linearized change of metric tensor associated with a displacement field $\eta_i\boldsymbol{a}^i$ of the middle surface S, as before. The 'new' functions $\gamma_{\alpha3}(\boldsymbol{\eta}, \boldsymbol{s})$ and $\rho_{\alpha\beta}^N(\boldsymbol{\eta}, \boldsymbol{s})$ are the covariant components of the linearized transverse shear strain tensor and of the Naghdi linearized change of curvature tensor associated with displacement and linearized rotation fields $\eta_i\boldsymbol{a}^i$ and $s_\alpha\boldsymbol{a}^\alpha$ of S; for a justification of these definitions, see, e.g., Bernadou (1994, Part I, Chapter 3).

The next existence and uniqueness result for the solution to the variational problem $\mathcal{P}_N^\varepsilon(\omega)$ is due to Bernadou, Ciarlet and Miara (1994, Theorem 3.1).

Theorem 12.1: Existence and uniqueness of solutions. Let ω be a domain in $\mathbb{R}^2$, let γ_0 be a subset of $\partial\omega$ with length $\gamma_0 > 0$, and let $\boldsymbol{\theta} \in C^3(\overline{\omega}; \mathbb{R}^3)$ be an injective mapping such that the two vectors $\boldsymbol{a}_\alpha = \partial_\alpha \boldsymbol{\theta}$ are linearly independent at all points of $\overline{\omega}$.

Then the associated Naghdi equations $\mathcal{P}_N^\varepsilon(\omega)$ have exactly one solution (which is also the unique solution to a minimization problem, the form of which should be clear).

Sketch of proof. Let

$$|(\boldsymbol{\eta}, \boldsymbol{s})| := \left\{ \sum_{\alpha,\beta} |\gamma_{\alpha\beta}(\boldsymbol{\eta})|_{0,\omega}^2 + \sum_\alpha |\gamma_{\alpha 3}(\boldsymbol{\eta}, \boldsymbol{s})|_{0,\omega}^2 + \sum_{\alpha,\beta} |\rho_{\alpha\beta}^N(\boldsymbol{\eta}, \boldsymbol{s})|_{0,\omega}^2 \right\}^{1/2},$$

$$\|(\boldsymbol{\eta}, \boldsymbol{s})\| := \left\{ \sum_i |\eta_i|_{0,\omega}^2 + \sum_\alpha |s_\alpha|_{0,\omega}^2 + |(\boldsymbol{\eta}, \boldsymbol{s})|^2 \right\}^{1/2},$$

$$\|(\boldsymbol{\eta}, \boldsymbol{s})\|_{1,\omega} := \left\{ \|\boldsymbol{\eta}\|_{1,\omega}^2 + \|\boldsymbol{s}\|_{1,\omega}^2 \right\}^{1/2},$$

where $\boldsymbol{\eta} = (\eta_i)$ and $\boldsymbol{s} = (s_\alpha)$.

(i) First, the Lemma of J. L. Lions, used as in Theorem 4.1, shows that there exists a constant c_0 such that

$$\|(\boldsymbol{\eta}, \boldsymbol{s})\|_{1,\omega} \le c_0 \|(\boldsymbol{\eta}, \boldsymbol{s})\|$$

for all $(\boldsymbol{\eta}, \boldsymbol{s}) \in \mathbf{H}^1(\omega) = (H^1(\omega))^5$.

(ii) Next, let $(\boldsymbol{\eta}, \boldsymbol{s}) \in \mathbf{H}^1(\omega)$ be such that

$$\gamma_{\alpha\beta}(\boldsymbol{\eta}) = \gamma_{\alpha 3}(\boldsymbol{\eta}, \boldsymbol{s}) = \rho_{\alpha\beta}^N(\boldsymbol{\eta}, \boldsymbol{s}) = 0 \text{ in } \omega.$$

These relations imply that $\eta_3 \in H^2(\omega)$ and that $\rho_{\alpha\beta}^N(\boldsymbol{\eta}, \boldsymbol{s}) = \rho_{\alpha\beta}(\boldsymbol{\eta})$ for such fields $(\boldsymbol{\eta}, \boldsymbol{s})$. Hence Theorem 4.3(a) shows that the vector field $\eta_i \boldsymbol{a}^i$ is a linearized rigid displacement of the surface $S = \boldsymbol{\theta}(\overline{\omega})$, in the sense that there exist two vectors $\widehat{\boldsymbol{c}}, \widehat{\boldsymbol{d}} \in \mathbb{R}^3$ such that

$$\eta_i(y)\boldsymbol{a}^i(y) = \widehat{\boldsymbol{c}} + \widehat{\boldsymbol{d}} \wedge \boldsymbol{\theta}(y) \text{ for all } y \in \overline{\omega}.$$

(iii) Let $(\boldsymbol{\eta}, \boldsymbol{s}) \in \mathbf{H}^1(\omega)$ be such that

$$\gamma_{\alpha\beta}(\boldsymbol{\eta}) = \gamma_{\alpha 3}(\boldsymbol{\eta}, \boldsymbol{s}) = \rho_{\alpha\beta}^N(\boldsymbol{\eta}, \boldsymbol{s}) = 0 \text{ in } \omega, \quad \eta_i = s_\alpha = 0 \text{ on } \gamma_0,$$

where $\gamma_0 \subset \gamma$ satisfies length $\gamma_0 > 0$. Then an argument similar to that in the proof of Theorem 4.3(b) shows that $(\boldsymbol{\eta}, \boldsymbol{s}) = (\mathbf{0}, \mathbf{0})$.

(iv) A proof by contradiction as in Theorem 4.4 shows that there exists a constant c such that another inequality of Korn's type on a general surface holds (compare with that in Theorem 4.4):

$$\|(\boldsymbol{\eta}, \boldsymbol{s})\|_{1,\omega} \leq c |(\boldsymbol{\eta}, \boldsymbol{s})|$$

for all $(\boldsymbol{\eta}, \boldsymbol{s}) \in \mathbf{V}_N(\omega)$, where

$$\mathbf{V}_N(\omega) := \{(\boldsymbol{\eta}, \boldsymbol{s}) = ((\eta_i), (s_\alpha)) \in \mathbf{H}^1(\omega) : \eta_i = s_\alpha = 0 \text{ on } \gamma_0\}.$$

(v) Finally, let $B_N^\varepsilon : \mathbf{V}_N(\omega) \times \mathbf{V}_N(\omega) \to \mathbb{R}$ denote the bilinear form defined by the left-hand side of the variational equations in problem $\mathcal{P}_N^\varepsilon(\omega)$. Then it is easily seen that there exists a constant c_N^ε such that

$$|(\boldsymbol{\eta}, \boldsymbol{s})|^2 \leq c_N^\varepsilon B_N^\varepsilon((\boldsymbol{\eta}, \boldsymbol{s}), (\boldsymbol{\eta}, \boldsymbol{s}))$$

for all $(\boldsymbol{\eta}, \boldsymbol{s}) \in \mathbf{V}_N(\omega)$. We thus conclude that the variational problem $\mathcal{P}_N^\varepsilon(\omega)$ has exactly one solution. □

Note that parts (ii) and (iii) of the above proof constitute another linearized rigid displacement lemma on a general surface, which is due to Coutris (1978).

The variational problem $\mathcal{P}_N^\varepsilon(\omega)$ is, at least formally, equivalent to a boundary value problem, which is given in Iosifescu (200x), where the *regularity* of its solution when $\gamma_0 = \gamma$ is also studied.

In the same manner that Blouza and Le Dret (1999) have generalized Theorem 11.1 to Koiter's equations for shells whose middle surface has little regularity (Theorem 11.6), Blouza (1997) has extended Theorem 12.1 to Naghdi's equations for shells whose middle surface has little regularity (the mapping $\boldsymbol{\theta}$ need only be in the space $W^{2,\infty}(\omega; \mathbb{R}^3)$).

Various asymptotic justifications of Naghdi's equations, including error estimates, are found in Lods and Mardare (1999, 2000a).

13. 'Shallow' shells

According to the definition justified via a formal analysis by Ciarlet and Paumier (1996) in the nonlinear case, then justified via a convergence theorem by Ciarlet and Miara (1992a) in the linear case, a shell is *shallow* if the deviation of its middle surface S^ε from a plane is of the order of the thickness, that is, if the surface S^ε can be written as $S^\varepsilon = \boldsymbol{\theta}^\varepsilon(\overline{\omega})$, with a mapping $\boldsymbol{\theta}^\varepsilon : \overline{\omega} \to \mathbb{R}^3$ of the form

$$\boldsymbol{\theta}^\varepsilon(y_1, y_2) = (y_1, y_2, \varepsilon\theta(y_1, y_2)) \text{ for all } (y_1, y_2) \in \overline{\omega},$$

and $\theta : \overline{\omega} \to \mathbb{R}$ is a sufficiently smooth function that is *independent of* ε; see Figure 13.1. This specific 'variation of the middle surface with ε' thus constitutes an additional assumption on the data, special to (linear and nonlinear) shallow shell theory.

Fig. 13.1. A shell is 'shallow' if, in its reference configuration, the deviation of its middle surface from a plane is (up to an additive constant)
of the order of the thickness of the shell

Like 'general' shells, linearly elastic 'shallow' shells are amenable to an
asymptotic analysis (as their thickness approaches zero) that also produces
'limit' two-dimensional equations. There are, however, crucial differences
between their analysis and that of 'general' shells.

First, *different scalings* are made at the *outset* of the asymptotic analysis on the *tangential* and *normal* components of the displacement field
and *different assumptions* are likewise made on the *tangential* and *normal*
components of the applied body force.

More specifically, another 'scaled unknown' $\boldsymbol{u}(\varepsilon) = (u_i(\varepsilon)) : \overline{\Omega} \to \mathbb{R}^3$ is
defined in this case by letting

$$u_\alpha^\varepsilon(x^\varepsilon) = \varepsilon u_\alpha(\varepsilon)(x) \text{ and } u_3^\varepsilon(x^\varepsilon) = u_3(\varepsilon)(x) \text{ for all } x^\varepsilon = \pi^\varepsilon x \in \overline{\Omega}^\varepsilon,$$

and it is assumed that the applied body forces are such that there exist
functions $f^i \in L^2(\Omega)$ independent of ε such that

$$f^{\alpha,\varepsilon}(x^\varepsilon) = \varepsilon f^\alpha(x) \text{ and } f^{3,\varepsilon}(x^\varepsilon) = \varepsilon^2 f^3(x) \text{ for all } x^\varepsilon = \pi^\varepsilon x \in \Omega^\varepsilon$$

(compare with Section 7.2). Note in passing that these scalings and assumptions are identical to those made in the asymptotic analysis of linearly
elastic *plates* (see Ciarlet (1997, Section 1.3)).

Making such scalings and assumptions on the data, Busse, Ciarlet and
Miara (1997) have shown how two-dimensional equations of a linearly elastic
shallow shell 'in curvilinear coordinates' can be given a rigorous justification
by means of a convergence theorem as the thickness goes to zero. We simply
list the limit equations that are found in this fashion, when they are stated
as a variational problem. Let

$$b^{\alpha\beta\sigma\tau,\varepsilon} := \frac{4\lambda^\varepsilon \mu^\varepsilon}{\lambda^\varepsilon + 2\mu^\varepsilon} \delta^{\alpha\beta} \delta^{\sigma\tau} + 2\mu^\varepsilon (\delta^{\alpha\sigma} \delta^{\beta\tau} + \delta^{\alpha\tau} \delta^{\beta\sigma}),$$

$$p^{i,\varepsilon} := \int_{-\varepsilon}^\varepsilon f^{i,\varepsilon} \, dx_3^\varepsilon, \quad q^{\alpha,\varepsilon} := \int_{-\varepsilon}^\varepsilon x_3^\varepsilon f^{\alpha,\varepsilon} \, dx_3^\varepsilon,$$

$$e_{\alpha\beta}^{sh,\varepsilon}(\boldsymbol{\eta}) := \frac{1}{2}(\partial_\beta \eta_\alpha + \partial_\alpha \eta_\beta) - \varepsilon \eta_3 \partial_{\alpha\beta}\theta$$

($\delta^{\alpha\beta}$ designates the Kronecker symbol), and let $a^{i,\varepsilon}$ designate the vectors of
the contravariant bases along the middle surface S^ε (like the middle surface

S^ε, they now depend on ε). Then the 'limit', de-scaled, vector field $\boldsymbol{\zeta}^\varepsilon = (\zeta_i^\varepsilon)$, where the functions $\zeta_i^\varepsilon : \bar{\omega} \to \mathbb{R}$ are the covariant components of the displacement field $\zeta_i^\varepsilon \boldsymbol{a}^{i,\varepsilon}$ of the middle surface S^ε, satisfies the following variational problem $\mathcal{P}_{sh}^\varepsilon(\varepsilon)$:

$$\boldsymbol{\zeta}^\varepsilon \in \mathbf{V}_K(\omega) := \{\boldsymbol{\eta} = (\eta_i) \in H^1(\omega) \times H^1(\omega) \times H^2(\omega) :$$
$$\eta_i = \partial_\nu \eta_3 = 0 \text{ on } \gamma_0\},$$
$$\int_\omega \left\{ \varepsilon b^{\alpha\beta\sigma\tau,\varepsilon} e_{\sigma\tau}^{sh,\varepsilon}(\boldsymbol{\zeta}^\varepsilon) e_{\alpha\beta}^{sh,\varepsilon}(\boldsymbol{\eta}) + \frac{\varepsilon^3}{3} b^{\alpha\beta\sigma\tau,\varepsilon} \partial_{\sigma\tau} \zeta_3^\varepsilon \partial_{\alpha\beta} \eta_3 \right\} \mathrm{d}y = \int_\omega p^{i,\varepsilon} \eta_i \, \mathrm{d}y$$

for all $\boldsymbol{\eta} = (\eta_i) \in \mathbf{V}_K(\omega)$.

Another major difference thus lies in the *outcome* of the asymptotic analysis: as evidenced by the equations given above, the 'limit' variational problem simultaneously includes 'membrane' and 'flexural' terms!

More precisely, even though it is still expressed in curvilinear coordinates, the variational problem $\mathcal{P}_{sh}^\varepsilon(\omega)$ resembles more the 'limit', de-scaled, two-dimensional problem of a linearly elastic plate (see Ciarlet (1997, Section 1.7)) than that of the shell! For the contravariant components of the metric tensor usually found in the two-dimensional elasticity tensor of a shell are now replaced by Kronecker deltas, the area element along the middle surface is replaced by $\mathrm{d}y$, and finally, the components of the linearized change of metric and change of curvature tensors are replaced by the functions $e_{\alpha\beta}^{sh,\varepsilon}(\boldsymbol{\eta})$ and $\partial_{\alpha\beta}\eta^3$, where neither the Christoffel symbols nor any components of the curvature tensor of S^ε are to be found.

Problem $\mathcal{P}_{sh}^\varepsilon(\omega)$ constitutes Novozhilov's model of a shallow shell, so named after Novozhilov (1959). These equations were given a first justification by Destuynder (1980) for special geometries.

As shown by Ciarlet and Miara (1992a) (see also Ciarlet (1997, Chapter 3)), the two-dimensional equations 'in Cartesian coordinates' of a linearly elastic shallow shell can likewise be justified by means of an asymptotic analysis of the three-dimensional equations. As expected, and shown by Andreoiu (1999), the 'limit' displacement fields found in either curvilinear or Cartesian coordinates, though not identical vector fields, are nevertheless 'essentially the same', that is, their components agree 'to within their first orders', once they are expressed in the same basis.

The asymptotic analysis of Busse, Ciarlet and Miara (1997) has been pursued substantially further by Andreoiu, Dauge and Faou (2000) and Andreoiu and Faou (200x), who showed how to construct expansions of the scaled unknown that yield error estimates of arbitrarily high order, thus generalizing analogous results of Destuynder (1981, Corollary 7) and Dauge and Gruais (1996, 1998) for plates. Such expansions comprise a 'polynomial' part of the form $\sum_{k=0}^p \varepsilon^k \boldsymbol{u}^k$ (as in a formal asymptotic expansion) and a 'boundary layer' part that compensates the violation of the boundary conditions by the polynomial part.

The asymptotic analysis of the corresponding eigenvalue problem has been carried out in Cartesian coordinates by Kesavan and Sabu (1999); there is no doubt that it could be similarly carried out in curvilinear coordinates.

The exponential nature of the boundary layers that arise in linearly elastic shallow shells is analysed in Pitkäranta, Matache and Schwab (2000).

Models of multi-layered, or composite, linearly elastic shallow shells, found in particular in hulls of sailboats, have been obtained by Kail (1994) by means of the method of formal asymptotic expansions.

Other definitions of 'shallowness' have been proposed, which often make explicit reference to the curvature of the middle surface. For instance, Destuynder (1985, Section 1) considers that a shell is 'shallow' if $\eta = \varepsilon^p$ for some $p \geq 2$, where the other 'small' parameter η is the ratio of the thickness 2ε to the smallest absolute value of the radii of curvature along the middle surface, $p = 2$ corresponding to Novozhilov's model. In this direction, see also Vekua (1965), Green and Zerna (1968, p. 400), Gordeziani (1974), Dikmen (1982, p. 158), Pitkäranta, Matache and Schwab (2000).

Acknowledgement

For the most part, this article is based on Chapters 1 to 7 from my book *Mathematical Elasticity, Volume III: Theory of Shells*, published in 2000 by North-Holland, Amsterdam. I am extremely grateful in this respect to Arjen Sevenster, who generously allowed me to use excerpts and figures from this book.

REFERENCES[1]

S. Agmon, A. Douglis and L. Nirenberg (1964), 'Estimates near the boundary for solutions of elliptic partial differential equations satisfying general boundary conditions, II', *Comm. Pure Appl. Math.* **17**, 35–92.

O. Alexandrescu (1994), 'Théorème d'existence pour le modèle bidimensionnel de coque non linéaire de W. T. Koiter', *C. R. Acad. Sci. Paris, Sér. I* **319**, 899–902.

C. Amrouche and V. Girault (1994), 'Decomposition of vector spaces and application to the Stokes problem in arbitrary dimension', *Czech. Math. J.* **44**, 109–140.

---

[1] This list of references is by no means exhaustive, as it only comprises references about recent, mathematical, developments of the 'asymptotic approach' to linear shell theory. A substantially more extensive bibliography, which includes in particular references about other possible approaches to shell theory and about its history, is found in Ciarlet (2000).

G. Andreoiu (1999), 'Comparaison entre modèles bidimensionnels de coques faible-ment courbées', *C. R. Acad. Sci. Paris, Sér. I* **329**, 339–342.

G. Andreoiu and E. Faou (200x), 'Complete asymptotics for shallow shells', *Asymptotic Anal.* To appear.

G. Andreoiu, M. Dauge and E. Faou (2000), 'Développements asymptotiques complets pour les coques faiblement courbées encastrées ou libres', *C. R. Acad. Sci. Paris, Sér. I* **330**, 523–528.

S. S. Antman (1995), *Nonlinear Problems of Elasticity*, Springer, Berlin.

J. M. Ball (1977), 'Convexity conditions and existence theorems in nonlinear elasticity', *Arch. Rational Mech. Anal.* **63**, 337–403.

K. J. Bathe (1996), *Finite Element Procedures*, Prentice-Hall, Englewood Cliffs, NJ.

M. Berger and B. Gostiaux (1992), *Géométrie Différentielle: Variétés, Courbes et Surfaces*, 2nd edn, Presses Universitaires de France, Paris.

M. Bernadou (1994), *Méthodes d'Eléments Finis pour les Coques Minces*, Masson, Paris. Translation: *Finite Element Methods for Thin Shell Problems*, Wiley, New York, 1995.

M. Bernadou and P. G. Ciarlet (1976), Sur l'ellipticité du modèle linéaire de coques de W.T. Koiter, in *Computing Methods in Applied Sciences and Engineering* (R. Glowinski and J. L. Lions, eds), Vol. 134 of *Lecture Notes in Economics and Mathematical Systems*, Springer, Heidelberg, pp. 89–136.

M. Bernadou, P. G. Ciarlet and B. Miara (1994), 'Existence theorems for two-dimensional linear shell theories', *J. Elasticity* **34**, 111–138.

L. Bers, F. John and M. Schechter (1964), *Partial Differential Equations*, Interscience Publishers, New York.

A. Blouza (1997), 'Existence et unicité pour le modèle de Naghdi pour une coque peu régulière', *C. R. Acad. Sci. Paris, Sér. I* **324**, 839–844.

A. Blouza and H. Le Dret (1999), 'Existence and uniqueness for the linear Koiter model for shells with little regularity', *Quart. Appl. Math.* **57**, 317–337.

A. Blouza and H. Le Dret (2000), An up-to-the boundary version of Friedrichs' lemma and applications to the linear Koiter shell model, Technical Report 00008, Laboratoire d'Analyse Numérique, Université Pierre et Marie Curie, Paris.

A. Blouza, F. Brezzi and C. Lovadina (1999), 'Sur la classification des coques linéairement élastiques', *C. R. Acad. Sci. Paris, Sér. I* **328**, 831–836.

P. Bolley and J. Camus (1976), 'Régularité pour une classe de problèmes aux limites elliptiques dégénérés variationnels', *C. R. Acad. Sci. Paris, Sér. A* **282**, 45–47.

W. Borchers and H. Sohr (1990), 'On the equations rot $v = g$ and div $u = f$ with zero boundary conditions', *Hokkaido Math. J.* **19**, 67–87.

S. C. Brenner and L. R. Scott (1994), *The Mathematical Theory of Finite Element Methods*, Springer, Berlin.

F. Brezzi and M. Fortin (1991), *Mixed and Hybrid Finite Element Methods*, Springer, Berlin.

B. Budiansky and J. L. Sanders (1967), On the 'best' first-order linear shell theory, in *Progress in Applied Mechanics, W. Prager Anniversary Volume*, MacMillan, New York, pp. 129–140.

S. Busse (1998), 'Asymptotic analysis of linearly elastic shells with variable thickness', *Rev. Roumaine Math. Pures Appl.* **43**, 553–590.

S. Busse, P. G. Ciarlet and B. Miara (1997), 'Justification d'un modèle bi-dimensionnel de coques 'faiblement courbées' en coordonnées curvilignes', *Math. Modelling Numer. Anal.* **31**, 409–434.

D. Caillerie (1996), 'Etude générale d'un type de problèmes raides et de perturbation singulière', *C. R. Acad. Sci. Paris, Sér. I* **323**, 835–840.

D. Caillerie and E. Sanchez-Palencia (1995*a*), 'A new kind of singular-stiff problems and application to thin elastic shells', *Math. Models Methods Appl. Sci.* **5**, 47–66.

D. Caillerie and E. Sanchez-Palencia (1995*b*), 'Elastic thin shells: asymptotic theory in the anisotropic and heterogeneous cases', *Math. Models Methods Appl. Sci.* **5**, 473–496.

D. Chapelle (1994), Personal communication.

P. G. Ciarlet (1978), *The Finite Element Method for Elliptic Problems*, North-Holland, Amsterdam.

P. G. Ciarlet (1988), *Mathematical Elasticity, Volume I: Three-Dimensional Elasticity*, North-Holland, Amsterdam.

P. G. Ciarlet (1991), Basic error estimates for elliptic problems, in *Handbook of Numerical Analysis* (P. G. Ciarlet and J. L. Lions, eds), Vol. II, North-Holland, Amsterdam, pp. 17–351.

P. G. Ciarlet (1993), Modèles bi-dimensionnels de coques: Analyse asymptotique et théorèmes d'existence, in *Boundary Value Problems for Partial Differential Equations and Applications* (J. L. Lions and C. Baiocchi, eds), Masson, Paris, pp. 61–80.

P. G. Ciarlet (1997), *Mathematical Elasticity, Volume II: Theory of Plates*, North-Holland, Amsterdam.

P. G. Ciarlet (2000), *Mathematical Elasticity, Volume III: Theory of Shells*, North-Holland, Amsterdam.

P. G. Ciarlet and V. Lods (1996*a*), 'On the ellipticity of linear membrane shell equations', *J. Math. Pures Appl.* **75**, 107–124.

P. G. Ciarlet and V. Lods (1996*b*), 'Asymptotic analysis of linearly elastic shells, I: Justification of membrane shell equations', *Arch. Rational Mech. Anal.* **136**, 119–161.

P. G. Ciarlet and V. Lods (1996*d*), 'Asymptotic analysis of linearly elastic shells: Generalized membrane shells', *J. Elasticity* **43**, 147–188.

P. G. Ciarlet and S. Mardare (2000), 'Sur les inégalités de Korn en coordonnées curvilignes', *C. R. Acad. Sci. Paris, Sér. I* **331**, 337–343.

P. G. Ciarlet and S. Mardare (200x), On Korn's inequalities in curvilinear coordinates. To appear.

P. G. Ciarlet and B. Miara (1992*a*), 'Justification of the two-dimensional equations of a linearly elastic shallow shell', *Comm. Pure Appl. Math.* **45**, 327–360.

P. G. Ciarlet and B. Miara (1992*b*), 'On the ellipticity of linear shell models', *Z. Angew. Math. Phys.* **43**, 243–253.

P. G. Ciarlet and J. C. Paumier (1996), 'A justification of the Marguerre–von Kármán equations', *Computational Mech.* **1**, 177–202.

P. G. Ciarlet and E. Sanchez-Palencia (1996), 'An existence and uniqueness theorem for the two-dimensional linear membrane shell equations', *J. Math. Pures
  Appl.* **75**, 51–67.
P. G. Ciarlet, V. Lods and B. Miara (1996), 'Asymptotic analysis of linearly elastic
  shells, II: Justification of flexural shell equations', *Arch. Rational Mech. Anal.*
  136, 163–190.
C. Collard and B. Miara (1999), 'Asymptotic analysis of the stresses in thin elastic
  shells', *Arch. Rational Mech. Anal.* **148**, 233–264.
R. Courant and D. Hilbert (1962), *Methods of Mathematical Physics*, Vol. II, Interscience Publishers, New York.
N. Coutris (1978), 'Théorème d'existence et d'unicité pour un problème de coque
  élastique dans le cas d'un modèle linéaire de P. M. Naghdi', *RAIRO Analyse
  Numérique* **12**, 51–57.
M. Dauge and I. Gruais (1996), 'Asymptotics of arbitrary order in thin elastic
  plates and optimal estimates for the Kirchhoff–Love model', *Asymptotic Anal.*
  13, 167–197.
M. Dauge and I. Gruais (1998), 'Asymptotics of arbitrary order for a thin elastic
  clamped plate, II: Analysis of the boundary layer terms', *Asymptotic Anal.*
  16, 99–124.
M. Delfour (1999), 'Characterization of the space of solutions of the membrane shell
  equation for arbitrary $C^{1,1}$ midsurfaces', *Control and Cybernetics* **28**, 481–501.
M. Delfour (200x), Tangential differential calculus and functional analysis on a $C^{1,1}$
  manifold, in *Differential-Geometric Methods in the Control of Partial Differential Equations* (R. Gulliver, W. Littman and R. Triggiani, eds), Vol. 268
  of *Contemporary Mathematics*, American Mathematical Society, Providence,
  RI. To appear.
P. Destuynder (1980), Sur une justification des modèles de plaques et de coques par
  les méthodes asymptotiques, PhD thesis, Université Pierre et Marie Curie,
  Paris.
P. Destuynder (1981), 'Comparaison entre les modèles tri-dimensionnels et bi-
  dimensionnels de plaques en élasticité', *RAIRO Analyse Numérique* **15**, 331–
  369.
P. Destuynder (1985), 'A classification of thin shell theories', *Acta Applicandae
  Mathematicae* **4**, 15–63.
P. Destuynder (1990), *Modélisation des Coques Minces Elastiques*, Masson, Paris.
J. Dieudonné (1968), *Eléments d'Analyse, Tome 1: Fondements de l'Analyse Mo-
  derne*, Gauthier-Villars, Paris. Translation: *Foundations of Modern Analysis*,
  Academic Press, New York, 1st edn 1960.
M. Dikmen (1982), *Theory of Thin Elastic Shells*, Pitman, Boston.
M. P. do Carmo (1976), *Differential Geometry of Curves and Surfaces*, Prentice-
  Hall, Englewood Cliffs, NJ.
G. Duvaut and J. L. Lions (1972), *Les Inéquations en Mécanique et en Physique*,
  Dunod, Paris. Translation: *Inequalities in Mechanics and Physics*, Springer,
  Berlin, 1976.
E. Faou (2000*a*), 'Elasticité linéarisée tridimensionnelle pour une coque mince:
  résolution en série formelle en puissances de l'épaisseur', *C. R. Acad. Sci.
  Paris, Sér. I* **330**, 415–420.

E. Faou (2000*b*), Développements asymptotiques dans les coques minces linéairement elastiques, PhD thesis, Université de Rennes.

I. N. Figueiredo and C. Leal (1998), 'Ellipticity of Koiter's and Naghdi's models for nonhomogeneous anisotropic shells', *Applicable Anal.* **70**, 75–84.

K. Genevey (1996), 'A regularity result for a linear membrane shell problem', *Math. Modelling Numer. Anal.* **30**, 467–488.

K. Genevey (1999), Justification of two-dimensional linear shell models by the use of Γ-convergence theory, in *CRM Proceedings and Lecture Notes*, Vol. 21, American Mathematical Society, Providence, pp. 185–197.

P. Gérard and E. Sanchez-Palencia (2000), 'Sensitivity phenomena for certain thin elastic shells with edges', *Math. Methods Appl. Sci.* **23**, 379–399.

K. Gerdes, A. M. Matache and C. Schwab (1998), 'Analysis of membrane locking in *hp*-FEM for a cylindrical shell', *Z. Angew. Math. Mech.* **78**, 663–686.

G. Geymonat and E. Sanchez-Palencia (1991), 'Remarques sur la rigidité infinitésimale de certaines surfaces elliptiques non régulières, non convexes et applications', *C. R. Acad. Sci. Paris, Sér. I* **313**, 645–651.

G. Geymonat and P. Suquet (1986), 'Functional spaces for Norton–Hoff materials', *Math. Methods Appl. Sci.* **8**, 206–222.

P. Giroud (1998), 'Analyse asymptotique de coques inhomogènes en élasticité linéarisée anisotrope', *C. R. Acad. Sci. Paris, Sér. I* **327**, 1011–1014.

R. Glowinski (1984), *Numerical Methods for Nonlinear Variational Problems*, Springer, Berlin.

D. G. Gordeziani (1974), 'On the solvability of some boundary value problems for a variant of the theory of thin shells', *Dokl. Akad. Nauk SSSR*. Translation: *Soviet Math. Dokl.* **15** (1974), 677–680.

A. E. Green and W. Zerna (1968), *Theoretical Elasticity*, 2nd edn, Oxford University Press.

H. Hakula (1997), High-order finite element tools for shell problems, PhD thesis, Helsinki University of Technology.

H. Hakula and J. Pitkäranta (1995), Pinched shells of revolution: Experiments on high-order FEM, in *Proceedings of the Third International Conference on Spectral and Higher-Order Methods* (ICOSAHOM'95) (A. V. Ilin and R. Scott, eds), pp. 193–201.

T. J. R. Hughes (1987), *The Finite Element Method: Linear Static and Dynamic Finite Element Analysis*, Prentice-Hall, Englewood Cliffs, NJ.

O. Iosifescu (200x), 'Regularity for Naghdi's shell equations', *Math. Mech. Solids*. To appear.

F. John (1965), 'Estimates for the derivatives of the stresses in a thin shell and interior shell equations', *Comm. Pure Appl. Math.* **18**, 235–267.

F. John (1971), 'Refined interior equations for thin elastic shells', *Comm. Pure Appl. Math.* **24**, 583–615.

R. Kail (1994), Modélisation asymptotique et numérique de plaques et coques stratifiées, PhD thesis, Université Pierre et Marie Curie, Paris.

P. Karamian (1998*a*), 'Réflexion des singularités dans les coques hyperboliques inhibées', *C. R. Acad. Sci. Paris, Sér. IIb* **326**, 609–614.

P. Karamian (1998*b*), 'Nouveaux résultats numériques concernant les coques minces hyperboliques inhibées: Cas du paraboloïde hyperbolique', *C. R. Acad. Sci. Paris, Sér. IIb* **326**, 755–760.

P. Karamian, J. Sanchez-Hubert and E. Sanchez-Palencia (2000), 'A model problem for boundary layers of thin elastic shells', *Math. Modelling Numer. Anal.* **34**, 1–30.

S. Kesavan and N. Sabu (1999), 'Two-dimensional approximation of eigenvalue problems in shallow shell theory', *Math. Mech. Solids* **4**, 441–460.

S. Kesavan and N. Sabu (2000), 'Two-dimensional approximation of eigenvalue problems in shell theory: Flexural shells', *Chinese Ann. Math.* **21B**, 1–16.

W. Klingenberg (1973), *Eine Vorlesung über Differentialgeometrie*, Springer, Berlin. Translation: *A Course in Differential Geometry*, Springer, Berlin, 1978.

W. T. Koiter (1960), A consistent first approximation in the general theory of thin elastic shells, in *Proceedings, IUTAM Symposium on the Theory of Thin Elastic Shells, Delft, August 1959*, North-Holland, Amsterdam, pp. 12–33.

W. T. Koiter (1966), 'On the nonlinear theory of thin elastic shells', *Proc. Kon. Ned. Akad. Wetensch.* **B69**, 1–54.

W. T. Koiter (1970), 'On the foundations of the linear theory of thin elastic shells', *Proc. Kon. Ned. Akad. Wetensch.* **B73**, 169–195.

P. Le Tallec (1994), Numerical methods for nonlinear three-dimensional elasticity, in *Handbook of Numerical Analysis*, Vol. III, North-Holland, Amsterdam, pp. 465–622.

D. Leguillon, J. Sanchez-Hubert and E. Sanchez-Palencia (1999), 'Model problem of singular perturbation without limit in the space of finite energy and its computation', *C. R. Acad. Sci. Paris, Sér. IIb* **327**, 485–492.

T. Lewiński and J. J. Telega (2000), *Plates, Laminates and Shells: Asymptotic Analysis and Homogenization*, World Scientific, Singapore.

J. L. Lions (1973), *Perturbations Singulières dans les Problèmes aux Limites et en Contrôle Optimal*, Vol. 323 of *Lecture Notes in Mathematics*, Springer, Berlin.

J. L. Lions and E. Sanchez-Palencia (1994), 'Problèmes aux limites sensitifs', *C. R. Acad. Sci. Paris, Sér. I* **319**, 1021–1026.

J. L. Lions and E. Sanchez-Palencia (1996), Problèmes sensitifs et coques élastiques minces, in *Partial Differential Equations and Functional Analysis: In Memory of Pierre Grisvard* (J. Céa, D. Chenais, G. Geymonat and J. L. Lions, eds), Birkhäuser, Boston, pp. 207–220.

J. L. Lions and E. Sanchez-Palencia (1997a), Sur quelques espaces de la théorie des coques et la sensitivité, in *Homogenization and Applications to Material Sciences* (D. Cioranescu, A. Damlamian and P. Donato, eds), Gakkotosho, Tokyo, pp. 271–278.

J. L. Lions and E. Sanchez-Palencia (1997b), Examples of sensitivity in shells with edges, in *Shells: Mathematical Modelling and Scientific Computing* (M. Bernadou, P. G. Ciarlet and J. M. Viaño, eds), Universidade de Santiago de Compostela, pp. 151–154.

J. L. Lions and E. Sanchez-Palencia (1998), Instabilities produced by edges in thin shells, in *Proceedings, IUTAM Symposium 'Variation of Domains and Free-Boundary Problems'* (P. Argoul, M. Fremond and Nguyen Quoc Son, eds), Kluwer Academic Publishers, Boston, pp. 277–284.

J. L. Lions and E. Sanchez-Palencia (2000), 'Sensitivity of certain constrained systems and application to shell theory', *J. Math. Pures Appl.* **79**, 821–838.

V. Lods and C. Mardare (1998*a*), 'The space of inextensional displacements for a partially clamped linearly elastic shell with an elliptic middle surface', *J. Elasticity* **51**, 127–144.

V. Lods and C. Mardare (1998*b*), 'Justification asymptotique des hypothèses de Kirchhoff–Love pour une coque encastrée linéairement élastique', *C. R. Acad. Sci. Paris, Sér. I* **326**, 909–912.

V. Lods and C. Mardare (1999), 'Une justification du modèle de coques de Naghdi', *C. R. Acad. Sci. Paris, Sér. I* **328**, 951–954.

V. Lods and C. Mardare (2000*a*), 'Estimations d'erreur entre le problème tridimensionnel de coque linéairement élastique et le modèle de Naghdi', *C. R. Acad. Sci. Paris, Sér. I* **330**, 157–162.

V. Lods and C. Mardare (2000*b*), 'Asymptotic justification of the Kirchhoff–Love assumptions for a linearly elastic clamped shell', *J. Elasticity* **58** 105–154.

E. Magenes and G. Stampacchia (1958), 'I problemi al contorno per le equazioni differenziali di tipo ellittico', *Ann. Scuola Norm. Sup. Pisa* **12**, 247–358.

C. Mardare (1998*a*), 'Asymptotic analysis of linearly elastic shells: Error estimates in the membrane case', *Asymptotic Anal.* **17**, 31–51.

C. Mardare (1998*b*), 'Two-dimensional models of linearly elastic shells: Error estimates between their solutions', *Math. Mech. Solids* **3**, 303–318.

C. Mardare (1998*c*), 'The generalized membrane problem for linearly elastic shells with hyperbolic or parabolic middle surface', *J. Elasticity* **51**, 145–165.

J. E. Marsden and T. J. R. Hughes (1983), *Mathematical Foundations of Elasticity*, Prentice-Hall, Englewood Cliffs, NJ.

B. Miara and E. Sanchez-Palencia (1996), 'Asymptotic analysis of linearly elastic shells', *Asymptotic Anal.* **12**, 41–54.

C. B. Morrey and L. Nirenberg (1957), 'On the analyticity of the solution of linear elliptic systems of partial differential equations', *Comm. Pure Appl. Math.* **10**, 271–290.

P. M. Naghdi (1963), Foundations of elastic shell theory, in *Progress in Solid Mechanics* (I. N. Sneddon and R. Hill, eds), Vol. 4, North-Holland, Amsterdam, pp. 1–90.

P. M. Naghdi (1972), The theory of shells and plates, in *Handbuch der Physik* (S. Flügge and C. Truesdell, eds), Vol. VIa/2, Springer, Berlin, pp. 425–640.

F. I. Niordson (1985), *Shell Theory*, North-Holland, Amsterdam.

V. V. Novozhilov (1959), *Thin Shell Theory*, Noordhoff, Groningen.

J. Pitkäranta and E. Sanchez-Palencia (1997), 'On the asymptotic behavior of sensitive shells with small thickness', *C. R. Acad. Sci. Paris, Sér. IIb* **325**, 127–134.

J. Pitkäranta, A. M. Matache and C. Schwab (2000), 'Fourier mode analysis of layers in shallow shell deformations'. Research Report SAM 99-18, ETH Zürich. To appear in *Comput. Methods Appl. Mech. Eng.*

O. Ramos (1995), Applications du calcul différentiel intrinsèque aux modèles bidimensionels linéaires de coques, PhD thesis, Université Pierre et Marie Curie, Paris.

J. E. Robert and J. M. Thomas (1991), Mixed and hybrid methods, in *Handbook of Numerical Analysis* (P. G. Ciarlet and J. L. Lions, eds), Vol. II, North-Holland, Amsterdam, pp. 523–633.

J. Sanchez-Hubert and E. Sanchez-Palencia (1997), *Coques Elastiques Minces: Propriétés Asymptotiques*, Masson, Paris.

E. Sanchez-Palencia (1980), *Nonhomogenous Media and Vibration Theory*, Springer, Berlin.

E. Sanchez-Palencia (1989a), 'Statique et dynamique des coques minces, I: Cas de flexion pure non inhibée', *C. R. Acad. Sci. Paris, Sér. I* **309**, 411–417.

E. Sanchez-Palencia (1989b), 'Statique et dynamique des coques minces, II: Cas de flexion pure inhibée: Approximation membranaire', *C. R. Acad. Sci. Paris, Sér. I* **309**, 531–537.

E. Sanchez-Palencia (1990), 'Passages à la limite de l'élasticité tri-dimensionnelle à la théorie asymptotique des coques minces', *C. R. Acad. Sci. Paris, Sér. II* **311**, 909–916.

E. Sanchez-Palencia (1992), 'Asymptotic and spectral properties of a class of singular–stiff problems', *J. Math. Pures Appl.* **71**, 379–406.

E. Sanchez-Palencia (1993), 'On the membrane approximation for thin elastic shells in the hyperbolic case', *Revista Matematica de la Universidad Complutense de Madrid* **6**, 311–331.

E. Sanchez-Palencia (1999), 'On sensitivity and related phenomena in thin shells which are not geometrically rigid', *Math. Models Methods Appl. Sci.* **9**, 139–160.

E. Sanchez-Palencia (2000), 'On a singular perturbation going out of the energy space', *J. Math. Pures Appl.* **79**, 591–602.

E. Sanchez-Palencia and J. Sanchez-Hubert (1998), 'Pathological phenomena in computation of thin elastic shells', *Trans. Canadian Soc. Mech. Eng.* **22**, 435–446.

J. L. Sanders (1959), An improved first-approximation theory for thin shells, NASA Report No. 24.

L. Schwartz (1966), *Théorie des Distributions*, Hermann, Paris.

S. Şlicaru (1997), 'On the ellipticity of the middle surface of a shell and its application to the asymptotic analysis of membrane shells', *J. Elasticity* **46**, 33–42.

S. Şlicaru (1998), Quelques résultats dans la théorie des coques linéairement elastiques à surface moyenne uniformément elliptique ou compacte sans bord, PhD thesis, Université Pierre et Marie Curie, Paris.

J. J. Stoker (1969), *Differential Geometry*, Wiley, New York.

L. Tartar (1978), Topics in Nonlinear Analysis, Technical Report 78.13, Université de Paris–Sud, Orsay.

J. J. Telega and T. Lewiński (1998a), 'Homogenization of linear elastic shells: Γ-convergence and duality, Part I: Formulation of the problem and the effective model', *Bull. Polish Acad. Sci., Technical Sci.* **46**, 1–9.

J. J. Telega and T. Lewiński (1998b), 'Homogenization of linear elastic shells: Γ-convergence and duality, Part II: Dual homogenization', *Bull. Polish Acad. Sci., Technical Sci.* **46**, 11–21.

I. N. Vekua (1962), *Generalized Analytic Functions*, Pergamon, New York.

I. N. Vekua (1965), 'Theory of thin shallow shells of variable thickness', *Acad. Nauk Gruzin. SSR Trudy Tbilissi Mat. Inst. Razmadze* **30**, 3–103.

Xiao Li-Ming (1998), 'Asymptotic analysis of dynamic problems for linearly elastic shells: Justification of equations for dynamic membrane shells', *Asymptotic Anal.* **17**, 121–134.

Xiao Li-Ming (1999), 'Existence and uniqueness of solutions to the dynamic equations for Koiter shells', *Appl. Math. Mech.* **20**, 801–806.

Xiao Li-Ming (200x *a*), 'Asymptotic analysis of dynamic problems for linearly elastic shells: Justification of the dynamic flexural shell equations', *Chinese Ann. Math.* To appear.

Xiao Li-Ming (200x *b*), 'Asymptotic analysis of dynamic problems for linearly elastic shells: Justification of the dynamic Koiter shell equations', *Chinese Ann. Math.* To appear.

Acta Numerica (2001), pp. 215–250

Some new results and current challenges in the finite element analysis of shells

Dominique Chapelle
INRIA-Rocquencourt, B.P. 105,
78153 Le Chesnay Cedex, France
E-mail: Dominique.Chapelle@inria.fr

This article, a companion to the article by Philippe G. Ciarlet on the mathematical modelling of shells also in this issue of *Acta Numerica*, focuses on numerical issues raised by the analysis of shells.

Finite element procedures are widely used in engineering practice to analyse the behaviour of shell structures. However, the concept of 'shell finite element' is still somewhat fuzzy, as it may correspond to very different ideas and techniques in various actual implementations. In particular, a significant distinction can be made between shell elements that are obtained via the discretization of shell models, and shell elements – such as the *general shell elements* – derived from 3D formulations using some kinematic assumptions, without the use of any shell theory. Our first objective in this paper is to give a unified perspective of these two families of shell elements. This is expected to be very useful as it paves the way for further thorough mathematical analyses of shell elements. A particularly important motivation for this is the understanding and treatment of the deficiencies associated with the analysis of *thin* shells (among which is the *locking* phenomenon). We then survey these deficiencies, in the framework of the *asymptotic behaviour* of shell models. We conclude the article by giving some detailed guidelines to numerically assess the performance of shell finite elements when faced with these pathological phenomena, which is essential for the design of improved procedures.

CONTENTS

1. Introduction

Finite element procedures are widely used to analyse the behaviour of shell structures in various areas of engineering (*e.g.*, in the automotive, aerospace and civil engineering industries), and such analyses have been conducted ever since the early development of finite element methods. Over time, however, and across the various disciplines involved, the concept of 'shell finite element' has referred to very different ideas and techniques, and this situation has resulted in an incredibly abundant literature and in a significant amount of confusion. In this respect, we can identify in the 'zoology' of shell finite elements a particularly distinct 'line of division' between finite element methods that result from the discretization of shell mathematical models (namely, essentially two-dimensional models with unknowns given on the mid-surface of the shell; see, *e.g.*, Ciarlet (2000)) and finite element techniques obtained from various other – usually engineering-rooted – considerations. Roughly speaking, the finite element methods derived from a shell model draw their justification from the assumed relevance of such models, the solution of which they can be shown to approximate within a certain accuracy using numerical analysis techniques; see Bernadou (1996). By contrast, other shell finite element methods are grounded in mechanical considerations used 'at the element level'; see, for example, Bathe (1996) and the references therein. Although they are usually numerically tested using various benchmark problems, what is typically lacking for such finite element procedures is the proof that the finite element solutions converge to some known limit solution when the mesh is being refined. Indeed, the characterization of such a limit solution would be extremely useful for evaluating the whole approach.

One of the earliest techniques used in shell analysis consists in simply combining plate bending and membrane behaviour in flat elements, the collection of which approximates the geometry of the actual shell structure. This is known as the *facet-shell element* approach. Examples of this category of elements can be found in Batoz and Dhatt (1992). In Bernadou, Ducatel and Trouvé (1988), a basic facet-shell procedure is analysed and proved to be a non-conforming discretization scheme of a shell model. However, it is also shown that the corresponding consistency error is non-converging, and this result is consistent with previous observations of the lack of convergence exhibited by these methods; see, *e.g.*, Irons and Ahmad (1980). Additional deficiencies associated with the facet-type approximation of the geometry are identified in Akian and Sanchez-Palencia (1992). Various further refinements of the facet-shell techniques have been introduced, but whether or not these developments actually counteract the above-mentioned deficiencies is not clear (and indeed not proven). Moreover, by construction the coarse approximation of the geometry involved in these procedures implies that, even if a successfully converging scheme were to be found, the accuracy of the approximation would be rather poor.

On the other hand, shell finite element procedures obtained by directly discretizing a shell model can be thoroughly analysed and explicit error estimates are then obtained under some well-identified conditions (pertaining to the regularity of the exact solutions, the type of numerical integration used, *etc.*); see Bernadou (1996) and the references therein. The major drawback of this approach is, however, that building a shell finite element procedure on a *set shell model* is clearly restrictive as to the variety of situations that can be analysed. In particular, most shell theories are based on linear elastic behaviour, and adapting the models to account for an arbitrary three-dimensional (3D) constitutive equation – or for general nonlinear behaviour – is not straightforward. Note that this drawback in fact also applies to facet shell elements.

By contrast, a third methodology has been developed that specifically allows for unrestricted versatility as regards shell modelling capabilities. This methodology is based on the idea of 'degenerating' a 3D solid finite element into a shell element by using some *kinematic assumptions* for describing the variation of displacements across the thickness of the shell structure. Since the only model used is the 3D model in consideration, general 3D constitutive laws can be employed. This is indeed why these finite elements are called *general shell elements* (and sometimes also 'degenerated solid elements'); see the seminal work of Ahmad, Irons and Zienkiewicz (1970), and also Bathe (1996), Bischoff and Ramm (1997) and Başar and Krätzig (1989), and the references therein, for later generalizations of these concepts. This approach is widely used in practice, and evidence of its effectiveness can be found abundantly in the literature; see, for example, Bathe (1996). However, because of its specific construction, which does not rely on any shell theory, a mathematical analysis of such a procedure is very difficult to achieve.

The first objective of this article is to present some recent results that give a unified perspective of general shell elements in the framework of shell models and their discretizations. More specifically, we show that we can identify a well-defined shell model that 'underlies' general shell elements, that is, such that the finite element solutions converge to the exact solution of this model when the mesh is refined. In addition, we compare this model to other classical shell models by means of an asymptotic analysis. This unification of concepts paves the way for further mathematical analyses of these highly attractive numerical procedures.

A particularly important motivation for a thorough mathematical analysis of shell finite elements is the understanding and the treatment of some serious numerical pathologies that are known to occur when the thickness of the shell structure is 'small' (to fix the ideas, these phenomena typically become very significant as soon as the ratio of the thickness over the other characteristic dimensions of the structure is of the order of 1%, which is a situation commonly encountered in practice). The sources of these difficulties, including the by-now classical *numerical locking*, are now well identified, and they are closely related to the complexity and diversity of the asymptotic behaviour exhibited by shells; see Chapelle and Bathe (1998). However, from a numerical analysis perspective, a finite element procedure with proven reliability with respect to variation of the thickness parameter is still to be found. On the other hand, the literature abounds with allegedly reliable or locking-free shell finite elements, but the assessment performed to reach this conclusion is often inadequate or incomplete, when not altogether irrelevant. The second main objective of this article is thus to survey the issue of the reliability of shell finite elements with respect to variations of the thickness parameter. Of course, this issue must be put into perspective with the variations of behaviour induced in the exact solutions by variations of the thickness, namely with the analysis of the *asymptotic behaviour* of shell models.

This article is organized in the following manner. In Section 2 we present the two families of shell finite elements that we want to analyse and compare, namely the finite elements obtained by discretizing shell models (Section 2.1), and the general shell elements (Section 2.2). Section 3 is then dedicated to the issue of the reliability of shell finite elements. We start by surveying the asymptotic behaviour of shell models (Section 3.1) before focusing on computational issues (Section 3.2), and we conclude in Section 3.3 by presenting some guidelines for assessing and improving the reliability of shell finite elements.

In all our discussions we focus on linear formulations. Indeed, the difficulties that we want to analyse are already present in linear problems, hence it is natural to treat them in this framework first. Another simplification that we use is to assume that the thickness is constant (equal to 2ε) over the whole shell structure. This assumption is made only in order to simplify the formulas and the discussions, and does not represent a restriction in our analysis; see Bathe and Chapelle (200x) for a general analysis with arbitrary thickness.

Our notation is based on the notation used in the companion article (Ciarlet 2001) (see also Ciarlet (2000)); hence we do not repeat the same definitions here and we only define the new notation (mainly related to finite element concepts). In addition, we use the symbol C to denote a generic positive constant that, unless otherwise stated, is independent of all the parameters appearing in the same equation (and in particular of the mesh parameter h and of the thickness parameter ε), with the classical convention that C may take different values at each occurrence (even when appearing several times in the same equation).

2. Two families of shell finite elements

2.1. Discretizations of classical shell models

As discussed in Ciarlet (2001), classical shell models are mathematical models in which the unknowns are defined on the mid-surface of the shell body or, equivalently, in the 2D reference domain ω from which the mid-surface is obtained via the mapping $\boldsymbol{\theta}$ (which we call the *chart*). In general, these models are based on kinematic assumptions that are used to describe the displacements of points located on material fibres that are orthogonal to the mid-surface in the original configuration. Roughly speaking, shell models can then be divided into two categories according to the kinematic assumption made.

(1) When the assumption is that any normal fibre remains straight and unstretched during the deformation, the displacements in the whole shell body are completely described by the data – on ω – of a displacement field and a rotation field. Namely, a displacement at any point in Ω^ε is given by

$$v(x_1, x_2, x_3) = \eta(x_1, x_2) + x_3 s(x_1, x_2), \qquad (2.1)$$

where s, as the vector that represents the effect of the rotation of the normal fibre, is a vector tangent to the mid-surface at the point of coordinates (x_1, x_2), *i.e.*, $s = s_\alpha a^\alpha$. We call this the Reissner–Mindlin kinematic assumption, and the shell models that are based on this assumption the Reissner–Mindlin-type models; see Reissner (1945) and Mindlin (1951). As an example of this category of shell models, we will consider the model summarized in Naghdi (1963), which we call the Naghdi model.

(2) When the assumption is that any normal fibre remains straight, unstretched *and normal to the deformed mid-surface* (the so-called Kirchhoff–Love kinematic assumption), the displacements can be described by a displacement field only, that is,

$$v(x_1, x_2, x_3) = \eta - x_3(\eta_{3,\alpha} + b_\alpha^\sigma \eta_\sigma) a^\alpha, \qquad (2.2)$$

where $\eta(x_1, x_2)$ is defined on ω. Note that, comparing (2.1) and (2.2), the additional assumption on the preserved orthogonality of fibres is equivalent to the rotation vector being given by

$$s_\alpha = -(\eta_{3,\alpha} + b_\alpha^\sigma \eta_\sigma). \qquad (2.3)$$

We will call these shell models the Kirchhoff–Love-type models. The shell model proposed in Koiter (1966) belongs to this category and will henceforth be used as an example (we refer to this model as the Koiter model).

An important difference between these two categories of models is the respective regularities of their solution spaces (which we also call the *displacement spaces*). Namely, for the Naghdi model the natural displacement space is

$$V_N = \{(\boldsymbol{\eta}, \boldsymbol{s}) \in [H^1(\omega)]^5\} \cap \mathcal{BC}, \tag{2.4}$$

and

$$V_K = \{\boldsymbol{\eta} \in [H^1(\omega)]^2 \times H^2(\omega)\} \cap \mathcal{BC} \tag{2.5}$$

for the Koiter model, where $\mathcal{BC}$ symbolically denotes the essential boundary conditions prescribed. Of course these boundary conditions must, in each case, be compatible with the nature of the functional space. Note that the discrepancy in the regularity of the transverse component of the displacement field (η_3) can be interpreted as the consequence of the additional constraint (2.3) acting on the solution space V_N where $\boldsymbol{s} \in [H^1(\omega)]^2$.

Clearly, in order to obtain well-posed variational formulations, some appropriate boundary conditions need to be prescribed in the displacement space. For the Naghdi and Koiter models (and indeed also for similar models from the above two categories), it can be shown that the corresponding bilinear forms are coercive and bounded on their respective displacement spaces provided that the boundary conditions are sufficient to prevent any rigid body motion of the shell body; see Bernadou and Ciarlet (1976) and Bernadou, Ciarlet and Miara (1994). Note that this holds in particular when the displacement field is set to zero on some part of the boundary $\partial\omega$ that does not correspond to a straight segment in the physical space. When the boundary conditions are compatible with the functional space considered and such that rigid body motions are prevented, we will say that we have *admissible boundary conditions*.

As a consequence, the discretization of classical shell models by conforming finite element methods is rather straightforward, at least in principle. The main difficulty lies in the required use of C^1-conforming finite elements for the transverse displacements in models of Kirchhoff–Love type. These issues are thoroughly addressed in Bernadou (1996). Let us emphasize that these finite element procedures are based on meshes that are constructed in the *reference domain* ω, and on computations that require an extensive use of the chart $\boldsymbol{\theta}$. Since conforming methods are used, *a priori* error estimates follow from interpolation bounds and are of the type

$$\|\boldsymbol{\zeta} - \boldsymbol{\zeta}_h\|_{[H^1(\omega)]^3} + \|\boldsymbol{r} - \boldsymbol{r}_h\|_{[H^1(\omega)]^2} \leq Ch^p, \tag{2.6}$$

for the Naghdi model, and

$$\|\boldsymbol{\zeta} - \boldsymbol{\zeta}_h\|_{[H^1(\omega)]^2 \times H^2(\omega)} \leq Ch^p, \tag{2.7}$$

for the Koiter model, where the constants C and the orders of convergence p depend on the regularity of the exact solution and on the finite element shape functions considered. We emphasize that these estimates are not independent of the thickness parameter ε, as the constants C in (2.6)–(2.7) above depend on ε for two reasons:

- the coercivity and continuity constants of the corresponding bilinear forms depend on ε;
- the regularity bounds are not independent of ε in general.

This issue will be further addressed in Section 3.2.

In practice, in spite of the extensive use of shell finite elements in engineering, the type of finite element procedure that we have been describing in this section is not very often encountered.

2.2. General shell elements

Most shell finite element procedures used in engineering practice fall into this category of *general shell elements*. Unlike the previously discussed procedures, general shell elements are not derived from a shell model, but instead from a 3D variational formulation. A typical general shell element procedure is constructed as follows.

(1) Consider a general 3D variational formulation posed on the 3D geometrical domain defined by the mapping

$$\Theta(x_1, x_2, x_3) = \theta(x_1, x_2) + x_3 a_3(x_1, x_2), \qquad (2.8)$$
$$(x_1, x_2, x_3) \in \omega \times] - \varepsilon, \varepsilon[,$$

and infer a modified variational problem from the *stress assumption*

$$\sigma^{33} \equiv 0. \qquad (2.9)$$

We symbolically denote this modified problem by

$$B^{3D}(u, v) = L^{3D}(v), \qquad \forall v \in V^{3D}, \qquad (2.10)$$

where B^{3D} and L^{3D} respectively represent the external and internal virtual works, and V^{3D} denotes the appropriate functional space, taking into account the essential boundary conditions.

(2) Consider a 2D mesh of ω given in the form of a set of points (the nodes) in ω and of a connectivity that defines the elements. Isoparametric elements are used; hence this defines a one-to-one mapping inside each element between the (x_1, x_2) coordinates and the local coordinates (ξ_1, ξ_2) of the form

$$\begin{pmatrix} x_1 \\ x_2 \end{pmatrix} = \sum_{i=1}^{k} \lambda_i(\xi_1, \xi_2) \begin{pmatrix} x_1^{(i)} \\ x_2^{(i)} \end{pmatrix}, \qquad (2.11)$$

where the functions λ_i are the shape functions associated with the k nodes of the element considered, and the quantities $(x_1^{(i)}, x_2^{(i)})$ denote the nodal coordinates. Of course, the shape functions depend on the polynomial order chosen and on the geometric type of the element (namely quadrilateral or triangular). Using this mapping, we define the interpolation operator $\mathcal{I}$ such that

$$\mathcal{I}(\phi)(x_1, x_2) = \sum_{i=1}^{k} \lambda_i(\xi_1, \xi_2) \, \phi(x_1^{(i)}, x_2^{(i)}) \qquad (2.12)$$

in every element of the mesh, for any continuous scalar or vector function ϕ.

(3) Consider displacement functions defined over Ω^ε that satisfy *Reissner–Mindlin kinematic assumptions at all the nodes* of the mesh, and are interpolated inside the elements using the above shape functions, namely

$$v(x_1, x_2, x_3) = \sum_{i=1}^{k} \lambda_i(\xi_1, \xi_2)(\eta^{(i)} + x_3 s^{(i)}), \qquad (2.13)$$

where $s^{(i)}$ corresponds to a rotation at node i, and hence satisfies

$$s^{(i)} \cdot a_3^{(i)} = 0. \qquad (2.14)$$

We call V_h^{3D} the space of functions of this type that satisfy the boundary conditions prescribed on the structure (note that boundary conditions can be prescribed here as they would be for a Reissner–Mindlin-type model).

(4) Consider the problem:
Find $u \in V_h^{3D}$ such that

$$B_h^{3D}(u, v) = L_h^{3D}(v), \qquad \forall v \in V_h^{3D}, \qquad (2.15)$$

where B_h^{3D} and L_h^{3D} are obtained from B^{3D} and L^{3D} by using, instead of the exact chart Θ, the approximation

$$\Theta_h(x_1, x_2, x_3) = \mathcal{I}(\theta)(x_1, x_2) + x_3 \mathcal{I}(a_3)(x_1, x_2), \qquad (2.16)$$

to be compared with (2.8).

The general shell element procedure consists in solving problem (2.15), hence the unknowns (degrees of freedom) are the 3 displacement components and the 2 rotation components at the nodes. Note that the degrees of freedom are similar to those of a classical shell finite element procedure obtained by discretizing a Reissner–Mindlin-type model. However, it appears from the above description that general shell elements have two key advantages.

- They are not obtained from a specific shell model, but from a *general* 3D formulation, hence they can be easily used for arbitrary 3D material laws and also for nonlinear formulations. This is indeed why these procedures are called *general* shell elements.
- As implied by the use of the approximate chart Θ_h, the derivation of the associated matrix problem does not require the data of the exact mid-surface chart θ, but only the position of the points and the value of the normal vectors at the nodes. This is extremely convenient in practice as most structures analysed are the result of design procedures (CAD systems) that do not provide these charts.

The connections between general shell elements and classical shell elements (*i.e.*, those obtained by discretizing classical shell models) have remained unclear for many years. Moreover, this issue has probably been a source of major misunderstanding between the engineering community (primarily using and developing general shell elements) and the numerical analysis community (considering only classical shell elements). These connections are now much better identified; see in particular Chapelle and Bathe (2000). Indeed, for a given 3D material law we can consider the problem:
Find $(\boldsymbol{\zeta}, \boldsymbol{r})$ in $\boldsymbol{V}_N$ such that

$$B^{3D}(\boldsymbol{\zeta} + x_3\boldsymbol{r}, \boldsymbol{\eta} + x_3\boldsymbol{s}) = L^{3D}(\boldsymbol{\eta} + x_3\boldsymbol{s}), \quad \forall(\boldsymbol{\eta}, \boldsymbol{s}) \in \boldsymbol{V}_N. \tag{2.17}$$

Clearly, this defines a shell model of Reissner–Mindlin type, since the unknowns are a displacement field and a rotation field, both being given on the mid-surface (or in the reference domain). This shell model is the natural candidate for being the model that underlies general shell elements, *i.e.*, the model that provides the solution to which finite element solutions converge when the mesh parameter h (namely the diameter of the largest element) tends to zero. With a view to analysing problem (2.15) as an internal approximation of problem (2.17), we define V_h as the scalar finite element space associated with the isoparametric discretization above, *i.e.*, a function ϕ is in V_h if and only if $\phi = \mathcal{I}(\phi)$. Note that $V_h \subset H^1(\omega)$. Then $v \in V_h^{3D}$ is equivalent to

$$v = \eta(x_1, x_2) + x_3\,\tilde{s}(x_1, x_2), \tag{2.18}$$

with the three conditions:

C1 $\boldsymbol{\eta} \cdot \hat{\boldsymbol{e}}_i \in V_h$ and $\tilde{\boldsymbol{s}} \cdot \hat{\boldsymbol{e}}_i \in V_h$ for $i = 1, 2, 3$;
C2 $\tilde{\boldsymbol{s}} \cdot \boldsymbol{a}_3 = 0$ at all the nodes of the mesh;
C3 $\boldsymbol{\eta}$ and $\tilde{\boldsymbol{s}}$ satisfy the proper boundary conditions.

Denoting by π the operator that projects a vector field defined over ω onto the tangential plane of the mid-surface at each point, that is,

$$\pi(\tilde{\boldsymbol{s}}) = \tilde{\boldsymbol{s}} - (\tilde{\boldsymbol{s}} \cdot \boldsymbol{a}_3)\boldsymbol{a}_3, \tag{2.19}$$

we define the finite element space V_{Nh} as the set of functions $(\boldsymbol{\eta}, \pi(\tilde{\boldsymbol{s}}))$ such that $(\boldsymbol{\eta}, \tilde{\boldsymbol{s}})$ satisfies the three conditions (C1, C2, C3) above. Clearly V_{Nh} is a (finite-dimensional) subspace of V_N. Note that the use of the projection operator is required for that purpose because we do not directly have $\tilde{\boldsymbol{s}}$ in the tangential plane at every point of ω. Using the equivalence

$$\pi(\tilde{\boldsymbol{s}}) = \boldsymbol{s} \Leftrightarrow \tilde{\boldsymbol{s}} = \mathcal{I}(\boldsymbol{s}), \tag{2.20}$$

valid for any $(\boldsymbol{\eta}, \pi(\tilde{\boldsymbol{s}}))$ in V_{Nh}, we can then reformulate the general shell element procedure as follows:
Find $(\boldsymbol{\zeta}, \boldsymbol{r})$ in V_{Nh} such that

$$B_h^{3D}(\boldsymbol{\zeta} + x_3\,\mathcal{I}(\boldsymbol{r}), \boldsymbol{\eta} + x_3\,\mathcal{I}(\boldsymbol{s})) = L_h^{3D}(\boldsymbol{\eta} + x_3\,\mathcal{I}(\boldsymbol{s})), \ \forall(\boldsymbol{\eta}, \boldsymbol{s}) \in V_{Nh}. \tag{2.21}$$

Now comparing this equation with the continuous problem (2.17), we can say that the general shell element procedure corresponds to a discretization scheme of the continuous problem with approximate bilinear and linear forms (see, e.g., Ciarlet (1978)), and we observe that consistency errors come from two sources: the approximation of the geometry corresponding to equation (2.16) and the presence of the interpolation operator in (2.21).

When choosing specific 3D formulations, the connection between the shell model represented by (2.17) and the corresponding general shell element procedure can be further analysed. In particular, when considering Hooke's constitutive law we obtain, as shown in Chapelle and Bathe (2000),

$$B^{3D}(\boldsymbol{u}, \boldsymbol{v}) = \int_{\Omega^\varepsilon} [\tfrac{1}{2} g^{\alpha\beta\sigma\tau} e_{\alpha\|\beta}(\boldsymbol{u}) e_{\sigma\|\tau}(\boldsymbol{v})$$
$$+ 4\mu g^{\alpha\beta} e_{\alpha\|3}(\boldsymbol{u}) e_{\beta\|3}(\boldsymbol{v})] \sqrt{g} \ dx, \tag{2.22}$$

with

$$g^{\alpha\beta\sigma\tau} = \tfrac{4\lambda\mu}{\lambda+2\mu} g^{\alpha\beta} g^{\sigma\tau} + 2\mu(g^{\alpha\sigma} g^{\beta\tau} + g^{\alpha\tau} g^{\beta\sigma}). \tag{2.23}$$

Note that we assume that the Lamé constants do not depend on ε. Taking into account the Reissner–Mindlin kinematic assumption, we have

$$\begin{cases} e_{\sigma\|\tau}(\boldsymbol{\eta} + x_3\boldsymbol{s}) = \gamma_{\sigma\tau}(\boldsymbol{\eta}) - x_3\,\rho_{\sigma\tau}^N(\boldsymbol{\eta}, \boldsymbol{s}) - (x_3)^2 \kappa_{\sigma\tau}(\boldsymbol{s}), \\ e_{\beta\|3}(\boldsymbol{\eta} + x_3\boldsymbol{s}) = \gamma_{\beta3}(\boldsymbol{\eta}, \boldsymbol{s}), \end{cases} \tag{2.24}$$

where we recall that $\gamma_{\sigma\tau}$, $\rho_{\sigma\tau}^N$ and $\gamma_{\beta3}$ respectively represent the components of the tensors of membrane strains, bending (or flexural) strains and shear strains in the Naghdi model, and $\kappa_{\sigma\tau}$ corresponds to a new tensor defined by

$$\kappa_{\sigma\tau}(\boldsymbol{s}) = \tfrac{1}{2}(b_\sigma^\alpha s_{\alpha|\tau} + b_\tau^\alpha s_{\alpha|\sigma}). \tag{2.25}$$

In addition, we consider the external virtual work given by

$$L^{3D}(\boldsymbol{v}) = \int_{\Omega^\varepsilon} \boldsymbol{f} \cdot \boldsymbol{v} \sqrt{g} \ dx, \tag{2.26}$$

i.e., $\boldsymbol{f}$ represents the applied distributed force, and we assume $\boldsymbol{f} \in [L^2(\Omega^\varepsilon)]^3$. We then have the following result, under standard assumptions regarding the mesh and the finite element shape functions used; see Chapelle and Bathe (2000).

Proposition 1. Provided that the boundary conditions are admissible, problem (2.17) has a unique solution. Moreover, this solution is the limit of the solutions of Problem (2.15) when the mesh parameter h tends to zero.

This result shows that (at least in the case of Hooke's law) the shell model given by (2.17) really is the mathematical model that underlies general shell element procedures. It is then important to investigate the relations between this shell model (which we call the *basic shell model*; see Bathe and Chapelle (200x)) and classical shell models. If we formally transform the expression (2.22) by truncating the expansions in powers of x_3 of the expressions appearing under the integral sign, taking

- the first-order approximation of the strains, namely

$$\begin{cases} e_{\sigma\|\tau}(\boldsymbol{\eta} + x_3 \boldsymbol{s}) \approx \gamma_{\sigma\tau}(\boldsymbol{\eta}) - x_3\, \rho^N_{\sigma\tau}(\boldsymbol{\eta}, \boldsymbol{s}), \\ e_{\beta\|3}(\boldsymbol{\eta} + x_3 \boldsymbol{s}) = \gamma_{\beta3}(\boldsymbol{\eta}, \boldsymbol{s}), \end{cases} \qquad (2.27)$$

- the zero-order approximation of all geometric coefficients, namely

$$g^{\alpha\beta\sigma\tau} \approx a^{\alpha\beta\sigma\tau}, \quad g^{\alpha\beta} \approx a^{\alpha\beta}, \quad \sqrt{g} \approx \sqrt{a}, \qquad (2.28)$$

we obtain exactly the bilinear form of the Naghdi model (with $c = 8$ in the equations recalled in Ciarlet (2001)); see Chapelle and Bathe (2000), and also Delfour (1998) where various combinations of truncations are considered and analysed. This suggests that some close connections exist between the basic model and the Naghdi model. We will further substantiate these connections in Section 3.1 by showing that the two models are *asymptotically equivalent* as regards their asymptotic solutions when the thickness parameter tends to zero.

As a conclusion, we can say that, although they do not seem (at first sight) to bear much resemblance to classical shell elements, general shell elements are – in fact – non-conforming discretization schemes of a shell model of Reissner–Mindlin type which is itself very closely connected to the Naghdi model. This is extremely important, as it allows us to analyse finite element procedures of both 'families' using similar tools. As a matter of fact, further advances in the analysis of some specific general shell procedures have already been obtained with this approach (*cf.*, *e.g.*, Malinen (2000)), in particular as regards the reliability of shell finite elements, which is the subject of our next section.

3. Computational reliability issues for thin shells

The essential motivation underlying the formulation of shell models and shell
finite elements is the reduction of the cost of the analysis (in particular the
computational cost), compared to the cost of a full 3D analysis, by instead
considering a 2D problem. This motivation is based on the assumption
(or the 'hope') that the accuracy of the solution obtained only depends
on criteria that prevail in *2D* analysis, that is, typically on the fineness of
the 2D mesh *regardless of the third dimension*, namely the thickness of the
structure. This means that we implicitly expect uniform convergence of the
finite element solution with respect to the thickness parameter, namely, error
bounds of the type

$$\frac{\|\zeta^{\varepsilon} - \zeta^{\varepsilon}_h\|_*}{\|\zeta^{\varepsilon}\|_*} \leq Ch^p, \tag{3.1}$$

where we now use ε as a superscript to indicate that the solutions depend
on this parameter, and where C (and also p) *should not depend on ε.* We
subscript the norm symbol with a '$*$' to signify that we do not necessarily
require that uniform convergence hold in the Sobolev space in which the
problem is originally set (for example $[H^1(\omega)]^2 \times H^2(\omega)$ for Kirchhoff–Love-
type models), but that we may tolerate uniform convergence in weaker norms
to be defined.

Unfortunately, it was soon recognized that standard finite element methods (such as the ones that we described in the previous section) do not
provide such uniform bounds in general. The reason for this will be analysed in Section 3.2. To that purpose it is necessary to start by surveying
the behaviour of the exact solution of the shell model when the thickness
parameter tends to zero, namely the asymptotic behaviour of shell models,
which is the objective of the forthcoming section. This analysis will, in particular, allow us to specify the norms in which we seek uniform estimates
for approximate solutions.

3.1. Asymptotic behaviour of shell models

Most classical shell models can be written in the following generic form:
Find $Z^{\varepsilon} \in V$ such that

$$\varepsilon^3 B_F(Z^{\varepsilon}, E) + \varepsilon B_M(Z^{\varepsilon}, E) = L^{\varepsilon}(E), \quad \forall E \in V. \tag{3.2}$$

The meaning of the symbols appearing in this formulation is as follows.

- Z^{ε}: the unknown solution, namely the displacement of the mid-surface
  for a Kirchhoff–Love-type model, or this displacement *and* the rotation
  of the normal fibre for a Reissner–Mindlin-type model.
- E: a corresponding test function that we call a 'displacement' (even
  though it also contains a rotation for a Reissner–Mindlin-type model).

- V: the solution space (we recall that the definition of this space takes into account the essential boundary conditions).

- B_F: a scaled representation of the bending (flexural) energy.

- B_M: a scaled representation of the membrane energy for a Kirchhoff–Love-type model, or of the membrane energy *and* the shear energy for a Reissner–Mindlin-type model.

- $L^\varepsilon(E)$: the external virtual work associated with E.

For example, for the Koiter model we have, as seen in Koiter (1966),

$$B_F^K(\zeta, \eta) = \tfrac{1}{3} \int_\omega a^{\alpha\beta\sigma\tau} \rho_{\alpha\beta}(\zeta)\rho_{\sigma\tau}(\eta)\sqrt{a}\ dy, \qquad (3.3)$$

$$B_M^K(\zeta, \eta) = \int_\omega a^{\alpha\beta\sigma\tau} \gamma_{\alpha\beta}(\zeta)\gamma_{\sigma\tau}(\eta)\sqrt{a}\ dy, \qquad (3.4)$$

and for the Naghdi model

$$B_F^N((\zeta, r), (\eta, s)) = \tfrac{1}{3} \int_\omega a^{\alpha\beta\sigma\tau} \rho_{\alpha\beta}^N(\zeta, r)\rho_{\sigma\tau}^N(\eta, s)\sqrt{a}\ dy, \qquad (3.5)$$

$$B_M^N((\zeta, r), (\eta, s)) = \qquad\qquad\qquad\qquad\qquad\qquad\qquad (3.6)$$
$$\int_\omega [a^{\alpha\beta\sigma\tau} \gamma_{\alpha\beta}(\zeta)\gamma_{\sigma\tau}(\eta) + c\mu a^{\alpha\beta} \gamma_{\alpha3}(\zeta, r)\gamma_{\beta3}(\eta, s)]\sqrt{a}\ dy.$$

We further emphasize that, in (3.2), *the expressions of the bilinear forms B_F and B_M do not depend on the thickness parameter ε.* We also introduced ε as a superscript in the right-hand side of the formulation because it is unlikely to obtain well-posed asymptotic behaviour while keeping the loading constant over the whole sequence of problems. More specifically, what we will be looking for in the asymptotic analysis is a scaling of the right-hand side in the form

$$L^\varepsilon(E) = \varepsilon^\rho G(E), \qquad (3.7)$$

where G is a function of V' *independent of ε* and ρ is a real number, for which the scaled external work $G(Z^\varepsilon)$ converges to a *finite and nonzero* limit when ε tends to zero. In this case, we will say that the given scaling provides *admissible asymptotic behaviour*. Of course, this is equivalent to requiring that the scaled internal work $\varepsilon^{3-\rho}B_F(Z^\varepsilon, Z^\varepsilon) + \varepsilon^{1-\rho}B_M(Z^\varepsilon, Z^\varepsilon)$ have a finite nonzero limit, since

$$\varepsilon^{3-\rho}B_F(Z^\varepsilon, Z^\varepsilon) + \varepsilon^{1-\rho}B_M(Z^\varepsilon, Z^\varepsilon) = G(Z^\varepsilon). \qquad (3.8)$$

We can then show the following result; see Blouza, Brezzi and Lovadina (1999), Baiocchi and Lovadina (200x) and Bathe and Chapelle (200x).

Proposition 2.   There is at most one exponent ρ that provides admissible asymptotic behaviour. In addition, if such a number exists we have

$$1 \le \rho \le 3. \tag{3.9}$$

Remark. In some cases the convergence may also be such that the sequence (Z^{ε}) tends to some limit in V, but we want to allow for more general situations in which a weaker convergence property may hold.

As discussed in Sanchez-Hubert and Sanchez-Palencia (1997) (see also Ciarlet (2001) for the Koiter model), classical shell models feature asymptotic behaviour that dramatically differs according to the contents of the subspace:

$$V_F = \{E \in V : B_M(E, E) = 0\}. \tag{3.10}$$

This subspace contains the displacements that have zero membrane energy (and also zero shear energy for Reissner–Mindlin-type models). For this reason, these displacements are called *pure-bending displacements* (since only the bending energy is nonzero). As a matter of fact, because of the severe kinematic constraints corresponding to zero membrane strains (3 equations versus 3 components of displacements), it may very well happen that

$$V_F = \{0\}. \tag{3.11}$$

In this case we say that *pure bending is inhibited.* This situation induces asymptotic behaviour very different from when pure bending is not inhibited. The existence of nonzero pure bending displacements is governed by the geometry of the mid-surface and the boundary conditions prescribed on the structure (we further analyse this issue below).

We start by considering the case of non-inhibited pure bending. It can then be shown that well-defined asymptotic behaviour exists for $\rho = 3$. In this case, the variational formulation can indeed be re-written as

$$B_F(Z^{\varepsilon}, E) + \frac{1}{\varepsilon^2} B_M(Z^{\varepsilon}, E) = G(E), \quad \forall E \in V. \tag{3.12}$$

Since ε is a small parameter, this equation can be interpreted as the *penalized form* of a problem in which the solution Z^0 is constrained to lie in the subspace V_F. Namely, the candidate limit problem for (3.12) is:
Find Z^0 in V_F such that

$$B_F(Z^0, E) = G(E), \quad \forall E \in V_F. \tag{3.13}$$

Note that this problem is well-posed, since it corresponds to the restriction to V_F, a closed subspace of V, of the minimization problem equivalent to the original variational formulation. We then have the following result; see Chenais and Paumier (1994), and also Pitkäranta (1992) and Sanchez-Palencia (1992).

Proposition 3. Assume that pure bending is not inhibited. Then, setting $\rho = 3$, $\boldsymbol{Z}^{\varepsilon}$ converges strongly in $\boldsymbol{V}$ to $\boldsymbol{Z}^{0}$, the solution of (3.13). In addition we have

$$\lim_{\varepsilon \to 0} \frac{1}{\varepsilon^2} B_M(\boldsymbol{Z}^{\varepsilon}, \boldsymbol{Z}^{\varepsilon}) = 0. \tag{3.14}$$

Equation (3.14) shows that, in the scaled deformation energy, the part corresponding to the membrane energy (and also the part corresponding to the shear energy when applicable) tends to zero, while the part that corresponds to bending deformations tends to a finite value. This is why this asymptotic behaviour is also called *bending-dominated behaviour*.

Note that, when pure bending is not inhibited, Proposition 3 shows that we have admissible asymptotic behaviour for the scaling $\rho = 3$ *provided that* $\boldsymbol{Z}^{0}$ *is nonzero.* If $\boldsymbol{Z}^{0} = \boldsymbol{0}$, that is, if and only if

$$G(\boldsymbol{E}) = 0, \quad \forall \boldsymbol{E} \in \boldsymbol{V}_F, \tag{3.15}$$

another admissible scaling (for $1 \le \rho < 3$) may exist. This, however, does not correspond to a 'physical situation' as small perturbations in the geometry (producing variations in $\boldsymbol{V}_F$) or in the loading lead to a violation of (3.15), hence to effects of smaller asymptotic order, namely of 'larger magnitude' (corresponding to bending-dominated behaviour, that is, $\rho = 3$). Therefore we do not further consider this situation; see Chapelle and Bathe (1998) for a more detailed analysis of this issue.

By contrast, when pure bending is inhibited (which is a situation more common than the previous one, although both may be encountered in practice – see Ciarlet (2001) and Bathe and Chapelle (200x)), the asymptotic behaviour is very different and more complex. We consider the tentative scaling $\rho = 1$. The variational formulation then gives

$$B_M(\boldsymbol{Z}^{\varepsilon}, \boldsymbol{E}) + \varepsilon^2 B_F(\boldsymbol{Z}^{\varepsilon}, \boldsymbol{E}) = G(\boldsymbol{E}), \quad \forall \boldsymbol{E} \in \boldsymbol{V}, \tag{3.16}$$

and we note that

$$\|\boldsymbol{E}\|_M = [B_M(\boldsymbol{E}, \boldsymbol{E})]^{\frac{1}{2}} \tag{3.17}$$

defines a norm over $\boldsymbol{V}$ which, of course, satisfies

$$\|\boldsymbol{E}\|_M \le C\|\boldsymbol{E}\|_V, \quad \forall \boldsymbol{E} \in \boldsymbol{V}, \tag{3.18}$$

but is not equivalent to the original norm (*cf.* examples of the Koiter and Naghdi models above). We call this new norm the *membrane energy norm*. We then recognize in (3.16) a *singular perturbation problem*. Defining $\boldsymbol{V}_M$ as the space obtained by completion of $\boldsymbol{V}$ for the membrane energy norm, we introduce the variational problem:
Find $\boldsymbol{Z}^M$ in $\boldsymbol{V}_M$ such that

$$B_M(\boldsymbol{Z}^M, \boldsymbol{E}) = G(\boldsymbol{E}), \quad \forall \boldsymbol{E} \in \boldsymbol{V}_M. \tag{3.19}$$

Clearly, this problem is well-posed *provided that* $G \in (V_M)'$. Note that this corresponds to a restriction on the loading since $(V_M)' \subset V'$. An equivalent way of writing this condition is

$$|G(E)| \le C[B_M(E, E)]^{\frac{1}{2}}, \quad \forall E \in V \tag{3.20}$$

since, by density considerations, this is equivalent to the same condition holding for any $E \in V_M$. Under this assumption we have admissible asymptotic behaviour as shown in the following proposition; see Lions (1973).

Proposition 4. Assume that pure bending is inhibited and that $G \in (V_M)'$. Then, setting $\rho = 1$, Z^ε converges strongly in V_M to Z^M, the solution of (3.19). In addition we have

$$\lim_{\varepsilon \to 0} \varepsilon^2 B_F(Z^\varepsilon, Z^\varepsilon) = 0. \tag{3.21}$$

Since, by (3.21), the bending part of the scaled energy tends to zero while the remaining part (the membrane part, and also the shear part for Reissner–Mindlin-type models) tends to a finite value, this asymptotic behaviour is called *membrane-dominated behaviour*. Note that (under the assumptions of the proposition) we always have admissible asymptotic behaviour since Z^M cannot be zero (unless, of course, G is zero).

We have therefore identified two very distinct categories of admissible asymptotic behaviour for classical shell models: bending-dominated behaviour and membrane-dominated behaviour. It is the contents of the subspace V_F that determine into which category a given problem falls. For each of these types of asymptotic behaviour, convergence of the solutions is obtained in a given norm, hence it is natural to seek uniform convergence of finite element solutions in the same norm (*cf.* the next section). Other types of asymptotic behaviour may exist and the above discussion shows that they can arise only when pure bending is inhibited and $G \notin (V_M)'$ (disregarding cases when (3.15) holds, for the above-mentioned reason). These situations are more complex than the two 'clear-cut' asymptotic behaviour types discussed, and they have not yet been completely elucidated, although significant advances have been made in their analysis; see in particular Pitkäranta and Sanchez-Palencia (1997) (*cf.* also Section 3.3).

We now consider the case of the basic shell model, namely the shell model that underlies general shell element. We recall that the general asymptotic analysis performed above is not directly applicable to this model since it cannot be written in the form (3.2). In particular the membrane and bending deformation energies are coupled in the term

$$\int_{\Omega^\varepsilon} \tfrac{1}{2} g^{\alpha\beta\sigma\tau} e_{\alpha\|\beta}(u) e_{\sigma\|\tau}(v) \sqrt{g} \, \mathrm{d}x, \tag{3.22}$$

as shown in (2.24). Another significant difference is that the loading acting on the structure is three-dimensional, taking the form

$$L^{(3D)}(v) = \int_{\Omega^\varepsilon} f^\varepsilon \cdot v \sqrt{g}\, dx. \tag{3.23}$$

In order to perform an asymptotic analysis consistent with the general one above, we consider the asymptotic assumption

$$f^\varepsilon = \varepsilon^{\rho-1} g, \tag{3.24}$$

recalling that the external work involves an integration over the thickness. Furthermore, we assume that g is sufficiently regular to allow the following expansion:

$$g(x_1, x_2, x_3) = g_0(x_1, x_2) + x_3 g_1(x_1, x_2) + (x_3)^2 g_2(x_1, x_2, x_3), \tag{3.25}$$

where g_0 and g_1 are in $L^2(\omega)$, while g_2 is a bounded function. Let $(\zeta^\varepsilon, r^\varepsilon)$ be the solution of the basic shell model, that is:
Find $(\zeta^\varepsilon, r^\varepsilon) \in V_N$ such that

$$B^{3D}(\zeta^\varepsilon + x_3 r^\varepsilon, \eta + x_3 s) = \tag{3.26}$$
$$\varepsilon^{\rho-1} \int_{\Omega^\varepsilon} g \cdot (\eta + x_3 s)\sqrt{g}\, dx, \quad \forall(\eta, s) \in V_N.$$

We also introduce the following Naghdi problem:
Find $(\tilde{\zeta}^\varepsilon, \tilde{r}^\varepsilon) \in V_N$ such that

$$\varepsilon^3 B_F^N((\tilde{\zeta}^\varepsilon, \tilde{r}^\varepsilon), (\eta, s)) + \varepsilon B_M^N((\tilde{\zeta}^\varepsilon, \tilde{r}^\varepsilon), (\eta, s)) = \tag{3.27}$$
$$\varepsilon^{\rho-1} G(\eta), \quad \forall(\eta, s) \in V_N,$$

where

$$G(\eta) = \int_\omega g_0 \cdot \eta \sqrt{a}\, dy, \tag{3.28}$$

and setting $c = 8$ in B_F^N; see Ciarlet (2001). Calling V_F^N the pure-bending displacement subspace relative to this Naghdi problem (3.27), and V_M^N the completed space for the membrane energy norm when pure bending is inhibited, we then have the following asymptotic properties (assuming the chart is smooth); see Chapelle and Bathe (2000).

Proposition 5. If pure bending is not inhibited in the Naghdi problem (3.27), then for the scaling $\rho = 3$ both sequences $(\zeta^\varepsilon, r^\varepsilon)$ and $(\tilde{\zeta}^\varepsilon, \tilde{r}^\varepsilon)$ converge in V_N to the same limit (ζ^0, r^0), which is the element of V_F^N that satisfies

$$B_F^N((\zeta^0, r^0), (\eta, s)) = G(\eta), \quad \forall(\eta, s) \in V_F^N. \tag{3.29}$$

Alternatively, if pure bending is inhibited and if $G \in (V_F^N)'$, then for the scaling $\rho = 1$ both sequences $(\zeta^\varepsilon, r^\varepsilon)$ and $(\tilde{\zeta}^\varepsilon, \tilde{r}^\varepsilon)$ converge in V_M^N to the same limit (ζ^M, r^M), which is the element of V_M^N that satisfies

$$B_M^N((\zeta^M, r^M), (\eta, s)) = G(\eta) \quad \forall (\eta, s) \in V_M^N. \tag{3.30}$$

Therefore, the solution of the basic shell model converges to the same limit solutions and under the exact same assumptions as the solution of the Naghdi problem (3.27). To summarize these properties, we will say that the basic shell model and the Naghdi model are 'asymptotically equivalent'.

We close this section by providing some guidelines for the analysis of the subspace of pure bending displacements, since the contents of this subspace have been shown to crucially determine the asymptotic behaviour of the shell problems. First of all we note that, for both Kirchhoff–Love-type and Reissner–Mindlin-type models, pure bending displacements have identically zero membrane strains. In addition, for a Reissner–Mindlin-type model pure bending displacements also have zero shear strains. However, the condition of zero shear strains can be written in the form of an explicit expression giving the rotation field η in terms of the displacement field s, namely

$$\eta_\alpha = -(s_{3,\alpha} + b_\alpha^\sigma s_\sigma). \tag{3.31}$$

By contrast, the condition of vanishing membrane strains

$$\gamma_{\alpha\beta}(\eta) = 0, \quad \forall \alpha, \beta = 1, 2, \tag{3.32}$$

make up an exactly determined differential system (three equations versus three unknowns) that must be satisfied by the displacement field only. Hence this is clearly the crucial condition that determines the contents of the pure bending displacement subspace, which is why elements of this subspace are also called *inextensional displacements* (we recall that the membrane strain tensor is the tensor of linearized change of metric of the mid-surface – see Koiter (1966)), even in the case of Reissner–Mindlin-type models. In addition, system (3.32) enjoys the following remarkable property (*cf.* Sanchez-Palencia (1989a, 1989b) and the references therein).

Proposition 6. The differential nature (elliptic, parabolic or hyperbolic) of system (3.32) is the same as the geometric nature of the mid-surface at the point in consideration. Furthermore, when such a concept is relevant (namely in the hyperbolic and – by extension – parabolic case) the characteristics of the system are also the asymptotic lines of the surface.

This property is extremely valuable for analysing the contents of the subspace of pure bending displacements. In particular it shows how the geometry (*i.e.*, the type of mid-surface) and the boundary conditions can affect the contents of the subspace (*cf.*, *e.g.*, Chapelle and Bathe (1998) for examples). We refer to Sanchez-Hubert and Sanchez-Palencia (1997) for a more detailed analysis of this issue.

3.2. Asymptotic reliability of shell finite elements

We recall that our objective is to obtain finite element procedures that behave uniformly well with respect to the thickness parameter, that is, for which the finite element solution $\boldsymbol{Z}_h^\varepsilon$ satisfies an estimate of the type

$$\frac{\|\boldsymbol{Z}^\varepsilon - \boldsymbol{Z}_h^\varepsilon\|_*}{\|\boldsymbol{Z}^\varepsilon\|_*} \leq Ch^k, \tag{3.33}$$

where the norm used remains to be specified. In the light of the above discussion regarding the asymptotic behaviour of shell models, it is natural to require that

(1) in a bending-dominated case, such a uniform estimate should hold in the norm of the displacement space $\boldsymbol{V}$;

(2) in a membrane-dominated case, such a uniform estimate should hold in the membrane energy norm.

We consider a standard finite element procedure obtained by discretizing the variational formulation (3.2), namely the problem:
Find $\boldsymbol{Z}_h^\varepsilon \in \boldsymbol{V}_h$ such that

$$\varepsilon^3 B_F(\boldsymbol{Z}_h^\varepsilon, \boldsymbol{E}) + \varepsilon B_M(\boldsymbol{Z}_h^\varepsilon, \boldsymbol{E}) = L^\varepsilon(\boldsymbol{E}), \quad \forall \boldsymbol{E} \in \boldsymbol{V}_h, \tag{3.34}$$

where $\boldsymbol{V}_h$ denotes the finite element displacement space used. For a membrane-dominated problem we – in fact – need to consider the scaled problem

$$B_M(\boldsymbol{Z}_h^\varepsilon, \boldsymbol{E}) + \varepsilon^2 B_F(\boldsymbol{Z}_h^\varepsilon, \boldsymbol{E}) = G(\boldsymbol{E}), \quad \forall \boldsymbol{E} \in \boldsymbol{V}_h. \tag{3.35}$$

We can then show that we have a uniform convergence property, as proved in Chapelle and Bathe (1998).

Proposition 7. Assume that there exist two interpolation operators $\mathcal{I}_h$ and $\mathcal{J}_h$, defined respectively in $\boldsymbol{V}_M$ and $\boldsymbol{V}$, and both with values in $\boldsymbol{V}_h$, such that

$$\begin{cases} \|\mathcal{I}_h(\boldsymbol{E})\|_M \leq C\|\boldsymbol{E}\|_M, & \forall \boldsymbol{E} \in \boldsymbol{V}_M, \\ \|\mathcal{I}_h(\boldsymbol{E})\|_{\boldsymbol{V}} \leq C\|\boldsymbol{E}\|_{\boldsymbol{V}}, & \forall \boldsymbol{E} \in \boldsymbol{V}, \\ \lim_{h\to 0} \|\boldsymbol{E} - \mathcal{I}_h(\boldsymbol{E})\|_M = 0, & \forall \boldsymbol{E} \in \boldsymbol{V}_M, \end{cases} \tag{3.36}$$

$$\begin{cases} \|\mathcal{J}_h(\boldsymbol{E})\|_{\boldsymbol{V}} \leq C\|\boldsymbol{E}\|_{\boldsymbol{V}}, & \forall \boldsymbol{E} \in \boldsymbol{V}, \\ \lim_{h\to 0} \|\boldsymbol{E} - \mathcal{J}_h(\boldsymbol{E})\|_{\boldsymbol{V}} = 0, & \forall \boldsymbol{E} \in \boldsymbol{V}. \end{cases} \tag{3.37}$$

Then, for any fixed $\varepsilon_{\max} > 0$ we have

$$\lim_{h\to 0} \sup_{\varepsilon\in]0,\varepsilon_{\max}]} \{\|\boldsymbol{Z}^\varepsilon - \boldsymbol{Z}_h^\varepsilon\|_M + \varepsilon\|\boldsymbol{Z}^\varepsilon - \boldsymbol{Z}_h^\varepsilon\|_{\boldsymbol{V}}\} = 0. \tag{3.38}$$

If we consider only the membrane energy norm in this uniform convergence result, this bound is weaker than an estimate of the type (3.33) in the membrane energy norm. However, this is clearly the best estimate that we can obtain without uniform regularity results on the exact solutions, and such regularity results are not available. We further point out that the interpolation assumptions (3.36) and (3.37) are rather unrestrictive. For example, the Clément operator satisfies these two sets of assumptions when V_M is a Sobolev space on which this operator is properly defined.

By contrast, in a bending-dominated situation, the problem to be considered is instead

$$B_F(Z_h^\varepsilon, E) + \frac{1}{\varepsilon^2} B_M(Z_h^\varepsilon, E) = G(E), \quad \forall E \in V_h. \tag{3.39}$$

The major difficulty associated with this type of finite element formulation is the classical *numerical locking* phenomenon. As is well known from other penalized formulations (*e.g.*, nearly incompressible elasticity, beams, plates), locking occurs at its worst in situations for which

$$V_h \cap V_F = \{0\}. \tag{3.40}$$

Indeed, Proposition 3 can now be applied with V_h instead of V. We thus have that, keeping the finite element subspace (in particular the mesh) fixed, when ε tends to zero the finite element solution Z_h^ε converges to the solution of the limit problem:
Find Z_h^0 in $V_h \cap V_F$ such that

$$B_F(Z_h^0, E) = G(E), \quad \forall E \in V_h \cap V_F, \tag{3.41}$$

and of course the solution of this problem is the zero displacement if (3.40) holds. Even though the thickness is always finite in practice, this implies that no uniform error bound of the type (3.33) holds in this case, and more specifically that, for a given mesh, the finite element solution gets 'smaller and smaller' when the thickness decreases (whereas the exact solution does not vanish). Hence, this is a purely numerical artefact that makes the structure appear stiffer as it gets thinner, which explains the terminology 'numerical locking'. The specific difficulty with shells is that, unlike for beams or plates, the pathological situation expressed by (3.40) is the common rule, as illustrated in the following result, for which we also give the proof because it is simple and illuminating as to how the geometry influences the asymptotic properties (*cf.* also Choï, Palma, Sanchez-Palencia and Vilarino (1998) and Sanchez-Hubert and Sanchez-Palencia (1997) for other results concerning (3.40)).

Proposition 8. Consider a regular hyperbolic shell fixed on some part of its boundary, and a finite element scheme, in the framework of (3.39), in which the displacement components η_1, η_2 and η_3 are approximated using continuous piecewise-polynomial functions. Assume that no element edge in the mesh is part of an asymptotic line. Then (3.40) holds.

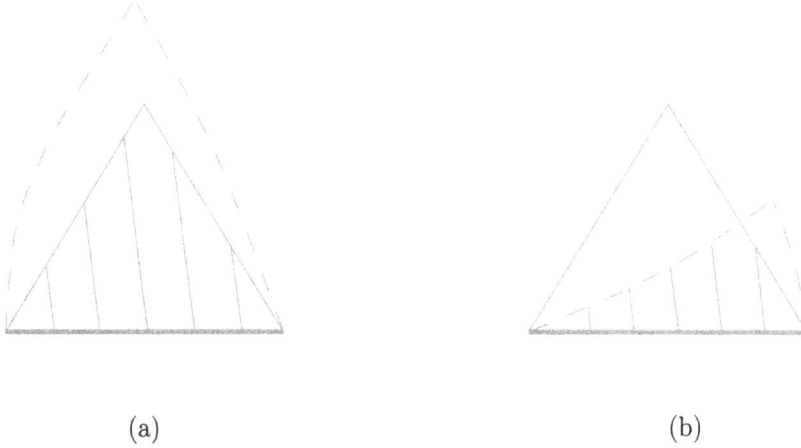

(a)   (b)

Fig. 3.1. Asymptotic lines and inhibited region for an element

Proof.   We will show that, for any inextensional displacement η, if η is zero on some edge of any element of the mesh, then η is identically zero over the whole element. This being granted, the result is immediately obtained by 'propagating' the zero displacements from the boundary conditions to the whole domain.

Consider an arbitrary element of the mesh, with η set to zero on any of its edges. Then we have two possibilities: *either*

(a) the lattice of asymptotic lines originating from this edge covers the element completely (as in Figure 3.1(a)) and the conclusion directly follows from Proposition 6;   *or*

(b) the element is only partially covered (Figure 3.1(b)), but in this case we still have $\eta_1 = \eta_2 = \eta_3 = 0$ on a part of the nonzero area of the element. Recalling that the displacement components are given by polynomial functions, this implies that they are zero over the whole element.   ☐

In order to make the statement of this property simpler, we have assumed that no single element edge corresponds to an asymptotic curve. It is obvious, however, that this result carries over to any case in which the same propagation technique can be applied, which is much more general. A practical consequence of this discussion is that, when using benchmark problems to detect locking phenomena in shells numerically, we should not align the mesh with the asymptotic lines of the mid-surface when such lines exist, in order to avoid inhibiting the locking mechanism identified in the above proof. This holds in particular for *cylindrical test problems* for which the (very strong) temptation to align the mesh along the axis should definitely

be resisted if one wants to draw conclusions of somewhat general significance (note that, in industrial computations, aligning meshes along lines of
specific geometric properties is impossible in general).

We further emphasize that Proposition 8 reveals a significant difference
between finite element methods for shells and for other structures. For
beams and plates, it is indeed always possible to avoid (3.40) by raising
the polynomial degrees of the discretization spaces; see in particular Arnold
(1981). Here our argument is independent of the polynomial degree, hence
locking always occurs using displacement-based finite element formulations.
But of course, for a given thickness, the errors in the solution may be ac-
ceptably small if the order of the polynomials is sufficiently high and the
mesh is sufficiently fine.

In order to circumvent the locking phenomenon, it is now classical to resort
to *mixed formulations*, in which an additional unknown – corresponding to
the 'delinquent' term of the energy – is introduced; see Brezzi and Fortin
(1991) and Bathe (1996). For example, for Reissner–Mindlin plates this
additional unknown corresponds to the shear stress. For shell formulations,
it is thus natural to introduce the membrane stresses (and the shear stresses
for Reissner–Mindlin-type models) as auxiliary unknowns. For instance,
for the Koiter shell model the additional unknown is the membrane stress
tensor, namely the tensor with contravariant components

$$m^{\alpha\beta} = \frac{1}{\varepsilon^2} a^{\alpha\beta\sigma\tau} \gamma_{\sigma\tau}(\boldsymbol{\zeta}). \tag{3.42}$$

If we symbolically denote by $\boldsymbol{\Sigma}^\varepsilon$ the group of additional stress unknowns,
and by $\boldsymbol{\Xi}$ the corresponding test function, the mixed formulation equivalent
to (3.12) then reads

$$\begin{cases} B_F(\boldsymbol{Z}^\varepsilon, \boldsymbol{E}) + M(\boldsymbol{E}, \boldsymbol{\Sigma}^\varepsilon) = G(\boldsymbol{E}), & \forall \boldsymbol{E} \in \boldsymbol{V}, \\ M(\boldsymbol{Z}^\varepsilon, \boldsymbol{\Xi}) - \varepsilon^2 D(\boldsymbol{\Sigma}^\varepsilon, \boldsymbol{\Xi}) = 0, & \forall \boldsymbol{\Xi} \in \boldsymbol{L}^2, \end{cases} \tag{3.43}$$

where the second equation expresses the stress–strain relationship for the
additional unknowns, and $\boldsymbol{L}^2$ denotes the product space obtained by taking
$L^2(\omega)$ for each component of the stress term. By construction, $D(\boldsymbol{\Xi}, \boldsymbol{\Xi})^{1/2}$
defines a norm which is equivalent to the L^2-norm, and the mixed bilinear
form M is continuous over $\boldsymbol{V} \times \boldsymbol{L}^2$; see, *e.g.*, Arnold and Brezzi (1997).
Then, both fields (displacements and stresses) are discretized in the finite
element procedure. In general, discontinuous shape functions are used for
the stresses, in order to allow the elimination of the stress degrees of freedom
at the element level. The resulting finite element formulation can then be
written as

$$B_F(\boldsymbol{Z}_h^\varepsilon, \boldsymbol{E}) + \frac{1}{\varepsilon^2} B_M^h(\boldsymbol{Z}_h^\varepsilon, \boldsymbol{E}) = G(\boldsymbol{E}), \quad \forall \boldsymbol{E} \in \boldsymbol{V}_h, \tag{3.44}$$

where B_M^h is a bilinear form that 'resembles B_M', but for which the constraint $B_M^h(\boldsymbol{E}, \boldsymbol{E}) = 0$ is 'more easily satisfied' for finite element displacements than $B_M(\boldsymbol{E}, \boldsymbol{E}) = 0$. Note that, from this point of view, this strategy is fairly close to the engineering-based idea of reduced integration. In fact, in some cases it has been possible to substantiate some previously proposed reduced-integration schemes using the theory of mixed methods (cf., e.g., Lyly (2000)). When the analysis of the mixed formulation can be achieved, it indeed provides a proof that uniform convergence with respect to ε holds (in the norm of $\boldsymbol{V}$). This analysis, however, relies on a crucial inf-sup condition which can be written in the general form

$$\sup_{\boldsymbol{E} \in V_h} \frac{M(\boldsymbol{E}, \boldsymbol{\Xi})}{\|\boldsymbol{E}\|_V} \geq \gamma \sup_{\boldsymbol{E} \in V} \frac{M(\boldsymbol{E}, \boldsymbol{\Xi})}{\|\boldsymbol{E}\|_V}, \qquad \forall \boldsymbol{\Xi} \in S_h, \tag{3.45}$$

where S_h denotes the discrete stress space and γ is a strictly positive constant independent of h and ε. Defining the seminorm

$$|\boldsymbol{\Xi}|_S = \sup_{\boldsymbol{E} \in V} \frac{M(\boldsymbol{E}, \boldsymbol{\Xi})}{\|\boldsymbol{E}\|_V}, \tag{3.46}$$

we have the equivalent – more classical – expression of the inf-sup condition

$$\inf_{\substack{\boldsymbol{\Xi} \in S_h \\ |\boldsymbol{\Xi}|_S \neq 0}} \sup_{\boldsymbol{E} \in V_h} \frac{M(\boldsymbol{E}, \boldsymbol{\Xi})}{|\boldsymbol{\Xi}|_S \|\boldsymbol{E}\|_V} \geq \gamma > 0. \tag{3.47}$$

This inf-sup condition requires a deep correlation between the displacement and stress finite element spaces in order to be fulfilled. For shell formulations, the problem of finding a couple of displacement/stress finite element spaces in order to satisfy the corresponding inf-sup condition is still open (cf. Arnold and Brezzi (1997) for an attempt in this direction).

In the case of nearly incompressible elasticity where locking also arises, owing to the kinematic constraint introduced by the incompressibility condition, a numerical procedure – called the inf-sup test – has been introduced in Chapelle and Bathe (1993) to test whether the inf-sup condition is satisfied without resorting to an actual mathematical proof. This procedure has been shown to give very reliable results (i.e., by clearly indicating whether the condition is satisfied or not) on numerous finite element procedures for which theoretical results where available; see also Iosilevich, Bathe and Brezzi (1997) for plate formulations. A similar test has been devised and used in Bathe, Iosilevich and Chapelle (2000) for shells. The MITC shell elements (MITC4, MITC9 and MITC16 elements, cf. Bathe and Dvorkin (1986) and Bucalem and Bathe (1993)), which are based on a mixed formulation, have passed this numerical testing procedure, whereas the corresponding standard finite elements (which are also called displacement-based methods to distinguish them from mixed methods) have all failed as expected. However, the absence of theoretical results pertaining to shell inf-sup conditions does not allow a thorough validation of this test at present. Moreover, two other reasons make it difficult to advocate the shell inf-sup test as strongly as for nearly incompressible elasticity.

(1) The dramatic dependence of the shell behaviour on the geometry (and the boundary conditions), which implies that many configurations should probably be considered in order to reach an acceptable level of confidence (only one configuration was used for nearly incompressible elasticity).

(2) The inf-sup condition that the test is based upon is in fact stronger than the inf-sup condition corresponding to the shell mixed formulation, which means that a given finite element procedure could satisfy the actual inf-sup condition and yet fail in the numerical test.

Given the considerable difficulty in proving an inf-sup condition for shell finite elements, another interesting idea, proposed and analysed in Bramble and Sun (1997), is to relax the inf-sup condition while tolerating some degree of non-uniformity in the convergence of the approximate solutions; namely, a mixed method is used for which a stability condition weaker than the inf-sup condition (3.45) is shown to hold, and a uniform error estimate is then obtained *under the condition*

$$h \leq C\sqrt{\varepsilon}. \tag{3.48}$$

This means that locking should not appear unless a very small thickness (or a coarse mesh) is used. However, a thorough numerical testing of this finite element procedure is lacking, as regards locking as well as other difficulties featured by mixed finite elements for shells (see below).

Finally, as regards the mathematical analysis of the locking of shell problems, to the best of our knowledge the only shell finite element method for which an analysis *valid for general geometries* has provided a uniform error estimate is the method given in Chapelle and Stenberg (1998). This method corresponds to a *mixed–stabilized* formulation in which additional stabilizing terms are added to the mixed formulation in order to bypass the inf-sup condition.

In the framework of shell models, mixed formulations feature another serious difficulty, which is that they are clearly tailored to the approximation of a problem in which a constraint is enforced by penalization. But for shell problems to fall into that category, according to our above discussion the asymptotic behaviour must be of the bending-dominated type. On the other hand, if the problem solved using a mixed method happens to be of the membrane-dominated type, instead of (3.44) we in fact need to consider

$$B_M^h(\boldsymbol{Z}_h^\varepsilon, \boldsymbol{E}) + \varepsilon^2 B_F(\boldsymbol{Z}_h^\varepsilon, \boldsymbol{E}) = G(\boldsymbol{E}), \quad \forall \boldsymbol{E} \in \boldsymbol{V}_h. \tag{3.49}$$

Hence, compared to equation (3.35) satisfied by the exact solution, the mixed formulation leads to a perturbation of the *leading term* of the formulation. Therefore, convergence (and in particular uniform convergence) cannot be expected in this case unless the *consistency* of B_M^h with respect to B_M is strictly controlled. This condition is very difficult to enforce, in particular because mixed methods are not designed to that purpose. As a matter of fact, for the above-mentioned mixed–stabilized scheme that works well in bending-dominated cases, uniform convergence is not obtained in membrane-dominated situations.

Therefore, it appears that we are 'trapped' in a dilemma caused by the dramatic discrepancy between the two major asymptotic states of shell problems. Displacement-based finite elements work well in membrane-dominated situations but are subject to numerical locking in bending-dominated situations, in general. On the other hand, if we try to 'fix' the bending-dominated behaviour by using mixed methods (or some kind of reduced integration strategy), there is a major risk of seriously deteriorating the numerical solution in membrane-dominated cases. As a matter of fact, we do not know of any shell finite element procedure that would have been proved to work well in both regimes.

3.3. Guidelines for assessing and improving the reliability of shell finite elements

Whereas the complete mathematical analysis required to design and justify a general uniformly convergent finite element method for shell problems still seems out of reach, we believe that it is crucial, in this reliability quest, to combine the insight provided by the analysis of the mathematical models and the elements of analysis available on their discretizations on the one hand, with extensive numerical testing on the other hand. In this interaction, *benchmarks* are both an essential interface and an invaluable means of assessing the performance and the reliability of shell finite element methods, hence also to improve these methods.

Of course, in order to provide a rigorous and meaningful assessment tool, benchmarks must be very carefully chosen, and properly used. To that purpose, some detailed guidelines are given in Chapelle and Bathe (1998), which rely on the following principles.

- A basic set of benchmarks must contain instances of the two fundamental categories of asymptotic behaviour of shell structures.
- Several geometries must be considered, since the asymptotic behaviour is highly sensitive on the geometry of the mid-surface (for example, a set of benchmarks only based on cylinders is not satisfactory).
- Numerical computations must be performed and reported upon for several values of h (since we are concerned with the whole convergence

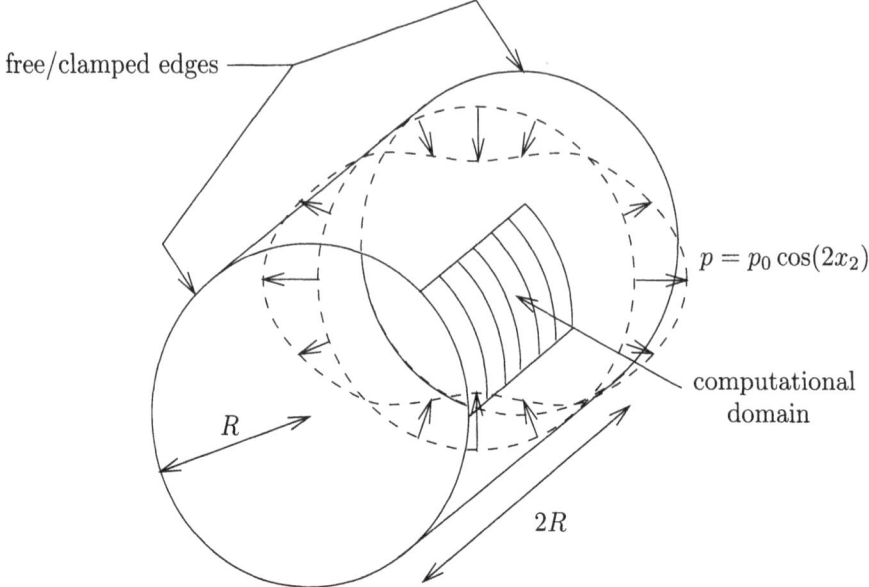

$$p = p_0 \cos(2x_2)$$

Fig. 3.2. Cylinder loaded by periodic pressure

behaviour and not only with the accuracy in one specific instance)
and for several values of ε (in order to assess the robustness of the
convergence with respect to this parameter).

In Chapelle and Bathe (1998), a basic set of benchmarks is also proposed
based on these principles. It consists of the following test problems.

Circular cylinder loaded by periodic pressure. (See Figure 3.2.) Depending on the boundary conditions we can obtain either one of the
two main asymptotic behaviour types (bending-dominated behaviour
if both ends are left free, well-posed membrane-dominated behaviour
if both ends are clamped). This problem was originally proposed in
Pitkäranta, Leino, Ovaskainen and Piila (1995) which also presented a
procedure for obtaining numerical solutions of arbitrary accuracy.[1]

Partly clamped hyperbolic paraboloid. The shell shown in Figure 3.3
is clamped on one side and loaded, *e.g.*, by self weight. It can be shown
that pure bending is not inhibited; *cf.* Chapelle and Bathe (1998). We
do not have an analytical solution for this problem (see Choï (1999)
where the exact solution of the limit problem is analysed), but we can
use reference solutions obtained with a very fine mesh to compute error
measures.

---

[1] The values of displacements and deformation energies given in Pitkäranta *et al.* (1995)
should be multiplied by $(1 - \nu^2)$ and $(1 - \nu^2)^2$, respectively.

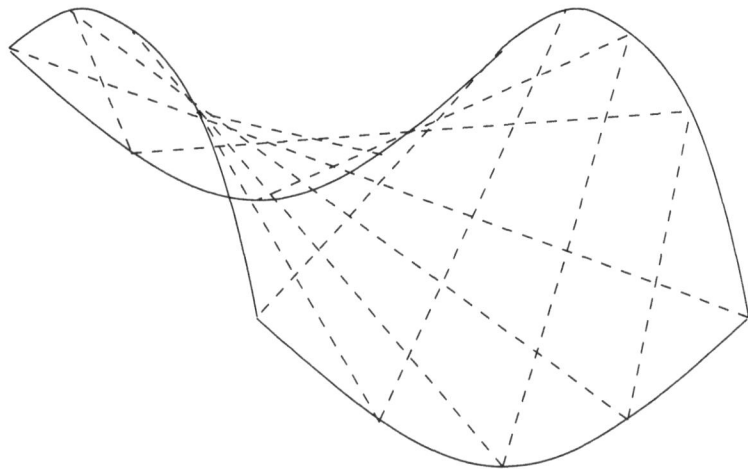

Fig. 3.3. Hyperbolic paraboloid

Clamped hemispherical cap. It can be shown that pure bending is inhibited for this problem (as a matter of fact, non-inhibited pure bending is obtained for a hemispherical cap only when the boundary is
completely free; see Sanchez-Hubert and Sanchez-Palencia (1997), and
also Chapelle and Bathe (1998)). If we choose an axisymmetric loading
such as the one shown in Figure 3.4, we can compare the results of the
shell analysis with the results of some (reliable) axisymmetric analysis
scheme.

We also emphasize that a deep insight into the behaviour of the exact solutions is an essential prerequisite in the assessment procedure, that is, in
the proper use of benchmarks. A particularly important feature of these
solutions is the development of boundary layers which vary in amplitude
and width when the thickness of the shell varies. If these boundary layers are not properly taken into account in the choice of the discretization
scheme (in particular by refining the mesh when and where appropriate),
they can strongly affect the convergence behaviour and induce wrong conclusions in the reliability assessment. This insight, available for plates (see,
e.g., Häggblad and Bathe (1990), Arnold and Falk (1996), Dauge and Yosibash (2000)), is now becoming available for shells also through some more
recent works (Pitkäranta, Matache and Schwab 1999, Karamian, Sanchez-
Hubert and Sanchez-Palencia 2000).

Once a good set of benchmarks has been selected, it is crucial to measure
the error made in the numerical computations by using proper error measures. We now review and evaluate some possible choices, including those
currently most widely used.

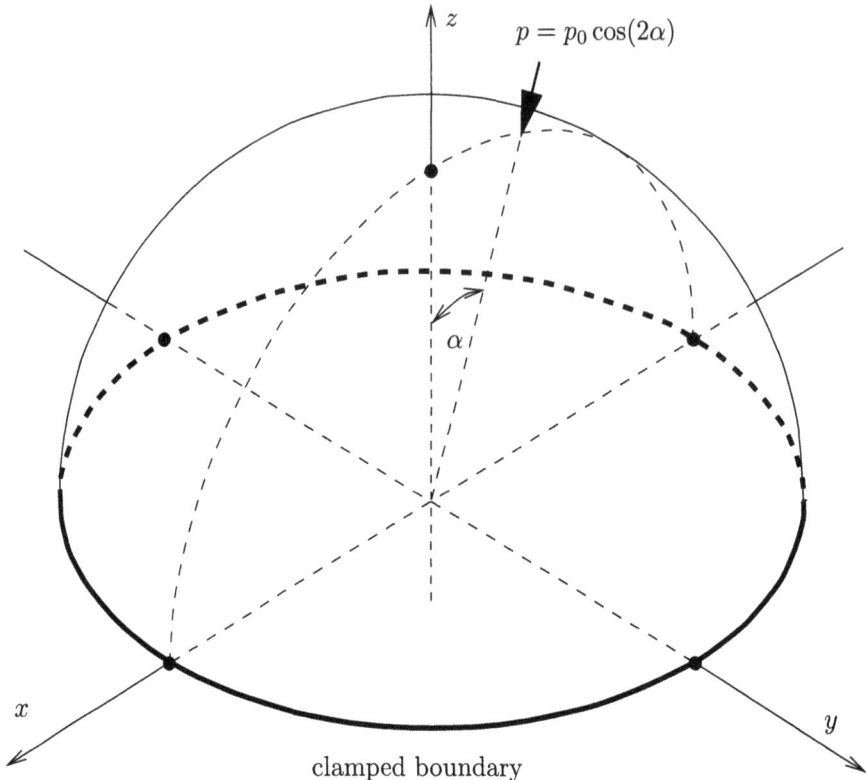

$$p = p_0 \cos(2\alpha)$$

clamped boundary

Fig. 3.4. Hemispherical shell under axisymmetric loading

(1) Pointwise displacements. This error measure is probably the most extensively used, especially in the engineering literature. It is, however, likely to be misleading, since the pointwise convergence of finite element methods is not guaranteed, in general.

(2) Energy norm, namely $[\varepsilon^2 B_F(\boldsymbol{E}, \boldsymbol{E}) + B_M(\boldsymbol{E}, \boldsymbol{E})]^{1/2}$ (note that we can scale the energy by any given power of ε since we are concerned with relative errors). This cannot be a good choice, since it is the norm in which the displacement-based finite element solution is the best approximation, whereas this solution locks in bending-dominated situations.

(3) Modified energy norm, namely $[\varepsilon^2 B_F(\boldsymbol{E}, \boldsymbol{E}) + B_M^h(\boldsymbol{E}, \boldsymbol{E})]^{1/2}$. This choice is advocated in Malinen and Pitkäranta (2000), in particular. Its major drawback, beside the fact that its use with the exact solution (or any reference solution) is delicate (since the B_M^h-form is usually tailored to one particular discretization), is that the use of a non-physical modification of the energy makes it difficult to interpret convergence in this norm.

(4) Energy variation, namely $|\mathcal{E} - \mathcal{E}_h|$ where $\mathcal{E}$ is the energy that corresponds to the exact (or reference) solution, namely

$$\mathcal{E} = \varepsilon^2 B_F(\mathbf{Z}^\varepsilon, \mathbf{Z}^\varepsilon) + B_M(\mathbf{Z}^\varepsilon, \mathbf{Z}^\varepsilon), \tag{3.50}$$

and $\mathcal{E}_h$ the energy of the finite element solution, namely

$$\mathcal{E}_h = \varepsilon^2 B_F(\mathbf{Z}_h^\varepsilon, \mathbf{Z}_h^\varepsilon) + B_M(\mathbf{Z}_h^\varepsilon, \mathbf{Z}_h^\varepsilon) \tag{3.51}$$

for displacement-based methods and

$$\mathcal{E}_h = \varepsilon^2 B_F(\mathbf{Z}_h^\varepsilon, \mathbf{Z}_h^\varepsilon) + B_M^h(\mathbf{Z}_h^\varepsilon, \mathbf{Z}_h^\varepsilon) \tag{3.52}$$

for mixed methods. For displacement-based methods, it can be shown that this measure equals the square of the error in the energy norm, but this does not hold for mixed methods, which makes the interpretation of this indicator very difficult (see also Malinen and Pitkäranta (2000)).

(5) Asymptotic convergence norms, namely – as discussed in the introduction of this section – the norms in which exact solutions are shown to converge to limit solutions, that is, the norm of the displacement space for bending-dominated problems and the membrane energy norm for membrane-dominated problems. For practical purposes the membrane energy norm can be directly computed using its definition (and the corresponding stiffness matrices). For the norm of the displacement space (typically a combination of Sobolev norms) it is more convenient to use an equivalent norm constructed with the deformation energy, for example $[L^2 B_F(\mathbf{E}, \mathbf{E}) + B_M(\mathbf{E}, \mathbf{E})]^{1/2}$ where L is a characteristic dimension of the structure (the diameter, for instance). Note that, when using general shell elements, these computations are not straightforward since the membrane and bending energies are combined in the basic shell model. However, as a consequence of our above discussion (cf. Section 3.1), since the Naghdi model is asymptotically equivalent we can use the corresponding energy expressions (and stiffness matrices) to compute adequate norms.

Using the norms in which asymptotic convergence is obtained for the exact solutions, it is then natural to seek finite element procedures for which uniformly decreasing errors can be measured on benchmarks such as those proposed above, when boundary layers (namely, the 'nonsmooth part' of the exact solutions) are properly resolved. Indeed, for a displacement-based finite element procedure used in a membrane-dominated case it can be shown that (see Chapelle and Bathe (1998))

$$\|\mathbf{Z}^\varepsilon - \mathbf{Z}_h^\varepsilon\|_M + \varepsilon\|\mathbf{Z}^\varepsilon - \mathbf{Z}_h^\varepsilon\|_V \tag{3.53}$$
$$\leq C \inf_{\mathbf{E} \in V_h} \{\|\mathbf{Z}^\varepsilon - \mathbf{E}\|_M + \varepsilon\|\mathbf{Z}^\varepsilon - \mathbf{E}\|_V\},$$

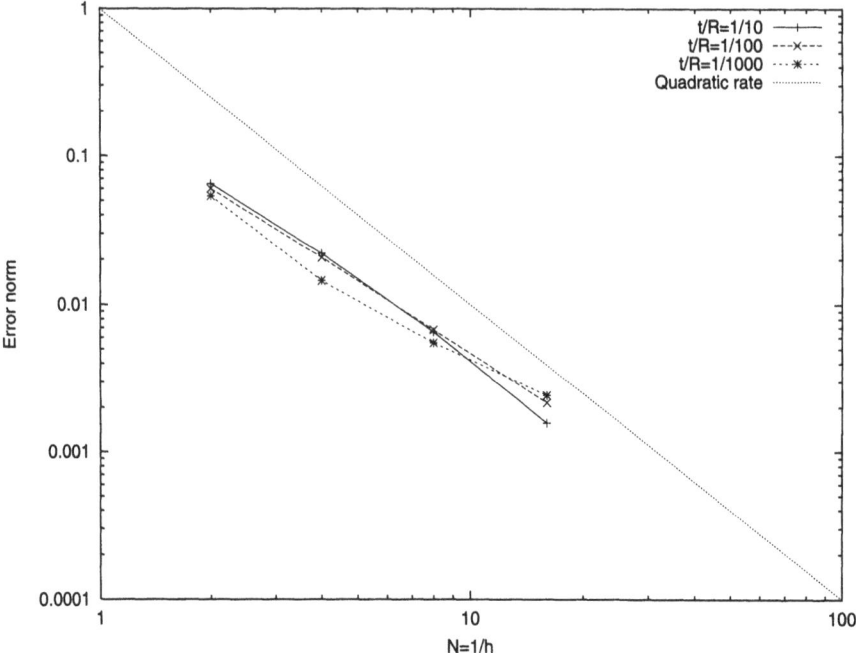

Fig. 3.5. Clamped cylinder: adapted mesh, membrane norm

which shows that a displacement-based finite element procedure is *uniformly optimal* with respect to ε for the norm $\|\cdot\|_M + \varepsilon\|\cdot\|_V$. Therefore, if the mesh is refined so that the boundary layers of a given test problem are properly resolved for this norm, uniform optimal convergence is obtained. This is illustrated in Figures 3.5–3.7, where the case of a clamped circular cylinder is considered (as described in Figure 3.2); see Chapelle, Oliveira and Bucalem (200x). The Naghdi model is used, and Q2 displacement-based elements are employed to discretize the problem. Figure 3.5 shows the convergence curves obtained for the membrane norm and for decreasing values of ε (t denotes the actual thickness of the shell, namely 2ε) when as many layers of elements are used within the layer of width $\sqrt{\varepsilon}$ along the boundaries (*i.e.*, the boundary layers) as in the rest of the domain. This can be shown to be the correct way to resolve the boundary layers; see Pitkäranta *et al.* (1995). We can see that the convergence rate is optimal for all values of the thickness, and the convergence behaviour is not particularly sensitive to the value of ε. By contrast, if no particular treatment of the boundary layers is applied, we obtain the convergence curves displayed in Figure 3.6 where some significant sensitivity with respect to ε is observed. Also, if the original norm of V is used instead of the membrane norm (Figure 3.7) we again detect some marked sensitivity. All these numerical results are clearly consistent with equation (3.53).

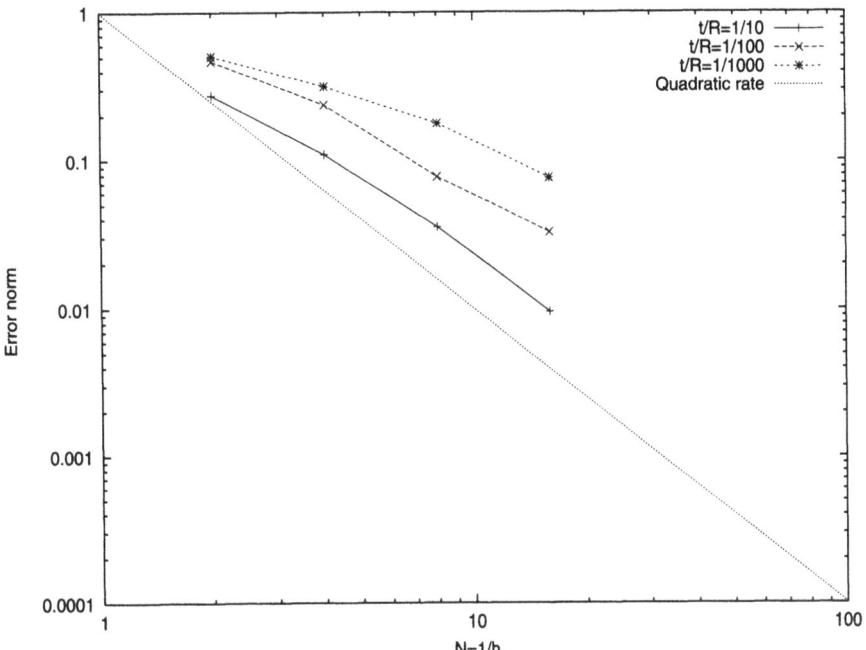

Fig. 3.6. Clamped cylinder: non-adapted mesh, membrane norm

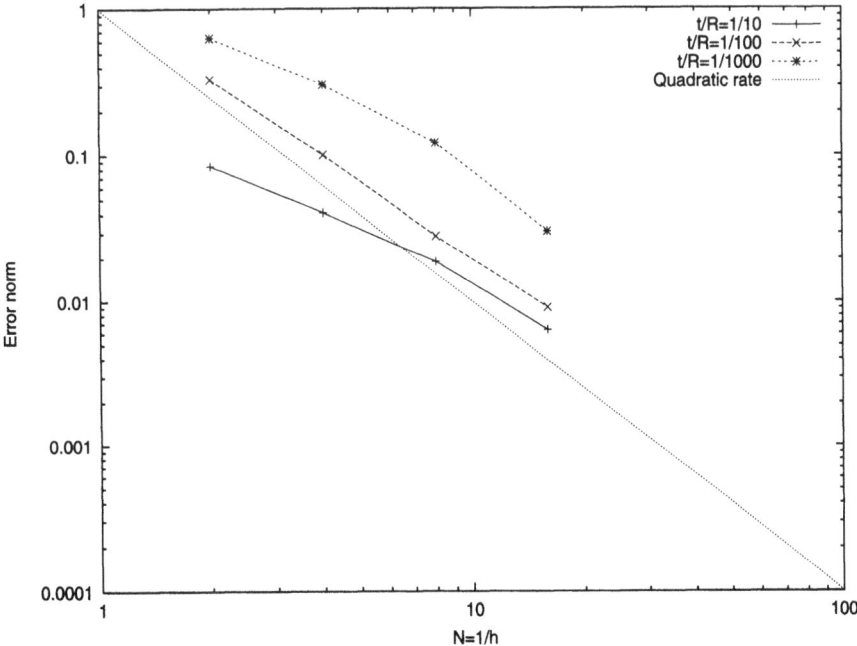

Fig. 3.7. Clamped cylinder: adapted mesh, original norm

Likewise, for bending-dominated problems classical results for mixed formulations show that, when the inf-sup condition is satisfied, we have the following optimal error bound (see Brezzi and Fortin (1991)):

$$\|Z^\varepsilon - Z_h^\varepsilon\|_V + |\Sigma^\varepsilon - \Sigma_h^\varepsilon|_S \le C \inf_{\substack{E \in V_h \\ \Xi \in S_h}} \{\|Z^\varepsilon - E\|_V + |\Sigma^\varepsilon - \Xi|_S\}. \quad (3.54)$$

Of course, as previously mentioned, we do not know of any shell finite element procedure that is proven to satisfy the inf-sup condition. Note, however, that a procedure which passes the inf-sup test on a given test problem automatically satisfies the estimate (3.54) *for this same test problem and on the sequence of meshes used for the inf-sup test* (as it is the inf-sup *values* that enter in the bounding constants). Moreover, the estimate (3.54) is very useful for determining how the meshes should be refined in order to ascertain uniformly converging errors in the numerical results obtained with finite element schemes that 'would satisfy' the inf-sup condition (hence locking-free finite elements), even though this inf-sup condition cannot be proven. Namely, the regularity of the stresses also needs to be considered to control the error in the seminorm $|\cdot|_S$ (note that this seminorm is always bounded by the L^2-norm – *cf.* Chapelle and Stenberg (1998)).

Finally, we emphasize that all the above-proposed benchmarks are based on problems for which a well-defined asymptotic limit exists, and more specifically for which the asymptotic behaviour is either membrane-dominated or bending-dominated. As previously discussed, other types of asymptotic behaviour (as defined in Section 3.1) do exist, and in particular in cases when pure bending is inhibited and '$G \notin (V_M)'$'; see in particular Pitkäranta and Sanchez-Palencia (1997) and Pitkäranta *et al.* (1995) for examples. In these cases, however, it does not seem possible to identify a norm for which an asymptotic limit exists, hence it is difficult to see what kind of uniform convergence could be sought in finite element solutions. This is why the use of such problems cannot be recommended to assess the reliability of finite element procedures. We point out that the 'Scordelis–Lo roof' (for which the geometry and the boundary conditions are defined in Figure 3.8, and the loading corresponds to self-weight), which appears to be one of the most widely used test problems in the engineering literature, belongs to this category, as it can be shown that it corresponds to a situation of inhibited pure bending with '$G \notin (V_M)'$'; see Chapelle and Bathe (1998) (in fact, it is the loading in the vicinity of the free boundaries which causes this difficulty). Nevertheless, some recent works have shown that in some cases (more specifically when the shell mid-surface is parabolic or hyperbolic) it is possible to split the solution into several components, namely a smooth part (in which the membrane energy is asymptotically dominant as in membrane-dominated situations) and a singular part, where the singular part consists of strong boundary layers or of 'generalized layers' that lie along asymp-

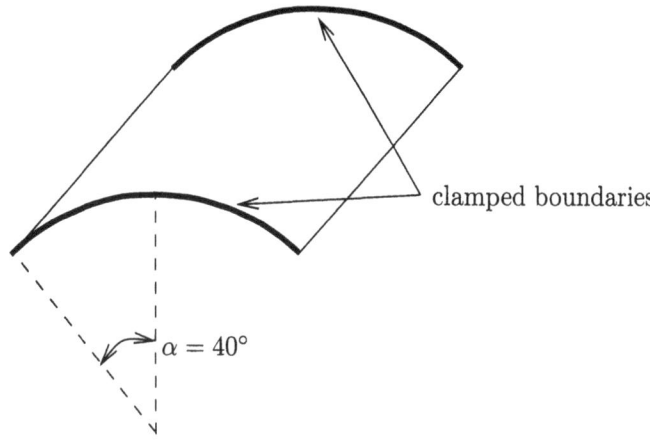

Fig. 3.8. Scordelis–Lo roof

totic lines of the surface; see Pitkäranta *et al.* (1999) and also Karamian *et al.* (2000). Moreover, the displacements that develop in these layers appear to be very similar to pure bending displacements. Even though more analysis is still needed, these recent results give hope that finite element procedures which would perform well (*i.e.*, converge uniformly) in bending-dominated and membrane-dominated situations would also perform well in these more complex situations, that is, converge uniformly for each of the above-mentioned components separately.

Acknowledgement

The author is particularly indebted to Professor Klaus-Jürgen Bathe, as many ideas and results presented in this article originated from a fruitful collaboration with him, which is attested by the joint publications listed in the references.

REFERENCES

S. Ahmad, B. M. Irons and O. C. Zienkiewicz (1970), 'Analysis of thick and thin shell structures by curved finite elements', *Internat. J. Numer. Methods Eng.* **2**, 419–451.

J. L. Akian and E. Sanchez-Palencia (1992), 'Approximation de coques élastiques minces par facettes plane: Phénomènes de blocage membranaire', *C. R. Acad. Sci. Paris, Sér. I* **315**, 363–369.

D. N. Arnold (1981), 'Discretization by finite elements of a model parameter dependent problem', *Numer. Math.* **37**, 405–421.

D. N. Arnold and F. Brezzi (1997), 'Locking-free finite element methods for shells', *Math. Comp.* **66**, 1–14.

D. N. Arnold and R. S. Falk (1996), 'Asymptotic analysis of the boundary layer for the Reissner–Mindlin plate model', *SIAM J. Math. Anal.* **27**, 486–514.

C. Baiocchi and C. Lovadina (200x), A shell classification by interpolation, To appear.

Y. Başar and W. B. Krätzig (1989), ' A consistent shell theory for finite deformations', *Acta Mech.* **76**, 73–87.

K. J. Bathe (1996), *Finite Element Procedures*, Prentice-Hall, Englewood Cliffs, NJ.

K. J. Bathe and D. Chapelle (200x), *The Finite Element Analysis of Shells*. To appear.

K. J. Bathe and E. N. Dvorkin (1986), 'A formulation of general shell elements: The use of mixed interpolation of tensorial components', *Internat. J. Numer. Methods Eng.* **22**, 697–722.

K. J. Bathe, A. Iosilevich and D. Chapelle (2000), 'An inf-sup test for shell finite elements', *Comput. Structures* **75**, 439–456.

J. L. Batoz and G. Dhatt (1992), *Modélisation des Structures par Eléments Finis, Vol. 3: Coques*, Hermes, Paris.

M. Bernadou (1996), *Finite Element Methods for Thin Shell Problems*, Wiley.

M. Bernadou and P. G. Ciarlet (1976), Sur l'ellipticité du modèle linéaire de coques de W. T. Koiter, in *Computing Methods in Applied Sciences and Engineering* (R. Glowinski and J. L. Lions, eds), Vol. 134 of *Lecture Notes in Economics and Mathematical Systems*, Springer, Heidelberg, pp. 89–136.

M. Bernadou, Y. Ducatel and P. Trouvé (1988), 'Approximation of a circular cylindrical shell by Clough–Johnson flat plate finite elements', *Numer. Math.* **52**, 187–217.

M. Bernadou, P. G. Ciarlet and B. Miara (1994), 'Existence theorems for two-dimensional linear shell theories', *J. Elasticity* **34**, 111–138.

M. Bischoff and E. Ramm (1997), 'Shear deformable shell elements for large strains and rotations', *Internat. J. Numer. Methods Eng.* **40**, 4427–4449.

A. Blouza, F. Brezzi and C. Lovadina (1999), 'Sur la classification des coques linéairement élastiques', *C. R. Acad. Sci. Paris, Sér. I* **328**, 831–836.

J. Bramble and T. Sun (1997), 'A locking-free finite element method for Naghdi shells', *J. Comput. Appl. Math.* **89**, 119–133.

F. Brezzi and M. Fortin (1991), *Mixed and Hybrid Finite Element Methods*, Springer.

M. L. Bucalem and K. J. Bathe (1993), 'Higher-order MITC general shell elements', *Internat. J. Numer. Methods Eng.* **36**, 3729–3754.

D. Chapelle and K. J. Bathe (1993), 'The inf-sup test', *Comput. Structures* **47**, 537–545.

D. Chapelle and K. J. Bathe (1998), 'Fundamental considerations for the finite element analysis of shell structures', *Comput. Structures* **66**, 19–36.

D. Chapelle and K. J. Bathe (2000), 'The mathematical shell model underlying general shell elements', *Internat. J. Numer. Methods Eng.* **48**, 289–313.

D. Chapelle, D. L. Oliveira, and M. L. Bucalem (200x), 'On the reliability of MITC shell finite elements based on Naghdi's model'. To appear.

D. Chapelle and R. Stenberg (1998), 'Stabilized finite element formulations for shells in a bending dominated state', *SIAM J. Numer. Anal.* **36**, 32–73.

D. Chenais and J. C. Paumier (1994), 'On the locking phenomenon for a class of elliptic problems', *Numer. Math.* **67**, 427–440.

D. Choï (1999), 'Computations on thin non-inhibited hyperbolic elastic shells: Benchmarks for membrane locking', *Math. Meth. Appl. Sci.* **22**, 1293–1321.

D. Choï, F. Palma, E. Sanchez-Palencia and M. Vilarino (1998), 'Membrane locking in the finite element computation of very thin elastic shells', *Modélisation Mathématique et Analyse Numérique* **32**, 131–152.

P. G. Ciarlet (1978), *The Finite Element Method for Elliptic Problems*, North-Holland.

P. G. Ciarlet (2000), *Mathematical Elasticity, Volume III: Theory of Shells*, North-Holland.

P. G. Ciarlet (2001), 'Mathematical modelling of linearly elastic shells', in *Acta Numerica*, Vol. 10, Cambridge University Press, pp. 103–214.

M. Dauge and Z. Yosibash (2000), 'Boundary layer realization in thin elastic 3D domains and 2D hierarchic plate models', *Internat. J. Solids Structures* **37**, 2443–2471.

M. Delfour (1998), Intrinsic differential geometric methods in the asymptotic analysis of linear thin shells, in *Boundaries, Interfaces and Transitions* (M. Delfour, ed.), *CRM Proceedings & Lecture Notes*, American Mathematical Society, Providence, RI, pp. 19–90.

B. Häggblad and K. J. Bathe (1990), 'Specifications of boundary conditions for Reissner/Mindlin plate bending finite elements', *Internat. J. Numer. Methods Eng.* **30**, 981–1011.

A. Iosilevich, K. J. Bathe and F. Brezzi (1997), 'On evaluating the inf-sup condition for plate bending elements', *Internat. J. Numer. Methods Eng.* **40**, 3639–3663.

B. M. Irons and S. Ahmad (1980), *Techniques of Finite Elements*, Wiley, London.

P. Karamian, J. Sanchez-Hubert and E. Sanchez-Palencia (2000), 'A model problem for boundary layers of thin elastic shells', *Math. Modelling Numer. Anal.* **34**, 1–30.

W. T. Koiter (1966), 'On the nonlinear theory of thin elastic shells', *Proc. Kon. Ned. Akad. Wetensch.* **B69**, 1–54.

J. L. Lions (1973), *Perturbations Singulières dans les Problèmes aux Limites et en Contrôle Optimal*, Springer.

M. Lyly (2000), 'On the connection between some linear triangular Reissner–Mindlin plate bending elements', *Numer. Math.* **85**, 77–107.

M. Malinen (2000), On geometrically incompatible bilinear shell elements and classical shell models, Research Report TKK-Lo-30, Helsinki University of Technology, Laboratory for Mechanics of Materials.

M. Malinen and J. Pitkäranta (2000), A benchmark study of reduced-strain shell finite elements: Quadratic schemes. *Internat. J. Numer. Methods Eng.* **48** 1637–1671.

R. Mindlin (1951), 'Influence of rotary inertia and shear on flexural motion of isotropic elastic plates', *J. Appl. Mech.* **18**, 31–38.

P. M. Naghdi (1963), Foundations of elastic shell theory, in *Progress in Solid Mechanics*, Vol. 4, North-Holland, pp. 1–90.

J. Pitkäranta (1992), 'The problem of membrane locking in finite element analysis of cylindrical shells', *Numer. Math.* **61**, 523–542.

J. Pitkäranta and E. Sanchez-Palencia (1997), 'On the asymptotic behaviour of sensitive shells with small thickness', *C. R. Acad. Sci. Paris, Sér. IIb* **325**, 127–134.

J. Pitkäranta, Y. Leino, O. Ovaskainen and J. Piila (1995), 'Shell deformation states and the finite element method: a benchmark study of cylindrical shells', *Comput. Methods Appl. Mech. Eng.* **128**, 81–121.

J. Pitkäranta, A. M. Matache and C. Schwab (1999), Fourier mode analysis of layers in shallow shell deformations, Research Report SAM 99-18, ETH Zürich. To appear in *Comput. Methods Appl. Mech. Eng.*

E. Reissner (1945), 'The effect of transverse shear deformation on the bending of elastic plates', *J. Appl. Mech.* **67**, A69–A77.

J. Sanchez-Hubert and E. Sanchez-Palencia (1997), *Coques Elastiques Minces: Propriétés Asymptotiques*, Masson, Paris.

E. Sanchez-Palencia (1989*a*), 'Statique et dynamique des coques minces, I: Cas de flexion pure non inhibée', *C. R. Acad. Sci. Paris, Sér. I* **309**, 411–417.

E. Sanchez-Palencia (1989*b*), 'Statique et dynamique des coques minces, II: Cas de flexion pure inhibée: Approximation membranaire', *C. R. Acad. Sci. Paris, Sér. I* **309**, 531–537.

E. Sanchez-Palencia (1992), 'Asymptotic and spectral properties of a class of singular-stiff problems', *J. Math. Pures Appl.* **71**, 379–406.

Acta Numerica (2001), pp. 251–312   © Cambridge University Press, 2001

Geometric aspects of the theory of Krylov subspace methods

Michael Eiermann[*] and Oliver G. Ernst[†]

Institut für Angewandte Mathematik II,

TU Bergakademie Freiberg,

09596 Freiberg, Germany

E-mail: `eiermann@math.tu-freiberg.de`

`ernst@math.tu-freiberg.de`

The development of Krylov subspace methods for the solution of operator equations has shown that two basic construction principles underlie the most commonly used algorithms: the orthogonal residual (OR) and minimal residual (MR) approaches. It is shown that these can both be formulated as techniques for solving an approximation problem on a sequence of nested subspaces of a Hilbert space, an abstract problem not necessarily related to an operator equation. Essentially all Krylov subspace algorithms result when these subspaces form a Krylov sequence. The well-known relations among the iterates and residuals of MR/OR pairs are shown to hold also in this rather general setting. We further show that a common error analysis for these methods involving the canonical angles between subspaces allows many of the known residual and error bounds to be derived in a simple and consistent manner. An application of this analysis to compact perturbations of the identity shows that MR/OR pairs of Krylov subspace methods converge q-superlinearly when applied to such operator equations.

[*] Partially supported by the Laboratoire d'Analyse Numérique et d'Optimisation of the Université des Sciences et Technologies de Lille.

[†] Partially supported by the University of Maryland Institute of Advanced Computer Studies and NSF grant no. ASC9704683.

CONTENTS

1. Overview

In the past three decades research on Krylov subspace techniques for solving linear systems of equations has brought forth a variety of algorithms and methods so large that even specialists in matrix computations have difficulties keeping up. This situation is all the more confusing for scientists whose main interests lie elsewhere when faced with choosing among the many approaches for solving a linear system, or merely attempting to obtain an overview of the methods and how they are related. This state of affairs applies not only to the methods themselves, but also to many theoretical results obtained time and again for each individual method.

It is our objective in this paper to develop the theory and algorithms on which all Krylov subspace methods are based through several layers of abstraction, proceeding from the most general to the most specific and leading to Krylov subspace methods in the form in which they are currently used. We have found several advantages to this approach. First, by obtaining each result in as general a setting as possible, it is easier to distinguish properties unique to Krylov subspace methods from those which these methods inherit as, say, projection methods or subspace correction methods. We will indeed see that many results on Krylov subspace methods hold in greater generality than usually stated, and that many computational elements of these methods can be translated with little modification to more general methods. Second, our approach emphasizes the common origin of all Krylov subspace methods and we have found this approach an elegant yet simple way of presenting this theory, tying together in a consistent and natural framework many results that are otherwise difficult to relate. Finally, we have found that this way of developing Krylov subspace theory offers a new and insightful perspective and is accessible to any reader familiar with only basic facts about inner product spaces.

In the literature on Krylov subspace techniques for solving linear systems of equations, two principal methods have emerged as the basis for most algorithms: these are the *minimal residual* (MR) and the *orthogonal residual* (OR) approaches. Both methods select an approximation to the solution of the linear system from a shifted Krylov space. The former does this in such a way that the resulting residual norm is minimized, whereas the latter chooses the approximation such that the associated residual is orthogonal to the Krylov space. The following is a list of the most widely used MR/OR pairs:

- the conjugate residual (CR)/conjugate gradient (CG) methods for Hermitian definite systems (Hestenes and Stiefel 1952)

- the minimal residual method (MINRES)/CG methods for Hermitian indefinite systems (Paige and Saunders 1975)

- the full orthogonalization method (FOM)/generalized minimal residual method (GMRES) for the non-Hermitian case (Saad 1981, Saad and Schultz 1986)

- the biconjugate gradient (BCG)/quasi-minimal residual (QMR) methods (QMR) for non-Hermitian problems (Lanczos 1952, Freund and Nachtigal 1991)

- conjugate gradients squared (CGS)/transpose-free QMR (TFQMR) for non-Hermitian problems (Sonneveld 1989, Freund 1993),

where the last two pairs of methods require us to define orthogonality with respect to a problem-dependent inner product, a matter which will be discussed in detail in Sections 4 and 5.

Expositions of these and other Krylov subspace methods can be found, for instance, in the monographs of Hageman and Young (1981), Axelsson (1994), Bruaset (1995), Fischer (1996), Saad (1996), Greenbaum (1997), Weiss (1997), and Meurant (1999). In addition, there are also several survey papers with differing emphases, among which we mention those of Ashby, Manteuffel and Saylor (1990), Freund, Golub and Nachtigal (1992), Gutknecht (1997), Golub and van der Vorst (1997) and van der Vorst and Saad (2000).

In what follows, $\mathcal{H}$ denotes a Hilbert space with inner product $(\cdot, \cdot)$ and associated norm $\| \cdot \|$. By $A : \mathcal{H} \to \mathcal{H}$, we always denote an invertible bounded linear operator. Krylov subspace methods for solving a linear operator equation

$$Ax = b \qquad (1.1)$$

begin with an initial approximation x_0 of $x = A^{-1}b$ and at each step $m = 1, 2, \ldots$ attempt to construct an improved approximation $x_m = x_0 + c_m$ by adding a correction c_m from the Krylov space

$$\mathcal{K}_m(A, r_0) := \operatorname{span}\{r_0, Ar_0, \ldots, A^{m-1}r_0\}$$

of order m with respect to A and the initial residual $r_0 = b - Ax_0$. The MR approach determines the correction c_m^{MR} in such a way that the associated residual is minimized, that is,

$$\|b - Ax_m^{MR}\| = \|r_0 - Ac_m^{MR}\| = \min_{c \in \mathcal{K}_m(A, r_0)} \|r_0 - Ac\|, \qquad (1.2a)$$

while the OR approach determines the correction from the Galerkin condition

$$b - Ax_m^{OR} = r_0 - Ac_m^{OR} \perp \mathcal{K}_m(A, r_0). \qquad (1.2b)$$

In contrast to the MR approximation, there are situations where the OR approximation may not exist or may not be uniquely determined and these are discussed below.

The Krylov subspace MR and OR methods are special cases of the more general *subspace correction methods*: given an arbitrary correction space $\mathcal{C}_m$ of dimension m, determine c_m^{MR} and c_m^{OR} such that

$$\|r_0 - Ac_m^{MR}\| = \min_{c \in \mathcal{C}_m} \|r_0 - Ac\| \qquad (1.3a)$$

and

$$r_0 - Ac_m^{OR} \perp \mathcal{V}_m, \qquad (1.3b)$$

respectively, for some suitable m-dimensional test space $\mathcal{V}_m$. Letting $\mathcal{W}_m := A\mathcal{C}_m$ denote the image of the correction space under A, we can reformulate (1.3a) and (1.3b) as the task of determining $w_m^{MR}, w_m^{OR} \in \mathcal{W}_m$ such that

$$\|r_0 - w_m^{MR}\| = \min_{w \in \mathcal{W}_m} \|r_0 - w\| \qquad (1.4a)$$

and

$$r_0 - w_m^{OR} \perp \mathcal{V}_m. \qquad (1.4b)$$

The solutions of (1.4a) and (1.4b) are, of course, given by the orthogonal projection $w_m^{MR} = P_{\mathcal{W}_m} r_0$ of r_0 onto $\mathcal{W}_m$ and the oblique projection $w_m^{OR} = P_{\mathcal{W}_m}^{\mathcal{V}_m} r_0$ of r_0 onto $\mathcal{W}_m$ orthogonal to $\mathcal{V}_m$, respectively. In view of (1.4a) and (1.4b), the MR and OR approaches consist of approximating $r_0 \in \mathcal{H}$ in the space $\mathcal{W}_m$ by its orthogonal and oblique projections onto $\mathcal{W}_m$.

In the following section we shall determine what can be said about the approximation of an arbitrary element in $\mathcal{H}$ by its orthogonal and oblique projections with respect to given spaces $\mathcal{W}$ and $\mathcal{V}$ of equal (finite) dimension. We then consider such approximations with respect to nested sequences of spaces $\{\mathcal{W}_m\}_{m \geq 0}$ and $\{\mathcal{V}_m\}_{m \geq 0}$ and show that all the well-known relations between Krylov subspace MR and OR approximations and their residuals already hold in this abstract setting of approximation by orthogonal and oblique projection. The source of these relations is thereby seen to lie in the properties of orthogonal and oblique projections on a nested sequence of subspaces, where the oblique projection is orthogonal to the previous error space (which we shall call residual space to emphasize its Krylov subspace correspondence). In particular, these relations are not restricted to Krylov spaces, or even to solving equations.

In the usual implementations, the Krylov MR/OR approximations are computed by solving a small least-squares problem or linear system, respectively, involving a tridiagonal or Hessenberg matrix at each step. This is done by maintaining a QR factorization of this Hessenberg matrix which is updated at each step using Givens rotations. Due to the Hessenberg structure, only one Givens rotation is required at each step, and the angles of these rotations can be used to characterize the OR and MR residual norms and also to express the iterates and residuals in terms of each other.

In our abstract setting, these relations are derived using only the notion of angles between the spaces which define the orthogonal and oblique projections. The sines and cosines occurring in these expressions are therefore not mere artifacts of the computational scheme for the MR and OR approximations, but have an intrinsic meaning. It is this intrinsic relationship between the orthogonal and oblique projections that is at the root of the often observed close relationship between MR and OR approximations.

We further discuss the relation between MR and quasi-minimal residual (QMR) approximations. The latter is also obtained by solving a least-squares problem in coordinate space, the only difference to the MR approximation being that these coordinates are with respect to a non-orthogonal basis. We show that QMR approximations become MR approximations if only the inner product is appropriately chosen. Again, the QMR approximation can be defined in our abstract setting and then has a structure identical to the MR approximation. This distinction between MR and QMR approximations via different inner products also extends to OR and quasi-orthogonal residual (QOR) approximations; in fact, we show that *essentially any* Krylov subspace method for solving a linear system can be classified as both an MR or OR method by appropriate choice of the inner product.

In Section 2 we cast the MR and OR approximations to the solution of an operator equation as abstract approximation problems or, equivalently, as an orthogonal and oblique projection method, respectively. In contrast to the orthogonal projection, there are situations in which the oblique projection, and hence the OR approximation, may not exist. A useful characterization of the oblique projection as the Moore–Penrose inverse of the product of two orthogonal projections leads to a canonical way of defining an OR approximation in case of such a breakdown. We characterize the conditions under which the oblique projection exists and relate the norm of the oblique projection to the angle between two subspaces.

Section 3 specializes the spaces characterizing the projections to two sequences of closely related nested spaces. It is shown that, under these general conditions, all the well-known relations among OR/MR pairs such as FOM/GMRES can be derived. These results also elucidate the theory behind residual smoothing techniques.

Section 4 explores the coordinate calculations required to compute the MR and OR approximations with respect to orthogonal and non-orthogonal bases of the underlying spaces. These involve solving least-squares problems and linear systems with a Hessenberg matrix. When a QR-factorization based on Givens rotations is used to solve these problems, the angles characterizing the Givens rotations are identified as the angles between these subspaces. Moreover, it is shown that methods based on non-orthogonal bases can be characterized as MR/OR methods with respect to a different, basis-dependent inner product. We further characterize the QMR and QOR approximations as oblique projections with respect to the original inner product and conclude with a characterization for when the QMR approximation coincides with the MR approximation.

Section 5 specializes the spaces yet further to Krylov subspaces and their images under the operator A. We recover familiar Krylov subspace algorithms and show that our framework covers all Krylov subspace methods.

In Section 6 we show how angles between subspaces may be used to derive known residual and error bounds for MR and OR methods and include an application to compact perturbations of the identity.

Final remarks and conclusions are given in Section 7.

2. Approximations, projections and angles

Here we consider two basic methods for approximating a vector $r_0 \in \mathcal{H}$ by an element w from a subspace $\mathcal{W} \subset \mathcal{H}$ and recall their well-known relations to projections and angles. Although the results in this and the following two sections make no reference to solving equation (1.1), we will, of course, ultimately make this connection. In order to avoid changing notation at that point, however, we adopt the notation and terminology of equation solving from the beginning. Thus we denote the (for now arbitrary) vector to be approximated by r_0, since this role will later be played by the residual vector associated with an initial guess x_0 for $A^{-1}b$. Similarly, we denote the associated approximation error $r_0 - w$ by r, as it will coincide with the residuals $r = b - A(x_0 + c) = r_0 - Ac$ of subsequent correction vectors c in the equation-solving context (i.e., w is then of the form $w = Ac$).

Given an arbitrary finite-dimensional subspace $\mathcal{W} \subset \mathcal{H}$ and an element $r_0 \in \mathcal{H}$, we define its MR approximation w^{MR} as the best approximation of r_0 from $\mathcal{W}$ and denote by r^{MR} the associated error

$$w^{\mathrm{MR}} := P_{\mathcal{W}} r_0, \qquad r^{\mathrm{MR}} := r_0 - w^{\mathrm{MR}} = (I - P_{\mathcal{W}}) r_0 \perp \mathcal{W},$$

where $P_{\mathcal{W}}$ is the orthogonal projection onto $\mathcal{W}$. The distance between r_0

and its best approximation $P_W r_0$ in W can be described in terms of angles between vectors and subspaces of $\mathcal{H}$. The angle $\measuredangle(x, y)$ between two non-zero elements $x, y \in \mathcal{H}$ is defined by the relation

$$\cos \measuredangle(x, y) := \frac{|(x, y)|}{\|x\| \, \|y\|},$$

which, in view of the Cauchy–Schwarz inequality, uniquely determines the number $\measuredangle(x, y) \in [0, \pi/2]$. We note here that the natural definition of the angle would replace the modulus in the numerator by the real part; our definition, however, is more appropriate for comparing subspaces (see also the discussion of this issue in Davis and Kahan (1970, p. 9)). Similarly, we define the angle between a nonzero vector $x \in \mathcal{H}$ and a subspace $\mathcal{U} \subset \mathcal{H}$, $\mathcal{U} \neq \{0\}$, as

$$\measuredangle(x, \mathcal{U}) := \inf_{0 \neq u \in \mathcal{U}} \measuredangle(x, u), \quad i.e., \quad \cos \measuredangle(x, \mathcal{U}) = \sup_{0 \neq u \in \mathcal{U}} \cos \measuredangle(x, u).$$

Further, we define the sine of this angle as

$$\sin \measuredangle(x, \mathcal{U}) := \sqrt{1 - \cos^2 \measuredangle(x, \mathcal{U})}.$$

The connection between angles and orthogonal projections is given in the following lemma, a proof of which can be found in Wedin (1983), for example.

Lemma 2.1. Let $\mathcal{U}$ be a finite-dimensional subspace of $\mathcal{H}$ and let $P_\mathcal{U}$ denote the orthogonal projection onto $\mathcal{U}$. For each $x \in \mathcal{H}$ we have

$$\measuredangle(x, \mathcal{U}) = \measuredangle(x, P_\mathcal{U} x) \tag{2.1}$$

and, as a consequence,

$$\|P_\mathcal{U} x\| = \|x\| \cos \measuredangle(x, \mathcal{U}), \tag{2.2}$$
$$\|(I - P_\mathcal{U}) x\| = \|x\| \sin \measuredangle(x, \mathcal{U}). \tag{2.3}$$

In light of this result, the distance between r_0 and its MR approximation w^{MR} may be expressed as

$$\|r^{\mathrm{MR}}\| = \|r_0 - w^{\mathrm{MR}}\| = \|(I - P_W) r_0\| = \|r_0\| \sin \measuredangle(r_0, W). \tag{2.4}$$

In order to define the OR approximation in this abstract setting we require a further finite-dimensional subspace $\mathcal{V} \subset \mathcal{H}$ to formulate the orthogonality constraint. The OR approximation $w^{\mathrm{OR}} \in W$ of r_0 is then defined by the requirement

$$w^{\mathrm{OR}} \in W, \qquad r_0 - w^{\mathrm{OR}} \perp \mathcal{V}.$$

Of course, since choosing $\mathcal{V} = W$ yields the MR approximation, the latter is just a special case of the OR approximation. We choose nonetheless to distinguish the two, both for historical reasons and for ease of exposition. Existence and uniqueness of w^{OR} are summarized in the following result.

Lemma 2.2. If V, W are subspaces of the Hilbert space $\mathcal{H}$ and $r_0 \in \mathcal{H}$, then

(a) there exists $w \in W$ such that $r_0 - w \perp V$ if and only if $r_0 \in W + V^\perp$;

(b) such a w is unique if and only if $W \cap V^\perp = \{0\}$.

Thus, a unique OR approximation is defined for any $r_0 \in \mathcal{H}$ whenever $\mathcal{H} = W \oplus V^\perp$. In this case w^{OR} is the *oblique* projection of r_0 onto W *orthogonal to* V (or, equivalently, along $V^\perp$), which we denote by $P_W^V : \mathcal{H} \to W$, and w^{OR} is characterized by

$$w^{OR} = P_W^V r_0, \qquad r^{OR} = r_0 - w^{OR} = (I - P_W^V) r_0 \perp V.$$

When it exists, the oblique projection P_W^V is given by the Moore–Penrose inverse $(P_V P_W)^+$ of $P_V P_W$ (cf. Wedin (1983)). This is established in the following two lemmas, the first of which describes the mapping properties of $(P_V P_W)^+$.

Lemma 2.3. Given two finite-dimensional subspaces V and W of the Hilbert space $\mathcal{H}$, let $S := (P_V P_W)^+$ denote the Moore–Penrose inverse of the product of the orthogonal projections onto W and V. Then S is a projection and we have

$$\text{range}(S) = W \cap (V + W^\perp), \tag{2.5}$$

$$\text{null}(S) = V^\perp + (W^\perp \cap V). \tag{2.6}$$

Proof. The proof of the projection property $S^2 = S$ follows along the lines of Greville (1974, Theorem 1). First, since $P := P_V P_W$ has finite rank, its Moore–Penrose pseudo-inverse $P^+ = S$ exists and satisfies $\text{range}(P^+) = \text{range}(P^*)$, from which it follows that

$$\text{range}(S) \subset \text{range}(P_W), \qquad \text{range}(S^*) \subset \text{range}(P_V),$$

and hence, by the idempotency of P_V and P_W,

$$P_W S = S, \qquad S P_V = S,$$

which together imply $S^2 = S P_V P_W S = P^+ P P^+ = P^+ = S$.

Since $\text{range}(S) = \text{range}((P_V P_W)^*) = \text{range}(P_W P_V)$ and analogously for the null space, it is sufficient to show (2.5) and (2.6) for the operator $P_W P_V$ instead of S. To derive (2.5), note that $w \in W$ lies in $\text{range}(P_W P_V)$ if and only if there exists $v \in V$ such that $v = w + w_\perp$ for some $w_\perp \in W^\perp$, which in turn is equivalent to $w \in W \cap (V + W^\perp)$. To see (2.6), note that any $x \in \mathcal{H}$ may be written as $x = (w + w_\perp) + v_\perp$ with $w \in W$, $w_\perp \in W^\perp$, $v_\perp \in V^\perp$, and $w + w_\perp \in V$. Thus, $x \in \text{null}(P_W P_V)$ if and only if $w = 0$ or, equivalently, $x \in V^\perp + (W^\perp \cap V)$. $\square$

Lemma 2.4. Given two finite-dimensional subspaces $\mathcal{V}$ and $\mathcal{W}$ of the Hilbert space $\mathcal{H}$ such that $\mathcal{H} = \mathcal{W} \oplus \mathcal{V}^{\perp}$, the oblique projection $P_{\mathcal{W}}^{\mathcal{V}}$ onto $\mathcal{W}$ orthogonal to $\mathcal{V}$ is given by

$$P_{\mathcal{W}}^{\mathcal{V}} = (P_{\mathcal{V}} P_{\mathcal{W}})^{+}. \tag{2.7}$$

Proof. By the previous lemma, S is a projection and, in view of $\mathcal{V} + \mathcal{W}^{\perp} = (\mathcal{V}^{\perp} \cap \mathcal{W})^{\perp}$, $\mathcal{W}^{\perp} \cap \mathcal{V} = (\mathcal{W} + \mathcal{V}^{\perp})^{\perp}$, and $\mathcal{H} = \mathcal{W} \oplus \mathcal{V}^{\perp}$, its range is $\mathcal{W}$ and its null space is $\mathcal{V}^{\perp}$, which together characterize it as $P_{\mathcal{W}}^{\mathcal{V}}$. □

Remark 2.5. We note that, while the left-hand side of (2.7) exists only under the condition $\mathcal{H} = \mathcal{W} \oplus \mathcal{V}^{\perp}$, the right-hand side is always defined. Thus, $(P_{\mathcal{V}} P_{\mathcal{W}})^{+} r_0$ can be viewed as a natural way of defining an OR approximation in those cases where this direct sum condition fails to hold, a situation sometimes referred to, in the context of Krylov subspace methods, as a *Galerkin breakdown*.

As is to be expected, the error $r^{\mathrm{OR}} = r_0 - P_{\mathcal{W}}^{\mathcal{V}} r_0$ depends on the angles between the subspaces $\mathcal{V}$ and $\mathcal{W}$, which we introduce as follows (*cf.* Golub and Van Loan (1996, Section 12.4.3)): Given two finite-dimensional subspaces $\mathcal{V}$ and $\mathcal{W}$ of $\mathcal{H}$, let $m := \min(\dim \mathcal{V}, \dim \mathcal{W})$. The *canonical* or *principal angles* $\{\theta_j\}_{j=1}^{m}$ between $\mathcal{V}$ and $\mathcal{W}$ are defined recursively by

$$\cos \theta_j := \max_{0 \neq v \in \mathcal{V}} \max_{0 \neq w \in \mathcal{W}} \frac{|(v, w)|}{\|v\| \|w\|} =: \frac{|(v_j, w_j)|}{\|v_j\| \|w_j\|}$$

subject to $v \perp v_1, \ldots, v_{j-1}$ and $w \perp w_1, \ldots, w_{j-1}$. We further define *the angle* between the spaces $\mathcal{V}$ and $\mathcal{W}$ as the largest canonical angle

$$\measuredangle(\mathcal{V}, \mathcal{W}) := \theta_m.$$

Remark 2.6. If $P_{\mathcal{V}} P_{\mathcal{W}} = \sum_{j=1}^{m} \sigma_j(\cdot, w_j) v_j$ is a singular value decomposition of $P_{\mathcal{V}} P_{\mathcal{W}}$, then the variational characterization of the singular values,

$$\sigma_j(P_{\mathcal{V}} P_{\mathcal{W}}) = \max_{v \in \mathcal{V}, w \in \mathcal{W}} \frac{|(P_{\mathcal{V}} P_{\mathcal{W}} w, v)|}{\|w\| \|v\|} =: \frac{|(P_{\mathcal{V}} P_{\mathcal{W}} w_j, v_j)|}{\|w_j\| \|v_j\|}$$

subject to $v \perp v_1, \ldots, v_{j-1}$, $w \perp w_1, \ldots, w_{j-1}$ for $j = 1, \ldots, m$ (*cf.* Björck and Golub (1973)), shows immediately that $\cos \theta_j = \sigma_j$. Furthermore, we note that, given any two orthonormal bases $\{v_j\}_{j=1}^{\dim \mathcal{V}}$ and $\{w_j\}_{j=1}^{\dim \mathcal{W}}$ of $\mathcal{V}$ and $\mathcal{W}$, then the cosines of the canonical angles are the singular values of the matrix of inner products $[(v_j, w_k)]_{j=1,\ldots,\dim \mathcal{V}, k=1,\ldots,\dim \mathcal{W}}$ (*cf.* Chatelin (1993)).

Remark 2.7. As a consequence of Remark 2.6, we see that $S = (P_{\mathcal{V}} P_{\mathcal{W}})^{+}$ can be written as

$$S = \sum_{j=1}^{m} \sigma_j^{+}(\cdot, v_j) w_j \quad \text{with} \quad \sigma_j^{+} := \begin{cases} 1/\sigma_j, & \text{if } \sigma_j > 0, \\ 0, & \text{otherwise.} \end{cases}$$

In particular, we have

$$\text{range}(S) = \text{span}\{w_j : \sigma_j > 0\}, \qquad \text{null}(S) = \text{span}\{v_j : \sigma_j > 0\}^\perp.$$

Thus, we have range$(S) = \mathcal{W}$ if and only if $\dim(\mathcal{W}) = m$ and $\sigma_j > 0$ for all $j = 1,\ldots,m$. Similarly, null$(S) = \mathcal{V}^\perp$ if and only if $\dim(\mathcal{V}) = m$ and $\sigma_j > 0$ for all $j = 1,\ldots,m$. Consequently, the oblique projection $P_{\mathcal{W}}^{\mathcal{V}}$ exists if and only if $\dim(\mathcal{V}) = \dim(\mathcal{W})$ and $\angle(\mathcal{V},\mathcal{W}) < \pi/2$.

Remark 2.8. Two further characterizations of the angle between two subspaces, $\mathcal{V}$ and $\mathcal{W}$, are given by

$$\angle(\mathcal{V},\mathcal{W}) = \max_{\substack{v\in\mathcal{V},\\v\neq 0}} \angle(v, P_{\mathcal{W}}^{\mathcal{V}}v)$$

(see Saad (1982)) and $\sin\angle(\mathcal{V},\mathcal{W}) = \|P_{\mathcal{V}} - P_{\mathcal{W}}\|$ (*cf.* Chatelin (1993, p. 5)).

Besides the relative position of the spaces $\mathcal{V}$ and $\mathcal{W}$, the error of the OR approximation also depends on the position of r_0 with respect to $\mathcal{V}$ and $\mathcal{W}$. In this generality, all we can do to bound the OR approximation error is determine the norm of the complementary projection $I - P_{\mathcal{W}}^{\mathcal{V}}$. For simplicity, since $\mathcal{H} = \mathcal{W}\oplus\mathcal{V}^\perp$ for finite-dimensional $\mathcal{V}$ and $\mathcal{W}$ implies $\dim\mathcal{V} = \dim\mathcal{W}$, we assume that both spaces have the same dimension $m < \infty$.

The following result was stated in Saad (1982).

Theorem 2.9. Given two m-dimensional subspaces $\mathcal{V},\mathcal{W}\subset\mathcal{H}$ of the Hilbert space $\mathcal{H}$ such that $\mathcal{H} = \mathcal{W}\oplus\mathcal{V}^\perp$, let $P_{\mathcal{W}}^{\mathcal{V}}:\mathcal{H}\to\mathcal{W}$ denote the oblique projection onto $\mathcal{W}$ orthogonal to $\mathcal{V}$. Then

$$\|I - P_{\mathcal{W}}^{\mathcal{V}}\| = \frac{1}{\cos\angle(\mathcal{V},\mathcal{W})}. \tag{2.8}$$

Proof. The proof follows from the fact that $\|P\| = \|I - P\|$ for any nontrivial projection operator P in a Hilbert space (*cf.* Kato (1960)). Thus, letting $\sigma_{\min}$ and $\sigma_{\max}$ denote the smallest and largest singular values of an operator, we have

$$\|I - (P_{\mathcal{V}}P_{\mathcal{W}})^+\| = \|(P_{\mathcal{V}}P_{\mathcal{W}})^+\| = \sigma_{\max}((P_{\mathcal{V}}P_{\mathcal{W}})^+) = \frac{1}{\sigma_{\min}(P_{\mathcal{V}}P_{\mathcal{W}})}$$

and the conclusion follows from Remark 2.6. $\qquad\square$

Without further assumptions on r_0 and the spaces $\mathcal{V}$ and $\mathcal{W}$, all we can say about the error of the OR approximation is

$$\|r^{\text{OR}}\| = \|r_0 - w^{\text{OR}}\| = \|(I - P_{\mathcal{W}}^{\mathcal{V}})r_0\| \leq \frac{\|r_0\|}{\cos\angle(\mathcal{V},\mathcal{W})}.$$

As an immediate consequence, noting that

$$r^{\text{OR}} = (I - P_{\mathcal{W}}^{\mathcal{V}})r_0 = (I - P_{\mathcal{W}}^{\mathcal{V}})(r_0 - w), \qquad \forall w\in\mathcal{W},$$

we obtain

$$\|r^{\mathrm{OR}}\| \leq \|I - P_{\mathcal{W}}^{\mathcal{V}}\| \inf_{w \in \mathcal{W}} \|r_0 - w\| = \|I - P_{\mathcal{W}}^{\mathcal{V}}\| \|r^{\mathrm{MR}}\|,$$

an estimate usually referred to as *Céa's lemma* in connection with projection methods (see, *e.g.*, Brenner and Scott (1994)), which, in view of (2.8), implies

$$\cos \angle(\mathcal{V}, \mathcal{W}) \|r^{\mathrm{OR}}\| \leq \|r^{\mathrm{MR}}\| \leq \|r^{\mathrm{OR}}\|.$$

3. Approximations from nested subspaces

Until now nothing further has been assumed to relate $\mathcal{V}$, $\mathcal{W}$ and r_0; hence we were only able to bound the error in the OR approximation in terms of the angle between the spaces $\mathcal{V}$ and $\mathcal{W}$. We will obtain more interesting results in this section by selecting a specific test space $\mathcal{V}$ which differs only slightly from the approximation space $\mathcal{W}$. This choice, however, is still general enough to cover all MR and OR methods, in particular Krylov subspace methods. We now investigate MR and OR approximations on nested sequences of subspaces, which is the setting in which these approximations are used by practical algorithms.

3.1. MR approximations

Consider a sequence of nested subspaces

$$\{0\} = \mathcal{W}_0 \subset \mathcal{W}_1 \subset \cdots \subset \mathcal{W}_{m-1} \subset \mathcal{W}_m \subset \cdots \qquad (3.1)$$

of $\mathcal{H}$ and assume for simplicity that $\dim \mathcal{W}_m = m$. Throughout this section, $\{w_1, \ldots, w_m\}$ will always denote an *ascending* orthonormal basis of $\mathcal{W}_m$, that is, one such that $\{w_1, \ldots, w_j\}$ forms a basis of $\mathcal{W}_j$ for every $j = 1, \ldots, m$. In terms of such a basis, the MR approximation, that is, the best approximation of $r_0 \in \mathcal{H}$ from $\mathcal{W}_m$, can be expressed as the truncated Fourier expansion

$$w_m^{\mathrm{MR}} = P_{\mathcal{W}_m} r_0 = \sum_{j=1}^{m} (r_0, w_j) w_j = w_{m-1}^{\mathrm{MR}} + (r_0, w_m) w_m, \qquad m \geq 1,$$

so that the norm of the associated error $r_m^{\mathrm{MR}} = r_0 - w_m^{\mathrm{MR}}$ is given by

$$\|r_m^{\mathrm{MR}}\|^2 = \|(I - P_{\mathcal{W}_m}) r_0\|^2 = \|r_0\|^2 - \sum_{j=1}^{m} |(r_0, w_j)|^2 \qquad (3.2)$$

$$= \|r_{m-1}^{\mathrm{MR}}\|^2 - |(r_0, w_m)|^2.$$

Relation (3.2) shows that no improvement in the MR approximation results whenever the direction in which $\mathcal{W}_{m-1}$ is enlarged is orthogonal to r_0. In other words,

$$\|r_m^{\text{MR}}\| < \|r_{m-1}^{\text{MR}}\| \quad \text{if and only if} \quad (r_0, w_m) \neq 0. \tag{3.3}$$

To relate the approximations on successive spaces, note that for $m \geq 1$ we have

$$w_m^{\text{MR}} = w_{m-1}^{\text{MR}} + P_{\mathcal{W}_m} r_{m-1}^{\text{MR}},$$

from which it follows that

$$r_m^{\text{MR}} = r_{m-1}^{\text{MR}} - P_{\mathcal{W}_m} r_{m-1}^{\text{MR}} = (I - P_{\mathcal{W}_m}) r_{m-1}^{\text{MR}}. \tag{3.4}$$

To express successive approximation errors in terms of angles, we note that (3.4) together with (2.3) yields

$$\|r_m^{\text{MR}}\| = \|(I - P_{\mathcal{W}_m}) r_{m-1}^{\text{MR}}\| = s_m \|r_{m-1}^{\text{MR}}\|, \tag{3.5}$$

where $s_m := \sin \angle(r_{m-1}^{\text{MR}}, \mathcal{W}_m)$. Note that the sine s_m is also given by (*cf.* (2.4))

$$s_m = \frac{\|r_m^{\text{MR}}\|}{\|r_{m-1}^{\text{MR}}\|} = \frac{\|(I - P_{\mathcal{W}_m}) r_0\|}{\|(I - P_{\mathcal{W}_{m-1}}) r_0\|} = \frac{\sin \angle(r_0, \mathcal{W}_m)}{\sin \angle(r_0, \mathcal{W}_{m-1})}, \tag{3.6}$$

that is, s_m is the sine of the angle between the previous approximation error and $\mathcal{W}_m$ or, equivalently, the quotient of the sines of the angles between r_0 and the current and previous approximation spaces. In order for the last three terms in (3.6) to make sense, we assume that $r_0 \notin \mathcal{W}_{m-1}$; otherwise, the approximation problem is solved exactly in the space $\mathcal{W}_{m-1}$ and the larger spaces no longer contribute toward improving the approximation.

In view of (3.2) and (3.5), the corresponding cosine is given by

$$c_m := \sqrt{1 - s_m^2} = \sqrt{1 - \frac{\|r_m^{\text{MR}}\|^2}{\|r_{m-1}^{\text{MR}}\|^2}} = \frac{|(r_0, w_m)|}{\|r_{m-1}^{\text{MR}}\|} \tag{3.7}$$

and we see that an equivalent way of stating (3.3) is

$$\|r_m^{\text{MR}}\| < \|r_{m-1}^{\text{MR}}\| \quad \text{if and only if} \quad c_m \neq 0. \tag{3.8}$$

An obvious induction applied to (3.5) leads to the error formula

$$\|r_m^{\text{MR}}\| = s_1 s_2 \cdots s_m \|r_0\|, \tag{3.9}$$

from which we see that the sequence of approximations will converge to r_0 if and only if the product of sines tends to zero. Moreover, if the numbers s_m themselves tend to zero, the convergence of the MR approximations is superlinear.

3.2. OR approximations

In order to define the OR approximations associated with the nested sequence $\{\mathcal{W}_m\}_{m \geq 0}$, we fix the sequence $\{\mathcal{V}_m\}_{m \geq 1}$ of spaces which define the orthogonality condition by setting

$$\mathcal{V}_m := \operatorname{span}\{r_0\} + \mathcal{W}_{m-1}, \qquad m = 1, 2, \ldots. \tag{3.10}$$

With this definition of $\mathcal{V}_m$, since the mth OR and MR approximations lie in $\mathcal{W}_m$, the corresponding approximation errors lie in $\mathcal{V}_{m+1}$. Since these errors turn out to be residual vectors in the context of subspace correction methods for solving $A\boldsymbol{x} = \boldsymbol{b}$, we shall refer to $\{\mathcal{V}_m\}_{m\geq 1}$ as the sequence of *residual spaces*.

We first investigate the question of when the OR approximation is well-defined. In view of Remark 2.7, this amounts to checking whether the angle between $\mathcal{V}_m$ and $\mathcal{W}_m$ is strictly less than $\pi/2$. As a consequence of the special choice (3.10) of $\mathcal{V}_m$, it turns out that this angle is the same as $\angle(r_{m-1}^{\mathrm{MR}}, \mathcal{W}_m)$.

Theorem 3.1. If the spaces $\mathcal{V}_m$ and $\mathcal{W}_m$ are related as in (3.10), then the largest canonical angle between them is given by

$$\angle(\mathcal{V}_m, \mathcal{W}_m) = \angle(r_{m-1}^{\mathrm{MR}}, \mathcal{W}_m). \tag{3.11}$$

Moreover, the remaining $m - 1$ canonical angles between $\mathcal{V}_m$ and $\mathcal{W}_m$ are zero.

Proof. Noting that $\{\boldsymbol{w}_1, \ldots, \boldsymbol{w}_{m-1}, \widehat{\boldsymbol{w}}_m\}$, with $\widehat{\boldsymbol{w}}_m = r_{m-1}^{\mathrm{MR}}/\|r_{m-1}^{\mathrm{MR}}\|$, is an orthonormal basis of $\mathcal{V}_m$, we obtain the cosine of the largest canonical angle as the smallest singular value of the matrix

$$\begin{bmatrix} (\boldsymbol{w}_1, \boldsymbol{w}_1) & \cdots & (\boldsymbol{w}_{m-1}, \boldsymbol{w}_1) & (\boldsymbol{w}_m, \boldsymbol{w}_1) \\ \vdots & & \vdots & \vdots \\ (\boldsymbol{w}_1, \boldsymbol{w}_{m-1}) & \cdots & (\boldsymbol{w}_{m-1}, \boldsymbol{w}_{m-1}) & (\boldsymbol{w}_m, \boldsymbol{w}_{m-1}) \\ (\boldsymbol{w}_1, \widehat{\boldsymbol{w}}_m) & \cdots & (\boldsymbol{w}_{m-1}, \widehat{\boldsymbol{w}}_m) & (\boldsymbol{w}_m, \widehat{\boldsymbol{w}}_m) \end{bmatrix} = \begin{bmatrix} I_{m-1} & 0 \\ 0 & \frac{(\boldsymbol{w}_m, r_0)}{\|r_{m-1}^{\mathrm{MR}}\|} \end{bmatrix}$$

(*cf.* Remark 2.6). Thus the smallest singular value is $|(\boldsymbol{w}_m, r_0)|/\|r_{m-1}^{\mathrm{MR}}\| = c_m$ (*cf.* (3.7)) and all remaining singular values are equal to one. $\square$

As an immediate consequence of Theorem 3.1, we obtain the following characterization of when the oblique projection $P_{\mathcal{W}_m}^{\mathcal{V}_m}$ is defined for our specific choice of $\mathcal{V}_m$.

Corollary 3.2. The OR approximation of an arbitrary $r_0 \in \mathcal{H}$ with respect to the sequence of spaces $\mathcal{V}_m$ and $\mathcal{W}_m$ as given by (3.1) and (3.10) is uniquely defined if and only if $(r_0, \boldsymbol{w}_m) \neq 0$, that is, if and only if the MR approximation improves as $\mathcal{W}_{m-1}$ is enlarged to $\mathcal{W}_m$.

Thus the degenerate case in which the OR approximation is not uniquely defined for all $r_0 \in \mathcal{H}$ and for which the MR approximation makes no progress is characterized by $c_m = 0$. We thus tacitly assume $(r_0, \boldsymbol{w}_m) \neq 0$ whenever mentioning the OR approximation of index m.

In many algorithms such as GMRES it is convenient to work also with an ascending basis $\{v_1, \ldots, v_m\}$ of $\mathcal{V}_m$. The following result, which will be used in Section 6.4, relates $\angle(r_0, w_m)$ with $\angle(v_{m+1}, w_m)$.

Lemma 3.3. If the subspaces $\mathcal{V}_m$ and $\mathcal{W}_m$ are related by (3.10) and $\{v_1, \ldots, v_m\}$ and $\{w_1, \ldots, w_m\}$ are ascending orthonormal bases of $\mathcal{V}_m$ and $\mathcal{W}_m$, respectively, then

$$|(v_{m+1}, w_m)| = s_m.$$

In the degenerate case $(r_0, w_m) = 0$, the vectors v_{m+1} and w_m must be collinear.

Proof. From $v_{m+1} \in \operatorname{span}\{r_0, w_1, \ldots, w_m\} \cap \operatorname{span}\{r_0, w_1, \ldots, w_{m-1}\}^\perp$ we conclude $v_{m+1} \in \operatorname{span}\{r_m^{\mathrm{MR}}, w_m\}$ and the assertion follows from the remaining requirements $\|v_{m+1}\| = 1$ and $v_{m+1} \perp r_0$: Using the notation from the proof of Theorem 3.1, we set $\hat{w}_{m+1} := r_m^{\mathrm{MR}} / \|r_m^{\mathrm{MR}}\|$ and note that $\{\hat{w}_m, w_m\}$ form an orthonormal basis of $\operatorname{span}\{r_m^{\mathrm{MR}}, w_m\}$, hence $v_{m+1} = \alpha \hat{w}_m + \beta w_m$ for some coefficients $\alpha, \beta \in \mathbb{C}$. Since $(r_m^{\mathrm{MR}}, r_0) = \|r_m^{\mathrm{MR}}\|^2$, orthogonality of v_{m+1} and r_0 yields

$$0 = (v_{m+1}, r_0) = \frac{\alpha}{\|r_m^{\mathrm{MR}}\|}(r_m^{\mathrm{MR}}, r_0) + \beta(w_m, r_0) = \alpha \|r_m^{\mathrm{MR}}\| + \beta(w_m, r_0).$$

The requirement that v_{m+1} have unit norm now gives

$$|\beta|^2 = \left(1 + \frac{|(w_m, r_0)|^2}{\|r_m^{\mathrm{MR}}\|^2}\right)^{-1} = \frac{\|r_m^{\mathrm{MR}}\|^2}{\|r_m^{\mathrm{MR}}\|^2 + |(w_m, r_0)|^2} = \frac{\|r_m^{\mathrm{MR}}\|^2}{\|r_{m-1}^{\mathrm{MR}}\|^2} = s_m^2.$$

Since $r_m^{\mathrm{MR}} \perp w_m$, we now obtain $|(v_{m+1}, w_m)| = |\beta| = s_m$.

If $(r_0, w_m) = 0$ then $|(v_{m+1}, w_m)| = 1$, i.e., $|(v_{m+1}, w_m)| = \|v_{m+1}\| \|w_m\|$, which means that these two vectors are collinear. $\square$

3.3. Relations between MR and OR approximations

Recall that $r_0 \in \mathcal{W}_{m-1}$ implies that the MR approximation with respect to the space $\mathcal{W}_{m-1}$ solves the approximation problem exactly. The same is true for the OR approximation, since $r_0 \in \mathcal{W}_{m-1}$ implies $r_0 - r_{m-1}^{\mathrm{OR}} \in \mathcal{W}_{m-1} \cap \mathcal{V}_{m-1}^\perp = \{0\}$, so that the OR approximation solves the problem exactly whenever the MR approximation does. In other words, the assumption $r_0 \notin \mathcal{W}_{m-1}$ is *equivalent* to saying that both r_{m-1}^{MR} and r_{m-1}^{OR} are not yet zero.

If we define $\tilde{w}_m := |(w_m, r_0)| \hat{w}_m / (r_0, w_m)$, where $\hat{w}_m = r_{m-1}^{\mathrm{MR}} / \|r_{m-1}^{\mathrm{MR}}\|$, then $(w_m, \tilde{w}_m) = c_m$ (cf. the proof of Theorem 3.1). Consequently, the sets $\{w_1, \ldots, w_m\}$ and $\{w_1, \ldots, w_{m-1}, \tilde{w}_m / c_m\}$ form a pair of *biorthonormal* bases of $\mathcal{W}_m$ and $\mathcal{V}_m$. This fact allows us to express the oblique projection which determines the OR approximation as the singular value expansion

$$P_{\mathcal{W}_m}^{\mathcal{V}_m} = \sum_{j=1}^{m-1} (\cdot, \boldsymbol{w}_j) \boldsymbol{w}_j + \frac{1}{c_m} (\cdot, \tilde{\boldsymbol{w}}_m) \boldsymbol{w}_m, \tag{3.12}$$

from which we derive the following expression for the difference of the OR and MR approximations:

$$\begin{aligned}
\boldsymbol{w}_m^{\mathrm{OR}} - \boldsymbol{w}_m^{\mathrm{MR}} &= (P_{\mathcal{W}_m}^{\mathcal{V}_m} - P_{\mathcal{W}_m}) \boldsymbol{r}_0 = \left[c_m^{-1} (\boldsymbol{r}_0, \tilde{\boldsymbol{w}}_m) - (\boldsymbol{r}_0, \boldsymbol{w}_m) \right] \boldsymbol{w}_m \\
&= \frac{\|\boldsymbol{r}_{m-1}^{\mathrm{MR}}\|^2 - |(\boldsymbol{r}_0, \boldsymbol{w}_m)|^2}{(\boldsymbol{w}_m, \boldsymbol{r}_0)} \boldsymbol{w}_m = \frac{\|\boldsymbol{r}_m^{\mathrm{MR}}\|^2}{(\boldsymbol{w}_m, \boldsymbol{r}_0)} \boldsymbol{w}_m.
\end{aligned} \tag{3.13}$$

In other words, since the spaces $\mathcal{V}_m$ and $\mathcal{V}_m$ are so closely related, the projection $P_{\mathcal{W}_m}^{\mathcal{V}_m}$ is simply a rank-one modification of $P_{\mathcal{W}_m}$ and this is the essential ingredient of the proof for the following familiar relations.

Theorem 3.4. Given an arbitrary element $\boldsymbol{r}_0 \in \mathcal{H}$, a nested sequence of subspaces $\mathcal{W}_m \subset \mathcal{H}$ of dimension m (cf. (3.1)) and a corresponding sequence of error spaces $\mathcal{V}_m$ as defined by (3.10), then the MR and OR approximations of $\boldsymbol{r}_0$ with respect to $\mathcal{W}_m$ and $\mathcal{V}_m$ satisfy

$$\|\boldsymbol{r}_m^{\mathrm{MR}}\| = s_m \|\boldsymbol{r}_{m-1}^{\mathrm{MR}}\|, \tag{3.14}$$

$$\|\boldsymbol{r}_m^{\mathrm{MR}}\| = s_1 s_2 \cdots s_m \|\boldsymbol{r}_0\|, \tag{3.15}$$

$$\|\boldsymbol{r}_m^{\mathrm{MR}}\| = c_m \|\boldsymbol{r}_m^{\mathrm{OR}}\|, \tag{3.16}$$

$$\|\boldsymbol{r}_m^{\mathrm{OR}}\| = s_1 s_2 \cdots s_m \|\boldsymbol{r}_0\| / c_m, \tag{3.17}$$

where $s_m = \sin \angle(\boldsymbol{r}_{m-1}^{\mathrm{MR}}, \mathcal{W}_m)$ and $c_m = \cos \angle(\boldsymbol{r}_{m-1}^{\mathrm{MR}}, \mathcal{W}_m)$.

Proof. Identities (3.14) and (3.15) are merely restatements of (3.5) and (3.9), which have already been proved. Next, from (3.13) we obtain

$$\boldsymbol{r}_m^{\mathrm{MR}} - \boldsymbol{r}_m^{\mathrm{OR}} = \boldsymbol{w}_m^{\mathrm{OR}} - \boldsymbol{w}_m^{\mathrm{MR}} \in \mathrm{span}\{\boldsymbol{w}_m\}, \tag{3.18}$$

and

$$\boldsymbol{w}_m^{\mathrm{OR}} = \boldsymbol{w}_m^{\mathrm{MR}} + \frac{\|\boldsymbol{r}_m^{\mathrm{MR}}\|^2}{(\boldsymbol{w}_m, \boldsymbol{r}_0)} \boldsymbol{w}_m = \boldsymbol{w}_{m-1}^{\mathrm{MR}} + \frac{\|\boldsymbol{r}_{m-1}^{\mathrm{MR}}\|^2}{(\boldsymbol{w}_m, \boldsymbol{r}_0)} \boldsymbol{w}_m, \tag{3.19}$$

where we have used (3.2) for the last equality. Since $\boldsymbol{r}_m^{\mathrm{MR}} \perp \boldsymbol{w}_m$, the Pythagorean identity and (3.2) yield

$$\|\boldsymbol{r}_m^{\mathrm{OR}}\|^2 = \left(1 + \frac{\|\boldsymbol{r}_m^{\mathrm{MR}}\|^2}{|(\boldsymbol{w}_m, \boldsymbol{r}_0)|^2} \right) \|\boldsymbol{r}_m^{\mathrm{MR}}\|^2 = \frac{\|\boldsymbol{r}_{m-1}^{\mathrm{MR}}\|^2}{|(\boldsymbol{w}_m, \boldsymbol{r}_0)|^2} \|\boldsymbol{r}_m^{\mathrm{MR}}\|^2,$$

which, in view of (3.7), gives the error formula

$$\|\boldsymbol{r}_m^{\mathrm{OR}}\| = \frac{\|\boldsymbol{r}_{m-1}^{\mathrm{MR}}\| \, \|\boldsymbol{r}_m^{\mathrm{MR}}\|}{|(\boldsymbol{w}_m, \boldsymbol{r}_0)|} = \frac{1}{c_m} \|\boldsymbol{r}_m^{\mathrm{MR}}\|$$

for the OR approximation, establishing (3.16) and (3.17).   □

In addition to the norm identities contained in Theorem 3.4, it is also possible to relate the MR and OR approximations and their errors, as the next theorem shows.

Theorem 3.5. Under the assumptions of Theorem 3.4, the MR and OR approximations and errors satisfy

$$w_m^{\mathrm{MR}} = s_m^2 w_{m-1}^{\mathrm{MR}} + c_m^2 w_m^{\mathrm{OR}}, \tag{3.20}$$

$$r_m^{\mathrm{MR}} = s_m^2 r_{m-1}^{\mathrm{MR}} + c_m^2 r_m^{\mathrm{OR}}, \tag{3.21}$$

$$\frac{w_m^{\mathrm{MR}}}{\|r_m^{\mathrm{MR}}\|^2} = \sum_{j=0}^{m} \frac{w_j^{\mathrm{OR}}}{\|r_j^{\mathrm{OR}}\|^2}, \tag{3.22}$$

$$\frac{r_m^{\mathrm{MR}}}{\|r_m^{\mathrm{MR}}\|^2} = \sum_{j=0}^{m} \frac{r_j^{\mathrm{OR}}}{\|r_j^{\mathrm{OR}}\|^2}, \tag{3.23}$$

$$\frac{1}{\|r_m^{\mathrm{MR}}\|^2} = \sum_{j=0}^{m} \frac{1}{\|r_j^{\mathrm{OR}}\|^2} = \frac{1}{\|r_{m-1}^{\mathrm{MR}}\|^2} + \frac{1}{\|r_m^{\mathrm{OR}}\|^2}. \tag{3.24}$$

Proof. From (3.19) and $w_m^{\mathrm{MR}} - w_{m-1}^{\mathrm{MR}} = (r_0, w_m) w_m$ we obtain

$$
\begin{aligned}
w_m^{\mathrm{OR}} &= w_m^{\mathrm{MR}} + \frac{\|r_m^{\mathrm{MR}}\|^2}{(w_m, r_0)} \frac{1}{(r_0, w_m)} (w_m^{\mathrm{MR}} - w_{m-1}^{\mathrm{MR}}) \\
&= w_m^{\mathrm{MR}} + \frac{\|r_m^{\mathrm{MR}}\|^2}{\|r_{m-1}^{\mathrm{MR}}\|^2} \frac{\|r_{m-1}^{\mathrm{MR}}\|^2}{|(r_0, w_m)|^2} (w_m^{\mathrm{MR}} - w_{m-1}^{\mathrm{MR}}) \\
&= w_m^{\mathrm{MR}} + \frac{s_m^2}{c_m^2} (w_m^{\mathrm{MR}} - w_{m-1}^{\mathrm{MR}})
\end{aligned}
$$

(*cf.* (3.14) and (3.7)), which implies the relationship (3.20) between the MR and OR approximations and, by way of $s_m^2 + c_m^2 = 1$, the corresponding relationship (3.21) between their errors.

Repeated application of these two formulas leads to

$$w_m^{\mathrm{MR}} = \sum_{j=0}^{m} \tau_{m,j}^2 w_j^{\mathrm{OR}} \qquad \text{and} \qquad r_m^{\mathrm{MR}} = \sum_{j=0}^{m} \tau_{m,j}^2 r_j^{\mathrm{OR}},$$

where $\tau_{m,0} := s_1 s_2 \ldots s_m$ and $\tau_{m,j} := c_j s_{j+1} \ldots s_m$ $(1 \le j \le m)$. Using (3.14) and (3.16) this can be simplified to

$$\tau_{m,j} = c_j \frac{\|r_{j+1}^{\mathrm{MR}}\| \, \|r_{j+2}^{\mathrm{MR}}\|}{\|r_j^{\mathrm{MR}}\| \, \|r_{j+1}^{\mathrm{MR}}\|} \cdots \frac{\|r_m^{\mathrm{MR}}\|}{\|r_{m-1}^{\mathrm{MR}}\|} = c_j \frac{\|r_m^{\mathrm{MR}}\|}{\|r_j^{\mathrm{MR}}\|} = \frac{\|r_m^{\mathrm{MR}}\|}{\|r_j^{\mathrm{OR}}\|},$$

and we obtain (3.22) as well as (3.23). Finally, since the errors r_j^{OR} are orthogonal, we have

$$\frac{1}{\|r_m^{\mathrm{MR}}\|^2} = \sum_{j=0}^{m} \frac{1}{\|r_j^{\mathrm{OR}}\|^2} = \sum_{j=0}^{m-1} \frac{1}{\|r_j^{\mathrm{OR}}\|^2} + \frac{1}{\|r_m^{\mathrm{OR}}\|^2} = \frac{1}{\|r_{m-1}^{\mathrm{MR}}\|^2} + \frac{1}{\|r_m^{\mathrm{OR}}\|^2},$$

which proves (3.24). Strictly speaking, this proves

$$\frac{1}{\|r_m^{\mathrm{MR}}\|^2} = \frac{1}{\|r_{m-1}^{\mathrm{MR}}\|^2} + \frac{1}{\|r_m^{\mathrm{OR}}\|^2}$$

only under the assumption that all OR approximations $w_1^{\mathrm{OR}}, \ldots, w_m^{\mathrm{OR}}$ exist. But this last equation is merely a reformulation of the Pythagorean identity,

$$1 = s_m^2 + c_m^2 = \frac{\|r_m^{\mathrm{MR}}\|^2}{\|r_{m-1}^{\mathrm{MR}}\|^2} + \frac{\|r_m^{\mathrm{MR}}\|^2}{\|r_m^{\mathrm{OR}}\|^2}$$

(cf. (3.14), (3.16)), requiring only the existence of w_m^{OR} (besides $r_0 \notin \mathcal{W}_m$).   □

Corollary 3.6. In view of

$$s_m = \frac{\|r_m^{\mathrm{MR}}\|}{\|r_{m-1}^{\mathrm{MR}}\|}, \qquad i.e., \quad c_m = \sqrt{1 - \frac{\|r_m^{\mathrm{MR}}\|^2}{\|r_{m-1}^{\mathrm{MR}}\|^2}},$$

an angle-free formulation of (3.16), (3.20) and (3.21) reads

$$\|r_m^{\mathrm{MR}}\| = \sqrt{1 - \frac{\|r_m^{\mathrm{MR}}\|^2}{\|r_{m-1}^{\mathrm{MR}}\|^2}} \, \|r_m^{\mathrm{OR}}\|,$$

$$w_m^{\mathrm{MR}} = w_m^{\mathrm{OR}} + \frac{\|r_m^{\mathrm{MR}}\|^2}{\|r_{m-1}^{\mathrm{MR}}\|^2} (w_{m-1}^{\mathrm{MR}} - w_m^{\mathrm{OR}}),$$

$$r_m^{\mathrm{MR}} = r_m^{\mathrm{OR}} + \frac{\|r_m^{\mathrm{MR}}\|^2}{\|r_{m-1}^{\mathrm{MR}}\|^2} (r_{m-1}^{\mathrm{MR}} - r_m^{\mathrm{OR}}).$$

Of course, the first of these identities, or its reformulation

$$\|r_m^{\mathrm{OR}}\| = \left(1 - \frac{\|r_m^{\mathrm{MR}}\|^2}{\|r_{m-1}^{\mathrm{MR}}\|^2}\right)^{-1/2} \|r_m^{\mathrm{MR}}\|,$$

only makes sense if w_m^{OR} is defined, which is equivalent to $\|r_m^{\mathrm{MR}}\| < \|r_{m-1}^{\mathrm{MR}}\|$. If $\|r_m^{\mathrm{MR}}\| \approx \|r_{m-1}^{\mathrm{MR}}\|$ then the factor $(1 - \|r_m^{\mathrm{MR}}\|^2/\|r_{m-1}^{\mathrm{MR}}\|^2)^{-1/2}$ will be large and, consequently, $\|r_m^{\mathrm{OR}}\| \gg \|r_m^{\mathrm{MR}}\|$. Conversely, if the MR approximation makes considerable progress in step m, then $(1 - \|r_m^{\mathrm{MR}}\|^2/\|r_{m-1}^{\mathrm{MR}}\|^2)^{-1/2} \approx 1$ and $\|r_m^{\mathrm{OR}}\| \approx \|r_m^{\mathrm{MR}}\|$. In the context of Krylov subspace methods, this observation is sometimes referred to as the *peak/plateau phenomenon* of MR/OR approximations (see, e.g., Cullum and Greenbaum (1996)).

Remark 3.7. We close this section by reconsidering the issue of the so-called Galerkin breakdown mentioned in Remark 2.5. Using the biorthonormal bases introduced in (3.12), we obtain the singular value expansion of $P_\mathcal{V}P_\mathcal{W}$ as

$$P_\mathcal{V}P_\mathcal{W} = \sum_{j=1}^{m-1}(\cdot,\boldsymbol{w}_j)\boldsymbol{w}_j + c_m(\cdot,\boldsymbol{w}_m)\widetilde{\boldsymbol{w}}_m.$$

If, in view of (2.7), one defines the OR approximation in the degenerate case $c_m = 0$ by $\boldsymbol{w}_m^{OR} := (P_\mathcal{V}P_\mathcal{W})^+\boldsymbol{r}_0$, then this leads to $\boldsymbol{w}_m^{OR} = \boldsymbol{w}_{m-1}^{MR} = \boldsymbol{w}_m^{MR}$. We thereby arrive at a natural extension of the definition of the OR approximation in the case of a Galerkin breakdown.

3.4. Smoothing algorithms

A *smoothing algorithm* transforms a given sequence $\{\boldsymbol{u}_m\} \subset \mathcal{W}_m$ of approximations to $\boldsymbol{r}_0$ into a new sequence $\{\hat{\boldsymbol{u}}_m\} \subset \mathcal{W}_m$ according to

$$\hat{\boldsymbol{u}}_m := (1-\alpha_m)\hat{\boldsymbol{u}}_{m-1} + \alpha_m\boldsymbol{u}_m \tag{3.25}$$

($m = 1,2,\ldots$, $\hat{\boldsymbol{u}}_0 := \boldsymbol{u}_0 = \boldsymbol{0}$). The associated approximation errors $\boldsymbol{r}_m = \boldsymbol{r}_0 - \boldsymbol{u}_m$ and $\hat{\boldsymbol{r}}_m = \boldsymbol{r}_0 - \hat{\boldsymbol{u}}_m$ then satisfy

$$\hat{\boldsymbol{r}}_m = (1-\alpha_m)\hat{\boldsymbol{r}}_{m-1} + \alpha_m\boldsymbol{r}_m.$$

The intention is that the errors of the transformed sequence should decrease 'more smoothly' than those of the original sequence. Ideally, we would like to have $\hat{\boldsymbol{u}}_m = \boldsymbol{w}_m^{MR}$ and we shall discuss two smoothing procedures which achieve this goal when applied to $\boldsymbol{u}_m = \boldsymbol{w}_m^{OR}$, that is, to the sequence of OR approximations.

In *minimal residual smoothing* (cf. Weiss (1994), Zhou and Walker (1994) or Gutknecht (1997, Section 17)), the parameter α_m in (3.25) is chosen to minimize the norm of the error $\hat{\boldsymbol{r}}_m = \hat{\boldsymbol{r}}_{m-1} - \alpha_m(\hat{\boldsymbol{r}}_{m-1} - \boldsymbol{r}_m)$ as a function of α_m. In other words, we seek the best approximation $\alpha_m(\hat{\boldsymbol{r}}_{m-1} - \boldsymbol{r}_m)$ from $\mathrm{span}\{\hat{\boldsymbol{r}}_{m-1} - \boldsymbol{r}_m\}$ to $\hat{\boldsymbol{r}}_{m-1}$, which is obtained for

$$\alpha_m^{MR} := \frac{(\hat{\boldsymbol{r}}_{m-1}, \hat{\boldsymbol{r}}_{m-1} - \boldsymbol{r}_m)}{\|\hat{\boldsymbol{r}}_{m-1} - \boldsymbol{r}_m\|^2}.$$

In an alternative smoothing procedure known as *quasi-minimal residual smoothing* (cf. Zhou and Walker (1994) or Gutknecht (1997, Section 17)), the parameter α_m is chosen as

$$\alpha_m^{QMR} := \frac{\tau_m^2}{\|\boldsymbol{r}_m\|^2} \quad \text{with } \tau_m \text{ such that} \quad \frac{1}{\tau_m^2} = \frac{1}{\tau_{m-1}^2} + \frac{1}{\|\boldsymbol{r}_m\|^2}, \ \tau_0 = \|\boldsymbol{r}_0\|.$$

It is easy to see by induction that

$$\tau_m^2 = \frac{1}{\sum_{j=0}^m 1/\|r_j\|^2}, \qquad i.e., \qquad \alpha_m^{\text{QMR}} = \frac{1/\|r_m\|^2}{\sum_{j=0}^m 1/\|r_j\|^2},$$

and therefore

$$\hat{u}_m = \frac{\sum_{j=0}^m u_j/\|r_j\|^2}{\sum_{j=0}^m 1/\|r_j\|^2} \qquad \text{as well as} \qquad \hat{r}_m = \frac{\sum_{j=0}^m r_j/\|r_j\|^2}{\sum_{j=0}^m 1/\|r_j\|^2}.$$

The last formula for $\hat{u}_m$ reveals the strategy behind quasi-minimal residual smoothing: $\hat{u}_m$ is a weighted sum of all previous iterates $u_0, u_1, \ldots, u_m$ with weights $(1/\|r_k\|^2)/(\sum_{j=0}^m(1/\|r_j\|^2))$ which are (relatively) large if u_k approximates r_0 well and (relatively) small if u_k is a poor approximation to r_0.

In general, that is, for an arbitrary sequence $\{u_m\} \in \mathcal{W}_m$, minimal and quasi-minimal residual smoothing will generate different 'smoothed' iterates $\hat{u}_m$. In the case of $u_m = w_m^{\text{OR}}$, however, these two methods are equivalent.

Proposition 3.8. If either the minimal residual or the quasi-minimal residual smoothing algorithm is applied to the sequence of OR approximations $\{w_m^{\text{OR}}\}$ for an element $r_0 \in \mathcal{H}$, then the resulting smoothed sequence consists of the MR approximations w_m^{MR} for r_0 and the associated smoothing parameters are given by $\alpha_m^{\text{MR}} = \alpha_m^{\text{QMR}} = c_m^2$.

Proof. An induction shows that minimal residual smoothing applied to $u_m = w_m^{\text{OR}}$ yields $\hat{u}_m = w_m^{\text{MR}}$ and that, in this case, $\alpha_m^{\text{MR}} = c_m^2$: the assertion is trivial for $m = 1$. Assuming $\hat{u}_{m-1} = h_{m-1}^{\text{MR}}$ for some $m \geq 2$, we see from (3.19) that

$$\hat{r}_{m-1} - r_m = r_{m-1}^{\text{MR}} - r_m^{\text{OR}} = \frac{\|r_{m-1}^{\text{MR}}\|^2}{(w_m, r_0)} w_m$$

and consequently, noting that $r_{m-1}^{\text{MR}} = r_m^{\text{MR}} + (r_0, w_m)w_m$ and $r_m^{\text{MR}} \perp w_m$,

$$(\hat{r}_{m-1}, \hat{r}_{m-1} - r_m) = \left((r_0, w_m)w_m, \frac{\|r_{m-1}^{\text{MR}}\|^2}{(w_m, r_0)} w_m \right) = \|r_{m-1}^{\text{MR}}\|^2.$$

From (3.7) and (3.20), there finally follows

$$\alpha_m^{\text{MR}} = \|r_{m-1}^{\text{MR}}\|^2 \frac{|(w_m, r_0)|^2}{\|r_{m-1}^{\text{MR}}\|^4} = \frac{|(w_m, r_0)|^2}{\|r_{m-1}^{\text{MR}}\|^2} = c_m^2 \qquad \text{and} \qquad \hat{u}_m = w_m^{\text{MR}}.$$

The analogous assertion for quasi-minimal residual smoothing follows, with (3.22) and (3.23), immediately from the orthogonality of the error vectors r_m^{OR}. $\qquad\square$

It should not come as a surprise that, in our setting, a one-dimensional minimization procedure such as minimal residual smoothing yields the best approximation w_m^{MR}, which is the global optimum on $\mathcal{W}_m$. Recall that w_m^{OR} is already 'nearly optimal' and needs to be corrected only in the direction of w_m.

4. Working with coordinates

The results of the previous sections more closely resemble the matrix formulation of familiar Krylov subspace methods once they are formulated in coordinates with respect to suitable bases of the spaces $\mathcal{W}_m$ and $\mathcal{V}_m$. Specifically, the relevant quantities may be represented in terms of bases of either space, and the situation is simplified if either of these bases is orthogonal. The distinction between which bases are used and whether these are orthogonal also results in three fundamental algorithmic approaches on which all Krylov subspace solvers and their generalizations are based. As we show in this section, these three approaches may all be expressed at the abstract level of the preceding two sections, and this serves to isolate the algorithmic features of these approximation methods from issues associated with their use for solving equations, for instance, by Krylov subspace methods.

We shall begin with the simplest formulation in terms of an ascending orthonormal basis of the sequence $\{\mathcal{W}_m\}$ followed by an approach which first generates an ascending orthonormal basis of $\{\mathcal{V}_m\}$ from a not necessarily orthogonal ascending basis of $\{\mathcal{W}_m\}$. It is seen that the familiar relation of these two bases via an unreduced upper Hessenberg matrix in the Krylov subspace context also holds in the abstract setting as a direct consequence of the definition of the sequence of residual spaces (3.10). The third basic approach allows for neither of the two ascending bases to be orthogonal, and it is shown how this case can be made to fit into the MR/OR framework by introducing a new, basis-dependent inner product.

Motivated by this last approach, we show that, when allowing for such a change of inner product, any sequence of approximations in a Hilbert space becomes a sequence of either MR or OR approximations.

4.1. Using an orthonormal basis of $\mathcal{W}_m$

If, for each $m \geq 1$, the vectors $\{w_1, \ldots, w_m\}$ form an orthonormal basis of $\mathcal{W}_m$, then each $w \in \mathcal{W}_m$ possesses the unique representation $w = W_m y$, in which W_m denotes the row vector $W_m := [w_1, \ldots, w_m]$ and $y \in \mathbb{C}^m$ is the coordinate vector of w with respect to this basis. The characterization $r_m^{\mathrm{MR}} = r_0 - W_m y_m^{\mathrm{MR}} \perp \mathcal{W}_m$ then immediately determines the coordinate vector y_m^{MR} of w_m^{MR} to be

$$y_m^{\mathrm{MR}} = [(r_0, w_1), \ldots, (r_0, w_m)]^\top. \tag{4.1}$$

The coordinate vector y_m^{OR} of the corresponding OR approximation w_m^{OR} is given by

$$y_m^{OR} = [(r_0, w_1), \ldots, (r_0, w_{m-1}), \|w_{m-1}^{MR}\|^2/(w_m, r_0)]^\top \qquad (4.2)$$

since $w_m^{OR} = w_{m-1}^{MR} + \|r_{m-1}^{MR}\|^2/(w_m, r_0)w_m$ (cf. (3.19)).

4.2. Using an orthonormal basis of $\mathcal{V}_m$

We now drop the orthogonality requirement on the vectors w_j, assuming only that the sequence of vectors $\{w_m\}$ forms an ascending basis of the sequence of approximation spaces $\{\mathcal{W}_m\}$. In the same manner, let $\{v_m\}$ form an ascending basis of the corresponding residual spaces space $\mathcal{V}_m$. Since $\mathcal{W}_m \subseteq \text{span}\{r_0\} + \mathcal{W}_m = \mathcal{V}_{m+1}$, we may represent each basis vector w_k as a linear combination of $v_1, \ldots, v_{k+1}$,

$$w_k = \sum_{j=1}^{k+1} \eta_{j,k} v_j, \qquad k = 1, \ldots, m,$$

or, more compactly, employing the row vector notation $W_m = [w_1, \ldots, w_m]$ and $V_m := [v_1, \ldots, v_m]$, $m = 1, 2, \ldots,$

$$W_m = V_{m+1}\tilde{H}_m = V_m H_m + [0, \ldots, 0, \eta_{m+1,m} v_{m+1}], \qquad (4.3)$$

where $\tilde{H}_m =: [\eta_{j,k}] \in \mathbb{C}^{(m+1)\times m}$ is an upper Hessenberg matrix and $H_m := [I_m \ 0]\tilde{H}_m$ is the square matrix obtained by deleting the last row of $\tilde{H}_m$. Note that as long as $r_0 \notin \mathcal{W}_m$, that is, $w_m \notin \mathcal{V}_m$, we have $\eta_{m+1,m} \neq 0$ and therefore $\text{rank}(\tilde{H}_m) = m$. If $r_0 \in \mathcal{W}_m$ for some index m, we let

$$L := \min\{m : r_0 \in \mathcal{W}_m\} \qquad (4.4)$$

be the smallest such index and note that $\mathcal{V}_{L+1} = \mathcal{V}_L = \text{span}\{v_1, \ldots, v_L\}$, that is, $W_L = V_L H_L$, implying $\text{rank}(H_L) = \text{rank}(\tilde{H}_L) = L$. If such an index does not exist, we set $L = \infty$. To avoid cumbersome notation we shall concentrate on the case of $L < \infty$, which is the most relevant for practical applications. We note, however, that all our conclusions can be proved in the general case.

For a given sequence $\{w_j\}_{j\geq 1}$, an *orthonormal* sequence of vectors $\{v_j\}_{j\geq 1}$ may be constructed recursively, starting with $v_1 := r_0/\|r_0\|$ and, in view of $\mathcal{V}_{m+1} = \mathcal{V}_m + \text{span}\{w_m\}$, successively orthogonalizing each w_m against the previously generated $v_1, \ldots, v_m$:

$$v_1 := r_0/\beta, \quad \beta := \|r_0\|,$$
$$v_{m+1} := \frac{(I - P_{\mathcal{V}_m})w_m}{\|(I - P_{\mathcal{V}_m})w_m\|}, \qquad m = 1, 2, \ldots, L - 1. \qquad (4.5)$$

Of course, this is simply the Gram–Schmidt orthogonalization procedure applied to the basis $\{r, w_1, \ldots, w_{m-1}\}$ of $\mathcal{V}_m$, and hence in this case the entries in the Hessenberg matrix introduced in (4.3) are given by

$$\eta_{j,m} = (w_m, v_j), \qquad j = 1, \ldots, m+1, \quad m \geq 1.$$

We also note that

$$\eta_{m+1,m} = (w_m, v_{m+1}) = \|(I - P_{\mathcal{V}_m})w_m\| \geq 0 \tag{4.6}$$

with equality holding if and only if $w_m \in \mathcal{V}_m$ or, equivalently, $r_0 \in \mathcal{W}_m$, that is, $m = L$.

We now turn to the determination of the coordinate vectors of the MR and OR approximations with respect to the basis $\{w_1, \ldots, w_m\}$. In the following lemma and in the remainder of the paper, we will employ the notation $u_1^{(m)}$ to denote the first unit coordinate vector of $\mathbb{C}^m$ and omit the superscript when the dimension is clear from the context.

Lemma 4.1. The coordinate vector $y_m^{MR} \in \mathbb{C}^m$ of the MR approximation w_m^{MR} with respect to the basis $\{w_1, \ldots, w_m\}$ is the solution of the least-squares problem

$$\|\beta u_1^{(m+1)} - \tilde{H}_m y\|_2 \to \min_{y \in \mathbb{C}^m}, \tag{4.7}$$

whereas the coordinate vector y_m^{OR} of the OR approximation solves the linear system of equations

$$H_m y = \beta u_1^{(m)}. \tag{4.8}$$

In short,

$$y_m^{MR} = \beta \tilde{H}_m^+ u_1^{(m+1)} \quad \text{and} \quad y_m^{OR} = \beta H_m^{-1} u_1^{(m)},$$

where $\tilde{H}_m^+ = (\tilde{H}_m^H \tilde{H}_m)^{-1} \tilde{H}_m^H$ is the Moore–Penrose pseudo-inverse of $\tilde{H}_m$.

Proof. The assertions of the lemma become obvious when the relevant quantities are represented in terms of the orthonormal basis $\{v_1, \ldots, v_{m+1}\}$ of $\mathcal{V}_{m+1}$. The vector r_0 to be approximated possesses the coordinate vector $\beta u_1 \in \mathbb{R}^{m+1}$ and the approximation space $\mathcal{W}_m = \text{span}\{w_1, \ldots, w_m\}$ is represented by the span of the columns of $\tilde{H}_m$. In other words, if $w \in \mathcal{W}_m$ has the coordinate vector y with respect to $\{w_1, \ldots, w_m\}$, then $r = r_0 - w \in \mathcal{V}_{m+1}$ has the coordinate vector $\beta u_1 - \tilde{H}_m y$ with respect to $\{v_1, \ldots, v_{m+1}\}$. More formally, for any $w = W_m y \in \mathcal{W}_m$ $(y \in \mathbb{C}^m)$,

$$r_0 - w = r_0 - W_m y = \beta v_1 - V_{m+1} \tilde{H}_m y = V_{m+1}(\beta u_1 - \tilde{H}_m y).$$

As the vectors $\{v_1, \ldots, v_{m+1}\}$ are orthonormal, it follows that

$$\|r_0 - w\| = \|\beta u_1 - \tilde{H}_m y\|_2$$

$(\|\cdot\|_2$ denoting the Euclidean norm in $\mathbb{C}^{m+1})$. Similarly, $r_0 - w \perp \mathcal{V}_m$

if and only if the first m components of $\beta u_1 - \tilde{H}_m y$ vanish, that is, if $\beta u_1^{(m)} - H_m y = 0$.   □

Remark 4.2. To determine y_m^{OR} using Lemma 4.1 we must of course assume that the linear system $H_m y = \beta u_1$ is solvable. But this is equivalent to our previous characterization of the existence of w_m^{OR}, namely that $c_m = \cos \measuredangle(r_{m-1}^{MR}, \mathcal{W}_m) \neq 0$ (cf. (3.8) and Corollary 3.2), which can be seen as follows. First, note that $u_1^{(m)}$ and the first $m - 1$ column vectors of H_m form a basis of $\mathbb{C}^m$ as long as $m \leq L$ (since $\eta_{j+1,j} \neq 0$ for $j = 1, 2, \ldots, L-1$). This implies that $H_m y = \beta u_1$ is consistent, that is, $u_1 \in \text{range}(H_m)$, if and only if H_m is nonsingular.

Next, recall from Remark 2.6 that c_m equals the smallest singular value of the matrix $[(v_j, \hat{w}_k)]_{j,k=1,2,\ldots,m}$, where $\{\hat{w}_1, \hat{w}_2, \ldots, \hat{w}_m\}$ is any orthonormal basis of $\mathcal{W}_m$. We select such an orthonormal basis and represent its elements as linear combinations in the original basis $\{w_1, w_2, \ldots, w_m\}$. In our row vector notation, this leads to a nonsingular matrix $T \in \mathbb{C}^{m \times m}$ with $[\hat{w}_1, \hat{w}_2, \ldots, \hat{w}_m] = [w_1, w_2, \ldots, w_m]T$. Now,

$$[(v_j, \hat{w}_k)] = T^H [(v_j, w_k)] = (H_m T)^H$$

and, consequently, the smallest singular value of $[(v_j, \hat{w}_k)]$ is positive if and only if H_m is nonsingular.

Remark 4.3. In view of (4.3) and the result of Lemma 4.1, the approximations w_m^{MR} and w_m^{OR} and their associated errors have the following representations in terms of the basis $\{v_1, \ldots, v_{m+1}\}$:

$$w_m^{MR} = V_{m+1} \tilde{H}_m \tilde{H}_m^+ \beta u_1^{(m+1)}, \qquad r_m^{MR} = V_{m+1} \left(I_{m+1} - \tilde{H}_m \tilde{H}_m^+ \right) \beta u_1^{(m+1)},$$

$$w_m^{OR} = V_{m+1} \tilde{H}_m H_m^{-1} \beta u_1^{(m)}, \qquad r_m^{OR} = V_{m+1} \left(\begin{bmatrix} I_m \\ 0 \end{bmatrix} - \tilde{H}_m H_m^{-1} \right) \beta u_1^{(m)}.$$

$$(4.9)$$

The last identity shows that the coordinate vector of r_m^{OR} has a particularly simple form. Introducing the notation $H_m^{-1} = \left[\eta_{j,k}^{[-1]} \right]$, we obtain, for $m < L$,

$$r_m^{OR} = \beta V_{m+1} \left(u_1^{(m+1)} - \tilde{H}_m H_m^{-1} u_1^{(m)} \right)$$

$$= \beta V_{m+1} \left(I_{m+1} - \tilde{H}_m [H_m^{-1} \ 0] \right) u_1^{(m+1)}$$

$$= \beta V_{m+1} \begin{bmatrix} 0 \\ -\eta_{m+1,m} \eta_{m,1}^{[-1]} \end{bmatrix} = -\beta \eta_{m+1,m} \eta_{m,1}^{[-1]} v_{m+1}.$$

The matrix $I_{m+1} - \tilde{H}_m [H_m^{-1} \ 0]$ represents $I - P_{\mathcal{W}_m}^{\mathcal{V}_m}$ restricted to $\mathcal{V}_{m+1}$ with respect to the orthonormal basis $\{v_1, \ldots, v_{m+1}\}$. The following lemma, which was recently obtained by Hochbruck and Lubich (1998), provides a simpler expression for this projection.

Lemma 4.4. Let the vector $\hat{w}_{m+1} \in \mathcal{V}_{m+1} \cap \mathcal{W}_m^\perp$ be defined by the condition $(v_{m+1}, \hat{w}_{m+1}) = 1$. Then, for all $v \in \mathcal{V}_{m+1}$, we have

$$(I - P_{\mathcal{W}_m}^{\mathcal{V}_m})v = (v, \hat{w}_{m+1})v_{m+1}. \tag{4.10}$$

Further, the coordinate vector $\hat{y}_{m+1}$ of $\hat{w}_{m+1}$ with respect to $\{v_1, \ldots, v_{m+1}\}$ has the form

$$\hat{y}_{m+1} = \begin{bmatrix} g_m \\ 1 \end{bmatrix}, \quad \text{where } g_m \text{ solves} \quad H_m^H g_m = -\eta_{m+1,m} u_m$$

and u_m denotes the last unit coordinate vector in $\mathbb{R}^m$.

Proof. On $\mathcal{V}_{m+1}$, the projection $I - P_{\mathcal{W}_m}^{\mathcal{V}_m}$ is characterized by the two properties

$$(I - P_{\mathcal{W}_m}^{\mathcal{V}_m})v = v \qquad \forall v \in \mathcal{V}_{m+1} \cap \mathcal{V}_m^\perp = \mathrm{span}\{v_{m+1}\},$$
$$(I - P_{\mathcal{W}_m}^{\mathcal{V}_m})v = 0 \qquad \forall v \in \mathcal{V}_{m+1} \cap \mathcal{W}_m,$$

that is, it is the oblique projection onto $\mathcal{V}_{m+1} \cap \mathcal{V}_m^\perp$ orthogonal to $\mathcal{V}_{m+1} \cap \mathcal{W}_m^\perp$, both of which are one-dimensional spaces of which $\{v_{m+1}\}$ and $\{\hat{w}_{m+1}\}$ are biorthonormal bases. Thus, (4.10) is the singular value expansion of $I - P_{\mathcal{W}_m}^{\mathcal{V}_m}$ restricted to $\mathcal{V}_{m+1}$.

To determine the coordinate vector $\hat{y}_{m+1}$, we first note that the requirement $(v_{m+1}, \hat{w}_{m+1}) = 1$ implies that its last component is equal to one. Furthermore, $\hat{w}_{m+1} \perp \mathcal{W}_m$ translates to $\hat{y}_{m+1} \in \mathrm{null}(\tilde{H}_m^H)$, since the columns of $\tilde{H}_m$ span the coordinate space of $\mathcal{W}_m$. Letting $g_m \in \mathbb{C}^m$ denote the first m components of $\hat{y}_{m+1}$ and recalling that $\eta_{m+1,m} > 0$, we obtain

$$0 = \tilde{H}_m^H \hat{y}_{m+1} = \begin{bmatrix} H_m^H & \eta_{m+1,m} u_m \end{bmatrix} \begin{bmatrix} g_m \\ 1 \end{bmatrix} = H_m^H g_m + \eta_{m+1,m} u_m. \quad \square$$

The representation (4.10) can be used to obtain another expression for the OR approximation error as follows: by virtue of the inclusion $\mathcal{W}_{m-1} \subset \mathcal{W}_m$, an arbitrary vector $w \in \mathcal{W}_{m-1}$ must lie in the nullspace of $I - P_{\mathcal{W}_m}^{\mathcal{V}_m}$. Furthermore, the difference $r_0 - w \in \mathrm{span}\{r_0\} + \mathcal{W}_{m-1} = \mathcal{V}_m$ has a representation $r_0 - w = V_m z$ with $z \in \mathbb{C}^m$. It follows that

$$r_m^{\mathrm{OR}} = (I - P_{\mathcal{W}_m}^{\mathcal{V}_m})r_0 = (I - P_{\mathcal{W}_m}^{\mathcal{V}_m})(r_0 - w) = (r_0 - w, \hat{w}_{m+1})v_{m+1}$$

$$= \left(V_{m+1} \begin{bmatrix} z \\ 0 \end{bmatrix}, V_{m+1} \begin{bmatrix} g_m \\ 1 \end{bmatrix} \right) v_{m+1} = (g_m^H z)\, v_{m+1}, \tag{4.11}$$

and therefore $\|r_m^{\mathrm{OR}}\| = |g_m^H z| \le \|g_m\|_2 \|z\|_2$ with equality holding if and only if g and z are collinear. At the same time, as g_m is fixed, equality must occur when $\|z\|_2 = \|r_0 - w\|$ is minimized among all $w \in \mathcal{W}_{m-1}$, which is the case for $w = w_{m-1}^{\mathrm{MR}}$. As a result, $\|r_m^{\mathrm{OR}}\| = \|g_m\|_2 \|r_{m-1}^{\mathrm{MR}}\|$ which, in view of (3.14) and (3.16), implies $\|g_m\|_2 = s_m/c_m$, an identity which could also have been derived directly from the definition of g_m.

The least-squares problem (4.7) can be solved with the help of a QR decomposition of the Hessenberg matrix $\tilde{H}_m$, which we write as

$$Q_m \tilde{H}_m = \begin{bmatrix} R_m \\ 0 \end{bmatrix}, \qquad (4.12)$$

with $Q_m \in \mathbb{C}^{(m+1)\times(m+1)}$ unitary ($Q_m^H Q_m = I_{m+1}$) and $R_m \in \mathbb{C}^{m\times m}$ upper triangular. Substituting (4.12) in (4.7) yields

$$\min_{y\in\mathbb{C}^m} \|\beta u_1 - \tilde{H}_m y\|_2 = \min_{y\in\mathbb{C}^m} \left\| Q_m^H \left(\beta Q_m u_1 - \begin{bmatrix} R_m \\ 0 \end{bmatrix} y \right) \right\|_2$$

$$= \min_{y\in\mathbb{C}^m} \left\| \beta Q_m u_1 - \begin{bmatrix} R_m \\ 0 \end{bmatrix} y \right\|_2$$

$$= \min_{y\in\mathbb{C}^m} \left\| \begin{bmatrix} \beta q_m - R_m y \\ \beta q_{m+1,1}^{(m)} \end{bmatrix} \right\|_2,$$

where $[q_m^\top, q_{m+1,1}^{(m)}]^\top$ ($q_m \in \mathbb{C}^m$) denotes the first column of Q_m. Since $\tilde{H}_m$ has full rank, R_m is nonsingular and the solution of the above least-squares problem is $y_m^{\mathrm{MR}} = \beta R_m^{-1} q_m$. The associated least-squares error is given by $\|r_m^{\mathrm{MR}}\| = \beta |q_{m+1,1}^{(m)}|$.

The following theorem identifies the angles between r_0 and $\mathcal{W}_m$ as well as those between the spaces $\mathcal{V}_m$ and $\mathcal{W}_m$ with quantities which occur in the first column of the matrix Q_m.

Theorem 4.5. If, for $m = 1, \ldots, L$, $Q_m = [q_{j,k}^{(m)}]_{j,k=1}^{m+1} \in \mathbb{C}^{(m+1)\times(m+1)}$ is the unitary matrix in the QR decomposition (4.12) of the Hessenberg matrix $\tilde{H}_m$ in (4.3), then

$$\sin \angle(r_0, \mathcal{W}_m) = \left| q_{m+1,1}^{(m)} \right|, \qquad (4.13)$$

$$\sin \angle(r_{m-1}^{\mathrm{MR}}, \mathcal{W}_m) = \sin \angle(\mathcal{V}_m, \mathcal{W}_m) = \left| q_{m+1,1}^{(m)}/q_{m,1}^{(m-1)} \right|. \qquad (4.14)$$

Proof. As mentioned earlier (cf. the proof of Lemma 4.1) the vector r_0 possesses the coordinates $\beta u_1^{(m+1)}$ with respect to the orthonormal basis $\{v_1, \ldots, v_{m+1}\}$ of $\mathcal{V}_{m+1}$, whereas $\mathcal{W}_m$ is represented by $\mathrm{range}(\tilde{H}_m) \subset \mathbb{C}^{m+1}$. This implies

$$\angle(r_0, \mathcal{W}_m) = \angle_2(\beta u_1, \mathrm{range}(\tilde{H}_m)) = \angle_2(u_1, \mathrm{range}(\tilde{H}_m)),$$

where the subscript 2 indicates that the last two angles are defined with respect to the Euclidean inner product on $\mathbb{C}^{m+1}$.

The vectors $[v_1, \ldots, v_{m+1}]Q_m^H$ form another orthonormal basis of $\mathcal{V}_{m+1}$

with respect to which r_0 possesses the coordinate vector $\beta Q_m u_1^{(m+1)} =:$
$\beta [q_m^\top, q_{m+1,1}^{(m)}]^\top$, that is, a multiple of the first column of Q_m, and $\mathcal{W}_m$ is
represented by $Q_m \tilde{H}_m y = \begin{bmatrix} R_m y \\ 0 \end{bmatrix}$ $(y \in \mathbb{C}^m)$, a subspace of $\mathbb{C}^{m+1}$ which we
identify with $\mathbb{C}^m$ because it consists of those vectors from $\mathbb{C}^{m+1}$ whose last
component equals zero. Consequently,

$$\angle(r_0, \mathcal{W}_m) = \angle_2(\beta [q_m^\top, q_{m+1,1}^{(m)}]^\top, \mathbb{C}^m) = \angle_2([q_m^\top, q_{m+1,1}^{(m)}]^\top, \mathbb{C}^m)$$

which, in view of (2.3), proves assertion (4.13). Formula (4.14) follows
directly from (3.6) and (3.11).   □

The matrix Q_m is usually constructed as a product of Givens rotations

$$Q_m = G_m \begin{bmatrix} G_{m-1} & 0 \\ 0 & 1 \end{bmatrix} \begin{bmatrix} G_{m-2} & 0 \\ 0 & I_2 \end{bmatrix} \cdots \begin{bmatrix} G_1 & 0 \\ 0 & I_{m-1} \end{bmatrix}$$

where, for $k = 1, 2, \ldots, m$,

$$G_k := \begin{bmatrix} I_{k-1} & 0 & 0 \\ 0 & \tilde{c}_k & \tilde{s}_k e^{-i\phi_k} \\ 0 & -\tilde{s}_k e^{i\phi_k} & \tilde{c}_k \end{bmatrix} \qquad (\tilde{c}_k, \tilde{s}_k \geq 0, \tilde{c}_k^2 + \tilde{s}_k^2 = 1, \phi_k \in \mathbb{R}).$$

We briefly explain how these rotations have to be chosen inductively. Assume that we have constructed $G_1, \ldots, G_{m-2}, G_{m-1}$ such that

$$\begin{bmatrix} G_{m-1} & 0 \\ 0 & 1 \end{bmatrix} \begin{bmatrix} G_{m-2} & 0 \\ 0 & I_2 \end{bmatrix} \cdots \begin{bmatrix} G_1 & 0 \\ 0 & I_{m-1} \end{bmatrix} \tilde{H}_m = \begin{bmatrix} R_{m-1} & r \\ 0 & \tau \\ 0 & \eta_{m+1,m} \end{bmatrix}.$$

For later use, we rewrite this identity in the form

$$\begin{bmatrix} Q_{m-1} & 0 \\ 0 & 1 \end{bmatrix} \begin{bmatrix} H_m \\ 0 \cdots 0 \ \eta_{m+1,m} \end{bmatrix} = \begin{bmatrix} R_{m-1} & r \\ 0 & \tau \\ 0 & \eta_{m+1,m} \end{bmatrix}. \qquad (4.15)$$

Now we set

$$\tilde{c}_m := \frac{|\tau|}{\sqrt{|\tau|^2 + \eta_{m+1,m}^2}}, \quad \tilde{s}_m := \frac{\eta_{m+1,m}}{\sqrt{|\tau|^2 + \eta_{m+1,m}^2}}, \qquad (4.16)$$

$$\phi_m := \arg(\eta_{m+1,m}) - \arg(\tau) = -\arg(\tau)$$

(recall $\eta_{m+1,m} \geq 0$) and verify by a simple calculation that

$$\begin{bmatrix} I_{m-1} & 0 & 0 \\ 0 & \tilde{c}_m & \tilde{s}_m e^{-i\phi_m} \\ 0 & -\tilde{s}_m e^{i\phi_m} & \tilde{c}_m \end{bmatrix} \begin{bmatrix} R_{m-1} & r \\ 0 & \tau \\ 0 & \eta_{m+1,m} \end{bmatrix} = \begin{bmatrix} R_{m-1} & r \\ 0 & \rho \\ 0 & 0 \end{bmatrix}$$

with $\rho = \sqrt{|\tau|^2 + \eta_{m+1,m}^2} \, e^{-i\phi_m}$.

That the quantities $\tilde{s}_m$ and $\tilde{c}_m$ are indeed the sines and cosines of the angles $\angle(r_{m-1}^{\mathrm{MR}}, \mathcal{W}_m) = \angle(\mathcal{V}_m, \mathcal{W}_m)$ can be easily seen as follows. Since $q_{m+1,1}^{(m)} = -\tilde{s}_m e^{i\phi_m} q_{m,1}^{(m-1)}$, by Theorem 4.5 we have

$$s_m = \angle(r_{m-1}^{\mathrm{MR}}, \mathcal{W}_m) = |q_{m+1,1}^{(m)}/q_{m,1}^{(m-1)}| = \tilde{s}_m.$$

When describing MR and OR approximations the alternate orthonormal basis of $\mathcal{V}_{m+1}$ which occurred in the proof of Theorem 4.5, and which was already employed by Paige and Saunders (1975), often proves useful: we thus define

$$\hat{V}_{m+1} := [\hat{v}_1^{(m+1)}, \ldots, \hat{v}_{m+1}^{(m+1)}] := V_{m+1} Q_m^H. \tag{4.17}$$

The notation $\hat{v}_1^{(m+1)}, \ldots, \hat{v}_{m+1}^{(m+1)}$ for these new basis vectors is not entirely appropriate, since, as the following proposition shows, all but the last do not change with the index m.

Proposition 4.6. We have

$$[\hat{v}_1^{(m+1)}, \ldots, \hat{v}_m^{(m+1)}, \hat{v}_{m+1}^{(m+1)}] = [\hat{v}_1, \ldots, \hat{v}_m, \tilde{v}_{m+1}],$$

where $\tilde{v}_1 = v_1$, and, for $m = 1, \ldots, L-1$,

$$\hat{v}_m = c_m \tilde{v}_m + s_m e^{i\phi_m} v_{m+1},$$
$$\tilde{v}_{m+1} = -s_m e^{-i\phi_m} \tilde{v}_m + c_m v_{m+1}.$$

The vectors $\{\hat{v}_1, \ldots, \hat{v}_m, \tilde{v}_{m+1}\}$ form an orthonormal basis of $\mathcal{V}_{m+1}$ such that $\{\hat{v}_1, \ldots, \hat{v}_m\}$ is a basis of $\mathcal{W}_m$. In addition,

$$w_m^{\mathrm{MR}} = \hat{V}_{m+1} \beta \begin{bmatrix} q_m \\ 0 \end{bmatrix} \quad \text{and} \quad r_m^{\mathrm{MR}} = \hat{V}_{m+1} \beta \begin{bmatrix} 0 \\ q_{m+1,1}^{(m)} \end{bmatrix}. \tag{4.18}$$

Proof. To prove the first two assertions, we observe

$$\hat{V}_{m+1} = V_{m+1} Q_m^H = [V_m, v_{m+1}] \begin{bmatrix} Q_{m-1}^H & 0 \\ 0 & 1 \end{bmatrix} G_m^H$$

$$= [\hat{V}_m, v_{m+1}] \begin{bmatrix} I_{m-1} & 0 & 0 \\ 0 & c_m & -s_m e^{-i\phi_m} \\ 0 & s_m e^{i\phi_m} & c_m \end{bmatrix}.$$

That the first m columns of $\hat{V}_{m+1}$ form a basis of the approximation space $\mathcal{W}_m$ follows from

$$W_m = V_{m+1} \tilde{H}_m = V_{m+1} Q_m^H \begin{bmatrix} R_m \\ 0 \end{bmatrix} = [\hat{v}_1, \ldots, \hat{v}_m] R_m,$$

which also implies (4.18). $\square$

We summarize the coordinate representations of the MR and OR errors in the following result.

Proposition 4.7. For the MR and OR errors we have

$$r_m^{\mathrm{MR}} = \beta\, q_{m+1,1}^{(m)}\, \tilde{v}_{m+1} = \beta \prod_{j=1}^{m} \left[-s_j e^{i\phi_j} \right] \tilde{v}_{m+1},$$

$$r_m^{\mathrm{OR}} = -\beta\, \frac{s_m}{c_m}\, e^{i\phi_m} q_{m,1}^{(m-1)}\, v_{m+1} = \frac{\beta}{c_m} \prod_{j=1}^{m} \left[-s_j e^{i\phi_j} \right] v_{m+1},$$

$$r_{m-1}^{\mathrm{MR}} - r_m^{\mathrm{OR}} = \frac{\beta}{c_m}\, q_{m,1}^{(m-1)}\, \hat{v}_m = \frac{\beta}{c_m} \prod_{j=1}^{m-1} \left[-s_j e^{i\phi_j} \right] \hat{v}_m.$$

Proof. The recursive definition of Q_m allows us to express its entries $q_{k,1}^{(m)}$ explicitly in terms of the quantities s_j and c_j:

$$q_{k,1}^{(m)} = c_k \prod_{j=1}^{k-1} \left[-s_j e^{i\phi_j} \right] \quad (1 \le k \le m), \qquad q_{m+1,1}^{(m)} = \prod_{j=1}^{m} \left[-s_j e^{i\phi_j} \right].$$

This, together with (4.18), proves the first identity.

Next, we recall from Remark 4.3 that $r_m^{\mathrm{OR}} = -\beta\eta_{m+1,m}\,\eta_{m,1}^{[-1]}\, v_{m+1}$. To eliminate $\eta_{m,1}^{[-1]}$ from this relation, we note that the matrix H_m possesses the QR decomposition (*cf.* (4.15))

$$Q_{m-1}H_m = \begin{bmatrix} R_{m-1} & r \\ 0 & \tau \end{bmatrix}, \quad \text{i.e.,} \quad H_m^{-1} = \begin{bmatrix} R_{m-1}^{-1} & \tilde{r} \\ 0 & 1/\tau \end{bmatrix} Q_{m-1}, \quad (4.19)$$

which implies $\eta_{m,1}^{[-1]} = q_{m,1}^{(m-1)}/\tau$. Since $\eta_{m+1,m}/\tau = e^{i\phi_m} s_m/c_m$ (*cf.* (4.16)) we conclude $\eta_{m,1}^{[-1]}\eta_{m+1,m} = q_{m,1}^{(m-1)} e^{i\phi_m} s_m/c_m$. This proves the second identity.

The desired representation of $r_{m-1}^{\mathrm{MR}} - r_m^{\mathrm{OR}}$ now follows from $\hat{v}_m = c_m\tilde{v}_m + s_m e^{i\phi_m} v_{m+1}$ (*cf.* Proposition 4.6). $\qquad\square$

We note that all relations of Theorem 3.4 connecting the MR and OR approaches can be easily obtained by manipulating the error representations of Proposition 4.7. Indeed, this is essentially how these relations are proven in the literature on Krylov subspace methods. The main difference there is that the occurring sines and cosines result from the Givens rotations needed to construct the QR decomposition of $\tilde{H}_m$, and they have not been identified as the sines and cosines of $\angle(r_0, \mathcal{W}_m)$.

Finally, we observe that the vector $\hat{w}_{m+1}$ introduced in Lemma 4.4 is given by $\tilde{v}_{m+1}/c_m$ in terms of the last vector in the Paige–Saunders basis, so that an equivalent formulation of (4.10) reads

$$\left(I - P_{\mathcal{W}_m}^{\mathcal{V}_m} \right) v = \frac{(v, \tilde{v}_{m+1})}{c_m}\, v_{m+1} \qquad (v \in \mathcal{V}_{m+1}).$$

4.3. Using arbitrary bases of $\mathcal{W}_m$ and $\mathcal{V}_m$

For practical computations it is desirable that the Hessenberg matrices $\tilde{H}_m$ introduced in (4.3) have small bandwidth. Indeed, if $\tilde{H}_m$ has only k non-vanishing diagonals, namely the ones with indices $-1, 0, 1, \ldots, k-2$ (we follow the standard notation according to which a diagonal has index k if its entries $\eta_{j,\ell}$ are characterized by $\ell - j = k$), then only k diagonals of the upper triangular matrices R_m are nonzero, namely those with indices $k = 0, 1, \ldots, k-1$. This follows easily from the fact that Q_m has lower Hessenberg form. This banded structure of the matrices R_m can then be used to derive k-term recurrence formulas for the coordinate vectors y_m in terms of $y_{m-1}, y_{m-2}, \ldots, y_{m-k+1}$ and for the approximations w_m in terms of $w_{m-1}, w_{m-2}, \ldots, w_{m-k+1}$. (This statement applies to both the MR and the OR approach.) The most important consequence of this observation is that, at each step, only the k previous approximations (or, in other implementations, the last k basis vectors) need to be stored, which means that in this case storage requirements do not increase with m. If we insist on choosing $v_1, \ldots, v_L$ as orthogonal vectors then the Hessenberg matrices will generally not have banded form. The main motivation for giving up orthogonality of the basis of $\mathcal{V}_m$ is therefore to constrain the bandwidth of H_m in order to keep storage requirements low.

As explained at the beginning of Section 4.2, no orthogonality conditions are required to derive the fundamental relationship (4.3)

$$W_m = V_{m+1}\tilde{H}_m = V_m H_m + [0, \ldots, 0, \eta_{m+1,m} v_{m+1}].$$

In this section, we assume only that $v_1, v_2, \ldots$ are linearly independent such that $\{v_1, \ldots, v_m\}$ constitutes a basis of $\mathcal{V}_m$ $(m = 1, 2, \ldots, L)$. Just as before, the jth column of the upper Hessenberg matrix $\tilde{H}_m \in \mathbb{C}^{(m+1)\times m}$ contains the coefficients of $w_j \in \mathcal{W}_j \subset \mathcal{V}_{m+1}$ with respect to the basis vectors $v_1, \ldots, v_{m+1}$. The difference is that, since now the vectors v_j need not be orthogonal, these coefficients can no longer be expressed in terms of the inner product with which $\mathcal{H}$ was originally endowed. We shall see below that we can recover the usual inner product representation by switching to another suitable inner product. Recall from the proof of Lemma 4.1 that, for each $w = W_m y \in \mathcal{W}_m$ $(y \in \mathbb{C}^m)$, the associated error in approximating r_0 is represented by

$$r = r_0 - w = r - W_m y = \beta v_1 - V_{m+1}\tilde{H}_m y = V_{m+1}(\beta u_1 - \tilde{H}_m y).$$

Minimizing the norm of r among all $w \in \mathcal{W}_m$ leads as above to the least squares problem

$$\min_{y \in \mathbb{C}^m} \left\| V_{m+1}(\beta u_1 - \tilde{H}_m y) \right\| = \min_{y \in \mathbb{C}^m} \| \beta u_1 - \tilde{H}_m y \|_v, \qquad (4.20)$$

in which now $\| \cdot \|_v$ denotes the norm induced on the coordinate space with respect to V_L by the inner product $(\cdot, \cdot)$ given on $\mathcal{H}$. More precisely, if we set for $\boldsymbol{x}, \boldsymbol{y} \in \mathbb{C}^L$

$$(\boldsymbol{x}, \boldsymbol{y})_v := (V_L \boldsymbol{x}, V_L \boldsymbol{y}) = \boldsymbol{y}^H M \boldsymbol{x}, \quad \text{where } M := [(v_j, v_k)] \in \mathbb{C}^{L \times L}, \quad (4.21)$$

then $(\cdot, \cdot)_v$ is an inner product on $\mathbb{C}^L$ with associated norm

$$\| \cdot \|_v = \sqrt{(\cdot, \cdot)_v}.$$

At this point one could proceed as in the algorithms which use an orthogonal basis, the only difference being that all inner products in the coordinate space now require knowledge of the Gram matrix M. In particular, the Givens rotations and the matrices Q_m in the QR factorization (4.12) must now be unitary with respect to the inner product $(\cdot, \cdot)_v$, that is, they must satisfy $Q_m^H M_m Q_m = I_m$, where $M_m \in \mathbb{C}^{(m+1) \times (m+1)}$ is the $(m+1)$st leading principal submatrix of M. The submatrices M_m, however, *cannot* be computed unless all basis vectors $v_1, v_2, \ldots, v_{m+1}$ are available, and thus any short recurrence for generating these would not result in any storage savings: this is precisely what we wish to avoid.

An alternative was originally proposed by Freund (1992): rather than solving the minimization problem (4.20), we instead solve

$$\min_{\boldsymbol{y} \in \mathbb{C}^m} \| \beta \boldsymbol{u}_1 - \tilde{H}_m \boldsymbol{y} \|_2,$$

and, if $\boldsymbol{y}_m^{\mathrm{QMR}} \in \mathbb{C}^m$ denotes the unique solution to this least-squares problem, regard

$$\boldsymbol{w}_m^{\mathrm{QMR}} := W_m \boldsymbol{y}_m^{\mathrm{QMR}}$$

as an approximation of $\boldsymbol{r}_0$. Adhering to the terminology introduced by Freund, we refer to this approach as the *quasi-minimal residual* (QMR) approach. Following the convention in the literature on Krylov subspace methods, we refer to the (coordinate) vector

$$\boldsymbol{s}_m^{\mathrm{QMR}} := \beta \boldsymbol{u}_1 - \tilde{H}_m \boldsymbol{y}_m^{\mathrm{QMR}} \in \mathbb{C}^{m+1}$$

as the *quasi-error* of $\boldsymbol{w}_m^{\mathrm{QMR}}$.

We note that, instead of changing the inner product in the coordinate space from $(\cdot, \cdot)_v$ to the Euclidean inner product, one could equivalently have replaced the given inner product $(\cdot, \cdot)$ on $\mathcal{V}_L \subseteq \mathcal{H}$ by

$$(\boldsymbol{v}, \boldsymbol{w})_V = (V_L \boldsymbol{x}, V_L \boldsymbol{y})_V =: \boldsymbol{y}^H \boldsymbol{x} \qquad \forall \boldsymbol{v} = V_L \boldsymbol{x}, \boldsymbol{w} = V_L \boldsymbol{y} \in \mathcal{V}_L, \quad (4.22)$$

and proceeded as in the MR algorithm of Section 4.2. The new inner product $(\cdot, \cdot)_V$ thus defined has the property that the basis vectors $\{v_1, \ldots, v_L\}$ are orthonormal. For related work on interpreting QMR approximations as MR approximations in a modified norm see Barth and Manteuffel (1994).

The following assertions are obvious if we keep in mind that, for $\boldsymbol{x}, \boldsymbol{y} \in \mathbb{C}^L$, we have the relations

$$
\begin{aligned}
(V_L\boldsymbol{x}, V_L\boldsymbol{y})_V &= (V_L\boldsymbol{x}, V_L M^{-1}\boldsymbol{y}) = (V_L M^{-1}\boldsymbol{x}, V_L\boldsymbol{z}) \\
&= (V_L M^{-1/2}\boldsymbol{x}, V_L M^{-1/2}\boldsymbol{y}), \\
(V_L\boldsymbol{x}, V_L\boldsymbol{y}) &= (V_L\boldsymbol{x}, V_L M\boldsymbol{y})_V = (V_L M\boldsymbol{x}, V_L\boldsymbol{y})_V \\
&= (V_L M^{1/2}\boldsymbol{x}, V_L M^{1/2}\boldsymbol{y})_V.
\end{aligned}
\tag{4.23}
$$

As a consequence, we obtain for instance the following result.

Theorem 4.8. The QMR iterates are the MR iterates with respect to the inner product $(\cdot, \cdot)_V$:

$$
\|\boldsymbol{r}_m^{\mathrm{QMR}}\|_V = \|\boldsymbol{r}_0 - \boldsymbol{w}_m^{\mathrm{QMR}}\|_V = \min_{\boldsymbol{w}\in\mathcal{W}_m} \|\boldsymbol{r}_0 - \boldsymbol{w}\|_V.
$$

In terms of the original norm on $\mathcal{H}$, the MR and QMR errors may be bounded by

$$
\|\boldsymbol{r}_m^{\mathrm{MR}}\| \leq \|\boldsymbol{r}_m^{\mathrm{QMR}}\| \leq \sqrt{\kappa_2(M_m)}\,\|\boldsymbol{r}_m^{\mathrm{MR}}\|,
\tag{4.24}
$$

in which $\kappa_2(M_m)$ denotes the (Euclidean) condition number of the (Hermitian positive definite) matrix M_m. Moreover, we have

$$
\|\boldsymbol{r}_m^{\mathrm{QMR}}\| \leq \lambda_{\max}^{1/2}(M_m)\|\boldsymbol{s}_m^{\mathrm{QMR}}\|.
$$

Proof. Only the second inequality in (4.24) remains to be proved. This follows immediately from

$$
\lambda_{\min}^{1/2}(M_m)\,\|\boldsymbol{y}\|_2 \leq \|\boldsymbol{y}\|_v \leq \lambda_{\max}^{1/2}(M_m)\,\|\boldsymbol{y}\|_2 \qquad \forall \boldsymbol{y}\in\mathbb{C}^{m+1}. \qquad \square
$$

In view of (4.24) the deviation of the QMR approach from the MR approach is bounded by the condition numbers $\kappa_2(M_m)$, that is, by the ratio of the extremal eigenvalues of M_m. The largest eigenvalue $\lambda_{\max}(M_m)$ is easily controlled: it merely requires choosing the basis vectors $\boldsymbol{v}_m$ to have unit length, that is, $\|\boldsymbol{v}_m\| = 1$ for all m, to ensure $\lambda_{\max}(M_m) \leq m+1$ (note that $\lambda_{\max}(M_m) \leq \|M_m\|_F := [\sum_{j,k=1}^{m+1}(\boldsymbol{v}_j, \boldsymbol{v}_k)^2]^{1/2} \leq [\sum_{j,k=1}^{m+1}\|\boldsymbol{v}_j\|\,\|\boldsymbol{v}_k\|]^{1/2})$. The crucial point is to construct the basis V_m in such a way that $\lambda_{\min}(M_m)$ does not approach zero (or does so only slowly).

Another immediate consequence of (4.23) is as follows.

Proposition 4.9. The QMR error vectors satisfy

$$
\boldsymbol{r}_m^{\mathrm{QMR}} \perp \mathcal{U}_m,
$$

where $\mathcal{U}_m := \{\boldsymbol{v} = V_{m+1}\boldsymbol{y} : \boldsymbol{y} = M_m^{-1}\tilde{H}_m\boldsymbol{z}$ for some $\boldsymbol{z}\in\mathbb{C}^m\}$ is an m-dimensional subspace of $\mathcal{V}_{m+1}$ ($\mathcal{U}_L = \mathcal{V}_L$ for $m = L$) and orthogonality is understood with respect to the original inner product $(\cdot, \cdot)$ on $\mathcal{H}$. Consequently,

$$
\boldsymbol{r}_m^{\mathrm{QMR}} = \left(I - P_{\mathcal{W}_m}^{\mathcal{U}_m}\right)\boldsymbol{r}_0,
$$

where $P_{\mathcal{W}_m}^{\mathcal{U}_m}$ denotes the oblique projection onto $\mathcal{W}_m$ orthogonal to $\mathcal{U}_m$.

Proof. Since the QMR approximations are merely the MR approximations with respect to $(\cdot, \cdot)_V$, their errors $r_m^{\mathrm{QMR}} \in \mathcal{V}_{m+1}$ are characterized by

$$r_m^{\mathrm{QMR}} \perp_V \mathcal{W}_m.$$

From this observation and (4.23) both assertions easily follow.

An equally simple proof results from the fact that

$$r_m^{\mathrm{QMR}} = V_{m+1}\left(\beta u_1 - \tilde{H}_m y_m^{\mathrm{QMR}}\right),$$

where y_m^{QMR} solves the least-squares problem $\|\beta u_1 - \tilde{H}_m y\|_2 \to \min$. In other words,

$$\beta u_1 - \tilde{H}_m y^{\mathrm{QMR}} \perp \mathrm{range}(\tilde{H}_m) \quad \text{or} \quad \beta u_1 - \tilde{H}_m y^{\mathrm{QMR}} \perp_v M^{-1}\mathrm{range}(\tilde{H}_m).$$

$\square$

Note that the orthogonal complement of $\mathcal{U}_m$ is given by $\mathcal{U}_m^\perp = \mathrm{span}\{r_m^{\mathrm{QMR}}\}$ $+\mathcal{V}_{m+1}^\perp$, that is, $\mathcal{U}_m^\perp \oplus \mathcal{V}_m = \mathcal{H}$ and the oblique projection $P_{\mathcal{W}_m}^{\mathcal{U}_m}$ exists.

We now briefly describe the analogue of the OR approach when using a non-orthogonal basis. In place of seeking $y \in \mathbb{C}^m$ such that

$$0 = \left(V_{m+1}[\beta u_1 - \tilde{H}_m y], v_j\right) = (\beta u_1 - \tilde{H}_m y, u_j)_v \quad (j = 1, 2, \ldots, m),$$

which would lead to $r - \mathcal{W}_m y \perp \mathcal{V}_m$ and thereby to a proper OR approximation, we instead determine a coordinate vector $y_m^{\mathrm{QOR}} \in \mathbb{C}^m$ to satisfy

$$0 = (\beta u_1 - \tilde{H}_m y_m^{\mathrm{QOR}}, u_j)_2 \quad (j = 1, 2, \ldots, m), \quad i.e., \quad H_m y_m^{\mathrm{QOR}} = \beta u_1 \quad (4.25)$$

(provided H_m is nonsingular) and thus obtain the corresponding approximants $w_m^{\mathrm{QOR}} := W_m y_m^{\mathrm{QOR}}$. In terms of the inner product $(\cdot, \cdot)_V$ on $\mathcal{V}_L$ these are characterized by

$$r_m^{\mathrm{QOR}} = r_0 - w_m^{\mathrm{QOR}} \perp_V \mathcal{V}_m$$

that is, the QOR iterates are the OR iterates with respect to the inner product $(\cdot, \cdot)_V$. By analogy with Proposition 4.9 we obtain the following result.

Proposition 4.10. The QOR errors satisfy

$$r_m^{\mathrm{QOR}} \perp \mathcal{J}_m,$$

where $\mathcal{J}_m := \{v = V_{m+1} y \in \mathcal{V}_{m+1} : y = M_m^{-1}[z^\top \ 0]^\top \text{ with } z \in \mathbb{C}^m\}$ is an m-dimensional subspace of $\mathcal{V}_{m+1}$ ($\mathcal{J}_L = \mathcal{V}_L$ for $m = L$) and orthogonality is understood with respect to the original inner product $(\cdot, \cdot)$ on $\mathcal{H}$. Consequently,

$$r_m^{\mathrm{QOR}} = \left(I - P_{\mathcal{W}_m}^{\mathcal{J}_m}\right) r_0,$$

where $P_{\mathcal{W}_m}^{\mathcal{T}_m}$ denotes the oblique projection onto $\mathcal{W}_m$ orthogonal to $\mathcal{T}_m$.

We know from Remark 4.2 that H_m being nonsingular is equivalent to $\mathcal{W}_m \oplus \mathcal{V}_m^{\perp_V} = \mathcal{H}$. But since, for every $h \in \mathcal{H}$, $h \perp_V \mathcal{V}_m$ if and only if $h \perp \mathcal{T}_m$, the oblique projection $P_{\mathcal{W}_m}^{\mathcal{T}_m}$ exists if H_m is nonsingular.

The simple observation that the QMR and QOR approximations are the MR and OR approximations, respectively, obtained when replacing the original inner product $(\cdot, \cdot)$ with the basis-dependent inner product $(\cdot, \cdot)_V$ implies that the assertions of the preceding sections, particularly those of Theorem 3.4 and Propositions 3.8 and 4.7, are valid for any pair of QMR/QOR methods. Note, however, that when formulating these results for QMR/QOR methods each occurrence of the original norm must be replaced by the $\|\cdot\|_V$-norm and that angles are understood to be defined with respect to $(\cdot, \cdot)_V$.

As an example, we mention that

$$w_m^{\mathrm{QMR}} = \hat{s}_m^2 \, w_{m-1}^{\mathrm{QMR}} + \hat{c}_m^2 \, w_m^{\mathrm{QMR}},$$

where

$$\hat{s}_m := \sin \angle_V(r_{m-1}^{\mathrm{QMR}}, \mathcal{W}_m), \quad \hat{c}_m := \cos \angle_V(r_{m-1}^{\mathrm{QMR}}, \mathcal{W}_m)$$

and $\angle_V(r_{m-1}^{\mathrm{QMR}}, \mathcal{W}_m)$ denotes the angle between r_{m-1}^{QMR} and $\mathcal{W}_m$ with respect to the inner product $(\cdot, \cdot)_V$.

We conclude our discussion of abstract QMR and QOR approximations with a comment on the effect of applying the smoothing procedures introduced in Section 3, namely minimal and quasi-minimal residual smoothing, to the QOR approximations. We first note that the two are no longer equivalent. Specifically, if we define

$$\alpha_m^{\mathrm{MR}} := \frac{(d_{m-1}^{\mathrm{QMR}}, d_{m-1}^{\mathrm{QMR}} - d_m^{\mathrm{QOR}})}{\|d_{m-1}^{\mathrm{QMR}} - d_m^{\mathrm{QOR}}\|^2} \tag{4.26}$$

then, in general, $w_m^{\mathrm{QMR}} \neq (1 - \alpha_m^{\mathrm{MR}}) w_{m-1}^{\mathrm{QMR}} + \alpha_m^{\mathrm{MR}} w_m^{\mathrm{QOR}}$ because the formula (4.26) for the smoothing parameter α_m^{MR} was derived in order that the errors of the smoothed approximations solve a local approximation problem with respect to the inner product $(\cdot, \cdot)$, which is different from the inner product $(\cdot, \cdot)_V$ which characterizes the QMR and QOR approximations. Minimal residual smoothing therefore does not lead from QOR to QMR.

It is an easy consequence of Remark 4.3 that the situation is different if we apply QMR smoothing. If we set

$$\alpha_m^{\mathrm{QMR}} = \frac{1/\|r_m^{\mathrm{QOR}}\|_V^2}{\sum_{j=0}^m 1/\|r_j^{\mathrm{QOR}}\|_V^2},$$

then $w_m^{\mathrm{QMR}} = (1 - \alpha_m^{\mathrm{QMR}}) w_{m-1}^{\mathrm{QMR}} + \alpha_m^{\mathrm{QMR}} w_m^{\mathrm{QOR}}$ does indeed hold (*cf.* Proposition 3.8). But since $r_m^{\mathrm{QOR}} = \gamma v_{m+1}$ for some $\gamma \in \mathbb{C}$, we have

$$\|r_m^{\mathrm{QOR}}\|_V = |\gamma| = \|r_m^{\mathrm{QOR}}\|/\|v_{m+1}\|$$

and consequently

$$\alpha_m^{\mathrm{QMR}} = \frac{\|v_{m+1}\|^2/\|r_m^{\mathrm{QOR}}\|^2}{\sum_{j=0}^m \|v_{j+1}\|^2/\|r_j^{\mathrm{QOR}}\|^2}.$$

If we make the usual assumption that the basis vectors v_j have unit length ($\|v_j\| = 1$ for all j), then

$$\alpha_m^{\mathrm{QMR}} = \frac{1/\|r_m^{\mathrm{QOR}}\|^2}{\sum_{j=0}^m 1/\|r_j^{\mathrm{QOR}}\|^2},$$

and the QMR and QOR approximants are related by exactly the formulas which hold for a proper MR/OR pair, namely,

$$w_m^{\mathrm{QMR}} = \frac{\sum_{j=0}^m w_j^{\mathrm{QOR}}/\|r_j^{\mathrm{QOR}}\|^2}{\sum_{j=0}^m 1/\|r_j^{\mathrm{QOR}}\|^2} \quad \text{and} \quad r_m^{\mathrm{QMR}} = \frac{\sum_{j=0}^m r_j^{\mathrm{QOR}}/\|r_j^{\mathrm{QOR}}\|^2}{\sum_{j=0}^m 1/\|r_j^{\mathrm{QOR}}\|^2}.$$

QMR smoothing applied to CGS and Bi-CGSTAB is discussed in Walker (1995). For smoothing techniques applied to the general class of Lanczos-type product methods, see Ressel and Gutknecht (1998).

The above analysis might lead one to believe that the QMR approximations will move steadily further away from the MR approximation at each step. The following observation due to Stewart (1998) shows that this is not necessarily the case, but that the QMR approximation may under certain conditions recover, regardless of how far it may have deviated from the (optimal) MR approximation in earlier steps.

Proposition 4.11. We have $w_m^{\mathrm{QMR}} = w_m^{\mathrm{MR}}$ if and only if $\tilde{v}_{m+1} \perp \mathcal{W}_m$, and $w_m^{\mathrm{QOR}} = w_m^{\mathrm{OR}}$ if and only if $v_{m+1} \perp \mathcal{V}_m$.

Proof. By Proposition 4.6, we have both $\mathcal{W}_m = \operatorname{span}\{\hat{v}_1, \ldots, \hat{v}_m\}$ and $r_m^{\mathrm{QMR}} \in \operatorname{span}\{\tilde{v}_{m+1}\}$. Since $r_m^{\mathrm{QMR}} = r_m^{\mathrm{MR}}$ if and only if $r_m^{\mathrm{QMR}} \perp \mathcal{W}_m$, the first assertion is proved. The analogous assertion for the QOR approximation follows from $r_m^{\mathrm{QOR}} \in \operatorname{span}\{v_{m+1}\}$; hence $r_m^{\mathrm{QOR}} \perp \mathcal{V}_m$ if and only if $v_{m+1} \perp \mathcal{V}_m$. □

4.4. Every method is an MR and an OR method

In Section 4.3 we saw how the QMR and QOR approximations can be reinterpreted as MR and OR approximations with respect to the basis-dependent inner product $(\cdot, \cdot)_V$. It turns out that an analogous interpretation is possible for *any* reasonable sequence of approximations $\{w_m\}$ to a given $r_0 \in \mathcal{H}$, in fact, as we will show, both as an MR *and* as an OR approximation.

In the remainder of this section $\{\mathcal{W}_m\}_{m=0}^{L}$ denotes any sequence of nested spaces with $\dim \mathcal{W}_m = m$ (in particular, $\mathcal{W}_0 = \{0\}$), and $\{\mathcal{V}_m\}_{m=1}^{L}$ denotes the associated sequence of error spaces $\mathcal{V}_m = \mathrm{span}\{r_0\} + \mathcal{W}_{m-1}$ relative to $r_0 \in \mathcal{H}$.

Theorem 4.12. Assume $\{h_m\}_{m=0}^{L}$ is a sequence of approximations to $r_0 \in \mathcal{H}$ such that $h_m \in \mathcal{W}_m$ and $h_L = r$. Then an inner product $(\cdot, \cdot)_V$ on $\mathcal{V}_L = \mathcal{W}_L$ such that

$$\|r_0 - h_m\|_V = \min_{w \in \mathcal{W}_m} \|r - w\|_V, \qquad m = 1, 2, \ldots, L-1,$$

exists if and only if $h_m \in \mathcal{W}_{m-1}$ implies $h_m = h_{m-1}$ for $m = 1, 2, \ldots, L-1$ or, in other words, if and only if

$$\text{either} \quad h_m \in \mathcal{W}_m \setminus \mathcal{W}_{m-1} \quad \text{or} \quad h_m = h_{m-1}, \qquad m = 1, 2, \ldots, L-1. \quad (4.27)$$

Proof. If the vectors $\{h_m\}_{m=1}^{L-1}$ are the best approximations to r_0 from $\mathcal{W}_m$ with respect to some inner product $(\cdot, \cdot)_V$, then, whenever h_m happens to lie also in $\mathcal{W}_{m-1}$, h_m must also be the best approximation to r_0 from $\mathcal{W}_{m-1}$, whereby $h_m = h_{m-1}$, which proves the necessity of (4.27).

Conversely, assuming that (4.27) is satisfied, we write

$$r_0 = (h_1 - h_0) + (h_2 - h_1) + \cdots + (h_L - h_{L-1})$$

and construct a basis $\{w_1, \ldots, w_L\}$ of $\mathcal{W}_L$ by setting

$$w_m := \begin{cases} h_m - h_{m-1}, & \text{if } h_m \in \mathcal{W}_m \setminus \mathcal{W}_{m-1}, \\ \text{an arbitrary vector from } \mathcal{W}_m \setminus \mathcal{W}_{m-1}, & \text{if } h_m = h_{m-1}. \end{cases}$$

Note that, for each m, $\{w_1, \ldots, w_m\}$ is a basis of $\mathcal{W}_m$. We further define the 'Fourier coefficients' α_m by

$$\alpha_m := \begin{cases} 1, & \text{if } h_m \in \mathcal{W}_m \setminus \mathcal{W}_{m-1}, \\ 0, & \text{if } h_m = h_{m-1}, \end{cases} \qquad m = 1, \ldots, L,$$

so that $r_0 = \alpha_1 w_1 + \alpha_2 w_2 + \cdots + \alpha_L w_L$ and

$$h_m = \alpha_1 w_1 + \alpha_2 w_2 + \cdots \alpha_m w_m, \qquad m = 1, \ldots, L,$$

that is, h_m is simply the truncated 'Fourier expansion' of r_0. Defining the inner product $(\cdot, \cdot)_V$ such that $\{w_1, \ldots, w_L\}$ are orthonormal then leads to the desired conclusion. ☐

The next theorem establishes the analogous result for the OR (or, more precisely, the QOR) approximation.

Theorem 4.13. If $\{h_m\}_{m=1}^{L}$ is a sequence of approximations to $r_0 \in \mathcal{H}$ such that $h_m \in \mathcal{W}_m$ and $h_L = r_0$, then an inner product $(\cdot, \cdot)_{\tilde{V}}$ on $\mathcal{V}_L = \mathcal{W}_L$ such that

$$r_0 - h_m \perp_{\tilde{V}} \mathcal{V}_m, \qquad m = 1, 2, \ldots, L-1,$$

exists if and only if $h_m \in \mathcal{W}_m \setminus \mathcal{W}_{m-1}$ for $m = 1, 2, \ldots, L-1$.

Proof. Assume that, for all $m = 1, 2, \ldots, L-1$, the vectors h_m are the OR approximations to r_0 from $\mathcal{W}_m$ with respect to some inner product $(\cdot, \cdot)_{\tilde{V}}$. If now, for some m, $h_m \in \mathcal{W}_{m-1}$, then $r_m := r_0 - h_m \in \text{span}\{r_0\} + \mathcal{W}_{m-1} = \mathcal{V}_m$, that is, $r_m \in \mathcal{V}_m \cap \mathcal{V}_m^{\perp} = \{0\}$, which implies $h_m = r_0$. But this is impossible unless $r_0 \in \mathcal{W}_m$, that is, $m = L$, and we have thus established that $h_m \in \mathcal{W}_m \setminus \mathcal{W}_{m-1}$ for $m = 1, \ldots, L-1$.

Conversely, since $h_m \in \mathcal{W}_m \setminus \mathcal{W}_{m-1}$ implies $r_m \in \mathcal{V}_{m+1} \setminus \mathcal{V}_m$ for $m = 1, \ldots, L-1$, we see that $\{r_0, r_1, \ldots, r_{L-1}\}$ ($h_0 = r_0$) is a basis of $\mathcal{V}_L$ such that, for every $m = 1, \ldots, L-1$, $\{r_0, r_1, \ldots, r_{m-1}\}$ is a basis of $\mathcal{V}_m$. Defining the inner product $(\cdot, \cdot)_{\tilde{V}}$ such that $\{r_0, \ldots, r_{L-1}\}$ is an orthogonal basis of $\mathcal{V}_L$ leads to the desired conclusion. $\square$

We have formulated these two theorems for the case of a sequence terminating with $h_L = r_0$, as this is the situation when solving linear equations in finite dimensions by MR and OR approximations, at least in the absence of rounding error. When the sequence of approximations does not terminate, we may proceed analogously to Theorem 4.12 with the difference that the inner product is then defined on the union of all error spaces $\mathcal{V}_m$ and we need not have convergence of the approximation to r_0 in the associated norm. Similar considerations apply for a formulation of Theorem 4.13 for a nonterminating approximation sequence.

In summary, we conclude that, by allowing the inner product to vary, the concept of MR and OR approximations becomes sufficiently general to include any reasonable sequence of approximations. Of course, this result is of a rather academic nature since an application in which such approximation problems arise often comes with a natural norm to be minimized. In view of this one might say that methods should not be compared on the grounds of whether they minimize a norm, but whether the norm being minimized is appropriate for the problem at hand. However, it does show that the given MR/OR framework includes all reasonable approximation schemes.

5. Krylov subspace methods and related algorithms

We now return to our original problem of approximating the solution of an operator equation (1.1) using MR and OR approximations. As mentioned in the introduction, this amounts to approximating the residual $r_0 = b - Ax_0$ of a given initial approximation x_0 in the sequence of nested approximation spaces $\mathcal{W}_m = A\mathcal{C}_m$, which are the images under A of a sequence of m-dimensional nested correction spaces $\mathcal{C}_m \subset \mathcal{H}$. In accordance with (3.10), the resulting residual spaces are $\mathcal{V}_m = \mathrm{span}\{r_0\} + A\mathcal{C}_{m-1}$, and thus both the approximation and residual spaces are determined by the sequence of correction spaces. In this equation-solving setting the termination index L defined in (4.4) becomes

$$
\begin{aligned}
L &= \min\{m : r_0 \in A\mathcal{C}_m\} = \min\{m : b = A(x_0 + c),\, c \in \mathcal{C}_m\} \\
&= \min\{m : A^{-1}b \in x_0 + \mathcal{C}_m\},
\end{aligned}
\tag{5.1}
$$

that is, the MR and OR methods terminate when the exact solution is found, at least in exact arithmetic. In general one has $r_0 \in \mathcal{C}_m$, so that a sufficient condition for termination is the A-invariance of the correction space. The angles which determine the rate of convergence (cf. (3.5), (3.11)) are now

$$
\angle(r_m^{\mathrm{MR}}, A\mathcal{C}_m) = \angle(\mathcal{V}_m, A\mathcal{C}_m) = \angle(\mathrm{span}\{r_0\} + A\mathcal{C}_{m-1}, A\mathcal{C}_m).
$$

We note that all statements about MR and OR approximations made in the previous sections immediately carry over to the associated equation-solving methods, in particular those of Theorems 3.4 and 3.5 as well as the results of Proposition 3.8 on minimal and quasi-minimal residual smoothing. Note also that, by the injectivity of A, relation (3.20) among the MR and OR approximations of r_0 implies

$$
c_m^{\mathrm{MR}} = s_m^2\, c_{m-1}^{\mathrm{MR}} + c_m^2\, c_m^{\mathrm{OR}}
$$

for the associated correction vectors c^{MR} and c^{OR}, and hence, by the identity $s_m^2 + c_m^2 = 1$, we obtain the analogous relation

$$
x_m^{\mathrm{MR}} = s_m^2\, x_{m-1}^{\mathrm{MR}} + c_m^2\, x_m^{\mathrm{OR}}
\tag{5.2}
$$

for the approximations of $A^{-1}b$.

Although in principle any nested sequence of correction spaces leads to the MR/OR methods discussed so far for solving (1.1), by far the most popular of such subspace correction methods employ *Krylov subspaces* as correction spaces, that is,

$$
\mathcal{C}_m = \mathcal{K}_m := \mathcal{K}_m(A, r_0) := \mathrm{span}\{r_0, Ar_0, \ldots, A^{m-1}r_0\}, \qquad m = 0, 1, \ldots .
$$

The name refers to a method introduced by Krylov (1931) for determining divisors of the minimal polynomial of an operator for the purpose of computing eigenvalues, in which such spaces were used (see also Householder (1964, Section 6.1)). In this case the residual spaces

$$
\mathcal{V}_m = \mathrm{span}\{r_0\} + A\mathcal{C}_{m-1} = \mathrm{span}\{r_0\} + A\mathcal{K}_{m-1}(A, r_0) = \mathcal{K}_m(A, r_0) = \mathcal{C}_m
$$

coincide with the correction spaces, so that Krylov subspace MR and OR methods are special cases of the abstract MR and OR methods described in

Section 3 with

$$\mathcal{C}_m = \mathcal{V}_m = \mathcal{K}_m(A, r_0), \qquad \mathcal{W}_m = A\mathcal{K}_m(A, r_0) = \mathcal{K}_m(A, Ar_0). \qquad (5.3)$$

The associated angles governing convergence in this case are thus

$$\angle(r_m^{\mathrm{MR}}, A\mathcal{K}_m) = \angle(\mathcal{K}_m, A\mathcal{K}_m), \qquad (5.4)$$

and the termination index L may now be characterized by

$$\begin{aligned}
L &= \min\{m : A^{-1}b \in x_0 + \mathcal{K}_m(A, r_0)\} \\
&= \min\{m : A^{-1}r_0 \in \mathcal{K}_m(A, r_0)\} \\
&= \min\{m : \mathcal{K}_m(A, r_0) = \mathcal{K}_{m+1}(A, r_0)\}.
\end{aligned} \qquad (5.5)$$

In the remainder of this section, we discuss some of the advantages of Krylov subspaces as correction spaces, review the most important examples of Krylov subspace algorithms, and recover some well-known results on Krylov subspace methods by specializing the abstract results of Sections 3 and 4.

Before turning to Krylov subspace methods, however, we point out that MR and OR methods which use non-Krylov correction spaces have received increasing attention recently. Methods of this type are the EN-method of Eirola and Nevanlinna (1989), the FGMRES method of Saad (1993), the GMRESR algorithm of van der Vorst and Vuik (1994), the augmented GMRES methods of Morgan (2000) and the GCROT method of de Sturler (1999). We refer to Eiermann, Ernst and Schneider (2000) for an overview and an analysis of these approaches.

5.1. Why Krylov subspaces?

The use of (shifted) Krylov spaces to construct approximate solutions to linear equations, at least implicitly, is as old as classical stationary iteration methods (see, e.g., Varga (1999)): given a splitting $A = M - N$ with M nonsingular, the induced stationary iteration

$$x_m = Tx_{m-1} + c, \qquad m = 1, 2, \ldots \qquad (5.6)$$

with $T = M^{-1}N$ and $c = M^{-1}b$ generates the approximations

$$x_m = x_0 + (I + T + \cdots + T^{m-1})r_0 \in x_0 + \mathcal{K}_m(T, r_0).$$

From this perspective, one can view Krylov subspace MR and OR methods as a cleverer strategy for choosing the approximations in $x_0 + \mathcal{K}_m$ or, equivalently, as techniques which accelerate the convergence of the stationary iterative method (5.6). This was the motivation in Varga (1999), where the term *semi-iterative methods* is used for this approach. In view of (5.5), a subtle difference between such stationary iterations and Krylov subspace MR and OR methods is that the latter terminate with the exact solution whenever the Krylov space becomes A-invariant, which is not the case for stationary methods or those based on Chebyshev recursions (Varga 1999).

The main reason for the prevalence of Krylov subspaces as correction spaces for MR/OR methods is the ease by which these can be generated, namely by multiplication of a vector by A in each step, an inexpensive operation whenever A is represented by a sparse or structured matrix or when the action of A can be implemented efficiently without reference to a matrix representation. Moreover, the fact that the correction and residual spaces coincide for Krylov subspace methods allows the same basis to be used for both in computations, which further reduces computing and storage requirements.

Whether or not Krylov spaces are well suited as correction spaces will, in view of (5.4) and the results of Section 3, depend on the behaviour of the sequence of angles $\angle(\mathcal{K}_m, A\mathcal{K}_m)$. There are classes of problems for which this behaviour is very favourable. An example where the angles actually tend to zero, which, in view of (3.15), implies superlinear convergence of the MR and OR approximants, is given by second-kind Fredholm equations (see Section 6.4). On the other hand, there are matrix problems of dimension n for which $\angle(\mathcal{K}_m, A\mathcal{K}_m) = \pi/2$ for $m = 1, 2, \ldots, n-1$, that is, no Krylov subspace method is able to improve the initial residual until the very last step. The convergence properties of Krylov subspace methods will be discussed in more detail in Section 6.

A great simplification in the analysis of Krylov subspace methods arises from the representation

$$\mathcal{K}_m(A, r_0) = \{q(A)r_0 : q \in \mathcal{P}_{m-1}\}, \qquad m = 1, 2, \ldots,$$

where $\mathcal{P}_m$ denotes the space of all complex polynomials of degree at most m. The linear map

$$\mathcal{P}_{m-1} \ni q \mapsto q(A)r_0 \in \mathcal{K}_m(A, r_0)$$

is thus always surjective but fails to be an isomorphism if and only if there exists a nonzero polynomial $q \in \mathcal{P}_{m-1}$ with $q(A)r_0 = 0$. If such a polynomial exists (e.g., if A has finite rank) then there also exists a (unique) monic polynomial $c = c_{A,r_0}$ of minimal degree with $c(A)r_0 = 0$ which is usually called the *minimal polynomial of r_0 with respect to A*. It is easy to see that the degree of c equals the smallest integer m such that $\mathcal{K}_m = \mathcal{K}_{m+1}$ and thus coincides with the index L introduced in (4.4) (*cf.* also (5.1) and (5.5)),

$$L = \min\{m : \mathcal{K}_m = \mathcal{K}_{m+1}\} = \min\{\deg q : q \text{ monic and } q(A)r_0 = 0\}. \tag{5.7}$$

In other words, $\mathcal{P}_{m-1}$ and $\mathcal{K}_m$ are isomorphic linear spaces if and only if $m \leq L$.

Since every vector $x \in x_0 + \mathcal{K}_m$ is of the form $x = x_0 + q_{m-1}(A)r_0$ for some $q_{m-1} \in \mathcal{P}_{m-1}$, the corresponding residual $r = b - Ax$ can be written as

$$r = r_0 - A q_{m-1}(A) r_0 = p_m(A) r_0, \quad \text{where } p_m(\zeta) := 1 - \zeta q_{m-1}(\zeta) \in \mathcal{P}_m.$$

Note that the *residual polynomial* p_m satisfies the normalization condition $p_m(0) = 1$. Characterizations of the residual polynomials which belong to the MR and OR iterates as well as their zeros were first given by Stiefel (1958) in the Hermitian case and by Freund (1992) in the non-Hermitian case; a concise presentation can be found in Eiermann *et al.* (2000).

Finally, we note that the MR and OR approximations of an arbitrary vector $r \in \mathcal{H}$ from a given sequence of nested spaces $\{W_m\}_{m=0}^{L}$, $\dim W_m = m$, can always be interpreted as Krylov subspace MR and OR approximations of the solution of a linear operator equation: indeed, relation (4.3) uniquely determines a linear operator A on the spaces $\{V_m\}_{m=0}^{L}$, and A has V_L as an invariant subspace. If A is then extended to an invertible operator on the entire space $\mathcal{H}$, then the MR and OR approximation sequences for r coincide with the corresponding Krylov subspace approximations for solving the equation $Ae = r$ with initial guess $e_0 = 0$.

As already mentioned at the beginning of this section, the results of Sections 2, 3 and 4 also hold for the equation-solving MR and OR methods. For the particular class of Krylov subspace methods many of these identities, in particular (3.21) and (5.2), have been derived separately for each method, for example in the papers of Brown (1991), Freund (1992), Gutknecht (1993a), and Cullum and Greenbaum (1996). In these works, however, the sines and cosines result from the Givens rotations needed to construct a QR decomposition of the Hessenberg matrix analogous to (4.3) in the course of a specific algorithm for computing the MR and OR approximations. Section 4, specifically Theorem 4.5, reveals the more fundamental significance of these rotation angles, namely as the angles between the subspaces $\mathcal{K}_m(A, r_0)$ and $A\mathcal{K}_m(A, r_0)$. Moreover, all these relations have been derived in the previous sections as properties of the abstract MR and OR approximation schemes on nested subspaces, of which equation solving based on Krylov subspaces is just one particular instance.

5.2. Algorithms based on orthonormal bases

The algorithms which have been proposed in the literature for calculating MR and OR approximations for solving linear equations are based on one of the coordinate representations introduced in Section 4. The novelty, in the context of equation-solving, is that one requires the coordinates of the correction vector in $\mathcal{C}_m$ as well as those of the residual approximation in W_m.

Using an orthonormal basis of W_m
The trivial construction of the MR/OR approximations in terms of an orthonormal basis of W_m described in Section 4.1 becomes a little more interesting

when solving equations. Given an ascending basis $C_m = [c_1, \ldots, c_m]$ of $\mathcal{C}_m$, the most direct approach is to orthonormalize the image sequence $\{Ac_j\}$ and then proceed as in Section 4.1. Because the spaces are nested, the orthonormalization results in a QR decomposition

$$AC_m = W_m R_m \tag{5.8}$$

with an upper triangular matrix $R_m \in \mathbb{C}^{m \times m}$ and a set of orthonormal basis vectors $W_m = [w_1, \ldots, w_m]$ of $\mathcal{W}_m$. The MR approximation x_m^{MR} of the solution of (1.1) with respect to $\mathcal{C}_m$ is given by $x_m^{\mathrm{MR}} = x_0 + C_m y_m^{\mathrm{MR}}$ with a coefficient vector $y_m^{\mathrm{MR}} \in \mathbb{C}^m$. By (4.1) the coefficient vector of $P_{\mathcal{W}_m} r_0$ with respect to W_m is $W_m^* r_0$, hence we must have $AC_m y_m^{\mathrm{MR}} = W_m W_m^* r_0$, which, in view of (5.8), leads to $y_m^{\mathrm{MR}} = R_m^{-1} W_m^* r_0$. If no Galerkin breakdown occurs at this step, that is, if $(r_0, w_m) \neq 0$, then the OR approximation $x_m^{\mathrm{OR}} = x_0 + C_m y_m^{\mathrm{OR}}$ is defined and may be computed following (4.2) by solving

$$R_m y_m^{\mathrm{OR}} = \begin{bmatrix} (r_0, w_1) \\ \vdots \\ (r_0, w_{m-1}) \\ \|r_{m-1}^{\mathrm{MR}}\|^2 / (w_m, r_0) \end{bmatrix}, \tag{5.9}$$

or by using (5.2) and noting $c_m = |(w_m, r_0)| / \|r_{m-1}^{\mathrm{MR}}\|$, $s_m = \sqrt{1 - c_m^2}$. The main computational expense of this algorithm lies in the orthogonalization process and the solution of a triangular system whenever the approximations are desired. In the Krylov subspace case (5.3), the correction spaces are $\mathcal{C}_1 = \mathrm{span}\{r_0\}$ and $\mathcal{C}_{m+1} = \mathrm{span}\{r_0\} + \mathcal{W}_m, m = 1, \ldots, L$. An obvious candidate for the new vector in the ascending basis of $\mathcal{C}_{m+1}$ is $c_{m+1} := w_m$, a vector which is constructed in the previous step by orthonormalizing Aw_{m-1} against the orthonormal basis of $\mathcal{W}_{m-1}$. This choice results in $C_{m+1} = [r_0, w_1, \ldots, w_m]$, so that no separate basis for the correction spaces is necessary. Although it appears to be the most straightforward implementation, this method was only recently proposed for computing MR approximations by Walker and Zhou (1994).

The more well-known *generalized conjugate residual* (GCR) algorithm, introduced by Eisenstat, Elman and Schultz (1983), can be derived from the fact that the image under A of the MR correction c_m^{MR} is the best approximation $W_m W_m^* r_0$ of r_0 from $\mathcal{W}_m$: if the basis C_m is taken to consist of the pre-images under A of a set of orthonormal basis vectors W_m of $\mathcal{W}_m$, then we obtain

$$c_m^{\mathrm{MR}} = A^{-1} W_m W_m^* r_0 = C_m W_m^* r_0.$$

In this case the coefficient vector y_m^{MR} of the MR correction with respect to C_m consists simply of the Fourier coefficients $W_m^* r_0$, that is, no triangular system needs to be solved. The corresponding OR coefficient vector is obtained, again in view of (4.2), by

$$y_m^{\mathrm{OR}} = [(r_0, w_1), \ldots, (r_0, w_{m-1}), \|r_{m-1}^{\mathrm{MR}}\|^2/(w_m, r_0)]^\top. \tag{5.10}$$

The associated residual vectors may be formed by using the same coefficient vectors, but with respect to the orthonormal basis W_m.

For the Krylov subspace GCR algorithm different ways of extending the bases C_m and W_m have been proposed in the literature. The older variant, introduced in Eisenstat *et al.* (1983), generates in step m the new basis vector w_m by orthonormalizing $A r_{m-1}$ against the previously generated orthonormal basis of W_{m-1}, while simultaneously updating r_{m-1} to obtain $c_m = A^{-1} w_m$. This approach has the drawback that it fails to extend the Krylov space whenever two consecutive MR approximations coincide, that is, when a Galerkin breakdown occurs. For this reason, Eisenstat *et al.* (1983) state that the algorithm should only be used for linear systems where A is *positive real*, which means that its Hermitian part $(A + A^*)/2$ is positive definite. As will be explained in Section 6.1, the algorithm may in fact be used provided $0 \notin W(A)$, a slightly more general criterion. An easy remedy for this deficiency is to extend the basis W_{m-1} by instead orthogonalizing $A w_{m-1}$ against W_{m-1}, which results in the Arnoldi process for $W_m = K_m(A, A r_0)$ for generating the w_m-sequence, while the c_m-sequence is again maintained such that $A c_m = w_m$. This observation is pointed out in Rozložník and Strakoš (1996), where many equivalent MR implementations are also compared with regard to their numerical stability.

The GCR algorithm belongs to the vast lineage of generalizations of the *conjugate gradient* and *conjugate residual methods* of Hestenes and Stiefel (1952), and we refer to Ernst (2000) and the monograph of Greenbaum (1997) for systematic surveys of this family of Krylov subspace methods.

Using an orthonormal basis of V_m

The equation-solving Krylov subspace MR/OR algorithms based on the abstract scheme of Section 4.2 are the *generalized minimum residual method* (GMRES) of Saad and Schultz (1986) and the *full orthogonalization method* (FOM) introduced by Saad (1981). These algorithms proceed by successively constructing an orthonormal basis $\{v_1, \ldots, v_m\}$ of $V_m = K_m(A, r_0)$ beginning with $V_1 = \mathrm{span}\{r_0\}$. We observe that this is exactly the Gram–Schmidt procedure introduced in (4.5) with $w_m = A v_m$, $m = 1, \ldots, L$, which in this setting is known as the *Arnoldi process*. In this case equation (4.3) reads

$$A V_m = V_{m+1} \tilde{H}_m = V_m H_m + [0, \ldots, 0, \eta_{m+1,m} v_{m+1}] \tag{5.11}$$

and the upper Hessenberg matrix $\tilde{H}_m$ is given by

$$\tilde{H}_m = [(A v_k, v_j)] \in \mathbb{C}^{(m+1) \times m}, \quad j = 1, \ldots, k+1, \quad k = 1, \ldots, m. \tag{5.12}$$

Equation (5.12) reveals that, in the Krylov subspace case, the Hessenberg matrix H_m has a closer connection to the operator A than merely expressing the basis W_m in terms of V_{m+1} as in equation (4.3): here it is also the *orthogonal section* of A onto $\mathcal{K}_m$, that is, it represents the linear map $P_{\mathcal{K}_m} A|_{\mathcal{K}_m} : \mathcal{K}_m \to \mathcal{K}_m$ with respect to the basis V_m. To compute the coordinate vector y_m^{MR} of the MR approximation $x_m^{MR} = x_0 + V_m y_m^{MR}$, we note that, in view of (5.11), the corresponding residual has the form $r_m^{MR} = r_0 - V_{m+1} \tilde{H}_m y_m^{MR} = V_{m+1}(\beta u_1 - \tilde{H}_m y_m^{MR})$, and hence the minimum residual condition again determines y_m^{MR} as the solution of the least-squares problem (4.7). Just as in Section 4.2, the GMRES algorithm uses a QR factorization of $\tilde{H}_m$ constructed from Givens rotations to solve this least-squares problem and thus, by Theorem 4.5, the rotation angles of these Givens rotations coincide with the angles (5.4) between the subspaces $\mathcal{K}_m(A, r_0)$ and $A\mathcal{K}_m(A, r_0)$.

The FOM implementation of the OR approach, which is also based on the Arnoldi process (*cf.* (5.11)), computes the coordinate vector $y_m^{OR} \in \mathbb{C}^m$ of the OR approximation $x_m^{OR} = x_0 + V_m y_m^{OR}$ just as in Lemma 4.1 as the solution of the linear system (4.8), which requires H_m that be nonsingular. We recall from Remark 4.2 that this condition is consistent with our previous result, namely that $\angle(\mathcal{K}_m, A\mathcal{K}_m) \neq \pi/2$.

The GMRES and FOM algorithms and the variants mentioned above implement the MR and OR approach for solving (1.1) in the case of general injective A. The method of choice for constructing an orthonormal basis of $\mathcal{V}_m = \mathcal{K}_m(A, r_0)$ is the Arnoldi process, of which there exist several modifications with varying degrees of stability in the presence of round-off error (see, *e.g.*, the discussions in Rozložník, Strakoš and Tůma (1996) and Greenbaum, Rozložník and Strakoš (1997)). Whenever A is self-adjoint, the Arnoldi process simplifies to the Hermitian Lanczos process (Lanczos 1950), with the result that the Hessenberg matrix in the fundamental relation (5.11) is tridiagonal, with all the computational benefits mentioned at the beginning of Section 4.3. In particular, the short recurrences for the basis vectors allow the approximations and residuals to be inexpensively obtained by clever update formulas. Well-known algorithms that follow this approach to implement the MR and OR approximations in the self-adjoint and positive definite case are the CR/CG algorithms of Hestenes and Stiefel (1952), and an MR/OR pair of algorithms for the self-adjoint but possibly indefinite case is MINRES/CG due to Paige and Saunders (1975).

5.3. Using a non-orthogonal basis

In this section we continue our discussion of Krylov subspace MR and OR algorithms for the iterative solution of (1.1) with MR/OR pairs such as QMR/BCG and TFQMR/CGS, which work with a non-orthogonal basis of the Krylov space. These represent examples of the MR/OR approximations discussed in Section 4.3 for solving equations using Krylov spaces.

To this end, let $v_1, v_2, \ldots, v_L \in \mathcal{H}$ be *any* set of ascending basis vectors, *i.e.*, linearly independent vectors such that $\{v_1, v_2, \ldots, v_m\}$ forms a basis of $\mathcal{V}_m = \mathcal{K}_m(A, r_0)$ for $m = 1, \ldots, L$ (in particular, $r_0 = \beta v_1$ for some $0 \neq \beta \in \mathbb{C}$). Such a basis leads naturally to a (quasi-)minimal residual as well as to a (quasi-)orthogonal residual method. Since, for every $1 \leq m \leq L$, $Av_m \in \mathrm{span}\{v_1, v_2, \ldots, v_{m+1}\}$, where we set $v_{L+1} = 0$, there exists an upper Hessenberg matrix $\tilde{H}_m \in \mathbb{C}^{(m+1) \times m}$ such that (*cf.* (4.3) and (5.11))

$$AV_m = V_m H_m + [0, \ldots, 0, \eta_{m+1,m} v_{m+1}] = V_{m+1} \tilde{H}_m. \tag{5.13}$$

As in (5.12), the mth column of $\tilde{H}_m$ contains the coefficients of $Av_m \in \mathcal{K}_{m+1}(A, r_0)$ with respect to the basis vectors $v_1, \ldots, v_{m+1}$. We are thus in the situation of Section 4.3 with $\mathcal{V}_m = \mathcal{K}_m(A, r_0)$ and $\mathcal{W}_m = A\mathcal{K}_m(A, r_0)$. Defining the auxiliary inner products $(\cdot, \cdot)_V$ on $\mathcal{V}_L$ and $(\cdot, \cdot)_v$ on the coordinate space $\mathbb{C}^L$ as in (4.22) and (4.21), respectively, we obtain the QMR approximation

$$x_m^{\mathrm{QMR}} := x_0 + V_m y_m^{\mathrm{QMR}} = x_0 + c_m^{\mathrm{QMR}}$$

of the solution $A^{-1}b$ by requiring that Ac_m^{QMR} be the MR approximation to r_0 with respect to the inner product $(\cdot, \cdot)_V$. Just as in Section 4.3, the coefficient vector $y_m^{\mathrm{QMR}} \in \mathbb{C}^m$ is characterized as the unique solution of the least-squares problem (4.7). Analogously, the associated QOR approximation

$$x_m^{\mathrm{QOR}} := x_0 + V_m y_m^{\mathrm{QOR}} = x_0 + c_m^{\mathrm{QOR}}$$

is obtained if Ac_m^{OR} is the OR approximation of r_0 with respect to the inner product $(\cdot, \cdot)_V$, that is, if the coordinate vector $y_m^{\mathrm{QOR}} \in \mathbb{C}^m$ satisfies the linear system of equations (4.25).

As stated in Section 4, the residuals of the QMR and QOR iterates are the errors of the MR and OR approximations of r_0 with respect to the inner product $(\cdot, \cdot)_V$, and therefore the results of Theorem 3.4 on the residual norms hold for the QMR and QOR residuals $r_m^{\mathrm{QMR}} = b - Ax_m^{\mathrm{QMR}}$ and $r_m^{\mathrm{QOR}} = b - Ax_m^{\mathrm{QOR}}$, albeit with respect to the norm $\|\cdot\|_V$. The same applies to the statements of Theorem 3.5 and Corollary 3.6 for the QMR and QOR approximations $w_m^{\mathrm{QMR}} = Ac_m^{\mathrm{QMR}}$ and $w_m^{\mathrm{QOR}} = Ac_m^{\mathrm{QOR}}$, where the angles are understood with respect to the inner product $(\cdot, \cdot)_V$. Finally, the bounds in Theorem 4.8 relating the two different norms as well as the assertions of Propositions 4.9, 4.10 and 4.11 all hold for the QMR and QOR residuals, respectively.

The motivation for using non-orthogonal bases in Krylov subspace MR and OR methods comes from the potential savings in storage and computation when using algorithms for generating a basis of $\mathcal{K}_m(A, r_0)$ which lead to a Hessenberg matrix in relation (5.13) with only a small number of non-zero bands. A tridiagonal Hessenberg matrix can be achieved using the non-Hermitian Lanczos process. The non-Hermitian variant of the Lanczos process is generally less expensive than the Arnoldi process, requiring in addition the generation of a basis of a Krylov space with respect to the adjoint A^* and thus storage for several additional vectors and, which can be expensive, multiplication by A^* at each step. The result are two bi-orthogonal, in general non-orthogonal, bases. The non-Hermitian Lanczos process may break down before the Krylov space becomes stationary, and in finite arithmetic this leads to numerical instabilities in the case of near-breakdowns. This problem is addressed by so-called look-ahead techniques for the non-Hermitian Lanczos process (*cf.* Freund, Gutknecht and Nachtigal (1993), Gutknecht (1997) and the references therein), which result in a pair of block-biorthogonal bases and a Hessenberg matrix which is as close to tridiagonal form as possible while maintaining stability.

Remark 5.1. We have mentioned that Krylov subspace QMR/QOR methods have the advantage of being able to work with an arbitrary basis of the Krylov space, in particular also with a non-orthogonal basis. As the bounds (4.24) show, how close the QMR iterates come to minimizing the residual in the original norm depends on the Euclidean condition number of the Gram matrices M_m, that is, on the largest and smallest singular value of $V_m = [v_1, \ldots, v_m]$ (interpreted as an operator from $\mathbb{C}^m$ to $\mathcal{K}_m(A, r_0)$). The largest singular value of V_m is bounded by $\sqrt{m+1}$ if the basis vectors are normalized with respect to the original norm. When the look-ahead Lanczos algorithm is used to generate the basis, bounds on the smallest singular value may be obtained depending on the look-ahead strategy being followed (Freund *et al.* 1993).

Examples: QMR/BCG and TFQMR/CGS
We now consider two specific examples of Krylov subspace QMR/QOR algorithms, the QMR method of Freund and Nachtigal (1991) and the BCG method due to Lanczos (1952) as well as the TFQMR and CGS methods due to Freund (1993) and Sonneveld (1989), respectively. As another QMR/QOR pair, we mention the QMRCGSTAB method of Chan, Gallopoulos, Simoncini, Szeto and Tong (1994) and BICGSTAB developed by van der Vorst (1992) and Gutknecht (1993*b*).

The QMR algorithm of Freund and Nachtigal proceeds exactly as described above, with the basis of the Krylov space generated by the look-ahead Lanczos algorithm. The QOR counterpart of Freund and Nachtigal's QMR is the BCG algorithm, the iterates of which are characterized by

$$x_m^{\text{BCG}} \in x_0 + \mathcal{K}_m(A, r_0), \qquad r_m^{\text{BCG}} \perp \mathcal{K}_m(A^*, \tilde{r}_0).$$

The algorithm proceeds by generating a basis of the Krylov space $\mathcal{K}_m(A^*, \tilde{r}_0)$, where $\tilde{r}_0$ is an arbitrary starting vector, simultaneously with a basis V_m of $\mathcal{K}_m(A, r_0)$ in such a way that these two bases are biorthogonal. If $y \in \mathbb{C}^m$ denotes the coefficient vector of the BCG approximation with respect to V_m, then the biorthogonality requirement

$$r_m^{\mathrm{BCG}} = V_{m+1}(\beta u_1 - \tilde{H}_m y) \perp \mathcal{K}_m(A^*, \tilde{r}_0),$$

implies $r_m^{\mathrm{BCG}} \perp_V V_m$, which is equivalent to $H_m y = \beta u_1$. The last equality identifies x_m^{BCG} as the mth QOR iterate.

To treat the TFQMR/CGS pair, recall that the residual of any Krylov subspace approximation can be expressed as $r_m = p_m(A)r_0$ in terms of a polynomial p_m of degree m satisfying $p_m(0) = 1$. Sonneveld (1989) defined the CGS iterate $x_m^{\mathrm{CGS}} \in x_0 + \mathcal{K}_{2m}(A, r_0)$ such that

$$r_m^{\mathrm{CGS}} = [p_m(A)]^2 r_0, \quad \text{where} \quad r_m^{\mathrm{BCG}} = p_m(A)r_0.$$

It is shown by Freund (1993) that $x_m^{\mathrm{CGS}} = x_0 + Y_{2m} z_{2m}$ for a coefficient vector $z_{2m} \in \mathbb{C}^{2m}$, where $\mathcal{K}_{2m}(A, r_0) = \operatorname{span}\{y_1, \ldots, y_{2m}\}$. Moreover, there exists a sequence $\{w_n\}$ such that $r_0 = \beta w_1$ and

$$AY_n = W_{n+1}\tilde{H}_n, \quad n = 1, \ldots, 2L.$$

In terms of this sequence, we find

$$r_m^{\mathrm{CGS}} = W_{2m+1}\left(\beta u_1 - \tilde{H}_{2m} z_{2m}\right),$$

where $H_{2m} z_{2m} = \beta u_1$, that is,

$$r_m^{\mathrm{CGS}} \perp_W \operatorname{span}\{w_1, \ldots, w_{2m}\} = \mathcal{K}_{2m}(A, r_0),$$

which identifies CGS as an OR method. The corresponding MR method is Freund's transpose-free QMR method TFQMR (cf. (Freund 1993, Freund 1994)), the iterates of which are defined as

$$x_n^{\mathrm{TFQMR}} = x_0 + Y_n z_n, \quad n = 1, \ldots, 2L,$$

where the coefficient vector $z_n \in \mathbb{C}^n$ solves the least-squares problem

$$\|\beta u_1 - \tilde{H}_n z_n\|_2 \to \min_{z \in \mathbb{C}^n}.$$

In other words,

$$\|r_n^{\mathrm{TFQMR}}\|_W = \min_{x \in x_0 + \mathcal{K}_n(A, r_0)} \|b - Ax\|_W.$$

Comparison of residuals

The availability of a sequence of vectors $\{\tilde{v}_j\}_{j \geq 1}$ which is biorthogonal to the Krylov basis $\{v_j\}$ permits a convenient representation of the QOR residual via (4.11). If

$$(v_j, \tilde{v}_k) = \delta_{jk} d_j, \quad j, k = 1, \ldots, L,$$

and if both sequences are normalized to one, then, since the MR-residual at step $m - 1$ lies in $\mathcal{W}_{m-1} \subset \mathcal{V}_m$, we have $r_{m-1}^{\mathrm{MR}} = \sum_{j=1}^{m} (r_{m-1}^{\mathrm{MR}}, \tilde{v}_j/d_j) v_j$. Inserting $r - w = r_{m-1}^{\mathrm{MR}}$ in (4.11) then yields

$$r_m^{\mathrm{QOR}} = (r_{m-1}^{\mathrm{MR}}, \hat{w}_{m+1})_V \, v_{m+1} = v_{m+1} \sum_{j=1}^{m} \frac{\bar{g}_j}{d_j} (r_{m-1}^{\mathrm{MR}}, \tilde{v}_j) v_j,$$

which, after taking norms and applying the Cauchy–Schwarz inequality, becomes

$$\|r_m^{\mathrm{QOR}}\| \le \|r_{m-1}^{\mathrm{MR}}\| \sum_{j=1}^{m} \frac{|g_j|}{|d_j|}.$$

This bound is a slightly improved version of a bound by Hochbruck and Lubich (1998).

6. Residual and error bounds

The usual residual and error bounds for Krylov subspace MR methods follow directly from the defining equation (1.2a),

$$\|r_m^{\mathrm{MR}}\| = \min_{c \in \mathcal{K}_m(A, r_0)} \|b - A(x_0 + c)\| = \min_{c \in \mathcal{K}_m(A, r_0)} \|r_0 - Ac\|,$$

and the close connection between Krylov subspaces and polynomials (see Section 5.1) which immediately leads to

$$\|r_m^{\mathrm{MR}}\| = \min_{q \in \mathcal{P}_{m-1}} \|r_0 - Aq(A)r_0\| = \min_{\substack{p \in \mathcal{P}_m \\ p(0)=1}} \|p(A)r_0\|,$$

that is, to

$$\frac{\|r_m^{\mathrm{MR}}\|}{\|r_0\|} \le \min_{\substack{p \in \mathcal{P}_m \\ p(0)=1}} \|p(A)\|.$$

If A is normal, the right-hand side represents a standard polynomial approximation problem,

$$\min_{\substack{p \in \mathcal{P}_m \\ p(0)=1}} \|p(A)\| = \sup_{\substack{p \in \mathcal{P}_m \\ p(0)=1}} \max_{\lambda \in \Lambda(A)} |p(\lambda)| \le \min_{\substack{p \in \mathcal{P}_m \\ p(0)=1}} \max_{\lambda \in \Omega} |p(\lambda)|, \qquad (6.1)$$

where $\Omega \subset \mathbb{C}$ is an arbitrary compact set which contains $\Lambda(A)$, the spectrum of A (of course, a useful bound will only be obtained when $0 \notin \Omega$). Most bounds, for instance, the standard bound for the conjugate gradient method, are derived in this way.

If A is not normal, then it is no longer true that its spectrum determines the convergence behaviour of a Krylov MR method (see Section 6.3).

Subsequent investigations into whether larger sets in the complex plane associated with A such as its field of values (Eiermann 1993) or pseudospectrum (Trefethen 1992) can predict convergence behaviour have to date failed to produce an analogous theoretical tool in the non-normal case. We refer to Embree (1999) for a detailed discussion of these issues.

In this section we will concentrate on bounds for the residual and error norms of Krylov subspace methods which are obtained by specializing Theorem 3.4. A similar approach in this direction has been taken by Saad (2000). Our goal is to relate properties of the operator A to the decay of the numbers s_m.

6.1. Residual bounds

Estimates based on the angles between $\mathcal{K}_m := \mathcal{K}_m(A, r_0)$ and $A\mathcal{K}_m$ naturally involve the field of values of A, which is defined by

$$W(A) := \left\{ \frac{(Av, v)}{(v, v)} : 0 \neq v \in \mathcal{H} \right\}.$$

Theorem 6.1. The residual with index m of a Krylov MR method satisfies

$$\frac{\|r_m^{\mathrm{MR}}\|}{\|r_0\|} \leq \prod_{j=1}^{m} \sqrt{1 - \nu_j(A)\tilde{\nu}_j(A^{-1})}, \tag{6.2}$$

where the quantities $\nu_j(A)$ and $\tilde{\nu}_j(A^{-1})$ are defined as

$$\nu_j(A) := \inf\{|z| : z \in W(A|_{\mathcal{S}_j})\},$$
$$\tilde{\nu}_j(A^{-1}) := \inf\{|z| : z \in W(A^{-1}|_{A\mathcal{S}_j})\}$$

and $\mathcal{S}_j \subseteq \mathcal{H}$ denotes a subspace which contains $\mathrm{span}\{r_{j-1}^{\mathrm{MR}}\} = \mathcal{K}_j \cap (A\mathcal{K}_{j-1})^{\perp}$.

Proof. From (3.15), the fact that $s_j = \sin \angle(r_{j-1}^{\mathrm{MR}}, A\mathcal{K}_j)$ as well as $r_{j-1}^{\mathrm{MR}} \in \mathcal{K}_j \cap (A\mathcal{K}_{j-1})^{\perp}$, we conclude

$$\frac{\|r_m^{\mathrm{MR}}\|}{\|r_0\|} = \prod_{j=1}^{m} s_j = \prod_{j=1}^{m} \sin \angle(r_{j-1}^{\mathrm{MR}}, A\mathcal{K}_j)$$

$$= \prod_{j=1}^{m} \left(1 - \sup_{v \in \mathcal{K}_j} \frac{|(r_{j-1}^{\mathrm{MR}}, Av)|^2}{\|r_{j-1}^{\mathrm{MR}}\|^2 \|Av\|^2} \right)^{1/2}$$

$$\leq \prod_{j=1}^{m} \left(1 - \frac{|(r_{j-1}^{\mathrm{MR}}, Ar_{j-1}^{\mathrm{MR}})|^2}{\|r_{j-1}^{\mathrm{MR}}\|^2 \|Ar_{j-1}^{\mathrm{MR}}\|^2} \right)^{1/2}$$

$$= \prod_{j=1}^{m} \left(1 - \left| \frac{(Ar_{j-1}^{\mathrm{MR}}, r_{j-1}^{\mathrm{MR}})}{(r_{j-1}^{\mathrm{MR}}, r_{j-1}^{\mathrm{MR}})} \right| \left| \frac{(r_{j-1}^{\mathrm{MR}}, Ar_{j-1}^{\mathrm{MR}})}{(Ar_{j-1}^{\mathrm{MR}}, Ar_{j-1}^{\mathrm{MR}})} \right| \right)^{1/2}$$

$$= \prod_{j=1}^{m} \left(1 - \left| \frac{(Ar_{j-1}^{\mathrm{MR}}, r_{j-1}^{\mathrm{MR}})}{(r_{j-1}^{\mathrm{MR}}, r_{j-1}^{\mathrm{MR}})} \right| \left| \frac{(A^{-1}Ar_{j-1}^{\mathrm{MR}}, Ar_{j-1}^{\mathrm{MR}})}{(Ar_{j-1}^{\mathrm{MR}}, Ar_{j-1}^{\mathrm{MR}})} \right| \right)^{1/2}$$

$$\leq \prod_{j=1}^{m} \left(1 - \inf_{s \in \mathcal{S}_j} \left| \frac{(As, s)}{(s, s)} \right| \inf_{t \in A\mathcal{S}_j} \left| \frac{(A^{-1}t, t)}{(t, t)} \right| \right)^{1/2}$$

$$= \prod_{j=1}^{m} \left(1 - \nu_j(A) \tilde{\nu}_j(A^{-1}) \right)^{1/2}. \qquad \qquad \square$$

For the choice $\mathcal{S}_j = \mathcal{H}$, we obtain the following simpler bound.

Corollary 6.2. The MR residual with index m satisfies

$$\frac{\|r_m^{\mathrm{MR}}\|}{\|r_0\|} \leq \left(1 - \nu(A)\nu(A^{-1}) \right)^{m/2}, \qquad (6.3)$$

where $\nu(A) := \inf\{|z| : z \in W(A)\}$.

Of course, the bound of Corollary 6.2 only yields a reduction provided $0 \notin W(A)$, which also implies $0 \notin W(A^{-1})$; see Horn and Johnson (1991, p. 66).

If A is a positive real matrix, that is, if its Hermitian part $H := (A+A^*)/2$ is positive definite, then $\nu(A) = \lambda_{\min}(H) > 0$ and

$$\nu(A^{-1}) = \min_{v \in \mathcal{H}} \frac{(A^{-1}v, v)}{(v, v)} = \min_{w \in \mathcal{H}} \frac{(w, Aw)}{(w, w)} \frac{(w, w)}{(Aw, Aw)} \geq \frac{\lambda_{\min}(H)}{\|A\|^2},$$

in view of which Corollary 6.2 yields a bound first given by Elman (1982):

$$\frac{\|r_m^{\mathrm{MR}}\|}{\|r_0\|} \leq \left(1 - \frac{\lambda_{\min}(H)^2}{\lambda_{\max}(A^T A)} \right)^{m/2} \qquad (6.4)$$

(for a collection of similar bounds, see Joubert (1994)). Since (see the proof of Theorem 6.1)

$$\frac{\|r_m^{\mathrm{MR}}\|}{\|r_{m-1}^{\mathrm{MR}}\|} = s_m \leq 1 - \nu(A)\nu(A^{-1}) < 1,$$

the residual norms of a Krylov MR method decrease strictly monotonically if A is positive real or, slightly more generally, if $W(A)$ is contained in any half-plane $\{z : \mathrm{Re}\,(e^{i\alpha}z) > 0\}$ with $\alpha \in \mathbb{R}$. Note that, in view of Corollary 3.2, this implies that Galerkin breakdowns can be excluded for OR methods applied to such systems.

Remark 6.3. If, in the derivation of the residual bound (6.2), one makes the cruder estimate

$$\frac{\|r_j^{\mathrm{MR}}\|^2}{\|r_{j-1}^{\mathrm{MR}}\|^2} \leq 1 - \frac{|(r_{j-1}^{\mathrm{MR}}, Ar_{j-1}^{\mathrm{MR}})|^2}{\|r_{j-1}^{\mathrm{MR}}\|^2 \|Ar_{j-1}^{\mathrm{MR}}\|^2} \leq 1 - \inf_{v \in \mathcal{H}} \frac{|(v, Av)|^2}{\|v\|^2 \|Av\|^2} =: \sin^2(\gamma(A)),$$

where $\gamma(A)$ is the largest angle between a nonzero vector $v \in \mathcal{H}$ and its image Av, one thus obtains the bound $\sin^m \gamma(A)$ on the residual reduction after m steps. The angle $\gamma(A)$ was introduced by Wielandt (1996) (see also Gustafson and Rao (1996)).

6.2. Error bounds

When solving a linear system (1.1) approximately using successive iterates x_m, the residual $r_m = b - Ax_m$ may be the only computable indication of the progress of the solution process. The quantity of primary interest, however, is the error $e_m = x - x_m = A^{-1} r_m$.

For Krylov subspace methods, we have $x_m \in x_0 + \mathcal{K}_m$, so that $e_m = e_0 - v$ for some $v \in \mathcal{K}_m$. Of course, the best one could do is to select this $v \in \mathcal{K}_m$ as the best approximation to e_0 from $\mathcal{K}_m(A, r_0) = A\mathcal{K}_m(A, e_0)$. This would correspond to computing the MR approximation of e_0 with respect to the sequence of spaces $\mathcal{W}_m = A\mathcal{K}_m(A, e_0)$, a process which would require knowledge of the initial error and hence the solution x. The relation between residuals and errors, however, allows us to bound the error of the MR approximation with respect to this best possible approximation.

Lemma 6.4. The error e_m^{MR} of the MR approximation satisfies

$$\|e_m^{\mathrm{MR}}\| \le \kappa(A) \inf_{v \in \mathcal{K}_m} \|e_0 - v\|, \tag{6.5}$$

where $\kappa(A) = \|A\| \, \|A^{-1}\|$ denotes the condition number of A.

Proof. We have

$$\|r_m^{\mathrm{MR}}\| = \min_{w \in A\mathcal{K}_m} \|r_0 - w\| = \min_{v \in \mathcal{K}_m} \|A(e_0 - v)\| \le \|A\| \min_{v \in \mathcal{K}_m} \|e_0 - v\|,$$

and thus the assertion follows from $\|e_m^{\mathrm{MR}}\| = \|A^{-1} r_m^{\mathrm{MR}}\| \le \|A^{-1}\| \, \|r_m^{\mathrm{MR}}\|$. □

Thus, the error of the MR approximation is within the condition number of A of the error of the best approximation to e_0 from the Krylov space. In view of the relation (3.16), this translates to the following bound for the OR error:

$$\|e_m^{\mathrm{OR}}\| \le \frac{\kappa(A)}{c_m} \inf_{v \in \mathcal{K}_m} \|e_0 - v\|. \tag{6.6}$$

However, a stronger bound can be obtained if the field of values of A is bounded away from the origin.

Theorem 6.5. If $\nu(A) = \inf\{|z| : z \in W(A)\} > 0$, then the Krylov OR error satisfies

$$\frac{\|e_m^{\mathrm{OR}}\|}{\|e_0\|} \le \frac{\|A\|}{\nu(A)} \inf_{v \in \mathcal{K}_m} \|e_0 - v\|.$$

Proof. From the characterization of the OR approximation we have $e_m^{OR} = e_0 - v$ for some $v \in \mathcal{K}_m(A, r_0)$ subject to

$$r_m^{OR} = Ae^{OR} \perp \mathcal{K}_m(A, r_0) \quad \Leftrightarrow \quad e_m^{OR} \perp A^*\mathcal{K}_m(A, r_0).$$

This means that the OR error is obtained as the error of an OR approximation of e_0 from the space $\mathcal{K}_m(A, r_0)$ orthogonal to $A^*\mathcal{K}_m(A, r_0)$. Thus, $e_m^{OR} = (I - P_{\mathcal{K}_m}^{A^*\mathcal{K}_m})e_0$ and Theorem 2.9 implies

$$\frac{\|e_m^{OR}\|}{\|e_0\|} \le \|I - P_{\mathcal{K}_m}^{A^*\mathcal{K}_m}\| = \frac{1}{\cos \angle(\mathcal{K}_m, A^*\mathcal{K}_m)}.$$

We bound the cosine of the largest canonical angle between $\mathcal{K}_m$ and $A^*\mathcal{K}_m$ by

$$\cos^2 \angle(\mathcal{K}_m, A^*\mathcal{K}_m) = \sup_{u\in\mathcal{K}_m} \sup_{v\in\mathcal{K}_m} \frac{|(u, A^*v)|^2}{\|u\|^2 \|A^*v\|^2}$$

$$\ge \sup_{v\in\mathcal{K}_m} \frac{|(v, A^*v)|^2}{\|v\|^2 \|A^*v\|^2} = \sup_{v\in\mathcal{K}_m} \frac{(Av, v)}{(v, v)} \frac{(A^*v, v)}{(A^*v, A^*v)}$$

$$\ge \inf_{v\in\mathcal{K}_m} \left|\frac{(Av, v)}{(v, v)}\right| \inf_{v\in\mathcal{K}_m} \left|\frac{(A^*v, v)}{(A^*v, A^*v)}\right| \ge \nu(A)\nu((A^*)^{-1})$$

$$\doteq \nu(A)\nu(A^{-1}) \ge \frac{\nu(A)^2}{\|A\|^2},$$

where the last inequality has already been used to prove (6.4). $\square$

If the Hermitian part $H := (A + A^*)/2$ of A is positive definite, that is, if A is positive real, then $(H\cdot, \cdot)$ is an inner product and thus defines a norm $\|\cdot\|_H$ on $\mathcal{H}$. The next theorem, which is due to Starke (1994), shows that the OR error measured in this norm is optimal up to a factor which depends on the skew-Hermitian part $S := (A - A^*)/2$ of A.

Theorem 6.6. If A is positive real with Hermitian and skew-Hermitian parts H and S, then the Krylov OR error satisfies

$$\|e_m^{OR}\|_H \le (1 + \rho(H^{-1}S)) \inf_{v\in\mathcal{K}_m} \|e_0 - v\|_H,$$

where $\rho(H^{-1}S)$ denotes the spectral radius of $H^{-1}S$.

Proof. Since $r_m^{OR} = Ae_m^{OR} \perp \mathcal{K}_m$, noting that $(Hv, v) = \mathrm{Re}\,(Av, v)$, $(Sv, v) = i\mathrm{Im}\,(Av, v)$ and $e_m^{OR} - e_0 \in \mathcal{K}_m$, we have

$$\|e_m^{OR}\|_H^2 = (He_m^{OR}, e_m^{OR})$$
$$\le |(Ae_m^{OR}, e_m^{OR})| = |(Ae_m^{OR}, e_0)| = |(Ae_m^{OR}, e_0 - v)|$$

for arbitrary $v \in \mathcal{K}_m$, and therefore

$$\|e_m^{OR}\|_H^2 \leq |(H e_m^{OR}, e_0 - v) + (S e_m^{OR}, e_0 - v)|.$$

The first term is bounded by $\|e_0 - v\|_H \|e_m^{OR}\|_H$, and for the second term we obtain

$$\begin{aligned} |(S e_m^{OR}, e_0 - v)| &= |(H^{1/2} H^{-1/2} S H^{-1/2} H^{1/2} e_m^{OR}, e_0 - v)| \\ &\leq \|H^{-1/2} S H^{-1/2} H^{1/2} e_m^{OR}\| \|e_0 - v\|_H \\ &\leq \|H^{-1/2} S H^{-1/2}\| \|e_m^{OR}\|_H \|e_0 - v\|_H \\ &= \rho(H^{-1} S) \|e_m^{OR}\|_H \|e_0 - v\|_H, \end{aligned}$$

which, together with the bound for the first term, yields the assertion. □

An immediate consequence of Theorem 6.6 is that, for a Hermitian positive definite operator A, the OR method, which simplifies to the well-known conjugate gradient method in this case, yields the smallest possible error in the A-norm.

Remark 6.7. In view of the remark preceding Lemma 6.4 that the best approximation of the initial error e_0 from the Krylov space $\mathcal{K}_m(A, r_0) = A\mathcal{K}_m(A, e_0)$ has the same structure as the best approximation of r_0 from $A\mathcal{K}(A, r_0)$ with r_0 replaced by e_0, the infimum in (6.5) and (6.6) may be bounded in an analogous manner to the MR residual in Theorem 6.1, Corollary 6.2 and Remark 6.3.

6.3. Quantities that determine the rate of convergence

We begin by recalling that, as a result of Theorem 3.4, the residual norm history of the MR and OR methods is completely determined by the sequence of angles between the spaces $\mathcal{V}_m$ and $\mathcal{W}_m$, which specialize to $\mathcal{K}_m(A, r_0)$ and $A\mathcal{K}_m(A, r_0)$, respectively, for Krylov subspace methods. In other words, convergence depends only on the sequence of these spaces and their relative position. Furthermore, Theorem 4.5 and the ensuing discussion revealed that the sines and cosines of these angles appear as the parameters of the Givens rotations in the recursive construction of the QR-factorizations (4.12) of the Hessenberg matrices $\tilde{H}_m$, $m = 1, \ldots, L$. As a consequence, we note that all the information regarding the progress of the solution algorithm is contained in the Q-factors. In particular, the residual norm history of the MR/OR approximations associated with the Arnoldi relations

$$A V_m = V_{m+1} \tilde{H}_m = V_{m+1} Q_m \begin{bmatrix} R_m \\ 0 \end{bmatrix}, \qquad m = 1, \ldots, L,$$

is independent of the matrices R_m. For later reference, we state this observation as follows.

Lemma 6.8. Let L denote the termination index (5.7), and

$$AV_L = V_L H_L = V_L Q_{L-1}^H R_L = \hat{V}_L R_L$$

the Arnoldi decomposition (5.11). Further, let $\tilde{R}_L$ be any nonsingular upper triangular matrix and define $\tilde{A}$ by

$$\tilde{A} V_L = V_L Q_{L-1}^H \tilde{R}_L = \hat{V}_L \tilde{R}_L.$$

Then A and $\tilde{A}$ are *MR-equivalent* in the sense that the Krylov MR method produces identical residual vectors r_m^{MR} for both systems $Ax = b$ and $\tilde{A}x = b$ provided the starting vectors x_0 and $\tilde{x}_0$ are chosen such that $b - Ax_0 = b - \tilde{A}\tilde{x}_0$.

Furthermore, Lemma 6.8 reveals that there is a particularly simple matrix $\tilde{A}$ which, for the same initial residual r_0, displays the identical MR/OR residual norm history, namely the unitary matrix obtained by setting R_L (and thus all $\{R_j\}_{j=1}^L$) equal to the identity. If $\mathcal{V}_L$ is a proper subspace of $\mathcal{H}$, $\tilde{A}$ may be made unique by extending it to be the identity on the complementary space. This observation, that for each linear system of equations there exists a linear system of equations with a unitary matrix for which MR/OR display the identical convergence behaviour, was first pointed out by Greenbaum and Strakoš (1994).

Lemma 6.8 is the point of departure for an approach for computing convergence bounds for GMRES given in Liesen (2000). This is obtained by applying the standard convergence bound (6.1) to the MR-equivalent matrix Q_{L-1}, which yields

$$\| r_m^{\mathrm{MR}} \| \le \min_{\substack{p \in \mathcal{P}_m \\ p(0)=1}} \max_{\lambda \in \Lambda(Q_{L-1})} |p(\lambda)|.$$

(We note that Liesen used somewhat more complicated MR-equivalent systems.) Since such a bound involves quantities not available until step L, which generally far exceeds the feasible number of iteration steps, Liesen suggests approximating the quantities in the above bound by those available at iteration step $m \ll L$. It is, however, clear that this works only under additional assumptions since, in general, Q_m need not contain any information about the progress of the MR iteration beyond step m. It is not difficult to show that any given sequence of residual norms $\|r_0\| \ge \|r_1^{\mathrm{MR}}\| \ge \cdots \ge \|r_m^{\mathrm{MR}}\|$ can be complemented in such a way that the iteration stagnates from step $m+1$ until step $L-1$, that is, $\|r_m^{\mathrm{MR}}\| = \cdots = \|r_{L-1}^{\mathrm{MR}}\|$.

As another consequence of Lemma 6.8, quantities such as the singular values of $\tilde{H}_m$, which coincide with those of R_m, can play no role in the rate of convergence.

Lemma 6.9. For any given linear system of equations (1.1), there is a linear system with coefficient matrix $\tilde{A}$ for which GMRES exhibits the identical convergence history, and for which $\tilde{A}$ has arbitrarily prescribed nonzero singular values.

Proof. Let L denote the termination index of A with respect to the given system as defined in (5.7). For any Krylov subspace method applied to this system, only the L singular values of $A|_{\mathcal{K}_L}$ are noticeable, hence A can be defined arbitrarily on $\mathcal{K}_L^\perp$. To prescribe the singular values of $A|_{\mathcal{K}_L}$, let $\tilde{R}_L$ denote an upper triangular $L \times L$ matrix possessing L arbitrary nonzero singular values. With $\hat{V}_L$ denoting the Paige–Saunders basis (4.17), the matrix $\tilde{A}$ by $\tilde{A}\hat{V}_L = \hat{V}_L\tilde{R}_L$ clearly possesses the same set of singular values and, by Proposition 6.8, $\tilde{A}$ is MR-equivalent to A. $\square$

By the same technique, we can prescribe, for fixed m, the singular values of $\tilde{H}_m$ since, in view of (4.12), these coincide with those of R_m.

For the singular values of the square Hessenberg matrices H_m there is a slight complication. From (4.19) we recall that a QR-factorization of H_m is given by

$$H_m = Q_{m-1}^H \begin{bmatrix} R_{m-1} & r \\ 0 & \tau \end{bmatrix}.$$

The mth plane rotation is determined so that the vector $[\tau \ \eta_{m+1,m}]^\top$ is rotated to the vector $[r_{m,m} \ 0]^\top$, where $r_{m,m}$ is the entry in the (m,m)-position of R_m (cf. (4.15)). This implies that $\tau = c_m r_{m,m}$. Thus, if we prescribe R_m to be a diagonal matrix, then the singular values of H_m are given by $|r_{1,1}|, \ldots, |r_{m-1,m-1}|, c_m|r_{m,m}|$ and thus can be selected arbitrarily. The only exception occurs when $c_m = 0$, that is, H_m is singular and clearly only $m - 1$ singular values can be chosen freely.

We remark that the same proof shows that, also in the case of an OR method, neither the singular values of A nor those of $\tilde{H}_m$ or H_m determine the convergence behaviour.

We conclude this subsection with a brief discussion of the role eigenvalues play for the convergence of MR methods. The bound (6.1) shows that the spectrum controls the convergence behaviour if A is normal. There are, however, examples of non-normal matrices which show that, in general, $\Lambda(A)$ may have no influence: Greenbaum, Strakoš and Ptak (1996) demonstrate that, for *any* nonincreasing finite sequence of positive real numbers $\rho_0 \geq \rho_1 \geq \cdots \geq \rho_{n-1}$ and *any* choice of (not necessarily distinct) nonzero complex numbers $\lambda_1, \lambda_2, \ldots, \lambda_n$, one can construct a matrix $A \in \mathbb{C}^{n \times n}$ and an initial residual r_0 with $\Lambda(A) = \{\lambda_1, \lambda_2, \ldots, \lambda_n\}$ and $\|r_m^{\mathrm{MR}}\| = \rho_m$ ($m = 0, 1, \ldots, n-1$). We illustrate this result by one of their striking examples. Any matrix

A in Frobenius form,

$$A = \begin{bmatrix} 0 & 0 & \cdots & 0 & -\alpha_0 \\ 1 & 0 & \cdots & 0 & -\alpha_1 \\ 0 & 1 & \cdots & 0 & -\alpha_2 \\ \vdots & & \ddots & \vdots & \vdots \\ 0 & 0 & \cdots & 1 & -\alpha_{n-1} \end{bmatrix} \in \mathbb{C}^{n \times n},$$

has $\zeta^n + \alpha_{n-1}\zeta^{n-1} + \cdots + \alpha_1\zeta + \alpha_0$ as its characteristic polynomial, so its eigenvalues can be arbitrarily prescribed. If we choose b and r_0 such that $r_0 = u_1$ is the first unit vector, then, for $m = 1, 2, \ldots, n - 1$, the approximation space $A\mathcal{K}_m(A, r_0)$ is the span of the unit vectors $u_2, u_3, \ldots, u_m$. The best approximation to r_0 from this space is obviously the null vector leading to $\|r_0\| = \|r_1^{\mathrm{MR}}\| = \cdots = \|r_{n-1}^{\mathrm{MR}}\| = 1$ independently of the chosen spectrum. In general it is therefore impossible to predict the convergence behaviour of an MR method such as GMRES (and of any other Krylov subspace method) on the basis of the eigenvalue distribution of A alone. Although this fact has been emphasized in several recent papers, it is still a widespread but nonetheless incorrect belief that spectral properties of the coefficient matrix (*i.e.*, without any additional assumptions on its departure from normality) determine the speed of convergence of GMRES.

6.4. An application: compact operators

Many applications such as the solution of elliptic boundary value problems by the integral equation method require the solution of second-kind Fredholm equations, that is, operator equations (1.1) in which A has the form $A = \lambda I + K$ with $\lambda \neq 0$ and $K : \mathcal{H} \to \mathcal{H}$ is a compact operator. The development of fast multiplication algorithms (*cf.* Greengard and Rokhlin (1987), Hackbusch and Nowak (1989)) has made Krylov subspace methods attractive as solution algorithms for discretizations of these problems, since they require only applications of the (discrete) operator to vectors. Moreover, as shown by Moret (1997) for GMRES and by Winther (1980) for CG, Krylov subspace methods converge q-superlinearly for operator equations involving compact perturbations of (multiples of) the identity.

The reason for this is that, for these operators, the sines s_m of the canonical angles between the Krylov space $\mathcal{K}_m$ and $A\mathcal{K}_m$ converge to zero. To show this, we recall a basic result on compact operators and orthonormal systems.

Theorem 6.10. Let $K : \mathcal{H} \to \mathcal{H}$ be a compact linear operator and $\{v_m\}_{m \geq 1} \subset \mathcal{H}$ be an orthonormal system. Then

$$\lim_{m \to \infty} (K v_m, v_{m+1}) = 0.$$

Proof. See, for example, Ringrose (1971). ☐

The next lemma, which is due to Moret (1997), gives a bound on the quantity $s_m = \sin \angle(\mathcal{K}_m, A\mathcal{K}_m)$.

Lemma 6.11. Let $\{v_j\}_{j\geq 1}$ be the Arnoldi basis defined in (5.11) of $\mathcal{K}_m$, where $A : \mathcal{H} \to \mathcal{H}$ possesses a bounded inverse. Then
$$s_m \leq \|A^{-1}\| \, (v_{m+1}, Av_m).$$

Proof. Let $\{w_1, \ldots, w_m\}$ denote an ascending orthonormal basis of $A\mathcal{K}_m$. Since $A^{-1}w_m \in \mathcal{K}_m$, we can write $(v_{m+1}, w_m) = (v_{m+1}, AP_{\mathcal{K}_m}A^{-1}w_m)$. Moreover, since $Av_j \in \mathcal{K}_m \perp v_{m+1}$ for $1 \leq j \leq m-1$, we have
$$|(v_{m+1}, AP_{\mathcal{K}_m}A^{-1}w_m)| = |(v_{m+1}, \sum_{j=1}^{m}(A^{-1}w_m, v_j)Av_j)|$$
$$= |(v_{m+1}, (A^{-1}w_m, v_m)Av_m)|$$
$$= |(A^{-1}w_m, v_m)(v_{m+1}, Av_m)|$$
$$\leq \|A^{-1}\| \, (Av_m, v_{m+1}).$$

Note that the modulus in $|(Av_m, v_{m+1})|$ can be omitted since $(Av_m, v_{m+1}) = \|(I - P_{v_m})Av_m\| \geq 0$. The assertion now follows from Lemma 3.3. ☐

Corollary 6.12. Let $A = \lambda I + K$ with $\lambda \neq 0$ and $K : \mathcal{H} \to \mathcal{H}$ compact, let $\{v_j\}_{j\geq 1}$ denote the Arnoldi basis of $\mathcal{K}_m$. Then the sines s_m of largest canonical angle between $\mathcal{K}_m$ and $A\mathcal{K}_m$ form a null sequence.

Proof. Lemma 6.11 and Theorem 6.10 yield
$$s_m \leq (v_{m+1}, Av_m)\|A^{-1}\| = (v_{m+1}, Kv_m)\|A^{-1}\| \to 0,$$
since A^{-1} is bounded whenever $\lambda \neq 0$. ☐

In particular, since $s_m \to 0$ implies that $s_m < 1$ for m sufficiently large, this means that the OR approximation is always defined except for possibly a finite number of indices. Moreover, as s_m is bounded away from one, c_m is accordingly bounded away from zero, hence the relation (3.16) also implies the q-superlinear convergence of the OR approximation. We summarize this result in the following theorem.

Theorem 6.13. Given $K : \mathcal{H} \to \mathcal{H}$ compact, $0 \neq \lambda \in \mathbb{C}$ and $b \in \mathcal{H}$, let $x_0 \in \mathcal{H}$ be an initial guess at the solution of (1.1) with $A = \lambda I + K$. Then the OR approximation with respect to the space $\mathcal{K}_m$ exists for all sufficiently large m. Moreover, the sequence of MR and OR approximations converge q-superlinearly.

We remark that the rate of superlinear convergence may be quantified in terms of the rate of decay of the singular values of K (see Moret (1997)). We also note, in view of (4.24), that this result applies to all MR/OR pairs of Krylov subspace methods including QMR/BCG, given a bound on the conditioning of the basis of the Krylov space being used. For bases generated by the look-ahead Lanczos method such bounds are guaranteed, for instance, by the implementation given in Freund (1993).

7. Conclusions and final remarks

We have presented a unifying framework for describing and analysing Krylov subspace methods by first introducing the abstract MR and OR approximation methods on nested sequences of subspaces, applying these to solving equations and then specializing further to the Krylov subspace setting. All known relations between MR/OR-type Krylov methods were shown to hold in the abstract formulation. In particular, the angles appearing in the Givens QR factorization of the Hessenberg matrix used in many Krylov subspace algorithms were identified as angles between the Krylov spaces and their images under A. Moreover, depending on whether orthogonal or non-orthogonal bases are employed, both MR/OR and QMR/QOR methods can be described and analysed in the same manner. Furthermore, we have shown that essentially all nested approximation schemes – and therefore also essentially all Krylov subspace methods – can be interpreted as QMR/QOR methods. The description of the algorithms in terms of angles was subsequently used to derive some of the previously known error and residual bounds.

Another benefit of the analysis in this paper is that, at least conceptually, it separates the issue of generating bases from the method for computing the approximations. Indeed, other algorithms besides Lanczos or Arnoldi could be used to generate the bases required for the MR and OR methods, but this is seldom done for lack of promising alternatives: Le Calvez and Saad (1999) introduce a nonstandard inner product and develop a QMR-like algorithm not based on the Lanczos algorithm.

Besides the basis-dependent inner product of QMR/QOR approximations, it can also be advantageous to use other fixed inner products other than the Euclidean inner product on the coordinate space. For an example of a non-standard inner product used in conjunction with GMRES see Starke (1997).

We have made no mention in this paper of preconditioning, which is indispensable for most problems of practical relevance; when preconditioning is accounted for, our results apply to the preconditioned system.

We believe that our approach provides a simple and intuitive way of describing Krylov subspace algorithms which simplifies many of the standard proofs and brings out the connections among the many algorithms in the literature.

Acknowledgements

The authors would like to thank Howard Elman for many helpful remarks and G. W. Stewart for his observation on QMR recovery.

REFERENCES

S. F. Ashby, T. A. Manteuffel and P. E. Saylor (1990), 'A taxonomy for conjugate gradient methods', *SIAM J. Numer. Anal.* **27**, 1542–1568.

O. Axelsson (1994), *Iterative Solution Methods*, Cambridge University Press, Cambridge.

T. Barth and T. A. Manteuffel (1994), Variable metric conjugate gradient algorithms, in *Advances in Numerical Methods for Large Sparse Sets of Linear Systems* (M. Natori and T. Nodera, eds), Vol. 10 of *Parallel Processing for Scientific Computing*, Keio University, Yokohama, Japan, pp. 165–188.

Å. Björck and G. H. Golub (1973), 'Numerical methods for computing angles between linear subspaces', *Math. Comp.* **27**, 579–594.

S. C. Brenner and L. R. Scott (1994), *The Mathematical Theory of Finite Element Methods*, Vol. 15 of *Texts in Applied Mathematics*, Springer, New York.

P. N. Brown (1991), 'A theoretical comparison of the Arnoldi and GMRES algorithms', *SIAM J. Sci. Statist. Comput.* **12**, 58–77.

A. M. Bruaset (1995), *A Survey of Preconditioned Iterative Methods*, Vol. 328 of *Pitman Research Notes in Mathematics*, Longman Scientific and Technical, Harlow.

T. F. Chan, E. Gallopoulos, V. Simoncini, T. Szeto and C. H. Tong (1994), 'A quasi-minimal residual variant of the Bi-CGSTAB algorithm for nonsymmetric systems', *SIAM J. Sci. Comput.* **15**, 338–347.

F. Chatelin (1993), *Eigenvalues of Matrices*, Wiley, New York.

J. Cullum and A. Greenbaum (1996), 'Relations between Galerkin and norm-minimizing iterative methods for solving linear systems.', *SIAM J. Matrix Anal. Appl.* **17**, 223–247.

C. Davis and W. M. Kahan (1970), 'The rotation of eigenvectors by a perturbation, III', *SIAM J. Numer. Anal.* **7**, 1–46.

M. Eiermann (1993), 'Fields of values and iterative methods', *Linear Algebra Appl.* **180**, 167–197.

M. Eiermann, O. G. Ernst and O. Schneider (2000), 'Analysis of acceleration strategies for restarted minimal residual methods', *J. Comput. Appl. Math.* **123**, 261–292.

T. Eirola and O. Nevanlinna (1989), 'Accelerating with rank-one updates', *Linear Algebra Appl.* **121**, 511–520.

S. C. Eisenstat, H. C. Elman and M. H. Schultz (1983), 'Variational iterative methods for nonsymmetric systems of linear equations', *SIAM J. Sci. Comput.* **20**, 345–357.

H. C. Elman (1982), Iterative methods for sparse, nonsymmetric systems of linear equations, PhD thesis, Yale University, Department of Computer Science.

M. Embree (1999), How descriptive are GMRES convergence bounds, Technical Report 99/08, Oxford University Computing Laboratory. Available from: http://web.comlab.ox.ac.uk/oucl/work/mark.embree/.

O. G. Ernst (2000), 'Residual-minimizing Krylov subspace methods for stabilized discretizations of convection-diffusion equations.', *SIAM J. Matrix Anal. Appl.* **21**, 1079–1101.

B. Fischer (1996), *Polynomial Based Iteration Methods for Symmetric Linear Systems*, Wiley-Teubner, Leipzig.

R. W. Freund (1992), 'Quasi-kernel polynomials and their use in non-Hermitian matrix iterations', *J. Comput. Appl. Math.* **43**, 135–158.

R. W. Freund (1993), 'A transpose-free quasi-minimal residual algorithm for non-Hermitian linear systems', *SIAM J. Sci. Comput.* **14**, 470–482.

R. W. Freund (1994), Transpose-free quasi-minimal residual methods for non-Hermitian linear systems, in *Recent Advances in Iterative Methods* (G. Golub, A. Greenbaum and M. Luskin, eds), Vol. 60 of *Mathematics and its Applications*, IMA, Springer, New York, pp. 69–94.

R. W. Freund and N. M. Nachtigal (1991), 'QMR: a quasi-minimal residual method for non-Hermitian linear systems', *Numer. Math.* **60**, 315–339.

R. W. Freund, G. H. Golub and N. M. Nachtigal (1992), Recent advances in iterative methods, in *Acta Numerica* (A. Iserles, ed.), Vol. 1, Cambridge University Press, Cambridge, pp. 57–100.

R. W. Freund, M. Gutknecht and N. M. Nachtigal (1993), 'An implementation of the look-ahead Lanczos algorithm for non-Hermitian matrices', *SIAM J. Sci. Comput.* **14**, 137–158.

G. H. Golub and H. A. van der Vorst (1997), Closer to the solution: iterative linear solvers, in *The State of the Art in Numerical Analysis* (I. S. Duff and G. A. Watson, eds), Clarendon Press, Oxford, pp. 63–92.

G. H. Golub and C. F. Van Loan (1996), *Matrix Computations*, 3rd edn, Johns Hopkins University Press.

A. Greenbaum (1997), *Iterative Methods for Solving Linear Systems*, Vol. 17 of *Frontiers in Applied Mathematics*, SIAM, Philadelphia, PA.

A. Greenbaum and Z. Strakoš (1994), Matrices that generate the same Krylov residual spaces, in *Recent Advances in Iterative Methods* (G. H. Golub, A. Greenbaum and M. Luskin, eds), Springer, New York, pp. 95–118.

A. Greenbaum, M. Rozložník and Z. Strakoš (1997), 'Numerical behaviour of the modified Gram-Schmidt GMRES implementation', *BIT* **37**, 706–719.

A. Greenbaum, Z. Strakoš and V. Ptak (1996), 'Any nonincreasing convergence curve is possible for GMRES', *SIAM J. Matrix Anal. Appl.* **17**, 465–469.

L. F. Greengard and V. Rokhlin (1987), 'A fast algorithm for particle simulations', *J. Comput. Phys.* **73**, 325–348.

T. N. E. Greville (1974), 'Solutions of the matrix equation $XAX = X$ and relations between oblique and orthogonal projectors', *SIAM J. Appl. Math.* **26**, 828–832.

K. E. Gustafson and D. K. M. Rao (1996), *Numerical Range: the Field of Values of Operators and Matrices*, Springer, New York.

M. H. Gutknecht (1993*a*), 'Changing the norm in conjugate gradient type algorithms', *SIAM J. Numer. Anal.* **30**, 40–56.

M. H. Gutknecht (1993*b*), 'Variants of BiCGStab for matrices with complex spectrum', *SIAM J. Sci. Statist. Comput.* **14**, 1020–1033.

M. H. Gutknecht (1997), Lanczos-type solvers for nonsymmetric linear systems of equations, in *Acta Numerica* (A. Iserles, ed.), Vol. 6, Cambridge University Press, Cambridge, pp. 271–397.

W. Hackbusch and Z. P. Nowak (1989), 'On the fast matrix multiplication in the boundary element method by panel clustering', *Numer. Math.* **54**, 463–491.

L. A. Hageman and D. M. Young (1981), *Applied Iterative Methods*, Academic Press, New York.

M. R. Hestenes and E. Stiefel (1952), 'Methods of conjugate gradients for solving linear systems', *J. Res. Nat. Bur. Standards* **49**, 409–436.

M. Hochbruck and C. Lubich (1998), 'Error analysis of Krylov methods in a nutshell', *SIAM J. Sci. Comput.* **19**, 695–701.

R. A. Horn and C. R. Johnson (1991), *Topics in Matrix Analysis*, Cambridge University Press, Cambridge, UK.

A. S. Householder (1964), *The Theory of Matrices in Numerical Analysis*, Dover, New York.

W. D. Joubert (1994), 'On the convergence behavior of the restarted GMRES algorithm for solving nonsymmetric linear systems', *Numer. Lin. Alg. Appl.* **1**, 427–447.

T. Kato (1960), 'Estimation of iterated matrices, with application to the von Neumann condition', *Numer. Math.* **2**, 22–29.

A. N. Krylov (1931), 'On the numerical solution of the equation by which the frequency of small oscillations is determined in technical problems', *Isz. Akad. Nauk SSSR Ser. Fiz.-Math.* **4**, 491–539.

C. Lanczos (1950), 'An iteration method for the solution of the eigenvalue problem of linear differential and integral operators', *J. Res. Nat. Bur. Standards* **45**, 255–282.

C. Lanczos (1952), 'Solution of systems of linear equations by minimized iterations', *J. Res. Nat. Bur. Standards* **49**, 33–53.

C. Le Calvez and Y. Saad (1999), 'Modified Krylov acceleration for parallel environments', *Appl. Numer. Math.* **30**, 191–212.

J. Liesen (2000), 'Computable convergence bounds for GMRES', *SIAM J. Matrix Anal. Appl.* **21**, 882–903.

G. Meurant (1999), *Computer Solution of Large Linear Systems*, Vol. 28 of *Studies in Mathematics and its Applications*, Elsevier, Amsterdam.

I. Moret (1997), 'A note on the superlinear convergence of GMRES', *SIAM J. Numer. Anal.* **34**, 513–516.

R. B. Morgan (2000), 'Implicitly restarted GMRES and Arnoldi methods for nonsymmetric systems of equations', *SIAM J. Matrix Anal. Appl.* **21**, 1112–1135.

C. C. Paige and M. A. Saunders (1975), 'Solution of sparse indefinite systems of linear equations', *SIAM J. Numer. Anal.* **12**, 617–629.

K. J. Ressel and M. H. Gutknecht (1998), 'QMR-smoothing for Lanczos-type product methods based on three-term recurrences', *SIAM J. Sci. Comput.* **19**, 55–73.

J. R. Ringrose (1971), *Compact Non-Self-Adjoint Operators*, Van Nostrand Reinhold Company, London.

M. Rozložník and Z. Strakoš (1996), Variants of the residual minimizing Krylov space methods, in *Proceedings of the XI. School 'Software and Algorithms of Numerical Mathematics', Zelezna Ruda, University of West Bohemia* (I. Marek, ed.), pp. 208–225.

M. Rozložník, Z. Strakoš and M. Tůma (1996), On the role of orthogonality in the GMRES method, in *SOFSEM'96: Theory and Practice of Informatics* (K. Jeffery, J. Kral and M. Bartosek, eds), Vol. 1175 of *Lecture Notes in Computer Science*, Springer, Berlin, pp. 409–416.

Y. Saad (1981), 'Krylov subspace methods for solving large unsymmetric linear systems', *Math. Comp.* **37**, 105–126.

Y. Saad (1982), 'The Lanczos biorthogonalization algorithm and other oblique projection methods for solving large unsymmetric systems', *SIAM J. Numer. Anal.* **19**, 470–484.

Y. Saad (1993), 'A flexible inner-outer preconditioned GMRES algorithm', *SIAM J. Sci. Comput.* **14**, 461–469.

Y. Saad (1996), *Iterative Methods for Sparse Linear Systems*, PWS Publishing, Boston.

Y. Saad (2000), 'Further analysis of minimum residual iterations', *Numer. Lin. Alg. Appl.* **7**, 67–93.

Y. Saad and M. H. Schultz (1986), 'GMRES: A generalized minimal residual algorithm for solving nonsymmetric linear systems', *SIAM J. Sci. Comput.* **7**, 856–869.

P. Sonneveld (1989), 'CGS, a fast Lanczos-type solver for nonsymmetric linear systems.', *SIAM J. Sci. Statist. Comput.* **10**, 36–52.

G. Starke (1994), 'Iterative methods and decomposition-based preconditioners for nonsymmetric elliptic boundary value problems', Habilitationsschrift, Universität Karlsruhe.

G. Starke (1997), 'Field-of-values analysis of preconditioned iterative methods for nonsymmetric elliptic problems', *Numer. Math.* **78**, 103–117.

G. W. Stewart (1998), *Matrix Algorithms, Vol. I: Basic Decompositions*, SIAM, Philadelphia, PA.

E. L. Stiefel (1958), 'Kernel polynomials in linear algebra and their numerical applications', *J. Res. Nat. Bur. Standards, Appl. Math. Ser.* **49**, 1–22.

E. de Sturler (1999), 'Truncation strategies for optimal Krylov subspace methods', *SIAM J. Numer. Anal.* **36**, 864–889.

L. N. Trefethen (1992), Pseudospectra of matrices, in *Numerical Analysis 1991* (D. F. Griffiths and G. A. Watson, eds), Longman, pp. 234–266.

R. S. Varga (1999), *Matrix Iterative Analysis*, 2nd edn, Springer, Berlin.

H. A. van der Vorst (1992), 'BICGSTAB: a fast and smoothly converging variant of Bi-CG for the solution of nonsymmetric linear systems', *SIAM J. Sci. Statist. Comput.* **13**, 631–644.

H. A. van der Vorst and Y. Saad (2000), 'Iterative solution of linear systems in the 20th century', *J. Comput. Appl. Math.* **123** 1–33.

H. A. van der Vorst and C. Vuik (1994), 'GMRESR: a family of nested GMRES methods', *Numer. Lin. Alg. Appl.* **1**, 369–386.

H. F. Walker (1995), 'Residual smoothing and peak/plateau behavior in Krylov subspace methods', *Appl. Numer. Math.* **19**, 279–286.

H. Walker and L. Zhou (1994), 'A simpler GMRES', *Numer. Lin. Alg. Appl.* **1**, 571–581.

P. Å. Wedin (1983), On angles between subspaces of a finite dimensional inner product space, in *Matrix Pencils* (B. Kågström and A. Ruhe, eds), Vol. 273 of *Lecture Notes in Mathematics*, Springer, New York, pp. 263–285.

R. Weiss (1994), 'Properties of generalized conjugate gradient methods', *Numer. Lin. Alg. Appl.* **1**, 45–63.

R. Weiss (1997), *Parameter-free Iterative Linear Solvers*, Akademie Verlag, Berlin.

H. Wielandt (1996), Topics in the analytic theory of matrices, in *Helmut Wielandt, Mathematical Works* (B. Huppert and H. Schneider, eds), Vol. 2, Walter de Gruyter, Berlin, pp. 271–352.

R. Winther (1980), 'Some superlinear convergence results for the conjugate gradient method', *SIAM J. Numer. Anal.* **17**, 14–17.

L. Zhou and H. F. Walker (1994), 'Residual smoothing techniques for iterative methods', *SIAM J. Sci. Comput.* **15**, 297–312.

Acta Numerica (2001), pp. 313–355

Data mining techniques

Markus Hegland

Centre for Mathematics and its Applications,
School of Mathematical Sciences,
Australian National University,
Canberra ACT 0200, Australia

E-mail: `Markus.Hegland@anu.edu.au`

Methods for knowledge discovery in data bases (KDD) have been studied for more than a decade. New methods are required owing to the size and complexity of data collections in administration, business and science. They include procedures for data query and extraction, for data cleaning, data analysis, and methods of knowledge representation. The part of KDD dealing with the analysis of the data has been termed data mining. Common data mining tasks include the induction of association rules, the discovery of functional relationships (classification and regression) and the exploration of groups of similar data objects in clustering. This review provides a discussion of and pointers to efficient algorithms for the common data mining tasks in a mathematical framework. Because of the size and complexity of the data sets, efficient algorithms and often crude approximations play an important role.

CONTENTS

1. Introduction

The following is an attempt at an introduction and review of some current data mining techniques. The main focus is on the computational aspects of data processing aspects, and not on statistics, data management or data retrieval. Data mining is a new and rapidly growing field. It draws ideas and resources from several disciplines, including machine learning, statistics, database research, high-performance computing and commerce. This explains the multifaceted and rapidly evolving nature of the data mining discipline. While there is a broad consensus that the abstract goal of data mining is to *discover new and useful information in databases*, this is where the consensus ends, and the means of achieving this goal are as diverse as the communities contributing. Thus any reasonably sized treatment of data mining techniques necessarily has to be selective, and perhaps biased towards a particular approach. Despite this, we hope that the following discussion will provide useful information for readers wishing to get some understanding of ideas and challenges underlying a selection of data mining techniques. This selection includes some of the most widely used data mining problems such as association rule mining, predictive models, and clustering. The focus is on fundamental concepts and on the challenges posed by data size, dimensionality, and data complexity. It is hoped that this overview gives the computational mathematician, in particular, some starting points for further exploration. We believe that data mining does provide many new and challenging questions for approximation theory, stochastic analysis, numerical analysis and parallel algorithm design.

Data mining techniques are used to find patterns in large datasets. A necessary property of algorithms capable of handling large and growing datasets is their scalability, or linear complexity with respect to the data size. Patterns in the database are described by relations between the attributes. In a sense, a relational database itself defines a pattern. However, the size of the relations makes it impossible to use them directly for further predictions or decisions. On the other hand, these relations only provide information about the available observations and cannot be directly applied to future observations. The power of generalization from specific observation is obtained from statistics and machine learning.

The variables, or attributes, considered here are assumed to be either continuous or categorical. However, more general data types are frequently analysed in data mining; see Bock and Diday (2000). The techniques discussed here are not based on sampling, and access every item in the full dataset.

The different disciplines are also reflected in different goals of data mining techniques (Ramakrishnan and Grama 1999):

Induction.
Find general patterns and equations that characterize the data.
Compression.
Reduce the complexity of the data, replace by simpler concepts.
Querying.
Find better ways to query data, *i.e.*, extract information and properties.
Approximation.
Find models that approximate the observations well.
Search.
Look for recurring patterns.

Furthermore, we may classify data mining algorithms according to three key elements (Ramakrishnan and Grama 1999):

Model representation.

Decision trees, regression functions and associations.

Data.

Continuous, time series, discrete, labelled, multimedia or nominal data.

Application areas.

Finance and economy, biology, web logs, web text mining.

The discovered patterns can be characterized according to accuracy and precision, expressiveness, interpretability, parsimony, 'surprisingness', 'interestingness', or 'actionability' (Ramakrishnan and Grama 1999).

Assume that the initial raw data are stored in a relational database, that is, a collection of tables or relations. Each table contains a sequence of records, each of which consists of a number of attribute values. These tables are typically used for transactional purposes, that is, for the management of a business. For example, a health insurance company would store a table containing information on all the doctors, another for the patients, and maybe a third for claims, containing pointers into the other two tables. Each record in these tables contains information about the individuals: for example, the patient table would contain the age, sex and location of a patient. In order to guarantee consistency of the information and avoid redundancy, the tables are normalized (Date 1995).

Assume now that in a data mining project the claims of patients are to be further examined. In particular, we might be interested in differences between claims for different doctors of a particular specialization. Thus a first step is to restructure the original database into a database where each new 'record' now contains many records from several tables and the data mining investigation is to compare these 'records' in order to find patterns and outliers.

In a next step, major characteristics of these records are extracted. In our health insurance example this might be the number of patients, the total claim or the service offered by a particular doctor. These characteristics may be *symbolic objects* (Bock and Diday 2000) including sets, intervals and frequency distributions, but in this discussion we will mainly contain our discussion to *features* which are either continuous variables or categorical. The choice of these features is very important and often the main reason for failure or success of a data mining project. If, in our example, the features do not provide information about doctors and patients, the analysis of their interrelations will not provide any interesting insights. One data mining task is indeed the identification of features containing information that can contribute to a particular research question. We will assume that the features have been chosen, and focus here instead on the algorithms used to analyse them. Thus, after these preliminary steps, we are left with one table or relation which is a sequence of feature vectors, and the data mining task is to explore and describe how these feature vectors are interrelated.

From a statistical perspective the data are a sequence of independent vectors described by a probability distribution. The goal of data mining is then to uncover interesting structures or aspects of the underlying probability distribution. Questions which are addressed are as follows.

- Are there areas that have a higher probability? This is addressed by clustering techniques and association rules.
- Can some of the variables be explained by others, and how well? This is addressed by classification and regression.
- Where are the areas where these functional relationships are better/worse? The analysis of areas of high misclassification rates and the residuals of regression address this.

The sizes of databases analysed are growing exponentially. In fact, it has been suggested that data grow at the same rate as computational resources, which, according to Moore's law, double every 18 months (Bell and Gray 1997). Today data collections in the megabyte range are very common. Gigabyte data collections are becoming available, including many business data collections, some of which easily extend into the terabyte range, particularly when the world wide web is involved. Further, data collections in the petabyte range are now emerging. The largest challenge is not so much the absolute size of the databases but their constant growth. Thus one of the largest challenges in computer science in general is the generation of systems which are capable of handling such growing datasets, that is, the systems need to be *scalable* (Bell and Gray 1997). Even if hardware and software performance scale at the same rate, the computational complexity may destroy overall scalability. Indeed, if one were to use an $\mathcal{O}(n^3)$ algorithm, then a typical data mining task in 10 years time would require roughly 10,000 times as long compared to a similar typical data mining task today, as both the data size (n) and the computational speed would have increased about 100-fold during this period, but the computational complexity would have increased by 10^6. Thus a typical data mining task may take one hour today, but in 10 years time a similar typical data mining task would take over one year due to the combined effects of data growth, increase in computing speed and $\mathcal{O}(n^3)$ complexity. Thus it is essential that all data mining algorithms have near-linear time complexity with data size ($\mathcal{O}(n \log(n))$ is usually acceptable).

The areas selected for the current discussion are association rules (Section 2), classification and regression (Section 3) and clustering (Section 5). The reader wishing to get further pointers to these and other areas of data mining is encouraged to search for relevant literature on the web, a useful guide being the book on data mining by Han and Kamber (2001). In addition to discussing a vast collection of data mining research it also provides a database perspective. Association rule discovery is perhaps data mining at its purest, and the techniques, while possibly motivated by earlier work in rule discovery, have evolved within the data mining community. As this area is the newest of the three covered, the mathematical foundations are less well studied. In contrast, there has been a lot of theoretical work in the area of classification in the machine learning community: in particular, we would like to point to the book by Devroye, Györfi and Lugosi (1996). While linear regression is widely studied in statistics, non-parametric regression is still a relatively new field, particularly in the case of high-dimensional data. Clustering techniques have been used for many years and the literature on clustering is quite formidable. We will only reveal the tip of the iceberg here.

2. Association rules

The motivation for association rules comes from market basket analysis (see Agrawal, Imielinski and Swami (1993)). A market basket is a collection of items purchased by a customer in one transaction. (In the case of retail shops this is the content of a shopping trolley.) Typically we are faced with a large number of possible items and market baskets which only cover a small proportion of all items. It is observed that items are purchased in groups, and the aim of the market basket analysis is to detect these groups. More specifically, we would like to be able to predict groups of items that occur in a market basket given that certain items are present. This information is then used for product placement in shops and on web pages.

The underlying data are modelled as a sequence of data records or *transactions* $\omega \in \Omega$. It is assumed that the records are distributed according to a distribution function $p : \Omega \to \mathbb{R}$. Rules are a very popular way to express knowledge about the data because they are comprehensible. Rules consist of 'if-then clauses' which depend on the data. The probabilities are normalized; in the case of a finite set Ω we have

$$P(\Omega) = \sum_{\omega \in \Omega} p(\omega) = 1. \tag{2.1}$$

In this discussion it is assumed that the probability is time-invariant; in practice the distributions may vary. Market basket analysis that takes this into account is discussed in Ramaswamy, Mahajan and Silbershatz (1998).

In the case of association rules, the records ω are subsets of a set of items $J = i_1, \ldots, i_m$, which is here assumed to be finite. *Association rules* are statements about itemsets $A \subset J$ and $B \subset J$; in particular, the rule

$$A \Rightarrow B$$

states that, for many transactions ω, we may conclude $B \subset \omega$ if we know that $A \subset \omega$. Such a rule is only useful if $A \subset \omega$, and we say that a transaction ω supports A in this case. The probability that a randomly selected transaction ω contains A is the *support* $s(A)$:

$$s(A) := P\left(\{\omega : A \subset \omega \subset J\}\right). \qquad (2.2)$$

In practice, the support is obtained from the database as the ratio of transactions ω supporting A to the total number of transactions N:

$$s(A) \approx \frac{N_A}{N}, \qquad (2.3)$$

where $N_A = |\{\omega_i : A \subset \omega_i\}|$ is the number of transactions supporting A, and $|X|$ denotes the cardinality of the set X.

Often, there is no distinction made in practice between the probability and this ratio of frequencies, which is then also denoted with $s(A)$.

The support of the association rule $A \Rightarrow B$ is defined by

$$s(A \Rightarrow B) := s(A \cup B), \qquad (2.4)$$

the probability that both A and B are subsets of a random ω. For association rule mining, the user imposes a minimal support s_0 for any rule that is further explored, that is, $s(A \Rightarrow B) > s_0$. This limits the number of spurious rules generated.

A second measure of the 'interestingness' of a rule is the *confidence*

$$c(A \Rightarrow B) := \frac{s(A \cup B)}{s(A)}, \qquad (2.5)$$

which is the (conditional) probability that $B \subset \omega$ given that $A \subset \omega$. In addition to the minimal support, the user selects a minimal confidence in order to prune out random rules. An association rule which satisfies both $s(A \Rightarrow B) \geq s_0$ and $c(A \Rightarrow B) \geq c_0$ is called a *strong association rule*. High confidence means that, in all the transactions containing A, a large proportion also contains B. However, this is only interesting if, in the subset of itemsets containing A, the itemsets also containing B are more frequent than in the full set. The ratio of these proportions is called the *lift* of the association rule:

$$\gamma(A \Rightarrow B) := \frac{c(A \Rightarrow B)}{s(B)} = \frac{s(A \cup B)}{s(A)s(B)}.$$

Note that $\gamma(A \Rightarrow B) = \gamma(B \Rightarrow A)$.

The co-occurrence of A and B in the transactions ω is further analysed in the *contingency table* (see Table 2.1), and it turns out that all the frequencies can be computed from N, N_A, N_B and $N_{A\cup B}$. The rule $A \Rightarrow B$ is irrelevant if the association of $A \subset \omega$ with $B \subset \omega$ only occurs owing to coincidence,

Table 2.1. Contingency table for $A \Rightarrow B$

	$A \subset \omega$	$A \not\subset \omega$	Σ
$B \subset \omega$	$N_{A \cup B}$	$N_B - N_{A \cup B}$	N_B
$B \not\subset \omega$	$N_A - N_{A \cup B}$	$N - N_A - N_B + N_{A \cup B}$	$N - N_B$
Σ	N_A	$N - N_A$	N

that is, the two events are statistically independent. This hypothesis can be tested using a χ^2 test statistic (see, $e.g.$, Dobson (1990)) because

$$X^2(A \Rightarrow B) = N \frac{(s(A \cup B) - s(A)s(B))^2}{s(A)s(B)(1 - s(A))(1 - s(B))}$$

can be seen to have a χ^2 distribution with 2 degrees of freedom if the supports are estimated with the frequencies. Note that all the measures so far could be expressed in terms of the support $s(A)$ of itemsets A.

If the number of transactions supporting N_A is small, the confidence estimate can become unreliable owing to large random fluctuations. In this case the *Laplace estimator*

$$c(A \Rightarrow B) \approx \frac{N_{A \cup B} + 1}{N_A + k},$$

where $1/k$ is an *a priori* guess of the confidence, may improve accuracy (Segal and Etioni 1994); an alternative is the Yates correction (Quinlan 1987) for an application to rule induction.

In addition to confidence, support, lift and χ^2, many other criteria have been used to describe the interestingness of an association rule. Many such measures depend on the application. However, the basic techniques rely on properties of support and confidence and thus these will be the only two measures discussed further here. Alternatives and further discussion of such measures can be found in Chen, Han and Yu (1996), Srikant and Agrawal (1995) and Silberschatz and Tuzhilin (1996).

The enumeration of all strong association rules is a considerable *search problem*. In the general case, the determination of optimal rules is known to be NP-hard (Morishita and Nakaya 2000). For the determination of strong association rules, however, efficient algorithms have been found. There are $\mathcal{O}(m^k)$ k-itemsets with a low number k of items from a total number of m items. Evaluating all the rules generated from these itemsets requires $\mathcal{O}(Nm^k)$ tests for support, which is not feasible computationally. Fortunately, not all the itemsets need to be checked, as most have a very small support. The foundation is the anti-monotonicity property of the support function. Even though the following properties follow readily, because of the importance of the application we formulate them as lemmata.

Lemma 1.   The function $s(A)$ is *anti-monotone*, that is, if $A \subset B$ then $s(A) \geq s(B)$ and furthermore $s(\emptyset) = 1$.

Proof.   The anti-monotonicity follows from the monotonicity of the probability P because, if $A \subset B$, then

$$\{\omega : A \subset \omega\} \supset \{\omega : B \subset \omega\}.$$

Furthermore, by equation (2.1), we have

$$s(\emptyset) = P(\{\omega : \emptyset \subset \omega\}) = 1. \qquad \square$$

Setting $B - A := \{i \in B : i \notin A\}$, we derive the following result from the definition.

Lemma 2.   If $B \subset A$ then $s(A \Rightarrow B) = s(A)$ and $c(A \Rightarrow B) = 1$. Furthermore, $s(A \Rightarrow B) = s(A \Rightarrow (B - A))$ and $c(A \Rightarrow B) = c(A \Rightarrow (B - A))$. In particular,

$$s(\emptyset \Rightarrow \emptyset) = 1 \tag{2.6}$$
$$s(\emptyset \Rightarrow A) = s(A \Rightarrow \emptyset) = s(A) \tag{2.7}$$
$$c(\emptyset \Rightarrow \emptyset) = 1 \tag{2.8}$$
$$c(\emptyset \Rightarrow A) = s(A) \tag{2.9}$$
$$c(A \Rightarrow \emptyset) = 1. \tag{2.10}$$

Proof.   Firstly, $s(A \Rightarrow B) = s(A \cup B) = s(A)$ and $c(A \Rightarrow B) = s(A \cup B)/s(A) = s(A)/s(A) = 1$.

Then $s(A \Rightarrow B) = s(A \cup B) = s(A \cup (B - A)) = s(A \Rightarrow (B - A)$ and $c(A \Rightarrow B) = s(A \cup B)/s(A) = s(A \cup (B - A))/s(A) = c(A \Rightarrow (B - A))$. $\square$

Basically, rules with $A \cap B \neq \emptyset$ do not do anything new, and so we only consider rules $A \Rightarrow B$ for which

$$A \cap B = \emptyset.$$

The association rule $\emptyset \Rightarrow \emptyset$ is the trivial rule, and the rule $\emptyset \Rightarrow A$, sometimes also written as '$\Rightarrow A$', can be identified with the itemset A, thus making the itemsets just a special case of an association rule.

An itemset A satisfying a condition $s(A) \geq s_0$ for a given s_0 is called a *frequent itemset*. (Note that this definition is not absolute: it depends on the choice of the parameter s_0.) The main property of frequent itemsets is as follows.

Lemma 3. (*a priori* property)   Every subset of a frequent itemset is frequent.

Proof. Let $A \subset J$ be frequent, that is, $s(A) \geq s_0$, and let $B \subset J$. Then, by the anti-monotonicity property we have $s(B) \geq s(A)$ and thus $s(B) \geq s_0$, and thus B is also a frequent itemset. □

The *a priori* property has been the strongest tool in the development of efficient association mining algorithms. Once the frequent itemsets are determined, finding strong association rules is very efficient.

Lemma 4. Once all the frequent itemsets and their supports are known, the strong association rules can be found without any further scans of the database.

Proof. By definition, for any strong association rule $A \Rightarrow B$ the itemset $A \cup B$ is frequent. By the *a priori* property, the itemset A is frequent too.

Thus all the the strong association rules can be found by searching through all partitions $C = A \cup B$ for all frequent itemsets C and selecting the ones for which the confidence $c(A \Rightarrow B) = s(C)/s(A) > c_0$. This does not require any further scans as both C and A are frequent and thus their supports $s(C)$ and $s(A)$ were evaluated earlier. □

So far, a general algorithm for the determination of strong association rules consists of two steps:

(1) find all frequent itemsets A and their support $s(A)$;
(2) from this determine the strong itemsets.

Only the first step requires scanning the database. Thus the first step is usually orders of magnitude more expensive than the second step. Most improvements of the algorithm focus on the first step; the second step is achieved simply by enumerating all possible rules from partitions of the frequent itemsets.

As the size $|\omega|$ of the transactions is finite, the size of the frequent itemsets $|A|$ is limited. Let a *k-itemset* be an itemset A with size $|A| = k$. Then it follows from the *a priori* property that there are many more frequent k-itemsets than there are frequent $(k+1)$-itemsets. A systematic search for the frequent itemsets may thus look for frequent itemsets by order of their size. Let the set of all frequent k-itemsets be denoted by L_k, that is,

$$L_k := \{A \subset J : s(A) \geq s_0, \quad |A| = k\},$$

in particular, $L_0 = \emptyset$. Assume in the following, that the items $i_k \in J$ are ordered (in an arbitrary but fixed way). Given this ordering we define the *join* $L_k * L_k$ of two sets of k-itemsets by

$$L_k * L_k = \{\{i_1, \ldots, i_{k+1}\} : \{i_1, \ldots, i_k\}, \{i_1, \ldots, i_{k-1}, i_{k+1}\} \in L_k\},$$

where the i_s are ordered, that is, $i_s \leq i_{s+1}$, for $s = 1, \ldots, k$.

This provides a first upper bound for the set of frequent k-itemsets.

Lemma 5. We have
$$L_{k+1} \subset L_k * L_k, \text{for } k \geq 0.$$

Proof. For any $\{i_1, \ldots, i_{k+1}\} \in L_{k+1}$, by the *a priori* property all sub-sets of size k are frequent k-itemsets; in particular, $\{i_1, \ldots, i_k\} \in L_k$ and $\{i_1, \ldots, i_{k-1}, i_{k+1}\} \in L_k$ and so $\{i_1, \ldots, i_{k+1}\} \in L_k * L_k$. □

This set can thus be easily constructed. While it may have much fewer than $\mathcal{O}(m^k)$ elements, it can still be large and may be pruned further using the *a priori* property. This *pruning step*, which does not require any database scanning, generates a *candidate itemset* C_{k+1} from the $L_k * L_k$ by

$$C_{k+1} := \{A \in L_k * L_k : \text{all } k\text{-itemsets } B \subset A \text{ satisfy } B \in L_k\}.$$

Application of the *a priori* property again gives

$$L_{k+1} \subset C_{k+1}, \quad k = 1, 2, \ldots.$$

This candidate itemset is the best we can get using the *a priori* property as any element contains only frequent itemsets.

Lemma 6. Any $B \subset A$ of any $A \in C_k$ is a frequent itemset.

Proof. By construction any k-itemset $B \subset A \in C_k$ is a frequent k-itemset. Any j-itemset with $j < k$ has to be by construction a subset of a frequent k itemset and is thus also frequent. □

Thus the *a priori* property provides ways to reduce the size of the *negative border*, that is, the points which are in C_k but not in L_k. While the supports of the itemsets in L_k are required for the determination of the confidence, the supports of the elements in the negative border are 'wasted', and so we would like to keep the size of $|C_k| - |L_k|$ to a minimum.

The *a priori algorithm* implements the repeated application of join, prune and evaluation of the support of itemsets to determine all itemsets. It is described in Algorithm 1, opposite.

The most expensive step is the scanning of the database. If we assume that determining whether a k-itemset is supported by a transaction ω requires $\mathcal{O}(k)$ operations, then the complexity of the *a priori* algorithm is $\mathcal{O}(N \sum_{k=1}^{\infty} |C_k| k)$. The performance of the *a priori* algorithm has been improved in numerous ways (see, *e.g.*, Han and Kamber (2001)), including

- reduction of the size of scans;
- reduction of the number of scans.

The reduction of the size of scans prunes transactions ω that do not contain frequent itemsets, using another consequence of the *a priori* property.

Lemma 7. Every item of a frequent $(k+1)$-itemset is contained in at least k frequent k-itemsets and thus in at least k candidate k-itemsets.

Algorithm 1   *a priori* algorithm

$C_1 := \{\{i\} : i \in J\}$ {set of all 1-itemsets}, $k = 1$
while $C_k \neq \emptyset$ **do**
  For all $A \in C_k$ determine $s(A)$ by scanning database.
  $L_k :=$ set of all frequent itemsets in C_k.
  Construct join $L_k * L_k$.
  Prune result by removing sets with infrequent subsets to get C_{k+1}.
  $k := k + 1$
end while
return all frequent itemsets $\bigcup L_k$

Proof.   Let i be an item of a frequent $(k+1)$-itemset A. Then A has exactly k subsets of size k that contain i. Because of the *a priori* property these subsets have to be frequent k-itemsets.

The set of candidate k-itemsets C_k contains all frequent k-itemsets because $L_{k+1} \subset C_{k+1}$, as derived in the proof of Lemma 5.   □

Using this lemma, the *a priori* algorithm is extended to contain a pruning of elements of transactions that are not in k different candidate k-itemsets. This pruning is done during the scanning step and no additional scanning is required. Note that whole transactions may be removed from the database and a substantial reduction in complexity may result. For more details and an implementation of this idea see Park, Chen and Yu (1995a), where a reduction of the number of counting steps based on hashing has been suggested.

Smaller association mining tasks may fit into memory and so multiple scans may not be necessary. However, for very large databases this is not realistic, and in this case the database has to be completely scanned from disk k times, where k is the maximal number of elements of the occurring frequent itemsets. A partitioning approach can reduce this number to two scans, which can substantially speed up the algorithm. For this let the database $D = (T_1, \ldots, T_N)$ be partitioned into p pairwise disjoint subsets D_i, that is,

$$D = D_1 \cup \cdots \cup D_p$$

such that $D_i \cap D_j = \emptyset$ for $i \neq j$. This is called a *partition of D*. We define the support of an itemset A in the partition D_j by

$$s_j(A) := \frac{|\{\omega \in D_j : A \subset \omega\}|}{|D_j|},$$

and an itemset A is said to be *frequent in D_j* if it satisfies

$$s_j(A) \geq s_0.$$

With this we get the following result.

Lemma 8. (Invariant partitioning property) Let $D_1, \ldots, D_p$ be a partition of D. Then each itemset that is frequent in D is frequent in at least one subset D_i.

Proof. Let A be an itemset that is frequent in D. Then

$$
s_0 \leq s(A) = \frac{|\{\omega \in D : A \subset \omega\}|}{|D|}
$$

$$
= \sum_{i=1}^{p} \frac{|D_i|}{|D|} \frac{|\{\omega \in D_i : A \subset \omega\}|}{|D_i|} \qquad \text{as } \{D_i\} \text{ are disjoint}
$$

$$
= \sum_{i=1}^{p} \frac{|D_i|}{|D|} s_i(A)
$$

$$
\leq \max_{1 \leq i \leq p} s_i(A).
$$

Thus A is least frequent in the set D_i in which it has maximal support. □

The modified *a priori* algorithm in a first scan gathers all the frequent itemsets and their supports in the partitions D_i which are chosen such that they fit comfortably in memory. Because of the invariant partitioning property the local frequent itemsets are potential frequent itemsets over the whole database. A second scan through the data returns the supports for all these potentially frequent itemsets over D. This can be done very efficiently, as the supports of many of these itemsets are already known for many partitions. See Savasere, Omieski and Navanthe (1995) for a further discussion of this approach. Other approaches to the reduction of the number of scans are found in Toivonen (1996) and Brin *et al.* (1997).

A common problem in the determination of strong association rules is that the items may be too specific to produce any rules of interest. In the case of market basket analysis, the brand names and quantities as typically represented by a bar code may not reveal anything interesting. Choosing a more general concept, such as a class of consumer goods, may lead to more interesting results. If items are included in the set J which refer to more general concepts the relationship between the items is modelled by a *directed acyclic graph with nodes in J*. This can represent multiple hierarchies or taxonomies. Of particular interest is the transitive closure of this graph. An edge (i_1, i_2) in this transitive closure means that the the item i_2 is a *generalization* of i_1, or i_1 is of type i_2. For example, i_1 could be rye bread, and i_2 could be bread, in the example of market basket analysis. The graph is defined by the application and is a type of domain knowledge or constraint on the data.

Using this we can generalize the concept of support. An *itemset B sup-ports an itemset A* if, for any element $i \in A$, either

- $i \in B$, or
- there is a $j \in B$ such that i is an ancestor of j.

We say that A generalizes B, or $A \leq B$. We see immediately that if $B \subset A$ then $B \leq A$. Also, the relationship $\leq$ is transitive owing to the transitivity of $\subset$, and it can be seen that this defines a partial ordering on Ω that extends the ordering given by $\subset$. The relationship with set operations is very close and we have the following.

Lemma 9.   For any itemsets A, B and C, we have

$$C \geq A \quad \text{and} \quad C \geq B \quad \text{if and only if} \quad C \geq A \cup B.$$

Proof. If $C \geq A$ and $C \geq B$ then each $i \in A \cup B$ is an element of one of A or B (or both) and so it is either an element of C or an ancestor of an element of C, which means that $A \cup B$ generalizes C.

Conversely, if $C \geq A \cup B$ then, as any element of A is also an element of $A \cup B$, it is either an element of C or an ancestor of an element of C. Thus $A \leq C$, and the same argument holds for B.   □

In the special case where $C = A$ the above lemma gives $A \geq A \cup B$ if $A \geq B$. Because $A \subset A \cup B$ we also have $A \leq A \cup B$. Thus the supports of the two itemsets are identical, and for our purposes they are thus indistinguishable. An example of such an itemset $A \cup B$ would contain both an item and an ancestor of this item. It is suggested that the itemsets are normalized to exclude this case, so that the only itemsets we will consider are modulo any ancestors of the elements in the itemset. We define $A \vee B$ to be the normalization of $A \cup B$ and we will still denote the normalized itemsets by A and B. The support of an itemset in this hierarchical context is now defined by

$$s(A) := P(\omega \in \Omega \mid A \leq \omega). \tag{2.11}$$

Note that the probability distribution has to be compatible with the hier-archy so that a distribution over the normalized itemsets can be defined. It follows that the support is also *anti-monotone*, that is, for $A \leq B$ we have $s(B) \leq s(A)$. A *frequent itemset* is defined as before by $s(A) \geq s_0$. The *a priori* property is obtained as above from the *a priori* algorithm. Further, frequent k-itemsets, the join operation, pruning and the candidate itemsets are straightforward generalizations. Normalization only has to be done explicitly in the *a priori* algorithm when generating the C_2, because all the itemsets are normalized automatically in the process (Srikant and Agrawal 1995).

Lemma 10. If s is defined as in equation (2.11) and C_2 is pruned such that it contains only normalized itemsets, then both C_k and L_k generated in the *a priori* algorithm contain only normalized itemsets.

Proof. By construction C_1, C_2 and L_1, L_2 are normalized. We use induction to show that the others are normalized as well. Assume now that $L_1, \ldots, L_k$ are normalized and $k \geq 2$. If there were an unnormalized set $A \in C_{k+1}$, then there would be two items $i_1, i_2 \in A$ such that i_1 is an ancestor of i_2. By Lemma 6, however, any pair of elements of a subset of C_{k+1} is frequent, that is, $\{i_1, i_2\} \in L_2$ and, because L_2 is normalized, i_1 can never be an ancestor of i_2 and thus C_{k+1} has to be normalized. $\qquad\square$

As a consequence of this lemma the *a priori* algorithm can easily be modified to include hierarchies. The importance of this approach becomes apparent for the case of *quantitative association rules*. In this case the transactions contain sets of real numbers. Typically, the probability of any particular combination of real numbers is zero for continuous probability distributions and thus there are no frequent itemsets in the original sense. However, if we include intervals as items we can get frequent itemsets again. We will not go into the theory of quantitative association rules here but instead we will show how hierarchies provide a framework for approximation of association rules, in particular for discretization in the case of quantitative association rules. Further discussion of the case of hierarchical association rules can be found in Agrawal and Srikant (1995), Mannila, Toivonen and Verkamo (1995), Mannila and Toivonen (1996) and Shintani and Kitsuregawa (2000).

In the following, hierarchy-based approximation will be discussed in the case of finite sets J but this can be generalized to more general cases. So, given a set of items J with a graph, we would like to understand how well a subset of items $J' \subset J$ covers the space of strong association rules. The set J' has to satisfy several properties: in particular, for any element in J there has to be at least an ancestor in J'. We can then approximate any item $i \in J$ by the closest ancestor $\pi(i) \in J'$. This mapping induces a mapping between itemsets on J and J' which is defined elementwise, and again denoted by π as

$$\pi(A) := \{\pi(i) : i \in J\}.$$

From this we get

$$\pi(A \cup B) = \pi(A) \cup \pi(B) \tag{2.12}$$

$$\pi(A) \leq A \tag{2.13}$$

$$\pi(A) \cup \pi(B) \leq A \cup B \tag{2.14}$$

$$s(A) \leq s(\pi(A)). \tag{2.15}$$

If we can also bound the support of this 'projection' from below by

$$s(\pi(A)) \leq K s(A)$$

for some $K \geq 1$, we say that the triple (J, J', K) is K-*complete*. In this case we can give a bound on the effect this approximation has on the confidence of rules. We assume that an association rule $A \Rightarrow B$ is approximated by $\pi(A) \Rightarrow \pi(B)$.

Lemma 11. If (J, J', π) is K-complete then

$$\frac{c(\pi(A) \Rightarrow \pi(B))}{c(A \Rightarrow B)} \in [1/K, K].$$

Proof. This bound was proved by Srikant and Agrawal (1996a), who also provide further analysis and a choice of J' for the case of quantitative association rules, where bins or intervals are used to approximate the itemsets.

$\square$

Association rule mining can be generalized to the analysis of sequences, which has been termed *sequence mining*. Typically, a sequence is defined as a sequence of itemsets and frequent k-sequences containing k items can be found with a variant of the *a priori* algorithm. Several algorithms have been discussed in Srikant and Agrawal (1996b), Agrawal and Srikant (1995), Oates, Schmill, Jensen and Cohen (1997) and Zaki (1998, 2000). Related to this is the discovery of frequent episodes (Mannila *et al.* 1995, Mannila and Toivonen 1996).

Ultimately, itemsets are described by predicates $A(\omega)$ defined on the transactions. In *multidimensional association rule mining* the algorithms are generalized for sets of predicates. Predicate sets are conjunctions of predicates and, using the *a priori* approach, frequent predicate sets are found from which if-then rules $A \Rightarrow B$ with precedent A and antecedent B can be derived. In general such *rule induction* is an area which has been of much interest in the machine learning community. The rule induction process selects rules that satisfy certain *constraints* depending on the dataset. These constraints relate to the quality or the interestingness of the rules. Typical data mining tasks are as follows.

- Find the set of all rules that satisfy the constraints. Note in particular that the consequent C is not determined either.
- For a given consequent and a given ordering of the rules, find the best antecedents A.

In general, finding best rules is an optimization problem that is known to be NP-hard (Morishita 1998). One heuristic assumes the availability of a LEARN-ONE-RULE subroutine which is capable of generating one rule from any dataset. After this is applied for the first time to the dataset all the true positives are removed and the LEARN-ONE-RULE is applied again. This greedy algorithm is the *sequential covering algorithm* (Mitchell 1997, p. 275).

A different approach is based on *beam search* and a very general algorithm for rule induction is suggested in Provost, Aronis and Buchanan (2000). At each node of a search tree, a set of possible specializations consisting of conjunctions with new predicates is chosen. Good efficiency is obtained by pruning off unpromising paths. This approach was also applied to association rule mining in Webb (2000). Classical rule induction techniques include RL (Clearwater and Provost 1990) or RIPPER (Cohen 1995). Parallel and distributed rule induction algorithms can be found in Hall, Chawla, Bowyer and Kegelmayer (2000), Williams (1990) and Hall, Chawla and Bowyer (1998).

Association rule mining is a recent development and has its origins in the data mining community. This is an area also undergoing constant change and development. At the core of the developments are the focus on frequent itemsets and the *a priori* property. Current research is dealing with the following questions:

- Which interestingness criteria produce the most useful rules?
- What are the most effective algorithms? Of interest are scalability with respect to data size, parallelism, higher dimensionality, disk access and memory hierarchies.
- How are complex data analysed? This problem includes multimedia and web data.

A good collection of pointers to the literature up to around 2000 can be found in Han and Kamber (2001).

3. Classification and regression

Functional relationships

$$Y = f(X_1, \ldots, X_d)$$

between the features Y and X_i form a powerful tool for summarizing aspects of the data in order to extract important data classes or to predict future data trends. The data is again modelled as a set of transactions $D = (\omega_1, \ldots, \omega_N)$ where $\omega_i \in \Omega$, but in addition it is assumed that we also know some features or random variables Y and X_i defined on Ω. If Y is categorical, the problem of determining the model f from the data is called classification, and in the case of continuous Y we speak of regression.

3.1. Classification

Classification is treated in depth in the machine learning literature; see, *e.g.*, Mitchell (1997). Both associations $A \Rightarrow B$ and functions $Y = f(X)$ provide relations between features of the data. The following is a comparison of some properties of associations and functions.

- The dependent variable Y is fixed for classification, while finding the consequent is part of association rule mining.
- The features of association rules are the predicates, which are Boolean, while in classification the features can take arbitrary types. The right-hand side Y of a classifier has values in a finite set $\{C_0, \ldots, C_m\}$ and the most commonly discussed case is $m = 2$, and in this case Y is basically a predicate.
- The performance of a classifier is judged by its accuracy, rather than by support and confidence.
- One function f summarizes all the data compared with a set of association rules.
- As in the case of association rule induction, the search for good classifiers often starts with a simple classifier, and successively more complex ones are considered until the desired performance is achieved.

In terms of the distribution of the probability the *misclassification rate* is defined by

$$L(f) = P(\{\omega : Y \neq f(X)\}).$$

This is an important measure of the quality of the classifier and is also called the Bayes risk. In the case where Y, and thus $f(X)$, is Boolean, the misclassification rate can be determined from the support defined in the previous section by

$$L(f) = s(Y) + s(f(X)) - 2s(Y \wedge f(X)).$$

This suggests that association rule technology can be applied to classification. Examples of this approach can be found in Dong and Li (1999), Li, Dong and Ramamohanrarao (2000), Lent, Swami and Widom (1997), Liu, Hsu and Ma (1998), and Meretakis and Wüthrich (1999).

3.2. Decision trees

Decision or classification trees use a partitioning of the domain Ω into hyper-rectangles parallel to the values of the features X_i. They can be defined recursively by

$$f(X) = \begin{cases} f_1(X), & \text{if } A(X) \text{ holds,} \\ f_2(X), & \text{else,} \end{cases}$$

where f_1 and f_2 are either constant or again decision trees. The predicate $A(X)$ defines the split. In most cases the predicate $A(X)$ only depends on one of the features X_i. Decision trees originated in machine learning and statistics (Breiman, Friedman, Olshen and Stone 1984, Quinlan 1986, Quinlan 1993). During the induction of the decision tree the data needs to be revisited for each new level of the tree. For N data points, a balanced tree will have $\mathcal{O}(\log(N))$ levels. As the tree building algorithm has to scan the data for each level, the average complexity of the tree building algorithm is $\mathcal{O}(N\log(N)m)$, where m is the number of attributes or features considered. Thus the algorithm is close to scalable in the number of data points, especially under the assumption that the levels of the tree are kept constant for growing data sizes.

The models from (small) decision trees are comprehensible and can quite easily be cast as standard database queries (which may be expressed in SQL, the standard for relational database queries). They can be generated rapidly and show good accuracy. These properties make them well suited to data mining. They can also be used as a starting point for the generation of rules. In fact, there is one rule per leaf of the tree, which is obtained by taking the conjunction of all the predicates $A(X)$ used to split the tree on a path from the leaf to the root. With the topmost split we get, in the case of Boolean f,

$$f(X) = A(X) \wedge f_1(X) \vee \overline{A(X)} \wedge f_2(X),$$

where $\overline{A(X)}$ denotes the negation of $A(X)$. This provides a recursion to build the rules. We start with constant values at the leaves. Then, let the $A_i^{(1)}(X) \Rightarrow B_i^{(1)}$ and $A_i^{(2)}(X) \Rightarrow B_i^{(2)}$ be the rules generated for both descendants of the node; then

$$A(X) \wedge A_i^{(1)}(X) \Rightarrow B_i^{(1)}, \quad \text{and} \quad \overline{A(X)} \wedge A_i^{(2)}(X) \Rightarrow B_i^{(2)}$$

are the rules for the top node. The antecedents are constants and do not change during the extraction process. They are used to initialize the rules for each leaf to True $\Rightarrow B_i^{(s)}$.

These initial rules, however, are often not very useful and may have low support, as the supports of the antecedents of the rules are mutually exclusive. Several steps are required to create rules that are useful for data mining from the initial rules (Quinlan 1993):

- Remove predicates in the rule which do not improve the rule. This pruning is often done based on a χ^2 test.
- Rules with overall low quality are removed and similar rules are merged.
- A default rule is required which covers all the cases not covered by other rules.

These steps do require further assessment of the quality of the generated rules and thus need further scanning of the data. As the pruning step may need to be repeated several times, the time spent on this data scanning may be substantial. See Quinlan (1987) for implementation of the rule induction from decision trees.

The basic algorithm for the construction of the decision tree consists of repeated assessment of the purity of the nodes in each partition and decisions on which partition to split into subpartitions and how; see Algorithm 2.

Algorithm 2	Basic decision tree algorithm

Partition(Ω)
if all the points in Ω are of the same class **then**
    return
else
    for each attribute A in Ω **do**
        evaluate all splits of A
    end for
    find best split $\Omega = \Omega_1 \cup \Omega_2$
    Partition(Ω_1)
    Partition(Ω_2)
end if

The evaluation of the splits is often based on an *impurity function* $\phi(p)$ based on the proportion p of transactions which are in the class of interest. If the proportion of class 1 cases is p in Ω, and p_1 and p_2 in the subsets $\Omega 1$ and $\Omega 2$ respectively, then the decrease of the impurity due to the splitting is modelled as

$$\phi(p) - s\phi(p_1) - (1 - s)\phi(p_2),$$

where s is the proportion of cases in Ω_1. Examples of impurity functions are:

- $\phi(p) = \min(p, (1 - p))$ (based on misclassification rate)
- $\phi(p) = 2p(1 - p)$ (Gini (Breiman *et al.* 1984))
- $\phi(p) = -p\log(p) - (1 - p)\log(1 - p)$ (entropy (Quinlan 1986))

Alternatively, the χ^2 value is used. The construction of the tree has two steps. First the tree is grown until each leaf is either pure or contains very few points. Then the tree is pruned again in order to reduce overfitting using an error estimator.

In order to handle the large and growing data sizes, scalable parallel algorithms are required. An example is the SPRINT algorithm (Shafer, Agrawal and Mehta 1996). The computing platform is a 'shared nothing' parallel computer (*e.g.*, a Beowulf cluster) where the processors have their own memories and disks and the data are distributed evenly over these local disks. The data records are assumed to be sequences of features $(x_1^{(s)}, \ldots, x_d^{(s)}, y^{(s)})$ for $s = 1, \ldots, N$ are stored with some redundancy in sorted *attribute lists* which for each attribute x_i consist of sequences $(i, x_s^{(i)}, y^{(i)})$ sorted on $x_s^{(i)}$. This allows the efficient evaluation of splits in the case of real attributes. This evaluation is done based on histograms of the values of y for each x_s, and can be done in parallel. Hash tables are used to associate the transaction record identifier i with both the processor and the partition, and the Gini index is used for splitting. Experiments demonstrate the scalability of this approach. In fact, SPRINT performs well with respect to three important criteria:

Scale-up. Fix the data per processor and study time as a function of the number of processors.

Speed-up. For a constant data size the time is studied as a function of p.

Size-up. The number of processors is fixed and the time is studied as a function of the data size n.

If the misclassification rate is determined from the dataset used for training, we get an estimate that is systematically too low. Better estimates are obtained from the *hold out* method, where the data are partitioned into training data for computation and test data for the estimation of the error. A popular alternative is *generalized cross-validation* or GCV (Golub, Heath and Wahba 1979), which emulates the hold-out technique without the reduction of the size of the training data.

3.3. Bayesian classifiers

The best classifier minimizes $L(f)$ and can be characterized in terms of the probability distribution as follows (see Devroye *et al.* (1996, p. 10)).

Theorem 1. Let $Y \in \{0, 1\}$ and

$$f^*(\mathbf{x}) = \begin{cases} 1, & \text{if } P(Y = 1|\mathbf{x}) \geq 0.5, \\ 0, & \text{else.} \end{cases}$$

Then $L(f^*) \leq L(f)$.

The classifier f^* minimizing $L(f)$ is called *Bayes' classifier*. The minimum $L^* = L(f^*)$ is the Bayes error and characterizes how difficult a certain classification problem is. Other measures have been considered in the literature but are seen to be closely related to the Bayes error (Devroye *et al.* 1996, p. 21ff).

In practice, the distribution of the random variables is unknown and so the classifier has to be determined from observations. The observations are modelled as a sequence $(\mathbf{X}_i, Y_i)_{i=1}^n$ of identically distributed independent random variables. Practical classifiers are then functions $f_n(\mathbf{x}; \mathbf{X}_1, Y_1, \ldots, \mathbf{X}_n, Y_n)$.

An alternative, more general definition of the Bayesian classifier (Duda and Hart 1973) is

$$f^*(\mathbf{x}) = \text{argmax}_y \, p(y|\mathbf{x})$$

where, as usual, $p(y|\mathbf{x}) = P(Y = y|\mathbf{X} = \mathbf{x})$ is the conditional probability of Y being in class y when $\mathbf{X} = \mathbf{x}$. Practical estimators for f^* are obtained by the introduction of estimates for the conditional probabilities $p(y|\mathbf{x})$ in the above formula. The classifiers obtained in this way are called *plug-in* *decisions* in the case of Boolean y and we have the following result (Devroye *et al.* 1996, p. 16).

Theorem 2. We have

$$P(f(X) \neq Y) \geq P(f^*(X) \neq Y) + 2E\,|\eta(\mathbf{X}) - \tilde{\eta}(\mathbf{X})|,$$

where $\eta(\mathbf{x}) = P(Y = 1|\mathbf{x})$, $\tilde{\eta}$ is the estimate of η, and $E(\cdot)$ denotes the expectation.

Many estimates of the conditional probability use *Bayes' formula*:

$$p(y|\mathbf{x}) = \frac{p(\mathbf{x}|y)p(y)}{p(\mathbf{x})},$$

where $p(y)$ is the probability of class y and $p(\mathbf{x})$ is the probability density function in the feature space in order to apply standard density estimators rather than have to estimate a family of probabilities.

In the case where the features are conditionally independent random variables we get

$$p(y|\mathbf{x}) = \frac{p(y)\amalg_{i=1}^{d}p(x_i|y)}{\sum_y p(y)\Pi_{i=1}^{d}p(x_i|y)}.$$

Techniques based on this assumption are called *naive Bayes* estimators. Langley, Iba and Thompson (1992) have demonstrated the effectiveness of naive Bayes in the case of independent Boolean features, and a newer analysis can be found in Langley and Sage (1999). It is also claimed that this approach works well even in cases where the features are not independent.

Note that f^* does not depend on the denominator. We obtain a *discriminant function* by removing the denominator and taking the logarithm

$$g(y|\mathbf{x}) = \log p(y) + \sum_{i=1}^{d} \log p(x_i|y),$$

and because of the monotonicity of the logarithm we get

$$f^b(\mathbf{x}) = \operatorname{argmax}_y g(y|\mathbf{x}).$$

The practical determination of f^b then uses estimates of the conditional probability densities $p(x_i|y)$ for all the variables x_i.

In the case of categorical x_i the estimate used for $p(x_i|y)$ is just the frequency, and for continuous features we often use a normal distribution for $p(x_i \mid y)$. In this case we get

$$g(y|\mathbf{x}) \approx k_y + \sum_{\text{categorical } x_i} k(x_i|y) + \sum_{\text{continuous } x_i} \frac{(x_i - \mu_i)^2}{2\sigma_i^2}.$$

There are two steps in the determination of $f(\mathbf{x})$:

(1) The determination of of the probability distributions. This step only has to be done once but requires the exploration of all the data points.

(2) The determination of the maximum of the discriminant function. This step has to be done for each evaluation of $f(\mathbf{x})$.

The determination of $p(y)$ requires n additions in order to count the frequencies of all classes. Only the c values of $p(y)$ need to be stored. In order to determine the distributions $p(x_i|y)$ for each feature x_i, the frequencies in the case of categorical x_i do require n additions; in the case of continuous variables and the normal distribution we require $2n$ additions and n multiplications to determine all the variances and means. Thus the total number of floating point operations required for the first step is

$$n + d_1 n + 3 d_2 n,$$

where d_1 is the number of categorical variables and d_2 the number of continuous variables (the dimension is thus $d = d_1 + d_2$).

The evaluation of $f(\mathbf{x})$ does require the computation of the discriminant function for all the c possible values of y, which requires cd additions for the sum and $5cd_2$ floating point operations to evaluate the terms of the second sum.

Thus the naive Bayes classifier is computationally very efficient, as

- the data only have to be read once and do not need to be kept in memory,
- the algorithm is scalable in the number of observations,
- the algorithm is linear in the number of dimensions, and
- in addition, the main operations of the learning stage can be done in parallel both over the test dataset and over the dimensions.

However, the naive Bayes classifier does rely on two assumptions: first, it is based on the probabilistic model of the observations, and second, it requires that the features are independent for each class. However, in practice, features are mostly correlated. While it has been observed that, even in examples with correlations, naive Bayes is often as effective as tree-based classifiers (Langley and Sage 1994), it would appear that dealing with the correlations may improve the performance even further. Domingos and Pazzani (1997) provide the region where Bayesian classifiers are optimal as well as further evidence for the good performance.

Langley and Sage (1994) introduce the *selective Bayesian classifier*, which is able to remove redundant or heavily correlated features. The algorithm starts with no attributes, then adds one at a time until no more improvement is found. As, at each step, the effect of all the remaining attributes has to be

Table 3.2. Costs in time and space of the naive and flexible
Bayes classifier

	Naive Bayes		Flexible Bayes	
	time	space	time	space
training on n cases	$\mathcal{O}(nk)$	$\mathcal{O}(k)$	$\mathcal{O}(nk)$	$\mathcal{O}(nk)$
evaluation	$\mathcal{O}(k)$		$\mathcal{O}(nk)$	

recomputed, the complexity of this method is of $\mathcal{O}(d^2)$. In order to estimate the errors we could consider cross-validation techniques or hold-out, but the authors have simply used the accuracy on the training set and obtained good results. They included attributes as long as they did not degrade accuracy. In view of the complexity of the selective Bayesian classifier we lose the linear dependence on the dimension, and we have to read the data several times, once for each feature included.

One limitation of the approach is the use of normal distributions for the conditional probabilities in the case of numerical variables. John and Langley (1995) propose an approach using kernel estimators for the densities, specifically

$$p(X_i = x_i | Y = y) = (nh)^{-1} \sum_j K \left(\frac{x - \mu_j}{h} \right).$$

This increases, in particular, the number of operations for the evaluation stage. It is shown that in most cases a significant improvement in the classification performance is obtained. The time and space costs are given in Table 3.2.

A different way to include dependencies of the features are the *Bayesian networks*. A Bayesian network is a *directed acyclic graph* (DAG) where all the vertices are random variables X_i such that X_i and all the variables that are not descendants of X_i or parents of X_i are conditionally independent given the parents of X_i. The main property of Bayesian networks is that the probabilities of all the random variables can be computed almost as if they were independent (Bender (1996, pp. 308–313), Heckerman, Geiger and Chickering (1995), Friedman, Geiger and Goldszmidt (1997)).

Theorem 3. If $(\mathbf{X}, E, P)$ is a Bayesian network, then

$$P(\mathbf{X}) = \prod_{P(C(X_i)) \neq 0} P(X_i | C(X_i)).$$

Conversely, if $(\mathbf{X}, E)$ is a DAG and if f_i is a nonnegative function such that

$\sum f_i(X_i) = 1$, then

$$P(\mathbf{X}) = \prod f_i(X_i)$$

defines a probability space for which $(\mathbf{X}, E, P)$ is a Bayesian network. Furthermore, $P(X_i|C(X_i))$ is either 0 or $f_i(X_i)$.

Here $C(X_i)$ denotes the parents (or direct causes) of X_i.

4. Regression

Regression refers in data mining to the determination of functions

$$Y = f(X_1, \ldots, X_m),$$

where the response variable Y is real-valued. The features X_i are assumed to be given and can have any type. In logistic regression, these models are also used to model decision functions or data density distributions. Parametric models, in particular linear models, are immensely popular in data mining, but non-parametric models are also widely used and will be the focus of the following discussion. Data mining applications to regression pose some additional challenges when compared to smaller statistical applications:

- The data mining datasets are very large, with millions of transactions. Furthermore, they are high-dimensional, including up to several thousand features.

- The models found should be understandable in the application domain.
- The data are usually collected for some other purpose, such as management of accounts. Thus carefully designed experiments are an exception in data mining studies.

This shows that data mining must pay close attention to computational efficiency and simple models are often preferred. Furthermore, even though the data can be very large, the limitations of the information in the data have to be carefully studied, and the effect of skewed or long-tail distributions has to be assessed.

As before, let $\omega \in \Omega$ denote a transaction and $p(\omega)$ be the probability distribution on Ω. The features X_i and Y are real functions on Ω, *i.e.*, random variables. The function f is assessed according to how well it fits the data, and often the expected squared error is used as a measure of the error:

$$E((Y - f(\mathbf{X}))^2) = \int_\Omega (Y(\omega) - f(\mathbf{X}(\omega)))^2 \, dp(\omega).$$

Of course, it is impossible to assess this error directly. Practical estimators use a special portion of the data selected as a test set to estimate the error by the sum of squares

$$R(f) := \frac{1}{N} \sum_{i=1}^{N} (y^{(i)} - f(\mathbf{x}^{(i)}))^2.$$

In order to avoid over-optimistic error rates, a different independent part of the data is used for the evaluation of the error. Alternatively, particularly in the case of small datasets, we use *generalized cross-validation*, or GCV (Golub *et al.* 1979).

The best estimator with respect to the exact mean squared error is the conditional probability

$$f(\mathbf{x}) = E(Y|\mathbf{X} = \mathbf{x}).$$

The straightforward determination of this using the average is usually not feasible, as it is unlikely that enough (or even one) data points are available for every possible value of $\mathbf{x}$ such that $f(\mathbf{x})$ may be estimated by the mean of the corresponding values of Y. Assume, for example, $m = 30$ categorical features with 6 possible values each. Then the total number of possible values for the independent variable is $6^{30} \approx 2 \cdot 10^{23}$, which is way beyond the number of observations available. Thus smoothing, simplified and parametric models have to be used.

A seemingly simple approach to regression is to quantify or discretize Y and then use a classification technique to approximate f. This approach can make use of all the classification techniques; however, in order to get good performance we may have to use a very high number of classes. Consider, for example, the approximation of a simple one-dimensional linear function. A linear model requires only two parameters but a classification technique may require hundreds of parameters. Thus real models may be much simpler than their approximating (discretized) classification models.

A large class of regression functions has the following form:

$$f(\mathbf{x}) = f_0 + \sum_i f_i(x_i) + \sum_{i,j} f_{i,j}(x_i, x_j) + \sum_{i,j,k} f_{i,j,k}(x_i, x_j, x_k) + \cdots,$$

where the level of interactions may be determined by the data but an upper bound is often supplied by the user. It is seen that one or two levels are often enough and hardly any cases are found with interactions higher than level 5. These are called *ANOVA decompositions*, because of the similarity to the interaction models in ANOVA.

4.1. Regression trees

Regression trees (Breiman *et al.* 1984) are based on the same ideas as decision trees. They do have the same advantages and, in fact, they may be used with logistic regression for the determination of classification probabilities. In contrast to techniques based on tensor products of one-dimensional spaces, regression trees are extremely successful in dealing with high dimensions as their complexity is proportional to the dimension. They have two shortcomings, however: they do not represent linear functions well and, more generally, they cannot approximate smooth functions accurately, as they are discontinuous, piecewise constant functions. This approach is described by Breiman *et al.* (1984).

The trees define, as in the case of classification, a partitioning of the space of the independent variable $\mathbf{x}$ and for each partition a function estimate is provided. The main questions in building the tree are how to split the domain into subdomains, how to prune the regression tree and how to associate functions to the leaves of the tree.

Addressing the last question first, we choose the function constant on all the subdomains Ω_i related to the leaves of the tree. Note that the mean

$$\overline{f}_j := \frac{1}{N_j} \sum_{\mathbf{x}^{(i)} \in \Omega_j} y^{(i)}$$

minimizes the sum of squared residuals

$$R_j(f) := \sum_{\mathbf{x}^{(i)} \in \Omega_j} (y^{(i)} - f(\mathbf{x}^{(i)}))^2$$

between constants and the data over the subdomain Ω_j, and, for all leaves Ω_j, we have the least squares approximation

$$\overline{f}(\mathbf{x}) = \overline{f}_j, \quad \mathbf{x} \in \Omega_j,$$

which is constant on Ω_j. Thus, once the partitioning of the domain is done, the determination of the 'best' function is straightforward. The splitting is selected based on the reduction of the sum of squares, or, in order to adapt the complexity of the model to the data, we use a penalized sum of squared residuals

$$R_\alpha(f) = R(f) + \alpha|L_f|,$$

where L_f is the number of leaves of f, that is, a complexity measure, possibly using a different dataset for its evaluation.

For every variable there are several splittings possible and we select the one that maximizes the decrease of error in ΔR. This splitting is continued until each of the leaf nodes contains at most $N_{\min}$ observations.

Typically, the resulting tree is substantially overfitting the data, and we would like to select a subtree that is less complex. Complexity is measured by the number of leaves of the tree, *i.e.*, $|L_f|$. Thus instead of minimizing the sum of squares $R'(f)$, we attempt to balance complexity with fit on the

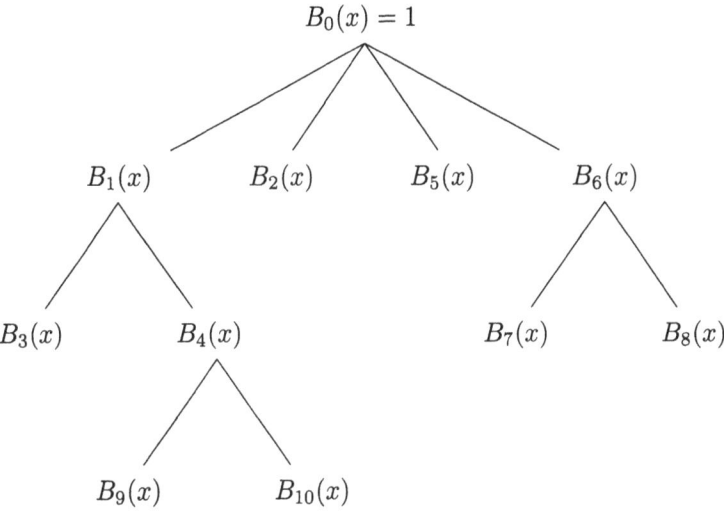

$$B_0(x) = 1$$

$B_1(x) \qquad B_2(x) \qquad B_5(x) \qquad B_6(x)$

$B_3(x) \qquad B_4(x) \qquad\qquad\qquad B_7(x) \qquad B_8(x)$

$B_9(x) \qquad B_{10}(x)$

Fig. 4.1. Hierarchy of MARS basis functions

training dataset by minimizing

$$R_\alpha(f) = R'(f) + \alpha|L_f|.$$

Note that on the right-hand side an independent data size has to be used in practice for the evaluation of $R(f)$. For an N-dimensional space Ω, an approximation defined by a tree with M leaves and N data points, this algorithm does require $\mathcal{O}(nNM)$ operations (Friedman 1991). Thus the procedure is scalable, that is, linear in the number of observations N.

4.2. Regression splines

Better approximation properties are obtained by piecewise polynomial functions. They are used in the MARS algorithm (*multivariate adaptive regression splines*): see Friedman (1991). Instead of explicitly generating a hierarchy of domains, MARS generates a hierarchy of basis functions that implicitly define a hierarchy of domains. An example of a basis function hierarchy is displayed in Figure 4.1. At the root of the tree is the constant basis function $B_0 \equiv 1$. At each 'partitioning' step two new children are generated:

$$B_{\text{child}_1} = B_{\text{parent}}(x)(x_j - \xi)_+ \quad \text{and} \quad B_{\text{child}_2} = B_{\text{parent}}(x)(-x_j + \xi)_+,$$

where $(z)_+$ denotes the usual truncated linear function. It is equal to z for $z > 0$ and equal to zero if $z < 0$. The parent, the variable x_j and the value ξ are all chosen such that the sum of squared residuals is minimized.

While each node can have several children, there are some rules for the generation of this tree:

- The depth of the tree is bounded, typically by a value of 5 or less. This is thought to be sufficient for practical purposes (Friedman 1991), and the bound is important for controlling the computational work required for the determination of the function values.

- Each variable x_j can only appear once in a factor of a basis function B_k. This guarantees that the function is piecewise multilinear.

The partitioning of the domain is defined such that on each partition the function is multilinear. Thus this partitioning does not have the same interpretative value as in the case of classification and is not recovered by the algorithm. However, the MARS method generates an ANOVA decomposition.

The computational complexity of the algorithm, some clever updating ideas having been applied, is shown in Friedman (1991) to be $\mathcal{O}\left(dN M_{\max}^4/L\right)$ where d is the dimension, N the number of data points, $M_{\max}$ the number of basis functions considered – some might not be used later because of pruning – and L is the number of levels of the tree. Thus the algorithm is scalable, that is, the complexity is proportional to the data size. However, the proportionality constant can be very large owing to the dependence on $\mathcal{O}(M^4)$, which limits the number of basis functions that can be used, and thus limits the approximation power of this approach. Another limitation is due to the fact that a greedy algorithm is used and the choice of basis functions may not be a global optimum. One problem with the usage of truncated powers is stability. Bakin, Hegland and Osborne (1998) suggest a variant of the MARS algorithm where the truncated powers are replaced by a hierarchy of B-splines. In addition, a parallel implementation is provided.

4.3. Additive models

In the case where the tree of basis functions generated by MARS contains two levels, namely, the root $B_0 \equiv 0$ and its children, a model of the following form is generated:

$$f(x_1,\ldots,x_d) = f_0 + \sum_{i=1}^{d} f_i(x_i).$$

The univariate components f_i of this additive model are piecewise linear in the case of MARS. Other commonly used additive models are based on smoothing splines and local parametric (polynomial) approximations (Hastie and Tibshirani 1990). A unified treatment for all these approximations reveals that in all cases additive models are scalable with respect to data size and, like the multivariate regression splines, conquer the curse of dimensionality. While additive models are computationally very competitive they also show good performance in practical applications. Like other regression models they provide good classification methods, especially if logistic regression is used.

If the probability distribution of the random variables $(X_1, \ldots, X_d, Y)$ which model the observations is known, the best (in the sense of expected squared error) approximation for the additive model is obtained when

$$f_i(x_i) = E\left(Y - f_0 - \sum_{\substack{k=1 \\ k \neq i}}^{d} f_k(X_k) \mid X_i = x_i\right) \qquad (4.1)$$

and $f_0 = E(Y)$. These equations do not have a unique solution, but uniqueness can easily be obtained if we introduce constraints like $E(f_i(X_i)) = 0$.

In practical algorithms, the estimates are approximated by smooth functions. Let $\mathbf{f}_i$ be the vector of function values $\left(f_i(x_i^{(k)})\right)_{k=1}^{n}$. Furthermore, let S_i be the matrix representing the mapping between the data and the smooth $\mathbf{f}_i$. The matrix S_i depends on the observations $x_i^{(1)}, \ldots, x_i^{(n)}$. Replacement of the estimation operator by the matrix S_i in equation (4.1) leads to

$$\begin{bmatrix} I & S_1 & \cdots & S_1 \\ S_2 & I & \cdots & S_2 \\ \vdots & \vdots & & \vdots \\ S_d & S_d & \cdots & I \end{bmatrix} \begin{bmatrix} \mathbf{f}_1 \\ \mathbf{f}_2 \\ \vdots \\ \mathbf{f}_d \end{bmatrix} = \begin{bmatrix} S_1 \mathbf{y} \\ S_2 \mathbf{y} \\ \vdots \\ S_d \mathbf{y} \end{bmatrix}.$$

If the eigenvalues of the S_i are in $(0, 1)$ and this linear system of equations is nonsingular, it can be seen that the Gauss–Seidel algorithm for this system converges. This method of determining the additive model is the *backfitting algorithm* (Hastie and Tibshirani 1990): it has complexity $\mathcal{O}(Nqd)$, where q denotes the number of iteration steps. The backfitting algorithm is very general and, in fact, is used even for nonlinear smoothers. For very large datasets, however, the algorithm becomes very costly – even though it is scalable in the data size and does not suffer from the curse of dimensionality – because it needs to revisit the data qd times.

The high cost of the solution of the previous linear system of equations resulted from the large size of its matrix. Smaller alternative systems are available for particular cases of smoothers S_i. For example, in the case of regression splines we can work with the system of normal equations. Consider the case of fitting with piecewise multilinear functions. These functions include the ones used in MARS and are, on each subdomain, linear combinations of products $x_{j_1} \cdots x_{j_k}$ where every variable x_j occurs at most once.

The basis functions of the full space of piecewise multilinear functions are products of hat functions of the form $b_1(x_1) \cdots b_d(x_d)$. For the full space, the normal matrix of the least squares fitting problem becomes sparse with nonzeros on 3^d of the diagonals. But the matrix is huge, being of order m^d if each of the d dimensions is discretized by m hat functions. This shows the computational advantage of the additive models where the dimension of the approximating function space is only md.

4.4. Radial basis functions

In order to explore further the challenge posed by the curse of dimensionality, we will investigate radial basis function approximation. In recent years, radial basis functions have received a lot of attention both theoretically and in applications. One of their outstanding features is that they are able to approximate high-dimensional functions very effectively. Thus they seem to be able to overcome the curse of dimensionality. In the case of real attributes $x \in \mathbb{R}^d$ a radial basis function is of the form

$$f(x) = \sum_{i=1}^{N} c_i \, \rho(\|x - x^{(i)}\|) + p(x),$$

where $x^{(i)}$ are the data points. Examples of the function ρ include Gaussians $\rho(r) = \exp(-\alpha r^2)$, powers and thin plate splines $\rho(r) = r^\beta$ and $\rho(r) = r^\beta \ln(r)$ (for even integers β only), multiquadrics $\rho(r) = (r^2 + c^2)^{\beta/2}$ and others. The function $p(x)$ is typically a polynomial of low degree and in many cases it is zero. The radial basis function approach may be generalized to metric spaces where the argument of ρ is replaced by the distance $d(x, x_i)$. Reviews of radial basis function research can be found in Dyn (1989), Powell (1992), and Buhmann (1993). Existence, uniqueness and approximation properties have been well studied.

The evaluation of $f(x)$ requires the computation of the distances between x and all the data points $x^{(i)}$. Thus the time required to compute one function value is $\mathcal{O}(dN)$; the complexity is linear in the number of attributes d and the curse of dimensionality has been overcome. However, if many function values need to be evaluated, this is still very expensive. Fast methods for evaluation of radial basis functions have been studied in Beatson, Goodsell and Powell (1996), Beatson and Light (1997), Beatson and Newsam (1992), and Beatson and Powell (1994). For example, a multipole method that reduces the complexity to $\mathcal{O}((m + N) \log(N))$ has been suggested in Beatson and Newsam (1992) for the evaluation of $f(x)$ for m values of x. For data mining applications, which have very large N, even this is still too expensive. What is required is an approximation for which the evaluation is independent of the data size N and does not suffer from the curse of dimensionality. In the following, we will revisit the determination of the function from the data points and see how the geometry of high-dimensional spaces influences the computational costs.

The vector of coefficients of the radial basis functions $\mathbf{c} = (c_1, \ldots, c_N)$ and the vector of coefficients of the polynomial term $\mathbf{d} = (d_1, \ldots, d_m)$ are determined, in the case of smoothing, by a linear system of equations of the form

$$\begin{bmatrix} A + \alpha I & P \\ P^T & 0 \end{bmatrix} \begin{bmatrix} \mathbf{c} \\ \mathbf{d} \end{bmatrix} = \begin{bmatrix} \mathbf{y} \\ 0 \end{bmatrix}. \tag{4.2}$$

The matrix $A = \left[\rho(\|x^{(i)} - x^{(j)}\|) \right]_{i,j=1\ldots n}$ has almost no zero elements for the case of thin plate splines and has to be treated as a dense matrix. However, the influence of the data points is local and it is mainly the observed points close to x that influence the value of $f(x)$. This locality is shared with nearest neighbour approximation techniques. However, in higher dimensions, points get more sparse. For example, Friedman (1994) observes that the expected distance of the nearest neighbour in a d-dimensional hypercube grows like $\mathcal{O}(N^{-1/d})$ with the dimension and that large numbers of data points are required to maintain a uniform coverage of the domain. In particular, a constant distance between a point and its nearest neighbour is obtained if $\log(N) = \mathcal{O}(d)$, that is, the number of points has to grow exponentially with the dimension. This is just another aspect of the curse of dimensionality.

If the function to be approximated is smooth enough, the number of points available may be sufficient even in high dimensions. But a computational difficulty appears which is related to the concentration of measure phenomenon (Talagrand 1996, Milman and Schechtman 1986, Ball 1997). The concentration of measure basically tells us that in high dimensions the neighbours of any point are concentrated close to a sphere around that point.

The effect of this on the computation is severe as the determination of a good approximant will require visiting a large number of neighbouring points for each evaluation point. In an attempt to decrease the computational work, a compactly supported radial basis function may be used. However, the support will have to be chosen such that, for a very large number of points, the values of the radial basis function $\rho(\|x^{(i)} - x^{(j)}\|)$ will be nonzero. Thus the linear system of equations (4.2) will have a substantial number of nonzeros which will ultimately render the solution computationally infeasible. This, however, does not mean that radial basis functions cannot be used in data mining. In fact, it has been found (Powell 1992) that the approximation order of radial basis functions increases with the dimension. Further, there are new iterative algorithms that can be used to solve (4.2), sometimes in $\mathcal{O}(n \log n)$ operations (Faul and Powell 1999). Therefore there is some evidence that radial basis functions do not suffer from the curse of dimensionality, although further work remains to be done.

5. Cluster analysis

Clustering refers to the task of grouping objects into classes of similar objects (Kaufman and Rousseeuw 1990). Cluster analysis is important in business, for the identification of customers with similar interests for which similar solutions may be found (market segmentation: see Berry and Linoff (1997, p. 47)), but also in science, where taxonomies of animals, stars or archaeological artefacts build the foundations for the understanding of evolution and development (Gordon 1981, p. 1). Cluster analysis has a long history both in statistical analysis and machine learning (Berry and Linoff 1997, Berson and Smith 1997, Gordon 1981, Hartigan 1975, Ripley 1996). Clusters of similar objects form one type of information discovered in data mining and can lead to new insights and suggest further actions. However, in order to be applicable for data mining, clustering algorithms need to have the following properties (see Han and Kamber (2001)):

- *scalable* in order to work on large and growing data sizes,
- capable of handling a large number of attributes of various types and complex data,
- capable of handling noisy data,
- insensitive to the ordering of the data, and
- able to represent clusters of arbitrary shapes.

The result of a cluster analysis is one or several subsets $\Omega' \subset \Omega$ of the set of all possible observations. Many clustering algorithms partition the domain Ω into (possibly disjoint) subdomains Ω_i such that

$$\Omega = \bigcup \Omega_i.$$

In data mining explorations, clustering is often combined with classification and rule detection in a 3-step procedure:

(1) determination of clusters,
(2) induction of a decision tree to predict the cluster label,
(3) extraction of rules relating to the clusters.

In order to find clusters, the dissimilarity or similarity of transactions or objects need to be assessed and most commonly, this is done with metrics on feature vectors:

- Minkowski distances (including Euclidean and Manhattan), or p-norms, for real-valued features, that is,

$$\rho(\omega_1, \omega_2) = \left(\sum_{i=1}^{d} |X_i(\omega_1) - X_i(\omega_2)|^q \right)^{1/q}$$

- simple matching for categorical (and binary) features, that is,

$$\rho(\omega_1, \omega_2) = |\{X_i : X_i(\omega_1) \neq X_i(\omega_2)\}|.$$

Often the distances are normalized or weighted differences are used. The four main types of clustering algorithm used in data mining are as follows.

Partitioning methods. If the number of clusters is known in advance, the data is partitioned into sets of similar data records. Starting with some initial partitioning (often random), data points are moved between clusters until the differences of objects within the clusters is small and the difference between elements of different clusters is large.

Hierarchical clustering methods. These methods provide a hierarchical partitioning of the data. Clusters are determined from the hierarchical partitioning.

There are two basic approaches. In the *agglomerative* approach an initial partitioning is given by first identifying each separate data point with one cluster, and then repeatedly joining clusters together to form successively larger clusters until only one cluster is left.

In the *divisive* approach all points are first assumed to be part of one big cluster. The partitions are then successively divided into smaller clusters until every point forms one separate cluster.

Density-based methods. Neighbouring elements are joined together using a local density measure. Often, only one scan through the database is required for this algorithm.

Model-based methods. Each cluster is modelled, for instance by a simple distribution function, and the best fit of these models to the data is determined.

Many clustering algorithms have an $\mathcal{O}(N^2)$ complexity, which results from the fact that the distances between all the points are computed. There are two major approximation techniques that address this computational problem: sampling (Kaufman and Rousseeuw 1990), where a subset of the data is selected to perform the algorithm, and binning, which is based on discretization of the domain (Wang, Yang and Muntz 1997). In both approaches the full dataset may be used for the evaluation of the result, particularly since this evaluation often only requires $\mathcal{O}(N)$ complexity. Both algorithms can lead to scalable algorithms, if, in the first case, only $\mathcal{O}(\sqrt{N})$ elements are chosen from the sample, and, in the second case, the bin sizes are not chosen to be too large and the dimension is low. Note that binning becomes less useful in higher dimensions, because it suffers from the curse of dimensionality.

5.1. Partitioning techniques

The aim of the clustering is to find subsets which are more pure than the original set in the sense that, on average, their elements are much more similar than the elements of the original domain. A partition $\Omega_1, \ldots, \Omega_k$ is represented by the *centroids* $z_1, \ldots, z_k$ such that

$$\mathbf{x} \in \Omega_i \iff \rho(\mathbf{x}, \mathbf{z}_i) \le \rho(\mathbf{x}, \mathbf{z}_j), \quad i, j = 1, \ldots, k.$$

The centroids define an *impurity measure* of the form

$$J(\mathbf{z}_1, \ldots, \mathbf{z}_p) = \frac{1}{N} \sum_{i=1}^{k} \sum_{\mathbf{x}^{(j)} \in \Omega_i} \rho(\mathbf{x}^{(j)}, \mathbf{z}_i) \tag{5.1}$$

$$= \frac{1}{N} \sum_{j=1}^{N} \min_{1 \le i \le k} \rho(\mathbf{x}^{(j)}, \mathbf{z}_i). \tag{5.2}$$

For simplicity we identify ω with the feature vector $\mathbf{x}$ here.

The two types of algorithm for partitioning differ in the way they estimate the centroid. In the k-means algorithm, the mean of the (real-valued) observations in Ω_i is used:

$$\mathbf{z}_i = \frac{1}{N_i} \sum_{\mathbf{x}^{(j)} \in \Omega_i} \mathbf{x}^{(j)},$$

where N_i denotes the number of data points in Ω_i. One disadvantage of the k-means approach is that the mean cannot be guaranteed to be close to any data point at all and the data are limited to real vectors. An alternative is the k-medoid approach, where the centroid is chosen to be the most central element of the set, *i.e.*,

$$\mathbf{z}_i = \mathbf{x}^{(s_i)}$$

such that

$$\sum_{\mathbf{x}^{(j)} \in \Omega_i} \rho(\mathbf{x}^{(j)}, \mathbf{x}^{(s_i)}) \le \sum_{\mathbf{x}^{(j)} \in \Omega_i} \rho(\mathbf{x}^{(j)}, \mathbf{x}^{(m)}), \quad \text{for all } \mathbf{x}^{(m)} \in \Omega_i.$$

k-means algorithm

The *k-means* algorithm (see Algorithm 3) was discussed in MacQueen (1967). It can be seen that the k-means algorithm cannot increase the function J, and, in fact, if any clusters are changed, J is reduced. As J is bounded from below it converges, and as a consequence the algorithm converges. It is also known that the k-means will always converge to a local minimum (Bottou and Bengio 1995). The k-means algorithm may be viewed as a variant of the EM algorithm (McLachlan and Krishnan 1996).

Algorithm 3 k-means algorithm

Select k arbitrary data points $\mathbf{z}_1, \ldots, \mathbf{z}_k$.
repeat
$\quad \Omega_i := \{\mathbf{x}^{(j)} : \rho(\mathbf{x}^{(j)}, \mathbf{z}_i) \leq \rho(\mathbf{x}^{(j)}, \mathbf{z}_s), \quad s = 1, \ldots, p \}$
$\quad \mathbf{z}_i := \frac{1}{|\Omega_i|} \sum_{\mathbf{x}^{(j)} \in \Omega_i} \mathbf{x}^{(j)}.$
until the $\mathbf{z}_i$ converge

Dhillon and Modha (2000) propose a parallel k-means algorithm. They also provide a careful analysis of the algorithm's computational complexity. There are two major steps in the algorithm: the determination of the distances between all the points, and the recalculation of the centroid. The determination of all the Euclidean distances between the N points and the k-cluster means requires $3Nkd$ floating point operations (one subtraction, square and one addition per component of all pairs). Finding the minimum for each point requires a total of kN comparisons; then we need to compute the new average for each cluster, which requires nd additions and kd divisions. In data mining the cost is usually dominated by the determination of all the distances, and thus the time is (Dhillon and Modha 2000)

$$T = \mathcal{O}(NkdI),$$

where I is the number of iterations.

For the parallel algorithm (shared nothing), the data are initially distributed over the discs of all the processors. Then each processor computes the distances of its elements to all cluster centres. This is done in parallel and so the most expensive computation gets a parallel speed-up of a factor of p (the number of processors). After the sums of all the elements are computed on all the processors, these sums are communicated, which requires an all-to-all communication with volume dk per iteration per processor. If this can be done in parallel, the time required is $\mathcal{O}(dkI\tau)$, where the time τ for reduction is typically of $\mathcal{O}(\sqrt{p})$. Thus the total time is

$$\mathcal{O}(NkdI/p) + \mathcal{O}(dkI\sqrt{p}).$$

Now the communication time is small compared to the computation time for large problems, that is, if

$$N \gg p^{3/2}\kappa,$$

where κ is the ratio of communication time to computation time. Typically, p is in the order of 100 and κ around 1000, thus data sizes of a million do qualify. A scalable k-means method has been discussed in Bradley, Fayyad and Reina (1998).

k-medoids algorithm

Determining a set of elements $\mathbf{x}^{(j_s)}$ which minimize $J(\mathbf{x}^{(j_1)}, \ldots, \mathbf{x}^{(j_k)})$ defines a discrete optimization problem. The search graph of this problem has as nodes all possible sets of centroids, *i.e.*, all possible $\binom{N}{k}$ combinations of the points $\mathbf{x}^{(i)}$. The edges of the search graph join any two sets of centroids $\{\mathbf{x}^{(j_1)}, \ldots, \mathbf{x}^{(j_k)}\}$ and $\{\mathbf{x}^{(i_1)}, \ldots, \mathbf{x}^{(i_k)}\}$ if they only differ in one centroid. The k-medoids algorithm starts at a random node and moves to the adjacent node for which the reduction in impurity is maximal. This descent approach requires the determination of J for all adjacent nodes in order to determine the best direction. The algorithm finds a local minimum.

The algorithm stores all the distances between data points and all the centres. From this the minimal distance between any data point and the centres can be determined in $\mathcal{O}(k(N-k))$ comparisons. Thus the complexity of the algorithm is $\mathcal{O}(k^2(N-k)^2)$. However, for many data points the minimum can actually be found without comparing the distance between the new centroid and the data points and all the earlier computed distances. This is based on the following simple observation.

Lemma 12. If $\rho_1, \ldots, \rho_{k+1} \in \mathbb{R}$ satisfy $\rho_1 > \min(\rho_1, \ldots, \rho_k)$ or $\rho_{k+1} \leq \rho_1$, we have

$$\min(\rho_2, \ldots, \rho_{k+1}) = \min(\min(\rho_1, \ldots, \rho_k), \rho_{k+1}).$$

Proof. The conditions on ρ_1 imply that including ρ_1 does not change the minimum in both cases, that is,

$$\min(\rho_2, \ldots, \rho_{k+1}) = \min(\rho_1, \ldots, \rho_{k+1}). \qquad \square$$

Applying this to the k-medoid algorithm, we see that each term in the sum for J does have a minimum over all the centroids. If one centroid is replaced by another, say, centroid 1 is replaced by centroid $k+1$, then we have to compute the minimum $\min(\rho(x^{(i)}, z_2), \ldots, \rho(x^{(i)} z_{k+1}))$. This can be computed from the previously determined minimum over the original centroids using Lemma 12, except in the case, not covered by the lemma, for which the centroid is just the one that is replaced, and the distance between the new centroid and the data point is larger than the distance between the removed centroid and the data point. This algorithm is the PAM (partitioning around medoids) algorithm: see Kaufman and Rousseeuw (1990).

The complexity is further reduced by repeated clustering on a random sample of the data and always selecting the cluster with the lowest value of J. This is implemented in the CLARA (clustering large applications) algorithm and leads to an algorithm that is scalable in the number of data points N. However, there are still very many search directions and a random selection is used in the CLARANS (clustering large applications based on randomized search) algorithm (Ng and Han 1994). This CLARANS algorithm is one of the early developments in clustering for data mining applications.

5.2. Hierarchical clustering

Hierarchical clustering algorithms transform a set of points with an associated dissimilarity metric into a tree structure known as a *dendrogram*. The nodes of the dendrogram correspond to sets of data points: for example, the leaves are single points. Like decision trees, dendrograms split the datasets at each node and the edges of the dendrogram correspond to set inclusions.

The two broad classes of hierarchical clustering algorithms are *agglomerative*, which start with single points and join them together whenever they are close, whereas *divisive* algorithms move from the top down and break the clusters up when they are dissimilar.

In contrast to decision trees, where the impurity of the nodes is determined through the value of a class attribute, in clustering the dissimilarity measure provides this measure of impurity.

An agglomerative hierarchical clustering algorithm consists of the following loop:

    Initialize all points as single clusters
    Determine all the dissimilarities between the clusters
    while There is more than one cluster **do**
      Merge the two clusters with the smallest dissimilarity
      Update the dissimilarities
    end while

It is important to assess the dissimilarity or impurity which is introduced through the joining of two partitions Ω_1 and Ω_2. The *single link* (or nearest neighbour) technique defines the dissimilarity as

$$\rho(A, B) = \min\{\rho(x, y) : x \in \Omega_1, y \in \Omega_2\}, \quad \text{for two partitions } \Omega_1 \text{ and } \Omega_2.$$

Other measures are used and they have a big effect on the computational load.

After the full generation of the dendrogram, it is pruned down to a desired level. The single link algorithm computes all the distances but does not require us to store them all. The classical SLINK (Sibson 1973) algorithm requires $\mathcal{O}(N^2)$ time and $\mathcal{O}(N)$ space.

In the case of a database distributed over p processors, Johnson and Kargupta (2000) present an algorithm for single-link clustering which has $\mathcal{O}(pN^2)$ time and $\mathcal{O}(pN)$ space complexity. The communication costs of the algorithm are $\mathcal{O}(N)$. The algorithm has the following three steps:

    Apply the hierarchical clustering algorithm at each site.
    Transmit the local dendrograms to the facilitator site.
    Generate the global dendrogram.

The local dendrograms are communicated together with the distances of each partition. Then an upper bound on the distance is obtained as the sum of the distances of the shortest path in each partition. This is an upper bound on the actual distance. We have

$$\rho(x_1, x_2) = \sqrt{\sum_{j=1}^{p}\left(\sum_{i\in P_j}(x_{1,i,j} - x_{i,2,j})^2\right)} \le \sqrt{\sum_{j=1}^{p}(\text{dist}_{\text{dendrogram }j}(x_1, x_2))}.$$

Basically, it uses the distance defined in the actual dendrograms. The merging algorithm is costly but the main advantage is that not all the data need to be communicated to one processor. For a parallel implementation of SLINK see Olson (1995).

There is a large variety of data mining clustering techniques including BIRCH (Zhang, Ramakrishnan and Livny 1996) and CURE (Guha, Rastogi and Shim 1998). A further class of clustering algorithms is based on density considerations, both using parametric models (Cheeseman and Stutz 1996) and non-parametric approaches in DBSCAN (Ester, Kriegel, Sander and Xu 1996).

Acknowledgement

The author would like to thank the editors and Brad Baxter (Imperial College, London), who carefully read the paper.

REFERENCES

R. Agrawal, T. Imielinski and T. Swami (1993), Mining association rules between sets of items in large databases, in *Proc. ACM–SIGMOD Conf. Management of Data*, ACM Press, pp. 207–216.

R. Agrawal and R. Srikant (1995), Mining sequential patterns, in *Proc. 11th Int. Conf. Data Engineering*, IEEE CS Press, Los Alamitos, CA, pp. 3–14.

S. Bakin, M. Hegland and M. Osborne (1998), Can MARS be improved with B-splines?, in *Computational Techniques and Applications: CTAC97* (B. J. Noye, M. D. Teubner and A. W. Gill, eds), World Scientific, pp. 75–82.

K. Ball (1997) 'An elementary introduction to modern convex geometry', in *Flavors of Geometry* (S. Levy, ed.), Cambridge University Press.

R. K. Beatson, G. Goodsell and M. J. D. Powell (1996), On multigrid techniques for thin plate spline interpolation in two dimensions, in *The Mathematics of Numerical Analysis* (Park City, UT, 1995), Vol. 32 of *Lectures in Appl. Math.*, American Mathematical Society, Providence, RI, pp. 77–97.

R. K. Beatson and W. A. Light (1997), Fast evaluation of radial basis functions: Methods for two-dimensional polyharmonic splines, *IMA J. Numer. Anal.*, **17** 343–372.

R. K. Beatson and G. N. Newsam (1992), Fast evaluation of radial basis functions, I, *Comput. Math. Appl.*, **24** 7–19.

R. K. Beatson and M. J. D. Powell (1994), An iterative method for thin plate spline interpolation that employs approximations to Lagrange functions, in *Numerical Analysis 1993*, Vol. 303 of *Pitman Res. Notes Math. Ser.*, Longman Sci. Tech., Harlow, pp. 17–39.

G. Bell and J. N. Gray (1997), The revolution yet to happen, in *Beyond Calculation* (P. J. Denning and R. M. Metcalfe, eds), Springer, pp. 5–32.

E. A. Bender (1996), *Mathematical Methods in Artificial Intelligence*, IEEE CS Press, Los Alamitos, CA.

M. J. A. Berry and G. Linoff (1997), *Data Mining Techniques: For Marketing, Sales and Customer Support*, Wiley.

A. Berson and S. J. Smith (1997), *Data Warehousing, Data Mining, and OLAP*, McGraw-Hill Series on Data Warehousing and Data Management, McGraw-Hill, New York.

H.-H. Bock and E. Diday, eds (2000), *Analysis of Symbolic Data*, Springer.

L. Bottou and Y. Bengio (1995), Convergence properties of the k-means algorithm, in *Adv. in Neural Info. Proc. Systems*, Vol. 7 (G. Tesauro and D. Touretzky, eds), MIT Press, Cambridge, MA, pp. 585–592.

P. Bradley, U. Fayyad and C. Reina (1998), Scaling clustering algorithms to large databases, in *Proc. 4th Int. Conf. KDD*, pp. 9–15.

L. Breiman, J. H. Friedman, R. A. Olshen and C. J. Stone (1984), *Classification and Regression Trees*, Wadsworth International Group, Blemont, CA.

S. Brin *et al.* (1997), Dynamic itemset counting and implication rules for market basket data, in *Proc. ACM–SIGMOD Int. Conf. Management of Data*, ACM Press, New York, pp. 255–264.

M. D. Buhmann (1993), New developments in the theory of radial basis function interpolation, in *Multivariate Approximation: From CAGD to Wavelets* (Santiago, 1992), Vol. 3 of *Ser. Approx. Decompos.*, World Scientific, River Edge, NJ, pp. 35–75.

P. Cheeseman and J. Stutz (1996), Bayesian classification (Autoclass): Theory and results, in *Advances in Knowledge Discovery and Data Mining* (U. M. Fayyad, G. Piatetsky-Shapiro, P. Smyth and R. Uthurusamy, eds), AAAI/MIT Press, Cambridge, MA, pp. 153–180.

M.-S. Chen, J. Han and P. S. Yu (1996), Data mining: An overview from a database perspective, *IEEE Trans. Knowledge and Data Engineering*, **8** 866–883.

S. Clearwater and F. Provost (1990), Rl4: A tool for knowledge-based induction.

W. W. Cohen (1995), Fast effective rule induction, in *Proc. 12th Int. Conf. Machine Learning*, Morgan Kaufmann, pp. 115–123.

C. J. Date (1995), *An Introduction to Database Systems*, Addison-Wesley, Reading, MA.

L. Devroye, L. Györfi and G. Lugosi (1996), *A Probabilistic Theory of Pattern Recognition*, Vol. 31 of *Applications of Mathematics*, Springer.

I. S. Dhillon and D. S. Modha (2000), A data-clustering algorithm on distributed memory multiprocessors, in *Large-Scale Parallel Data Mining* (M. J. Zaki and C.-T. Ho, eds), Springer, pp. 245–260.

A. J. Dobson (1990), *An Introduction to Generalized Linear Models*, Chapman and Hall, London. Second edn of *Introduction to Statistical Modelling*.

P. Domingos and M. Pazzani (1997), On the optimality of the simple Bayesian classifier under zero–one loss, *Machine Learning*, **29** 103–130.

G. Dong and J. Li (1999), Efficient mining of emerging patterns: Discovering trends and differences, in *Proc. 1999 Int. Conf. Knowledge Discovery and Data Mining* (KDD'99), ACM Press, pp. 43–52.

R. O. Duda and P. E. Hart (1973), *Pattern Classification and Scene Analysis*, Wiley, New York.

N. Dyn (1989), *Interpolation and Approximation by Radial and Related Functions*, Vol. 1, Academic Press, pp. 211–234.

M. Ester, H.-P. Kriegel, J. Sander and X. Xu (1996), A density-based algorithm for discovering clusters in large spatial databases, in *Proc. 1996 Int. Conf. Knowledge Discovery and Data Mining* (KDD'96), AAAI Press, pp. 226–231.

A. C. Faul and M. J. D. Powell (1999) Proof of convergence of an iterative technique for thin plate spline interpolation in two dimensions, *Adv. Comput. Math.* **11** 183–192.

J. H. Friedman (1991), Multivariate adaptive regression splines, *The Annals of Statistics*, **19** 1–141.

J. H. Friedman (1994), Flexible metric nearest neighbor classification, Technical Report, Department of Statistics, Stanford University.

N. Friedman, D. Geiger and M. Goldszmidt (1997), Bayesian network classifiers, *Machine Learning*, **29** 131–163.

G. Golub, M. Heath and G. Wahba (1979), Generalized cross validation as a method for choosing a good ridge parameter, *Technometrics*, **21** 215–224.

A. D. Gordon (1981), *Classification*, Vol. 82 of *Monographs on Statistics and Applied Probability*, Chapman and Hall, London.

S. Guha, R. Rastogi and K. Shim (1998), CURE: An efficient clustering algorithm for large databases, in *Proc. ACM–SIGMOD Int. Conf. Management of Data*, ACM Press, pp. 73–84.

L. Hall, N. Chawla and K. Bowyer (1998), Decision tree learning on very large data sets, in *Int. Conf. Systems, Man and Cybernetics*, IEEE Press, pp. 2579–2584,

L. O. Hall, N. Chawla, K. W. Bowyer and W. P. Kegelmayer (2000), Learning rules from distributed data, in *Large-Scale Parallel Data Mining* (M. J. Zaki and C.-T. Ho, eds), Springer, pp. 211–220.

J. Han and M. Kamber (2001), *Data Mining, Concepts and Techniques*, Morgan Kaufmann.

J. A. Hartigan (1975), *Clustering Algorithms*, Wiley Series in Probability and Mathematical Statistics, Wiley, New York/London/Sydney.

T. J. Hastie and R. J. Tibshirani (1990), *Generalized Additive Models*, Vol. 43 of *Monographs on Statistics and Applied Probability*, Chapman and Hall.

D. Heckerman, D. Geiger and D. E. Chickering (1995), Learning Bayesian networks, *Machine Learning*, **20** 197–243.

G. H. John and P. Langley (1995), Estimating continuous distributions in Bayesian classifiers, in *Proc. 11th Conference on Uncertainty in Artificial Intelligence* (P. Besnard and S. Hanks, eds), Morgan Kaufmann, pp. 338–345.

E. L. Johnson and H. Kargupta (2000), Collective, hierarchical clustering from distributed heterogeneous data, in *Large-Scale Parallel Data Mining* (M. J. Zaki and C.-T. Ho, eds), Springer, pp. 221–244.

L. Kaufman and P. J. Rousseeuw (1990), *Finding Groups in Data: An Introduction to Cluster Analysis*, Wiley, New York.

P. Langley, W. Iba and K. Thompson (1992), An analysis of Bayesian classifiers, in *Proc. 10th Nat. Conf. Artificial Intelligence* (W. Swartout, ed.), AAAI Press, pp. 223–228.

P. Langley and S. Sage (1994), Induction of selective Bayesian classifiers, in *Proc. 10th Conf. Uncertainty in Artificial Intelligence* (R. L. Mantaras and D. Poole, eds), Morgan Kaufmann, pp. 399–406.

P. Langley and S. Sage (1999), Tractable average-case analysis of naive Bayesian classifiers, in *Proc. 16th Int. Conf. Machine Learning*, Bled, Slovenia, Morgan Kaufmann, pp. 220–228,

B. Lent, A. Swami and J. Widom (1997), Clustering association rules, in *Proc. 1997 Int. Conf. Data Engineering* (ICDE'97), IEEE CS Press, pp. 220–231.

J. Li, G. Dong and K. Ramamohanrarao (2000), Making use of the most expressive jumping emerging patterns for classification, in *Proc. 2000 Pacific–Asia Conf. Knowledge Discovery and Data Mining* (PAKDD'00), Springer, pp. 220–232.

B. Liu, W. Hsu and Y. Ma (1998), Integrating classification and association rule mining, in *Proc. 1998 Int. Conf. Knowledge Discovery and Data Mining* (KDD'98), AAAI Press, pp. 80–86.

J. MacQueen (1967), Some methods for classification and analysis of multivariate observations, in *Proc. 5th Berkeley Sympos. Math. Statist. and Probability* (1965/66), Vol. I: *Statistics*, University of California Press, Berkeley, CA, pp. 281–297.

G. J. McLachlan and T. Krishnan (1996), *The EM Algorithm and Extensions*, Wiley.

H. Mannila, H. Toivonen and A. I. Verkamo (1995), Discovering frequent episodes in sequences, in *Proc. 1st Int. Conf. Knowledge Discovery Databases and Data Mining* (KDD'95), AAAI Press, pp. 210–215.

H. Mannila and H. Toivonen (1996), Discovering generalized episodes using minimal occurrences, in *Proc. 2nd Int. Conf. Knowledge Discovery Databases and Data Mining* (KDD'96), AAAI Press, pp. 146–151.

D. Meretakis and B. Wüthrich (1999), Extending naive Bayes classifiers using long itemsets, in *Proc. 1999 Int. Conf. Knowledge Discovery and Data Mining* (KDD'99), ACM Press, pp. 165–174.

V. D. Milman and G. Schechtman (1986) *Asymptotic Theory of Finite Dimensional Normed Spaces*, Vol. 1200 of *Lecture Notes in Mathematics*, Springer.

T. M. Mitchell (1997), *Machine Learning*, McGraw-Hill.

S. Morishita (1998), On classification and regression, in *Proc. 1st Int. Conf. Discovery Science*, Vol. 1532, Springer, pp. 40–57.

S. Morishita and A. Nakaya (2000), Parallel branch-and-bound graph search for correlated association rules, in *Large-Scale Parallel Data Mining* (M. J. Zaki and C.-T. Ho, eds), Springer, pp. 127–144.

R. Ng and J. Han (1994), Efficient and effective clustering methods for spatial data mining, in *Proc. 20th Int. Conf. Very Large Data Bases*, Morgan Kaufmann, pp. 144–155.

T. Oates, M. D. Schmill, D. Jensen and P. R. Cohen (1997), A family of algorithms for finding temporal structure in data, in *Preliminary Papers of the 6th Int. Workshop on AI and Statistics*, Society for Artificial Intelligence and Statistics, pp. 371–378.

C. Olson (1995), Parallel algorithms for hierarchical clustering, *Parallel Computing*, **8** 1313–1325.

J. S. Park, M. S. Chen and P. S. Yu (1995*a*), An effective hash-based algorithm for mining association rules, in *Proc. 1995 ACM–SIGMOD Int. Conf. Management of Data* (SIGMOD'95), ACM Press, pp. 175–186.

J. S. Park, M. S. Chen and P. S. Yu (1995*b*), Efficient parallel data mining for association rules, in *Proc. 4th Int. Conf. Information and Knowledge Management*, ACM Press, pp. 31–36.

M. J. D. Powell (1992), The theory of radial basis function approximation in 1990, in *Advances in Numerical Analysis, Vol. II* (Lancaster, 1990), Oxford Sci. Publ., Oxford University Press, New York, pp. 105–210.

F. Provost, J. Aronis and B. Buchanan (2000), Rule-space search for knowledge-based discovery, CIIO Working Paper Number #IS 99-0012, Stern School of Business, New York University, NY. Available from:
http://www.stern.nyu.edu/ fprovost/

J. R. Quinlan (1986), Induction of decision trees, *Machine Learning*, **1** 81–106.

J. R. Quinlan (1987), Generating production rules from decision trees, in *Proc. 10th Int. Conf. Artificial Intelligence* (IJCAI-87), Morgan Kaufmann, pp. 304–307.

J. R. Quinlan (1993), *C4.5: Programs for Machine Learning*, Morgan Kaufmann.

N. Ramakrishnan and A. Y. Grama (1999), Data mining: From serendipity to science, *Computer*, **32**(8) 34–37.

S. Ramaswamy, S. Mahajan and A. Silbershatz (1998), On the discovery of interesting patterns in association rules, in *Proc. 24th Int. Conf. Very Large Data Bases*, Morgan Kaufmann, pp. 368–379.

B. D. Ripley (1996), *Pattern Recognition and Neural Networks*, Cambridge University Press, Cambridge.

A. Savasere, E. Omieski and S. Navanthe (1995), An efficient algorithm for mining association rules in large databases, in *Proc. 21st Int. Conf. Very Large Data Bases*, Morgan Kaufmann, pp. 432–444.

R. Segal and O. Etioni (1994), Learning decision lists using homogeneous rules, in *Proc. 12th Nat. Conf. Artificial Intelligence*, AAAI Press, pp. 619–625.

J. Shafer, R. Agrawal and M. Mehta (1996), SPRINT: A scalable parallel classifier for data mining, in *Proc. 1996 Int. Conf. Very Large Data Bases* (VLDB'96), Morgan Kaufmann, pp. 544–555.

T. Shintani and M. Kitsuregawa (2000), Parallel generalized association rule mining on large scale PC clusters, in *Large-Scale Parallel Data Mining* (M. J. Zaki and C.-T. Ho, eds), Springer, pp. 145–160.

R. Sibson (1973), SLINK: An optimally efficient algorithm for the single-link cluster method, *The Computer Journal*, **16** 30–34.

A. Silberschatz and A. Tuzhilin (1996), What makes patterns interesting in knowledge discovery systems, *IEEE Trans. Knowledge and Data Engineering*, **8** 970–974.

R. Srikant and R. Agrawal (1995), Mining generalized association rules, in *Proc. 1995 Int. Conf. Very Large Data Bases* (VLDB'95), pp. 407–419.

R. Srikant and R. Agrawal (1996 a), Mining quantitative association rules in large relational tables, in *Proc. 1996 ACM–SIGMOD Int. Conf. Management of Data* (SIGMOD'96), ACM Press, pp. 1–12.

R. Srikant and R. Agrawal (1996 b), Mining sequential patterns: Generalizations and performance improvements, in *5th Int. Conf. Extending Database Technology*, Vol. 1057 of *Lecture Notes in Computer Science*, Springer, pp. 3–17.

M. Talagrand (1996), A new look at independence, *Ann. Prob.*, **23** 1–37.

H. Toivonen (1996), Sampling large databases for association rules, in *Proc. 22nd Int. Conf. Very Large Data Bases*, Morgan Kaufmann, pp. 134–145.

W. Wang, J. Yang and R. Muntz (1997), STING: A statistical information grid approach to spatial data mining, in *Proc. 1997 Int. Conf. Very Large Data Bases* (VLDB'97), Morgan Kaufmann, pp. 186–195.

G. Webb (2000), Efficient search for association rules, in *Proc. of the 6th ACM–SIGKDD Int. Conf. Knowledge Discovery and Data Mining*, ACM Press, pp. 99–107.

G. Williams (1990), Inducing and combining multiple decision trees, PhD thesis, Australian National University.

M. J. Zaki (1998), Efficient enumeration of frequent sequences, in *7th Int. Conf. Information and Knowledge Management*, ACM Press, pp. 68–75.

M. J. Zaki (2000), Parallel sequence mining on shared-memory machines, in *Large-Scale Parallel Data Mining* (M. J. Zaki and C.-T. Ho, eds), Springer, pp. 161–189.

M. J. Zaki and C.-T. Ho, eds (2000), *Large-Scale Parallel Data Mining*, Vol. 1759 of *Lecture Notes in Artificial Intelligence*, Springer.

T. Zhang, R. Ramakrishnan and M. Livny (1996), BIRCH: An efficient data clustering method for very large databases, in *Proc. 1996 ACM–SIGMOD Int. Conf. Management of Data* (SIGMOD'96), ACM Press, pp. 103–114.

Acta Numerica (2001), pp. 357–514

Discrete mechanics and variational integrators

J. E. Marsden and M. West

Control and Dynamical Systems 107-81,

Caltech, Pasadena, CA 91125-8100, USA

E-mail: `marsden@cds.caltech.edu`

`mwest@cds.caltech.edu`

This paper gives a review of integration algorithms for finite dimensional mechanical systems that are based on discrete variational principles. The variational technique gives a unified treatment of many symplectic schemes, including those of higher order, as well as a natural treatment of the discrete Noether theorem. The approach also allows us to include forces, dissipation and constraints in a natural way. Amongst the many specific schemes treated as examples, the Verlet, SHAKE, RATTLE, Newmark, and the symplectic partitioned Runge–Kutta schemes are presented.

CONTENTS

PART ONE
Discrete variational mechanics

1.1. Introduction

This paper gives a survey of the variational approach to discrete mechanics and to mechanical integrators. This point of view is not confined to conservative systems, but also applies to forced and dissipative systems, so is useful for control problems (for instance) as well as traditional conservative problems in mechanics. As we shall show, the variational approach gives a comprehensive and unified view of much of the literature on both discrete mechanics as well as integration methods for mechanical systems and we view these as closely allied subjects.

Some of the important topics that come out naturally from this method are symplectic–energy-momentum methods, error analysis, constraints, forcing, and Newmark algorithms. Besides giving an account of methods such as these, we connect these techniques to other recent and exciting developments, including the PDE setting of multisymplectic spacetime integrators (also called AVI, or asynchronous variational integrators), which are being used for problems such as nonlinear wave equations and nonlinear shell dynamics. In fact, one of our points is that by basing the integrators on fundamental mechanical concepts and methods from the outset, one eases the way to other areas, such as continuum mechanics and systems with forcing and constraints.

In the last few years this subject has grown to be very large and an active area of research, with many points of view and many topics. We shall focus here on our own point of view, namely the variational view. Naturally we must omit a number of important topics, but include several of our own. We do make contact with some, but not all, of other topics in the final part of this article and in the brief history below.

As in standard mechanics, some things are easier from a Hamiltonian perspective and others are easier from a Lagrangian perspective. Regarding symplectic integrators from both viewpoints gives greater insight into their properties and derivations. We have tried to give a balanced perspective in this article.

We will assume that the configuration manifold is finite-dimensional. This means that at the outset, we will deal with the context of ordinary differential equations. However, as we have indicated, our approach is closely tied with the variational spacetime multisymplectic approach, which is the approach that is suitable for the infinite-dimensional, PDE context, so an investment in the methodology of this article eases the transition to the corresponding PDE context.

One of the simple, but important ideas in discrete mechanics is easiest to say from the Lagrangian point of view. Namely, consider a mechanical system with configuration manifold Q. The velocity phase space is then TQ and the Lagrangian is a map $L : TQ \to \mathbb{R}$. In discrete mechanics, the starting point is to replace TQ with $Q \times Q$ and we regard, intuitively, two nearby points as being the discrete analogue of a velocity vector.

There is an important note about constraints that we would like to say at the outset. Recall from basic geometric mechanics (as in Marsden and Ratiu (1999) for instance) that specifying a constraint manifold Q means that one may already have specified constraints: for example, Q may already be a submanifold of a linear space that is specified by constraints. However, when constructing integrators in Section 2.1 we will take Q to be linear, although this is only for simplicity. One way of handling a nonlinear Q is to embed it within a linear space and use the theory of constrained systems: this point of view is developed in Section 3. This approach has computational advantages, but we will also discuss implementations of variational integrators on arbitrary configuration manifolds Q.

1.1.1. History and literature

Of course, the variational view of mechanics goes back to Euler, Lagrange and Hamilton. The form of the variational principle most important for continuous mechanics we use in this article is due, of course, to Hamilton (1834). We refer to Marsden and Ratiu (1999) for additional history, references and background on geometric mechanics.

There have been many attempts at the development of a discrete mechanics and corresponding integrators that we will not attempt to survey in any systematic fashion. The theory of discrete variational mechanics in the form we shall use it (that uses $Q \times Q$ for the discrete analogue of the velocity phase space) has its roots in the optimal control literature of the 1960s: see, for example, Jordan and Polak (1964), Hwang and Fan (1967) and Cadzow (1970). In the context of mechanics early work was done, often independently, by Cadzow (1973), Logan (1973), Maeda (1980, 1981a, 1981b), and Lee (1983, 1987), by which point the discrete action sum, the discrete Euler–Lagrange equations and the discrete Noether's theorem were clearly understood. This theory was then pursued further in the context of integrable systems in Veselov (1988, 1991) and Moser and Veselov (1991), and in the context of quantum mechanics in Jaroszkiewicz and Norton (1997a, 1997b) and Norton and Jaroszkiewicz (1998).

The variational view of discrete mechanics and its numerical implementation is further developed in Wendlandt and Marsden (1997a) and (1997b) and then extended in Kane, Marsden and Ortiz (1999a), Marsden, Pekarsky and Shkoller (1999a, 1999b), Bobenko and Suris (1999a, 1999b) and Kane,

Marsden, Ortiz and West (2000). The beginnings of an extension of these ideas to a nonsmooth framework is given in Kane, Repetto, Ortiz and Marsden (1999b), and is carried further in Fetecau, Marsden, Ortiz and West (2001).

Other discretizations of Hamilton's principle are given in Mutze (1998), Cano and Lewis (1998) and Shibberu (1994). Other versions of discrete mechanics (not necessarily discrete Hamilton's principles) are given in (for instance) Itoh and Abe (1988), Labudde and Greenspan (1974, 1976a, 1976b), and MacKay (1992).

Of course, there have been many works on symplectic integration, largely done from other points of view than that developed here. We will not attempt to survey this in any systematic fashion, as the literature is simply too large with too many points of view and too many intricate subtleties. We give a few highlights and give further references in the body of the paper. For instance, we shall connect the variational view with the generating function point of view that was begun in De Vogelaére (1956). Generating function methods were developed and used in, for example, Ruth (1983), Forest and Ruth (1990) and in Channell and Scovel (1990). See also Berg, Warnock, Ruth and Forest (1994), and Warnock and Ruth (1991, 1992). For an overview of symplectic integration, see Sanz-Serna (1992b) and Sanz-Serna and Calvo (1994). Qualitative properties of symplectic integration of Hamiltonian systems are given in Gonzalez, Higham and Stuart (1999) and Cano and Sanz-Serna (1997). Long-time energy behaviour for oscillatory systems is studied in Hairer and Lubich (2000). Long-time behaviour of symplectic methods for systems with dissipation is given in Hairer and Lubich (1999). A numerical study of preservation of adiabatic invariants is given in Reich (1999b) and Shimada and Yoshida (1996). Backward error analysis is studied in Benettin and Giorgilli (1994), Hairer (1994), Hairer and Lubich (1997) and Reich (1999a). Other ideas connected to the above literature include those of Baez and Gilliam (1994), Gilliam (1996), Gillilan and Wilson (1992). For other references see the large literature on symplectic methods in molecular dynamics, such as Schlick, Skeel et al. (1999), and for various applications, see Hardy, Okunbor and Skeel (1999), Leimkuhler and Skeel (1994), Barth and Leimkuhler (1996) and references therein.

A single-step variational idea that is relevant for our approach is given in Ortiz and Stainier (1998), and developed further in Radovitzky and Ortiz (1999), and Kane et al. (1999b, 2000).

Direct discretizations on the Hamiltonian side, where one discretizes the Hamiltonian and the symplectic structure, are developed in Gonzalez (1996b) and (1996a) and further in Gonzalez (1999) and Gonzalez et al. (1999). This is developed and generalized much further in McLachlan, Quispel and Robidoux (1998) and (1999).

Finally, we mention that techniques of geometric integration in the sense of preserving manifold or Lie group structures, as given in Budd and Iserles (1999), Iserles, Munthe-Kaas and Zanna (2000) and references therein, presumably could, and probably should, be combined with the techniques described herein for a more efficient treatment of certain classes of constraints in mechanical systems. Such an enterprise is for the future.

1.1.2. A simplified introduction

In this section we give a brief overview of how discrete variational mechanics can be used to derive variational integrators. We begin by reviewing the derivation of the Euler–Lagrange equations, and then show how to mimic this process on a discrete level.

For concreteness, consider the Lagrangian system $L(q, \dot{q}) = \frac{1}{2}\dot{q}^T M \dot{q} - V(q)$, where M is a symmetric positive-definite mass matrix and V is a potential function. We work in $\mathbb{R}^n$ or in generalized coordinates and will use vector notation for simplicity, so $q = (q^1, q^2, \ldots, q^n)$. In the standard approach of Lagrangian mechanics, we form the action function by integrating L along a curve $q(t)$ and then compute variations of the action while holding the endpoints of the curve $q(t)$ fixed. This gives

$$\delta \int_0^T L\big(q(t), \dot{q}(t)\big)\, \mathrm{d}t = \int_0^T \left[\frac{\partial L}{\partial q} \cdot \delta q + \frac{\partial L}{\partial \dot{q}} \cdot \delta \dot{q} \right] \mathrm{d}t$$

$$= \int_0^T \left[\frac{\partial L}{\partial q} - \frac{\mathrm{d}}{\mathrm{d}t}\left(\frac{\partial L}{\partial \dot{q}} \right) \right] \cdot \delta q\, \mathrm{d}t,$$

where we have used integration by parts and the condition $\delta q(T) = \delta q(0) = 0$. Requiring that the variations of the action be zero for all δq implies that the integrand must be zero for each time t, giving the well-known Euler–Lagrange equations

$$\frac{\partial L}{\partial q}(q, \dot{q}) - \frac{\mathrm{d}}{\mathrm{d}t}\left(\frac{\partial L}{\partial \dot{q}}(q, \dot{q}) \right) = 0.$$

For the particular form of the Lagrangian chosen above, this is just

$$M\ddot{q} = -\nabla V(q),$$

which is Newton's equation: mass times acceleration equals force. It is well known that the system described by the Euler–Lagrange equations has many special properties. In particular, the flow on state space is symplectic, meaning that it conserves a particular two-form, and if there are symmetry actions on phase space then there are corresponding conserved quantities of the flow, known as momentum maps.

We will now see how discrete variational mechanics performs an analogue of the above derivation. Rather than taking a position q and velocity $\dot{q}$, consider now two positions q_0 and q_1 and a time-step $h \in \mathbb{R}$. These positions should be thought of as being two points on a curve at time h apart, so that $q_0 \approx q(0)$ and $q_1 \approx q(h)$.

We now consider a discrete Lagrangian $L_d(q_0, q_1, h)$, which we think of as approximating the action integral along the curve segment between q_0 and q_1. For concreteness, consider the very simple approximation to the integral $\int_0^T L\,dt$ given by using the rectangle rule[1] (the length of the interval times the value of the integrand with the velocity vector replaced by $(q_1 - q_0)/h$):

$$L_d(q_0, q_1, h) = h\left[\left(\frac{q_1 - q_0}{h}\right)^T M\left(\frac{q_1 - q_0}{h}\right) - V(q_0)\right].$$

Next consider a discrete curve of points $\{q_k\}_{k=0}^N$ and calculate the discrete action along this sequence by summing the discrete Lagrangian on each adjacent pair. Following the continuous derivation above, we compute variations of this action sum with the boundary points q_0 and q_N held fixed. This gives

$$\delta \sum_{k=0}^{N-1} L_d(q_k, q_{k+1}, h)$$

$$= \sum_{k=0}^{N-1} \left[D_1 L_d(q_k, q_{k+1}, h) \cdot \delta q_k + D_2 L_d(q_k, q_{k+1}, h) \cdot \delta q_{k+1}\right]$$

$$= \sum_{k=1}^{N-1} \left[D_2 L_d(q_{k-1}, q_k, h) + D_1 L_d(q_k, q_{k+1}, h)\right] \cdot \delta q_k,$$

where we have used a discrete integration by parts (rearranging the summation) and the fact that $\delta q_0 = \delta q_N = 0$. If we now require that the variations of the action be zero for any choice of δq_k, then we obtain the *discrete Euler–Lagrange equations*

$$D_2 L_d(q_{k-1}, q_k, h) + D_1 L_d(q_k, q_{k+1}, h) = 0,$$

which must hold for each k. For the particular L_d chosen above, we compute

$$D_2 L_d(q_{k-1}, q_k, h) = M\left(\frac{q_k - q_{k-1}}{h}\right)$$

$$D_1 L_d(q_k, q_{k+1}, h) = -\left[M\left(\frac{q_{k+1} - q_k}{h}\right) + h\nabla V(q_k)\right],$$

and so the discrete Euler–Lagrange equations are

$$M\left(\frac{q_{k+1} - 2q_k + q_{k-1}}{h^2}\right) = -\nabla V(q_k).$$

[1] As we shall see later, more sophisticated quadrature rules lead to higher-order accurate integrators.

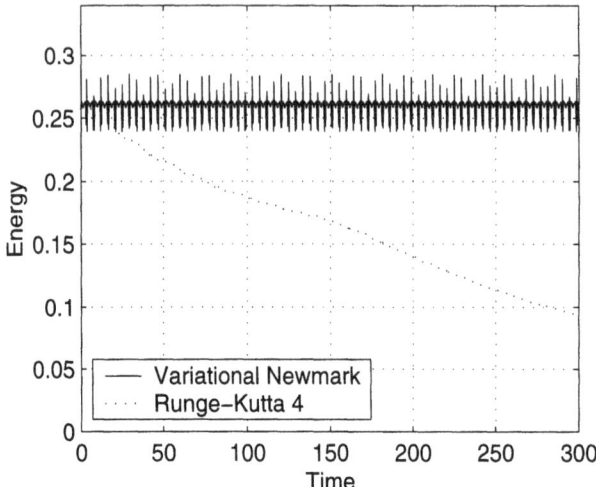

Fig. 1. Energy computed with variational Newmark
(solid line) and Runge–Kutta (dashed line). Note that
the variational method does not artificially dissipate
energy

This is clearly a discretization of Newton's equations, using a simple finite difference rule for the derivative.

If we take initial conditions (q_0, q_1) then the discrete Euler–Lagrange equations define a recursive rule for calculating the sequence $\{q_k\}_{k=0}^N$. Regarded in this way, they define a map $F_{L_d} : (q_k, q_{k+1}) \mapsto (q_{k+1}, q_{k+2})$ which we can think of as a one-step integrator for the system defined by the continuous Euler–Lagrange equations.

Indeed, as we will see later, many standard one-step methods can be derived by such a procedure. An example of this is the well-known Newmark method, which for the parameter settings $\gamma = \frac{1}{2}$ and $\beta = 0$ is derived by choosing the discrete Lagrangian

$$L_d(q_0, q_1, h) = h\left[\left(\frac{q_1 - q_0}{h}\right)^T M \left(\frac{q_1 - q_0}{h}\right) - \left(\frac{V(q_0) + V(q_1)}{2}\right)\right].$$

If we use this variational Newmark method to simulate a model system and plot the energy versus time, then we obtain a graph like that in Figure 1. For comparison, this graph also shows the energy curve for a simulation with a standard stable method such as RK4 (the common fourth-order Runge–Kutta method).

The system being simulated here is purely conservative and so there should be no loss of energy over time. The striking aspect of this graph is that while the energy associated with a standard method decays due to numerical damping, for the Newmark method the energy error remains bounded. This may be understood by recognizing that the integrator is symplectic, that is, it preserves the same two-form on state space as the true system.

In fact, *all variational integrators have this property*, as it is a consequence of the variational method of derivation, just as it is for continuous Lagrangian systems. In addition, they will also have the property of conserving momentum maps arising from symmetry actions, again due to the variational derivation. To understand this behaviour more deeply, however, we must first return to the beginning and consider in more detail the geometric structures underlying both continuous and discrete variational mechanics.

Of course, such sweeping statements as above have to be interpreted and used with great care, as in the precise statements in the text that follows. For example, if the integration step size is too large, then sometimes energy can behave very badly, even for a symplectic integrator (see, for example, Gonzalez and Simo (1996)). It is likewise well known that energy conservation does not guarantee accuracy (Ortiz 1986).

1.2. Background: Lagrangian mechanics

1.2.1. *Basic definitions*

Consider a *configuration manifold* Q, with associated *state space* given by the tangent bundle TQ, and a *Lagrangian* $L : TQ \to \mathbb{R}$.

Given an interval $[0, T]$, define the *path space* to be

$$\mathcal{C}(Q) = \mathcal{C}([0, T], Q) = \{q \colon [0, T] \to Q \mid q \text{ is a } C^2 \text{ curve}\}$$

and the *action map* $\mathfrak{G} : \mathcal{C}(Q) \to \mathbb{R}$ to be

$$\mathfrak{G}(q) \equiv \int_0^T L(q(t), \dot{q}(t)) \, \mathrm{d}t. \tag{1.2.1}$$

It can be proved that $\mathcal{C}(Q)$ is a smooth manifold (Abraham, Marsden and Ratiu 1988), and $\mathfrak{G}$ is as smooth as L.

The tangent space $T_q\mathcal{C}(Q)$ to $\mathcal{C}(Q)$ at the point q is the set of C^2 maps $v_q : [0, T] \to TQ$ such that $\pi_Q \circ v_q = q$, where $\pi_Q : TQ \to Q$ is the canonical projection.

Define the *second-order submanifold* of $T(TQ)$ to be

$$\ddot{Q} \equiv \{w \in T(TQ) \mid T\pi_Q(w) = \pi_{TQ}(w)\} \subset T(TQ)$$

where $\pi_{TQ} : T(TQ) \to TQ$ and $\pi_Q : TQ \to Q$ are the canonical projections. $\ddot{Q}$ is simply the set of second derivatives $\mathrm{d}^2q/\mathrm{d}t^2(0)$ of curves $q : \mathbb{R} \to Q$, which are elements of the form $((q, \dot{q}), (\dot{q}, \ddot{q})) \in T(TQ)$.

Theorem 1.2.1. Given a C^k Lagrangian L, $k \geq 2$, there exists a unique C^{k-2} mapping $D_{\mathrm{EL}}L : \ddot{Q} \to T^*Q$ and a unique C^{k-1} one-form Θ_L on TQ, such that, for all variations $\delta q \in T_q \mathcal{C}(Q)$ of $q(t)$, we have

$$\mathbf{d}\mathfrak{G}(q) \cdot \delta q = \int_0^T D_{\mathrm{EL}}L(\ddot{q}) \cdot \delta q \, \mathrm{d}t + \Theta_L(\dot{q}) \cdot \hat{\delta q}\Big|_0^T, \qquad (1.2.2)$$

where

$$\hat{\delta q}(t) = \left(\left(q(t), \frac{\partial q}{\partial t}(t)\right), \left(\delta q(t), \frac{\partial \delta q}{\partial t}(t)\right)\right).$$

The mapping $D_{\mathrm{EL}}L$ is called the *Euler–Lagrange map* and has the coordinate expression

$$(D_{\mathrm{EL}}L)_i = \frac{\partial L}{\partial q^i} - \frac{\mathrm{d}}{\mathrm{d}t}\frac{\partial L}{\partial \dot{q}^i}.$$

The one-form Θ_L is called the *Lagrangian one-form* and in coordinates is given by

$$\Theta_L = \frac{\partial L}{\partial \dot{q}^i}\mathrm{d}q^i. \qquad (1.2.3)$$

Proof.   Computing the variation of the action map gives

$$\mathbf{d}\mathfrak{G}(q) \cdot \delta q = \int_0^T \left[\frac{\partial L}{\partial q^i}\delta q^i + \frac{\partial L}{\partial \dot{q}^i}\frac{\mathrm{d}}{\mathrm{d}t}\delta q^i\right]\mathrm{d}t$$

$$= \int_0^T \left[\frac{\partial L}{\partial q^i} - \frac{\mathrm{d}}{\mathrm{d}t}\frac{\partial L}{\partial \dot{q}^i}\right] \cdot \delta q^i \, \mathrm{d}t + \left[\frac{\partial L}{\partial \dot{q}^i}\delta q^i\right]_0^T$$

using integration by parts, and the terms of this expression can be identified as the Euler–Lagrange map and the Lagrangian one-form.   □

1.2.2. Lagrangian vector fields and flows

The *Lagrangian vector field* $X_L : TQ \to T(TQ)$ is a second-order vector field on TQ satisfying

$$D_{\mathrm{EL}}L \circ X_L = 0 \qquad (1.2.4)$$

and the *Lagrangian flow* $F_L : TQ \times \mathbb{R} \to TQ$ is the flow of X_L (we shall ignore issues related to global versus local flows, which are easily dealt with by restricting the domains of the flows). We shall write $F_L^t : TQ \to TQ$ for the map F_L at the frozen time t.

For an arbitrary Lagrangian, equation (1.2.4) may not uniquely define the vector field X_L and hence the flow map F_L may not exist. For now we will assume that L is such that these objects exist and are unique, and in Section 1.4.3 we will see under what conditions this is true.

A curve $q \in \mathcal{C}(Q)$ is said to be a *solution of the Euler–Lagrange equations* if the first term on the right-hand side of (1.2.2) vanishes for all variations $\delta q \in T_q \mathcal{C}(Q)$. This is equivalent to $(q, \dot{q})$ being an integral curve of X_L, and means that q must satisfy the *Euler–Lagrange equations*

$$\frac{\partial L}{\partial q^i}(q, \dot{q}) - \frac{\mathrm{d}}{\mathrm{d}t}\left(\frac{\partial L}{\partial \dot{q}^i}(q, \dot{q})\right) = 0 \tag{1.2.5}$$

for all $t \in (0, T)$.

1.2.3. Lagrangian flows are symplectic

Define the *solution space* $\mathcal{C}_L(Q) \subset \mathcal{C}(Q)$ to be the set of solutions of the Euler–Lagrange equations. As an element $q \in \mathcal{C}_L(Q)$ is an integral curve of X_L, it is uniquely determined by the initial condition $(q(0), \dot{q}(0)) \in TQ$ and we can thus identify $\mathcal{C}_L(Q)$ with the space of initial conditions TQ.

Defining the *restricted action map* $\hat{\mathfrak{G}} : TQ \to \mathbb{R}$ to be

$$\hat{\mathfrak{G}}(v_q) = \mathfrak{G}(q), \quad q \in \mathcal{C}_L(Q) \text{ and } (q(0), \dot{q}(0)) = v_q,$$

we see that (1.2.2) reduces to

$$\begin{aligned}
\mathbf{d}\hat{\mathfrak{G}}(v_q) \cdot w_{v_q} &= \Theta_L(\dot{q}(T))((F_L^T)_*(w_{v_q})) - \Theta_L(v_q)(w_{v_q}) \\
&= ((F_L^T)^*(\Theta_L))(v_q)(w_{v_q}) - \Theta_L(v_q)(w_{v_q})
\end{aligned} \tag{1.2.6}$$

for all $w_{v_q} \in T_{v_q}(TQ)$. Taking a further derivative of this expression, and using the fact that $\mathbf{d}^2\hat{\mathfrak{G}} = 0$, we obtain

$$(F_L^T)^*(\Omega_L) = \Omega_L,$$

where $\Omega_L = \mathbf{d}\Theta_L$ is the *Lagrangian symplectic form*, given in coordinates by

$$\Omega_L(q, \dot{q}) = \frac{\partial^2 L}{\partial q^i \partial \dot{q}^j}\mathbf{d}q^i \wedge \mathbf{d}q^j + \frac{\partial^2 L}{\partial \dot{q}^i \partial \dot{q}^j}\mathbf{d}\dot{q}^i \wedge \mathbf{d}q^j.$$

1.2.4. Lagrangian flows preserve momentum maps

Suppose that a Lie group G, with Lie algebra $\mathfrak{g}$, acts on Q by the (left or right) action $\Phi : G \times Q \to Q$. Consider the tangent lift of this action to $\Phi^{TQ} : G \times TQ \to TQ$ given by $\Phi_g^{TQ}(v_q) = T(\Phi_g) \cdot v_q$, which is

$$\Phi^{TQ}(g, (q, \dot{q})) = \left(\Phi^i(g, q), \frac{\partial \Phi^i}{\partial q^j}(g, q)\dot{q}^j\right).$$

For $\xi \in \mathfrak{g}$ define the *infinitesimal generators* $\xi_Q : Q \to TQ$ and $\xi_{TQ} : TQ \to T(TQ)$ by

$$\xi_Q(q) = \frac{\mathrm{d}}{\mathrm{d}g}\left(\Phi_g(q)\right) \cdot \xi,$$

$$\xi_{TQ}(v_q) = \frac{\mathrm{d}}{\mathrm{d}g}\left(\Phi_g^{TQ}(v_q)\right) \cdot \xi.$$

In coordinates these are given by

$$\xi_Q(q) = \left(q^i, \frac{\partial \Phi^i}{\partial g^m}(e,q)\xi^m \right),$$

$$\xi_{TQ}(q,\dot q) = \left(q^i, \dot q^i, \frac{\partial \Phi^i}{\partial g^m}(e,q)\xi^m, \frac{\partial^2 \Phi^i}{\partial g^m \partial q^j}(e,q)\dot q^j \xi^m \right).$$

We now define the *Lagrangian momentum map* $J_L : TQ \to \mathfrak{g}^*$ to be

$$J_L(v_q) \cdot \xi = \Theta_L \cdot \xi_{TQ}(v_q). \tag{1.2.7}$$

It can be checked that an equivalent expression for J_L is

$$J_L(v_q) \cdot \xi = \left\langle \frac{\partial L}{\partial \dot q}, \xi_Q(q) \right\rangle,$$

where $\partial L/\partial \dot q$ represents the Legendre transformation, discussed shortly. This equation is convenient for computing momentum maps in examples: see Marsden and Ratiu (1999).

The traditional linear and angular momenta are momentum maps, with the linear momentum $J_L : T\mathbb{R}^n \to \mathbb{R}^n$ arising from the additive action of $\mathbb{R}^n$ on itself, and the angular momentum $J_L : T\mathbb{R}^n \to \mathfrak{so}(n)^*$ coming from the action of $SO(n)$ on $\mathbb{R}^n$.

An important property of momentum maps is *equivariance*, which is the condition that the following diagram commutes.

$$\begin{array}{ccc} TQ & \xrightarrow{\;J_L\;} & \mathfrak{g}^* \\ {\scriptstyle \Phi_g^{TQ}}\Big\downarrow & & \Big\downarrow{\scriptstyle \mathrm{Ad}^*_{g^{-1}}} \\ TQ & \xrightarrow[\;J_L\;]{} & \mathfrak{g}^* \end{array} \tag{1.2.8}$$

In general, Lagrangian momentum maps are not equivariant, but we give here a simple sufficient condition for this property to be satisfied. Recall that a map $f : TQ \to TQ$ is said to be *symplectic* if $f^*\Omega_L = \Omega_L$. If, furthermore, f is such that $f^*\Theta_L = \Theta_L$, then f is said to be a *special symplectic map*. Clearly a special symplectic map is also symplectic, but the converse does not hold.

Theorem 1.2.2. Consider a Lagrangian system $L : TQ \to \mathbb{R}$ with a left action $\Phi : G \times Q \to Q$. If the lifted action $\Phi^{TQ} : G \times TQ \to TQ$ acts by special canonical transformations, then the Lagrangian momentum map $J_L : TQ \to \mathfrak{g}^*$ is equivariant.

Proof. Observing that $(\Phi_g^{TQ})^{-1} = \Phi_{g^{-1}}^{TQ}$, we see that equivariance is equivalent to

$$J_L(v_q) \cdot \xi = J_L \circ T\Phi_{g^{-1}}(v_q) \cdot \mathrm{Ad}_{g^{-1}}\xi.$$

We now compute the right-hand side of this expression to give

$$J_L \circ \Phi_{g^{-1}}^{TQ}(v_q) \cdot \mathrm{Ad}_{g^{-1}}\xi = \langle \Theta_L(\Phi_{g^{-1}}^{TQ}(v_q)), (\mathrm{Ad}_{g^{-1}}\xi)_{TQ}(\Phi_{g^{-1}}^{TQ}(v_q)) \rangle$$

$$= \langle \Theta_L(\Phi_{g^{-1}}^{TQ}(v_q)), T(\Phi_{g^{-1}}^{TQ}) \cdot \xi_{TQ}(v_q) \rangle$$

$$= \langle ((\Phi_{g^{-1}}^{TQ})^* \Theta_L)(v_q), \xi_{TQ}(v_q) \rangle$$

$$= \langle \Theta_L(v_q), \xi_{TQ}(v_q) \rangle,$$

which is just $J_L(v_q) \cdot \xi$, as desired. Here we used the identity $(\mathrm{Ad}_g\xi)_M = \Phi_{g^{-1}}^* \xi_M$ (Marsden and Ratiu 1999) to pass from the first to the second line.

$\square$

A Lagrangian $L : TQ \to \mathbb{R}$ is said to be *invariant* under the lift of the action $\Phi : G \times Q \to Q$ if $L \circ \Phi_g^{TQ} = L$ for all $g \in G$, and in this case the group action is said to be a *symmetry* of the Lagrangian. Differentiating this expression implies that the Lagrangian is *infinitesimally invariant*, which is the statement $\mathbf{d}L \cdot \xi_{TQ} = 0$ for all $\xi \in \mathfrak{g}$.

Observe that if L is invariant then this implies that Φ^{TQ} acts by special symplectic transformations, and so the Lagrangian momentum map is equivariant. To see this, we write $L \circ \Phi_g^{TQ} = L$ in coordinates to obtain $L(\Phi_g(q), \partial_q\Phi_g(q) \cdot \dot{q}) = L(q, \dot{q})$, and now differentiating this with respect to $\dot{q}$ in the direction δq gives

$$\frac{\partial L}{\partial \dot{q}}(\Phi_g(q), \partial_q\Phi_g(q) \cdot \dot{q}) \cdot \partial_q\Phi_g(q) \cdot \delta q = \frac{\partial L}{\partial \dot{q}}(q, \dot{q}) \cdot \delta q.$$

But the left- and right-hand sides are simply $(\Phi_g^{TQ})^*\Theta_L$ and Θ_L, respectively, evaluated on $((q, \dot{q}), (\delta q, \delta\dot{q}))$, and thus we have $(\Phi_g^{TQ})^*\Theta_L = \Theta_L$.

We will now show that, when the group action is a symmetry of the Lagrangian, then the momentum maps are preserved by the Lagrangian flow. This result was originally due to Noether (1918), using a technique similar to the one given below.

Theorem 1.2.3. (Noether's theorem)   Consider a Lagrangian system $L : TQ \to \mathbb{R}$ which is invariant under the lift of the (left or right) action $\Phi : G \times Q \to Q$. Then the corresponding Lagrangian momentum map $J_L : TQ \to \mathfrak{g}^*$ is a conserved quantity of the flow, so that $J_L \circ F_L^t = J_L$ for all times t.

Proof.   The action of G on Q induces an action of G on the space of paths $\mathcal{C}(Q)$ by pointwise action, so that $\Phi_g : \mathcal{C}(Q) \to \mathcal{C}(Q)$ is given by $\Phi_g(q)(t) = \Phi_g(q(t))$. As the action is just the integral of the Lagrangian, invariance of L implies invariance of $\mathfrak{G}$ and the differential of this gives

$$\mathbf{d}\mathfrak{G}(q) \cdot \xi_{\mathcal{C}(Q)}(q) = \int_0^T \mathbf{d}L \cdot \xi_{TQ} \, dt = 0.$$

Invariance of $\mathfrak{G}$ also implies that Φ_g maps solution curves to solution curves and thus $\xi_{\mathcal{C}(Q)}(q) \in T_q\mathcal{C}_L$, which is the corresponding infinitesimal version. We can thus restrict $\mathbf{d}\mathfrak{G} \cdot \xi_{\mathcal{C}(Q)}$ to the space of solutions $\mathcal{C}_L$ to obtain

$$0 = \hat{\mathfrak{G}}(v_q) \cdot \xi_{TQ}(v_q) = \Theta_L(\dot{q}(T)) \cdot \xi_{TQ}(\dot{q}(T)) - \Theta_L(v_q) \cdot \xi_{TQ}(v_q).$$

Substituting in the definition of the Lagrangian momentum map J_L, however, shows that this is just $0 = J_L(F_L^T(v_q)) \cdot \xi - J_L(v_q) \cdot \xi$, which gives the desired result. □

We have thus seen that conservation of momentum maps is a direct consequence of the invariance of the variational principle under a symmetry action. The fact that the symmetry maps solution curves to solution curves will extend directly to discrete mechanics.

In fact, only infinitesimal invariance is needed for the momentum map to be conserved by the Lagrangian flow, as a careful reading of the above proof will show. This is because it is only necessary that the Lagrangian be invariant in a neighbourhood of a given trajectory, and so the global statement of invariance is stronger than necessary.

1.3. Discrete variational mechanics: Lagrangian viewpoint

Take again a configuration manifold Q, but now define the *discrete state space* to be $Q \times Q$. This contains the same amount of information as (is locally isomorphic to) TQ. A *discrete Lagrangian* is a function $L_d : Q \times Q \to \mathbb{R}$.

To relate discrete and continuous mechanics it is necessary to introduce a *time-step* $h \in \mathbb{R}$, and to take L_d to depend on this time-step. For the moment, we will take $L_d : Q \times Q \times \mathbb{R} \to \mathbb{R}$, and will neglect the h dependence except where it is important. We shall come back to this point later when we discuss the context of time-dependent mechanics and adaptive algorithms. However, the idea behind this was explained in the introduction.

Construct the increasing sequence of times $\{t_k = kh \mid k = 0, \ldots, N\} \subset \mathbb{R}$ from the time-step h, and define the *discrete path space* to be

$$\mathcal{C}_d(Q) = \mathcal{C}_d(\{t_k\}_{k=0}^N, Q) = \{q_d \colon \{t_k\}_{k=0}^N \to Q\}.$$

We will identify a discrete trajectory $q_d \in \mathcal{C}_d(Q)$ with its image $q_d = \{q_k\}_{k=0}^N$, where $q_k = q_d(t_k)$. The *discrete action map* $\mathfrak{G}_d : \mathcal{C}_d(Q) \to \mathbb{R}$ is defined by

$$\mathfrak{G}_d(q_d) = \sum_{k=0}^{N-1} L_d(q_k, q_{k+1}).$$

As the discrete path space $\mathcal{C}_d$ is isomorphic to $Q \times \cdots \times Q$ ($N+1$ copies), it can be given a smooth product manifold structure. The discrete action $\mathfrak{G}_d$ clearly inherits the smoothness of the discrete Lagrangian L_d.

The tangent space $T_{q_d}\mathcal{C}_d(Q)$ to $\mathcal{C}_d(Q)$ at q_d is the set of maps v_{q_d} : $\{t_k\}_{k=0}^N \to TQ$ such that $\pi_Q \circ v_{q_d} = q_d$, which we will denote by $v_{q_d} = \{(q_k, v_k)\}_{k=0}^N$.

The discrete object corresponding to $T(TQ)$ is the set $(Q \times Q) \times (Q \times Q)$. We define the *projection operator* π and the *translation operator* σ to be

$$\pi : ((q_0, q_1), (q_0', q_1')) \mapsto (q_0, q_1),$$
$$\sigma : ((q_0, q_1), (q_0', q_1')) \mapsto (q_0', q_1').$$

The *discrete second-order submanifold* of $(Q \times Q) \times (Q \times Q)$ is defined to be

$$\ddot{Q}_d \equiv \{w_d \in (Q \times Q) \times (Q \times Q) \mid \pi_1 \circ \sigma(w_d) = \pi_2 \circ \pi(w_d)\},$$

which has the same information content as (is locally isomorphic to) $\ddot{Q}$. Concretely, the discrete second-order submanifold is the set of pairs of the form $((q_0, q_1), (q_1, q_2))$.

Theorem 1.3.1. Given a C^k discrete Lagrangian L_d, $k \geq 1$, there exists a unique C^{k-1} mapping $D_{\mathrm{DEL}}L_d : \ddot{Q}_d \to T^*Q$ and unique C^{k-1} one-forms $\Theta_{L_d}^+$ and $\Theta_{L_d}^-$ on $Q \times Q$, such that, for all variations $\delta q_d \in T_{q_d}\mathcal{C}(Q)$ of q_d, we have

$$\mathbf{d}\mathfrak{G}_d(q_d) \cdot \delta q_d = \sum_{k=1}^{N-1} D_{\mathrm{DEL}}L_d((q_{k-1}, q_k), (q_k, q_{k+1})) \cdot \delta q_k$$

$$+ \Theta_{L_d}^+(q_{N-1}, q_N) \cdot (\delta q_{N-1}, \delta q_N) - \Theta_{L_d}^-(q_0, q_1) \cdot (\delta q_0, \delta q_1). \quad (1.3.1)$$

The mapping $D_{\mathrm{DEL}}L_d$ is called the *discrete Euler–Lagrange map* and has coordinate expression

$$D_{\mathrm{DEL}}L_d((q_{k-1}, q_k), (q_k, q_{k+1})) = D_2 L_d(q_{k-1}, q_k) + D_1 L_d(q_k, q_{k+1}).$$

The one-forms $\Theta_{L_d}^+$ and $\Theta_{L_d}^-$ are called the *discrete Lagrangian one-forms* and in coordinates are

$$\Theta_{L_d}^+(q_0, q_1) = D_2 L_d(q_0, q_1)\mathbf{d}q_1 = \frac{\partial L_d}{\partial q_1^i} dq_1^i, \qquad (1.3.2a)$$

$$\Theta_{L_d}^-(q_0, q_1) = -D_1 L_d(q_0, q_1)\mathbf{d}q_0 = -\frac{\partial L_d}{\partial q_0^i} dq_0^i. \qquad (1.3.2b)$$

Proof. Computing the derivative of the discrete action map gives

$$\mathbf{d}\mathfrak{G}_d(q_d) \cdot \delta q_d = \sum_{k=0}^{N-1} [D_1 L_d(q_k, q_{k+1}) \cdot \delta q_k + D_2 L_d(q_k, q_{k+1}) \cdot \delta q_{k+1}]$$

$$= \sum_{k=1}^{N-1} [D_1 L_d(q_k, q_{k+1}) + D_2 L_d(q_{k-1}, q_k)] \cdot \delta q_k$$

$$+ D_1 L_d(q_0, q_1) \cdot \delta q_0 + D_2 L_d(q_{N-1}, q_N) \cdot \delta q_N$$

using a discrete integration by parts (rearrangement of the summation). Identifying the terms with the discrete Euler–Lagrange map and the discrete Lagrangian one-forms now gives the desired result.   □

Unlike the continuous case, in the discrete case there are two one-forms that arise from the boundary terms. Observe, however, that $\mathbf{d}L_d = \Theta^+_{L_d} - \Theta^-_{L_d}$ and so using $\mathbf{d}^2 = 0$ shows that

$$\mathbf{d}\Theta^+_{L_d} = \mathbf{d}\Theta^-_{L_d}.$$

This will be reflected below in the fact that there is only a single discrete two-form, which is the same as the continuous situation and is important for symplecticity.

1.3.1. Discrete Lagrangian evolution operator and mappings

A *discrete evolution operator* X plays the same role as a continuous vector field, and is defined to be any map $X : Q \times Q \to (Q \times Q) \times (Q \times Q)$ satisfying $\pi \circ X = \mathrm{id}$. The discrete object corresponding to the flow is the *discrete map* $F : Q \times Q \to Q \times Q$ defined by $F = \sigma \circ X$. In coordinates, if the discrete evolution operator maps $X : (q_0, q_1) \mapsto (q_0, q_1, q'_0, q'_1)$, then the discrete map will be $F : (q_0, q_1) \mapsto (q'_0, q'_1)$.

We will be mainly interested in discrete evolution operators which are *second-order*, which is the requirement that $X(Q \times Q) \subset \ddot{Q}_d$. This implies that they have the form $X : (q_0, q_1) \mapsto (q_0, q_1, q_1, q_2)$, and so the corresponding discrete map will be $F : (q_0, q_1) \mapsto (q_1, q_2)$. We now consider the particular case of a discrete Lagrangian system.

The *discrete Lagrangian evolution operator* X_{L_d} is a second-order discrete evolution operator satisfying

$$D_{\mathrm{DEL}}L_d \circ X_{L_d} = 0$$

and the *discrete Lagrangian map* $F_{L_d} : Q \times Q \to Q \times Q$ is defined by $F_{L_d} = \sigma \circ X_{L_d}$.

As in the continuous case, the discrete Lagrangian evolution operator and discrete Lagrangian map are not well-defined for arbitrary choices of discrete Lagrangian. We will henceforth assume that L_d is chosen so as to make these structures well-defined, and in Section 1.5 we will give a condition on L_d which ensures that this is true.

A discrete path $q_d \in C_d(Q)$ is said to be a *solution of the discrete Euler–Lagrange equations* if the first term on the right-hand side of (1.3.1) vanishes for all variations $\delta q_d \in T_{q_d}C_d(Q)$. This means that the points $\{q_k\}$ satisfy $F_{L_d}(q_{k-1}, q_k) = (q_k, q_{k+1})$ or, equivalently, that they satisfy the *discrete Euler–Lagrange equations*

$$D_2 L_d(q_{k-1}, q_k) + D_1 L_d(q_k, q_{k+1}) = 0, \quad \text{for all } k = 1, \dots, N-1. \quad (1.3.3)$$

1.3.2. Discrete Lagrangian maps are symplectic

Define the *discrete solution space* $C_{L_d}(Q) \subset C_d(Q)$ to be the set of solutions of the discrete Euler–Lagrange equations. Since an element $q_d \in C_{L_d}(Q)$ is formed by iteration of the map F_{L_d}, it is uniquely determined by the initial condition $(q_0, q_1) \in Q \times Q$. We can thus identify $C_{L_d}(Q)$ with the space of initial conditions $Q \times Q$.

Defining the *restricted discrete action map* $\hat{\mathfrak{G}}_d : Q \times Q \to \mathbb{R}$ to be

$$\hat{\mathfrak{G}}_d(q_0, q_1) = \mathfrak{G}_d(q_d); \quad q_d \in C_{L_d}(Q) \text{ and } (q_d(t_0), q_d(t_1)) = (q_0, q_1),$$

we see that (1.3.1) reduces to

$$
\begin{aligned}
\mathbf{d}\hat{\mathfrak{G}}_d(v_d) \cdot w_{v_d} &= \Theta_{L_d}^+(F_{L_d}^{N-1}(v_d))((F_{L_d}^{N-1})_*(w_{v_d})) - \Theta_{L_d}^-(v_d)(w_{v_d}) \\
&= ((F_{L_d}^{N-1})^*(\Theta_{L_d}^+))(v_d)(w_{v_d}) - \Theta_{L_d}^-(v_d)(w_{v_d}) \quad (1.3.4)
\end{aligned}
$$

for all $w_{v_d} \in T_{v_d}(Q \times Q)$ and $v_d = (q_0, q_1) \in Q \times Q$. Taking a further derivative of this expression, and using the fact that $\mathbf{d}^2\hat{\mathfrak{G}}_d = 0$, we obtain

$$(F_{L_d}^{N-1})^*(\Omega_{L_d}) = \Omega_{L_d}$$

where $\Omega_{L_d} = \mathbf{d}\Theta_{L_d}^+ = \mathbf{d}\Theta_{L_d}^-$ is the *discrete Lagrangian symplectic form*, with coordinate expression

$$\Omega_{L_d}(q_0, q_1) = \frac{\partial^2 L_d}{\partial q_0^i \partial q_1^j} \mathbf{d}q_0^i \wedge \mathbf{d}q_1^j.$$

This argument also holds if we take any subinterval of $0, \ldots, N$ and so the statement is true for any number of steps of F_{L_d}. For a single step we have $(F_{L_d})^*\Omega_{L_d} = \Omega_{L_d}$.

Given a map $f : Q \times Q \to Q \times Q$, we will say that f is *discretely symplectic* if $f^*\Omega_{L_d} = \Omega_{L_d}$. The above calculations thus prove that the discrete Lagrangian map F_{L_d} is discretely symplectic, just as we saw in the last section that the Lagrangian flow map is symplectic on TQ.

1.3.3. Discrete Noether's theorem

Consider the (left or right) action $\Phi : G \times Q \to Q$ of a Lie group G on Q, with infinitesimal generator as defined in Section 1.2. This action can be lifted to $Q \times Q$ by the product $\Phi_g^{Q \times Q}(q_0, q_1) = (\Phi_g(q_0), \Phi_g(q_1))$, which has an *infinitesimal generator* $\xi_{Q \times Q} : Q \times Q \to T(Q \times Q)$ given by

$$\xi_{Q \times Q}(q_0, q_1) = (\xi_Q(q_0), \xi_Q(q_1)).$$

The two *discrete Lagrangian momentum maps* $J_{L_d}^+, J_{L_d}^- : Q \times Q \to \mathfrak{g}^*$ are

$$J_{L_d}^+(q_0, q_1) \cdot \xi = \Theta_{L_d}^+ \cdot \xi_{Q \times Q}(q_0, q_1),$$

$$J_{L_d}^-(q_0, q_1) \cdot \xi = \Theta_{L_d}^- \cdot \xi_{Q \times Q}(q_0, q_1).$$

Using the expressions for $\Theta_{L_d}^{\pm}$ allows the discrete momentum maps to be alternatively written as

$$J_{L_d}^+(q_0, q_1) \cdot \xi = \langle D_2 L_d(q_0, q_1), \xi_Q(q_1) \rangle,$$
$$J_{L_d}^-(q_0, q_1) \cdot \xi = \langle -D_1 L_d(q_0, q_1), \xi_Q(q_0) \rangle,$$

which are computationally useful formulations.

As in the continuous case, it is interesting to consider when the discrete momentum maps are equivariant. This is the conditions

$$J_{L_d}^+ \circ \Phi_g^{Q \times Q} = \mathrm{Ad}_{g^{-1}}^* \circ J_{L_d}^+,$$
$$J_{L_d}^- \circ \Phi_g^{Q \times Q} = \mathrm{Ad}_{g^{-1}}^* \circ J_{L_d}^-.$$

In general these equations will not be satisfied; however, there is a simple sufficient condition, similar to the condition in the continuous case.

Recall that we have defined a map $f : Q \times Q \to Q \times Q$ to be discretely symplectic if $f^* \Omega_{L_d} = \Omega_{L_d}$. We now define f to be a *special discrete symplectic map* if $f^* \Theta_{L_d}^+ = \Theta_{L_d}^+$ and $f^* \Theta_{L_d}^- = \Theta_{L_d}^-$. This clearly means that f is also discretely symplectic, but the reverse is not true.

Theorem 1.3.2. Take a discrete Lagrangian system $L_d : Q \times Q \to \mathbb{R}$ with a (left or right) group action $\Phi : G \times Q \to Q$. If the product lifted action $\Phi^{Q \times Q} : G \times Q \times Q \to Q \times Q$ acts by special discrete symplectic maps, then the discrete Lagrangian momentum maps are equivariant.

Proof. The proof used in Theorem 1.2.2 for the continuous case can also be used here, with $J_{L_d}^+$ and $J_{L_d}^-$ being considered separately. □

If the lifted action only preserves one of $\Theta_{L_d}^+$ or $\Theta_{L_d}^-$ then only the corresponding momentum map will necessarily be equivariant.[2]

If a discrete Lagrangian $L_d : Q \times Q \to \mathbb{R}$ is such that $L_d \circ \Phi_g^{Q \times Q} = L_d$ for all $g \in G$, then L_d is said to be *invariant* under the lifted action, and Φ is said to be a *symmetry* of the discrete Lagrangian. Note that invariance implies *infinitesimal invariance*, which is $\mathbf{d}L_d \cdot \xi_{Q \times Q} = 0$ for all $\xi \in \mathfrak{g}$. Also note that

$$\mathbf{d}L_d = \Theta_{L_d}^+ - \Theta_{L_d}^-,$$

and so when L_d is infinitesimally invariant under the lifted action the two discrete momentum maps are equal. In such cases we will use the notation $J_{L_d} : Q \times Q \to \mathfrak{g}^*$ for the unique single *discrete Lagrangian momentum map*.

---

[2] As in the continuous case, equivariance plays an important role in reduction theory and, in the Hamiltonian context, equivariance guarantees that the momentum map is Poisson, which is often useful.

Note that invariance of L_d under the lifted action implies that $\Phi_g^{Q \times Q}$ is a special discrete symplectic map. This can be seen by differentiating $L_d \circ \Phi_g^{Q \times Q} = L_d$ with respect to q_1 to obtain

$$D_2 L_d \big(\Phi_g^{Q \times Q}(q_0, q_1) \big) \cdot \partial_q \Phi_g(q_1) \cdot \delta q_1 = D_2 L_d(q_0, q_1) \cdot \delta q_1,$$

and observing that the left- and right-hand sides are just $(\Phi_g^{Q \times Q})^* \Theta_{L_d}^+$ and $\Theta_{L_d}^+$, respectively, applied to $(q_0, q_1, \delta q_0, \delta q_1)$. Hence $(\Phi_g^{Q \times Q})^* \Theta_{L_d}^+ = \Theta_{L_d}^+$, and a similar calculation gives the result for $\Theta_{L_d}^-$.

We now give the discrete analogue of Noether's theorem, Theorem 1.2.3, which states that momentum maps of symmetries are constants of the motion.

Theorem 1.3.3. (Discrete Noether's theorem)   Consider a given discrete Lagrangian system $L_d : Q \times Q \to \mathbb{R}$ which is invariant under the lift of the (left or right) action $\Phi : G \times Q \to Q$. Then the corresponding discrete Lagrangian momentum map $J_{L_d} : Q \times Q \to \mathfrak{g}^*$ is a conserved quantity of the discrete Lagrangian map $F_{L_d} : Q \times Q \to Q \times Q$, so that $J_{L_d} \circ F_{L_d} = J_{L_d}$.

Proof.   We will use the same idea as in the proof of the continuous Noether's theorem, based on the fact that the variational principle is invariant under the symmetry action.

Begin by inducing an action of G on the discrete path space $\mathcal{C}_d(Q)$ by using the pointwise action. Then

$$\mathbf{d}\mathfrak{G}_d(q_d) \cdot \xi_{\mathcal{C}_d(Q)}(q_d) = \sum_{k=0}^{N-1} \mathbf{d}L_d \cdot \xi_{Q \times Q},$$

and so the space of solutions $\mathcal{C}_{L_d}(Q)$ of the discrete Euler–Lagrange equations is invariant under the lifted action of G, and the discrete Lagrangian map $F_{L_d} : Q \times Q \to Q \times Q$ commutes with the lifted action $\Phi_g : Q \times Q \to Q \times Q$.

Identifying $\mathcal{C}_{L_d}(Q)$ with the space of initial conditions $Q \times Q$ and using equation (1.3.4) gives

$$\mathbf{d}\mathfrak{G}_d(q_d) \cdot \xi_{\mathcal{C}(Q)}(q_d) = \mathbf{d}\hat{\mathfrak{G}}_d(q_0, q_1) \cdot \xi_{Q \times Q}(q_0, q_1)$$

$$= \big((F_{L_d}^N)^* (\Theta_{L_d}^+) - \Theta_{L_d}^- \big)(q_0, q_1) \cdot \xi_{Q \times Q}(q_0, q_1).$$

For symmetries the left-hand side is zero, and so we have

$$(\Theta_{L_d}^+ \cdot \xi_{Q \times Q}) \circ F_{L_d}^N = \Theta_{L_d}^- \cdot \xi_{Q \times Q},$$

which is simply the statement of preservation of the discrete momentum map, given that for symmetry actions there is only a single momentum map and that the above argument holds for all subintervals, including a single time-step.   □

As in the continuous case, only infinitesimal invariance of the discrete Lagrangian is actually required for the discrete momentum maps to be conserved. This is due to the fact that only local invariance is used in the proof above, and global invariance is not necessary.

Note that if G is not a symmetry of L_d then the two discrete momentum maps will not be equal, and it is precisely the difference $J_{L_d}^+ - J_{L_d}^-$ which describes the evolution of either momentum map during the time-step. To see this, define

$$J_{L_d}^\triangle(q_k, q_{k+1}) = J_{L_d}^+(q_k, q_{k+1}) - J_{L_d}^-(q_k, q_{k+1})$$

and observe that the discrete Euler–Lagrange equations imply

$$J_{L_d}^+(q_{k-1}, q_k) = J_{L_d}^-(q_k, q_{k+1}).$$

Combining the two above expressions shows that the two discrete momentum maps evolve according to

$$J_{L_d}^+(q_k, q_{k+1}) = J_{L_d}^+(q_{k-1}, q_k) + J_{L_d}^\triangle(q_k, q_{k+1}),$$
$$J_{L_d}^-(q_k, q_{k+1}) = J_{L_d}^-(q_{k-1}, q_k) + J_{L_d}^\triangle(q_{k-1}, q_k).$$

Clearly, if L_d is invariant then $J_{L_d}^\triangle = 0$, and so the momentum maps are equal and they are conserved. If not, then these equations describe how the momentum maps evolve.

1.4. Background: Hamiltonian mechanics

1.4.1. Hamiltonian mechanics

We will only concern ourselves here with the case of a phase space that is the cotangent bundle of a configuration manifold. Although some of the elegance and power of the Hamiltonian formalism is lost in this restriction, it is simpler for our purposes, and of course is the most important case for applications.

Consider then a configuration manifold Q, and define the *phase space* to be the cotangent bundle T^*Q. The *Hamiltonian* is a function $H : T^*Q \to \mathbb{R}$. We will take local coordinates on T^*Q to be (q, p).

Define the *canonical one-form* Θ on T^*Q by

$$\Theta(p_q) \cdot u_{p_q} = \langle p_q, T\pi_{T^*Q} \cdot u_{p_q} \rangle, \tag{1.4.1}$$

where $\pi_{T^*Q} : T^*Q \to Q$ is the standard projection and $\langle \cdot, \cdot \rangle$ denotes the natural pairing between vectors and covectors. In coordinates, $\Theta(q, p) = p_i dq^i$. The *canonical two-form* Ω on T^*Q is defined to be

$$\Omega = -d\Theta,$$

which has coordinate expression $\Omega(q, p) = \mathbf{d}q^i \wedge \mathbf{d}p^i$. The pair (T^*Q, Ω) is an example of a *symplectic manifold* and a mapping $F : T^*Q \to T^*Q$ is said to be *canonical* or *symplectic* if $F^*\Omega = \Omega$. If $F^*\Theta = \Theta$ then F is said to be a *special symplectic map*, which clearly implies that it is also symplectic. Note that a particular case of special symplectic maps is given by cotangent lifts of maps $Q \to Q$, which automatically preserve the canonical one-form on T^*Q (see Marsden and Ratiu (1999) for details).

Given a Hamiltonian H, define the corresponding *Hamiltonian vector field* X_H to be the unique vector field on T^*Q satisfying

$$\mathbf{i}_{X_H}\Omega = \mathbf{d}H. \tag{1.4.2}$$

Writing $X_H = (X_q, X_p)$ in coordinates, we see that the above expression is

$$-X_{p_i}\mathbf{d}q^i + X_{q^i}\mathbf{d}p_i = \frac{\partial H}{\partial q^i}\mathbf{d}q^i + \frac{\partial H}{\partial p_i}\mathbf{d}p_i,$$

which gives the familiar *Hamilton's equations* for the components of X_H, namely

$$X_{q^i}(q, p) = \frac{\partial H}{\partial p_i}(q, p), \tag{1.4.3a}$$

$$X_{p_i}(q, p) = -\frac{\partial H}{\partial q^i}(q, p). \tag{1.4.3b}$$

The *Hamiltonian flow* $F_H : T^*Q \times \mathbb{R} \to T^*Q$ is the flow of the Hamiltonian vector field X_H. Note that, unlike the Lagrangian situation, the Hamiltonian vector field X_H and flow map F_H are always well-defined for any Hamiltonian.

For any fixed $t \in \mathbb{R}$, the flow map $F_H^t : T^*Q \to T^*Q$ is symplectic, as can be seen by differentiating to obtain

$$\left.\frac{\partial}{\partial t}\right|_{t=0} (F_H^t)^*\Omega = \mathcal{L}_{X_H}\Omega = \mathbf{d}\mathbf{i}_{X_H}\Omega + \mathbf{i}_{X_H}\mathbf{d}\Omega$$

$$= \mathbf{d}^2 H - \mathbf{i}_{X_H}\mathbf{d}^2\Theta = 0,$$

where we have used Cartan's magic formula $\mathcal{L}_X\alpha = \mathbf{d}\mathbf{i}_X\alpha + \mathbf{i}_X\mathbf{d}\alpha$ for the Lie derivative and the fact that $\mathbf{d}^2 = 0$.

1.4.2. Hamiltonian form of Noether's theorem

Consider a (left or right) action $\Phi : G \times Q \to Q$ of G on Q, as in Section 1.2. The cotangent lift of this action is $\Phi^{T^*Q} : G \times T^*Q \to T^*Q$ given by $\Phi_g^{T^*Q}(p_q) = \Phi_{g^{-1}}^*(p_q)$, which in coordinates is

$$\Phi^{T^*Q}(g, (q, p)) = \left((\Phi_g^{-1})^i(q), p_j \frac{\partial \Phi_g^j}{\partial q^i}(q)\right).$$

This has the corresponding *infinitesimal generator* $\xi_{T^*Q} : T^*Q \to T(T^*Q)$ defined by

$$\xi_{T^*Q}(p_q) = \frac{\mathrm{d}}{\mathrm{d}g}\left(\Phi_g^{T^*Q}(p_q)\right) \cdot \xi$$

which has coordinate form

$$\xi_{T^*Q}(q,p) = \left(q^i, p_i, -\left[\left(\frac{\partial \Phi}{\partial q}\right)^{-1}\right]_j^i \frac{\partial \Phi^j}{\partial g^m}\xi^m, \right.$$

$$\left. p_j \frac{\partial^2 \Phi^j}{\partial q^i \partial g^m}\xi^m - p_j \frac{\partial^2 \Phi_j}{\partial q^i \partial q^j}\left[\left(\frac{\partial \Phi}{\partial q}\right)^{-1}\right]_k^j \frac{\partial \Phi^k}{\partial g^m}\xi^m\right),$$

where the derivatives of Φ are all evaluated at (e, q).

The *Hamiltonian momentum map* $J_H : T^*Q \to \mathfrak{g}^*$ is defined by

$$J_H(p_q) \cdot \xi = \Theta(p_q) \cdot \xi_{T^*Q}(p_q).$$

For each $\xi \in \mathfrak{g}$ we define $J_H^\xi : T^*Q \to \mathbb{R}$ by $J_H^\xi(p_q) = J_H(p_q) \cdot \xi$, which has the expression $J_H^\xi = i_{\xi_{T^*Q}}\Theta$. Note that the Hamiltonian map is also given by the expression

$$J_H(p_q) \cdot \xi = \langle p_q, \xi_Q(q) \rangle,$$

which is useful for computing it in applications.

Writing the requirement for equivariance of a Hamiltonian momentum map gives the equation

$$J_H \circ \Phi_g^{T^*Q} = \mathrm{Ad}_{g^{-1}}^* \circ J_H.$$

Unlike the Lagrangian setting, however, cotangent lifted actions are *always* special symplectic maps, and so we have $(\Phi_g^{T^*Q})^*\Theta = \Theta$ irrespective of the Hamiltonian. This gives the following result.

Theorem 1.4.1. Consider a Hamiltonian system $H : T^*Q \to \mathbb{R}$ with a (left or right) group action $\Phi : G \times Q \to Q$. Then the Hamiltonian momentum map $J_H : T^*Q \to \mathfrak{g}^*$ is always equivariant with respect to the cotangent lifted action $\Phi^{T^*Q} : G \times T^*Q \to T^*Q$.

Proof. Once again, we can use exactly the same proof as for Theorem 1.2.2 in the continuous case. The only difference is that H need not be restricted to ensure that the lifted action is a special symplectic map. □

A Hamiltonian $H : T^*Q \to \mathbb{R}$ is said to be *invariant* under the cotangent lift of the action $\Phi : G \times Q \to Q$ if $H \circ \Phi_g^{T^*Q} = H$ for all $g \in G$, in which case the action is said to be a *symmetry* for the Hamiltonian. The derivative of this expression implies that such a Hamiltonian is also *infinitesimally invariant*, which is the requirement $\mathbf{d}H \cdot \xi_{T^*Q} = 0$ for all $\xi \in \mathfrak{g}$, although the converse is not generally true.

Theorem 1.4.2. (Hamiltonian Noether's theorem) Let $H : T^*Q \to \mathbb{R}$ be a Hamiltonian which is invariant under the lift of the (left or right) action $\Phi : G \times Q \to Q$. Then the corresponding Hamiltonian momentum map $J_H : T^*Q \to \mathfrak{g}^*$ is a conserved quantity of the flow; that is, $J_H \circ F_H^t = J_H$ for all times t.

Proof. Recall that $(\Phi_g^{T^*Q})^* \Theta = \Theta$ for all $g \in G$ as the action is a cotangent lift, and hence $\mathcal{L}_{\xi_{T^*Q}} \Theta = 0$. Now computing the derivative of J_H^ξ in the direction given by the Hamiltonian vector field X_H gives

$$
\begin{aligned}
\mathbf{d}J_H^\xi \cdot X_H &= \mathbf{d}(\mathbf{i}_{\xi_{T^*Q}} \Theta) \cdot X_H \\
&= \mathcal{L}_{\xi_{T^*Q}} \Theta \cdot X_H - \mathbf{i}_{\xi_{T^*Q}} \mathbf{d}\Theta \cdot X_H \\
&= -\mathbf{i}_{X_H} \Omega \cdot \xi_{T^*Q} \\
&= -\mathbf{d}H \cdot \xi_{T^*Q} = 0
\end{aligned}
$$

using Cartan's magic formula $\mathcal{L}_X \alpha = \mathbf{d}\mathbf{i}_X \alpha + \mathbf{i}_X \mathbf{d}\alpha$ and (1.4.2). As F_H^t is the flow map for X_H this gives the desired result. $\square$

Noether's theorem still holds even if the Hamiltonian is only infinitesimally invariant, as it is only this local statement which is used in the proof.

1.4.3. Legendre transforms

To relate Lagrangian mechanics to Hamiltonian mechanics we define the *Legendre transform* or *fibre derivative* $\mathbb{F}L : TQ \to T^*Q$ by

$$
\mathbb{F}L(v_q) \cdot w_q = \frac{\mathrm{d}}{\mathrm{d}\epsilon}\bigg|_{\epsilon=0} L(v_q + \epsilon w_q),
$$

which has coordinate form

$$
\mathbb{F}L : (q, \dot{q}) \mapsto (q, p) = \left(q, \frac{\partial L}{\partial \dot{q}}(q, \dot{q}) \right).
$$

If the fibre derivative of L is locally an isomorphism then we say that L is *regular*, and if it is a global isomorphism then L is said to be *hyperregular*. We will generally assume that we are working with hyperregular Lagrangians.

The *fibre derivative* of a Hamiltonian is the map $\mathbb{F}H : T^*Q \to TQ$ defined by

$$
\alpha_q \cdot \mathbb{F}H(\beta_q) = \frac{\mathrm{d}}{\mathrm{d}\epsilon}\bigg|_{\epsilon=0} H(\beta_q + \epsilon \alpha_q),
$$

which in coordinates is

$$
\mathbb{F}H : (q, p) \mapsto (q, \dot{q}) = \left(q, \frac{\partial H}{\partial p}(q, p) \right).
$$

Similarly to the situations for Lagrangians, we say that H is *regular* if $\mathbb{F}H$ is a local isomorphism, and that H is *hyperregular* if $\mathbb{F}H$ is a global isomorphism.

The canonical one- and two-forms and the Hamiltonian momentum maps are related to the Lagrangian one- and two-forms and the Lagrangian momentum maps by pullback under the fibre derivative, so that

$$\Theta_L = (\mathbb{F}L)^*\Theta, \quad \Omega_L = (\mathbb{F}L)^*\Omega, \quad \text{and} \quad J_L = (\mathbb{F}L)^*J_H.$$

If we additionally relate the Hamiltonian to the Lagrangian by

$$H(q, p) = \mathbb{F}L(q, \dot{q}) \cdot \dot{q} - L(q, \dot{q}), \tag{1.4.4}$$

where (q, p) and $(q, \dot{q})$ are related by the Legendre transform, then the Hamiltonian and Lagrangian vector fields and their associated flow maps will also be related by pullback to give

$$X_L = (\mathbb{F}L)^*X_H; \quad F_L^t = (\mathbb{F}L)^{-1} \circ F_H^t \circ \mathbb{F}L.$$

In coordinates this means that Hamilton's equations (1.4.3) are equivalent to the Euler–Lagrange equations (1.3.3). To see this, we compute the derivatives of (1.4.4) to give

$$\frac{\partial H}{\partial q}(q, p) = p \cdot \frac{\partial \dot{q}}{\partial q} - \frac{\partial L}{\partial q}(q, \dot{q}) - \frac{\partial L}{\partial \dot{q}}(q, \dot{q})\frac{\partial \dot{q}}{\partial q}$$

$$= \frac{\partial L}{\partial q}(q, \dot{q}) \tag{1.4.5a}$$

$$= -\frac{\mathrm{d}}{\mathrm{d}t}\left(\frac{\partial L}{\partial \dot{q}}(q, \dot{q})\right)$$

$$= -\dot{p}, \tag{1.4.5b}$$

$$\frac{\partial H}{\partial \dot{q}}(q, p) = \dot{q} + p \cdot \frac{\partial \dot{q}}{\partial p} - \frac{\partial L}{\partial \dot{q}}(q, \dot{q})\frac{\partial \dot{q}}{\partial p}$$

$$= \dot{q}, \tag{1.4.5c}$$

where $p = \mathbb{F}L(q, \dot{q})$ defines $\dot{q}$ as a function of (q, p).

A similar calculation to the above also shows that if L is hyperregular and H is defined by (1.4.4) then H will also be hyperregular and the fibre derivatives will satisfy $\mathbb{F}H = (\mathbb{F}L)^{-1}$. The converse statement also holds (see Marsden and Ratiu (1999) for more details).

The above relationship between the Hamiltonian and Lagrangian flows can be summarized by the following commutative diagram, where we recall that the symplectic forms and momentum maps are also preserved under each map.

$$
\begin{array}{ccc}
TQ & \xrightarrow{\ F_L^t\ } & TQ \\
{\scriptstyle \mathbb{F}L}\downarrow & & \downarrow{\scriptstyle \mathbb{F}L} \\
T^*Q & \xrightarrow[\ F_H^t\ ]{} & T^*Q
\end{array}
\tag{1.4.6}
$$

One consequence of this relationship between the Lagrangian and Hamiltonian flow maps is a condition for when the Lagrangian vector field and flow map are well-defined.

Theorem 1.4.3. Given a Lagrangian $L : TQ \to \mathbb{R}$, the Lagrangian vector field X_L, and hence the Lagrangian flow map F_L, are well-defined if and only if the Lagrangian is regular.

Proof. This can be seen by relating the Hamiltonian and Lagrangian settings with $\mathbb{F}L$, or by computing the Euler–Lagrange equations in coordinates to give

$$0 = D_1 L(q, \dot{q}) - \frac{\mathrm{d}}{\mathrm{d}t} D_2 L(q, \dot{q})$$
$$= D_1 L(q, \dot{q}) - D_1 D_2 L(q, \dot{q}) \cdot \dot{q} - D_2 D_2 L(q, \dot{q}) \cdot \ddot{q}.$$

Thus, $\ddot{q}$ is well-defined as a function of $(q, \dot{q})$ if and only if $D_2 D_2 L$ is invertible, which by the implicit function theorem is equivalent to $\mathbb{F}L$ being locally invertible. $\qquad \square$

1.4.4. Generating functions

As with Hamiltonian mechanics, a useful general context for discussing canonical transformations and generating functions is that of symplectic manifolds. Here we limit ourselves, as above, to the case of T^*Q with the canonical symplectic form Ω.

Let $F : T^*Q \to T^*Q$ be a transformation from T^*Q to itself and let $\Gamma(F) \subset T^*Q \times T^*Q$ be the graph of F. Consider the one-form on $T^*Q \times T^*Q$ defined by

$$\hat{\Theta} = \pi_2^* \Theta - \pi_1^* \Theta.$$

where $\pi_i : T^*Q \times T^*Q$ are the projections onto the two components. The corresponding two-form is then

$$\hat{\Omega} = -\mathbf{d}\hat{\Theta} = \pi_2^* \Omega - \pi_1^* \Omega.$$

Denoting the inclusion map by $i_F : \Gamma(F) \to T^*Q \times T^*Q$, we see that we have the identities

$$\pi_1 \circ i_F = \pi_1|_{\Gamma(F)}, \quad \text{and} \quad \pi_2 \circ i_F = F \circ \pi_1 \text{ on } \Gamma(F).$$

Using these relations, we have

$$i_F^* \hat{\Omega} = i_F^* (\pi_2^* \Omega - \pi_1^* \Omega)$$
$$= (\pi_2 \circ i_F)^* \Omega - (\pi_1 \circ i_F)^* \Omega$$
$$= (\pi_1|_{\Gamma(F)})^* (F^* \Omega - \Omega).$$

Using this last equality, it is clear that F *is a canonical transformation if and only if* $i_F^* \hat{\Omega} = 0$ or, equivalently, if and only if $\mathbf{d}(i_F^* \hat{\Theta}) = 0$. By the

Poincaré lemma, this last statement is equivalent to there existing, at least locally, a function $S : \Gamma(F) \to \mathbb{R}$ such that $i_F^* \hat{\Theta} = \mathbf{d}S$. Such a function S is known as the *generating function* of the symplectic transformation F. Note that S is not unique.

The generating function S is specified on the graph $\Gamma(F)$, and so can be expressed in any local coordinate system on $\Gamma(F)$. The standard choices, for coordinates (q_0, p_0, q_1, p_1) on $T^*Q \times T^*Q$, are any two of the four quantities q_0, p_0, q_1 and p_1; note that $\Gamma(F)$ has the same dimension as T^*Q.

1.4.5. Coordinate expression

We will be particularly interested in the choice (q_0, q_1) as local coordinates on $\Gamma(F)$, and so we give the coordinate expressions for the above general generating function derivation for this particular case. This choice results in generating functions of the so-called *first kind* (Goldstein 1980).

Consider a function $S : Q \times Q \to \mathbb{R}$. Its differential is

$$\mathbf{d}S = \frac{\partial S}{\partial q_0} \mathbf{d}q_0 + \frac{\partial S}{\partial q_1} \mathbf{d}q_1.$$

Let $F : T^*Q \to T^*Q$ be the canonical transformation generated by S. In coordinates, the quantity $i_F^* \hat{\Theta}$ is

$$i_F^* \hat{\Theta} = -p_0 \mathbf{d}q_0 + p_1 \mathbf{d}q_1,$$

and so the condition $i_F^* \hat{\Theta} = \mathbf{d}S$ reduces to the equations

$$p_0 = -\frac{\partial S}{\partial q_0}(q_0, q_1), \tag{1.4.7a}$$

$$p_1 = \frac{\partial S}{\partial q_1}(q_0, q_1), \tag{1.4.7b}$$

which are an implicit definition of the transformation $F : (q_0, p_0) \mapsto (q_1, p_1)$. From the above general theory, we know that such a transformation is automatically symplectic, and that all symplectic transformations have such a representation, at least locally.

Note that there is not a one-to-one correspondence between symplectic transformations and real-valued functions on $Q \times Q$, because for some functions the above equations either have no solutions or multiple solutions, and so there is no well-defined map $(q_0, p_0) \mapsto (q_1, p_1)$. For example, taking $S(q_0, q_1) = 0$ forces p_0 to be zero, and so there is no corresponding map φ. In addition, one has to be careful about the special case of generating the identity transformation, as was noted in Channell and Scovel (1990) and Ge and Marsden (1988). As we will see later, this situation is identical to the existence of solutions to the discrete Euler–Lagrange equations, and, as in that case, we will assume for now that we choose generating functions and time-steps so that the equations (1.4.7) do indeed have solutions.

1.5. Discrete variational mechanics: Hamiltonian viewpoint

1.5.1. Discrete Legendre transforms

Just as the standard Legendre transform maps the Lagrangian state space TQ to the Hamiltonian phase space T^*Q, we can define *discrete Legendre transforms* or *discrete fibre derivatives* $\mathbb{F}^+L_d, \mathbb{F}^-L_d : Q \times Q \to T^*Q$, which map the discrete state space $Q \times Q$ to T^*Q. These are given by

$$\mathbb{F}^+L_d(q_0, q_1) \cdot \delta q_1 = D_2 L_d(q_0, q_1) \cdot \delta q_1,$$
$$\mathbb{F}^-L_d(q_0, q_1) \cdot \delta q_0 = -D_1 L_d(q_0, q_1) \cdot \delta q_0,$$

which can be written

$$\mathbb{F}^+L_d : (q_0, q_1) \mapsto (q_1, p_1) = (q_1, D_2 L_d(q_0, q_1)),$$
$$\mathbb{F}^-L_d : (q_0, q_1) \mapsto (q_0, p_0) = (q_0, -D_1 L_d(q_0, q_1)).$$

If both discrete fibre derivatives are locally isomorphisms (for nearby q_0 and q_1), then we say that L_d is *regular*. We will generally assume that we are working with regular discrete Lagrangians. In some special cases, such as if Q is a vector space, it may be that both discrete fibre derivatives are global isomorphisms. In that case we say that L_d is *hyperregular*.

Using the discrete fibre derivatives it can be seen that the canonical one- and two-forms and Hamiltonian momentum maps are related to the discrete Lagrangian forms and discrete momentum maps by pullback, so that

$$\Theta^{\pm}_{L_d} = (\mathbb{F}^{\pm}L_d)^*\Theta, \quad \Omega_{L_d} = (\mathbb{F}^{\pm}L_d)^*\Omega, \quad \text{and} \quad J^{\pm}_{L_d} = (\mathbb{F}^{\pm}L_d)^*J_H.$$

When the discrete momentum maps arise from a symmetry action, the pullback of the Hamiltonian momentum map by either discrete Legendre transform gives the unique discrete momentum map $J_{L_d} = (\mathbb{F}^{\pm}L_d)^*J_H$.

In the continuous case there is a particular relationship between a Lagrangian and a Hamiltonian so that the corresponding vector fields and flow maps are related by pullback under the Legendre transform. Indeed, we rarely consider pairs of Lagrangian and Hamiltonian systems which are not related in this way. In the discrete case a similar relationship exists, as will be shown in Section 1.6.

Unlike the continuous case, however, we will generally be interested in discrete Lagrangian systems that do not exactly correspond to a given Hamiltonian system. In this case, the symplectic structures and momentum maps are related by pullback under the discrete Legendre transforms, but the flow maps are not. As we will see later, this is a reflection of the fact that discrete Lagrangian systems can be regarded as symplectic-momentum integrators. The relationship between the energies of a discrete Lagrangian system and a Hamiltonian system is investigated in Part 4.

1.5.2. Momentum matching

The discrete fibre derivatives also permit a new interpretation of the discrete Euler–Lagrange equations. To see this, we introduce the notation

$$p^+_{k,k+1} = p^+(q_k, q_{k+1}) = \mathbb{F}^+ L_d(q_k, q_{k+1}),$$
$$p^-_{k,k+1} = p^-(q_k, q_{k+1}) = \mathbb{F}^- L_d(q_k, q_{k+1}),$$

for the momentum at the two endpoints of each interval $[k, k+1]$. Now observe that the discrete Euler–Lagrange equations are

$$D_2 L_d(q_{k-1}, q_k) = -D_1 L_d(q_k, q_{k+1}),$$

which can be written as

$$\mathbb{F}^+ L_d(q_{k-1}, q_k) = \mathbb{F}^- L_d(q_k, q_{k+1}), \tag{1.5.1}$$

or simply

$$p^+_{k-1,k} = p^-_{k,k+1}.$$

That is, the discrete Euler–Lagrange equations are simply enforcing the condition that the momentum at time k should be the same when evaluated from the lower interval $[k-1, k]$ or the upper interval $[k, k+1]$. This means that along a solution curve there is a unique momentum at each time k, which we denote by

$$p_k = p^+_{k-1,k} = p^-_{k,k+1}.$$

A discrete trajectory $\{q_k\}_{k=0}^N$ in Q can thus also be regarded as either a trajectory $\{(q_k, q_{k+1})\}_{k=0}^{N-1}$ in $Q \times Q$ or, equivalently, as a trajectory $\{(q_k, p_k)\}_{k=0}^N$ in T^*Q.

It will be useful to note that (1.5.1) can be written as

$$\mathbb{F}^+ L_d = \mathbb{F}^- L_d \circ F_{L_d}. \tag{1.5.2}$$

A consequence of viewing the discrete Euler–Lagrange equations as a matching of momenta is that it gives a condition for when the discrete Lagrangian evolution operator and discrete Lagrangian map are well-defined.

Theorem 1.5.1. Given a discrete Lagrangian system $L_d : Q \times Q \to \mathbb{R}$, the discrete Lagrangian evolution operator X_{L_d} and the discrete Lagrange map F_{L_d} are well-defined if and only if $\mathbb{F}^- L_d$ is locally an isomorphism. The discrete Lagrange map is well-defined and invertible if and only if the discrete Lagrangian is regular.

Proof. Given $(q_0, q_1) \in Q \times Q$, the point $q_2 \in Q$ required to satisfy

$$X_{L_d}(q_0, q_1) = (q_0, q_1, q_1, q_2)$$

is defined by equation (1.5.1), and so q_2 is uniquely defined as a function of

q_0 and q_1 if and only if $\mathbb{F}^- L_d$ is locally an isomorphism. From the definition of F_{L_d} it is well-defined if and only if X_{L_d} is.

The above argument only implies that F_{L_d} is well-defined as a map, however, meaning that it can be applied to map forward in time. For it to be invertible, equation (1.5.1) shows that it is necessary and sufficient for $\mathbb{F}^+ L_d$ also to be a local isomorphism, which is equivalent to regularity of L_d. $\quad\square$

1.5.3. Discrete Hamiltonian maps

Using the discrete fibre derivatives also enables us to push the discrete Lagrangian map $F_{L_d} : Q \times Q \to Q \times Q$ forward to T^*Q. We define the *discrete Hamiltonian map* $\tilde{F}_{L_d} : T^*Q \to T^*Q$ by $\tilde{F}_{L_d} = \mathbb{F}^{\pm} L_d \circ F_{L_d} \circ (\mathbb{F}^{\pm} L_d)^{-1}$. The fact that the discrete Hamiltonian map can be equivalently defined with either discrete Legendre transform is a consequence of the following theorem.

Theorem 1.5.2. The following diagram commutes.

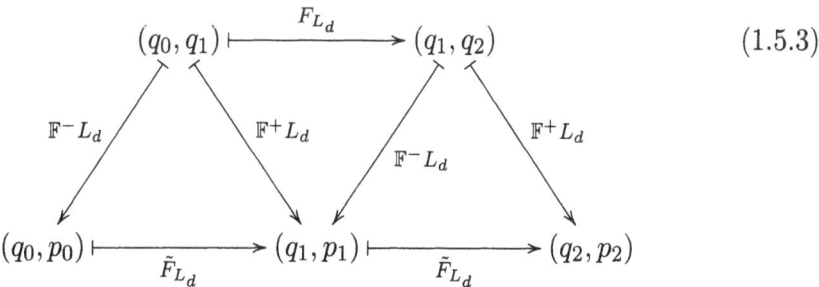

$$(1.5.3)$$

Proof. The central triangle is simply (1.5.2). Assume that we define the discrete Hamiltonian map by $\tilde{F}_{L_d} = \mathbb{F}^+ L_d \circ F_{L_d} \circ (\mathbb{F}^+ L_d)^{-1}$, which gives the right-hand parallelogram. Replicating the right-hand triangle on the left-hand side completes the diagram. If we choose to use the other discrete Legendre transform then the reverse argument applies. $\quad\square$

Corollary 1.5.3. The following three definitions of the discrete Hamiltonian map,

$$\tilde{F}_{L_d} = \mathbb{F}^+ L_d \circ F_{L_d} \circ (\mathbb{F}^+ L_d)^{-1},$$
$$\tilde{F}_{L_d} = \mathbb{F}^- L_d \circ F_{L_d} \circ (\mathbb{F}^- L_d)^{-1},$$
$$\tilde{F}_{L_d} = \mathbb{F}^+ L_d \circ (\mathbb{F}^- L_d)^{-1},$$

are equivalent and have coordinate expression $\tilde{F}_{L_d} : (q_0, p_0) \mapsto (q_1, p_1)$, where

$$p_0 = -D_1 L_d(q_0, q_1), \qquad\qquad (1.5.4a)$$
$$p_1 = D_2 L_d(q_0, q_1). \qquad\qquad (1.5.4b)$$

Proof. The equivalence of the three definitions can be read directly from the diagram in Theorem 1.5.2.

The coordinate expression for $\tilde{F}_{L_d} : (q_0, p_0) \mapsto (q_1, p_1)$ can be readily seen from the definition $\tilde{F}_{L_d} = \mathbb{F}^+ L_d \circ (\mathbb{F}^- L_d)^{-1}$. Taking initial condition $(q_0, p_0) \in T^*Q$ and setting $(q_0, q_1) = (\mathbb{F}^- L_d)^{-1}(q_0, p_0)$ implies that $p_0 = -D_1 L_d(q_0, q_1)$, which is (1.5.4a). Now, letting $(q_1, p_1) = \mathbb{F}^+ L_d(q_0, q_1)$ gives $p_1 = D_2 L_d(q_0, q_1)$, which is (1.5.4b). $\qquad\square$

As the discrete Lagrangian map preserves the discrete symplectic form and discrete momentum maps on $Q \times Q$, the discrete Hamiltonian map will preserve the pushforwards of these structures. As we saw above, however, these are simply the canonical symplectic form and canonical momentum maps on T^*Q, and so the discrete Hamiltonian map is symplectic and momentum-preserving.

We can summarize the relationship between the discrete and continuous systems in the following diagram, where the dashed arrows represent the discretization.

$$
\begin{array}{ccc}
TQ, F_L & \dashrightarrow & Q \times Q, F_{L_d} \\
{\scriptstyle \mathbb{F}L}\big\downarrow & & \big\downarrow{\scriptstyle \mathbb{F}L_d} \\
T^*Q, F_H & \dashrightarrow & T^*Q, \tilde{F}_{L_d}
\end{array}
\qquad (1.5.5)
$$

1.5.4. Discrete Lagrangians are generating functions

As we have seen above, a discrete Lagrangian is a real-valued function on $Q \times Q$ which defines a map $\tilde{F}_{L_d} : T^*Q \to T^*Q$. In fact, a discrete Lagrangian is simply a generating function of the first kind for the map $\tilde{F}_{L_d}$, in the sense defined in Section 1.4. This is seen by comparing the coordinate expression (1.5.4) for the discrete Hamiltonian map with the expression (1.4.7) for the map generated by a generating function of the first kind.

1.6. Correspondence between discrete and continuous mechanics

We will now define a particular choice of discrete Lagrangian which gives an *exact* correspondence between discrete and continuous systems. To do this, we must firstly recall the following fact.

Theorem 1.6.1. Consider a regular Lagrangian L for a configuration manifold Q, two points $q_0, q_1 \in Q$ and a time $h \in \mathbb{R}$. If $\|q_1 - q_0\|$ and $|h|$ are sufficiently small then there exists a unique solution $q : \mathbb{R} \to Q$ of the Euler–Lagrange equations for L satisfying $q(0) = q_0$ and $q(h) = q_1$.

Proof. See Marsden and Ratiu (1999). $\qquad\square$

For some regular Lagrangian L we now define the *exact discrete Lagrangian* to be

$$L_d^E(q_0, q_1, h) = \int_0^h L(q_{0,1}(t), \dot{q}_{0,1}(t)) \, dt \qquad (1.6.1)$$

for sufficiently small h and close q_0 and q_1. Here $q_{0,1}(t)$ is the unique solution of the Euler–Lagrange equations for L which satisfies the boundary conditions $q_{0,1}(0) = q_0$ and $q_{0,1}(h) = q_1$, and whose existence is guaranteed by Theorem 1.6.1.

We will now see that with this exact discrete Lagrangian there is an exact correspondence between the discrete and continuous systems. To do this, we will first establish that there is a special relationship between the Legendre transforms of a regular Lagrangian and its corresponding exact discrete Lagrangian. This result will also prove that exact discrete Lagrangians are automatically regular.

Lemma 1.6.2. A regular Lagrangian L and the corresponding exact discrete Lagrangian L_d^E have Legendre transforms related by

$$\mathbb{F}^+ L_d^E(q_0, q_1, h) = \mathbb{F}L(q_{0,1}(h), \dot{q}_{0,1}(h)),$$
$$\mathbb{F}^- L_d^E(q_0, q_1, h) = \mathbb{F}L(q_{0,1}(0), \dot{q}_{0,1}(0)),$$

for sufficiently small h and close $q_0, q_1 \in Q$.

Proof. We begin with $\mathbb{F}^- L_d^E$ and compute

$$\mathbb{F}^- L_d^E(q_0, q_1, h) = -\int_0^h \left[\frac{\partial L}{\partial q} \cdot \frac{\partial q_{0,1}}{\partial q_0} + \frac{\partial L}{\partial \dot{q}} \cdot \frac{\partial \dot{q}_{0,1}}{\partial q_0} \right] dt$$

$$= -\int_0^h \left[\frac{\partial L}{\partial q} - \frac{d}{dt} \frac{\partial L}{\partial \dot{q}} \right] \cdot \frac{\partial q_{0,1}}{\partial q_0} \, dt - \left[\frac{\partial L}{\partial \dot{q}} \cdot \frac{\partial q_{0,1}}{\partial q_0} \right]_0^h,$$

using integration by parts. The fact that $q_{0,1}(t)$ is a solution of the Euler–Lagrange equations shows that the first term is zero. To compute the second term we recall that $q_{0,1}(0) = q_0$ and $q_{0,1}(h) = q_1$, so that

$$\frac{\partial q_{0,1}}{\partial q_0}(0) = \mathrm{Id} \qquad \text{and} \qquad \frac{\partial q_{0,1}}{\partial q_0}(h) = 0.$$

Substituting these into the above expression for $\mathbb{F}^- L_d^E$ now gives

$$\mathbb{F}^- L_d^E(q_0, q_1, h) = \frac{\partial L}{\partial \dot{q}}(q_{0,1}(0), \dot{q}_{0,1}(0)),$$

which is simply the definition of $\mathbb{F}L(q_{0,1}(0), \dot{q}_{0,1}(0))$.

The result for $\mathbb{F}^+ L_d^E$ can be established by a similar computation. $\square$

Since $(q_{0,1}(h), \dot{q}_{0,1}(h)) = F_L^h(q_{0,1}(0), \dot{q}_{0,1}(0))$, Lemma 1.6.2 is equivalent to the following commutative diagram.

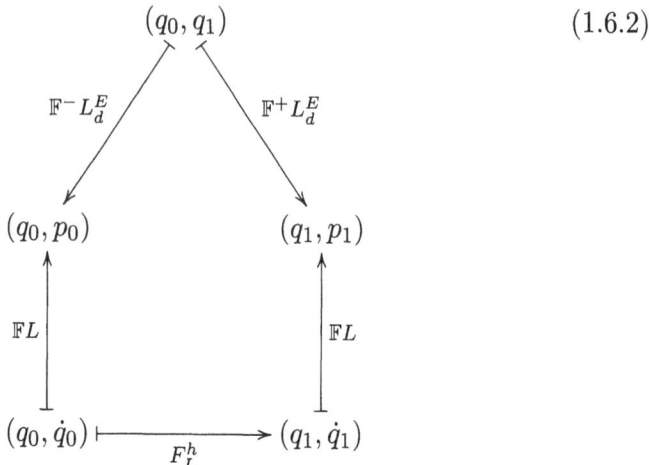

$$(1.6.2)$$

Combining this diagram with (1.4.6) and (1.5.3) gives the following commutative diagram for the exact discrete Lagrangian.

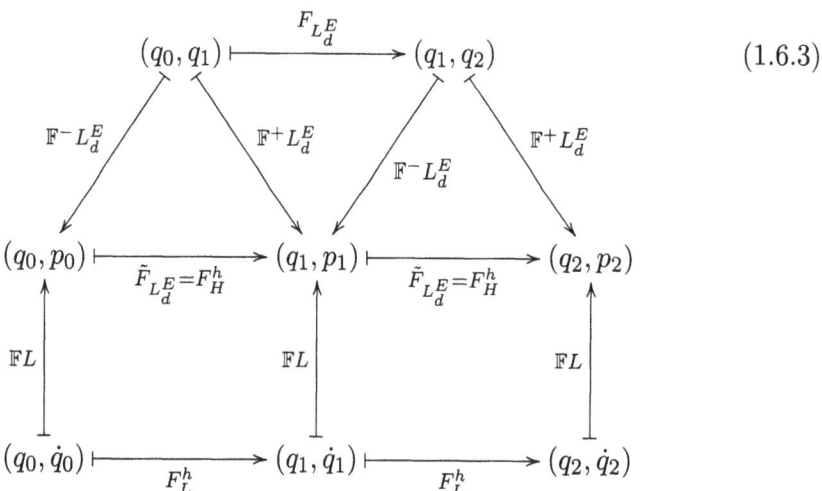

$$(1.6.3)$$

This proves the following theorem.

Theorem 1.6.3. Consider a regular Lagrangian L, its corresponding exact discrete Lagrangian L_d^E, and the pushforward of both the continuous and discrete systems to T^*Q, yielding a Hamiltonian system with Hamiltonian H and a discrete Hamiltonian map $\tilde{F}_{L_d^E}$, respectively. Then, for a sufficiently small time-step $h \in \mathbb{R}$, the Hamiltonian flow map equals the pushforward discrete Lagrangian map:

$$F_H^h = \tilde{F}_{L_d^E}.$$

This theorem is a statement about the time evolution of the system, and can also be interpreted as saying that the diagram (1.5.5) commutes with the dashed arrows understood as samples at times $\{t_k\}_{k=0}^N$, rather than merely as discretizations.

We can also interpret the equivalence of the discrete and continuous systems as a statement about trajectories. On the Lagrangian side, this gives the following theorem.

Theorem 1.6.4. Take a series of times $\{t_k = kh, \ k = 0, \ldots, N\}$ for a sufficiently small time-step h, and a regular Lagrangian L and its corresponding exact discrete Lagrangian L_d^E. Then solutions $q : [0, t_N] \to Q$ of the Euler–Lagrange equations for L and solutions $\{q_k\}_{k=0}^N$ of the discrete Euler–Lagrange equations for L_d^E are related by

$$q_k = q(t_k) \text{ for } k = 0, \ldots, N, \tag{1.6.4a}$$
$$q(t) = q_{k,k+1}(t) \text{ for } t \in [t_k, t_{k+1}]. \tag{1.6.4b}$$

Here the curves $q_{k,k+1} : [t_k, t_{k+1}] \to Q$ are the unique solutions of the Euler–Lagrange equations for L satisfying $q_{k,k+1}(kh) = q_k$ and $q_{k,k+1}((k+1)h) = q_{k+1}$.

Proof. The main non-obvious issue is smoothness. Let $q(t)$ be a solution of the Euler–Lagrange equations for L and define $\{q_k\}_{k=0}^N$ by (1.6.4a). Now the discrete Euler–Lagrange equations at time k are simply a matching of discrete Legendre transforms, as in (1.5.1), but by construction and Lemma 1.6.2 both sides of this expression are equal to $\mathbb{F}L(q(t_k), \dot{q}(t_k))$. We thus see that $\{q_k\}_{k=0}^N$ is a solution of the discrete Euler–Lagrange equations.

Conversely, let $\{q_k\}_{k=0}^N$ be a solution of the discrete Euler–Lagrange equations for L_d^E and define $q : [0, t_N] \to Q$ by (1.6.4b). Clearly $q(t)$ is C^2 and a solution of the Euler–Lagrange equations on each open interval (t_k, t_{k+1}), and so we must only establish that it is also C^2 at each t_k, from which it will follow that it is C^2 and a solution on the entire interval $[0, t_N]$.

At time t_k the discrete Euler–Lagrange equations in the form (1.5.1) together with Lemma 1.6.2 reduce to

$$\mathbb{F}L(q_{k-1,k}(t_k), \dot{q}_{k-1,k}(t_k)) = \mathbb{F}L(q_{k,k+1}(t_k), \dot{q}_{k,k+1}(t_k)),$$

and, as $\mathbb{F}L$ is a local isomorphism (due to the regularity of L), we see that $q(t)$ is C^1 on $[0, t_N]$. The regularity of L also implies that

$$\ddot{q}(t) = (D_2 D_2 L)^{-1}(D_1 L - D_1 D_2 L \cdot \dot{q}(t))$$

on each open interval (t_k, t_{k+1}), and as the right-hand side only depends on $q(t)$ and $\dot{q}(t)$ this expression is continuous at each t_k, giving that $q(t)$ is indeed C^2 on $[0, t_N]$. $\qquad\square$

To summarize, given Lagrangian and Hamiltonian systems with the Legendre transform mapping between them, the symplectic forms and momentum maps are always related by pullback under $\mathbb{F}L$. If, in addition, L and H satisfy the special relationship (1.4.4), then the flow maps and energy functions will also be related by pullback.

Exactly the same statements hold for the relationship between a discrete Lagrangian system and a Hamiltonian system. However, when discussing continuous systems we almost always assume that L and H are related by (1.4.4), whereas for discrete systems we generally do not assume that L_d and L or H are related by (1.6.1). This is because we are interested in using the discrete mechanics to derive integrators, and the exact discrete Lagrangian is generally not computable.

1.7. Background: Hamilton–Jacobi theory

1.7.1. Generating function for the flow

As discussed in Section 1.4, it is a standard result that the flow map F_H^t of a Hamiltonian system is a canonical map for each fixed time t. From the generating function theory, it must therefore have a generating function $S(q_0, q_1, t)$. We will now derive a partial differential equation which S must satisfy.

Consider first the time-preserving extension of F_H to the map

$$\hat{F}_H : T^*Q \times \mathbb{R} \to T^*Q \times \mathbb{R}, \quad (p_q, t) \mapsto (F_H^t(p_q), t).$$

Let $\pi_{T^*Q} : T^*Q \times \mathbb{R} \to T^*Q$ be the projection, and define the *extended canonical one-form* and the *extended canonical two-form* to be

$$\Theta_H = i_{T^*Q}^* \Theta - i_{T^*Q}^* H \wedge \mathrm{d}t,$$
$$\Omega_H = -\mathrm{d}\Theta_H = i_{T^*Q}^* \Omega - i_{T^*Q}^* \mathrm{d}H \wedge \mathrm{d}t.$$

We now calculate

$$T\hat{F}_H \cdot \left(\delta p_q, \delta t \right) = \left(TF_H^t \cdot \delta p_q + \frac{\partial}{\partial t} F_H^t(p_q) \cdot \delta t, \delta t \right)$$
$$= (TF_H^t \cdot \delta p_q + X_H \circ F_H^t \cdot \delta t, \delta t),$$

using that F_H^t is the flow map of the vector field X_H, and so

$$\hat{F}_H^* \Omega_H = (i_{T^*Q} \circ \hat{F}_H)^* \Omega - ((i_{T^*Q} \circ \hat{F}_H)^* \mathrm{d}H) \wedge (\hat{F}_H^* \mathrm{d}t)$$
$$= i_{T^*Q}^* (F_H^t)^* \Omega + (i_{T^*Q}^* (F_H^t)^* \mathrm{d}H) \wedge \mathrm{d}t - ((i_{T^*Q} \circ \hat{F}_H)^* \mathrm{d}H) \wedge \mathrm{d}t$$
$$= (i_{T^*Q})^* (F_H^t)^* \Omega = i_{T^*Q}^* \Omega$$

as F_H^t preserves Ω for fixed t. This identity essentially states that the extended flow map pulls back the extended symplectic form to the standard symplectic form.

Consider now the space $T^*Q \times \mathbb{R} \times T^*Q$ and the projection $\pi_1 : T^*Q \times \mathbb{R} \times T^*Q \to T^*Q \times \mathbb{R}$ onto the first two components and $\pi_2 : T^*Q \times \mathbb{R} \times T^*Q \to T^*Q \times \mathbb{R}$ onto the last two components. Define the one-form

$$\hat{\Theta} = \pi_2^* \Theta_H - \pi_1^* i_{T^*Q}^* \Theta,$$

and let the corresponding two-form be

$$\hat{\Omega} = -\mathbf{d}\hat{\Theta} = \pi_2^* \Omega_H - \pi_1^* i_{T^*Q}^* \Omega.$$

The flow map of the Hamiltonian system acts as $F_H : T^*Q \times \mathbb{R} \to T^*Q$ and so the graph of F_H is a subset $\Gamma(F_H) \subset T^*Q \times \mathbb{R} \times T^*Q$. Denote the inclusion map by $i_{F_H} : \Gamma(F_H) \to T^*Q \times \mathbb{R} \times T^*Q$. We now observe that

$$\pi_1 \circ i_{F_H} = \pi_1|_{\Gamma(F_H)},$$
$$\pi_2 \circ i_{F_H} = \hat{F}_H \circ \pi_1 \text{ on } \Gamma(F_H),$$

and using these relations calculate

$$i_{F_H}^* \hat{\Omega} = i_{F_H}^* \pi_2^* \Omega_H - i_{F_H}^* \pi_1^* i_{T^*Q}^* \Omega$$
$$= (\pi_2 \circ i_{F_H})^* \Omega_H - (\pi_1 \circ i_{F_H})^* i_{T^*Q}^* \Omega$$
$$= (\pi_1|_{\Gamma(F_H)})^* (\hat{F}_H^* \Omega_H - i_{T^*Q}^* \Omega)$$
$$= 0.$$

We have thus established that $\mathbf{d}(i_{F_H}^* \hat{\Theta}) = 0$ and so, by the Poincaré lemma, there must locally exist a function $S : \Gamma(F_H) \to \mathbb{R}$ so that $i_{F_H}^* \hat{\Theta} = \mathbf{d}S$. It is clear that restricting the above derivations to a section with fixed t simply reproduces the earlier derivation of generating functions for symplectic maps, and so the restriction $S^t : \Gamma(F_H^t) \to \mathbb{R}$ is a generating function for the map $F_H^t : T^*Q \to T^*Q$. The additional information contained in the statement $i_{F_H}^* \hat{\Theta} = \mathbf{d}S$ dictates how S depends on t.

1.7.2. Hamilton–Jacobi equation

As for the case of general generating functions discussed in Section 1.4 we will now choose a particular set of coordinates on ΓF_H and investigate the implications of $i_{F_H}^* \hat{\Theta} = \mathbf{d}S$.

Consistent with our earlier choice, we will take coordinates (q_0, q_1, t) for $\Gamma(F_H)$ and thus regard the generating function as a map $S : Q \times Q \times \mathbb{R} \to \mathbb{R}$. The differential is thus

$$\mathbf{d}S = \frac{\partial S}{\partial q_0} \mathbf{d}q_0 + \frac{\partial S}{\partial q_1} \mathbf{d}q_1 + \frac{\partial S}{\partial t} \mathbf{d}t,$$

and we also get

$$\hat{\Theta} = -p_0 \mathbf{d}q_0 + p_1 \mathbf{d}q_1 - H(q_1, p_1) \mathbf{d}t,$$

so the condition $i_{F_H}^* \hat{\Theta} = dS$ is

$$p_0 = -\frac{\partial S}{\partial q_0}(q_0, q_1, t),$$

$$p_1 = \frac{\partial S}{\partial q_1}(q_0, q_1, t),$$

$$H\left(q_1, \frac{\partial S}{\partial q_1}(q_0, q_1, t)\right) = \frac{\partial S}{\partial t}(q_0, q_1, t).$$

The first two equations are simply the standard relations which implicitly specify the map F_H^t from the generating function S^t. The third equation specifies the time-dependence of S and is known as the *Hamilton–Jacobi PDE*, and can be regarded as a partial-differential equation to be solved for S.

To fully specify the Hamilton–Jacobi PDE it is necessary also to provide boundary conditions. As it is first-order in t, it is clear that specifying S as a function of q_0 and q_1 at some time t will define the solution in a neighbourhood of that time. This is equivalent to specifying the map generated by S at some time, up to an arbitrary function of t. Taking this to be the flow map for some fixed time, we see that the unique solution of the Hamilton–Jacobi PDE must be the flow map for nearby t.

1.7.3. *Jacobi's solution*

While it is possible in principle to solve the Hamilton–Jacobi PDE directly for S, it is generally nonlinear and a closed form solution is not normally possible. By 1840, however, Jacobi had realized that the solution is simply the action of the trajectory joining q_0 and q_1 in time t: see Jacobi (1866). This is known as *Jacobi's solution*,

$$S(q_0, q_1, t) = \int_0^t L(q_{0,1}(\tau), \dot{q}_{0,1}(\tau)) \, d\tau, \qquad (1.7.1)$$

where $q_{0,1}(t)$ is a solution of the Euler–Lagrange equations for L satisfying the boundary conditions $q(0) = q_0$ and $q(t) = q_1$, and where L and H are related by the Legendre transform (assumed to be regular). This can be proved in the same way as Lemma 1.6.2.

1.8. Discrete variational mechanics: Hamilton–Jacobi viewpoint

As was discussed in Section 1.5, a discrete Lagrangian can be regarded as the generating function for the discrete Hamiltonian map $\tilde{F}_{L_d} : T^*Q \to T^*Q$. We then showed in Section 1.6 that there is a particular choice of discrete Lagrangian, the so-called *exact* discrete Lagrangian, which exactly generates the flow map F_H of the corresponding Hamiltonian system. From the development of Hamilton–Jacobi theory in Section 1.7, it is clear that this exact discrete Lagrangian must be a solution of the Hamilton–Jacobi equation. In fact, as can be seen by comparing the definitions given in equations (1.6.1) and (1.7.1), the exact discrete Lagrangian is precisely Jacobi's solution of the Hamilton–Jacobi equation.

To summarize, discrete Lagrangian mechanics can be regarded as a variational Lagrangian derivation of the standard generating function and Hamilton–Jacobi theory. Discrete Lagrangians generate symplectic transformations, and given a Lagrangian or Hamiltonian system, one can construct the exact discrete Lagrangian which solves the Hamilton–Jacobi equation, and this will then generate the exact flow of the continuous system.

PART TWO
Variational integrators

2.1. Introduction

We now turn our attention to considering a discrete Lagrangian system as an approximation to a given continuous system. That is, the discrete system is an integrator for the continuous system.

As we have seen, discrete Lagrangian maps preserve the symplectic structure and so, regarded as integrators, they are necessarily symplectic. Furthermore, generating function theory shows that any symplectic integrator for a mechanical system can be regarded as a discrete Lagrangian system, a fact we state here as a theorem.

Theorem 2.1.1. If the integrator $F : T^*Q \times \mathbb{R} \to T^*Q$ is symplectic then there exists[3] a discrete Lagrangian L_d whose discrete Hamiltonian map $\tilde{F}_{L_d}$ is F.

Proof. As shown above in Section 1.4, any symplectic transformation locally has a corresponding generating function, which is then a discrete Lagrangian for the method, as discussed in Section 1.5.4.   □

In addition, if the discrete Lagrangian inherits the same symmetry groups as the continuous system, then the discrete system will also preserve the corresponding momentum maps. As an integrator, it will thus be a so-called *symplectic–momentum integrator*.

Just as with continuous mechanics, we have seen that discrete variational mechanics has both a variational (Lagrangian) and a generating function (Hamiltonian) interpretation. These two viewpoints are complementary and both give insight into the behaviour and derivation of useful integrators.

However, the above theorem is not literally used in the construction of variational integrators, but is rather used as the first steps in obtaining inspiration. We will obtain much deeper insight from the variational principle itself and this is, in large part, what sets variational methods apart from standard symplectic methods.

Symplectic integrators have traditionally been approached from a Hamiltonian viewpoint and there is much existing literature treating this topic (see, for example, Hairer, Nørsett and Wanner (1993), Hairer and Wanner (1996), MacKay (1992) and Sanz-Serna (1992a)). In this paper, we concentrate on the analysis of symplectic methods from the variational viewpoint, and we reinterpret many standard concepts from ODE integration theory in this light.

---

[3] The discrete Lagrangian may exist only locally, as is the case with generating functions, as was discussed in Section 1.4.

It is also important to distinguish the two ways in which we can derive variational or generating function integrators. First, we can attempt to approximately solve the Hamilton–Jacobi PDE for a given system, such as by taking power series expansions of the generating function. This was used in some of the earliest derivations of symplectic integrators (such as De Vogelaére (1956) and Channell and Scovel (1990)). Second, the method we advocate involves trying to approximate the known Jacobi's solution to the Hamilton–Jacobi PDE: that is, we construct discrete Lagrangians that approximate the exact discrete Lagrangian. This approach is powerful not only because of the coherent and unifying underlying theory that reveals the beautiful geometry underlying discrete mechanics, but also because it leads to practical integrators.

In this section we will assume that Q, and thus also TQ and T^*Q, is a finite-dimensional vector space with an inner product $\langle \cdot, \cdot \rangle$ and corresponding norm $\| \cdot \|$. In the case that it is not a vector space, we can embed Q within a vector space and use the theory of constrained discrete systems developed below in Section 3.4 and discussed further in Section 3.5.2.

A word of caution: we must be careful about imagining that we can simply pick a coordinate chart and apply the vector space methods described below in such a chart. Doing so indiscriminately can lead to coordinate-dependent integrators that can be unattractive theoretically as well as impractical: for instance, using Euler angles for rigid body integrators has the difficulty that we may spend most of our computational time switching coordinate systems. See, for instance, Wisdom, Peale and Mignard (1984), Leimkuhler and Patrick (1996) and related papers. For some special classes of configuration manifolds, however, such as when Q is a Lie group, there may be particular global coordinate systems that can be used for this purpose.

We will also frequently consider integrators for Lagrangian systems of the form $L(q, \dot{q}) = \frac{1}{2}\dot{q}^T M \dot{q} - V(q)$. When dealing with such systems, we will always assume that M is a positive-definite symmetric mass matrix, so that $\mathbb{F}L(q, \dot{q}) = M\dot{q}$ and thus that L is regular.

2.1.1. Implementation of variational integrators

Although the distinction between the discrete Lagrangian map $F_{L_d} : Q \times Q \times \mathbb{R} \to Q \times Q$ and its pushforward $\tilde{F}_{L_d} : T^*Q \times \mathbb{R} \to T^*Q$ is important geometrically, for implementation purposes the two maps are essentially the same. This is because of the observation made in Section 1.5.2 that the discrete Euler–Lagrange equations that define F_{L_d} can be interpreted as matching of momenta between adjacent intervals.

In other words, given a trajectory $q_0, q_1, q_2, \ldots, q_{k-1}, q_k$ the map $F_{L_d} : Q \times Q \times \mathbb{R} \to Q \times Q$ calculates q_{k+1} according to

$$D_2 L_d(q_{k-1}, q_k, h) = -D_1 L_d(q_k, q_{k+1}, h).$$

If we now take $p_k = D_2 L_d(q_{k-1}, q_k, h)$ for each k, then this equation is simply

$$p_k = -D_1 L_d(q_k, q_{k+1}, h), \qquad (2.1.1)$$

which, together with the next update

$$p_{k+1} = D_2 L_d(q_k, q_{k+1}, h), \qquad (2.1.2)$$

defines the pushforward map $\tilde{F}_{L_d} : T^*Q \times \mathbb{R} \to T^*Q$. Another way to think of this is that the p_k are merely storing the values $D_2 L_d(q_{k-1}, q_k, h)$ from the last step.

For this reason it is typically easier to implement a variational integrator as the single step map $\tilde{F}_{L_d}$, as this also provides a simple method of initialization from initial values $(q_0, p_0) \in T^*Q$. Many discrete Lagrangians have pushforward maps that are simple to implement. For example, $\tilde{F}_{L_d}$ may be explicit, or it may be a Runge–Kutta method or other integrator type with standard implementation techniques.

In the general case when no special form is apparent, however, the equations (2.1.1) and (2.1.2) must be solved directly. The update $(q_k, p_k) \mapsto (q_{k+1}, p_{k+1})$ thus involves first solving the implicit equation (2.1.1) for q_{k+1} and then evaluating the explicit update (2.1.2) to give p_{k+1}.

To solve the implicit equation (2.1.1) we must typically use an iterative technique such as Newton's method. This involves computing a first guess $q_{k+1,0}$ for q_{k+1}, such as $q_{k+1,0} = 2q_k - q_{k-1}$, and then computing the sequence of approximations $q_{k+1,n}$, $n = 1, 2, \ldots$ until they converge to the solution value q_{k+1}. For Newton's method, the iteration rule is given by

$$q^i_{k+1,n+1} = q^i_{k+1,n} - A^{ij}\left[p^j_k + \frac{\partial L_d}{\partial q^j_0}(q_0, q_1, h)\right],$$

where A^{ij} is the inverse of the matrix

$$A_{ij} = \frac{\partial^2 L_d}{\partial q^i_0 \partial q^j_1}(q_0, q_1, h).$$

In the case that the Lagrangian has a simple form, such as $L(q, \dot{q}) = \frac{1}{2}\dot{q}^T M \dot{q} - V(q)$, then we can use an initial guess based on p_k, such as $q_{k+1,0} = q_k + h M^{-1} p_k$.

While the Newton's method outlined above typically experiences very fast convergence, it is also expensive to have to recompute A^{ij} at each iteration of the method. For this reason, it is typical to use an approximation to this matrix which can be held constant for all iterations of Newton's method. See Hairer *et al.* (1993) for details of this approach for Runge–Kutta methods.

2.1.2. Equivalence of integrators

Given two discrete Lagrangians L^1_d and L^2_d, we would like to know whether the integrators they generate are the same. Here it will be important to distinguish between the discrete Lagrangian maps $Q \times Q \to Q \times Q$ and the discrete Hamiltonian maps $T^*Q \to T^*Q$. We assume that we are dealing with regular discrete Lagrangians, so that the corresponding maps are well-defined.

We say that L_d^1 is (strongly) *equivalent* to L_d^2 if their discrete *Hamiltonian* maps are equal, so that $\tilde{F}_{L_d^1} = \tilde{F}_{L_d^2}$. Using the expression $\tilde{F}_{L_d^1} = \mathbb{F}^+ L_d^1 \circ (\mathbb{F}^- L_d^1)^{-1}$, we see that if L_d^1 and L_d^2 are equivalent then their discrete Legendre transforms will be equal. This implies that the difference $L_d^{\Delta} = L_d^1 - L_d^2$ must be a function of h only. That is, $L_d^{\Delta}(q_0, q_1, h) = f(h)$ for some function f. This is clearly also a sufficient condition, as well as being necessary.

We define L_d^1 to be *weakly equivalent* to L_d^2 if their discrete *Lagrangian* maps $F_{L_d^1}$ and $F_{L_d^2}$ are equal. A sufficient (and presumably necessary) condition for this to be true is that their difference $L_d^{\Delta} = L_d^1 - L_d^2$ is a *null discrete Lagrangian*; that is, the discrete Euler–Lagrange equations for L_d^{Δ} are satisfied by any triplet (q_0, q_1, q_2). This terminology follows that of the continuous case, as in, for example, Oliver and Sivaloganathan (1988).

If L_d^{Δ} is a null discrete Lagrangian, then $D_2 L_d^{\Delta}(q_0, q_1, h)$ cannot depend on q_0 and $D_1 L_d^{\Delta}(q_1, q_2, h)$ cannot depend on q_2. Furthermore, these two derivatives must be the negative of each other for all q_1. We thus have that L_d^{Δ} is a null discrete Lagrangian if and only if it is of the form $L_d^{\Delta}(q_0, q_1, h) = f(q_1, h) - f(q_0, h)$ for some function f.

Using the above calculations, it is clear that strong equivalence implies weak equivalence of discrete Lagrangians. For variational integrators, weak equivalence is thus in some sense the more fundamental notion. Intuitively, if two integrators give solutions which have the same positions q_k for all time, but different momenta p_k at each step, then we would like to regard the methods as being essentially the same. This is exactly weak equivalence.

2.2. Background: Error analysis

In this section we consider a numerical method $F : T^*Q \times \mathbb{R} \to T^*Q$ which approximates the flow $F_H : T^*Q \times \mathbb{R} \to T^*Q$ of a given Hamiltonian vector field X_H. Error analysis is concerned with difference between an exact trajectory and a discrete trajectory.[4]

---

[4] The reader should be cautioned that in many circumstances, such as the integration of chaotic or complex systems, it may make little sense to imagine accurately computing an exact, but highly unstable, individual trajectory. Instead, we often want to accurately compute robust quantities, such as statistical measures that are insensitive to modelling errors and dynamical sensitivities. This vision, as important as it is, awaits a theory.

2.2.1. Local error and method order

An integrator F of X_H is said to be of *order* r if there exist an open set $U \subset T^*Q$ and constants $C_l > 0$ and $h_l > 0$ so that

$$\|F(q, p, h) - F_H(q, p, h)\| \leq C_l h^{r+1} \qquad (2.2.1)$$

for all $(q, p) \in U$ and $h \leq h_l$. The expression on the left-hand side of this inequality is known as the *local error*, and if a method has order of at least 1 then it is said to be *consistent*.

2.2.2. Global error and convergence

Having defined the error after one step, we now consider the error after many steps. The integrator F of X_H is said to be *convergent of order r* if there exist an open set $U \subset T^*Q$ and constants $C_g > 0$, $h_g > 0$ and $T_g > 0$ so that

$$\|(F)^N(q, p, h) - F_H(q, p, T)\| \leq C_g h^r,$$

where $h = T/N$, for all $(q, p) \in U$, $h \leq h_g$ and $T \leq T_g$. The expression on the left-hand side is the *global error* at time T.

For one-step methods such as we consider here, convergence follows from a local error bound on the method and a Lipschitz bound on X_H.

Theorem 2.2.1. Suppose that the integrator F for X_H is of order r on the open set $U \subset T^*Q$ with local error constant C_l, and assume that $L > 0$ is such that

$$\left\| \frac{\partial X_H}{\partial(q, p)} \right\| \leq L$$

on U. Then the method is consistent on U with global error constant C_g given by

$$C_g = \frac{C_l}{L} \left(e^{LT_g} - 1 \right)$$

Proof. See, for example, Hairer *et al.* (1993). ☐

2.2.3. Order calculation

Given an integrator F for X_H, the order can be calculated by expanding both the true flow F_H and the integrator F in a Taylor series in h and then comparing terms. If the terms agree up to order r then the method will be of order r.

Here we explicitly write the first few terms of the Taylor series for the true flow for a Hamiltonian of the form $H(q, p) = \frac{1}{2} p^T M^{-1} p + V(q)$. The corresponding Hamiltonian vector field X_H is

$$\dot{q} = M^{-1} p, \qquad (2.2.2a)$$
$$\dot{p} = -\nabla V(q), \qquad (2.2.2b)$$

and so the flow $(q(h), p(h)) = F_H(q_0, p_0, h)$ has the expansion

$$q(h) = q_0 + hM^{-1}p_0 - \frac{1}{2}h^2M^{-1}\nabla V(q_0) + \mathcal{O}(h^3), \tag{2.2.3a}$$

$$p(h) = p_0 - h\nabla V(q_0) - \frac{1}{2}h^2\nabla^2 V(q_0)M^{-1}p_0 + \mathcal{O}(h^3). \tag{2.2.3b}$$

We will see below an example of using this to calculate the order of a simple class of methods.

2.3. Variational error analysis

Rather than considering how closely the trajectory of F matches the exact trajectory given by F_H, we can alternatively consider how closely a discrete Lagrangian matches the ideal discrete Lagrangian given by the action. As we have seen in Section 1.6, if the discrete Lagrangian is equal to the action, then the corresponding discrete Hamiltonian map $\tilde{F}_{L_d}$ will exactly equal the flow F_H. We now investigate what happens when this is only an approximation.

2.3.1. Local variational order

Recall that the exact discrete Lagrangian (1.6.1) is defined by

$$L_d^E(q_0, q_1, h) = \int_0^h L(q, \dot{q}) \, \mathrm{d}t,$$

where $q(t)$ is the solution of the Euler–Lagrange equations satisfying $q(0) = q_0$ and $q(h) = q_1$.

We say that a given discrete Lagrangian L_d is of *order* r if there exist an open subset $U_v \subset TQ$ with compact closure and constants $C_v > 0$ and $h_v > 0$ so that

$$\|L_d(q(0), q(h), h) - L_d^E(q(0), q(h), h)\| \le C_v h^{r+1} \tag{2.3.1}$$

for all solutions $q(t)$ of the Euler–Lagrange equations with initial condition $(q, \dot{q}) \in U_v$ and for all $h \le h_v$.

2.3.2. Discrete Legendre transform order

The discrete Legendre transforms $\mathbb{F}^+ L_d$ and $\mathbb{F}^- L_d$ of a discrete Lagrangian L_d are said to be of *order* r if there exists an open subset $U_f \subset T^*Q$ with compact closure and constants $C_f > 0$ and $h_f > 0$ so that

$$\|\mathbb{F}^+ L_d(q(0), q(h), h) - \mathbb{F}^+ L_d^E(q(0), q(h), h)\| \le C_f h^{r+1}, \tag{2.3.2a}$$

$$\|\mathbb{F}^- L_d(q(0), q(h), h) - \mathbb{F}^- L_d^E(q(0), q(h), h)\| \le C_f h^{r+1}, \tag{2.3.2b}$$

for all solutions $q(t)$ of the Euler–Lagrange equations with initial condition $(q, \dot{q}) \in U_f$ and for all $h \le h_f$.

The relationship between the orders of a discrete Lagrangian, its discrete Legendre transforms and its discrete Hamiltonian map is given in the following fundamental theorem.

Theorem 2.3.1. Given a regular Lagrangian L and corresponding Hamiltonian H, the following are equivalent for a discrete Lagrangian L_d:

(1) the discrete Hamiltonian map for L_d is of order r,

(2) the discrete Legendre transforms of L_d are of order r,

(3) L_d is equivalent to a discrete Lagrangian of order r.

Proof. Begin by assuming that L_d is equivalent to a discrete Lagrangian of order r, and we will show that the discrete Legendre transforms are of order r. From Section 2.1.2 we know that equivalent discrete Lagrangians have the same discrete Legendre transforms, and we may thus assume without loss that L_d is itself of order r. Now note that (2.3.1) is equivalent to there existing a function $e_v : T^*Q \times \mathbb{R} \to T^*Q$ so that

$$L_d(q(0), q(h), h) = L_d^E(q(0), q(h), h) + h^{r+1} e_v(q(0), q(h), h)$$

with $\|e_v(q(0), q(h), h)\| \leq C_v$ on U_v. Also, from Theorem 1.6.1 it is clear that we can parametrize the set U_v by either the initial condition $(q, \dot{q})$ or by the endpoints $(q(0), q(h))$.

Taking derivatives of the above expression with respect to $q(h)$ gives

$$\mathbb{F}^+ L_d(q(0), q(h), h) = \mathbb{F}^+ L_d^E(q(0), q(h), h) + h^{r+1} D_2 e_v(q(0), q(h), h),$$

and as e_v is smooth and bounded on the compact set $\mathrm{cl}(U_v)$, so too is $D_2 e_v$, giving (2.3.2a). Taking derivatives with respect to $q(0)$ now shows that the discrete Legendre transforms of L_d are of order r.

Now assume that $\mathbb{F}^+ L_d$ and $\mathbb{F}^- L_d$ are of order r, and set

$$e_v(q(0), q(h), h) = \frac{1}{h^{r+1}} \left[L_d(q(0), q(h), h) - L_d^E(q(0), q(h), h) \right].$$

Taking derivatives with respect to $q(0)$ and $q(h)$ and using (2.3.2) shows that $\|D_1 e_v\| \leq C_f$ and $\|D_2 e_v\| \leq C_f$ on $\mathrm{cl}(U_f)$, which is compact. This then implies that $e_v(\cdot, \cdot, h)$ is itself locally bounded in its first two arguments, and so there exists a function $d(h)$ and a constant C_v such that $\|e_v(q(0), q(h), h) - d(h)\| \leq C_v$. This then proves that $L_d(q(0), q(h), h) - d(h)$ has variational order r, and so L_d is equivalent to a discrete Lagrangian of order r.

We will now show the equivalence of the discrete Hamiltonian map being of order r and the discrete Legendre transforms being of order r. To do this we will make use of the following fact, which is a consequence of the implicit function theorem.

Assume that we have smooth functions related by

$$f_1(x, h) = g_1(x, h) + h^{r+1} e_1(x, h),$$
$$f_2(y, h) = g_2(y, h) + h^{r+1} e_2(y, h),$$

with e_1 and e_2 bounded on some compact sets. Then we have

$$f_2(f_1(x, h), h) = g_2(g_1(x, h), h) + h^{r+1} e_{12}(x, h), \tag{2.3.3a}$$
$$f_1^{-1}(y, h) = g_1^{-1}(y, h) + h^{r+1} \bar{e}_1(y, h), \tag{2.3.3b}$$

for some functions $e_{12}(x, h)$ and $\bar{e}_1(y, h)$ bounded on compact sets.

Now assume that $\mathbb{F}^+ L_d$ and $\mathbb{F}^- L_d$ are of order r and use Corollary 1.5.3 to write

$$\tilde{F}_{L_d} = \mathbb{F}^+ L_d \circ (\mathbb{F}^- L_d)^{-1},$$
$$\tilde{F}_{L_d^E} = \mathbb{F}^+ L_d^E \circ (\mathbb{F}^- L_d^E)^{-1}.$$

Equation (2.3.3) gives the existence of a bounded function e_l such that

$$\tilde{F}_{L_d}(q(0), q(h), h) = \tilde{F}_{L_d^E}(q(0), q(h), h) + h^{r+1} e_l(q(0), q(h), h),$$

and thus we see that the discrete Hamiltonian map is of order r.

Finally, assume that $\tilde{F}_{L_d^E}$ is of order r, and observe that

$$(\mathbb{F}^- L_d)^{-1}(q_0, p_0) = (q_0, \pi_Q \circ \tilde{F}_{L_d}(q_0, p_0)),$$

so (2.3.3) implies (2.3.2b). But now we recall from (1.5.2) that

$$\mathbb{F}^+ L_d = \tilde{F}_{L_d} \circ \mathbb{F}^- L_d,$$

and together with (2.3.3a) this gives (2.3.2a), showing that the discrete Legendre transforms are of order r.   □

2.3.3. *Variational order calculation*

Given a discrete Lagrangian, its order can be calculated by expanding the expression for $L_d(q(0), q(h), h)$ in a Taylor series in h and comparing this to the same expansion for the exact Lagrangian. If the series agree up to r terms then the discrete Lagrangian is of order r.

We explicitly evaluate the first few terms of the expansion of the exact discrete Lagrangian to give

$$L_d^E(q(0), q(h), h) = hL(q, \dot{q}) + \frac{1}{2} h^2 \left(\frac{\partial L}{\partial q}(q, \dot{q}) \cdot \dot{q} + \frac{\partial L}{\partial \dot{q}}(q, \dot{q}) \cdot \ddot{q} \right) + \mathcal{O}(h^3),$$
$$\tag{2.3.4}$$

where $q = q(0)$, $\dot{q} = \dot{q}(0)$ and so forth. Higher derivatives of $q(t)$ are determined by the Euler–Lagrange equations.

Example 2.3.2. An illustrative class of discrete Lagrangian is given by

$$L_d^\alpha(q_0, q_1, h) = hL\left((1 - \alpha)q_0 + \alpha q_1, \frac{q_1 - q_0}{h}\right)$$

for some parameter $\alpha \in [0, 1]$. Calculating the expansion in h gives

$$L_d^\alpha(q(0), q(h), h) = hL(q, \dot{q}) + \frac{1}{2}h^2\left(2\alpha\frac{\partial L}{\partial q}(q, \dot{q}) \cdot \dot{q} + \frac{\partial L}{\partial \dot{q}}(q, \dot{q}) \cdot \ddot{q}\right) + \mathcal{O}(h^3).$$

Comparing this to the expansion (2.3.4) for the exact discrete Lagrangian shows that the method is second-order if and only if $\alpha = 1/2$; otherwise it is only consistent.

Calculating the discrete Hamiltonian map for $L(q, \dot{q}) = \frac{1}{2}\dot{q}^T M\dot{q} - V(q)$ gives the integrator $\tilde{F}_{L_d}^\alpha : (q_0, p_0) \mapsto (q_1, p_1)$ defined implicitly by the relations

$$\frac{q_1 - q_0}{h} = M^{-1}(\alpha p_0 + (1 - \alpha)p_1), \tag{2.3.5a}$$

$$\frac{p_1 - p_0}{h} = -\nabla V((1 - \alpha)q_0 + \alpha q_1). \tag{2.3.5b}$$

Note that this method is explicit for $\alpha = 0$ or $\alpha = 1$ and that it is simply the midpoint rule for $\alpha = 1/2$. Expanding (2.3.5) in h gives

$$q_1 = q_0 + hM^{-1}p_0 - (1 - \alpha)h^2\nabla V(q_0) + \mathcal{O}(h^3),$$
$$p_1 = p_0 - h\nabla V(q_0) - \alpha h^2\nabla^2 V(q_0)M^{-1}p_0 + \mathcal{O}(h^3),$$

and comparing this to the expansion (2.2.3) of the true flow shows that the method is second-order if and only if $\alpha = 1/2$, and otherwise it is only consistent.

The local error and the discrete Lagrangian error thus agree, as expected.
$\Diamond$

Example 2.3.3. As the expansions of discrete Lagrangians are linear in L_d, if we take the symmetrized discrete Lagrangian

$$L_d^{\text{sym},\alpha} = \frac{1}{2}L_d^\alpha + \frac{1}{2}L_d^{1-\alpha},$$

then the expansion will agree with that of the exact discrete Lagrangian up to terms of order h^2, so it gives a method that is second-order for any α. $\Diamond$

2.4. The adjoint of a method and symmetric methods

For a one-step method $F : T^*Q \times \mathbb{R} \to T^*Q$ the *adjoint method* is $F^* : T^*Q \times \mathbb{R} \to T^*Q$ defined by

$$(F^*)^h \circ F^{-h} = \text{Id}; \tag{2.4.1}$$

that is, $(F^*)^h = (F^{-h})^{-1}$. The method F is said to be *self-adjoint* if $F^* = F$. Note that we always have $F^{**} = F$.

Given a discrete Lagrangian $L_d : Q \times Q \times \mathbb{R} \to \mathbb{R}$, we define the *adjoint discrete Lagrangian* to be $L_d^* : Q \times Q \times \mathbb{R} \to \mathbb{R}$ defined by

$$L_d^*(q_0, q_1, h) = -L_d(q_1, q_0, -h). \tag{2.4.2}$$

The discrete Lagrangian L_d is said to be *self-adjoint* if $L_d^* = L_d$. Note that $L_d^{**} = L_d$ for any L_d.

Theorem 2.4.1. If the discrete Lagrangian L_d has discrete Hamiltonian map $\tilde{F}_{L_d}$ then the adjoint L_d^* of the discrete Lagrangian has discrete Hamiltonian map equal to the adjoint map, so that $\tilde{F}_{L_d^*} = \tilde{F}_{L_d}^*$. If the discrete Lagrangian is self-adjoint then the method is self-adjoint. Conversely, if the method is self-adjoint then the discrete Lagrangian is equivalent to a self-adjoint discrete Lagrangian.

Proof. Consider discrete Lagrangians L_d and L_d^* and the corresponding discrete Hamiltonian maps $\tilde{F}_{L_d}$ and $\tilde{F}_{L_d^*}$. For $\tilde{F}_{L_d}$ and $\tilde{F}_{L_d^*}$ to be adjoint, the definition (2.4.1) requires that $\tilde{F}_{L_d}(q_0, p_0, -h) = (q_1, p_1)$ and $\tilde{F}_{L_d^*}(q_1, p_1, h) = (q_0, p_0)$ for all (q_0, p_0). In terms of the generating functions this is

$$\begin{aligned}
p_0 &= -D_1 L_d(q_0, q_1, -h), \\
p_1 &= D_2 L_d(q_0, q_1, -h), \\
p_1 &= -D_1 L_d^*(q_1, q_0, h), \\
p_0 &= D_2 L_d^*(q_1, q_0, h).
\end{aligned} \tag{2.4.3}$$

Equating the expressions for p_0 and p_1 shows that this, in turn, is equivalent to

$$\begin{aligned}
-D_1 L_d(q_0, q_1, -h) &= D_2 L_d^*(q_1, q_0, h), \\
D_2 L_d(q_0, q_1, -h) &= D_1 L_d^*(q_1, q_0, h).
\end{aligned} \tag{2.4.4}$$

Now, if L_d and L_d^* are mutually adjoint, then the definition (2.4.2) implies (2.4.4) and so (2.4.3), thus establishing that $\tilde{F}_{L_d}$ and $\tilde{F}_{L_d^*}$ must also be mutually adjoint, which is written $\tilde{F}_{L_d^*} = \tilde{F}_{L_d}^*$. Note that this implies that $\tilde{F}_{L_d^*}^* = \tilde{F}_{L_d}$.

If L_d is self-adjoint and so $L_d = L_d^*$, then this immediately gives that $\tilde{F}_{L_d} = \tilde{F}_{L_d}^*$ and so $\tilde{F}_{L_d}$ is also self-adjoint.

Conversely, if $\tilde{F}_{L_d}$ and $\tilde{F}_{L_d^*}$ are adjoint, then (2.4.1) implies (2.4.3) which implies (2.4.4). As this simply states that the derivatives of L_d and L_d^* with respect to q_0 and q_1 satisfy the requirement (2.4.2) for adjointness it follows that L_d and L_d^* are mutually adjoint up to the addition of a function of h. Symmetry of $\tilde{F}_{L_d}$ thus implies symmetry of L_d up to a function of h, and so L_d is equivalent to a self-adjoint discrete Lagrangian.   □

2.4.1. Exact discrete Lagrangian is self-adjoint

It is easy to verify that the exact discrete Lagrangian (1.6.1) is self-adjoint. This can be done either directly from the definition (2.4.2), or by realizing that the exact flow map F_H generated by L_d^E satisfies (2.4.1), and then using Theorem 2.4.1.

2.4.2. Order of adjoint methods

To relate the expansions of L_d and its adjoint in terms of h, it is necessary to work with the modified form

$$L_d^*\big(q(-h/2), q(h/2), h\big) = -L_d\big(q(h/2), q(-h/2), -h\big),$$

which can be used in the same way as $L_d^*(q(0), q(h), h) = -L_d(q(h), q(0), -h)$. From this it is clear that the expansion of L_d^* is the negative of the expansion of L_d with h replaced by $-h$. In other words, if L_d has expansion

$$L_d(h) = hL_d^{(1)} + \frac{1}{2}h^2 L_d^{(2)} + \frac{1}{6}h^3 L_d^{(3)} + \cdots$$

then L_d^* will have expansion

$$L_d^*(h) = -(-h)L_d^{(1)} - \frac{1}{2}(-h)^2 L_d^{(2)} - \frac{1}{6}(-h)^3 L_d^{(3)} - \cdots$$
$$= hL_d^{(1)} - \frac{1}{2}h^2 L_d^{(2)} + \frac{1}{6}h^3 L_d^{(3)} - \cdots$$

and so the series agree on odd terms and are opposite on even terms.

This shows that the order of the adjoint discrete Lagrangian L_d^* is the same as the order of L_d. Furthermore, if L_d is self-adjoint, then all the even terms in its expansion must be zero, showing that self-adjoint discrete Lagrangians are necessarily of even order (the first nonzero term, which is $r+1$, must be odd).

These same conclusions can be also be reached by working with the discrete Hamiltonian map, and showing that its adjoint has the same order as it, and that it is of even order whenever it is self-adjoint. Theorems 2.4.1 and 2.3.1 then give the corresponding statements for the discrete Lagrangians.

Example 2.4.2. Perhaps the simplest example of adjoint discrete Lagrangians is the pair

$$L_d(q_0, q_1, h) = hL\left(q_0, \frac{q_1 - q_0}{h}\right),$$
$$L_d^*(q_0, q_1, h) = hL\left(q_1, \frac{q_1 - q_0}{h}\right),$$

which clearly satisfy (2.4.2). For a Lagrangian of the form $L = \frac{1}{2}\dot{q}^T M\dot{q} -$

$V(q)$ these two discrete Lagrangians produce the methods $\tilde{F}_{L_d}$ and $\tilde{F}_{L_d^*}$ given by

$$\tilde{F}_{L_d} \begin{cases} q_1 = q_0 + hM^{-1}p_1, \\ p_1 = p_0 - h\nabla V(q_0), \end{cases}$$

$$\tilde{F}_{L_d^*} \begin{cases} q_1 = q_0 + hM^{-1}p_0, \\ p_1 = p_0 - h\nabla V(q_1). \end{cases}$$

In the terminology of Hairer (1998) these are the two types of symplectic Euler. We can now explicitly compute:

$$(\tilde{F}_{L_d^*})^h \circ (\tilde{F}_{L_d})^{(-h)}(q_0, p_0) = \tilde{F}_{L_d^*}(q_0 + hM^{-1}p_1, p_0 - h\nabla V(q_0), h)$$
$$= (q_0, p_0),$$

which shows that $\tilde{F}_{L_d}$ and $\tilde{F}_{L_d^*}$ are indeed mutually adjoint.   ◇

Example 2.4.3. The discrete Lagrangians in the previous example are just L_d^α for $\alpha = 0$ and $\alpha = 1$, respectively. Extending this gives $(L_d^\alpha)^* = L_d^{1-\alpha}$, which shows that the midpoint rule (given by $\alpha = 1/2$) is self-adjoint. From this it is also clear that the symmetrized versions $L_d^{\text{sym},\alpha}$ are self-adjoint for all α.   ◇

2.5. Composition methods

We now consider how to combine several discrete Lagrangians together to obtain a new discrete Lagrangian with higher order, or some other desirable property. The resulting discrete Hamiltonian map will be the composition of the maps of the component discrete Lagrangians. References on composition methods include Yoshida (1990), Qin and Zhu (1992), McLachlan (1995) and Murua and Sanz-Serna (1999).

The strength of the composition methodology can be illustrated by a few simple examples. Given a one-step method $F : T^*Q \times \mathbb{R} \to T^*Q$ with corresponding adjoint F^*, then the method $\hat{F}^h = F^{h/2} \circ (F^*)^{h/2}$ will be self-adjoint and have order at least equal to that of F. Furthermore, for a self-adjoint method F with order r, which we recall must be even, the method $\hat{F}^h = F^{\gamma h} \circ F^{(1-2\gamma)h} \circ F^{\gamma h}$ with the constant $\gamma = (2 - 2^{1/(r+1)})^{-1}$ will have order $r + 2$. This thus provides a simple way to derive methods of arbitrarily high order starting from a given low-order method. See the above references for details and more complicated examples.

Consider now discrete Lagrangians L_d^i and time-step fractions γ^i for $i = 1, \ldots, s$ satisfying $\sum_{i=1}^s \gamma^i = 1$. Note that the γ^i may each be positive or negative. We now give three equivalent interpretations of composition discrete Lagrangians.

2.5.1. Multiple steps

Begin by taking a discrete trajectory $\{q_k\}_{k=0}^N$, dividing each step (q_k, q_{k+1}) into s substeps $(q_k = q_k^0, q_k^1, q_k^2, \ldots, q_k^s = q_{k+1})$. Rather than using the same discrete Lagrangian for each step, as we have previously always assumed, we will now use the different L_d^i on each substep in turn.

This is equivalent to taking the discrete action sum to be

$$\mathfrak{G}_d\big(\{(q_k = q_k^0, \ldots, q_k^s = q_{k+1})\}_{k=1}^N\big) = \sum_{k=0}^N \sum_{i=1}^s L_d^i(q_k^{i-1}, q_k^i, \gamma^i h). \quad (2.5.1)$$

The discrete Euler–Lagrange equations, resulting from requiring that this be stationary, pair neighbouring discrete Lagrangians together to give

$$D_2 L_d^i(q_k^{i-1}, q_k^i, \gamma^i h) + D_1 L_d^{i+1}(q_k^i, q_k^{i+1}, \gamma^{i+1} h) = 0, \quad (2.5.2a)$$
$$i = 1, \ldots, s-1,$$
$$D_2 L_d^s(q_k^{s-1}, q_k^s, \gamma^s h) + D_1 L_d^1(q_{k+1}^0, q_{k+1}^1, \gamma^1 h) = 0, \quad (2.5.2b)$$

where the steps are joined with $q_k^s = q_{k+1}^0$.

Considering the L_d^i as generating functions for the discrete Hamiltonian maps $\tilde{F}_{L_d^i}$ shows that this is simply taking a step with $\tilde{F}_{L_d^1}$ of length $\gamma^1 h$, followed by a step with $\tilde{F}_{L_d^2}$ of length $\gamma^2 h$, and so on. The map over the entire time-step is thus the composition of the maps

$$\tilde{F}_{L_d^s}^{\gamma^s h} \circ \cdots \circ \tilde{F}_{L_d^2}^{\gamma^2 h} \circ \tilde{F}_{L_d^1}^{\gamma^1 h}.$$

2.5.2. Single step, multiple substeps

We now combine the discrete Lagrangians on each step into one multipoint discrete Lagrangian defined by

$$L_d(q_k^0, q_k^1, \ldots, q_k^s, h) = \sum_{i=1}^s L_d^i(q_k^{i-1}, q_k^i, \gamma^i h), \quad (2.5.3)$$

and we define the discrete action sum over the entire trajectory to be

$$\mathfrak{G}_d\big(\{(q_k = q_k^0, \ldots, q_k^s = q_{k+1})\}_{k=1}^N\big) = \sum_{k=0}^N L_d(q_k^0, q_k^1, \ldots, q_k^s, h), \quad (2.5.4)$$

which is clearly equal to (2.5.1).

Requiring that $\mathfrak{G}_d$ be stationary gives the extended set of discrete Euler–Lagrange equations

$$D_i L_d(q_k^0, q_k^1, \ldots, q_k^s, h) = 0 \qquad i = 2, \ldots, s \quad (2.5.5a)$$
$$D_{s+1} L_d(q_k^0, q_k^1, \ldots, q_k^s, h) + D_1 L_d(q_{k+1}^0, q_{k+1}^1, \ldots, q_{k+1}^s, h) = 0, \quad (2.5.5b)$$

which are equivalent to (2.5.2a) and (2.5.2b), respectively.

2.5.3. Single step

Finally, we form a standard discrete Lagrangian which is equivalent to the above methods. Set the *composition discrete Lagrangian* to be

$$L_d(q_k, q_{k+1}, h) = \operatorname*{ext}_{(q_k^1, \dots, q_k^{s-1})} L_d(q_k = q_k^0, q_k^1, q_k^2, \dots, q_k^{s-1}, q_k^s = q_{k+1}, h)$$

$$(2.5.6)$$

which is the multipoint discrete Lagrangian defined above, evaluated on the trajectory within the step which solves (2.5.5a).

Note that the derivatives of this discrete Lagrangian satisfy

$$D_1 L_d(q_k, q_{k+1}, h) = D_1 L_d(q_k, q_k^1, q_k^2, \dots, q_k^{s-1}, q_{k+1}, h)$$

$$+ \sum_{i=1}^{s-1} D_i L_d(q_k, q_k^1, q_k^2, \dots, q_k^{s-1}, q_{k+1}, h) \cdot \frac{\partial q_k^i}{\partial q_k}$$

$$= D_1 L_d(q_k, q_k^1, q_k^2, \dots, q_k^{s-1}, q_{k+1}, h)$$

$$= D_1 L_d^1(q_k, q_k^1, \gamma^1 h)$$

using (2.5.5a), and similarly

$$D_2 L_d(q_k, q_{k+1}, h) = D_{s+1} L_d(q_k, q_k^1, q_k^2, \dots, q_k^{s-1}, q_{k+1}, h)$$

$$= D_2 L_d^s(q_k^{s-1}, q_{k+1}, \gamma^s h).$$

This gives the following theorem.

Theorem 2.5.1. Take discrete Lagrangians L_d^i and time-step fractions γ^i for $i = 1, \dots, s$ satisfying $\sum_{i=1}^s \gamma^i = 1$. Define the composition discrete Lagrangian L_d by (2.5.6). Then the discrete Hamiltonian map $\tilde{F}_{L_d}$ is

$$\tilde{F}_{L_d}^h = \tilde{F}_{L_d^s}^{\gamma^s h} \circ \cdots \circ \tilde{F}_{L_d^2}^{\gamma^2 h} \circ \tilde{F}_{L_d^1}^{\gamma^1 h}$$

formed by the composition of the discrete Hamiltonian maps for each L_d^i.

Proof. The equations that define $\tilde{F}_{L_d}$ are

$$p_k = -D_1 L_d(q_k, q_{k+1}, h) = -D_1 L_d^1(q_k, q_k^1, \gamma^1 h),$$

$$p_{k+1} = D_2 L_d(q_k, q_{k+1}, h) = D_2 L_d^s(q_k^{s-1}, q_{k+1}, \gamma^s h),$$

together with (2.5.5a), which is equivalent to (2.5.2a), which we write as

$$p_k^i = D_2 L_d^i(q_k^{i-1}, q_k^i, \gamma^i h) = -D_1 L_d^{i+1}(q_k^i, q_k^{i+1}, \gamma^{i+1} h)$$

for $i = 1, \ldots, s - 1$. Setting $p_k^0 = p_k$ and $p_k^s = p_{k+1}$, we can group these to give

$$p_k^{i-1} = -D_1 L_d^i(q_k^{i-1}, q_k^i, \gamma^i h),$$
$$p_k^i = D_2 L_d^i(q_k^{i-1}, q_k^i, \gamma^i h),$$

for $i = 1, \ldots, s$, which are the definition of $\tilde{F}_{L_d^s}^{\gamma^s h} \circ \cdots \circ \tilde{F}_{L_d^2}^{\gamma^2 h} \circ \tilde{F}_{L_d^1}^{\gamma^1 h}$, thus giving the required equivalence. □

2.6. Examples of variational integrators

In this section we will consider a number of standard symplectic methods and show how to write them as variational integrators. Recall that we are assuming that Q is a linear space with inner product $\langle \cdot, \cdot \rangle$ and corresponding norm $\| \cdot \|$. We will always assume that the Lagrangian $L : TQ \rightarrow \mathbb{R}$ is regular, so that it has a corresponding Hamiltonian $H : T^*Q \rightarrow \mathbb{R}$. In addition, we will sometimes consider the Lagrangian composed of a kinetic and a potential energy, so that it is of the form $L(q, \dot{q}) = \frac{1}{2}\dot{q}^T M \dot{q} - V(q)$ where M is a positive-definite symmetric matrix.

2.6.1. Midpoint rule

Given a Hamiltonian system $H : T^*Q \rightarrow \mathbb{R}$, the *midpoint rule* is an integrator $F^h : (q_0, p_0) \mapsto (q_1, p_1)$. Setting $z_0 = (q_0, p_0)$ and $z_1 = (q_1, p_1)$ the map is defined implicitly by the relation

$$\frac{z_1 - z_0}{h} = X_H \left(\frac{z_1 + z_0}{2} \right),$$

where X_H is the Hamiltonian vector field. Writing the two components separately gives

$$\frac{q_1 - q_0}{h} = \frac{\partial H}{\partial p} \left(\frac{q_1 + q_0}{2}, \frac{p_1 + p_0}{2} \right), \tag{2.6.1a}$$

$$\frac{p_1 - p_0}{h} = -\frac{\partial H}{\partial p} \left(\frac{q_1 + q_0}{2}, \frac{p_1 + p_0}{2} \right). \tag{2.6.1b}$$

The symplectic nature of the midpoint rule is often explained by using the Cayley transform (this remark is due, as far as we know, to Krishnaprasad and J. C. Simo; see, for example, Austin, Krishnaprasad and Wang (1993), Simo, Tarnow and Wong (1992) and Simo and Tarnow (1992), and related papers). See Marsden (1999) for an exposition of this method.

   To write the midpoint rule as a variational integrator, assume that H is regular and that L is the corresponding regular Lagrangian defined by (1.4.4). Define the discrete Lagrangian

$$L_d^{\frac{1}{2}}(q_0, q_1, h) = hL \left(\frac{q_1 + q_0}{2}, \frac{q_1 - q_0}{h} \right).$$

Evaluating the expressions (1.5.4) for the discrete Hamiltonian map gives

$$p_0 = -\frac{h}{2}\frac{\partial L}{\partial q}\left(\frac{q_1 + q_0}{2}, \frac{q_1 - q_0}{h}\right) + \frac{\partial L}{\partial \dot{q}}\left(\frac{q_1 + q_0}{2}, \frac{q_1 - q_0}{h}\right),$$

$$p_1 = \frac{h}{2}\frac{\partial L}{\partial q}\left(\frac{q_1 + q_0}{2}, \frac{q_1 - q_0}{h}\right) + \frac{\partial L}{\partial \dot{q}}\left(\frac{q_1 + q_0}{2}, \frac{q_1 - q_0}{h}\right),$$

and subtracting and adding these two equations produces

$$\frac{p_1 - p_0}{h} = \frac{\partial L}{\partial q}\left(\frac{q_1 + q_0}{2}, \frac{q_1 - q_0}{h}\right), \qquad (2.6.2)$$

$$\frac{p_1 + p_0}{2} = \frac{\partial L}{\partial \dot{q}}\left(\frac{q_1 + q_0}{2}, \frac{q_1 - q_0}{h}\right).$$

The second of these equations is simply the statement that

$$\left(\frac{q_1 + q_0}{2}, \frac{p_1 + p_0}{2}\right) = \mathbb{F}L\left(\frac{q_1 + q_0}{2}, \frac{q_1 - q_0}{h}\right),$$

and so using (1.4.5a) shows that (2.6.2) is equivalent to (2.6.1b), while (1.4.5c) gives (2.6.1a).

For regular Lagrangian systems, the midpoint discrete Lagrangian $L_d^{1/2}$ thus has discrete Hamiltonian map which is the midpoint rule on T^*Q for the corresponding Hamiltonian system.

2.6.2. Störmer–Verlet

The Verlet method (Verlet 1967) (also known as Störmer's rule) was originally formulated for molecular dynamics problems and remains popular in that field. The derivation of Verlet as a variational integrator is in Wendlandt and Marsden (1997a) and is implicitly in Gillilan and Wilson (1992) as well.

Verlet is usually written for systems of the form $L(q, \dot{q}) = \frac{1}{2}\dot{q}^T M \dot{q} - V(q)$, and was originally formulated as a map $Q \times Q \to Q \times Q$ given by $(q_k, q_{k+1}) \mapsto (q_{k+1}, q_{k+2})$ with

$$q_{k+1} = 2q_k - q_{k-1} + h^2 a_k,$$

where we use the notation $a_k = M^{-1}(-\nabla V(q_k))$. As can be readily seen, this is just the discrete Lagrangian map $F_{L_d} : Q \times Q \to Q \times Q$ for either of

$$L_d^0(q_0, q_1, h) = hL\left(q_0, \frac{q_1 - q_0}{h}\right),$$

$$L_d^1(q_0, q_1, h) = hL\left(q_1, \frac{q_1 - q_0}{h}\right),$$

or indeed any affine combination of these two. In particular, consider the symmetric version

$$L_d(q_0, q_1, h) = \frac{1}{2} h L \left(q_0, \frac{q_1 - q_0}{h} \right) + \frac{1}{2} h L \left(q_1, \frac{q_1 - q_0}{h} \right),$$

which gives Verlet as the corresponding F_{L_d}. Pushing this forward to T^*Q with $\mathbb{F}^{\pm} L_d$ now gives $\tilde{F}_{L_d} : T^*Q \to T^*Q$ defined by (1.5.4). Evaluating these yields

$$p_k = M \left(\frac{q_{k+1} - q_k}{h} \right) + \frac{1}{2} h \nabla V(q_k),$$

$$p_{k+1} = M \left(\frac{q_{k+1} - q_k}{h} \right) - \frac{1}{2} h \nabla V(q_{k+1}).$$

Now we subtract the first equation from the second and solve the first equation for q_{k+1} to obtain

$$q_{k+1} = q_k + h M^{-1} p_k + \frac{1}{2} h^2 M^{-1} (-\nabla V(q_k)),$$

$$p_{k+1} = p_k + h \left(\frac{-\nabla V(q_k) - \nabla V(q_{k+1})}{2} \right),$$

which is the so-called *velocity* Verlet method (Swope, Andersen, Berens and Wilson 1982, Allen and Tildesley 1987) written on T^*Q. Using the Legendre transform $\mathbb{F}L(q, \dot{q}) = (q, M\dot{q})$ this can also be mapped to TQ.

We thus see that velocity Verlet will preserve the canonical two-form Ω on T^*Q, and as L_d is invariant under linear symmetries of the potential, Verlet will also preserve quadratic momentum maps such as linear and angular momentum.

2.6.3. Newmark methods

The Newmark family of integrators, originally given in Newmark (1959), are widely used in structural dynamics codes. They are usually written (see, for example, Hughes (1987)) for the system $L = \frac{1}{2} \dot{q}^T M \dot{q} - V(q)$ as maps $TQ \to TQ$ given by $(q_k, \dot{q}_k) \mapsto (q_{k+1}, \dot{q}_{k+1})$ satisfying the implicit relations

$$q_{k+1} = q_k + h\dot{q}_k + \frac{h^2}{2} \left[(1 - 2\beta) a(q_k) + 2\beta a(q_{k+1}) \right], \qquad (2.6.3a)$$

$$\dot{q}_{k+1} = \dot{q}_k + h \left[(1 - \gamma) a(q_k) + \gamma a(q_{k+1}) \right], \qquad (2.6.3b)$$

$$a(q) = M^{-1}(-\nabla V(q)), \qquad (2.6.3c)$$

where the parameters $\gamma \in [0, 1]$ and $\beta \in [0, \frac{1}{2}]$ specify the method. It is simple to check that the method is second-order if $\gamma = 1/2$ and first-order otherwise, and that it is generally explicit only for $\beta = 0$.

The $\beta = 0$, $\gamma = 1/2$ case is well known to be symplectic (see, for example, Simo *et al.* (1992)) with respect to the canonical symplectic form Ω_L on TQ. This can be easily seen from the fact that this method is simply the pullback by $\mathbb{F}L$ of the discrete Hamiltonian map for $L_d^{\text{sym},\alpha}$ with $\alpha = 0$ or $\alpha = 1$. Note that this method is the same as velocity Verlet.

It is also well known (for example, Simo *et al.* (1992)) that the Newmark algorithm with $\beta \neq 0$ does not preserve the *canonical* symplectic form. Nonetheless, based on a remark by Suris, it can be shown (Kane *et al.* 2000) that the Newmark method with $\gamma = 1/2$ and any β can be generated from a discrete Lagrangian. To see this, we introduce the map $\varphi^\beta : Q \times Q \to TQ$ defined by

$$\varphi^\beta(q_k, q_{k+1}) = \left(q_k, \left[\frac{q_{k+1} - q_k}{h} \right] - \frac{h}{2}\left[(1 - 2\beta)a(q_k) + 2\beta a(q_{k+1}) \right] \right).$$

Pulling the Newmark map back by φ^β to a map $Q \times Q \to Q \times Q$ gives the map $(q_k, q_{k+1}) \mapsto (q_{k+1}, q_{k+2})$ where

$$\frac{q_{k+2} - 2q_{k+1} + q_k}{h^2} = \beta a(q_{k+2}) + (1 - 2\beta)a(q_{k+1}) + \beta a(q_k). \qquad (2.6.4)$$

A straightforward calculation now shows that this is in fact the discrete Lagrange map $F_{L_d^\beta}$ for the discrete Lagrangian

$$L_d^\beta(q_0, q_1, h) = h\frac{1}{2}\left(\frac{\eta^\beta(q_1) - \eta^\beta(q_0)}{h} \right)^T M \left(\frac{\eta^\beta(q_1) - \eta^\beta(q_0)}{h} \right) - h\tilde{V}(\eta^\beta(q_0)),$$

where we have introduced the map $\eta^\beta : Q \to Q$ defined by

$$\eta^\beta(q) = q - \beta h^2 M^{-1}\nabla V(q)$$

and the modified potential function $\tilde{V} : Q \to \mathbb{R}$ satisfying $\nabla\tilde{V} \circ \eta^\beta = \nabla V$, which will exist for small h.

This result shows that the Newmark method for $\gamma = 1/2$ is the pullback of the discrete Hamiltonian map $\tilde{F}_{L_d^\beta}$ by the map $\mathbb{F}^+ L_d^\beta \circ (\varphi^\beta)^{-1}$. As the discrete Hamiltonian map preserves the canonical symplectic form on T^*Q, this means that Newmark preserves the two-form $[\mathbb{F}^+ L_d^\beta \circ (\varphi^\beta)^{-1}]^*\Omega$ on TQ. Note that this is not the canonical two-form Ω_L on TQ, but this is enough to explain the otherwise inexplicably good long-time behaviour of $\gamma = 1/2$ Newmark for nonlinear problems.

An alternative and independent method of analysing the symplectic members of Newmark has been given by Skeel, Zhang and Schlick (1997), including an interesting nonlinear analysis in Skeel and Srinivas (2000). This is based on the observation that if we define the map $\bar{\eta}^\beta : TQ \to TQ$ by

$$\bar{\eta}^\beta(q, v) = \left(\eta^\beta(q), v \right)$$

then the pushforward of the Newmark method by $\bar{\eta}^\beta$ is given by $(x_k, v_k) \mapsto$
(x_{k+1}, v_{k+1}), where

$$x_{k+1} = x_k + hv_k + \frac{1}{2}h^2 a_k, \tag{2.6.5a}$$

$$v_{k+1} = v_k + \frac{1}{2}h(a_k + a_{k+1}), \tag{2.6.5b}$$

$$a_k = a(x_k + \beta h^2 a_k). \tag{2.6.5c}$$

This map can be shown to be symplectic with respect to the canonical
two-form Ω_L on TQ, and so Newmark will preserve the two-form $(\bar{\eta}^\beta)^*\Omega_L$
on TQ.

To summarize, we have the following commutative diagram, where the
map $\tilde{F}_{L_d^\beta}$ preserves the canonical two-form Ω on T^*Q, the map (2.6.5) preserves the Lagrange two-form Ω_L on TQ, and we have set $\gamma = 1/2$ in the
Newmark equation (2.6.3).

$$
\begin{array}{ccccccc}
T^*Q & \xleftarrow{\;\mathbb{F}^+ L_d^\beta\;} & Q \times Q & \xrightarrow{\;\varphi^\beta\;} & TQ & \xrightarrow{\;\bar{\eta}^\beta\;} & TQ \\
\Big\downarrow{\scriptstyle \tilde{F}_{L_d^\beta}} & & \Big\downarrow{\scriptstyle F_{L_d^\beta}\;(2.6.4)} & & \Big\downarrow{\scriptstyle (2.6.3)} & & \Big\downarrow{\scriptstyle (2.6.5)} \\
T^*Q & \xleftarrow{\;\mathbb{F}^+ L_d^\beta\;} & Q \times Q & \xrightarrow{\;\varphi^\beta\;} & TQ & \xrightarrow{\;\bar{\eta}^\beta\;} & TQ
\end{array}
$$

2.6.4. Explicit symplectic partitioned Runge–Kutta methods

Symplectic integrators which are explicit partitioned Runge–Kutta methods
were first used by Ruth (1983) and Forest and Ruth (1990), who constructed
them as a composition of steps, each one generated by a generating function
of the third kind. Using the same idea shows that these methods are also
variational, at least for Hamiltonians with kinetic energy of the form $T(p) =
1/2p^T M^{-1}p$ for some constant mass matrix M.

It can be shown (Hairer *et al.* 1993) that explicit symplectic partitioned
Runge–Kutta methods can always be written as the composition of a number
of steps of the method $F^{a,b} : T^*Q \times \mathbb{R} \to T^*Q$ given by

$$q_1 = q_0 + ahM^{-1}p_0,$$
$$p_1 = p_0 - bh\nabla V(q_1),$$

and of its adjoint method $(F^{a,b})^*$, with each step having different parameters
(a, b). Furthermore, it is simple to check that these can be chosen so that
all steps have nonzero a.

We now see, however, that the method $F^{a,b}$ is the discrete Hamiltonian map for the discrete Lagrangian $L_d^{a,b}$ given by

$$L_d^{a,b}(q_0, q_1, h) = h\left[b\frac{1}{2}\left(\frac{q_1 - q_0}{h}\right)^T M\left(\frac{q_1 - q_0}{h}\right) - \frac{1}{a}V(q_1)\right],$$

and from Theorem 2.4.1 it is clear that $(F^{a,b})^*$ is the discrete Hamiltonian map of the adjoint discrete Lagrangian $(L_d^{a,b})^*$.

We can thus form a composition discrete Lagrangian as in Theorem 2.5.1 whose discrete Hamiltonian map is the composition of the $F^{a,b}$ and $(F^{a,b})^*$, and is therefore the explicit symplectic partitioned Runge–Kutta method.

2.6.5. Symplectic partitioned Runge–Kutta methods

Partitioned Runge–Kutta methods are a class of integrators about which much is known and which generalize standard Runge–Kutta methods. The symplectic members of Runge–Kutta were first identified by Lasagni (1988), Sanz-Serna (1988) and Suris (1989). Symplectic partitioned Runge–Kutta methods appeared in Sanz-Serna (1992a) and Suris (1990). Good general references are Hairer *et al.* (1993) and Hairer and Wanner (1996). See also Geng (1995, 2000), Sofroniou and Oevel (1997) and Oevel and Sofroniou (1997) for order conditions and derivations. An explicit construction has been given by Suris (1990) for the discrete Lagrangian which generates any symplectic partitioned Runge–Kutta method. We summarize this derivation below.

Recall that a partitioned Runge–Kutta method for the regular Lagrangian system L is a map $T^*Q \times \mathbb{R} \to T^*Q$ specified by the coefficients b_i, a_{ij}, $\tilde{b}_i$, $\tilde{a}_{ij}$ for $i, j = 1, \ldots, s$, and defined by $(q_0, p_0) \mapsto (q_1, p_1)$ for

$$q_1 = q_0 + h\sum_{j=1}^{s} b_j \dot{Q}_j, \quad p_1 = p_0 + h\sum_{j=1}^{s} \tilde{b}_j \dot{P}_j, \tag{2.6.6a}$$

$$Q_i = q_0 + h\sum_{j=1}^{s} a_{ij} \dot{Q}_j, \quad P_i = p_0 + h\sum_{j=1}^{s} \tilde{a}_{ij} \dot{P}_j, \quad i = 1, \ldots, s, \tag{2.6.6b}$$

$$P_i = \frac{\partial L}{\partial \dot{q}}(Q_i, \dot{Q}_i), \quad \dot{P}_i = \frac{\partial L}{\partial q}(Q_i, \dot{Q}_i), \quad i = 1, \ldots, s, \tag{2.6.6c}$$

where the points (Q_i, P_i) are known as the internal stages. In the special case that $a_{ij} = \tilde{a}_{ij}$ and $b_i = \tilde{b}_i$ then a partitioned Runge–Kutta method is said to be simply a Runge–Kutta method.

It is well known that the method is symplectic (that is, it preserves the canonical symplectic form Ω on T^*Q) if the coefficients satisfy

$$b_i \tilde{a}_{ij} + \tilde{b}_j a_{ji} = b_i \tilde{b}_j, \quad i, j = 1, \ldots, s, \tag{2.6.7a}$$

$$b_i = \tilde{b}_i, \quad i = 1, \ldots, s. \tag{2.6.7b}$$

We now assume that we have coefficients satisfying (2.6.7) and write a discrete Lagrangian that generates the corresponding symplectic partitioned Runge–Kutta method. Given points $(q_0, q_1) \in Q \times Q$, we can regard (2.6.6) as implicitly defining p_0, p_1, Q_i, P_i, $\dot{Q}_i$ and $\dot{P}_i$ for $i = 1, \ldots, s$. Taking these to be so defined as functions of (q_0, q_1), we construct a discrete Lagrangian

$$L_d(q_0, q_1, h) = h \sum_{i=1}^{s} b_i L(Q_i, \dot{Q}_i). \tag{2.6.8}$$

It can now be shown (Suris 1990) that the corresponding discrete Hamiltonian map is exactly the map $(q_0, p_0) \mapsto (q_1, p_1)$, which is the symplectic partitioned Runge–Kutta method. Nonsymplectic partitioned Runge–Kutta methods will clearly not have a corresponding discrete Lagrangian formulation.

Theorem 2.6.1. The discrete Hamiltonian map generated by the discrete Lagrangian (2.6.8) is a symplectic partitioned Runge–Kutta method.

Proof. To check that the discrete Hamiltonian map defined by L_d is indeed the partitioned Runge–Kutta method specified by (2.6.6), we need only check that equations (1.5.4) are satisfied. We compute

$$\frac{\partial L_d}{\partial q_0}(q_0, q_1) = (\Delta t) \sum_{i=1}^{s} b_i \left[\frac{\partial L}{\partial q} \cdot \frac{\partial Q_i}{\partial q_0} + \frac{\partial L}{\partial \dot{q}} \cdot \frac{\partial \dot{Q}_i}{\partial q_0} \right]$$

$$= (\Delta t) \sum_{i=1}^{s} b_i \left[\dot{P}_i \cdot \frac{\partial Q_i}{\partial q_0} + P_i \cdot \frac{\partial \dot{Q}_i}{\partial q_0} \right],$$

using the definitions for P_i and $\dot{P}_i$ in (2.6.6). Differentiating the definition for Q_i in (2.6.6b) and substituting in this and the definition of P_i in (2.6.6b) now gives

$$\frac{\partial L_d}{\partial q_0}(q_0, q_1) = (\Delta t) \sum_{i=1}^{s} b_i \left[\dot{P}_i \cdot \left(I + (\Delta t) \sum_{j=1}^{s} a_{ij} \frac{\partial \dot{Q}_j}{\partial q_0} \right) \right.$$

$$\left. + \left(p_0 + (\Delta t) \sum_{j=1}^{s} \tilde{a}_{ij} \dot{P}_j \right) \cdot \frac{\partial \dot{Q}_i}{\partial q_0} \right]$$

$$= (\Delta t) \sum_{i=1}^{s} b_i \left[\dot{P}_i + p_0 \cdot \frac{\partial \dot{Q}_i}{\partial q_0} \right]$$

$$+ (\Delta t)^2 \sum_{i=1}^{s} \sum_{j=1}^{s} (b_i \tilde{a}_{ij} + b_j a_{ji}) \dot{P}_j \cdot \frac{\partial \dot{Q}_i}{\partial q_0},$$

and we can use the symplecticity identities (2.6.7) to obtain

$$\frac{\partial L_d}{\partial q_0}(q_0, q_1) = p_0 \cdot \left[(\Delta t) \sum_{i=1}^{s} b_i \frac{\partial \dot{Q}_i}{\partial q_0} \right] + (\Delta t) \sum_{i=1}^{s} b_i \dot{P}_i$$

$$+ (\Delta t) \sum_{j=1}^{s} b_j \dot{P}_j \cdot \left[(\Delta t) \sum_{i=1}^{s} b_i \frac{\partial \dot{Q}_i}{\partial q_0} \right]$$

$$= -p_0,$$

where we have differentiated the expression for q_1 in (2.6.6a) to obtain the identity

$$(\Delta t) \sum_{i=1}^{s} b_i \frac{\partial \dot{Q}_i}{\partial q_0} = -I.$$

This thus establishes that the first equation of (1.5.4) is satisfied.

Differentiating L_d with respect to q_1 and following a similar argument to that above gives the second part of (1.5.4), and shows that the discrete Hamiltonian map $\tilde{F}L_d$ generated by the discrete Lagrangian (2.6.8) is indeed the symplectic partitioned Runge–Kutta method.   □

This construction thus provides a proof of the well-known fact that the restrictions (2.6.7) on the coefficients mean that the partitioned Runge–Kutta method is symplectic, as discrete Hamiltonian maps always preserve the canonical symplectic form. In addition, the linear nature of the definition of the discrete Lagrangian (2.6.8) means that it will inherit linear symmetries of the Lagrangian L, which thus proves the standard result that partitioned Runge–Kutta methods preserve quadratic momentum maps.

Another way to regard the above derivation is to say that we have written down a generating function of the first kind for the symplectic partitioned Runge–Kutta map. A generalization of this is given in Jalnapurkar and Marsden (200x), where it is shown how to construct generating functions of arbitrary type for any given symplectic partitioned Runge–Kutta method.

The above construction has also been generalized to multisymplectic partial differential equations in West (2001), thereby obtaining multisymplectic product partitioned Runge–Kutta methods (Reich 2000).

2.6.6. Galerkin methods

To obtain accurate variational integrators, the results in Section 2.3 show that the discrete Lagrangian should approximate the action over short trajectory segments. One way to do this practically is to use polynomial approximations to the trajectories and numerical quadrature to approximate the integral. This can be shown to be equivalent both to a class of continuous

Galerkin methods and to a subset of symplectic partitioned Runge–Kutta methods.

This approach is related to the Continuous Galerkin and Discontinuous Galerkin methods, as in Estep and French (1994), Hulme (1972a, 1972b) and Thomée (1997). These methods differ in the precise choice of function space (continuous or discontinuous) and whether the position and velocity components are projected separately or the velocity projection is given by the lift of a position projection.

We know that a discrete Lagrangian should be an approximation

$$L_d(q_0, q_1, h) \approx \operatorname*{ext}_{q \in \mathcal{C}([0,h],Q)} \mathfrak{G}(q),$$

where $\mathcal{C}([0, h], Q)$ is the space of trajectories $q : [0, h] \to Q$ with $q(0) = q_0$ and $q(h) = q_1$, and $\mathfrak{G} : \mathcal{C}(0, h) \to \mathbb{R}$ is the action (1.2.1).

To approximate this quantity, we choose the particular finite-dimensional approximation $\mathcal{C}^s([0, h], Q) \subset \mathcal{C}([0, h], Q)$ of the trajectory space given by

$$\mathcal{C}^s([0, h], Q) = \{q \in \mathcal{C}([0, h], Q) \mid q \text{ is a polynomial of degree } s\},$$

and we approximate the action integral with numerical quadrature to give an approximate action $\mathfrak{G}^s : \mathcal{C}([0, h], Q) \to \mathbb{R}$ by

$$\mathfrak{G}^s(q) = h \sum_{i=1}^{s} b_i L\big(q(c_i h), \dot{q}(c_i h)\big), \tag{2.6.9}$$

where $c_i \in [0, 1]$, $i = 1, \ldots, s$ are a set of quadrature points and b_i are the associated maximal order weights. We now set the Galerkin discrete Lagrangian to be

$$L_d(q_0, q_1, h) = \operatorname*{ext}_{q \in \mathcal{C}^s([0,h],Q)} \mathfrak{G}^s(q), \tag{2.6.10}$$

which can be practically evaluated. This procedure, of course, is simply performing Galerkin projection of the weak form of the ODE onto the space of piecewise polynomial trajectories. Furthermore, as we will show below, the resulting integrator is a symplectic partitioned Runge–Kutta method.

To make the above equations explicit, choose control times $0 = d_0 < d_1 < d_2 < \cdots < d_{s-1} < d_s = 1$ and control points $q_0^0 = q_0, q_0^1, q_0^2, \ldots, q_0^{s-1}, q_0^s = q_1$. These uniquely define the degree s polynomial $q_d(t; q_0^\nu, h)$ which passes through each q_0^ν at time $d_\nu h$, that is, $q_d(d_\nu h) = q_0^\nu$ for $\nu = 0, \ldots, s$. Letting $\tilde{l}_{\nu,s}(t)$ denote the Lagrange polynomials associated with the d_ν, we can express $q_d(t; q_0^\nu, h)$ as

$$q_d(\tau h; q_0^\nu, h) = \sum_{\nu=0}^{s} q_0^\nu \tilde{l}_{\nu,s}(\tau). \tag{2.6.11}$$

For $q_d(t; q_0^\nu, h)$ to be a critical point of the discrete action (2.6.9) we must have stationarity with respect to variations in q_0^ν for $\nu = 1, \ldots, s - 1$. Differentiating (2.6.9) and (2.6.11) implies that we have

$$0 = h \sum_{i=1}^{s} b_i \left[\frac{\partial L}{\partial q}(c_i h) \tilde{l}_{\nu,s}(c_i) + \frac{1}{h} \frac{\partial L}{\partial \dot{q}}(c_i h) \dot{\tilde{l}}_{\nu,s}(c_i) \right] \qquad (2.6.12)$$

for each $\nu = 1, \ldots, s - 1$, where we denote $\frac{\partial L}{\partial q}(c_i h) = \frac{\partial L}{\partial q}(q_d(c_i h), \dot{q}_d(c_i h))$ and similarly for the other expressions.

The integration scheme $(q_0, p_0) \mapsto (q_1, p_1)$ generated by the Galerkin discrete Lagrangian (2.6.10) is now given implicitly by the relations

$$-p_0 = \frac{\partial L_d}{\partial q_0}(q_0, q_1, h), \qquad p_1 = \frac{\partial L_d}{\partial q_1}(q_0, q_1, h).$$

Evaluating these expressions and restating (2.6.12) gives the set of equations

$$E(0) : -p_0 = h \sum_{i=1}^{s} b_i \left[\frac{\partial L}{\partial q}(c_i h) \tilde{l}_{0,s}(c_i) + \frac{1}{h} \frac{\partial L}{\partial \dot{q}}(c_i h) \dot{\tilde{l}}_{0,s}(c_i) \right],$$

$$E(\nu) : \quad 0 = h \sum_{i=1}^{s} b_i \left[\frac{\partial L}{\partial q}(c_i h) \tilde{l}_{\nu,s}(c_i) + \frac{1}{h} \frac{\partial L}{\partial \dot{q}}(c_i h) \dot{\tilde{l}}_{\nu,s}(c_i) \right], \quad \nu = 1, \ldots, s - 1,$$

$$E(s) : \quad p_1 = h \sum_{i=1}^{s} b_i \left[\frac{\partial L}{\partial q}(c_i h) \tilde{l}_{s,s}(c_i) + \frac{1}{h} \frac{\partial L}{\partial \dot{q}}(c_i h) \dot{\tilde{l}}_{s,s}(c_i) \right],$$

which define the discrete Hamiltonian map $(q_0, p_0) \mapsto (q_1, p_1)$.

The above Galerkin discrete Lagrangian can also be interpreted as a function of several points, in a similar way to the composition discrete Lagrangians discussed in Section 2.5. Essentially, we choose a set of interior points which act as a parametrization of the space of degree s polynomials mapping $[0, h]$ to Q.

More precisely, we form the multipoint discrete Lagrangian

$$L_d(q_0^0, q_0^1, \ldots, q_0^s, h) = \mathfrak{G}^s\big(q_d(t; q_0^\nu, h)\big),$$

where we recall that $q_d(t; q_0^\nu, h)$ is the unique polynomial of degree s passing through q_0^ν at time $d_\nu h$ and $\mathfrak{G}^s$ is defined by (2.6.9). This multipoint discrete Lagrangian is the analogue of the discrete Lagrangian (2.5.3). The appropriate discrete action is then

$$\mathfrak{G}_d\big(\{(q_k = q_k^0, \ldots, q_k^s = q_{k+1})\}_{k=1}^{N}\big) = \sum_{k=0}^{N} L_d(q_k^0, q_k^1, \ldots, q_k^s, h),$$

and the corresponding discrete Euler–Lagrange equations are given by (2.5.5). Clearly, the discrete Lagrangian defined by extremizing the above multipoint L_d with respect to the interior points q_0^ν for $\nu = 1, \ldots, s$ is just the original Galerkin discrete Lagrangian (2.6.10), and the extended discrete

Euler–Lagrange equations are thus equivalent to $E(\nu)$ above for $\nu = 0, \ldots, s$. This follows in the same way as the proof of Theorem 2.5.1.

We will now see that these Galerkin variational integrators can be realized as particular examples of Runge–Kutta or partitioned Runge–Kutta schemes.

Theorem 2.6.2. Take a set of quadrature points c_i with corresponding maximal order weights b_i and let L_d be the corresponding Galerkin discrete Lagrangian (2.6.10). Then the integrator generated by this discrete Lagrangian is the partitioned Runge–Kutta scheme defined by the coefficients

$$b_i = \tilde{b}_i = \int_0^1 l_{i,s}(\rho)\mathrm{d}\rho,$$

$$a_{ij} = \int_0^{c_i} l_{j,s}(\rho)\mathrm{d}\rho,\qquad\qquad (2.6.13)$$

$$\tilde{a}_{ij} = \tilde{b}_j\left(1 - \frac{a_{ji}}{b_i}\right),$$

where the $l_{i,s}(\rho)$ are the Lagrange polynomials associated with the c_i.

Proof. Given (q_0, p_0), (q_1, p_1) and q_0^ν satisfying $E(\nu)$, $\nu = 0, \ldots, s$, we will show that they also satisfy the partitioned Runge–Kutta equations (2.6.6) written for a Lagrangian system with coefficients given by (2.6.13). We restate the defining equations here for reference:

$$q_1 = q_0 + h\sum_{j=1}^{s} b_j \dot{Q}_j, \quad p_1 = p_0 + h\sum_{j=1}^{s} \tilde{b}_j \dot{P}_j, \qquad\qquad (2.6.14\mathrm{a})$$

$$Q_i = q_0 + h\sum_{j=1}^{s} a_{ij} \dot{Q}_j, \quad P_i = p_0 + h\sum_{j=1}^{s} \tilde{a}_{ij} \dot{P}_j, \quad i = 1, \ldots, s, \quad (2.6.14\mathrm{b})$$

$$P_i = \frac{\partial L}{\partial \dot{q}}(Q_i, \dot{Q}_i), \qquad \dot{P}_i = \frac{\partial L}{\partial q}(Q_i, \dot{Q}_i), \qquad i = 1, \ldots, s. \quad (2.6.14\mathrm{c})$$

We will show that these equations are satisfied by the discrete Hamiltonian map.

Set $\dot{Q}_i = \dot{q}_d(c_i h; q_0^\nu, h)$ so that $\dot{q}_d(\tau h; q_0^\nu, h) = \sum_{j=1}^{s} \dot{Q}_j l_{j,s}(\tau)$. Integrating this expression and using the fact that $q_d(0; q_0^\nu, h) = q_0$ gives

$$q_d(\tau h; q_0^\nu, h) = q_0 + h\sum_{j=1}^{s} \dot{Q}_j \int_0^\tau l_j(\rho)\mathrm{d}\rho.$$

Setting $Q_i = q_d(c_i h; q_0^\nu, h)$ and using $q_1 = q_d(h; q_0^\nu, h)$ now gives the first parts of (2.6.14a) and (2.6.14b) for Q_i and q_1. Now define P_i and $\dot{P}_i$ according to (2.6.14c).

Until this point we have not made use of the relations $E(\nu)$. We will now begin to do so by forming the sum of the $E(\nu)$, $\nu = 0, \ldots, s$. This gives

$$p_1 - p_0 = h \sum_{i=1}^{s} b_i \left[\frac{\partial L}{\partial q}(c_i h) \sum_{\nu=0}^{s} \tilde{l}_{\nu,s}(c_i) + \frac{1}{h} \frac{\partial L}{\partial \dot{q}}(c_i h) \sum_{\nu=0}^{s} \dot{\tilde{l}}_{\nu,s}(c_i) \right].$$

However, the Lagrange polynomials $\tilde{l}_{\nu,s}(\tau)$ sum to the identity function, and therefore the sum of their derivatives must be zero. We thus recover the second part of (2.6.14a) for p_1.

Note that the $\tilde{l}_{\nu,s+1}$ are a set of $s+1$ independent polynomials of degree s and thus are a basis for $\mathcal{P}_s$, the space of polynomials of degree s. For each $j = 1, \ldots, s$ the polynomial $l_{j,s}$ is of degree $s-1$ and so has an integral of degree s. This implies that there exist coefficients m_ν^j such that

$$\sum_{\nu=0}^{s} m_\nu^j \tilde{l}_{\nu,s+1}(\tau) = \int_0^\tau l_{j,s}(\rho) d\rho - b_j.$$

Differentiating this expression with respect to τ and evaluating it at $\tau = 0$ and $\tau = 1$ gives the following three identities:

$$\sum_{\nu=0}^{s} m_\nu^j \dot{\tilde{l}}_{\nu,s+1}(\tau) = l_{j,s}(\tau),$$

$$m_s^j = \sum_{\nu=0}^{s} m_\nu^j \tilde{l}_{\nu,s+1}(1) = \int_0^1 l_{j,s}(\rho) d\rho - b_j = 0,$$

$$m_0^j = \sum_{\nu=0}^{s} m_\nu^j \tilde{l}_{\nu,s+1}(0) = -b_j.$$

If we now form the sum $\sum_{\nu=0}^{s} m_\nu^j E(\nu)$ and make use of the above identities, we obtain

$$b_j p_0 = h \sum_{i=1}^{s} b_i \left[\dot{P}_i \left(\int_0^{c_i} l_{j,s}(\rho) d\rho - b_j \right) + \frac{1}{h} P_i l_{j,s}(c_i) \right]$$

$$= h \sum_{i=1}^{s} \dot{P}_i \left[b_i (a_{ij} - b_j) \right] + b_j P_j,$$

which can be rearranged to give the second part of (2.6.14b) for P_i.   □

If the $\hat{a}_{ij}$ in Theorem 2.6.2 are equal to the a_{ij}, then the method is clearly the special case of a Runge–Kutta method, rather than the general partitioned Runge–Kutta case. Note that the definition of the $\hat{a}_{ij}$ in (2.6.13) is simply a rearrangement of the requirement (2.6.7a), and so the partitioned Runge–Kutta methods equivalent to the Galerkin variational integrators are naturally symplectic, as is clear from the symplectic nature of variational integrators in general. In addition, the additive structure of the Galerkin discrete Lagrangian means that L_d will inherit linear symmetries of L, so Noether's theorem recovers the well-known fact that the partitioned Runge–Kutta methods will preserve quadratic momentum maps.

A particularly elegant symplectic Runge–Kutta method is the *collocation Gauss–Legendre rule*. In the present derivation this results simply from taking the quadrature points c_i to be those given by the Gauss–Legendre quadrature, which is the highest-order quadrature for a given number of points. The c_i produced in this manner are all strictly between 0 and 1.

If the system being integrated is stiff then better numerical performance results from having $c_s = 1$, making the integrator stiffly accurate (Hairer and Wanner 1996). If we also wish to preserve the symmetry of the discrete Lagrangian, then it is natural to seek the c_i giving the highest order quadrature rule while enforcing $c_0 = 0$ and $c_s = 1$. This is the so-called Lobatto quadrature, and the Galerkin variational integrator generated in this way is the *standard Lobatto IIIA–IIIB partitioned Runge–Kutta method*.

PART THREE
Forcing and constraints

3.1. Background: Forced systems

Lagrangian and Hamiltonian systems with external forcing arise in many different contexts. Particular examples include control forces from actuators, dissipation and friction, and loading on mechanical systems. As we will see below, when integrating such systems it is important to take account of the geometric structure to avoid spurious numerical artifacts. One way to do this is by extending the discrete variational framework to include forcing, which we will now do.

3.1.1. Forced Lagrangian systems

A *Lagrangian force* is a fibre-preserving map $f_L : TQ \rightarrow T^*Q$ over the identity, which we write in coordinates as

$$f_L : (q, \dot{q}) \mapsto (q, f_L(q, \dot{q})).$$

Given such a force, it is standard to modify Hamilton's principle, seeking stationary points of the action, to the *Lagrange–d'Alembert principle*, which seeks curves $q \in \mathcal{C}(Q)$ satisfying

$$\delta \int_0^T L(q(t), \dot{q}(t)) \, dt + \int_0^T f_L(q(t), \dot{q}(t)) \cdot \delta q(t) \, dt = 0, \qquad (3.1.1)$$

where the δ represents variations vanishing at the endpoints. Using integration by parts shows that this is equivalent to the *forced Euler–Lagrange equations*, which have coordinate expression

$$\frac{\partial L}{\partial q}(q, \dot{q}) - \frac{d}{dt}\left(\frac{\partial L}{\partial \dot{q}}(q, \dot{q})\right) + f_L(q, \dot{q}) = 0. \qquad (3.1.2)$$

Note that these are the same as the standard Euler–Lagrange equations (1.2.5) with the forcing term added.

3.1.2. Forced Hamiltonian systems

A *Hamiltonian force* is a fibre-preserving map $f_H : T^*Q \rightarrow T^*Q$ over the identity. Given such a force, we define the corresponding horizontal one-form f_H' on T^*Q by

$$f_H'(p_q) \cdot u_{p_q} = \langle f_H(p_q), T\pi_Q \cdot u_{p_q} \rangle,$$

where $\pi_Q : T^*Q \rightarrow Q$ is the projection. This expression is reminiscent of the definition (1.4.1) of the canonical one-form Θ on T^*Q, and in coordinates it is $f_H'(q, p) \cdot (\delta q, \delta p) = f_H(q, p) \cdot \delta q$, so the one-form is clearly horizontal.

The *forced Hamiltonian vector field* X_H is now defined to satisfy

$$\mathbf{i}_{X_H}\Omega = \mathbf{d}H - f'_H$$

and in coordinates this gives the well-known *forced Hamilton's equations*

$$X_q(q,p) = \frac{\partial H}{\partial q}(q,p), \tag{3.1.3a}$$

$$X_p(q,p) = -\frac{\partial H}{\partial p}(q,p) + f_H(q,p), \tag{3.1.3b}$$

which are the same as the standard Hamilton's equations (1.4.3) with the forcing term added to the momentum equation.

3.1.3. Legendre transform with forces

Given a Lagrangian L, we can take the standard Legendre transform $\mathbb{F}L : T^*Q \to TQ$ and relate Hamiltonian and Lagrangian forces by

$$f_L = f_H \circ \mathbb{F}L.$$

If we also have a Hamiltonian H related to L by the Legendre transform according to (1.4.4) then it can be shown that the forced Euler–Lagrange equations and the forced Hamilton's equations are equivalent. That is, if X_L and X_H are the forced Lagrangian and Hamiltonian vector fields, respectively, then $(\mathbb{F}L)^*(X_H) = X_L$.

3.1.4. Noether's theorem with forcing

We now consider the effect of forcing on the evolution of momentum maps that arise from symmetries of the Lagrangian $L : TQ \to \mathbb{R}$. Let $\Phi : G \times Q \to Q$ be a symmetry of L and let the Lagangian momentum map $J_L : TQ \to \mathfrak{g}^*$ be as defined in Section 1.2.4.

Evaluating the left-hand side of (3.1.1) for a variation of the form $\delta q(t) = \xi_Q(q(t))$ gives

$$\int_0^T \mathbf{d}L \cdot \xi_{TQ}\, dt + \int_0^T f_L \cdot \xi_Q\, dt = \int_0^T f_L \cdot \xi_Q\, dt,$$

as L is assumed to be invariant. Using integration by parts as in the derivation of the forced Euler–Lagrange equations, we see that the above expression is equal to

$$\int_0^T \left[\frac{\partial L}{\partial q}(q,\dot{q}) - \frac{d}{dt}\left(\frac{\partial L}{\partial \dot{q}}(q,\dot{q}) \right) + f_L(q,\dot{q}) \right] + \Theta_L \cdot \xi_{TQ}\big|_0^T$$

$$= (J_L \circ F_L^T)(q(0),\dot{q}(0)) \cdot \xi - J_L(q(0),\dot{q}(0)) \cdot \xi,$$

and so equating these two statements of (3.1.1) gives

$$[(J_L \circ F_L^T)(q(0), \dot{q}(0)) - J_L(q(0), \dot{q}(0))] \cdot \xi = \int_0^T f_L(q(t), \dot{q}(t)) \cdot \xi_Q(q(t)) \, dt.$$
$$(3.1.4)$$

This equation describes the evolution of the momentum map from time 0 to time T, and shows that forcing will generally alter the momentum map. In the special case that the forcing is orthogonal to the group action, the above derivation shows that Noether's theorem will still hold.

Theorem 3.1.1. (Forced Noether's theorem)   Consider a Lagrangian system $L : TQ \to \mathbb{R}$ with forcing $f_L : TQ \to T^*Q$ and a symmetry action $\Phi : G \times Q \to Q$ such that $\langle f_L(q, \dot{q}), \xi_Q(q) \rangle = 0$ for all $(q, \dot{q}) \in TQ$ and all $\xi \in \mathfrak{g}$. Then the Lagrangian momentum map $J_L : TQ \to \mathfrak{g}^*$ will be preserved by the flow, so that $J_L \circ F_L^t = J_L$ for all t.

A similar result can also be derived for Hamiltonian systems, either by taking the Legendre transform of a regular forced Lagrangian system, or by working directly on the Hamiltonian side as in Section 1.4. For more details on the relationship between momentum maps and forcing see Bloch, Krishnaprasad, Marsden and Ratiu (1996*b*).

Note that, for nonzero forcing, the Lagrangian and Hamiltonian flows do not preserve the symplectic two-form. This can be seen by calculating $d\hat{\omega}$ as was done in Section 1.2.3, and realizing that it contains a term with the integral of the force which does not vanish except when $f_L = 0$.

3.2. Discrete variational mechanics with forces

3.2.1. Discrete Lagrange–d'Alembert principle

As with other discrete structures, we take two *discrete Lagrangian forces* $f_d^+, f_d^- : Q \times Q \to T^*Q$, which are fibre-preserving in the sense that $\pi_Q \circ f_d^\pm = \pi_Q^\pm$, and which thus have coordinate expressions

$$f_d^+(q_0, q_1) = (q_1, \bar{f}_d^+(q_0, q_1)),$$
$$f_d^-(q_0, q_1) = (q_0, \bar{f}_d^-(q_0, q_1)).$$

We combine the two discrete forces to give a single one-form $f_d : Q \times Q \to T^*(Q \times Q)$ defined by

$$f_d(q_0, q_1) \cdot (\delta q_0, \delta q_1) = f_d^+(q_0, q_1) \cdot \delta q_1 + f_d^-(q_0, q_1) \cdot \delta q_0. \qquad (3.2.1)$$

As with discrete Lagrangians, the discrete forces will also depend on the time-step h, which is important when relating the discrete and continuous mechanics. Given such forces, we modify the discrete Hamilton's principle, following Kane *et al.* (2000), to the *discrete Lagrange–d'Alembert principle*, which seeks discrete curves $\{q_k\}_{k=0}^{N}$ that satisfy

$$\delta \sum_{k=0}^{N-1} L_d(q_k, q_{k+1}) + \sum_{k=0}^{N-1} \left[f_d^-(q_k, q_{k+1}) \cdot \delta q_k + f_d^+(q_k, q_{k+1}) \cdot \delta q_{k+1} \right] = 0$$

(3.2.2)

for all variations $\{\delta q_k\}_{k=0}^{N}$ vanishing at the endpoints. This is equivalent to the *forced discrete Euler–Lagrange equations*

$$D_2 L_d(q_{k-1}, q_k) + D_1 L_d(q_k, q_{k+1}) + f_d^+(q_{k-1}, q_k) + f_d^-(q_k, q_{k+1}) = 0, \quad (3.2.3)$$

which are the same as the standard discrete Euler–Lagrange equations (1.3.3) with the discrete forces added. These implicitly define the *forced discrete Lagrangian map* $f_d : Q \times Q \to Q \times Q$.

3.2.2. Discrete Legendre transforms with forces

Although in the continuous case we used the standard Legendre transform for systems with forcing, in the discrete case it is necessary to take the *forced discrete Legendre transforms* to be

$$\mathbb{F}^{f+}L_d : (q_0, q_1) \mapsto (q_1, p_1) = (q_1, D_2 L_d(q_0, q_1) + f_d^+(q_0, q_1)), \qquad (3.2.4a)$$

$$\mathbb{F}^{f-}L_d : (q_0, q_1) \mapsto (q_0, p_0) = (q_0, -D_1 L_d(q_0, q_1) - f_d^-(q_0, q_1)). \qquad (3.2.4b)$$

Using these definitions and the forced discrete Euler–Lagrange equations (3.2.3), we can see that the corresponding *forced discrete Hamiltonian map* $\tilde{F}_{L_d} = \mathbb{F}^{f\pm}L_d \circ F_{L_d} \circ (\mathbb{F}^{f\pm}L_d)^{-1}$ is given by the map $\tilde{F}_{L_d} : (q_0, p_0) \mapsto (q_1, p_1)$, where

$$p_0 = -D_1 L_d(q_0, q_1) - f_d^-(q_0, q_1), \qquad (3.2.5a)$$

$$p_1 = D_2 L_d(q_0, q_1) + f_d^+(q_0, q_1), \qquad (3.2.5b)$$

which is the same as the standard discrete Hamiltonian map (1.5.4) with the discrete forces added.

3.2.3. Discrete Noether's theorem with forcing

Consider a group action $\Phi : G \times Q \to Q$ and assume that the discrete Lagrangian $L_d : Q \times Q \to \mathbb{R}$ is invariant under the lifted product action, as in Section 1.3.3. We can now calculate (3.2.2) in the direction of a variation $\delta q_k = \xi_Q(q_k)$ to give

$$\sum_{k=0}^{N-1} \mathbf{d}L_d(q_k, q_{k+1}) \cdot \xi_{Q \times Q}(q_k, q_{k+1}) + \sum_{k=0}^{N-1} f_d(q_k, q_{k+1}) \cdot \xi_{Q \times Q}(q_k, q_{k+1})$$

$$= \sum_{k=0}^{N-1} f_d(q_k, q_{k+1}) \cdot \xi_{Q \times Q}(q_k, q_{k+1}),$$

or we can use a discrete integration by parts to obtain the alternative expression

$$\sum_{k=1}^{N-1} [D_2 L_d(q_{k-1}, q_k) + D_1 L_d(q_k, q_{k+1}) + f_d^+(q_{k-1}, q_k) + f_d^-(q_k, q_{k+1})] \cdot \xi_Q(q_k)$$

$$+ [D_2 L_d(q_{N-1}, q_N) + f_d^+(q_{N-1}, q_N)] \cdot \xi_Q(q_N)$$

$$+ [D_1 L_d(q_0, q_1) + f_d^-(q_0, q_1)] \cdot \xi_Q(q_0)$$

$$= \mathbb{F}^{f+} L_d(q_{N-1}, q_N) \cdot \xi_Q(q_N) - \mathbb{F}^{f-} L_d(q_0, q_1) \cdot \xi_Q(q_0).$$

We now consider how the discrete momentum map should be defined in the presence of forcing, as there is a choice between the expressions (1.2.7) involving $\Theta_{L_d}^{\pm}$ and the expressions

$$J_{L_d}^{f+}(q_0, q_1) \cdot \xi = \langle \mathbb{F}^{f+} L_d(q_0, q_1), \xi_Q(q_1) \rangle, \qquad (3.2.6a)$$

$$J_{L_d}^{f-}(q_0, q_1) \cdot \xi = \langle \mathbb{F}^{f-} L_d(q_0, q_1), \xi_Q(q_1) \rangle, \qquad (3.2.6b)$$

which are based on the discrete Legendre transforms. In the unforced discrete case and in the continuous case both with and without forcing, these expressions are equal to the definition based on Θ_L and so the question does not arise. For a discrete system, however, consideration of the forced exact discrete Lagrangian defined below shows that (3.2.6) are the correct definitions. Given this, we can equate the above two forms of (3.2.2) to obtain

$$[J_{L_d}^{f+} \circ F_{L_d}^{N-1} - J_{L_d}^{f-}](q_0, q_1) \cdot \xi = \sum_{k=0}^{N-1} f_d(q_k, q_{k+1}) \cdot \xi_{Q \times Q}(q_k, q_{k+1}),$$

which describes the evolution of the discrete momentum map. If the discrete forces are orthogonal to the group action, so that $\langle f_d, \xi_{Q \times Q} \rangle = 0$ for all $\xi \in \mathfrak{g}$, then we have

$$0 = \langle \mathbf{d} L_d + f_d, \xi_{Q \times Q} \rangle = J_{L_d}^{f+} - J_{L_d}^{f-},$$

and thus the two discrete Lagrangian momentum maps are equal. Denoting this unique map by $J_{L_d}^f : Q \times Q \to \mathfrak{g}^*$, we see that the momentum map evolution equation gives a forced Noether's theorem for discrete mechanics.

Theorem 3.2.1. (Discrete forced Noether's theorem) Consider a discrete Lagrangian system $L_d : Q \times Q \to \mathbb{R}$ with discrete forces $f_d^+, f_d^- : Q \times Q \to T^*Q$ and a symmetry action $\Phi : G \times Q \to Q$ such that $\langle f_d, \xi_{Q \times Q} \rangle = 0$ for all $\xi \in \mathfrak{g}$. Then the discrete Lagrangian momentum map $J_{L_d}^f : Q \times Q \to \mathfrak{g}^*$ will be preserved by the discrete Lagrangian evolution map, so that $J_{L_d}^f \circ F_{L_d} = J_{L_d}^f$.

With the above definition of the discrete Lagrangian momentum map in the presence of forcing, we see that it will be the pullback of the Hamiltonian momentum map under the forced discrete Legendre transforms, and so the discrete forced Noether's theorem can also be stated for the forced discrete Hamiltonian map $\tilde{F}_{L_d}$ with the canonical momentum map $J_H : T^*Q \to \mathfrak{g}^*$.

As in the continuous case, a similar calculation to that given above shows that the discrete symplectic form will not be preserved in the presence of forcing.

3.2.4. Exact discrete forcing

In the unforced case, we have seen that the discrete Lagrangian should approximate the continuous action over the time-step. When forces are added, this must be modified so that the discrete Lagrange–d'Alembert principle (3.2.2) approximates the continuous expression (3.1.1).

Given a Lagrangian $L : TQ \to \mathbb{R}$ and a Lagrangian force $f_L : TQ \to T^*Q$, we define the *exact forced discrete Lagrangian* $L_d^E : Q \times Q \times \mathbb{R} \to \mathbb{R}$ and the *exact discrete forces* $f_d^{E+}, f_d^{E-} : Q \times Q \times \mathbb{R} \to T^*Q$ to be

$$L_d^E(q_0, q_1, h) = \int_0^h L(q(t), \dot{q}(t))\, dt, \tag{3.2.7a}$$

$$f_d^{E+}(q_0, q_1, h) = \int_0^h f_L(q(t), \dot{q}(t)) \cdot \frac{\partial q(t)}{\partial q_1}\, dt, \tag{3.2.7b}$$

$$f_d^{E-}(q_0, q_1, h) = \int_0^h f_L(q(t), \dot{q}(t)) \cdot \frac{\partial q(t)}{\partial q_0}\, dt, \tag{3.2.7c}$$

where $q : [0, h] \to Q$ is the solution of the forced Euler–Lagrange equations (3.1.2) for L and f_L satisfying the boundary conditions $q(0) = q_0$ and $q(h) = q_1$.

Note that this exact discrete Lagrangian is not the same as that for the unforced system with Lagrangian L, as the curves $q(t)$ are different. In other words, the exact discrete Lagrangian depends on both the continuous Lagrangian and the continuous forces, as do the discrete forces.

Given these definitions of the exact discrete quantities and the forced discrete Legendre transforms, it is easy to check that the forced version of Lemma 1.6.2 holds, and thus so too do forced versions of Theorems 1.6.4 and 1.6.3, showing the equivalence of the exact discrete system to the continuous systems. This is of particular interest because it shows that the variational error analysis developed in Section 2.3 can also be extended to the case of forced systems in the obvious way, and that there will be a forced version of Theorem 2.3.1.

Note that, if $\Phi : G \times Q \to Q$ is a symmetry of L such that $\langle f_L(q, \dot{q}), \xi_Q(q) \rangle = 0$, so the forced Noether's theorem holds, then the exact discrete forces will satisfy $\langle f_d, \xi_{Q \times Q} \rangle = 0$ and so the forced discrete Noether's theorem will also hold, as we would expect. This shows that (3.2.6) are the correct choice for the definition of the discrete Lagrangian momentum maps in the presence of forcing.

3.2.5. Integration of forced systems

To simulate a given forced Lagrangian or Hamiltonian system, we can choose a discrete Lagrangian and discrete forces to approximate the exact quantities given above, and then consider the resulting discrete system as an integrator for the continuous problem. We now give some simple examples of how to effect this.

Example 3.2.2. The natural discrete forces for the discrete Lagrangian L_d^α given in Example 2.3.2 are

$$ f_d^{\alpha+}(q_0, q_1, h) = \alpha h f_L \left((1-\alpha)q_0 + \alpha q_1, \frac{q_1 - q_0}{h} \right), $$

$$ f_d^{\alpha-}(q_0, q_1, h) = (1-\alpha)h f_L \left((1-\alpha)q_0 + \alpha q_1, \frac{q_1 - q_0}{h} \right). $$

For $L = \frac{1}{2}\dot{q}^T M \dot{q} - V(q)$, the discrete Hamiltonian map is then

$$ \frac{q_1 - q_0}{h} = M^{-1} \left(\alpha p_0 + (1-\alpha)p_1 \right), $$

$$ \frac{p_1 - p_0}{h} = -\nabla V \left((1-\alpha)q_0 + \alpha q_1 \right) $$
$$ + f_H \left((1-\alpha)q_0 + \alpha q_1, \alpha p_0 + (1-\alpha)p_1 \right), $$

which is the same as the unforced map (2.3.5) with the Hamiltonian force $f_H = (\mathbb{F}L)^{-1} \circ f_L$ added to the momentum equation. For $\alpha = 1/2$ this is once again simply the midpoint rule.   ◇

A particularly interesting class of Lagrangian forces $f_L : TQ \rightarrow T^*Q$ consists of those forces that satisfy

$$ \langle f_L(q, \dot{q}), (q, \dot{q}) \rangle < 0, $$

for all $(q, \dot{q}) \in TQ$. Such forces are said to be (*strongly*) *dissipative*. This terminology can be justified by computing the time evolution of the energy $E_L : TQ \rightarrow \mathbb{R}$ along a solution of the forced Euler–Lagrange equations to give

$$ \frac{d}{dt} E_L(q(t), \dot{q}(t)) = \frac{d}{dt} \left(\frac{\partial L}{\partial \dot{q}} \right) \cdot \dot{q} + \frac{\partial L}{\partial \dot{q}} \cdot \ddot{q} - \frac{d}{dt} L $$
$$ = \left(\frac{\partial L}{\partial q} + f_L \right) \cdot \dot{q} + \frac{\partial L}{\partial \dot{q}} \cdot \ddot{q} - \frac{\partial L}{\partial q} \cdot \dot{q} - \frac{\partial L}{\partial \dot{q}} \cdot \ddot{q} $$
$$ = f_L(q(t), \dot{q}(t)) \cdot \dot{q}(t). $$

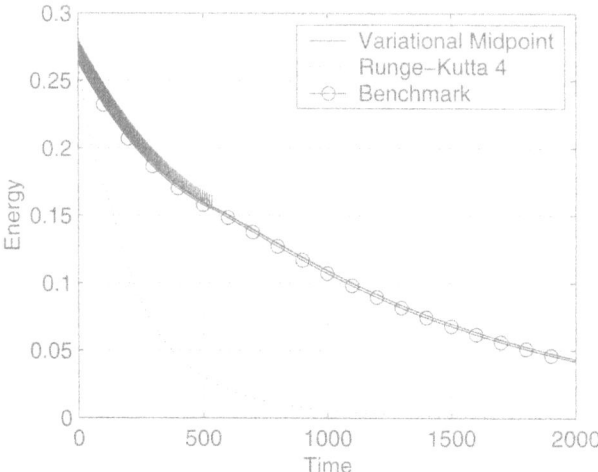

Fig. 2. Energy of a dissipative system computed with
variational midpoint and Runge–Kutta. Note the accurate energy behaviour of the variational method

We thus see that dissipative forces are those for which the energy of the
system always decreases. If we only have $f_L \cdot \dot{q} \leq 0$ then the force is said to
be *weakly dissipative*.

Because the discrete Euler–Lagrange equations do not, in general, conserve energy, it is unlikely that, without some time-step adaptation, there is
a discrete analogue of this result.

Example 3.2.3. As an example of a dissipative system, consider the move-
ment of a unit mass particle in the plane with radial potential $V(q) =
\|q\|^2(\|q\|^2 - 1)^2$ and forcing $f_L(q, \dot{q}) = -10^{-3}\dot{q}$. For this force we have
$f_L \cdot \dot{q} = -10^{-3}\|\dot{q}\|^2 \leq 0$.

In Figure 2 we plot the energy behaviour of the L_d^{α} method with $\alpha = 1/2$
for this system. For comparison, we also plot an extremely accurate benchmark trajectory, showing the true energy of the system, and the trajectory
of the standard fourth-order Runge–Kutta method.

Observe that the variational method dissipates energy due to the discrete
forces added to the Euler–Lagrange equations, but this energy dissipation
is of the correct amount to accurately track the true energy. In contrast,
non-conservative methods such as the Runge–Kutta integrator used here
artificially dissipate energy.

These effects are of particular importance when the amount of forcing or
dissipation in the system is small compared to the magnitude of the conservative dynamics and the time period of integration. For an investigation

of the long time behaviour of symplectic methods applied to systems with dissipation, see Hairer and Lubich (1999). ◇

Example 3.2.4. (Composition methods) Consider a sequence of discrete Lagrangians L_d^i, discrete forces f_d^{i+}, f_d^{i-} and time-step fractions γ^i for $i = 1, \ldots, s$ satisfying $\sum_{i=1}^s \gamma^i = 1$. Then we can form a composition discrete Lagrangian L_d and composition discrete forces f_d^+, f_d^- in a similar way to the procedures in Section 2.5.

Given points q_0 and q_s, define q_i for $i = 1, \ldots, s-1$ to satisfy the forced discrete Euler–Lagrange equations (3.2.3) along the sequence $q_0, q_1, \ldots, q_s$. Regarding the q_i as functions of q_0 and q_1, we now define the *composition discrete Lagrangian* and *composition discrete forces* by

$$L_d(q_0, q_1, h) = \sum_{i=1}^s L_d(q_{i-1}, q_i, \gamma^i h),$$

$$f_d^+(q_0, q_1, h) = f_d^{s+}(q_{s-1}, q_s, \gamma^s h)$$
$$+ \sum_{i=1}^{s-1} \left(f_d^{i+}(q_{i-1}, q_i, \gamma^i h) + f_d^{i-}(q_i, q_{i+1}, \gamma^{i+1} h) \right) \cdot \frac{\partial q_i}{\partial q_1},$$

$$f_d^-(q_0, q_1, h) = f_d^{1-}(q_0, q_1, \gamma^1 h)$$
$$+ \sum_{i=1}^{s-1} \left(f_d^{i+}(q_{i-1}, q_i, \gamma^i h) + f_d^{i-}(q_i, q_{i+1}, \gamma^{i+1} h) \right) \cdot \frac{\partial q_i}{\partial q_0}.$$

With these definitions it can be shown, using a similar derivation to that in Section 2.5, that the forced discrete Hamiltonian map for L_d and f_d^+, f_d^- is the composition of the individual forced discrete Hamiltonian maps, so that

$$\tilde{F}_{L_d}^h = \tilde{F}_{L_d^s}^{\gamma^s h} \circ \tilde{F}_{L_d^{s-1}}^{\gamma^{s-1} h} \circ \cdots \circ \tilde{F}_{L_d^1}^{\gamma^1 h}.$$

In forming composition methods it is often useful to use a sequence consisting of copies of a method together with its adjoint. It is thus worth noting that the adjoint of a discrete Lagrangian and discrete forces is given by

$$L_d^*(q_0, q_1, h) = -L_d(q_1, q_0, -h),$$
$$f_d^{*+}(q_0, q_1, h) = -f_d^-(q_1, q_0, -h),$$
$$f_d^{*-}(q_0, q_1, h) = -f_d^+(q_1, q_0, -h).$$

The discrete Hamiltonian map of the adjoint discrete Lagrangian and adjoint discrete forces will be the adjoint map of the original discrete Hamiltonian map. Observe that the exact discrete Lagrangian and exact discrete forces (3.2.7) are self-adjoint. ◇

Example 3.2.5. (Symplectic partitioned Runge–Kutta methods)
Recall that the discrete Lagrangian (2.6.8) given by

$$L_d(q_0, q_1, h) = h \sum_{i=1}^{s} b_i L(Q_i, \dot{Q}_i)$$

generates symplectic partitioned Runge–Kutta methods. Reasonable choices of corresponding discrete forces are

$$f_d^+(q_0, q_1, h) = h \sum_{i=1}^{s} b_i f_L(Q_i, \dot{Q}_i) \cdot \frac{\partial Q_i}{\partial q_0}, \qquad (3.2.8a)$$

$$f_d^-(q_0, q_1, h) = h \sum_{i=1}^{s} b_i f_L(Q_i, \dot{Q}_i) \cdot \frac{\partial Q_i}{\partial q_1}, \qquad (3.2.8b)$$

which approximate the exact forces (3.2.7b) and (3.2.7c) in the same way that L_d approximates the exact discrete Lagrangian (3.2.7a).

With these choices of discrete forces, it can be shown that the discrete Hamiltonian map defined by (3.2.5) is exactly a partitioned Runge–Kutta method for the forced Hamiltonian system (3.1.3). ◊

In most of the other examples of variational integrators discussed above, discrete forces can be chosen in a natural way so that the discrete Hamiltonian maps give the expected integrator for the forced Hamiltonian system. In particular, this can be done for the symplectic Newmark methods (see Kane *et al.* (2000)). We can also use alternative splitting-style methods to include forcing (see Kane *et al.* (2000) for details).

3.3. Background: Constrained systems

A particularly elegant way to study many systems is to consider them as a constrained version of some larger system. This can be appealing for both theoretical reasons and, as we shall see, also on numerical grounds. Here we will only consider so-called *holonomic constraints*, which are constraints on the configuration manifold of a system.

More precisely, if we have a Lagrangian or Hamiltonian system with configuration manifold Q, we consider a *constraint function* $\phi : Q \to \mathbb{R}^d$ and constrain the dynamics to the *constraint submanifold* $N = \phi^{-1}(0) \subset Q$. Here we will always assume that $0 \in \mathbb{R}^d$ is a regular point of ϕ, so that N is truly a submanifold of Q (Abraham *et al.* 1988).

Observe that, if $i : N \to Q$ is the embedding map, then $Ti : TN \to TQ$ provides a canonical way to embed TN in TQ and we will thus regard TN as a submanifold of TQ. There is, however, no canonical way to embed the cotangent bundle T^*N in T^*Q, a fact which has important consequences for the development of constrained Hamiltonian dynamics. We will see below that, in the special case when we have a regular Lagrangian or Hamiltonian, we can use this additional structure to provide a canonical embedding.

As in other areas of mechanics, we may consider constrained systems from both the Hamiltonian and the Lagrangian viewpoint. We will concentrate on the variational approach, however, as it is this formulation which readily extends to the discrete setting. The primary tool for constrained optimization problems is the Lagrange multiplier theorem, which we recall here (see Abraham *et al.* (1988) for the proof).

Theorem 3.3.1.   Consider a smooth manifold $\mathcal{C}$ and a function $\Phi : \mathcal{C} \to V$ mapping to some inner product space V, such that $0 \in V$ is a regular point of Φ. Set $\mathcal{D} = \Phi^{-1}(0) \subset \mathcal{C}$. Given a function $\mathfrak{G} : \mathcal{C} \to \mathbb{R}$, define $\bar{\mathfrak{G}} : \mathcal{C} \times V \to \mathbb{R}$ by $\bar{\mathfrak{G}}(q, \lambda) = \mathfrak{G}(q) - \langle \lambda, \Phi(q) \rangle$. Then the following are equivalent:

(1) $q \in \mathcal{D}$ is an extremum of $\mathfrak{G}|_{\mathcal{D}}$;
(2) $(q, \lambda) \in \mathcal{C} \times V$ is an extremum of $\bar{\mathfrak{G}}$.

3.3.1. Constrained Lagrangian systems

Given a Lagrangian system specified by a configuration manifold Q and a Lagrangian $L : TQ \to \mathbb{R}$, consider the holonomic constraint $\phi : Q \to \mathbb{R}^d$ and the corresponding constraint submanifold $N = \phi^{-1}(0) \subset Q$. Now TN is a submanifold of TQ, and so we may restrict L to $L^N = L|_{TN}$. We are interested in the relationship of the dynamics of L^N on TN to the dynamics of L on TQ.

To consider this, we will make use of the following convenient notation. Assume that we are working on a given time interval $[0, T] \subset \mathbb{R}$, and that we have fixed endpoints $q_0, q_T \in N \subset Q$. Now set $\mathcal{C}(Q) = \mathcal{C}([0, T], Q; q_0, q_T)$ to be the space of smooth functions $q : [0, T] \to Q$ satisfying $q(0) = q_0$ and $q(T) = q_T$, and $\mathcal{C}(N)$ to be the corresponding space of curves in N. Similarly, we set $\mathcal{C}(\mathbb{R}^d) = \mathcal{C}([0, T], \mathbb{R}^d)$ to be curves $\lambda : [0, T] \to \mathbb{R}^d$ with no boundary conditions. In general $\mathcal{C}(P)$ is the space of curves from $[0, T]$ to the manifold P with the appropriate boundary conditions.

Theorem 3.3.2.   Given a Lagrangian system $L : TQ \to \mathbb{R}$ with holonomic constraint $\phi : Q \to \mathbb{R}^d$, set $N = \phi^{-1}(0) \subset Q$ and $L^N = L|_{TN}$. Then the following are equivalent:

(1) $q \in \mathcal{C}(N)$ extremizes $\mathfrak{G}^N$ and hence solves the Euler–Lagrange equations for L^N;
(2) $q \in \mathcal{C}(Q)$ and $\lambda \in \mathcal{C}(\mathbb{R}^d)$ satisfy the *constrained Euler–Lagrange equations*

$$\frac{\partial L}{\partial q^i}(q(t), \dot{q}(t)) - \frac{\mathrm{d}}{\mathrm{d}t}\left(\frac{\partial L}{\partial \dot{q}^i}(q(t), \dot{q}(t)) \right) = \left\langle \lambda(t), \frac{\partial \phi}{\partial q^i}(q(t)) \right\rangle, \quad (3.3.1a)$$

$$\phi(q(t)) = 0; \quad (3.3.1b)$$

(3) $(q, \lambda) \in C(Q \times \mathbb{R}^d)$ extremizes $\bar{\mathfrak{G}}(q, \lambda) = \mathfrak{G}(q) - \langle \lambda, \Phi(q) \rangle$ and hence solves the Euler–Lagrange equations for the *augmented Lagrangian* $\bar{L}$: $T(Q \times \mathbb{R}^d) \to \mathbb{R}$ defined by

$$\bar{L}(q, \lambda, \dot{q}, \dot{\lambda}) = L(q, \dot{q}) - \langle \lambda, \phi(q) \rangle.$$

Proof. We make use of the Lagrange multiplier theorem, Theorem 3.3.1. To do so, we prepare the following definitions. The full space is $C = C(Q)$ and the function to be extremized is the action $\mathfrak{G} : C(Q) \to \mathbb{R}$. Take $V = C(\mathbb{R}^d)$ with the L_2 inner product and define the constraint function $\Phi : C \to V$ by $\Phi(q)(t) = \phi(q(t))$. Clearly $\Phi(q) = 0$, and hence $\phi(q(t)) = 0$ for all $t \in [0, T]$, if and only if $q \in C(N)$. We thus obtain that the constraint submanifold is $\mathcal{D} = \Phi^{-1}(0) = C(N)$.

Condition (1) simply means that $q \in C(N) = \mathcal{D}$ is an extremum of the action for L^N, which is readily seen to be the standard action restricted to $C(N)$. Thus $q \in \mathcal{D}$ is an extremum of $\mathfrak{G}|_{\mathcal{D}}$ and so, by the Lagrange multiplier theorem, this is equivalent to $(q, \lambda) \in C \times V$ being an extremum of $\bar{\mathfrak{G}}(q, \lambda) = \mathfrak{G}(q) - \langle \lambda, \Phi(q) \rangle$.

Now $C \times V = C(Q) \times C(\mathbb{R}^d)$ and so it can be identified with $C(Q \times \mathbb{R}^d)$. Furthermore, we see that $\bar{\mathfrak{G}} : C(Q \times \mathbb{R}^d) \to \mathbb{R}$ is

$$\begin{aligned}
\bar{\mathfrak{G}}(q, \lambda) &= \mathfrak{G}(q) - \langle \lambda, \Phi(q) \rangle \\
&= \int_0^T L(q(t), \dot{q}(t))\, dt - \int_0^T \langle \lambda(t), \Phi(q)(t) \rangle\, dt \\
&= \int_0^T [L(q(t), \dot{q}(t)) - \langle \lambda(t), \phi(q(t)) \rangle]\, dt,
\end{aligned}$$

which is simply the action for the augmented Lagrangian $\bar{L}(q, \lambda, \dot{q}, \dot{\lambda}) = L(q, \dot{q}) - \langle \lambda, \phi(q) \rangle$. As $(q, \lambda) \in C(Q \times \mathbb{R}^d)$ must extremize this action, we see that it is a solution of the Euler–Lagrange equations for $\bar{L}$, which is statement (3).

Finally, we extremize $\bar{\mathfrak{G}}$ by solving $d\bar{\mathfrak{G}} = 0$ to obtain the Euler–Lagrange equations. The standard integration by parts argument gives (3.3.1a) for variations with respect to q, and variations with respect to λ imply (3.3.1b), and thus we have equivalence to statement (2). □

If $i : N \to Q$ is the embedding, then by differentiating $L^N = L \circ Ti$ with respect to $\dot{q}$ we see that

$$\frac{\partial L^N}{\partial \dot{q}}(v_q) \cdot w_q = \frac{\partial L}{\partial \dot{q}}\big(Ti(v_q)\big) \cdot Ti \cdot w_q, \qquad (3.3.2)$$

which means that if L is regular then so is L^N and shows that the following diagram commutes.

$$TQ|_N \xrightarrow{\ \mathbb{F}L\ } T^*Q|_N \qquad\qquad (3.3.3)$$

$$Ti \uparrow \qquad\qquad \downarrow T^*i$$

$$TN \xrightarrow[\ \mathbb{F}L^N\ ]{} T^*N$$

Using this together with the fact that $\pi_Q \circ Ti = i \circ \pi_N$ for the projections $\pi_Q : TQ \to Q$ and $\pi_N : TN \to N$, we compute the pullback of the Lagrange one-form Θ_L on TQ to be

$$
\begin{aligned}
\big((Ti)^*\Theta_L\big)(v_q) \cdot \delta v_q &= \big\langle \mathbb{F}L\big(Ti(v_q)\big), T\pi_Q \circ T(Ti) \cdot \delta v_q \big\rangle \\
&= \big\langle \mathbb{F}L\big(Ti(v_q)\big), Ti \circ T\pi_N \cdot \delta v_q \big\rangle \\
&= \big\langle \mathbb{F}L^N(v_q), T\pi_N \cdot \delta v_q \big\rangle,
\end{aligned}
$$

and thus we see that $(Ti)^*\Theta_L = \Theta_{L^N}$, and so

$$(Ti)^*\Omega_L = \Omega_{L^N}. \qquad\qquad (3.3.4)$$

Using the projection $T^*i : T^*Q \to T^*N$ we can reinterpret statement (2) of Theorem 3.3.2. Observe that the span of the $\nabla\phi^i$, $i = 1, \ldots, d$ is exactly the null space of T^*i, and so (3.3.1) is equivalent to

$$(T^*i)_{q(t)}\left[\frac{\partial L}{\partial q}(q(t), \dot q(t)) - \frac{\mathrm{d}}{\mathrm{d}t}\left(\frac{\partial L}{\partial \dot q}(q(t), \dot q(t)) \right) \right] = 0. \qquad\qquad (3.3.5)$$

The above relationships hold for any Lagrangian L, irrespective of regularity. Also note that, although there is a canonical projection $T^*i : T^*Q \to T^*N$, there is no corresponding canonical embedding of T^*N into T^*Q. We will see below that when L is regular we can use the Legendre transform to define such an embedding.

3.3.2. Constrained Hamiltonian systems: Augmented approach

One can consider the Hamiltonian formulation of constrained systems by either working on the augmented space $T^*(Q \times \mathbb{R}^d)$, or working directly on T^*N, which gives the Dirac theory of constraints. We consider the former option first.

Given a Hamiltonian $H : T^*Q \to \mathbb{R}$, we define the *augmented Hamiltonian* to be

$$\bar H(q, \lambda, p, \pi) = H(q, p) + \langle \lambda, \phi(q) \rangle,$$

where π is the conjugate variable to λ. We now consider the *primary constraint set* $\Pi \subset T^*(Q \times \mathbb{R}^d)$ defined by $\pi = 0$. Pulling Ω back to Π gives the degenerate two-form Ω^Π, and the augmented Hamiltonian vector field $\bar X_{\bar H}$ is defined by

$$i_{\bar X_{\bar H}} \Omega^\Pi = \mathrm{d}\bar H,$$

which in coordinates is the set of *constrained Hamilton's equations*

$$\bar{X}_{q^i}(q,\lambda,p,\pi) = \frac{\partial H}{\partial p_i}, \tag{3.3.6a}$$

$$\bar{X}_{p_i}(q,\lambda,p,\pi) = -\frac{\partial H}{\partial q^i} - \left\langle \lambda, \frac{\partial \phi}{\partial q^i}(q) \right\rangle, \tag{3.3.6b}$$

$$\phi(q) = 0, \tag{3.3.6c}$$

where there is no λ equation owing to the degeneracy of Ω^Π. Note that for nonregular H these equations will not, in general, uniquely define the vector field $\bar{X}_{\bar{H}}$.

Consider now a regular Lagrangian L and its corresponding Hamiltonian H. Observe that the augmented Lagrangian $\bar{L}$ is degenerate, owing to the lack of dependence on $\dot{\lambda}$, and that the primary constraint manifold Π is exactly the image of $\mathbb{F}\bar{L}$. The augmented Hamiltonian and Lagrangian satisfy the equation $\bar{H} \circ \mathbb{F}\bar{L} = E_{\bar{L}}$, but this does not uniquely specify $\bar{H}$ since $\mathbb{F}\bar{L}$ need not be invertible. Nonetheless, it is simple to check that the constrained Hamilton's equations given above are equivalent to the constrained Euler–Lagrange equations (3.3.1) when we neglect the π component.

3.3.3. Constrained Hamiltonian systems: Dirac theory

As an alternative to working on the augmented space $T^*(Q \times \mathbb{R}^d)$, we can directly compare the dynamics of the constrained system on T^*N with those on T^*Q. The general form of this is the *Dirac theory of constraints* (Marsden and Ratiu 1999), but here we use only the simple case of holonomic constraints on cotangent bundles.

The main problem with this approach is that there is no canonical way to embed T^*N within T^*Q. For now we will assume that we have an embedding $\eta : T^*N \to T^*Q$ such that $\pi_Q \circ \eta = i \circ \pi_N$ and $\eta^*\Omega = \Omega^N$, where Ω and Ω^N are the canonical two-forms on T^*Q and T^*N respectively, and we will see below how to construct η given a regular Hamiltonian or Lagrangian.

Given a Hamiltonian $H : T^*Q \to \mathbb{R}$, we define $H^N : T^*N \to \mathbb{R}$ by $H^N = H \circ \eta$. The constrained Hamiltonian vector field $X_{H^N} : T^*N \to T(T^*N)$ is then defined by

$$i_{X_{H^N}}\Omega^N = dH^N.$$

Taking $\pi_\Omega : T(T^*Q) \to T(T^*N)$ to be the projection operator determined by using Ω to define the orthogonal complement of $T\eta \cdot T(T^*N) \subset T(T^*Q)$, leads us to the following simple relationship between the Hamiltonian vector field X_H and the constrained vector field X_{H^N}.

Theorem 3.3.3. Consider a Hamiltonian system $H : T^*Q \to \mathbb{R}$ and the corresponding constrained system $H^N : T^*N \to \mathbb{R}$ as defined above. Then

$$X_{H^N} = \pi_\Omega \cdot X_H \circ \eta.$$

Proof. We have that $\eta^*\Omega = \Omega^N$. Take an arbitrary $V^N \in T(T^*N)$ and compute

$$\mathbf{i}_{(\pi_\Omega \cdot X_H \circ \eta)}\Omega^N \cdot V^N = \Omega(T\eta \cdot \pi_\Omega \cdot X_H, T\eta \cdot V^N)$$
$$= \Omega(X_H, T\eta \cdot V^N)$$
$$= \mathbf{d}H \cdot T\eta \cdot V^N$$
$$= \mathbf{d}H^N \cdot V^N$$
$$= \mathbf{i}_{X_{H^N}}\Omega^N \cdot V^N,$$

where we used the fact that $(Id - T\eta \cdot \pi_\Omega) \cdot X_H$ is Ω-orthogonal to the set $T\eta \cdot T(T^*N)$. Finally, the fact that Ω^N is nondegenerate gives the desired equivalence.   □

3.3.4. Legendre transforms

Until this point we have assumed that we are using any symplectic embedding $\eta : T^*N \to T^*Q$ covering the embedding $i : N \to Q$. We now consider a hyperregular Hamiltonian H and the corresponding hyperregular Lagrangian L. Recall that *hyperregularity* of H, for example, means that $\mathbb{F}H$ is not only a local diffeomorphism (equivalent to regularity), but is a *global* diffeomorphism. Of course, if we only have regularity, these constructions may be done locally. We will show that, using this additional structure, there is a canonical way to construct η.

To do this, begin from either a hyperregular Lagrangian L or a hyperregular Hamiltonian H, and construct the corresponding L or H, which is necessarily hyperregular as well and has $\mathbb{F}L = (\mathbb{F}H)^{-1}$. This implies that L^N and H^N are also hyperregular.

We now define $\eta : T^*N \to T^*Q$ by requiring that the following diagram commutes, where $i : N \to Q$ is the embedding as before.

$$\begin{array}{ccc} TQ|_N & \xrightarrow{\mathbb{F}L} & T^*Q|_N \\ Ti\uparrow & & \uparrow\eta \\ TN & \xrightarrow[\mathbb{F}L^N]{} & T^*N \end{array} \qquad (3.3.7)$$

Clearly $\pi_Q \circ \eta = i \circ \pi_N$, and from (3.3.4) we see that $\eta^*\Omega = \Omega^N$, and so η gives a symplectic embedding of T^*N in T^*Q. Note that, although T_qN is a linear subset of T_qQ, the map η is in general not linear and so T_q^*N is not a linear subspace of T_q^*Q. It is true, however, that $T_{p_q}(T^*N)$ is a linear subspace of $T_{p_q}(T^*Q)$.

Regarding T^*N as a submanifold of T^*Q by means of η, we have the natural embedding $T\eta : T(T^*N) \to T(T^*Q)$ and so we can regard X_{H^N} as a vector field on $\eta(T^*N)$. Using canonical coordinates (q^i, p_i) on T^*Q we can derive a simple coordinate representation of this vector field:

$$\dot{q} = \frac{\partial H}{\partial p},$$

$$\dot{p} = -\frac{\partial H}{\partial q} - \lambda^T \nabla \phi(q),$$

$$\phi(q) = 0.$$

These equations are clearly equivalent to (3.3.6) above if we neglect the π variable there.

Consider the projection operator $\pi_{\Omega_L} : T(TQ) \to T(TN)$ defined by the Ω_L-orthogonal complement to $T(TN)$ regarded as a subspace of $T(TQ)$ by the map TTi. As $\Omega_L = (\mathbb{F}L)^*\Omega$, elements of $T(T^*Q)$ which are Ω-orthogonal pull back under $\mathbb{F}L$ to elements of $T(TQ)$ which are Ω_L-orthogonal. It follows that $T\mathbb{F}L^n \circ \pi_{\Omega_L} = \pi_\Omega \circ T\mathbb{F}L$. In addition, observe that, as both the constrained and unconstrained systems are regular, we obtain $X_L = (\mathbb{F}L)^*X_H$ and $X_{LN} = (\mathbb{F}L^N)^*X_{HN}$. Combining this with the statement of Theorem 3.3.3 and regarding TN and T^*N as submanifolds of TQ and T^*Q, respectively, gives the following commutative diagram.

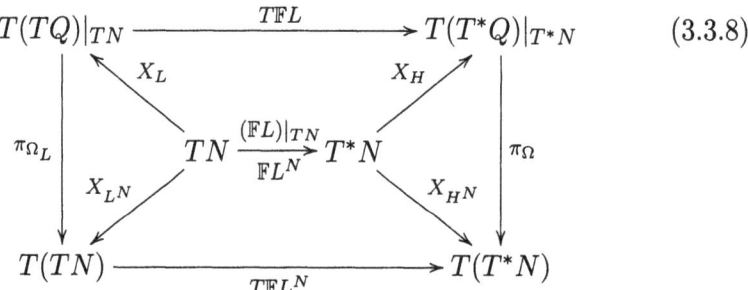

$$(3.3.8)$$

This establishes that $X_{LN} = \pi_{\Omega_L} \circ X_L \circ Ti$, which is the Lagrangian analogue of Theorem 3.3.3. Note that this only holds for regular Lagrangians, whereas the Hamiltonian result does not require regularity.

A special case of hyperregular systems is when we have a Riemannian metric $\langle\!\langle \cdot, \cdot \rangle\!\rangle$ on Q and the Lagrangian is of the form

$$L(v_q) = \frac{1}{2} \langle\!\langle v_q, v_q \rangle\!\rangle - V \circ \pi_Q(v_q)$$

for a potential function $V : Q \to \mathbb{R}$. Computing the Legendre transform gives

$$\mathbb{F}L(v_q) \cdot w_q = \langle\!\langle v_q, w_q \rangle\!\rangle = v_q^T M(q) w_q,$$

where we introduce the symmetric positive definite mass matrix $M(q)$ as the coordinate representation of the metric. In coordinates, the Legendre transform is thus $p = M(q)\dot{q}$, and we see that the Legendre transform is linear in $\dot{q}$ and so $\eta(T_q^* N)$ is a linear subspace of $T_q^* Q$ at each $q \in N$. Note that the constrained subspaces can be expressed as

$$TN = \{(q, \dot{q}) \in TQ \mid \phi(q) = 0 \text{ and } \nabla\phi \cdot \dot{q} = 0\}, \tag{3.3.9}$$

$$\eta(T^* N) = \{(q, p) \in T^* Q \mid \phi(q) = 0 \text{ and } \nabla\phi \cdot M^{-1}(q)\dot{q} = 0\}. \tag{3.3.10}$$

We define the projection map $\mathbb{P} : T^* Q|_N \rightarrow \eta(T^* N)$ by $\mathbb{P} = \eta \circ T^* i$, and as it must satisfy $\mathbb{P}(\nabla\phi^m) = 0$ for each $m = 1, \ldots, d$ we can calculate the coordinate expression to be

$$\mathbb{P} = I - (\nabla\phi)^T [(\nabla\phi) M^{-1} (\nabla\phi)^T]^{-1} (\nabla\phi) M^{-1}, \tag{3.3.11}$$

where I is the $n \times n$ identity matrix and $\nabla\phi$ is the $d \times n$ matrix $[\nabla\phi(q)]_{mi} = \frac{\partial\phi^m}{\partial q^i}$, and all quantities are evaluated at $q \in N$.

Another way to derive this expression is to define an induced Riemannian metric on $T^* Q$ by $\langle\langle p_q, r_q \rangle\rangle = \langle p_q, \mathbb{F}H(r_q)\rangle$, which has coordinate expression $p^T M^{-1}(q) r$. The projection $\mathbb{P}$ is then the projection onto the orthogonal subspace to the span of $\{\nabla\phi^m\}$ in the inner product given by this metric.

In this case, note that $\mathbb{P} : TQ|_N \rightarrow TN$, and so $T\mathbb{P} : T(TQ|_N) \rightarrow T(TN)$. However, observe that

$$T(TQ|_N) = \{w \in T(TQ) \mid T\pi_Q(w) \in TN\},$$

and as X_L is a second-order vector field, it satisfies $T\pi_Q \circ X_L = id$, and so we have that $X_L(v_q) \in T(TQ|_N)$ for all $v_q \in TN$. In particular, we can now show that, on the intersection of their domains, $\pi_{\Omega_L} = T\mathbb{P}$, which gives an explicit expression for the Lagrangian projection operator. This development is closely related to the expression of forces of constraint in terms of the second fundamental form (see Marsden and Ratiu (1999), Section 8.4).

3.3.5. Conservation properties

As we have seen above, the constrained systems on TN and $T^* N$ defined by $L^N = L \circ Ti$ and $H^N = H \circ \eta$, respectively, are standard Lagrangian or Hamiltonian systems and so have the usual conservation properties.

In particular, the constrained Lagrangian system $L^N : TN \rightarrow \mathbb{R}$ will have a flow map that preserves the symplectic two-form $\Omega_{L^N} = (Ti)^* \Omega_L$, and the constrained Hamiltonian system $H^N : T^* N \rightarrow \mathbb{R}$ preserves the canonical two-form $\Omega^N = \eta^* \Omega$ on $T^* N$. For (hyper)regular systems, the Lagrangian and Hamiltonian two-forms are related by the Legendre transforms on both the constrained and unconstrained levels, so that $\Omega_L = (\mathbb{F}L)^* \Omega$ and $\Omega_{L^N} = (\mathbb{F}L^N)^* \Omega^N$.

Suppose that we have a group action $\Phi : G \times Q \rightarrow Q$ that leaves N invariant, that is, there is a restricted action $\Phi^N : G \times N \rightarrow N$ satisfying $i \circ \Phi^N = \Phi \circ i$. It is now a simple matter to check that the infinitesimal generators are related by

$$\xi_Q \circ i = Ti \circ \xi_N,$$
$$\xi_{TQ} \circ Ti = T(Ti) \circ \xi_{TN},$$
$$\xi_{T^*Q} \circ \eta = T\eta \circ \xi_{T^*N},$$

and so the momentum maps satisfy

$$J_{LN} = J_L \circ Ti,$$
$$J_{HN} = J_H \circ \eta.$$

Since Noether's theorem holds for both the constrained and unconstrained systems, the above relationship shows that essentially the same momentum map is preserved on both levels. Note that if the group action does not leave the constraint submanifold N invariant, however, then in general it is not possible to define J_{LN} or J_{HN} and there will be no constrained Noether's theorems.

3.4. Discrete variational mechanics with constraints

We now consider a discrete Lagrangian system $L_d : Q \times Q \to \mathbb{R}$ with the holonomic constraint $\phi : Q \to \mathbb{R}^d$ and corresponding constraint submanifold $N = \phi^{-1}(0) \subset Q$. As in the continuous case, the fact that $N \times N$ is naturally a submanifold of $Q \times Q$ means that we can restrict the discrete Lagrangian to $L_d^N = L_d|_{N \times N}$ to obtain a discrete Lagrangian system on $N \times N$. More precisely, we define the embedding $i^{N \times N} : N \times N \to Q \times Q$ by $i^{N \times N}(q_0, q_1) = (i(q_0), i(q_1))$.

To relate the dynamics of L_d^N to that of L_d, it is useful to introduce the notation for discrete trajectories corresponding to that used in the continuous case. Given times $\{0, h, 2h, \ldots, Nh = T\}$ and endpoints $q_0, q_T \in N$ we set $\mathbb{C}_d(Q) = \mathbb{C}_d(\{0, h, 2h, \ldots, Nh\}, Q; q_0, q_T)$ to be the set of discrete trajectories $q_d : \{0, h, 2h, \ldots, Nh\} \to Q$ satisfying $q_d(0) = q_0$ and $q_d(Nh) = q_T$, and $\mathbb{C}_d(N)$ to be the corresponding set of discrete trajectories in N.

Similarly, we denote by $\mathcal{C}_d(\mathbb{R}^d) = \mathcal{C}_d(\{h, 2h, \ldots, (N-1)h\}, \mathbb{R}^d)$ the set of maps $\lambda_d : \{h, 2h, \ldots, (N-1)h\} \to \mathbb{R}^d$ with no boundary conditions. We will see below why we do not include the boundary points 0 and Nh. In general, $\mathcal{C}_d(P)$ is the space of maps from $\{0, h, 2h, \ldots, Nh\}$ to the manifold P, and we identify such maps with their images, and write $q_d = \{q_k\}_{k=0}^N$ for $k = 0, 1, 2, \ldots, N$, and similarly for $\lambda_d = \{\lambda_k\}_{k=0}^N$.

3.4.1. Constrained discrete variational principle

As we have do not use vector fields to define the dynamics in the discrete case, and so cannot project such objects onto the constraint manifold, we turn instead to constraining the variational principle. The following theorem gives the result of this procedure.

Theorem 3.4.1. Given a discrete Lagrangian system $L_d : Q \times Q \to \mathbb{R}$ with holonomic constraint $\phi : Q \to \mathbb{R}^d$, set $N = \phi^{-1}(0) \subset Q$ and $L_d^N = L_d|_{N \times N}$. Then the following are equivalent:

(1) $q_d = \{q_k\}_{k=0}^N \in C_d(N)$ extremizes $\mathfrak{G}_d^N = \mathfrak{G}_d|_{N \times N}$ and hence solves the discrete Euler–Lagrange equations for L_d^N;

(2) $q_d = \{q_k\}_{k=0}^N \in C_d(Q)$ and $\lambda_d = \{\lambda_k\}_{k=1}^{N-1} \in C_d(\mathbb{R}^d)$ satisfy the *constrained discrete Euler–Lagrange equations*

$$D_2 L_d(q_{k-1}, q_k) + D_1 L_d(q_k, q_{k+1}) = \langle \lambda_k, \nabla \phi(q_k) \rangle, \qquad (3.4.1a)$$
$$\phi(q_k) = 0; \qquad (3.4.1b)$$

(3) $(q_d, \lambda_d) = \{(q_k, \lambda_k)\}_{k=0}^N \in C_d(Q \times \mathbb{R}^d)$ extremizes $\bar{\mathfrak{G}}_d(q_d, \lambda_d) = \mathfrak{G}_d(q_d) - \langle \lambda_d, \Phi_d(q_d) \rangle_{l_2}$ and hence solves the discrete Euler–Lagrange equations for either of the *augmented discrete Lagrangians* $\bar{L}_d^+, \bar{L}_d^- : (Q \times \mathbb{R}^d) \times (Q \times \mathbb{R}^d) \to \mathbb{R}$ defined by

$$\bar{L}_d^+(q_k, \lambda_k, q_{k+1}, \lambda_{k+1}) = L_d(q_k, q_{k+1}) - \langle \lambda_{k+1}, \phi(q_{k+1}) \rangle,$$
$$\bar{L}_d^-(q_k, \lambda_k, q_{k+1}, \lambda_{k+1}) = L_d(q_k, q_{k+1}) - \langle \lambda_k, \phi(q_k) \rangle.$$

Proof. The proof of Theorem 3.3.2 in the continuous case can be almost directly applied in the discrete case.

We take the full space to be $C_d = C_d(Q)$ and the function we are extremizing is the discrete action $\mathfrak{G}_d : C_d(Q) \to \mathbb{R}$. The constraint is specified by setting $V_d = C_d(\mathbb{R}^d)$ with the l_2 inner product, and defining the constraint function $\Phi_d : C_d \to V_d$ by $\Phi_d(q_d)(kh) = \phi(q_d(kh)) = \phi(q_k)$. Thus $q_d \in C_d(N)$ if and only if $\phi(q_k) = 0$ for all k, and hence if and only if $\Phi_d(q_d) = 0$. The constraint submanifold is therefore $\mathcal{D}_d = \Phi_d^{-1}(0) = C_d(N)$.

As in the continuous case, statement (1) means that $q_d \in C_d(N) = \mathcal{D}_d$ is an extremum of the action for L_d^N, which is the full action restricted to $C_d(N)$. From the Lagrange multiplier theorem (Theorem 3.3.1), $q_d \in \mathcal{D}_d$ being an extremum of $\mathfrak{G}_d|_{\mathcal{D}_d}$ is equivalent to $(q_d, \lambda_d) \in C_d \times V_d$ being an extremum of $\bar{\mathfrak{G}}_d(q_d, \lambda_d) = \mathfrak{G}_d(q_d) - \langle \lambda_d, \Phi_d(q_d) \rangle$. Computing, this gives

$$\bar{\mathfrak{G}}_d(q_d, \lambda_d) = \mathfrak{G}_d(q_d) - \langle \lambda_d, \Phi_d(q_d) \rangle$$
$$= \sum_{k=0}^{N-1} L_d(q_k, q_{k+1}) - \sum_{k=1}^{N-1} \langle \lambda_d(kh), \Phi_d(q_d)(kh) \rangle.$$

Extremizing this function with respect to q_d now gives (3.4.1a), and extremizing with respect to λ_d recovers (3.4.1b). We therefore have equivalence to statement (2).

As we only extremize with respect to the internal points, and hold the boundary terms fixed, we may extend $C_d(\mathbb{R}^d)$ to include λ_0 and λ_N. We now identify $C_d \times V_d = C_d(Q) \times C_d(\mathbb{R}^d)$ with the space $C_d(Q \times \mathbb{R}^d) =$

$\mathcal{C}_d(\{0, h, 2h, \ldots, Nh\}, Q \times \mathbb{R}^d)$, and group the terms in the above expression for $\bar{\mathfrak{G}}_d$ to give two alternative functions $\bar{\mathfrak{G}}_d^+, \bar{\mathfrak{G}}_d^- : \mathcal{C}_d(Q \times \mathbb{R}^d) \to \mathbb{R}$ defined by

$$\bar{\mathfrak{G}}_d^+ = \sum_{k=0}^{N-1} [L_d(q_k, q_{k+1}) - \langle \lambda_{k+1}, \phi(q_{k+1}) \rangle],$$

$$\bar{\mathfrak{G}}_d^- = \sum_{k=0}^{N-1} [L_d(q_k, q_{k+1}) - \langle \lambda_k, \phi(q_k) \rangle],$$

which have the same extrema as $\bar{\mathfrak{G}}_d$ when the boundary terms are held fixed. Identifying the terms in the summations as the augmented discrete Lagrangians $\bar{L}_d^+$ and $\bar{L}_d^-$, respectively, gives equivalence to statement (3). $\square$

Note that in Theorem 3.4.1 one can actually take any convex combination of $\bar{L}_d^+$ and $\bar{L}_d^-$, although this will not substantially alter the result.

We may also use the projection operator $T^*i : T^*Q|_N \to T^*N$ to act on statement (2) of Theorem 3.4.1, showing that (3.4.1) is equivalent to

$$(T^*i)_{q_k} [D_2 L_d(q_{k-1}, q_k) + D_1 L_d(q_k, q_{k+1})] = 0. \tag{3.4.2}$$

This is the counterpart of the continuous equation (3.3.5).

3.4.2. Augmented Hamiltonian viewpoint

Just as in the continuous case, one can either work on the augmented space $T^*(Q \times \mathbb{R}^d)$ or directly on the constrained space T^*N.

The problem with trying to form the augmented discrete Hamiltonian maps $\bar{L}_d^{\pm}$ is the same as in this continuous case, namely the fact that the augmented discrete Lagrangians $\bar{L}_d^{\pm}$ are necessarily degenerate. Nonetheless, we will *define* the discrete Hamiltonian map $\tilde{F}_{\bar{L}_d^-} : (q_0, \lambda_0, p_0, \pi_0) \mapsto (q_1, \lambda_1, p_1, \pi_1)$ by the equations

$$p_0 = -D_1 L_d(q_0, q_1) + \langle \lambda_0, \nabla \phi(q_0) \rangle, \tag{3.4.3a}$$

$$\pi_0 = \phi(q_0), \tag{3.4.3b}$$

$$p_1 = D_2 L_d(q_0, q_1), \tag{3.4.3c}$$

$$\pi_1 = 0. \tag{3.4.3d}$$

Restricting to the same primary constraint set $\Pi \subset T^*(Q \times \mathbb{R}^d)$ as in the continuous case, we see that these equations are the equivalent to (3.4.1) together with the requirement $\phi(q_k) = 0$ and hence $q_d \in \mathcal{C}_d(N)$, that is, they are equivalent to statement (2) in Theorem 3.4.1.

Note that the evolution of λ is not well-defined, as in the continuous case, so that (3.4.3) do not define a map $\Pi \to \Pi$, that is, λ_0 is not a free initial condition, as it will be determined by (q_0, p_0). Note that constructing the alternative map $\tilde{F}_{\bar{L}_d^+}$ does not give a well-defined forward map in general. In fact, to map forward in time it is necessary to use $\tilde{F}_{\bar{L}_d^-}$ as defined above, while $\tilde{F}_{\bar{L}_d^+}$ can be used to map backward in time.

3.4.3. Direct Hamiltonian viewpoint

Alternatively, one can neglect the augmented space and directly relate T^*N and T^*Q. To do so, we differentiate $L_d^N = L_d \circ i^{N \times N}$ with respect to q_0 and q_1 to obtain the discrete equivalents of (3.3.2), thus establishing that the following diagrams commute.

$$\begin{array}{ccc} T^*Q|_N \xleftarrow{\mathbb{F}^- L_d} Q \times Q|_{\{q_0 \in N\}} & Q \times Q|_{\{q_1 \in N\}} \xrightarrow{\mathbb{F}^+ L_d} T^*Q|_N & (3.4.4) \\ \uparrow{\scriptstyle T^*i} \qquad \uparrow{\scriptstyle i^{N \times N}} & {\scriptstyle i^{N \times N}}\uparrow \qquad \qquad \downarrow{\scriptstyle T^*i} & \\ T^*N \xleftarrow{\mathbb{F}^- L_d^N} N \times N & N \times N \xrightarrow{\mathbb{F}^+ L_d^N} T^*N & \end{array}$$

We will henceforth assume that L_d is regular, which means that L_d^N is also regular and that the discrete Hamiltonian maps $\tilde{F}_{L_d}$ and $\tilde{F}_{L_d^N}$ are well-defined. Combining the above diagrams with the expressions $\tilde{F}_{L_d} = \mathbb{F}^+ L_d \circ (\mathbb{F}^- L_d)^{-1} : T^*Q \to T^*Q$ and $\tilde{F}_{L_d^N} = \mathbb{F}^+ L_d^N \circ (\mathbb{F}^- L_d^N)^{-1} : T^*N \to T^*N$ gives the following commutative diagram.

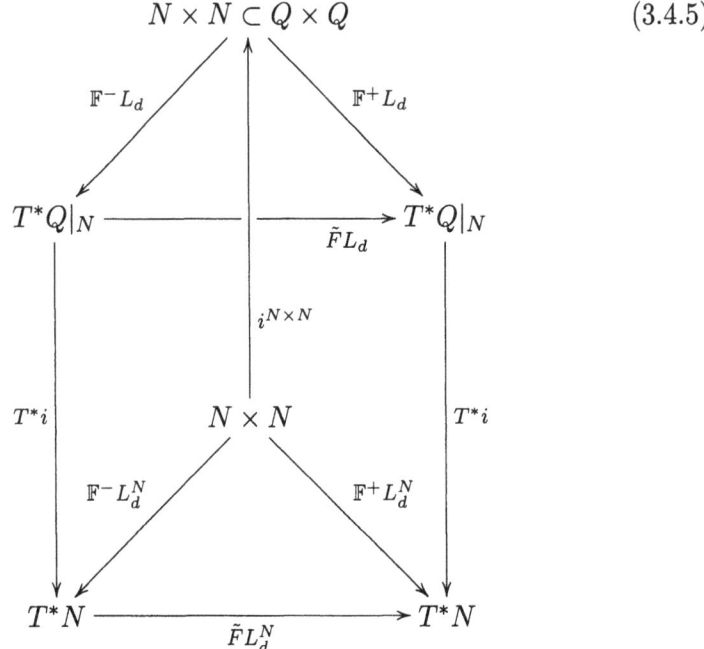

$$(3.4.5)$$

This proves the following theorem.

Theorem 3.4.2. Consider a regular discrete Lagrangian system $L_d : Q \times Q \to \mathbb{R}$ and the constrained system $L_d^N : N \times N \to \mathbb{R}$ defined by $L_d^N = L_d \circ i^{N \times N}$. Then the discrete Hamiltonian map $\tilde{F}_{L_d^N} : T^*N \to T^*N$ has the following equivalent formulations:

(1) $\tilde{F}_{L_d^N} : (q_0, p_0) \mapsto (q_1, p_1)$ for $(q_0, p_0), (q_1, p_1) \in T^*N$ satisfying

$$p_0 = -D_1 L_d^N(q_0, q_1), \tag{3.4.6a}$$
$$p_1 = D_2 L_d^N(q_0, q_1); \tag{3.4.6b}$$

(2) $\tilde{F}_{L_d^N} : (q_0, p_0) \mapsto (q_1, p_1)$ for $(q_0, p_0), (q_1, p_1) \in T^*N$ satisfying

$$p_0 = (T^*i)_{q_0}(-D_1 L_d \circ i^{N \times N}(q_0, q_1)), \tag{3.4.7a}$$
$$p_1 = (T^*i)_{q_1}(D_2 L_d \circ i^{N \times N}(q_0, q_1)); \tag{3.4.7b}$$

(3) $\tilde{F}_{L_d^N} : \eta(T^*N) \mapsto \eta(T^*N)$ for $(q_0, p_0) \in \eta(T^*N)$ and $(q_1, p_1) \in T^*Q$ satisfying

$$p_0 = \mathbb{P}_{q_0}(-D_1 L_d \circ i^{N \times N}(q_0, q_1)), \tag{3.4.8a}$$
$$p_1 = \mathbb{P}_{q_1}(D_2 L_d \circ i^{N \times N}(q_0, q_1)), \tag{3.4.8b}$$
$$\phi(q_1) = 0. \tag{3.4.8c}$$

Here $\eta : T^*N \to T^*Q$ is any symplectic embedding covering the identity, so that $\pi_Q \circ \eta = i \circ \pi_N$, and $\mathbb{P} : T^*Q|_N \to \eta(T^*N)$ is the map defined by $\mathbb{P} = \eta \circ T^*i$.

This theorem is the discrete analogue of Theorem 3.3.3, and shows how the unconstrained Hamiltonian equations are related to the constrained equations. If we further assume that η is defined by (3.3.7) for some regular Lagrangian L with corresponding Hamiltonian H, then we can use the fact that the null space of $\mathbb{P}$ is the span of the $\nabla\phi^m$ and introduce Lagrange multipliers to write (3.4.8) as

$$p_0 = -D_1 L_d(q_0, q_1) + (\lambda^{(0)})^T \nabla\phi(q_0), \tag{3.4.9a}$$
$$p_1 = D_2 L_d(q_0, q_1) - (\lambda^{(1)})^T \nabla\phi(q_1), \tag{3.4.9b}$$

$$\phi(q_1) = 0, \tag{3.4.9c}$$
$$\nabla\phi(q_1) \cdot \frac{\partial H}{\partial p}(q_1, p_1) = 0, \tag{3.4.9d}$$

defining a map from $(q_0, p_0) \in \eta(T^*N)$ to $(q_1, p_1) \in T^*Q$ which will satisfy $(q_1, p_1) \in \eta(T^*N)$. Here the arbitrary signs on the Lagrange multipliers have been chosen to correspond to the signs for discrete forces in (3.2.5).

Now consider the special case when Q is a Riemannian manifold with metric $\langle\!\langle \cdot, \cdot \rangle\!\rangle$ having coordinate representation $M(q)$ and η is defined by (3.3.7) for a Lagrangian with kinetic energy given by the metric. As we have seen in Section 3.3.4 above, $\eta(T^*N)$ and $\mathbb{P}$ are now given explicitly by (3.3.10) and (3.3.11), respectively. Using this, we can write (3.4.8) as

$$p_0 = -\left(I - (\nabla\phi)^T\left[(\nabla\phi)M^{-1}(\nabla\phi)^T\right]^{-1}(\nabla\phi)M^{-1}\right)D_1L_d(q_0, q_1),$$
$$\tag{3.4.10a}$$

$$p_1 = \left(I - (\nabla\phi)^T\left[(\nabla\phi)M^{-1}(\nabla\phi)^T\right]^{-1}(\nabla\phi)M^{-1}\right)D_2L_d(q_0, q_1),$$
$$\tag{3.4.10b}$$

$$\phi(q_1) = 0, \tag{3.4.10c}$$

where $\nabla\phi$ and M are evaluated at q_0 or q_1 as appropriate.

3.4.4. Conservation properties

A constrained discrete Lagrangian system on $N \times N$ and an unconstrained system on $Q \times Q$ will clearly preserve the standard discrete symplectic two-forms $\Omega_{L_d^N}$ and Ω_{L_d}, respectively. Now define the projections $\pi_Q^1 : Q \times Q \to Q$ and $\pi_N^1 : N \times N \to N$ onto the first components of $Q \times Q$ and $N \times N$. Observe that $\pi_Q^1 \circ i^{N \times N} = i \circ \pi_N^1$ and, together with the left-hand diagram in (3.4.4), a similar calculation to that preceding equation (3.3.4) will now establish that $\Theta_{L_d^N}^- = (i^{N \times N})^* \Theta_{L_d}^-$. Using the same idea for $\Theta_{L_d}^+$ and taking the exterior derivative of these expressions shows that the constrained and unconstrained discrete one- and two-forms are related by

$$\Theta_{L_d^N}^+ = (i^{N \times N})^* \Theta_{L_d}^+, \qquad \Theta_{L_d^N}^- = (i^{N \times N})^* \Theta_{L_d}^-, \qquad \Omega_{L_d^N} = (i^{N \times N})^* \Omega_{L_d}.$$

Pushing all of these structures forward with the discrete Legendre transforms shows that the constrained discrete Hamiltonian map $\tilde{F}_{L_d^N}$, regarded as acting either on T^*N or $\eta(T^*N)$, preserves the canonical two-form Ω^N, while $\tilde{F}_{L_d}$ naturally preserves Ω.

If we further consider a symmetry action $\Phi : G \times Q \to Q$ which leaves N invariant, so that it covers an action $\Phi^N : G \times N \to N$, then the infinitesimal generators are related by

$$\xi_Q \circ i = Ti \circ \xi_N, \tag{3.4.11a}$$
$$\xi_{Q \times Q} \circ i^{N \times N} = T(i^{N \times N}) \circ \xi_{N \times N}. \tag{3.4.11b}$$

Using now the above relations between the constrained and unconstrained symplectic one-forms, we have that the momentum maps for the product action will be related by

$$J_{L_d^N}^+ = J_{L_d}^+ \circ i^{N \times N}, \tag{3.4.12a}$$
$$J_{L_d^N}^- = J_{L_d}^- \circ i^{N \times N}. \tag{3.4.12b}$$

If the group action is a symmetry of the Lagrangian then these momentum maps are equal and Noether's theorem holds on both the constrained and unconstrained levels with this unique momentum map.

3.4.5. Constrained exact discrete Lagrangians

The exact discrete Lagrangian for a constrained system is not simply the standard exact discrete Lagrangian restricted to the constraint submanifold, as that would be the action along an unconstrained trajectory. Instead, the constrained exact discrete Lagrangian is the action of the constrained system, evaluated along the trajectory which lies on the constraint submanifold: that is,

$$L_d^{N,E}(q_0, q_1, h) = \int_0^h L^N(q_{0,1}(t), \dot{q}_{0,1}(t)) \, dt, \qquad (3.4.13)$$

where $q : [0, h] \to N$ is the solution of the Euler–Lagrange equations for $L^N : TN \to \mathbb{R}$ which satisfies $q(0) = q_0$ and $q(h) = q_1$. As this discrete Lagrangian is defined on $N \times N \times \mathbb{R}$, it satisfies

$$\mathbb{F}^- L_d^{N,E}(q_0, q_1, h)) = \mathbb{F}L^N(q_{0,1}(0), \dot{q}_{0,1}(0)),$$
$$\mathbb{F}^+ L_d^{N,E}(q_0, q_1, h)) = \mathbb{F}L^N(q_{0,1}(h), \dot{q}_{0,1}(h)).$$

We would like, however, to define a function on $Q \times Q \times \mathbb{R}$ whose restriction to $N \times N \times \mathbb{R}$ would give $L_d^{N,E}$. Without introducing additional structure, however, there is no canonical way to do so. Indeed, let $L_d^{Q,E} : Q \times Q \times \mathbb{R} \to \mathbb{R}$ be any smooth extension of $L_d^{N,E}$. Then from (3.3.3), (3.4.4) and the above relations we have immediately that

$$(T^*i)_{q_0}\left(\mathbb{F}^- L_d^{Q,E}(q_0, q_1, h)\right) = (T^*i)_{q_0}\left(\mathbb{F}L(q_{0,1}(0), \dot{q}_{0,1}(0))\right),$$
$$(T^*i)_{q_1}\left(\mathbb{F}^+ L_d^{Q,E}(q_0, q_1, h)\right) = (T^*i)_{q_1}\left(\mathbb{F}L(q_{0,1}(h), \dot{q}_{0,1}(h))\right),$$

which is a constrained version of Lemma 1.6.2. The equivalence of the discrete and continuous systems now follows as in Section 1.6.

Note that this means that the order of accuracy of a discrete Lagrangian constrained to $N \times N$ will not, in general, be the same as the order of accuracy on $Q \times Q$: that is, if $L_d : Q \times Q \times \mathbb{R} \to \mathbb{R}$ approximates the action on Q to some particular order, then the restriction $L_d^N = L_d|_{N \times N}$ will typically approximate the action of constrained solutions in N to some different order. Indeed, to derive high-order discrete Lagrangians for a constrained system, it is necessary to take account of the constraints in defining L_d^N, since a high-order L_d will typically restrict to only a first- or second-order L_d^N.

3.5. Constrained variational integrators

In this section we consider implementing the integration of a mechanical system with constraints. First we review standard geometric methods, and then we turn to variational integrators.

3.5.1. Constrained geometric integration

There are a number of standard approaches to the numerical integration of constrained mechanical systems. These include working in local coordinates on the constraint submanifold (for example, see Bobenko and Suris (1999*b*) in the case of Lie groups, or Leimkuhler and Reich (1994)), solving a modified system on the containing space which has the constraint submanifold as a stable invariant set (for example, see Leimkuhler and Reich (1994)), and methods based in the containing space which explicitly enforce the constraints. Constrained mechanical systems are particular examples of differential algebraic systems, and many of the techniques for the numerical integration of such systems can also be applied in the mechanical setting (see Hairer and Wanner (1996) and Ascher and Petzold (1998)).

Unless the system under consideration has a particularly simple structure, working in local coordinates on the constraint submanifold suffers from a number of problems, including the fact that changing charts during the integration is not smooth, which breaks many of the nice properties of geometric integrators. In addition, local coordinate computations can be very expensive, and the equations can be very complicated, making the integrator difficult to code. For all of these reasons, it is often preferable to use integration techniques based on the containing space.

There are a number of different approaches to this, with representative samples being Gonzalez (1999), Seiler (1999, 1998*a*, 1998*b*), McLachlan and Scovel (1995) and Brasey and Hairer (1993). For a good overview of this area see Hairer (200x).

3.5.2. Variational integrators for constrained systems

Here we consider a constrained discrete Lagrangian system as an integrator for a continuous system. Given a continuous system $L : TQ \to \mathbb{R}$ and a constraint submanifold $N \subset Q$ defined by $N = \phi^{-1}(0)$ for some $\phi : Q \to \mathbb{R}^d$, we would like a discrete Lagrangian $L_d : Q \times Q \times \mathbb{R} \to \mathbb{R}$ so that its restriction to $N \times N \times \mathbb{R}$ approximates the exact constrained discrete Lagrangian (3.4.13). The order of this approximation is related to the order of the resulting integrator.

Given such an L_d, we can now use any of the equivalent formulations of the constrained Euler–Lagrange equations from Section 3.4.3 to obtain an integrator. As in the unconstrained case, we can regard such an integrator as defined on the product $N \times N$ or on the corresponding cotangent bundle, although the latter interpretation is typically simpler for implementation purposes.

To be explicit, we will henceforth assume that the given continuous system is regular, so that we have equivalent Lagrangian and Hamiltonian representations, and that the containing manifold Q is linear, so that it is isomorphic to $\mathbb{R}^n$. We will use (3.4.9) to define the constrained discrete Hamiltonian map $\tilde{F}_{L_d^N}$ regarded as mapping $\eta(T^*N)$ to $\eta(T^*N)$, where we recall that $\eta(T^*N)$ is the embedding of T^*N in T^*Q defined by

$$\eta(T^*N) = \left\{ (q,p) \in T^*Q \mid \phi(q) = 0 \text{ and } \nabla\phi(q) \cdot \frac{\partial H}{\partial p}(q,p) = 0 \right\}. \quad (3.5.1)$$

As we are now treating the discrete Lagrangian as the approximation to the exact system, it will be dependent upon a time-step h and thus have the form $L_d(q_0, q_1, h)$. Given this, we may rescale the Lagrange multipliers in (3.4.9) by h so that the constraint terms appear in the same way as discrete forces, allowing them to be interpreted as discrete forces of constraint. This gives

$$p_0 = -D_1 L_d(q_0, q_1) + h(\lambda^{(0)})^T \nabla\phi(q_0), \quad (3.5.2a)$$
$$p_1 = D_2 L_d(q_0, q_1) - h(\lambda^{(1)})^T \nabla\phi(q_1), \quad (3.5.2b)$$
$$\phi(q_1) = 0, \quad (3.5.2c)$$
$$\nabla\phi(q_1) \cdot \frac{\partial H}{\partial p}(q_1, p_1) = 0. \quad (3.5.2d)$$

To use these equations as an integrator, we must take an initial condition $(q_0, p_0) \in \eta(T^*N)$, so that q_0 and p_0 satisfy the conditions given by (3.5.1). The $2n + 2d$ system (3.5.2) must then be solved implicitly to find (q_1, p_1) and the accompanying Lagrange multipliers. Iterating this process gives the integrated trajectory.

Although this is generally the simplest way to implement a variational integrator, note that if the Lagrangian has a special form, such as being composed of kinetic and potential terms, then we could also use one of the other equivalent expressions of the discrete Hamiltonian map given previously. Alternatively, we could also choose to work directly on $N \times N$ and to use (3.4.1) as an integrator mapping each pair (q_k, q_{k+1}) to (q_{k+1}, q_{k+2}).

Using the above theory, we recall that any such methods will always be symplectic, and if the discrete Lagrangian inherits the symmetries of the continuous system, then the integrator will also conserve the corresponding momentum maps.

To implement a constrained variational integrator, it is of course necessary to choose a particular discrete Lagrangian. We give below a number of ways in which this can be done and we explicitly evaluate the defining equations (3.5.2) in several cases.

3.5.3. Low-order methods

Given a low-order discrete Lagrangian, such as L_d^α given in Example 2.3.2, one can simply restrict it to $N \times N$ to obtain an integrator for the constrained system. As N will generally not be convex, the points $(1 - \alpha)q_0 + \alpha q_1$ will not be in N when q_0 and q_1 are. If the Lagrangian on N is the restriction of a smooth Lagrangian on Q, then this will not matter for sufficiently small stepsizes.

For a Lagrangian which is not defined off N, or which varies quickly compared to the stepsize, it is important to only evaluate L and its derivatives on N. Perhaps the simplest examples of such methods are given by L_d^0 and L_d^1, which give constrained versions of the symplectic Euler methods.

3.5.4. SHAKE and RATTLE

As we saw in Section 2.6.2, the Verlet algorithm is the discrete Lagrangian map $F_{L_d} : Q \times Q \to Q \times Q$ generated by the discrete Lagrangian

$$L_d(q_0, q_1, h) = \frac{1}{2}hL\left(q_0, \frac{q_1 - q_0}{h}\right) + \frac{1}{2}hL\left(q_1, \frac{q_1 - q_0}{h}\right), \qquad (3.5.3)$$

where we assume that the continuous system has the form $L(q, \dot{q}) = \frac{1}{2}\dot{q}^T M\dot{q} - V(q)$. To form a constrained version of this method, we can simply restrict L_d to $N \times N$ and calculate the constrained discrete Euler–Lagrange equations (3.4.1). These give

$$M\left(\frac{q_{k+1} - 2q_k + q_{k-1}}{h}\right) + h\nabla V(q_k) + (\lambda_k)^T \nabla \phi(q_k) = 0,$$

$$\phi(q_{k+1}) = 0,$$

which is known as the SHAKE algorithm. This was first proposed by Ryckaert, Ciccotti and Berendsen (1977) as a constrained version of Verlet.

A constrained version of the velocity Verlet integrator, RATTLE, was given by Anderson (1983). This was later shown by Leimkuhler and Skeel (1994) to be a symplectic integrator on T^*N. In fact, RATTLE *is simply the constrained discrete Hamiltonian map* $\tilde{F}_{L_d^N} : T^*N \to T^*N$ *associated to the discrete Lagrangian* (3.5.3). To see this, we calculate the coordinate expressions of (3.5.2) with $L(q, \dot{q}) = \frac{1}{2}\dot{q}^T M\dot{q} - V(q)$ to give

$$p_k = M\left(\frac{q_{k+1} - q_k}{h}\right) + \frac{1}{2}h\nabla V(q_k) + (\lambda_k^{(0)})^T \nabla \phi(q_k),$$

$$p_{k+1} = M\left(\frac{q_{k+1} - q_k}{h}\right) - \frac{1}{2}h\nabla V(q_{k+1}) + (\lambda_k^{(1)})^T \nabla \phi(q_{k+1}),$$

$$0 = \phi(q_{k+1}),$$

$$0 = \nabla \phi(q_{k+1}) M^{-1} p_{k+1}.$$

Now we subtract the first equation from the second and solve the first equation for q_{k+1} to obtain

$$q_{k+1} = q_k + hM^{-1}p_k + \frac{1}{2}h^2 M^{-1}(-\nabla V(q_k)) + \frac{1}{2}h^2 M^{-1}(\lambda_k^{(0)})^T \nabla \phi(q_k),$$

$$p_{k+1} = p_k + h\left(\frac{-\nabla V(q_k) - \nabla V(q_{k+1})}{2}\right)$$

$$+ h\left(\frac{(\lambda_k^{(0)})^T \nabla \phi(q_k) + (\lambda_k^{(1)})^T \nabla \phi(q_{k+1})}{2}\right),$$

$$0 = \phi(q_{k+1}),$$

$$0 = \nabla \phi(q_{k+1}) M^{-1} p_{k+1},$$

where we are assuming

$$(\nabla \phi)_{ij}(q) = \frac{\partial \phi^i}{\partial q^j}$$

and where we have scaled $\lambda_k^{(0)}$ and $\lambda_k^{(1)}$ by $-\frac{1}{2}$. This is exactly the RATTLE method.

This integrator is also the 2-stage member of the Lobatto IIIA–IIIB family (Jay 1996, 1999), which is discussed further below.

To summarize, the integrators known as Verlet, velocity Verlet, SHAKE and RATTLE are all derived from the discrete Lagrangian (3.5.3). Verlet is the discrete Lagrangian map $F_{L_d} : Q \times Q \to Q \times Q$, velocity Verlet is the discrete Hamiltonian map $\tilde{F}_{L_d} : T^*Q \to T^*Q$, SHAKE is the constrained discrete Lagrangian map $F_{L_d^N} : N \times N \to N \times N$, and RATTLE is the constrained discrete Hamiltonian map $\tilde{F}_{L_d^N} : T^*N \to T^*N$.

Thus, the variational formulation shows the natural connection between these methods, and proves in a unified way that they all conserve both the symplectic structure and quadratic momentum maps, as linear symmetries of V will be inherited by L_d.

3.5.5. Composition methods

To construct high-order integrators for a constrained system, a simple low-order constraint-preserving method can be used in a composition rule, as in Section 2.5 (Reich 1996). This approach has the advantage that the resulting method will inherit properties such as symplecticity from the base method, and will necessarily preserve the constraint.

Composing discrete Lagrangians extends directly to constrained systems. Given discrete Lagrangians L_d^i and time-step fractions γ^i for $i = 1, \ldots, s$, we can use any of the three interpretations of the composition L_d from Section 2.5. For the multiple steps method or the single step, multiple substeps method, the correct constraint to impose is that all the points q_k^i lie on the constraint submanifold. This implies that the single step constrained composition discrete Lagrangian should be defined as

$$L_d(q_k, q_{k+1}, h) = \operatorname*{ext}_{q_k^i \in N} L_d(q_k, q_k^i, q_{k+1}, h),$$

which denotes the extreme value of the multipoint discrete Lagrangian over the set of interior points in the constraint submanifold N. The constrained discrete Hamiltonian map for this L_d will then be the composition of the constrained discrete Hamiltonian maps of the component L_d^i.

When composing non-self-adjoint methods, it is common to use a sequence including both the methods themselves and their adjoints. For this reason, it is worth noting that the adjoint of a constrained discrete Lagrangian is equal to the constrained version of the adjoint, that is, $(L_d^*)^N = (L_d^N)^*$. Furthermore, the associated constrained discrete Hamiltonian maps are adjoint as integrators.

3.5.6. Constrained symplectic partitioned Runge–Kutta methods

For a Hamiltonian system $H : T^*Q \to \mathbb{R}$ with holonomic constraint $\phi : Q \to \mathbb{R}^d$, a *constrained partitioned Runge–Kutta method* is a map $T^*N \to T^*N$ specified by $(q_0, p_0) \mapsto (q_1, p_1)$ where

$$q_1 = q_0 + h \sum_{j=1}^{s} b_j \dot{Q}_j, \quad p_1 = p_0 + h \sum_{j=1}^{s} \tilde{b}_j \dot{P}_j, \tag{3.5.4a}$$

$$Q_i = q_0 + h \sum_{j=1}^{s} a_{ij} \dot{Q}_j, \quad P_i = p_0 + h \sum_{j=1}^{s} \tilde{a}_{ij} \dot{P}_j, \qquad i = 1, \dots, s, \tag{3.5.4b}$$

$$\dot{Q}_i = \frac{\partial H}{\partial p}(Q_i, P_i), \qquad \dot{P}_i = -\frac{\partial H}{\partial q}(Q_i, P_i) - \Lambda_i^T \nabla \phi(Q_i), \quad i = 1, \dots, s, \tag{3.5.4c}$$

$$0 = \phi(Q_i), \qquad 0 = \nabla \phi(q_1) \cdot \frac{\partial H}{\partial p}(q_1, p_1), \qquad i = 1, \dots, s. \tag{3.5.4d}$$

In addition, it is necessary to place some restrictions on the coefficients to ensure that these equations do in fact define a map on T^*N. We begin by imposing the requirement (2.6.7) of symplecticity to give

$$b_i \tilde{a}_{ij} + \tilde{b}_j a_{ji} = b_i \tilde{b}_j, \qquad i, j = 1, \dots, s,$$
$$b_i = \tilde{b}_i, \qquad i = 1, \dots, s.$$

We also require that the method be *stiffly accurate*: that is, $a_{si} = b_i$ for $i = 1, \dots, s$. This means that $q_1 = Q_s$, and hence $q_1 \in N$. Further requiring that $b_i \neq 0$ for $i = 1, \dots, s$ implies that $\tilde{a}_{is} = 0$ for each $i = 1, \dots, s$.

To ensure that the system is not over-determined, we set $a_{1i} = 0$ for $i = 1, \ldots, s$ and so obtain $q_0 = Q_1$. Requiring that $b_i \neq 0$ for $i = 1, \ldots, s$ now implies that $\tilde{a}_{i1} = \tilde{b}_i$ for $i = 1, \ldots, s$. Given that we start from $(q_0, p_0) \in T^*N$ we thus have that $\phi(Q_1) = \phi(q_0) = 0$ is immediately satisfied.

With these restrictions, (3.5.4) is a system of $s(4n + d) + 2n$ equations for the same number of unknowns, defining a map $\eta(T^*N) \to \eta(T^*N)$. It can be shown (Jay 1996) that this is a well-defined symplectic map on T^*N. Such methods are a particular example of the SPARK methods of Jay (1999), and the subset of these methods which are explicit have been analysed for constrained systems by Reich (1997).

To see how such constrained symplectic partitioned Runge–Kutta methods can be derived variationally, we proceed in a similar fashion to the unconstrained case in Section 2.6.5. Given $(q_0, q_1) \in Q \times Q$, we implicitly define $\bar{p}_0, \bar{p}_1, \bar{Q}_i, \bar{P}_i, \dot{\bar{Q}}_i, \dot{\bar{P}}_i$ for $i = 1, \ldots, s$, and $\bar{\Lambda}_i$ for $i = 2, \ldots, (s-1)$ by the equations

$$q_1 = q_0 + h \sum_{j=1}^{s} b_j \dot{\bar{Q}}_j, \qquad\qquad \bar{p}_1 = \bar{p}_0 + h \sum_{j=1}^{s} \tilde{b}_j \dot{\bar{P}}_j, \qquad (3.5.5a)$$

$$\bar{Q}_i = q_0 + h \sum_{j=1}^{s} a_{ij} \dot{\bar{Q}}_j, \qquad \bar{P}_i = \bar{p}_0 + h \sum_{j=1}^{s} \tilde{a}_{ij} \dot{\bar{P}}_j, \quad i = 1, \ldots, s,$$
$$(3.5.5b)$$

$$\dot{\bar{Q}}_i = \frac{\partial H}{\partial p}(\bar{Q}_i, \bar{P}_i), \qquad\qquad\qquad i = 1, \ldots, s,$$
$$(3.5.5c)$$

$$\dot{\bar{P}}_i = -\frac{\partial H}{\partial q}(\bar{Q}_i, \bar{P}_i) - \bar{\Lambda}_i^T \nabla \phi(\bar{Q}_i), \quad 0 = \phi(\bar{Q}_i), \qquad i = 2, \ldots, (s-1),$$
$$(3.5.5d)$$

$$\dot{\bar{P}}_1 = -\frac{\partial H}{\partial q}(\bar{Q}_1, \bar{P}_1), \qquad\qquad \dot{\bar{P}}_s = -\frac{\partial H}{\partial q}(\bar{Q}_s, \bar{P}_s). \qquad (3.5.5e)$$

This is a system of $4sn + (s-2)d$ equations in the same number of variables and the restrictions on the coefficients ensure that it will have a solution for sufficiently small h.

This subset of the equations (3.5.4) was chosen from the fact that $\bar{Q}_1 = q_0$ and $\bar{Q}_s = q_1$, so it is necessary to relax the constraints on these two points. Having done so, the same number of Lagrange multipliers must also then be disregarded. Given these definitions of the various quantities in terms of q_0 and q_1 we define the discrete Lagrangian $L_d : Q \times Q \times \mathbb{R} \to \mathbb{R}$ by

$$L_d(q_0, q_1, h) = h \sum_{i=1}^{s} b_i L(\bar{Q}_i, \dot{\bar{Q}}_i), \qquad (3.5.6)$$

where we assume that the coefficients satisfy all of the previous requirements. For a given continuous system (L or H) this is not the same as
the corresponding expression (2.6.8) in the unconstrained case, as the equations defining $\dot{Q}_i$ and $\ddot{Q}_i$ have been modified here to take account of the
constraints. We now show that the constrained discrete Hamiltonian map
corresponding to (3.5.6) is indeed the constrained symplectic partitioned
Runge–Kutta method.

Theorem 3.5.1. The constrained discrete Hamiltonian map for the discrete Lagrangian (3.5.6) is exactly the integrator defined by the constrained
symplectic partitioned Runge–Kutta equations (3.5.4).

Proof. Differentiating $\phi(\bar{Q}_i) = 0$ for $i = 2, \ldots, s - 1$ gives

$$\nabla\phi(\bar{Q}_i) \cdot \frac{\partial \bar{Q}_i}{\partial q_0} = 0, \qquad i = 2, \ldots, s - 1,$$

and using this together with the definitions (3.5.5) and the same argument
as in Theorem 2.6.1 shows that

$$\frac{\partial L_d}{\partial q_0} = -\bar{p}_0, \qquad\qquad \frac{\partial L_d}{\partial q_1} = \bar{p}_1.$$

We now consider a given initial condition $(q_0, p_0) \in T^*N$ and recall that the
discrete Hamiltonian map will give $(q_1, p_1) \in T^*N$ which satisfy (3.5.2). To
see the relation of this mapping to the symplectic partitioned Runge–Kutta
map, we make the following change of variables:

$$\begin{aligned}
Q_i &= \bar{Q}_i, & P_i &= \bar{P}_i, & i &= 1, \ldots, s, \\
\dot{Q}_i &= \dot{\bar{Q}}_i, & & & i &= 1, \ldots, s, \\
\Lambda_i &= \bar{\Lambda}_i, & \dot{P}_i &= \dot{\bar{P}}_i, & i &= 2, \ldots, s - 1, \\
\tilde{b}_1 \Lambda_1 &= \lambda^{(0)}, & \dot{P}_1 &= \dot{\bar{P}}_1 - \Lambda_1^T \nabla\phi(Q_1), \\
\tilde{b}_s \Lambda_s &= \lambda^{(1)}, & \dot{P}_s &= \dot{\bar{P}}_s - \Lambda_s^T \nabla\phi(Q_s).
\end{aligned}$$

Recalling that the coefficients are such that $Q_1 = q_0$ and $Q_s = q_1$, we now
see that (3.5.2c) and (3.5.2d), together with the restrictions (3.5.5d) on $\bar{Q}_i$,
give the conditions (3.5.4d) on the non-overbar quantities.

Furthermore, (3.5.2a) and (3.5.2b) give

$$\bar{p}_0 = p_0 - h\tilde{b}_1 \Lambda_1^T \nabla\phi(Q_1),$$
$$\bar{p}_1 = p_1 + h\tilde{b}_s \Lambda_s^T \nabla\phi(Q_s).$$

Substituting these definitions into the equations (3.5.5) and using the fact
that $\tilde{a}_{is} = 0$ and $\tilde{a}_{i1} = \tilde{b}_i$ for $i = 1, \ldots, s$ now shows that the non-overbar
quantities satisfy (3.5.4a), (3.5.4b) and (3.5.4c). We thus have that the

discrete Hamiltonian map $(q_0, p_0) \mapsto (q_1, p_1)$ on $\eta(T^*N)$ is identical to the constrained symplectic Runge–Kutta map.   □

3.5.7. Constrained Galerkin methods

With the insight gained from the definition of the constrained exact discrete Lagrangian (3.4.13) it is simple to extend the Galerkin discrete Lagrangians of Section 2.6.6 to include holonomic constraints.

In the particular example of polynomial trajectory approximations and numerical quadrature, the definition (2.6.10) of the Galerkin discrete Lagrangian should be modified to

$$L_d(q_0, q_1, h) = \underset{\substack{q \in \mathcal{C}^s([0,h],Q) \\ \phi(q(c_i h))=0}}{\mathrm{ext}} \mathfrak{G}^s(q), \qquad (3.5.7)$$

where $\phi : Q \to \mathbb{R}$ is the constraint function. This constrains the intermediate trajectories to intersect the constraint submanifold at each quadrature point. For such methods it is typically reasonable to require that $c_0 = 0$ and $c_s = 1$, so that the endpoints q_0 and q_1 also satisfy the constraint.

Evaluating the constrained discrete Euler–Lagrange equations for (3.5.7) shows that the associated discrete Hamiltonian map is a constrained symplectic partitioned Runge–Kutta method, in the sense of the preceding section and of Jay (1999). In particular, choosing the quadrature rule to be Lobatto quadrature results in the constrained Lobatto IIIA–IIIB method of Jay (1999).

3.6. Background: Forced and constrained systems

We now consider Lagrangian and Hamiltonian systems with *both* external forcing and holonomic constraints. The formulations and equations for such systems are straightforward combinations of the material in the preceding sections for systems with only forces or only constraints. For this reason, we will simply state the results without proof.

As before, we assume that we have a system on the unconstrained configuration manifold Q, and a holonomic constraint function $\phi : Q \to \mathbb{R}^d$ so that the constraint manifold is $N = \phi^{-1}(0) \subset Q$. The inclusion map is denoted $i : N \to Q$, and we have the natural lifts $Ti : TN \to TQ$ and $T^*i : T^*Q \to T^*N$.

3.6.1. Lagrangian systems

Given a Lagrangian force $f_L : TQ \to T^*Q$, we restrict it to $f_L^N = T^*i \circ f_L \circ Ti : TN \to T^*N$, which is then a Lagrangian force on TN. Taking the Lagrange–d'Alembert principle and restricting to the space of constrained curves gives the following theorem.

Theorem 3.6.1. Given a Lagrangian system $L : TQ \to \mathbb{R}$ with Lagrangian force $f_L : TQ \to T^*Q$ and holonomic constraint $\phi : Q \to \mathbb{R}^d$, set $N = \phi^{-1}(0) \subset Q$, $f_L^N = T^*i \circ f_L \circ Ti$, and $L^N = L|_{TN}$. Then the following are equivalent:

(1) $q \in \mathcal{C}(N)$ satisfies the Lagrange–d'Alembert principle for L^N and f_L^N and hence solves the forced Euler–Lagrange equations;

(2) $q \in \mathcal{C}(Q)$ and $\lambda \in \mathcal{C}(\mathbb{R}^d)$ satisfy the *forced constrained Euler–Lagrange equations*

$$\frac{\partial L}{\partial q^i}(q(t), \dot{q}(t)) - \frac{\mathrm{d}}{\mathrm{d}t}\left(\frac{\partial L}{\partial \dot{q}^i}(q(t), \dot{q}(t))\right) + f_L(q(t), \dot{q}(t))$$

$$= \left\langle \lambda(t), \frac{\partial \phi}{\partial q^i}(q(t)) \right\rangle, \quad (3.6.1a)$$

$$\phi(q(t)) = 0; \quad (3.6.1b)$$

(3) $(q, \lambda) \in \mathcal{C}(Q \times \mathbb{R}^d)$ satisfies the Lagrange–d'Alembert principle, and hence solves the forced Euler–Lagrange equations, for $\bar{L} : T^*(Q \times \mathbb{R}^d) \to \mathbb{R}$ and $\bar{f}_L : T(Q \times \mathbb{R}^d) \to T^*(Q \times \mathbb{R}^d)$ defined by

$$\bar{L}(q, \lambda, \dot{q}, \dot{\lambda}) = L(q, \dot{q}) - \langle \lambda, \phi(q) \rangle,$$

$$\bar{f}_L(q, \lambda, \dot{q}, \dot{\lambda}) = \pi_Q^* \circ f_L(q, \dot{q}),$$

where $\pi_Q : Q \times \mathbb{R}^d \to Q$ is the projection.

One can also project (3.6.1a) with $T^*i : T^*Q \to T^*N$ to obtain a system without λ, as in Section 3.3.

Observe that in the forced constrained Euler–Lagrange equations (3.6.1) the forcing and Lagrange multiplier terms enter in same way. For this reason, the Lagrange multiplier term is sometimes referred to as the *forces of constraint*, and we can regard it as being a force which is constructed exactly so that the solution is kept on the constraint submanifold N.

3.6.2. Hamiltonian systems

Following the development of the unforced constrained case, we can move to the Hamiltonian framework by either taking the Legendre transform of the degenerate augmented system, or by working directly on T^*N.

The former approach takes a Hamiltonian force $f_H : T^*Q \to T^*Q$ and forms the augmented Hamiltonian force $\bar{f}_H : T^*(Q \times \mathbb{R}^d) \to T^*(Q \times \mathbb{R}^d)$ by $\bar{f}_H(q, \lambda, p, \pi) = \pi_Q^* \circ f_H(q, p)$. The forced constrained Hamiltonian vector field $\bar{X}_H$ on the primary constraint set Π is defined by

$$i_{\bar{X}_H}\Omega^\Pi = \mathrm{d}\bar{H} - \bar{f}_H'$$

where $\bar{H}$ and Ω^Π are as before, and $\bar{f}'_H$ is the horizontal one-form on $T^*(Q \times \mathbb{R}^d)$ corresponding to $\bar{f}_H$. In coordinates this gives the *forced constrained Hamilton equations*

$$X_{q^i}(q, \lambda, p, \pi) = \frac{\partial H}{\partial p^i},$$

$$X_{p^i}(q, \lambda, p, \pi) = -\frac{\partial H}{\partial q^i} + f_H(q, p) - \left\langle \lambda, \frac{\partial \phi}{\partial q^i}(q) \right\rangle,$$

$$\phi(q) = 0.$$

Alternatively, we can directly relate the unconstrained Hamiltonian system to the constrained system as in Section 3.3.3. To do this, we must choose a symplectic embedding $\eta : T^*N \to T^*Q$, which we will assume covers the embedding $i : N \to Q$. Given such a map, we now define the *constrained Hamiltonian force* $f^N_H : T^*N \to T^*N$ by $f^N_H = T^*i \circ f_H \circ \eta$ and we let $f^{N'}_H$ be the corresponding horizontal one-form on T^*N. We assume that all other structures are as in Section 3.3.3, so that the constrained Hamiltonian is $H^N = H \circ \eta$.

The forced constrained Hamiltonian vector field X_{H^N} and the forced unconstrained Hamiltonian vector field X_H are now defined by

$$i_{X_{H^N}} \Omega^N = dH^N - f^{N'}_H,$$

$$i_{X_H} \Omega = dH - f'_H.$$

Denoting the Ω-orthogonal projection to $\eta(T^*N)$ by $\pi_\Omega : T^*Q \to T^*N$, we can show that the projection of the forced unconstrained vector field is just the forced constrained vector field.

Theorem 3.6.2. Consider a Hamiltonian system $H : T^*Q \to \mathbb{R}$ with forcing $f_H : T^*Q \to T^*Q$ and constraint submanifold $N \subset Q$ and let the constrained system $H^N : T^*N \to \mathbb{R}$ and $f^N_H : T^*N \to T^*N$ be defined as above. Then $X_{H^N} = \pi_\Omega \cdot X_H \circ \eta$.

Proof. We can use essentially the same proof as for Theorem 3.3.3 in the unforced case. The only additional requirement is to check that the one-form $f^{N'}_H$ is the pullback under η of f'_H, so that $f'_H(\eta(p_q)) \cdot T\eta \cdot V^N = f^{N'}_H(p_q) \cdot V^N$.

To see this, we recall that η covers the identity and so $\pi_Q \circ \eta = i \circ \pi_N$. Using the derivative of this expression we calculate

$$f^{N'}_H(p_q) \cdot V^N = \langle T^*i \circ f_H \circ \eta(p_q), T\pi_N \cdot V^N \rangle$$

$$= \langle f_H \circ \eta(p_q), Ti \circ T\pi_N \cdot V^N \rangle$$

$$= \langle f_H \circ \eta(p_q), T\pi_Q \circ T\eta \cdot V^N \rangle$$
$$= (\eta^*(f'_H))(p_q) \cdot V^N,$$

which can then be used to modify the proof of Theorem 3.3.3, to obtain the desired result. □

3.6.3. Legendre transforms

Given a regular Lagrangian system and the corresponding regular Hamiltonian system, we have seen in Section 3.3.4 that the standard Legendre transforms provide a canonical way to construct a map $\eta : T^*N \to T^*Q$ and so to regard T^*N as a submanifold of T^*Q.

Furthermore, as we saw in Section 3.1.3, the forced Lagrangian and Hamiltonian vector fields are related by the standard Legendre transform, so this will hold for both the constrained and unconstrained systems. Note that our definitions of constrained Lagrangian and Hamiltonian forces commute with the Legendre transform, so that if $f_L = f_H \circ \mathbb{F}L$ then $f_L^N = f_H^N \circ \mathbb{F}L^N$. This can be seen by recalling that $\eta \circ \mathbb{F}L^N = \mathbb{F}L \circ Ti$ and using the definitions of the constrained forces.

We thus have that the constrained and unconstrained forced vector fields on both the Lagrangian and Hamiltonian sides are related by projection and Legendre transforms, which fully commute. In particular, we can write the projected vector field on the Hamiltonian side in coordinates to give

$$\dot{q} = \frac{\partial H}{\partial p},$$
$$\dot{p} = -\frac{\partial H}{\partial q} - \lambda^T \nabla \phi(q) + f_H(q, p),$$
$$\phi(q) = 0.$$

In the special case when the Hamiltonian depends quadratically on p then this projection is induced by the metric given on T^*Q by the kinetic energy, as in Section 3.3.4 above.

3.6.4. Conservation properties

Given a group action $\Phi : G \times Q \to Q$, we have seen in Section 3.4.4 that if Φ leaves N invariant then it can be restricted to an action Φ^N on N and the infinitesimal generators of this restricted action are related by projection to the generators of the action on Q. This then shows that the momentum maps of the constrained systems are just the appropriate restrictions of the unconstrained momentum maps.

In addition, from Section 3.1.4 we know that if the Lagrangian is invariant under the group action and the forces are orthogonal to the action, then

Noether's theorem will still hold. In the constrained setting, observe that we have

$$\langle f_L^N(v_q), \xi_N(q) \rangle = \langle T^*i \circ f_L \circ Ti(v_q), \xi_N(q) \rangle$$
$$= \langle f_L \circ Ti(v_q), Ti \cdot \xi_N(q) \rangle$$
$$= \langle f_L \circ Ti(v_q), \xi_Q \circ i(q) \rangle,$$

and so if f_L is orthogonal to ξ_Q then the constrained force f_L^N will also be orthogonal to the constrained infinitesimal generator ξ_N. This gives us the following Noether's theorem.

Theorem 3.6.3. (Forced constrained Noether's theorem) Consider a Lagrangian system $L : TQ \to \mathbb{R}$ with constraint submanifold $N \subset Q$, forcing $f_L : TQ \to T^*Q$ and a symmetry action $\Phi : G \times Q \to Q$ such that $\langle f_L(q, \dot{q}), \xi_Q(q) \rangle = 0$ for all $(q, \dot{q}) \in TQ$ and $\xi \in \mathfrak{g}$. Then the constrained Lagrangian momentum map $J_{LN} : TN \to \mathfrak{g}^*$ will be preserved by the forced constrained Lagrangian flow.

Of course, it is only necessary that the constrained force be orthogonal to the group action on the constraint submanifold and that the reduced action be a symmetry of the constrained Lagrangian. The above theorem simply gives sufficient conditions for this in terms of the unconstrained quantities.

3.7. Discrete variational mechanics with forces and constraints

We now combine the previous results for forced and constrained systems to consider discrete Lagrangian systems with *both* forcing and constraints. The definitions and results are the expected combinations of the special cases of only forcing or only constraints, and so we will not give detailed proofs.

3.7.1. Lagrangian viewpoint

Given discrete Lagrangian forces $f_d^+, f_d^- : Q \times Q \to T^*Q$, we form the restrictions $f_d^{N+}, f_d^{N-} : N \times N \to T^*N$ by $f_d^{N\pm} = T^*i \circ f_d^{\pm} \circ i^{N \times N}$, which are then discrete Lagrangian forces on N. As in the continuous Lagrangian case, we now take the discrete Lagrange–d'Alembert principle from Section 3.4 and constrain it to N, thus obtaining the following theorem.

Theorem 3.7.1. Given discrete Lagrangian system $L_d : Q \times Q \to \mathbb{R}$ with discrete Lagrangian forces $f_d^+, f_d^- : Q \times Q \to T^*Q$ and holonomic constraint $\phi : Q \to \mathbb{R}^d$, set $N = \phi^{-1}(0) \subset Q$, $f_d^{N\pm} = T^*i \circ f_d^{\pm} \circ i^{N \times N}$, and $L_d^N = L_d|_{Q \times Q}$.

Then the following are equivalent:

(1) $q_d = \{q_k\}_{k=0}^N \in C_d(N)$ satisfies the discrete Lagrange–d'Alembert principle for L_d^N, f_d^{N+} and f_d^{N-}, and hence solves the forced discrete Euler–Lagrange equations;

(2) $q_d = \{q_k\}_{k=0}^N \in C_d(Q)$ and $\lambda_d = \{\lambda_k\}_{k=1}^{N-1} \in C_d(\mathbb{R}^d)$ satisfy the *forced constrained discrete Euler–Lagrange equations*

$$D_2 L_d(q_{k-1}, q_k) + D_1 L_d(q_k, q_{k+1})$$
$$+ f_d^+(q_{k-1}, q_k) + f_d^-(q_k, q_{k+1}) = \langle \lambda_k, \nabla\phi(q_k)\rangle, \quad (3.7.1a)$$
$$\phi(q_k) = 0; \qquad\qquad\qquad (3.7.1b)$$

(3) $(q_d, \lambda_d) = \{(q_k, \lambda_k)\}_{k=0}^N \in C_d(Q \times \mathbb{R}^d)$ satisfies the discrete Lagrange–d'Alembert principle, and hence solves the forced discrete Euler–Lagrange equations, for either of $\bar{L}_d^+, \bar{L}_d^- : (Q \times \mathbb{R}^d) \times (Q \times \mathbb{R}^d) \to \mathbb{R}$ defined by

$$\bar{L}_d^+(q_k, \lambda_k, q_{k+1}, \lambda_{k+1}) = L_d(q_k, q_{k+1}) - \langle \lambda_{k+1}, \phi(q_{k+1})\rangle,$$
$$\bar{L}_d^-(q_k, \lambda_k, q_{k+1}, \lambda_{k+1}) = L_d(q_k, q_{k+1}) - \langle \lambda_k, \phi(q_k)\rangle,$$

with the discrete Lagrangian forces $\bar{f}_d^+, \bar{f}_d^- : (Q \times \mathbb{R}^d) \times (Q \times \mathbb{R}^d) \to T^*(Q \times \mathbb{R}^d)$ defined by

$$\bar{f}_d^+(q_k, \lambda_k, q_{k+1}, \lambda_{k+1}) = \pi_Q^* \circ f_d^+(q_k, q_{k+1}),$$
$$\bar{f}_d^-(q_k, \lambda_k, q_{k+1}, \lambda_{k+1}) = \pi_Q^* \circ f_d^-(q_k, q_{k+1}),$$

where $\pi_Q : Q \times \mathbb{R}^d \to Q$ is the projection.

Using the canonical projection operator $T^*i : T^*Q \to T^*N$, we can also write (3.7.1) without the Lagrange multipliers.

3.7.2. Discrete Hamiltonian maps

We first consider the augmented approach to constructing a discrete Hamiltonian map, despite the lack of regularity. The *forced augmented discrete Hamiltonian map* $\tilde{F}_{\bar{L}_d^-} : (q_0, \lambda_0, p_0, \pi_0) \mapsto (q_1, \lambda_1, p_1, \pi_1)$ is defined by the equations

$$p_0 = -D_1 L_d(q_0, q_1) - f_d^-(q_0, q_1) + \langle \lambda_0, \nabla\phi(q_0)\rangle, \qquad (3.7.2a)$$
$$\pi_0 = \phi(q_0), \qquad\qquad\qquad (3.7.2b)$$
$$p_1 = D_2 L_d(q_0, q_1) + f_d^+(q_0, q_1), \qquad\qquad (3.7.2c)$$
$$\pi_1 = 0. \qquad\qquad\qquad (3.7.2d)$$

Restricting to the primary constraint set $\Pi \subset T^*(Q \times \mathbb{R}^d)$ now shows that these equations are equivalent to the forced constrained discrete Euler–Lagrange equations (3.7.1) together with the constraint $\phi(q_k) = 0$. As before, the evolution of λ is not well-defined.

Rather than considering the augmented systems, we can also directly relate the constrained and unconstrained systems. Here we must use the forced discrete Legendre transforms (3.2.4), which we recall are

$$\mathbb{F}^{f+}L_d : (q_0, q_1) \mapsto (q_1, p_1) = (q_1, D_2 L_d(q_0, q_1) + f_d^+(q_0, q_1)),$$
$$\mathbb{F}^{f-}L_d : (q_0, q_1) \mapsto (q_0, p_0) = (q_0, -D_1 L_d(q_0, q_1) - f_d^-(q_0, q_1)).$$

These depend on both the discrete Lagrangian and discrete forces. From (3.4.4) we have the relations

$$D_2 L_d^N = T^* i \circ D_2 L_d \circ i^{N \times N},$$
$$-D_1 L_d^N = T^* i \circ (-D_1 L_d) \circ i^{N \times N},$$

and, combining these with the definitions of the constrained discrete forces f_d^{N+} and f_d^{N-}, we have the following commutative diagrams, where the discrete Legendre transforms are those which include the forcing.

$$
\begin{array}{ccc}
T^*Q|_N \xleftarrow{\mathbb{F}^{f-}L_d} Q \times Q|_{\{q_0 \in N\}} & \qquad & Q \times Q|_{\{q_1 \in N\}} \xrightarrow{\mathbb{F}^{f+}L_d} T^*Q|_N \\
{\scriptstyle T^*i}\Big\downarrow \qquad \qquad \Big\uparrow{\scriptstyle i^{N \times N}} & \qquad & {\scriptstyle i^{N \times N}}\Big\uparrow \qquad \qquad \Big\downarrow{\scriptstyle T^*i} \\
T^*N \xleftarrow[\mathbb{F}^{f-}L_d^N]{} N \times N & \qquad & N \times N \xrightarrow[\mathbb{F}^{f+}L_d^N]{} T^*N
\end{array}
\qquad (3.7.3)
$$

This is the equivalent of (3.4.4) in the unforced case, and using this we now have the equivalent of diagram (3.4.5) for the forced discrete Legendre transforms, proving the following theorem.

Theorem 3.7.2. Consider a regular discrete Lagrangian system $L_d : Q \times Q \to \mathbb{R}$ with constraint submanifold $N \subset Q$ and forcing $f_d^+, f_d^- : Q \times Q \to T^*Q$. Then the forced constrained discrete Hamiltonian map $\tilde{F}_{L_d^N} : T^*N \to T^*N$ has the following equivalent formulations:

(1) $\tilde{F}_{L_d^N} : (q_0, p_0) \mapsto (q_1, p_1)$ for $(q_0, p_0), (q_1, p_1) \in T^*N$ satisfying

$$p_0 = -D_1 L_d^N(q_0, q_1) - f_d^{N-}(q_0, q_1), \qquad (3.7.4a)$$
$$p_1 = D_2 L_d^N(q_0, q_1) + f_d^{N+}(q_0, q_1); \qquad (3.7.4b)$$

(2) $\tilde{F}_{L_d^N} : (q_0, p_0) \mapsto (q_1, p_1)$ for $(q_0, p_0), (q_1, p_1) \in T^*N$ satisfying

$$p_0 = (T^*i)_{q_0}\left((-D_1 L_d - f_d^-) \circ i^{N \times N}(q_0, q_1)\right), \qquad (3.7.5a)$$
$$p_1 = (T^*i)_{q_1}\left((D_2 L_d + f_d^+) \circ i^{N \times N}(q_0, q_1)\right); \qquad (3.7.5b)$$

(3) $\tilde{F}_{L_d^N} : \eta(T^*N) \mapsto \eta(T^*N)$ for $(q_0, p_0) \in \eta(T^*N)$ and $(q_1, p_1) \in T^*Q$
    satisfying

$$p_0 = \mathbb{P}_{q_0}\left((-D_1 L_d - f_d^-) \circ i^{N \times N}(q_0, q_1)\right), \qquad (3.7.6a)$$

$$p_1 = \mathbb{P}_{q_1}\left((D_2 L_d + f_d^+) \circ i^{N \times N}(q_0, q_1)\right), \qquad (3.7.6b)$$

$$\phi(q_1) = 0. \qquad (3.7.6c)$$

Here $\eta : T^*N \to T^*Q$ is any symplectic embedding covering the identity,
so that $\pi_Q \circ \eta = i \circ \pi_N$, and $\mathbb{P} : T^*Q|_N \to \eta(T^*N)$ is the map defined by
$\mathbb{P} = \eta \circ T^*i$.

These equations are clearly the combination of the constrained equations
from Theorem 3.4.2 with the forced equations (3.2.5).

Now assume that η is constructed from the Legendre transforms of some
regular Lagrangian according to (3.3.7). Introducing Lagrange multipliers
allows us to rewrite (3.7.6) as

$$p_0 = -D_1 L_d(q_0, q_1) - f_d^-(q_0, q_1) + (\lambda^{(0)})^T \nabla \phi(q_0),$$

$$\hspace{10cm} (3.7.7a)$$

$$p_1 = D_2 L_d(q_0, q_1) + f_d^+(q_0, q_1) - (\lambda^{(1)})^T \nabla \phi(q_1), \quad (3.7.7b)$$

$$\phi(q_1) = 0, \qquad (3.7.7c)$$

$$\nabla \phi(q_1) \cdot \frac{\partial H}{\partial p}(q_1, p_1) = 0, \qquad (3.7.7d)$$

where (q_0, p_0) are in $\eta(T^*N)$. As before, we have chosen the signs on the Lagrange multipliers to correspond with the conventions of the discrete forces.

This form of the forced constrained discrete Hamiltonian map shows
clearly that one can interpret the Lagrange multiplier terms as *discrete
forces of constraint*. That is, the additional terms due to the constraints
enter the equations in exactly the same way as the forcing terms. Indeed,
the constraint terms can be regarded as forces which have exactly the correct
action to keep the discrete trajectory on the constraint submanifold N.

If we are working with a particular form of Lagrangian, such as one involving a quadratic kinetic energy, then we can explicitly write the projection
form of the discrete Hamiltonian map as was done in Section 3.4.3.

3.7.3. Exact forced constrained discrete Lagrangian

Given a Lagrangian system with forces and constraints, we can combine the
ideas from Sections 3.2.4 and 3.4.5 to define the appropriate exact discrete
Lagrangian and exact discrete forces.

Begin by considering the constrained system $L^N : TN \to \mathbb{R}$ with constrained force $f_L^N : TN \to T^*N$. Recall that the exact forced discrete

Lagrangian $L_d^{N,E} : N \times N \times \mathbb{R}$ is the action (3.2.7a) along a solution of the forced Euler–Lagrange equations, and that the exact discrete forces $f_d^{N,E+}, f_d^{N,E-} : N \times N \times \mathbb{R} \to T^*N$ are the integrals of the forces (3.2.7b), (3.2.7c) along the variations of such a solution.

Having constructed these functions on $N \times N \times \mathbb{R}$, we take any smooth extension to functions $L_d^{Q,E} : Q \times Q \times \mathbb{R}$ and $f_d^{Q,E+}, f_d^{Q,E-} : Q \times Q \times \mathbb{R} \to \mathbb{R}$, as in Section 3.4.5. The same argument as used there now shows that

$$(T^*i)_{q_0}\left(\mathbb{F}^{f-}L_d^{Q,E}(q_0,q_1,h)\right) = (T^*i)_{q_0}\left(\mathbb{F}L(q_{0,1}(0),\dot{q}_{0,1}(0))\right),$$

$$(T^*i)_{q_1}\left(\mathbb{F}^{f+}L_d^{Q,E}(q_0,q_1,h)\right) = (T^*i)_{q_1}\left(\mathbb{F}L(q_{0,1}(h),\dot{q}_{0,1}(h))\right),$$

for all $q_0, q_1 \in N$ and the corresponding solutions $q : [0,h] \to N$ of the forced constrained Euler–Lagrange equations.

Using the above definitions, it is clear that to derive high-order discrete Lagrangians and discrete forces in the presence of constraints, both the discrete Lagrangian and the discrete forces will have to depend upon the continuous Lagrangian, the continuous forces and also the constraints. We will see examples of this below.

3.7.4. Noether's theorem

Consider a group action $\Phi : G \times Q \to Q$ and assume that it leaves N invariant, so that it restricts to $\Phi^N : G \times N \to N$. In the presence of forcing we saw in Section 3.2.3 that it is necessary to use the forced Legendre transforms to define the discrete momentum maps by (3.2.6). For the unconstrained system this gives

$$J_{L_d}^{f+}(q_0,q_1) \cdot \xi = \left\langle \mathbb{F}^{f+}L_d(q_0,q_1), \xi_Q(q_1)\right\rangle, \qquad (3.7.8a)$$

$$J_{L_d}^{f-}(q_0,q_1) \cdot \xi = \left\langle \mathbb{F}^{f-}L_d(q_0,q_1), \xi_Q(q_1)\right\rangle, \qquad (3.7.8b)$$

while the constrained forced momentum maps are

$$J_{L_d^N}^{f+}(q_0,q_1) \cdot \xi = \left\langle \mathbb{F}^{f+}L_d^N(q_0,q_1), \xi_N(q_1)\right\rangle, \qquad (3.7.9a)$$

$$J_{L_d^N}^{f-}(q_0,q_1) \cdot \xi = \left\langle \mathbb{F}^{f-}L_d^N(q_0,q_1), \xi_N(q_1)\right\rangle. \qquad (3.7.9b)$$

Recalling that the forced discrete Legendre transforms satisfy (3.7.3), we can use the relations (3.4.11) between the constrained and unconstrained infinitesimal generators to show that

$$J_{L_d^N}^{f+} = J_{L_d}^{f+} \circ i^{N \times N}, \qquad (3.7.10a)$$

$$J_{L_d^N}^{f-} = J_{L_d}^{f-} \circ i^{N \times N}, \qquad (3.7.10b)$$

which is the forced equivalent of (3.4.12). If the group action is a symmetry of the discrete Lagrangian then these momentum maps will be equal. In general Noether's theorem does not hold in the presence of forcing, except in the special case when the forces are orthogonal to the group action. We will now see how this occurs in the presence of constraints.

Recall that, given discrete forces f_d^+ and f_d^-, we can construct a one-form f_d on $Q \times Q$ by (3.2.1), which gives

$$f_d^N(q_0, q_1) \cdot (\delta q_0, \delta q_1) = f_d^{N+}(q_0, q_1) \cdot \delta q_1 + f_d^{N-}(q_0, q_1) \cdot \delta q_0,$$

$$f_d(q_0, q_1) \cdot (\delta q_0, \delta q_1) = f_d^+(q_0, q_1) \cdot \delta q_1 + f_d^-(q_0, q_1) \cdot \delta q_0,$$

and so we have the relation $f_d^N = T^*(i^{N \times N}) \circ f_d \circ i^{N \times N}$. Using this, we compute

$$\langle f_d^N(q_0, q_1), \xi_{N \times N}(q_0, q_1) \rangle = \langle T^*(i^{N \times N}) \circ f_d \circ i^{N \times N}(q_0, q_1), \xi_{N \times N}(q_0, q_1) \rangle$$

$$= \langle f_d \circ i^{N \times N}(q_0, q_1), T(i^{N \times N}) \circ \xi_{N \times N}(q_0, q_1) \rangle$$

$$= \langle f_d \circ i^{N \times N}(q_0, q_1), \xi_{Q \times Q} \circ i^{N \times N}(q_0, q_1) \rangle,$$

where we have used the fact that $\xi_{Q \times Q} \circ i^{N \times N} = T(i^{N \times N}) \circ \xi_{N \times N}$. This shows that if f_d is orthogonal to $\xi_{Q \times Q}$, so that $\langle f_d, \xi_{Q \times Q} \rangle = 0$, then f_d^N will be orthogonal to $\xi_{N \times N}$. We thus have a Noether's theorem in this case.

Theorem 3.7.3. (Discrete forced constrained Noether's theorem)
Consider a discrete Lagrangian system $L_d : Q \times Q \to \mathbb{R}$ with constraint submanifold $N \subset Q$, discrete forces $f_d^+, f_d^- : Q \times Q \to T^*Q$ and a symmetry action $\Phi : G \times Q \to Q$ such that $\langle f_d, \xi_{Q \times Q} \rangle = 0$ for all $\xi \in \mathfrak{g}$. Then the constrained Lagrangian momentum map $J_{L_d^N}^f : N \times N \to \mathfrak{g}^*$ is preserved by the forced constrained discrete Hamiltonian map.

As in the continuous case with forcing and constraints, this only provides a sufficient condition as it is enough to just have orthogonality and invariance on N.

3.7.5. Variational integrators with forces and constraints

Consider a Lagrangian system $L : TQ \to \mathbb{R}$ with a constraint submanifold $N \subset Q$ specified by $N = \phi^{-1}(0)$ for some $\phi : Q \to \mathbb{R}^d$ and a Lagrangian force $f_L : TQ \to T^*Q$. We would now like to construct a discrete Lagrangian $L_d : Q \times Q \to \mathbb{R}$ and discrete forces $f_d^+, f_d^- : Q \times Q \to T^*Q$ which approximate an extension of the exact discrete Lagrangian and exact forces. The discrete Hamiltonian map will then be an integrator for the continuous system.

We will assume here that the Lagrangian is regular, so that it has an equivalent Hamiltonian formulation, and also that Q is linear and isomorphic to $\mathbb{R}^n$. Regularity of the Lagrangian also provides a canonical embedding $\eta : T^*N \to T^*Q$, and we will use the Lagrange multiplier formulation (3.7.7) of the forced constrained Hamiltonian map. As in Section 3.5.2, we will rescale the Lagrange multipliers by the time-step to give

$$p_0 = -D_1 L_d(q_0, q_1) - f_d^-(q_0, q_1) + h(\lambda^{(0)})^T \nabla \phi(q_0),$$
$$(3.7.11a)$$

$$p_1 = D_2 L_d(q_0, q_1) + f_d^+(q_0, q_1) - h(\lambda^{(1)})^T \nabla \phi(q_1),$$
$$(3.7.11b)$$

$$\phi(q_1) = 0, \tag{3.7.11c}$$

$$\nabla \phi(q_1) \cdot \frac{\partial H}{\partial p}(q_1, p_1) = 0, \tag{3.7.11d}$$

where the initial condition (q_0, p_0) is in $\eta(T^*N)$, and we solve over $(q_1, p_1) \in T^*Q$. The last two equations ensure that the solution (q_1, p_1) will also lie in $\eta(T^*N)$. Of course, we could also use one of the alternative formulations from Theorem 3.7.2 or we could use the forced constrained discrete Euler–Lagrange equations (3.7.1) and work directly on $N \times N$.

To construct discrete Lagrangians and discrete forces we can use any of the techniques discussed previously. Here we give a few examples.

Example 3.7.4. (Low-order methods) For a low-order discrete Lagrangian and discrete forces, such as the L_d^α and $f_d^{\alpha,\pm}$ from Example 3.2.2, we can simply restrict them to $N \times N$, as in Section 3.5.3. This yields a simple method that remains on the constraint manifold and includes the forcing. $\Diamond$

Example 3.7.5. (Composition methods) As we have seen in several examples already, composition methods provide a particularly elegant method to construct high-order methods from a given low-order integrator. In the case of systems with both forcing and constraints, the appropriate composed discrete forces and discrete Lagrangians are given by the combination of the definitions for the forced and constrained cases. $\Diamond$

Example 3.7.6. (Symplectic partitioned Runge–Kutta methods) Combining the definitions of the discrete forces (3.2.8) with the constrained formulation of the discrete Lagrangian (3.5.6), we arrive at discrete forces and a discrete Lagrangian for which the discrete Hamiltonian map is a constrained symplectic partitioned Runge–Kutta method with forcing. $\Diamond$

PART FOUR
Time-dependent mechanics

4.1. Introduction

In this part we address the issues of nonautonomous systems and discrete energies, which are closely related. In doing so, we rederive the symplectic–energy–momentum-conserving integrators of Kane *et al.* (1999*a*) in a purely variational way. We stress, however, that the theory developed here has many applications, aside from deriving such integrators. These applications include a deeper understanding of the behaviour of the symplectic–momentum integrators discussed in previous parts, as well as apparently unrelated areas such as nonsmooth variational mechanics (Fetecau *et al.* 2001).

The basic methodology used here is that of variational mechanics and variational discretizations. Unlike the standard discrete variational mechanics discussed in earlier parts of this work, however, we extend the framework to include time variations in addition to the usual configuration variable variations, as in Lee (1983) and (1987).

For continuous Lagrangian dynamics, these extra variations, which produce conservation of energy, do not contribute any new information to the Euler–Lagrange equations. In the discrete setting, however, we see that we obtain an extra equation which exactly ensures preservation of a quantity we can identify as the discrete energy. In this way, both the definition of the discrete energy and the fact that it is preserved arise naturally from the variational principle.

This same approach is also used in variational multisymplectic mechanics, where a configuration is regarded as a section of a fibre bundle over spacetime and we distinguish between vertical (configuration) variations and horizontal (spacetime) variations. In that theory the nature of the dual state space and fully covariant momentum maps can be properly defined. The interested reader is referred to Gotay, Isenberg and Marsden (1997) and Marsden, Patrick and Shkoller (1998), as here we use only the limited subset of the formalism which is sufficient for our purposes. To keep the exposition as simple and direct as possible, we have had to compromise with the general theory just a little: for example, in discussing momentum maps, one should really use the *affine dual* rather than the *linear dual* as we do here. This distinction becomes crucial in the multisymplectic, or PDE context.

We also investigate the links between the discrete variational mechanics and Hamilton–Jacobi theory. In particular, we see that requiring the discrete and continuous energies to be equal is exactly the Hamilton–Jacobi PDE, which proves that the symplectic–energy–momentum variational integrators will only conserve the continuous energy exactly if they also exactly integrate the continuous flow.

This result is clearly consistent with the well-known theorem of Ge and Marsden (1988) and shows the advantage of identifying the correct discrete energy. While we cannot hope to conserve the continuous energy for an arbitrary non-integrable system, we can exactly conserve the corresponding discrete quantity.

This is similar to the situation with the Newmark integrator, discussed in Kane *et al.* (2000), where it was long thought that the integrator did not conserve the momenta, but where the problem was simply that the wrong discretization of the momenta was being used. Once the correct discretization was chosen, it was seen that the algorithm was actually momentum-conserving (and indeed symplectic).

4.2. Background: Extended Lagrangian mechanics

4.2.1. Basic definitions

Consider a *configuration manifold* Q and the *time space* $\mathbb{R}$. Define the *extended configuration manifold* to be $\bar{Q} = \mathbb{R} \times Q$ and the corresponding *extended state space* to be $\mathbb{R} \times TQ$. Take an *extended Lagrangian* $L : \mathbb{R} \times TQ \to \mathbb{R}$.

The *extended path space* is

$$\bar{C} = \{c : [a_0, a_f] \to \bar{Q} \mid c \text{ is a } C^2 \text{ curve and } c'_t(a) > 0\},$$

and we denote the two components of c by $c(a) = (c_t(a), c_q(a))$. Given a path $c(a)$, we define the *initial time* $t_0 = c_t(a_0)$ and *final time* $t_f = c_t(t_f)$ and we form the *associated curve* $q : [t_0, t_f] \to Q$ by

$$q(t) = c_q(c_t^{-1}(t)).$$

It is simple to check that two paths $c^1(a)$ and $c^2(a)$ have the same associated curve if and only if they are reparametrizations of each other: that is, if there exists a smooth monotone increasing isomorphism $h : [a_0, a_f] \to [a_0, a_f]$ such that $c^1 \circ h = c^2$.

Given an associated curve, it will be useful to define

$$\hat{q}(t) = \big(t, q(t), \dot{q}(t)\big) \in \mathbb{R} \times TQ,$$
$$\hat{\hat{q}}(t) = \big((t, q(t), \dot{q}(t)), (1, \dot{q}(t), \ddot{q}(t))\big) \in T(\mathbb{R} \times TQ).$$

Now define the *extended action map* $\bar{\mathfrak{G}} : \bar{C} \to \mathbb{R}$ to be

$$\bar{\mathfrak{G}}(c) = \int_{t_0}^{t_f} L(\hat{\hat{q}}) \, \mathrm{d}t, \qquad (4.2.1)$$

where $q(t)$ is the associated curve to $c(a)$ and t_0 and t_f are the initial and final times, respectively, for c. Differentiating $q(t) = c_q(c_t^{-1}(t))$ gives

$$\dot{q}(t) = \frac{c_q'(c_t^{-1}(t))}{c_t'(c_t^{-1}(t))}, \tag{4.2.2}$$

where $\dot{q}$ denotes the derivative with respect to t and c' denotes the derivative with respect to a. With this, $\bar{\mathfrak{G}}$ can be written as

$$\bar{\mathfrak{G}}(c) = \int_{a_0}^{a_f} L\left(c_t(a), c_q(a), \frac{c_q'(a)}{c_t'(a)}\right) c_t'(a)\, \mathrm{d}a. \tag{4.2.3}$$

The tangent space $T_c\bar{\mathcal{C}}$ to the extended path space $\bar{\mathcal{C}}$ at the path c is the set of all C^2 maps $\delta c : [a_0, a_f] \to T\bar{Q}$ such that $\pi_{\bar{Q}} \circ \delta c = c$, where $\pi_{\bar{Q}} : T\bar{Q} \to \bar{Q}$ is the canonical projection map.

Define the *extended second-order submanifold* of $T(\mathbb{R} \times TQ)$ to be

$$\ddot{Q} = \{ w \in T(\mathbb{R} \times TQ) \mid T\pi_Q(w) = \pi_{TQ}(w) \} \subset T(\mathbb{R} \times TQ), \tag{4.2.4}$$

where $\pi_Q : \mathbb{R} \times TQ \to Q$ and $\pi_{TQ} : T(\mathbb{R} \times TQ) \to TQ$ are the canonical projections. $\ddot{Q}$ is the set of points of the form $((t, q, q'), (t', q', q''))$, where the third and fifth entries are equal.

4.2.2. Variations of the action

Now that we have defined the spaces we are working on and have formed the action function, we are ready to derive the Euler–Lagrange equations of motion and prove a number of facts about solutions of these equations. The basic methodology here is variational, that is, we derive the equations and symplectic forms from taking variations of the action with respect to the path. To begin this process, we state the following fundamental theorem.

Theorem 4.2.1. Given a C^k extended Lagrangian $L : \mathbb{R} \times TQ \to \mathbb{R}$, $k \geq 2$, there exists a unique C^{k-2} mapping $\bar{D}_{\mathrm{EL}}L : \ddot{Q} \to T^*\bar{Q}$ and a unique C^{k-1} one-form $\bar{\Theta}_L$ on the extended state space $\mathbb{R} \times TQ$, such that for all variations $\delta c \in T_c\bar{\mathcal{C}}$ of $c \in \bar{\mathcal{C}}$ we have

$$\mathbf{d}\bar{\mathfrak{G}}(c) \cdot \delta c = \int_{t_0}^{t_f} \bar{D}_{\mathrm{EL}}L(\hat{q}) \cdot \delta c \, \mathrm{d}t + \bar{\Theta}_L(\hat{q}) \cdot \hat{\delta c}\Big|_{t_0}^{t_f}, \tag{4.2.5}$$

where

$$\hat{\delta c}(a) = \frac{\partial}{\partial \epsilon}\Big|_{\epsilon=0} \left(c_t^\epsilon(a), c_q^\epsilon(a), \frac{(c_q^\epsilon)'(a)}{(c_t^\epsilon)'(a)} \right),$$

and $c^\epsilon(a)$ is such that $\delta c(a) = \frac{\partial}{\partial \epsilon}\big|_{\epsilon=0} c^\epsilon(a)$ and $q(t)$ is the associated curve to $c(a)$.

The one-form $\bar{\Theta}_L$ is called the *extended Lagrangian one-form* and the mapping $\bar{D}_{\mathrm{EL}}L$ is called the *extended Euler–Lagrange map*. In coordinates they have the form

$$\bar{D}_{\mathrm{EL}}L(c'') = \left[\frac{\partial L}{\partial q^i} - \frac{\mathrm{d}}{\mathrm{d}t}\left(\frac{\partial L}{\partial \dot{q}^i}\right)\right]\mathrm{d}c_q^i + \left[\frac{\partial L}{\partial t} + \frac{\mathrm{d}}{\mathrm{d}t}\left(\frac{\partial L}{\partial \dot{q}^i}\dot{q}^i - L\right)\right]\mathrm{d}c_t,$$

$$(4.2.6)$$

$$\bar{\Theta}_L(c') = \left[\frac{\partial L}{\partial \dot{q}^i}\right]\mathrm{d}c_q^i - \left[\frac{\partial L}{\partial \dot{q}^i}\dot{q}^i - L\right]\mathrm{d}c_t \qquad (4.2.7)$$

where the various quantities are evaluated at either t or $a = c_t^{-1}(t)$ as appropriate.

Proof. Take the derivative of equation (4.2.3) in the direction δc to obtain

$$\mathrm{d}\bar{\mathfrak{G}}(c)\cdot\delta c = \int_{a_0}^{a_f}\left[\frac{\partial L}{\partial t}\delta c_t + \frac{\partial L}{\partial q}\delta c_q + \frac{\partial L}{\partial \dot{q}}\left(\frac{\delta c_q'}{c_t'} - \frac{c_q'\delta c_t'}{(c_t')^2}\right)\right]c_t'\,\mathrm{d}a$$

$$+ \int_{a_0}^{a_f} L\delta c_t'\,\mathrm{d}a$$

$$= \int_{a_0}^{a_f}\left[\frac{\partial L}{\partial q}c_t' - \frac{\mathrm{d}}{\mathrm{d}a}\frac{\partial L}{\partial \dot{q}}\right]\delta c_q\,\mathrm{d}a$$

$$+ \int_{a_0}^{a_f}\left[\frac{\partial L}{\partial t}c_t' + \frac{\mathrm{d}}{\mathrm{d}a}\left(\frac{\partial L}{\partial \dot{q}}\frac{c_q'}{c_t'} - L\right)\right]\delta c_t\,\mathrm{d}a$$

$$+ \left[\frac{\partial L}{\partial \dot{q}}\delta c_q\right]_{a_0}^{a_f} + \left[-\left(\frac{\partial L}{\partial \dot{q}}\frac{c_q'}{c_t'} - L\right)\delta c_t\right]_{a_0}^{a_f},$$

where integration by parts has been used on the $\delta c_q'$ and $\delta c_t'$ terms.

Now change coordinates so that the integrals are taken with respect to t rather than a. In doing this, use the facts that $\mathrm{d}t = c_t'(a)\,\mathrm{d}a$ and $\frac{\mathrm{d}}{\mathrm{d}a} = c_t'(a)\frac{\mathrm{d}}{\mathrm{d}t}$. This gives the desired expression (4.2.5) with the Euler–Lagrange derivative (4.2.6) and the Lagrangian one-form (4.2.7). $\qquad\square$

The fact that the extended Euler–Lagrange operator and extended Lagrangian one-form are functions only of the associated curve $q(t)$ and not of the full path $c(a)$ is a reflection of the fact that the extended action $\bar{\mathfrak{G}}$ is itself only a function of $q(t)$. This will be very important in what follows.

4.2.3. Euler–Lagrange equations

Hamilton's principle of critical action now seeks those paths $c \in \bar{C}$ which are critical points of the action. More precisely, define the *space of solutions* $\bar{C}_L \subset \bar{C}$ to be all those paths c that satisfy $\mathrm{d}\bar{\mathfrak{G}}(c)\cdot\delta c = 0$ for all variations $\delta c \in T_c\bar{C}$ which are zero at the boundary points a_0 and a_f.

Using equation (4.2.5) it is clear that c is a solution if and only if the extended Euler–Lagrange map (4.2.6) is zero on the associated curve to c for all $t \in (t_0, t_f)$. For any time t this statement is

$$\bar{D}_{\mathrm{EL}} L(\hat{\bar{q}}) = 0 \tag{4.2.8}$$

and is called the *extended Euler–Lagrange equations*.

The fact that the extended Euler–Lagrange map depends only on the associated curve, and not on the path c itself, means that any other paths which have the same associated curve will be solutions if and only if c is a solution. As remarked earlier, this is equivalent to the paths being reparametrizations of each other. We can thus group the solution space $\bar{C}_L$ into equivalence classes of paths, each of which correspond to a single associated curve that satisfies the extended Euler–Lagrange equations. We will use this identification later when deriving an extended Lagrangian vector field and flow map, and in proving the conservation properties of these structures.

Of course, for an arbitrary Lagrangian the extended Euler–Lagrange equations will not necessarily uniquely define the associated curve, either. For the remainder of this section, however, we will assume that L is chosen such that $q(t)$ is uniquely determined, and in Section 4.4.4 we will consider conditions under which this is true.

Considering the expression of the extended Euler–Lagrange equations in more detail, we can break it up into the two components to give

$$\frac{\partial L}{\partial q^i} - \frac{\mathrm{d}}{\mathrm{d}t}\left(\frac{\partial L}{\partial \dot{q}^i}\right) = 0, \tag{4.2.9a}$$

$$\frac{\partial L}{\partial t} + \frac{\mathrm{d}}{\mathrm{d}t}\left(\frac{\partial L}{\partial \dot{q}^i}\dot{q}^i - L\right) = 0. \tag{4.2.9b}$$

In fact, only the first component of the Euler–Lagrange equations (4.2.9) is necessary, as it implies the second. To see this, consider an associated curve q satisfying (4.2.9a) for all $t \in (t_0, t_f)$. We now compute the second component (4.2.9b) of the Euler–Lagrange equations to be

$$\frac{\partial L}{\partial t} + \frac{\mathrm{d}}{\mathrm{d}t}\left(\frac{\partial L}{\partial \dot{q}^i}\dot{q}^i - L\right) = \frac{\partial L}{\partial t} + \frac{\mathrm{d}}{\mathrm{d}t}\left(\frac{\partial L}{\partial \dot{q}^i}\right)\dot{q}^i + \frac{\partial L}{\partial \dot{q}^i}\ddot{q}^i - \frac{\mathrm{d}L}{\mathrm{d}t}$$

$$= \left[\frac{\partial L}{\partial t} + \frac{\partial L}{\partial q^i}\dot{q}^i + \frac{\partial L}{\partial \dot{q}^i}\ddot{q}^i\right] - \frac{\mathrm{d}L}{\mathrm{d}t}$$

$$= 0,$$

where we used (4.2.9a) to pass from the first to the second line. The space of solutions $\bar{C}_L$ is thus the space of paths whose associated curves satisfy (4.2.9a) for all t.

This redundancy in the equations is a reflection of the fact that they only depend on the associated curve, and so the equations cannot determine the map $c_t(a)$. There must therefore be some functional dependency in the system, as is explicitly shown above.

The second part of the extended Euler–Lagrange equation (4.2.9) is actually the statement of energy evolution for an extended Lagrangian system. The *energy* $E_L : \mathbb{R} \times TQ \to \mathbb{R}$ is defined to be

$$E_L(t, q, \dot{q}) = \frac{\partial L}{\partial \dot{q}}(t, q, \dot{q}) \cdot \dot{q} - L(t, q, \dot{q}) \tag{4.2.10}$$

and is constant along solutions if L does not explicitly depend on time. This is clear from the fact that the second part (4.2.9b) of the extended Euler–Lagrange equations can be written as

$$\frac{\partial L}{\partial t} + \frac{\mathrm{d} E_L}{\mathrm{d} t} = 0.$$

Another reason that only the first part (4.2.9a) of the extended Euler–Lagrange equations is needed is that it already has the energy evolution built into it, making the second part redundant. If it did not, then the second part would ensure energy conservation (for L time-independent) or the correct energy change (for L time-varying). This will be especially important when we consider discrete Lagrangian systems.

Note that the definition of the energy function (4.2.10) also allows us to write the extended Lagrangian one-form in the compact notation

$$\bar{\Theta}_L = \frac{\partial L}{\partial \dot{q}} \mathrm{d} q - E_L \mathrm{d} t, \tag{4.2.11}$$

where we use (q, t) to refer to the two components of c. This expression for $\bar{\Theta}_L$ will be useful when we consider the corresponding discrete object.

4.2.4. Lagrangian vector fields and flow maps

As we have already seen, the solution paths are only uniquely defined up to reparametrizations in a. Equivalently, this means that only the associated curve is uniquely defined by the extended Euler–Lagrange equations. For this reason, there is no unique vector field on $T\bar{Q}$ for which solution paths are integral curves, but there is a unique vector field on $\mathbb{R} \times TQ$ for which the associated curves are integral curves, assuming that the time evolution is fixed as the identity.

More precisely, define *extended Lagrangian vector field* $\bar{X}_L : \mathbb{R} \times TQ \to T(\mathbb{R} \times TQ)$ to be the unique second-order vector field satisfying

$$\bar{D}_{\mathrm{EL}} L \circ \bar{X}_L = 0,$$
$$T\pi_{\mathbb{R}} \circ \bar{X}_L(t, q, \dot{q}) = (t, 1),$$

where $\pi_{\mathbb{R}} : \mathbb{R} \times TQ \to \mathbb{R}$ is the projection.

The *extended Lagrangian flow* $\bar{F}_L : \mathbb{R} \times (\mathbb{R} \times TQ) \to \mathbb{R} \times TQ$ is the flow map of $\bar{X}_L$, and we write $\bar{F}_L^h : \mathbb{R} \times TQ \to \mathbb{R} \times TQ$ for some fixed time h.

4.2.5. *Conservation of the extended symplectic form*

Using the definition of $\bar{F}_L$, it is clear that setting $q(t) = \pi_Q \circ \bar{F}_L^{t-t_0}(t_0, q_0, \dot{q}_0)$ for some initial condition $(t_0, q_0, \dot{q}_0) \in \mathbb{R} \times TQ$ means that $q(t)$ will satisfy the extended Euler–Lagrange equations and will thus be an element of $\bar{C}_L$. Also, given any $q(t) \in \bar{C}_L$, it will satisfy $q(t) = \pi_Q \circ \bar{F}_L^{t-t_0}(t_0, q(t_0), \dot{q}(t_0))$.

Choosing an elapsed time $h \in \mathbb{R}$, we may thus define the *restricted extended action map* $\hat{\mathfrak{G}} : \mathbb{R} \times TQ \to \mathbb{R}$ to be

$$\hat{\mathfrak{G}}^h(t_0, q_0, \dot{q}_0) = \bar{\mathfrak{G}}(c),$$

where $c \in \mathcal{C}_L(Q)$ is any solution satisfying

$$\left(c_t(a_0), c_q(a_0), \frac{c_q'(a_0)}{c_t'(a_0)} \right) = (t_0, q_0, \dot{q}_0),$$
$$c_t(a_f) = t_0 + h.$$

We now wish to calculate the derivative of this expression. Begin by considering a variation derived from a family of initial conditions by

$$((t_0, q_0, \dot{q}_0), (\delta t_0, \delta q_0, \delta \dot{q}_0)) = \left. \frac{\partial}{\partial \epsilon} \right|_{\epsilon=0} (t_0^\epsilon, q_0^\epsilon, \dot{q}_0^\epsilon) \in T(\mathbb{R} \times TQ),$$

and let $c^\epsilon \in \bar{C}_L$ be a corresponding family of solutions satisfying

$$\left(c_t^\epsilon(a_0), c_q^\epsilon(a_0), \frac{(c_q^\epsilon)'(a_0)}{(c_t^\epsilon)'(a_0)} \right) = (t_0^\epsilon, q_0^\epsilon, \dot{q}_0^\epsilon),$$
$$c_t^\epsilon(a_f) = t_0^\epsilon + h.$$

Observe that we thus have

$$\bar{F}_L^{t-t_0^\epsilon}(t_0^\epsilon, q_0^\epsilon, \dot{q}_0^\epsilon) = \left(c_t^\epsilon(a), c_q^\epsilon(a), \frac{(c_q^\epsilon)'(a)}{(c_t^\epsilon)'(a)} \right), \qquad a = (c_t^\epsilon)^{-1}(t),$$

and so

$$\hat{\delta c}(a_0) = \left. \frac{\partial}{\partial \epsilon} \right|_{\epsilon=0} \bar{F}_L^0(t_0^\epsilon, q_0^\epsilon, \dot{q}_0^\epsilon) = (\delta t_0, \delta q_0, \delta \dot{q}_0),$$

$$\hat{\delta c}(a_f) = \left. \frac{\partial}{\partial \epsilon} \right|_{\epsilon=0} \bar{F}_L^h(t_0^\epsilon, q_0^\epsilon, \dot{q}_0^\epsilon) = T\bar{F}_L^h(t_0, q_0, \dot{q}_0) \cdot (\delta t_0, \delta q_0, \delta \dot{q}_0).$$

To calculate $\mathbf{d}\hat{\mathfrak{G}}$ we now use equation (4.2.5) for the derivative of the extended action together with the above expressions for the variations and the fact that the extended Euler–Lagrange map is zero on solutions to give

$$\mathbf{d}\bar{\mathfrak{G}}^h(t_0, q_0, \dot{q}_0) \cdot (\delta t_0, \delta q_0, \delta \dot{q}_0) = \bar{\Theta}_L\big(\bar{F}_L^h(t_0, q_0, \dot{q}_0)\big) \cdot T\bar{F}_L^h \cdot (\delta t_0, \delta q_0, \delta \dot{q}_0)$$
$$- \bar{\Theta}_L(t_0, q_0, \dot{q}_0) \cdot (\delta t_0, \delta q_0, \delta \dot{q}_0),$$

which implies that

$$\mathbf{d}\bar{\mathfrak{G}}^h = (\bar{F}_L^h)^* \bar{\Theta}_L - \bar{\Theta}_L. \tag{4.2.12}$$

Taking a further derivative of (4.2.12), and using the fact that $\mathbf{d}^2 = 0$, now gives conservation of the extended symplectic structure

$$(F_L^h)^* \Omega_L = \Omega_L,$$

where $\Omega_L = -\mathbf{d}\Theta_L$ is the *extended Lagrangian symplectic form* on $\mathbb{R} \times TQ$.

4.2.6. *Conservation of momentum maps*

Consider a Lie group G which acts on $\bar{Q}$ by the (left or right) action $\Phi^{\bar{Q}}$: $G \times \bar{Q} \to \bar{Q}$ in such a way that it covers an action $\Phi^t : G \times \mathbb{R} \to \mathbb{R}$, which is assumed to be monotone ($\partial_t \Phi^t(t) \neq 0$). In coordinates this is

$$\Phi_g^{\bar{Q}}(t, q) = \big(\Phi^t(t), \Phi^q(t, q)\big).$$

The tangent lift of this action to $\Phi^{T\bar{Q}} : G \times T\bar{Q} \to T\bar{Q}$ is defined by $\Phi_g^{T\bar{Q}} = T\Phi_g$, and the lift to the extended state space $\Phi^{\mathbb{R} \times T\bar{Q}} : G \times (\mathbb{R} \times T\bar{Q}) \to \mathbb{R} \times T\bar{Q}$ is defined by

$$\Phi_g^{\mathbb{R} \times TQ}(t, q, \dot{q}) = \left(\Phi_g^t(t), \Phi_g^q(t, q), \frac{\partial_t \Phi_g^q(t, q) + \partial_q \Phi_g^q(t, q) \cdot \dot{q}}{\partial_t \Phi_g^t(t)} \right).$$

This is defined so that, if a path $c(a)$ is transformed by the pointwise action of $\Phi_g^{T\bar{Q}}$, then the associated curve and its first derivative will transform pointwise with $\Phi_g^{\mathbb{R} \times TQ}$.

The *infinitesimal generators* corresponding to these actions are $\xi_{\bar{Q}} : \bar{Q} \to T(\bar{Q})$, $\xi_{T\bar{Q}} : T\bar{Q} \to T(T\bar{Q})$ and $\xi_{\mathbb{R} \times TQ} : \mathbb{R} \times TQ \to T(\mathbb{R} \times TQ)$, given by

$$\xi_{\bar{Q}}(t, q) = \frac{\mathrm{d}}{\mathrm{d}g}\big(\Phi_g^{\bar{Q}}(t, q)\big) \cdot \xi,$$

$$\xi_{T\bar{Q}}(t, q, \delta t, \delta q) = \frac{\mathrm{d}}{\mathrm{d}g}\big(\Phi_g^{T\bar{Q}}(t, q, \delta t, \delta q)\big) \cdot \xi,$$

$$\xi_{\mathbb{R} \times TQ}(t, q, \dot{q}) = \frac{\mathrm{d}}{\mathrm{d}g}\big(\Phi_g^{\mathbb{R} \times TQ}(t, q, \dot{q})\big) \cdot \xi.$$

The *extended Lagrangian momentum map* $\bar{J}_L : \mathbb{R} \times TQ \to \mathfrak{g}^*$ is now defined by

$$\bar{J}_L(t, q, \dot{q}) \cdot \xi = \bar{\Theta}_L \cdot \xi_{\mathbb{R} \times TQ}(t, q, \dot{q}).$$

Denoting the two components of $\xi_{\bar{Q}}$ by $\xi_{\bar{Q}}(t,q) = (\xi^t_{\bar{Q}}(t,q), \xi^q_{\bar{Q}}(t,q))$, it can be checked that an equivalent expression for $\bar{J}_L$ is

$$\bar{J}_L(t,q,\dot{q}) \cdot \xi = \left\langle \frac{\partial L}{\partial \dot{q}}, \xi^q_{\bar{Q}}(t,q) \right\rangle - E_L(t,q,\dot{q})\xi^t_{\bar{Q}}(t,q),$$

which is useful in applications as it does not involve the lifted action.

If $\Phi^{\bar{Q}}$ acts by extended special symplectic maps, so that $(\Phi^{\bar{Q}}_g)^*\bar{\Theta}_L = \bar{\Theta}_L$, then it can be proved that the lifted action on the extended state space is equivariant, so that $\bar{J}_L \circ \Phi^{\mathbb{R} \times TQ}_g = \mathrm{Ad}^*_{g^{-1}} \circ \bar{J}_L$. The proof of this is essentially identical to that of Theorem 1.2.2.

In the autonomous setting, invariance of the Lagrangian implies invariance of the action. As the action is an integral with respect to time, however, and extended group actions can reparametrize time, this is no longer sufficient. Instead, we say that the group action is a *symmetry* if the one-form Ldt is preserved, so that

$$(\Phi^{\mathbb{R} \times TQ}_g)^*(Ldt) = Ldt.$$

In fact, this same condition ensures that the group action is by extended special symplectic maps, and hence the momentum maps are equivariant. To see this, we differentiate the above expression with respect to $\dot{q}$ in the direction δq to obtain

$$\frac{\partial L}{\partial \dot{q}} \circ \Phi^{\mathbb{R} \times TQ}_g \cdot \partial_q \Phi^q_g \cdot \delta q = \frac{\partial L}{\partial \dot{q}} \cdot \delta q,$$

and substituting this into the expression for $(\Phi^{\mathbb{R} \times TQ}_g)^*(\bar{\Theta}_L)(t,q,\dot{q}) \cdot (\delta t, \delta q, \delta \dot{q})$ we can rearrange to obtain $\bar{\Theta}_L \cdot (\delta t, \delta q, \delta \dot{q})$.

Note that in the language of multisymplectic mechanics we have required that the Lagrangian density be invariant, and the extended Lagrangian momentum map defined above is the fully covariant Lagrangian momentum map. See Gotay *et al.* (1997) for more details.

Theorem 4.2.2. (Extended Noether's theorem)   Consider an extended Lagrangian system $L : \mathbb{R} \times TQ \to \mathbb{R}$ with a (left or right) symmetry action $\Phi : G \times \bar{Q} \to \bar{Q}$. Then the corresponding (left or right) extended Lagrangian momentum map $\bar{J}_L : \mathbb{R} \times TQ \to \mathfrak{g}^*$ is a conserved quantity of the flow, so that $\bar{J}_L \circ \bar{F}^h_L = \bar{J}_L$ for all elapsed times h.

Proof.   First define the lifted action on paths $\Phi^{\bar{C}} : G \times \bar{C} \to \bar{C}$ by $\Phi^{\bar{C}}_g(c)(a) = \Phi^{T\bar{Q}}_g(c(a))$. If two paths c_1 and c_2 have the same associated curve q, then the transformed paths $\bar{c}_1 = \Phi^{\bar{C}}_g(c_1)$ and $\bar{c}_2 = \Phi^{\bar{C}}_g(c_2)$ will also map to a single associated curve $\bar{q}$. We can thus define the lifted action of G on the set of associated curves by

$$\Phi^{\bar{C}}_g(q) = \Phi^{\mathbb{R} \times TQ}_g \circ q \circ (\Phi^t_g)^{-1},$$

so that if q is the associated curve to c, then $\Phi_g^{\bar{C}}(q)$ is the associated curve to $\Phi_g^{\bar{C}}(c)$. The infinitesimal generators corresponding to the action on the space of paths is $\xi_{\bar{C}} : \bar{C} \to T\bar{C}$ defined by

$$\xi_{\bar{C}}(c) = \frac{\mathrm{d}}{\mathrm{d}g}\left(\Phi_g^{\bar{C}}(c)\right) \cdot \xi.$$

If the one-form Ldt is invariant, then taking an integral of this along an associated curve shows that the action $\bar{\mathfrak{G}} : \bar{C} \to \mathbb{R}$ is also invariant, which is $\bar{\mathfrak{G}} \circ \Phi_g^{\bar{C}} = \bar{\mathfrak{G}}$.

We thus have that solution curves map to solution curves and so $\xi_{\bar{C}}(c) \in T_c\bar{C}$. Restricting the infinitesimal invariance of the action to the space of solutions gives

$$0 = \mathbf{d}\bar{\mathfrak{G}}(c) \cdot \xi_{\bar{C}}(c) = \left.\bar{\Theta}_L(\hat{\dot{q}}) \cdot \hat{\delta c}\right|_{t_0}^{t_f},$$

where

$$\hat{\delta c}(a) = \frac{\mathrm{d}}{\mathrm{d}g}\left(\Phi_g^t(c_t(a)), \Phi_g^q(c_t(a), c_q(a)), \frac{\partial_a \Phi_g^q(c_t(a), c_q(a))}{\partial_a \Phi_g^t(c_t(a))}\right) \cdot \xi$$

$$= \frac{\mathrm{d}}{\mathrm{d}g}\Phi_g^{\mathbb{R} \times TQ}(\hat{\dot{q}}(t)) \cdot \xi \quad \text{where } t = c_t(a)$$

$$= \xi_{\mathbb{R} \times TQ}(\hat{\dot{q}}(t)).$$

Recall now that, for an initial condition $(t_0, q_0, \dot{q}_0)$ with corresponding solution $q(t)$, we have $\hat{\dot{q}}(t_0) = (t_0, q_0, \dot{q}_0)$ and $\hat{\dot{q}}(t_f) = \bar{F}_L^h(t_0, q_0, \dot{q}_0)$. Combining this with the above expressions now gives

$$0 = \bar{\Theta}_L\left(\bar{F}_L^h(t_0, q_0, \dot{q}_0)\right) \cdot \xi_{\mathbb{R} \times TQ}\left(\bar{F}_L^h(t_0, q_0, \dot{q}_0)\right)$$

$$- \bar{\Theta}_L(t_0, q_0, \dot{q}_0) \cdot \xi_{\mathbb{R} \times TQ}(t_0, q_0, \dot{q}_0)$$

which is exactly the statement $\bar{J}_L \circ \bar{F}_L^h = \bar{J}_L$. $\square$

An interesting example of an extended momentum map arises for the time translation action of $\mathbb{R}$ on $\bar{Q}$. The corresponding momentum map is $\bar{J}_L(t, q, \dot{q}) = -E_L(t, q, \dot{q})\mathrm{d}\xi$ and if the Lagrangian is time-independent then Noether's theorem recovers the statement of energy conservation.

4.3. Discrete variational mechanics: Lagrangian viewpoint

We now turn to constructing a discrete time-dependent Lagrangian mechanics. As in the continuous case, the basic idea is to use a variational principle to derive the equations of motion, which then automatically guarantees conservation of discrete symplectic forms and discrete momentum maps.

The difference from standard discrete variational mechanics is that we work here with time-varying systems and take variations with respect to time, and not only with respect to the configuration variables. This gives us two components to the discrete Euler–Lagrange equations, the second of which ensures the preservation of a discrete energy. This is different to the continuous case, where the first component of the Euler–Lagrange equations implies the second. In the discrete setting, both are independent and necessary.

4.3.1. Basic definitions

Consider once again a configuration manifold Q, time space $\mathbb{R}$ and extended configuration manifold $\bar{Q} = \mathbb{R} \times Q$. Define the *extended discrete Lagrangian state space* to be $\bar{Q} \times \bar{Q}$ and take an *extended discrete Lagrangian* $L_d : \bar{Q} \times \bar{Q} \to \mathbb{R}$.

The *extended discrete path space* is

$$\bar{\mathcal{C}}_d = \{c : \{0, \ldots, N\} \to \bar{Q} \mid c_t(k+1) > c_t(k) \text{ for all } k\},$$

and as before we denote the two components of c by $c(k) = (c_t(k), c_q(k))$. Given a discrete path c, we form the *associated discrete curve*

$$q : \{c_t(0), \ldots, c_t(N)\} \to Q$$

by

$$q(c_t(k)) = c_q(k),$$

and we denote the images of the time component of c by $t_k = c_t(k)$ and the images of the configuration component by $q_k = q(c_t(k)) = c_q(k)$. An extended discrete path is thus a sequence $\{(t_k, q_k)\}_{k=0}^N$ of points $(t_k, q_k) \in \bar{Q}$.

Observe that, unlike the autonomous situation, the extended discrete Lagrangian state space $\bar{Q} \times \bar{Q} = \mathbb{R} \times Q \times \mathbb{R} \times Q$ is not locally isomorphic to the extended Lagrangian state space $\mathbb{R} \times TQ$, as it has one extra dimension. Regarding a discrete Lagrangian system as an integrator, this is a reflection of the fact that nonautonomous discrete Lagrangian systems require adaptive time-stepping, which implies that it is necessary to keep track of both the current time and the current time-step.

In the case when the system being modelled is indeed autonomous, one could take the discrete system to be $Q \times \mathbb{R} \times Q$, with each element (q_0, h, q_1) consisting of two points q_0 and q_1 and a time-step h. As we will see later, however, this is not essentially different from the more general case of (t_0, q_0, t_1, q_1), and so we will only consider this more general formulation.

The *extended discrete action sum* $\bar{\mathfrak{G}}_d : \mathcal{C}_d \to \mathbb{R}$ is given by

$$\bar{\mathfrak{G}}_d(c) = \sum_{k=0}^{N-1} L_d(c(k), c(k+1)), \tag{4.3.1a}$$

$$\bar{\mathfrak{G}}_d(c) = \sum_{k=0}^{N-1} L_d(t_k, q_k, t_{k+1}, q_{k+1}). \tag{4.3.1b}$$

Equation (4.3.1a) is the discrete equivalent of (4.2.3) and equation (4.3.1b) is the equivalent of (4.2.1).

The tangent space $T_c\bar{C}_d$ to the discrete path space $\bar{C}_d$ at the path c is the set of all maps $\delta c : \{0, \ldots, N\} \to T\bar{Q}$ such that $\pi_{\bar{Q}} \circ \delta c = c$.

Consider the space $(\bar{Q} \times \bar{Q}) \times (\bar{Q} \times \bar{Q})$ and define the *extended projection operator* π and the *extended translation operator* σ by

$$\pi : ((c_0, c_1), (c_0', c_1')) \mapsto (c_0, c_1),$$
$$\sigma : ((c_0, c_1), (c_0', c_1')) \mapsto (c_0', c_1').$$

We define the *extended discrete second-order submanifold* of $(\bar{Q}\times\bar{Q})\times(\bar{Q}\times\bar{Q})$ to be

$$\ddot{\bar{Q}}_d = \{w_d \in (\bar{Q} \times \bar{Q}) \times (\bar{Q} \times \bar{Q}) \mid \pi_1 \circ \sigma(w_d) = \pi_2 \circ \pi(w_d)\},$$

where $\pi_1, \pi_2 : \bar{Q} \times \bar{Q} \to \bar{Q}$ are the projections onto the first and second components, respectively. $\ddot{\bar{Q}}_d$ is thus the set of points of the form $((t_0, q_0, t_1, q_1), (t_1, q_1, t_2, q_2))$, with the second and third values in $\bar{Q}$ being equal.

In contrast to the autonomous case, the extended discrete second-order submanifold $\ddot{\bar{Q}}_d$ is not isomorphic to the extended continuous second-order submanifold $\ddot{\bar{Q}}$. This will be reflected in the choice of initial conditions for a solution curve, as we will see below.

4.3.2. Variations of the discrete action

As in the continuous case, the discrete equations of motion and conservation laws are derived from a stationarity principle: that is, we take variations of the discrete action sum with respect to the discrete path, as stated in the following theorem.

Theorem 4.3.1. Given a C^k extended discrete Lagrangian $L_d : \bar{Q} \times \bar{Q} \to \mathbb{R}$, $k \geq 1$, there exists a unique C^{k-1} mapping $\bar{D}_{\text{DEL}}L_d : \ddot{\bar{Q}}_d \to T^*\bar{Q}$ and unique C^{k-1} one-forms $\bar{\Theta}_{L_d}^-$ and $\bar{\Theta}_{L_d}^+$ on the discrete Lagrangian state space $\bar{Q} \times \bar{Q}$, such that, for all variations $\delta c \in T_c\bar{C}_d$ of $c \in \bar{C}_d$, we have

$$\mathbf{d}\bar{\mathfrak{G}}_d(c) \cdot \delta c = \sum_{k=1}^{N-1} \bar{D}_{\text{DEL}}L_d(t_{k-1}, q_{k-1}, t_k, q_k, t_{k+1}, q_{k+1}) \cdot (\delta q_k, \delta t_k)$$
$$+ \bar{\Theta}_{L_d}^+(t_{N-1}, q_{N-1}, t_N, q_N) \cdot (\delta t_{N-1}, \delta q_{N-1}, \delta t_N, \delta q_N)$$
$$- \bar{\Theta}_{L_d}^-(t_0, q_0, t_1, q_1) \cdot (\delta t_0, \delta q_0, \delta t_1, \delta q_1), \qquad (4.3.2)$$

where we write $t_k = c_t(k)$ and $q_k = c_q(k)$ and similarly for the variations. The map $\bar{D}_{\text{DEL}}L_d$ is called the *extended discrete Euler–Lagrange map* and the one-forms $\bar{\Theta}_{L_d}^+$ and $\bar{\Theta}_{L_d}^-$ are the *extended discrete Lagrangian one-forms*.

In coordinates these have the expressions

$$\bar{D}_{\text{DEL}}L_d(t_{k-1}, q_{k-1}, t_k, q_k, t_{k+1}, q_{k+1}) \tag{4.3.3}$$
$$= \big[D_4 L_d(t_{k-1}, q_{k-1}, t_k, q_k) + D_2 L_d(t_k, q_k, t_{k+1}, q_{k+1})\big] dq$$
$$+ \big[D_3 L_d(t_{k-1}, q_{k-1}, t_k, q_k) + D_1 L_d(t_k, q_k, t_{k+1}, q_{k+1})\big] dt,$$

$$\bar{\Theta}^+_{L_d}(t_k, q_k, t_{k+1}, q_{k+1}) \tag{4.3.4}$$
$$= D_4 L_d(t_k, q_k, t_{k+1}, q_{k+1}) dq_{k+1} + D_3 L_d(t_k, q_k, t_{k+1}, q_{k+1}) dt_{k+1},$$

$$\bar{\Theta}^-_{L_d}(t_k, q_k, t_{k+1}, q_{k+1}) \tag{4.3.5}$$
$$= -D_2 L_d(t_k, q_k, t_{k+1}, q_{k+1}) dq_k - D_1 L_d(t_k, q_k, t_{k+1}, q_{k+1}) dt_k.$$

Proof. Taking the derivative of (4.3.1a) at c in the direction δc yields

$$\mathbf{d}\bar{\mathfrak{G}}_d(c) \cdot \delta c = \sum_{k=0}^{N-1} \big[D_1 L_d(c(k), c(k+1)) \cdot \delta c(k)$$
$$+ D_2 L_d(c(k), c(k+1)) \cdot \delta c(k+1) \big]$$
$$= \sum_{k=1}^{N-1} \big[D_2 L_d(c(k-1), c(k)) + D_1 L_d(c(k), c(k+1)) \big] \cdot \delta c(k)$$
$$+ D_2 L_d(c(N-1), c(N)) \cdot \delta c(N)$$
$$+ D_1 L_d(c(0), c(1)) \cdot \delta c(0),$$

where we have rearranged and grouped terms to isolate internal and boundary terms, as integration by parts achieved in the continuous setting.

Now we change notations to $(t_k, q_k) = (c_t(k), c_q(k)) = c(k)$, and similarly for the variations, and we obtain the desired expressions. □

4.3.3. Discrete Euler–Lagrange equations

We now apply the discrete Hamilton's principle, and seek discrete paths $c \in \bar{C}_d$ which are critical points of the discrete action. That is, we define the *discrete space of solutions* $\bar{C}_{L_d} \subset \bar{C}_d$ to be all those paths which satisfy $\mathbf{d}\bar{\mathfrak{G}}_d(c) \cdot \delta c = 0$ for all variations $\delta c \in T_c \bar{C}_d$ which are zero at the boundary points 0 and N.

As in the continuous case, it is clear from (4.3.2) that c is a solution if and only if the discrete Euler–Lagrange derivative is zero at all points other than the endpoints 0 and N. This statement at k reads

$$\bar{D}_{\text{DEL}} L_d(t_{k-1}, q_{k-1}, t_k, q_k, t_{k+1}, q_{k+1}) = 0 \tag{4.3.6}$$

and is known as the *extended discrete Euler–Lagrange equations*. It can be separated into the configuration and time components to give

$$D_4 L_d(t_{k-1}, q_{k-1}, t_k, q_k) + D_2 L_d(t_k, q_k, t_{k+1}, q_{k+1}) = 0, \tag{4.3.7a}$$
$$D_3 L_d(t_{k-1}, q_{k-1}, t_k, q_k) + D_1 L_d(t_k, q_k, t_{k+1}, q_{k+1}) = 0. \tag{4.3.7b}$$

Unlike the continuous case, however, paths which satisfy the first component (4.3.7a) of the Euler–Lagrange equations do not automatically satisfy
the second component (4.3.7b) as well: that is, both components contribute
restrictions on the space of solutions, and both are necessary. This is because the extended discrete action is a function of the entire extended path,
not just of some discrete associated curve.

The interpretation of the discrete Euler–Lagrange equations and the discrete Lagrangian one-forms can be aided by defining the *discrete energies*
to be

$$E_{L_d}^+(t_k, q_k, t_{k+1}, q_{k+1}) = -D_3 L_d(t_k, q_k, t_{k+1}, q_{k+1}),$$
$$E_{L_d}^-(t_k, q_k, t_{k+1}, q_{k+1}) = D_1 L_d(t_k, q_k, t_{k+1}, q_{k+1}).$$
(4.3.8)

With these definitions we see that the second component of the discrete
Euler–Lagrange equations (4.3.7b) is simply

$$E_{L_d}^+(t_{k-1}, q_{k-1}, t_k, q_k) = E_{L_d}^-(t_k, q_k, t_{k+1}, q_{k+1}),$$
(4.3.9)

and reflects the evolution of the discrete energy.

If the discrete Lagrangian is time-invariant, so that $L_d(t_k + \tau, q_k, t_{k+1} + \tau, q_{k+1}) = L_d(t_k, q_k, t_{k+1}, q_{k+1})$ for all $\tau \in \mathbb{R}$, then taking derivatives with
respect to τ shows that $D_1 L_d + D_3 L_d = 0$, and thus

$$E_{L_d}^-(t_k, q_k, t_{k+1}, q_{k+1}) = E_{L_d}^+(t_k, q_k, t_{k+1}, q_{k+1}).$$
(4.3.10)

Combining this with the time component of the Euler–Lagrange equations
(4.3.7b) written as (4.3.9) then yields

$$E_{L_d}^+(t_k, q_k, t_{k+1}, q_{k+1}) = E_{L_d}^+(t_{k-1}, q_{k-1}, t_k, q_k),$$
$$E_{L_d}^-(t_k, q_k, t_{k+1}, q_{k+1}) = E_{L_d}^-(t_{k-1}, q_{k-1}, t_k, q_k),$$

which shows that time invariance of the discrete Lagrangian leads to conservation of the discrete energies. This is a special case of the extended
discrete Noether's theorem, as we will see below. If the Lagrangian is not
time-invariant, then the equivalent of equation (4.3.10) will indicate how the
discrete energies evolve.

4.3.4. Extended discrete Lagrangian evolution operator and mappings

The *extended discrete Lagrangian evolution operator* $\bar{X}_{L_d}$ is the second-order
discrete evolution operator defined by

$$\bar{D}_{\mathrm{DEL}} L_d \circ \bar{X}_{L_d} = 0.$$

If this expression uniquely determines $\bar{X}_{L_d}$ then we will say that the extended discrete Lagrangian is *discretely well-posed*. In this case, the *extended
discrete Lagrangian map* $\bar{F}_{L_d} : \bar{Q} \times \bar{Q} \to \bar{Q} \times \bar{Q}$ is specified by $\bar{F}_{L_d} = \sigma \circ \bar{X}_{L_d}$.

For the rest of this section we will assume that we are working with an extended discrete Lagrangian which is discretely well-posed. Note that this is equivalent to the extended discrete Euler–Lagrange equations having unique solutions for t_2 and q_2 in terms of the other variables. Later we will investigate under what conditions L_d is in fact discretely well-posed.

Note that here, and in the material which follows, we are implicitly making use of the restriction on the extended discrete path space that $t_{k+1} > t_k$ for each k. This rules out spurious solutions which do not uniformly increase in time.

An extended discrete path $c_d \in \bar{C}_d$ is a solution of the extended discrete Euler–Lagrange equations if the points $\{(t_k, q_k)\}_{k=0}^N$ satisfy

$$\bar{F}_{L_d}(t_{k-1}, q_{k-1}, t_k, q_k) = (t_k, q_k, t_{k+1}, q_{k+1})$$

for each $k = 1, \dots, N-1$. That is, the extended discrete Lagrangian map defines a trajectory forward in time.

4.3.5. Extended discrete Lagrangian maps are symplectic

The extended discrete space of solutions $\bar{C}_{L_d}$ defined above can, by means of the extended discrete Lagrangian map, be identified with the space of initial conditions $\bar{Q} \times \bar{Q}$. Restricting the extended discrete action to this space gives the *restricted extended discrete action* $\hat{\mathfrak{G}}_d : \bar{Q} \times \bar{Q} \to \mathbb{R}$ defined by

$$\hat{\mathfrak{G}}_d(t_0, q_0, t_1, q_1) = \bar{\mathfrak{G}}_d(c_d),$$

where $c_d \in \bar{C}_{L_d}$ is the solution of the extended discrete Euler–Lagrange equations satisfying $c(0) = (t_0, q_0)$ and $c(1) = (t_1, q_1)$. We now use (4.3.2) to calculate

$$\mathbf{d}\hat{\mathfrak{G}}_d = (\bar{F}_{L_d}^{N-1})^*(\Theta_{L_d}^+) - \Theta_{L_d}^-,$$

and so taking another derivative and using the fact that $\mathbf{d}^2 = 0$ gives the conservation law

$$(\bar{F}_{L_d}^{N-1})^*(\Omega_{L_d}) = \Omega_{L_d},$$

where the *extended discrete Lagrangian two-form* Ω_{L_d} is defined by

$$\Omega_{L_d} = -\mathbf{d}\Theta_{L_d}^+ = -\mathbf{d}\Theta_{L_d}^-.$$

Taking any subinterval of $0, \dots, N$ and using the same argument gives the same conservation law for any number of steps of $\bar{F}_{L_d}$.

4.3.6. Extended discrete Noether's theorem

Take a Lie group G with a (left or right) action $\Phi^{\bar{Q}}$ on $\bar{Q}$, and consider the lift to the extended discrete state space $\bar{Q} \times \bar{Q}$ defined by the product action

$\Phi_g^{\bar{Q}\times\bar{Q}}(t_0, q_0, t_1, q_1) = (\Phi_g^{\bar{Q}}(t_0, q_0), \Phi_g^{\bar{Q}}(t_0, q_0))$. The corresponding *infinitesimal generator* is then $\xi_{\bar{Q}\times\bar{Q}} : \bar{Q} \times \bar{Q} \to T(\bar{Q} \times \bar{Q})$, defined by

$$\xi_{\bar{Q}\times\bar{Q}}(t_0, q_0, t_1, q_1) = (\xi_{\bar{Q}}(t_0, q_0), \xi_{\bar{Q}}(t_1, q_1))$$

and the *extended discrete Lagrangian momentum maps* $\bar{J}_{L_d}^+, \bar{J}_{L_d}^- : \bar{Q} \times \bar{Q} \to \mathfrak{g}^*$ are

$$\bar{J}_{L_d}^+(t_0, q_0, t_1, q_1) \cdot \xi = \bar{\Theta}_{L_d}^+ \cdot \xi_{\bar{Q}\times\bar{Q}}(t_0, q_0, t_1, q_1),$$
$$\bar{J}_{L_d}^-(t_0, q_0, t_1, q_1) \cdot \xi = \bar{\Theta}_{L_d}^- \cdot \xi_{\bar{Q}\times\bar{Q}}(t_0, q_0, t_1, q_1).$$

Equivalent expressions for these are

$$\bar{J}_{L_d}^+(t_0, q_0, t_1, q_1) \cdot \xi$$
$$= \langle D_4 L_d(t_0, q_0, t_1, q_1), \xi_{\bar{Q}}^q(t_1, q_1) \rangle + D_3 L_d(t_0, q_0, t_1, q_1)\xi_{\bar{Q}}^t(t_1),$$
$$\bar{J}_{L_d}^-(t_0, q_0, t_1, q_1) \cdot \xi$$
$$= \langle -D_2 L_d(t_0, q_0, t_1, q_1), \xi_{\bar{Q}}^q(t_0, q_0) \rangle - D_1 L_d(t_0, q_0, t_1, q_1)\xi_{\bar{Q}}^t(t_0).$$

If the lifted action on $\bar{Q} \times \bar{Q}$ acts by extended special discrete symplectic maps, which requires $(\Phi_g^{\bar{Q}\times\bar{Q}})^*(\bar{\Theta}_{L_d}^+) = \bar{\Theta}_{L_d}^+$ and $(\Phi_g^{\bar{Q}\times\bar{Q}})^*(\bar{\Theta}_{L_d}^-) = \bar{\Theta}_{L_d}^-$, then the extended discrete Lagrangian momentum maps will be equivariant: that is,

$$\bar{J}_{L_d}^+ \circ \Phi_g^{\bar{Q}\times\bar{Q}} = \mathrm{Ad}_{g^{-1}}^* \circ \bar{J}_{L_d}^+,$$
$$\bar{J}_{L_d}^- \circ \Phi_g^{\bar{Q}\times\bar{Q}} = \mathrm{Ad}_{g^{-1}}^* \circ \bar{J}_{L_d}^-.$$

This can be proved in the same way as Theorem 1.2.2.

An extended discrete Lagrangian $L_d : \bar{Q} \times \bar{Q} \to \mathbb{R}$ is said to be *invariant* under the lifted group action if $L_d(g \cdot (t_0, q_0), g \cdot (t_1, q_1)) = L_d(t_0, q_0, t_1, q_1)$ for all g and all t_0, q_0, t_1 and q_1, and then G is said to be *symmetry* of L_d. If a discrete Lagrangian is invariant then it is necessarily *infinitesimally invariant*, which is the requirement that $\mathbf{d}L_d \cdot \xi_{\bar{Q}\times\bar{Q}} = 0$ for all $\xi \in \mathfrak{g}$.

In the case that L_d is infinitesimally invariant, the fact that $\mathbf{d}L_d = \bar{\Theta}_{L_d}^+ - \bar{\Theta}_{L_d}^-$ implies that the two extended discrete momentum maps are equal, and we will denote the single *extended discrete Lagrangian momentum map* by $\bar{J}_{L_d} : \bar{Q} \times \bar{Q} \to \mathfrak{g}^*$.

As the discrete Lagrangian plays an analogous role to the continuous action, it is not surprising that for the extended continuous Noether's theorem we had to require invariance of the one-form $L\mathbf{d}t$, which gives invariance of the action, while for the discrete version action invariance is directly implied by invariance of the extended discrete Lagrangian itself.

If the extended discrete Lagrangian is invariant, then the lifted group action acts by extended special discrete symplectic maps. This can be seen in a similar way to the corresponding autonomous statement, by differentiating

$L_d \circ \Phi_g^{\bar{Q} \times \bar{Q}}(t_0, q_0, t_1, q_1) = L_d(t_0, q_0, t_1, q_1)$ with respect to t_1 and q_1 and using the resulting identities to show $(\Phi_g^{\bar{Q} \times \bar{Q}})^* (\bar{\Theta}_{L_d}^+) = \bar{\Theta}_{L_d}^+$, and similarly for $\bar{\Theta}_{L_d}^+$.

Theorem 4.3.2. (Extended discrete Noether's theorem)   Consider an extended discrete Lagrangian system $L_d : \bar{Q} \times \bar{Q} \to \mathbb{R}$ which is invariant under the lift of the (left or right) group action $\Phi^{\bar{Q}} : G \times \bar{Q} \to \bar{Q}$. Then the corresponding extended discrete Lagrangian momentum map $\bar{J}_{L_d} : \bar{Q} \times \bar{Q} \to \mathfrak{g}^*$ is a conserved quantity of the extended discrete Lagrangian map $\bar{F}_{L_d} : \bar{Q} \times \bar{Q} \to \bar{Q} \times \bar{Q}$, in the sense that $\bar{J}_{L_d} \circ \bar{F}_{L_d} = \bar{J}_{L_d}$.

Proof.   The extended discrete Noether's theorem can be proved in exactly the same way as the standard discrete Noether's theorem (Theorem 1.3.3): that is, we compute

$$0 = \mathbf{d}\bar{\mathfrak{G}}_d(t_0, q_0, t_1, q_1) \cdot \xi_{\bar{Q} \times \bar{Q}}(t_0, q_0, t_1, q_1)$$
$$= ((\bar{F}_{L_d}^{N-1})^* (\bar{\Theta}_{L_d}^+) - \bar{\Theta}_{L_d}^-)(t_0, q_0, t_1, q_1) \cdot \xi_{\bar{Q} \times \bar{Q}}(t_0, q_0, t_1, q_1),$$

which gives the desired expression, using the fact that we may take N to be arbitrary.   □

As in earlier Noether's theorems, we have required more than is actually necessary for the proof. In particular, the result will still hold if L_d is only infinitesimally invariant.

Example 4.3.3.   If we have a discrete Lagrangian which is autonomous, in the sense that it is invariant with respect to the additive action of $\mathbb{R}$ on the time component of $\bar{Q}$, then the extended discrete Noether's theorem recovers the statement of discrete energy conservation.

To see this, consider the group action $g \cdot (t, q) = (t + g, q)$ for $g \in \mathbb{R}$, which has infinitesimal generator $\xi_{\bar{Q}}(t, q) = ((t, q), (\xi, 0))$ for $\xi \in \mathfrak{g}^* \cong \mathbb{R}$. Using this, we compute the extended discrete Lagrangian momentum map to be $\bar{J}_{L_d} = -E_{L_d}^+ dt_1$, or equivalently $\bar{J}_{L_d} = -E_{L_d}^- dt_0$. Noether's theorem thus gives $E_{L_d}^+(k, k+1) = E_{L_d}^+(k-1, k)$, or $E_{L_d}^-(k, k+1) = E_{L_d}^-(k-1, k)$, which are the statements of discrete energy conservation. This is directly analogous to the continuous case, where energy conservation is also recovered by Noether's theorem in the case when L is autonomous.   ◇

A difference from the continuous case arises because the proof of the extended discrete Noether's theorem relied upon the fact that both components of the extended discrete Euler–Lagrange equations (4.3.7) are satisfied at intermediate (non-boundary) points, unlike the continuous case where only the first component was required to be satisfied. Weakening this requirement to only insist on (4.3.7a) being satisfied means that full extended discrete Noether's theorem will only be true in the case that the symmetry action is autonomous and time-preserving, which shows that the discrete energy would not then be preserved.

4.3.7. Autonomous discrete Lagrangians

The previous example is also interesting because this is the case in which one could formulate a discrete theory using a state space of $Q \times \mathbb{R} \times Q$ with elements (q_0, h, q_1) consisting of two points q_0 and q_1 and a time-step h. In such a theory there would be only a single discrete energy E_{L_d} rather than $E_{L_d}^+$ and $E_{L_d}^-$. Given such a discrete Lagrangian $L_d'(q_0, h, q_1)$, however, we can also simply define an extended discrete Lagrangian by

$$L_d(t_0, q_0, t_1, q_1) = L_d'(q_0, t_1 - t_0, q_1)$$

and proceed to use the extended discrete Lagrangian theory derived in this section. Such an L_d will automatically have the time translation symmetry, and so the statement of infinitesimal invariance $\mathbf{d}L_d \cdot \xi_{\bar{Q} \times \bar{Q}} = 0$ can be seen to be just

$$
\begin{aligned}
0 &= \mathbf{d}L_d(t_0, q_0, t_1, q_1) \cdot (\xi, 0, \xi, 0) \\
&= D_1 L_d(t_0, q_0, t_1, q_1) \cdot \xi + D_3 L_d(t_0, q_0, t_1, q_1) \cdot \xi \\
&= \left(E_{L_d}^-(t_0, q_0, t_1, q_1) - E_{L_d}^+(t_0, q_0, t_1, q_1) \right) \cdot \xi,
\end{aligned}
$$

and so the two discrete energies in this case are equal. This is just the single discrete energy $E_{L_d} = E_{L_d}^\pm$, which would appear in a theory based on $Q \times \mathbb{R} \times Q$.

This same calculation can also be carried through for the discrete one- and two-forms and the other discrete structures, and shows that a $Q \times \mathbb{R} \times Q$ theory would not be essentially different from that based on $\bar{Q} \times \bar{Q}$.

4.4. Background: Extended Hamiltonian mechanics

As in the autonomous setting, the Hamiltonian picture of mechanics is complementary to the Lagrangian, and additional insight can be gained for variational integrators if the connection is made with Hamiltonian mechanics and, in particular, with extended Hamilton–Jacobi theory.

4.4.1. Basic definitions

Consider a configuration manifold Q, time space $\mathbb{R}$ and extended configuration manifold $\bar{Q} = \mathbb{R} \times Q$. Define the *extended phase space* to be $\mathbb{R} \times T^*Q$, that is, the usual phase space augmented with time, and consider an *extended Hamiltonian* $H : \mathbb{R} \times T^*Q \to \mathbb{R}$.

Given an extended Hamiltonian H, we define the *extended canonical one-form* $\bar{\Theta}_H$ and the *extended canonical two-form* or *extended symplectic* *form* $\bar{\Omega}_H$ to be one- and two-forms respectively on the extended phase space $\mathbb{R} \times T^*Q$ given by

$$\bar{\Theta}_H(t, q, p) = p_i dq^i - H(t, q, p) dt, \tag{4.4.1}$$

$$\bar{\Omega}_H(t, q, p) = -d\bar{\Theta}_H = dq^i \wedge dp_i + dH(t, q, p) \wedge dt. \tag{4.4.2}$$

If $\phi : \mathbb{R} \times T^*Q \to \mathbb{R} \times T^*Q$ is a map from the extended phase space with Hamiltonian H_0 to the extended phase space with Hamiltonian H_1, then we say that ϕ is *symplectic* or *canonical* if ϕ pulls the extended symplectic form back to the appropriate other extended symplectic form, so that

$$\phi^*(\bar{\Omega}_{H_1}) = \bar{\Omega}_{H_0}.$$

4.4.2. Hamiltonian vector fields and flow maps

For an extended Hamiltonian $H : \mathbb{R} \times T^*Q$ we define the *extended Hamiltonian vector field* $\bar{X}_H$ on the extended phase space $\mathbb{R} \times T^*Q$ to be the unique vector field with unit time flow satisfying

$$i_{\bar{X}_H} \bar{\Omega}_{II} = 0. \tag{4.4.3}$$

Note that this uniquely defines $\bar{X}_H$ for any H. Writing $\bar{X}_H = (1, X_q, X_p)$ in coordinates, we see that the above expression is

$$i_{\bar{X}_H} \bar{\Omega}_H = \left[\frac{\partial H}{\partial q} \cdot X_q + \frac{\partial H}{\partial p} \cdot X_p \right] dt + \left[-X_p - \frac{\partial H}{\partial q} \right] dq + \left[X_q - \frac{\partial H}{\partial p} \right] dp,$$

and requiring that this be zero gives the familiar *extended Hamilton's equations*

$$X_q(t, q, p) = \frac{\partial H}{\partial p}(t, q, p), \tag{4.4.4a}$$

$$X_p(t, q, p) = -\frac{\partial H}{\partial q}(t, q, p). \tag{4.4.4b}$$

Notice that the time component of (4.4.3) does not contribute an equation, because the other two equations automatically imply that

$$\frac{\partial H}{\partial q} \cdot X_q + \frac{\partial H}{\partial p} \cdot X_p = \frac{\partial H}{\partial q} \cdot \left(\frac{\partial H}{\partial p} \right) + \frac{\partial H}{\partial p} \cdot \left(-\frac{\partial H}{\partial q} \right) = 0,$$

which is equivalent to the fact that time variations in the Lagrangian case do not give additional constraints beyond the Euler–Lagrange equations.

The time component of (4.4.3) is a statement about the time evolution of the Hamiltonian. Computing

$$\frac{dH}{dt} = \frac{\partial H}{\partial t} \cdot X_t + \frac{\partial H}{\partial q} \cdot X_q + \frac{\partial H}{\partial p} \cdot X_p = \frac{\partial H}{\partial t},$$

we see that the derivative of the Hamiltonian with respect to time simply reflects the explicit dependence of H on t. If this is zero, so that the extended Hamiltonian is time-independent, then H is constant along the flow of $\bar{X}_H$.

We denote the *extended flow map* of the extended Hamiltonian vector field $\bar{X}_H$ by $\bar{F}_H : \mathbb{R} \times (\mathbb{R} \times T^*Q) \to \mathbb{R} \times T^*Q$, defined as

$$\bar{F}_H^h(t_0, q_0, p_0) = \bar{F}_H(h, t_0, q_0, p_0) = (t_0 + h, q_1, p_1).$$

The flow map $\bar{F}_H^h$ for any fixed h is a symplectic map from $\mathbb{R} \times T^*Q$ to itself, as can be readily seen by taking the time derivative to obtain

$$\left.\frac{\partial}{\partial h}\right|_{h=0} (\bar{F}_H^h)^* \bar{\Omega}_H = \mathcal{L}_{\bar{X}_H} \bar{\Omega}_H$$

$$= \mathbf{di}_{\bar{X}_H} \bar{\Omega}_H + \mathbf{i}_{\bar{X}_H} \mathbf{d}\bar{\Omega}_H$$

$$= 0,$$

where we have used Cartan's magic formula $\mathcal{L}_X \alpha = \mathbf{di}_X \alpha + \mathbf{i}_X \mathbf{d}\alpha$, and then the facts that $\mathbf{i}_{\bar{X}_H} \bar{\Omega}_H = 0$ is the definition of the extended Hamiltonian vector field, and that $\mathbf{d}^2 = 0$ implies $\mathbf{d}\bar{\Omega}_H = -\mathbf{d}^2\bar{\Theta}_H = 0$.

4.4.3. Extended Hamiltonian Noether's theorem

Consider a (left or right) action $\Phi : G \times \bar{Q} \to \bar{Q}$ of G on the extended configuration manifold $\bar{Q}$, as in Section 4.2.6. The lift of this action to the extended phase space $\mathbb{R} \times T^*Q$ is denoted by $\Phi^{\mathbb{R} \times T^*Q} : G \times (\mathbb{R} \times T^*Q) \to \mathbb{R} \times T^*Q$ and is defined by

$$\left\langle \Phi_g^{\mathbb{R} \times T^*Q}(t, q, p), (t, q, \dot{q}) \right\rangle = \left\langle (t, q, p), (\Phi_g^t(t), \Phi_g^q(t, q), \partial_q \Phi_g^q(t, q) \cdot \dot{q}) \right\rangle.$$

This has the corresponding *infinitesimal generator* $\xi_{\mathbb{R} \times T^*Q} : \mathbb{R} \times T^*Q \to T(\mathbb{R} \times T^*Q)$ defined by

$$\xi_{\mathbb{R} \times T^*Q}(t, q, p) = \frac{\mathrm{d}}{\mathrm{d}g}\left(\Phi_g^{\mathbb{R} \times T^*Q}(t, q, p) \right) \cdot \xi.$$

The *extended Hamiltonian momentum map* $\bar{J}_H : \mathbb{R} \times T^*Q \to \mathfrak{g}^*$ is defined by

$$\bar{J}_H(t, q, p) \cdot \xi = \bar{\Theta}_H(t, q, p) \cdot \xi_{\mathbb{R} \times T^*Q}(t, q, p).$$

For each $\xi \in \mathfrak{g}$ we define $\bar{J}_H^\xi : \mathbb{R} \times T^*Q \to \mathbb{R}$ by $\bar{J}_H^\xi(t, q, p) = \bar{J}_H(t, q, p) \cdot \xi$, which has expression $\bar{J}_H^\xi = \mathbf{i}_{\xi_{\mathbb{R} \times T^*Q}} \bar{\Theta}_H$. Note that the Hamiltonian map is also given by the expression

$$\bar{J}_H(t, q, p) \cdot \xi = \left\langle (t, q, p), \xi_{\bar{Q}}^q(t, q) \right\rangle - H(t, q, p)\xi_{\bar{Q}}^t(t, q),$$

where the two components of $\xi_{\bar{Q}}$ are denoted by $\xi_{\bar{Q}} = (\xi_{\bar{Q}}^t, \xi_{\bar{Q}}^q)$.

As in the case of the extended Lagrangian system, invariance of the Hamiltonian is not the correct requirement to ensure conservation of the momentum maps. Instead, we define a group action to be a *symmetry* if its lift acts by extended special symplectic maps, which is the requirement that $(\Phi_g^{\mathbb{R} \times T^* Q})^* \bar{\Theta}_H = \bar{\Theta}_H$ for all $g \in G$.

With this requirement the same proof as in Theorem 1.2.2 shows that the extended Hamiltonian momentum map is equivariant, in the sense that $\bar{J}_H \circ \bar{T}^* \Phi_g = \mathrm{Ad}^*_{g^{-1}} \circ \bar{J}_H$.

Theorem 4.4.1. (Extended Hamiltonian Noether's theorem)   Let $H : \mathbb{R} \times T^* Q \to \mathbb{R}$ be an extended Hamiltonian system that is invariant under the lift of the (left or right) action $\Phi : G \times \bar{Q} \to \bar{Q}$. Then the corresponding extended Hamiltonian momentum map $\bar{J}_H : \mathbb{R} \times T^* Q \to \mathfrak{g}^*$ is a conserved quantity of the flow, so that $\bar{J}_H \circ F_H^h = \bar{J}_H$ for all times h.

Proof.   By assumption we have that $(\bar{T}^* \Phi_g)^* \bar{\Theta}_H = \bar{\Theta}_H$ for all $g \in G$ and hence $\mathcal{L}_{\xi_{\mathbb{R} \times T^* Q}} \bar{\Theta}_H = 0$. Now, computing the derivative of $\bar{J}_H^\xi$ in the direction given by the extended Hamiltonian vector field $\bar{X}_H$ gives

$$
\begin{aligned}
\mathrm{d}\bar{J}_H^\xi \cdot \bar{X}_H &= \mathrm{d}(\mathrm{i}_{\xi_{\mathbb{R} \times T^* Q}} \bar{\Theta}_H) \cdot \bar{X}_H \\
&= \mathcal{L}_{\xi_{\mathbb{R} \times T^* Q}} \bar{\Theta}_H \cdot \bar{X}_H - \mathrm{i}_{\xi_{\mathbb{R} \times T^* Q}} \mathrm{d}\bar{\Theta}_H \cdot \bar{X}_H \\
&= -\mathrm{i}_{\bar{X}_H} \bar{\Omega}_H \cdot \xi_{\mathbb{R} \times T^* Q} \\
&= 0
\end{aligned}
$$

using Cartan's magic formula $\mathcal{L}_X \alpha = \mathrm{d}\mathrm{i}_X \alpha + \mathrm{i}_X \mathrm{d}\alpha$ and (4.4.3). As $\bar{F}_H^h$ is the flow map for $\bar{X}_H$ this gives the desired result.   ☐

Observe that in this extended case the assumption that $\bar{\Theta}_H$ is preserved by the lifted group action plays the same role as invariance of H did in the autonomous case, when invariance of Θ was a result of the fact that the action was a lift.

4.4.4. Extended Legendre transform

To relate the extended Lagrangian picture developed previously to that of extended Hamiltonian mechanics we define the *extended Legendre transform* $\bar{\mathbb{F}}L : \mathbb{R} \times TQ \to \mathbb{R} \times T^* Q$ to be

$$
\bar{\mathbb{F}}L : (t, q, \dot{q}) \mapsto (t, q, p) = \left(t, q, \frac{\partial L}{\partial \dot{q}}(t, q, \dot{q}) \right).
$$

If $\bar{\mathbb{F}}L$ is a local isomorphism then we say that L is *regular*, and if is a global isomorphism then we say that L is *hyperregular*.

Using the definitions, it is simple to check that the Lagrangian and Hamiltonian symplectic forms and momentum maps are related by

$$
\bar{\Theta}_L = (\bar{\mathbb{F}}L)^* \bar{\Theta}_H, \quad \bar{\Omega}_L = (\bar{\mathbb{F}}L)^* \bar{\Omega}_H, \quad \text{and} \quad \bar{J}_L = (\bar{\mathbb{F}}L)^* \bar{J}_H.
$$

Given a regular extended Lagrangian L, we define the associated extended Hamiltonian by

$$H = E_L \circ (\bar{\mathbb{F}}L)^{-1},$$

which in coordinates is

$$H(t, q, p) = \mathbb{F}L(t, q, \dot{q}) \cdot \dot{q} - L(t, q, \dot{q}),$$

since (t, q, p) and $(t, q, \dot{q})$ are related by the extended Legendre transform. With this additional assumption on H and L we can use calculations similar to those in Section 1.4.3 to see that the Hamiltonian and Lagrangian vector fields and flow maps are related by

$$\bar{X}_L = (\bar{\mathbb{F}}L)^* \bar{X}_H, \quad \bar{F}_L^h = (\bar{\mathbb{F}}L)^{-1} \circ \bar{F}_H^h \circ \bar{\mathbb{F}}L.$$

The definition of regularity of an extended Lagrangian can also be used to give a characterization of when the Euler–Lagrange equations uniquely define solution curves.

Theorem 4.4.2. Given an extended Lagrangian $L : \mathbb{R} \times TQ \to \mathbb{R}$, the extended Lagrangian vector field $\bar{X}_L$, and hence the extended flow map $\bar{F}_L$, is well-defined if and only if the Lagrangian is regular.

Proof. The same coordinate proof as is given for Theorem 1.4.3 also works here. $\square$

4.4.5. Extended generating functions

Consider two extended phase spaces with Hamiltonians H_0 and H_1 and extended canonical two-forms $\bar{\Omega}_{H_0} = -\mathbf{d}\bar{\Theta}_{H_0}$ and $\bar{\Omega}_{H_1} = -\mathbf{d}\bar{\Theta}_{H_1}$ respectively, and let $\varphi : \mathbb{R} \times T^*Q \to \mathbb{R} \times T^*Q$ be a time-dependent transformation. Set $\Gamma(\varphi) \subset (\mathbb{R} \times T^*Q) \times (\mathbb{R} \times T^*Q)$ to be the graph of φ. Consider the one-form on $(\mathbb{R} \times T^*Q) \times (\mathbb{R} \times T^*Q)$ defined by

$$\bar{\Theta}_{H_0, H_1} = \pi_2^* \bar{\Theta}_{H_0} - \pi_1^* \bar{\Theta}_{H_1},$$

where $\pi_i : (\mathbb{R} \times T^*Q) \times (\mathbb{R} \times T^*Q) \to \mathbb{R} \times T^*Q$ are the projections. The corresponding two-form is then

$$\bar{\Omega}_{H_0, H_1} = -\mathbf{d}\bar{\Theta}_{H_0, H_1} = -\pi_2^* \bar{\Omega}_{H_0} + \pi_1^* \bar{\Omega}_{H_1}.$$

Denoting the inclusion map by $i_\varphi : \Gamma(\varphi) \to (\mathbb{R} \times T^*Q) \times (\mathbb{R} \times T^*Q)$, we see that we have the identities

$$\pi_1 \circ i_\varphi = \pi_1|_{\Gamma(\varphi)}, \tag{4.4.5}$$

$$\pi_2 \circ i_\varphi = \varphi \circ \pi_1 \text{ on } \Gamma(\varphi). \tag{4.4.6}$$

Using these relations we have

$$
\begin{aligned}
i_\varphi^* \bar{\Omega}_{H_0,H_1} &= i_\varphi^*(\pi_1^* \bar{\Omega}_{H_0} - \pi_2^* \bar{\Omega}_{H_1}) \\
&= (\pi_1 \circ i_\varphi)^* \bar{\Omega}_{H_0} - (\pi_2 \circ i_\varphi^* \bar{\Omega}_{H_1} \\
&= (\pi_1|_{\Gamma(\varphi)})^* (\bar{\Omega}_{H_0} - \varphi^* \bar{\Omega}_{H_1}),
\end{aligned}
$$

where we have used (4.4.5) and (4.4.6).

This last relation shows that φ is an extended symplectic transformation if and only if $i_\varphi^* \bar{\Omega}_{H_0,H_1} = 0$ or, equivalently, if and only if $\mathbf{d}(i_\varphi^* \bar{\Theta}_{H_0,H_1}) = 0$. By the Poincaré lemma, this last statement is equivalent to there existing a function $\bar{S} : \Gamma(\varphi) \to \mathbb{R}$ so that, locally, $i_\varphi^* \bar{\Theta}_{H_0,H_1} = \mathbf{d}\bar{S}$. Such a function $\bar{S}$ is known as the *extended generating function* of the extended symplectic transformation φ. Note that $\bar{S}$ is not unique, as it is only defined up to a constant.

The extended generating function $\bar{S}$ is specified on the graph $\Gamma(\varphi)$, and so can be expressed in any local coordinate system on $\Gamma(\varphi)$.

4.4.6. Coordinate expression

We will be particularly interested in the choice (q_0, q_1, t_0) as local coordinates on $\Gamma(\varphi)$, and so we give the coordinate expressions for the above general extended generating function derivation for this particular case. This choice results in extended generating functions of the so-called *first kind*.

Consider a function $\bar{S} : Q \times Q \times \mathbb{R} \to \mathbb{R}$. Now the differential will be

$$
\mathbf{d}\bar{S} = \frac{\partial \bar{S}}{\partial q_0}\mathbf{d}q_0 + \frac{\partial \bar{S}}{\partial q_1}\mathbf{d}q_1 + \frac{\partial \bar{S}}{\partial t_0}\mathbf{d}t_0.
$$

Let $\varphi : \mathbb{R} \times T^*Q \to \mathbb{R} \times T^*Q$ be the extended symplectic transformation generated by $\bar{S}$. In coordinates, the quantity $i_\varphi^* \bar{\Theta}_{H_0,H_1}$ is

$$
i_\varphi^* \bar{\Theta}_{H_0,H_1} = \left(-p_0 - H_1\frac{\partial t_1}{\partial q_0}\right)\mathbf{d}q_0 + \left(p_1 - H_1\frac{\partial t_1}{\partial q_1}\right)\mathbf{d}q_1 + \left(H_0 - H_1\frac{\partial t_1}{\partial t_0}\right)\mathbf{d}t_0
$$

and so the condition $i_\varphi^* \bar{\Theta}_{H_0,H_1} = \mathbf{d}\bar{S}$ reduces to the equations

$$
\frac{\partial \bar{S}}{\partial q_0} = -p_0 - H_1\frac{\partial t_1}{\partial q_0}, \tag{4.4.7a}
$$

$$
\frac{\partial \bar{S}}{\partial q_1} = p_1 - H_1\frac{\partial t_1}{\partial q_1}, \tag{4.4.7b}
$$

$$
\frac{\partial \bar{S}}{\partial t_0} = H_0 - H_1\frac{\partial t_1}{\partial t_0}, \tag{4.4.7c}
$$

which are an implicit definition of the transformation $\varphi : (t_0, q_0, p_0) \mapsto (t_1, q_1, p_1)$. From the above general theory, we know that such a transformation is automatically symplectic in an extended sense, and that all extended symplectic transformations have such a representation.

Observe that for some choices of $\bar{S}$ the above equations do not define a single map φ, as was also the case for autonomous generating functions and the extended discrete Euler–Lagrange equations. As before, we will assume for now that $\bar{S}$ is chosen so that there is a single well-defined map φ which satisfies (4.4.7), and we will investigate this issue further in Sections 4.5.4 and 4.9.4.

4.5. Discrete variational mechanics: Hamiltonian viewpoint

4.5.1. Extended discrete Legendre transforms

The fact that the extended discrete state space $\bar{Q} \times \bar{Q}$ is larger than the extended state space $\mathbb{R} \times TQ$ is particularly important when it comes to defining extended discrete Legendre transforms. We will see this below, where the Legendre transform to the Hamiltonian phase space $\mathbb{R} \times T^*Q$ will not be a local isomorphism, and so it will be necessary to define another map to $\mathbb{R} \times (\mathbb{R} \times T^*Q)$ in order to push the extended discrete Lagrangian map forward.

We begin by defining the *extended discrete Legendre transforms* $\bar{\mathbb{F}}^{\pm}L_d : \bar{Q} \times \bar{Q} \to \mathbb{R} \times T^*Q$ to be

$$\bar{\mathbb{F}}^+ L_d(t_0, q_0, t_1, q_1) = (t_1, q_1, D_4 L_d(t_0, q_0, t_1, q_1)),$$
$$\bar{\mathbb{F}}^- L_d(t_0, q_0, t_1, q_1) = (t_0, q_0, -D_2 L_d(t_0, q_0, t_1, q_1)).$$

As $\bar{Q} \times \bar{Q}$ is larger than $\mathbb{R} \times T^*Q$, these maps cannot be even local isomorphisms. If they are both onto, however, then we say that the extended discrete Lagrangian is *regular*, and if they are both globally onto then we say that L_d is *hyperregular*, which will typically require that Q be a linear space.

Note that, in general, symplectic forms and momentum maps do not pull back to their discrete counterparts. If a momentum map arises from a vertical action, however, then we essentially reduce to the autonomous case and it can be seen that the associated extended Hamiltonian momentum map will indeed pull back to the extended discrete momentum maps.

4.5.2. Momentum and energy matching

Just as we earlier defined the discrete energies, we can also define the discrete momenta to be the image of the extended discrete Legendre transforms: that is, we set

$$p^+_{k,k+1} = p^+(t_k, q_k, t_{k+1}, q_{k+1}) = D_4 L_d(t_k, q_k, t_{k+1}, q_{k+1}),$$
$$p^-_{k,k+1} = p^-(t_k, q_k, t_{k+1}, q_{k+1}) = -D_2 L_d(t_k, q_k, t_{k+1}, q_{k+1}),$$

or equivalently

$$\bar{\mathbb{F}}^+ L_d(t_k, q_k, t_{k+1}, q_{k+1}) = (t_{k+1}, q_{k+1}, p^+_{k,k+1}),$$
$$\bar{\mathbb{F}}^- L_d(t_k, q_k, t_{k+1}, q_{k+1}) = (t_k, q_k, p^-_{k,k+1}).$$

In other words, $p^+_{k,k+1}$ is the discrete momentum at the right endpoint of the interval $[t_k, t_{k+1}]$, while $p^-_{k,k+1}$ is the discrete momentum at the left endpoint. We also introduce the notation

$$E^+_{k,k+1} = E^+_{L_d}(t_k, q_k, t_{k+1}, q_{k+1}),$$
$$E^-_{k,k+1} = E^-_{L_d}(t_k, q_k, t_{k+1}, q_{k+1}).$$

Using these definitions, we see that the extended discrete Euler–Lagrange equations can be written

$$p^+_{k-1,k} = p^-_{k,k+1},$$
$$E^+_{k-1,k} = E^-_{k,k+1},$$

and thus can be interpreted as a matching of momenta and energies at each time t_k. This is a generalization of the autonomous case, where only the discrete momenta are matched.

The definitions above also allow us to write the discrete Lagrangian one-forms in the more compact form

$$\bar{\Theta}^+_{L_d}(k, k+1) = p^+_{k,k+1}dq_{k+1} - E^+_{k,k+1}dt_{k+1},$$
$$\bar{\Theta}^-_{L_d}(k, k+1) = p^-_{k,k+1}dq_k - E^-_{k,k+1}dt_k,$$

which makes the analogy to the continuous extended Lagrangian one-form (4.2.11) even more apparent.

Note that, unlike the continuous case, regularity of an extended discrete Lagrangian is not sufficient to ensure that the extended discrete Euler–Lagrange equations have unique solutions, or indeed any solutions at all. We will investigate this issue further in Section 4.9.4.

4.5.3. Extended discrete Hamiltonian maps

The extended discrete Legendre transforms defined above clearly cannot be used to push the extended discrete Lagrangian map $\bar{F}_{L_d} : \bar{Q} \times \bar{Q} \to \bar{Q} \times \bar{Q}$ forward to the Hamiltonian phase space, as these Legendre transforms are not injective. Another way of saying this is that we need to augment the Hamiltonian phase space with time-step information to give a well-defined map.

This results in the space $(\mathbb{R} \times T^*Q) \times \mathbb{R}$ where an element (t, q, p, h) is interpreted as being a point (t, q, p) in phase space together with a time-step h. Given a sequence $\{(t_k, q_k, p_k, h_k)\}_{k=0}^N$, we regard h_k as being the time-step $t_{k+1} - t_k$.

Now define the map $\tilde{\mathbb{F}}^- L_d : \bar{Q} \times \bar{Q} \to (\mathbb{R} \times T^*Q) \times \mathbb{R}$ by

$$\tilde{\mathbb{F}}^- L_d : (t_0, q_0, t_1, q_1) \mapsto (t_0, q_0, p_0, h_0) = (t_0, q_0, -D_2 L_d(t_0, q_0, t_1, q_1), t_1 - t_0)$$

and define the *extended discrete Hamiltonian map* $\tilde{F}_{L_d} : (\mathbb{R} \times T^*Q) \times \mathbb{R} \to (\mathbb{R} \times T^*Q) \times \mathbb{R}$ to be

$$\tilde{F}_{L_d} : (t_0, q_0, p_0, h_0) \mapsto (t_1, q_1, p_1, h_1),$$
$$\tilde{F}_{L_d} = (\tilde{\mathbb{F}}^- L_d) \circ \bar{F}_{L_d} \circ (\tilde{\mathbb{F}}^- L_d)^{-1},$$

which is equivalent to the following commutative diagram.

$$
\begin{array}{ccc}
(t_0, q_0, t_1, q_1) & \xrightarrow{\bar{F}_{L_d}} & (t_1, q_1, t_2, q_2) \\
{\scriptstyle \tilde{\mathbb{F}}^- L_d} \Big\uparrow & & \Big\downarrow {\scriptstyle \tilde{\mathbb{F}}^- L_d} \\
(t_0, q_0, p_0, h_0) & \xrightarrow[\tilde{F}_{L_d}]{} & (t_1, q_1, p_1, h_1)
\end{array}
$$

Using this we can define the map $\tilde{\mathbb{F}}^+ L_d : \bar{Q} \times \bar{Q} \to \mathbb{R} \times T^*Q$ by the following commutative diagram.

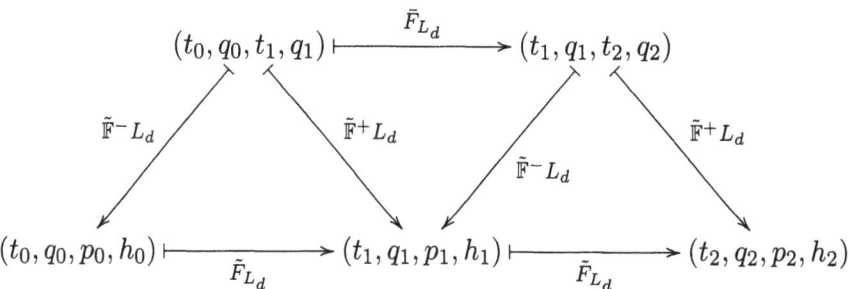

Now that we have a discrete system on an extension of the Hamiltonian phase space, it is natural to ask what structure is preserved by the evolution map. Unlike the autonomous case, the pushforward of the extended discrete symplectic form does not give any canonical structure on $(\mathbb{R} \times T^*Q) \times \mathbb{R}$. Of course, it does define a two-form there, which will be preserved by $\tilde{F}_{L_d}$, and we will see below one way in which this is related to $\bar{\Omega}_H$ on $\mathbb{R} \times T^*Q$.

An alternative to mapping to a point $(t, q, p) \in \mathbb{R} \times T^*Q$ and an associated time-step h would be to map to (t, q, p, E), where E is a discrete energy. This is reminiscent of the structure found in the continuous formulation of multisymplectic mechanics, but such discrete systems will often fail to behave well, as is further investigated in Section 4.9.4.

4.5.4. Extended discrete Lagrangians are extended generating functions

Although the extended discrete Hamiltonian map is not a map on the Hamiltonian phase space, we will see that a particular restriction of it is in fact

generated by a restriction of the extended discrete Lagrangian, in the sense
of Section 4.4.5.

Given $L_d : \bar{Q} \times \bar{Q} \to \mathbb{R}$, choose a fixed time-step $h \in \mathbb{R}$ and define the
restriction of the extended discrete Hamiltonian map to $\varphi^h : \mathbb{R} \times T^*Q \to$
$\mathbb{R} \times T^*Q$ by

$$\varphi^h : (t_0, q_0, p_0) \mapsto (t_1, q_1, p_1),$$

where $(t_1, q_1, p_1, h_1) = \tilde{F}_{L_d}(t_0, q_0, p_0, h)$. This provides an implicit definition
of (t_0, q_0, t_1, q_1) in terms of either (t_0, q_0, p_0) or (t_1, q_1, p_1). Given this, we
define the Hamiltonians H_0^h and H_1^h to be

$$H_0^h(t_0, q_0, p_0) = E_{L_d}^-(t_0, q_0, t_1, q_1),$$
$$H_1^h(t_1, q_1, p_1) = E_{L_d}^+(t_0, q_0, t_1, q_1),$$

and the extended generating function $\bar{S}^h : Q \times Q \times \mathbb{R} \to \mathbb{R}$ to be

$$\bar{S}^h(q_0, q_1, t_0) = L_d(t_0, q_0, t_1, q_1).$$

With these definitions we have the following result.

Theorem 4.5.1. Take a fixed time-step $h \in \mathbb{R}$ and let the functions $\bar{S}^h$,
φ^h, H_0^h and H_1^h be defined as above. Then the map φ^h from the space
$\mathbb{R} \times T^*Q$ with Hamiltonian H_0^h to the space $\mathbb{R} \times T^*Q$ with Hamiltonian H_1^h
is generated by the extended generating function $\bar{S}^h$.

Proof. Observe that the definition of φ^h implies that $t_1 = t_0 + h$. Computing the derivatives of $\bar{S}^h$ and using the definitions above we thus obtain

$$\frac{\partial \bar{S}}{\partial q_0} = D_2 L_d(t_0, q_0, t_1, q_1) = -p_0,$$

$$\frac{\partial \bar{S}}{\partial q_1} = D_4 L_d(t_0, q_0, t_1, q_1) = p_1,$$

$$\frac{\partial \bar{S}}{\partial t_0} = D_1 L_d(t_0, q_0, t_1, q_1) + D_3 L_d(t_0, q_0, t_1, q_1)$$

$$= H_0^h(t_0, q_0, p_0) - H_1^h(t_1, q_1, p_1).$$

However, these are simply the equations (4.4.7) which define the map gen-
erated by $\bar{S}^h$, and so φ^h must be this map. $\square$

As maps generated by extended generating functions must be symplectic,
this construction shows that $(\varphi^h)^* \Omega_{H_1^h} = \Omega_{H_0^h}$. This provides a way in which
the discrete Hamiltonian map $\tilde{F}_{L_d}$ can be viewed as preserving canonical
structures on $\mathbb{R} \times T^*Q$.

4.6. Correspondence between discrete and continuous mechanics

We will now investigate the choice of extended discrete Lagrangian which gives an exact correlation between a continuous and a discrete system, for which we need the following result.

Theorem 4.6.1. Consider a regular extended Lagrangian $L : \mathbb{R} \times TQ \to \mathbb{R}$, two points $q_0, q_1 \in Q$ and two times $t_0, t_1 \in \mathbb{R}$. If $\|q_1 - q_0\|$ and $|t_1 - t_0|$ are sufficiently small then there exists a unique solution $q : [t_0, t_1] \to Q$ of the Euler–Lagrange equations for L satisfying $q(t_0) = q_0$ and $q(t_1) = q_1$.

Proof. Essentially the same proof as that given in Marsden and Ratiu (1999), Section 7.4, for the autonomous case holds in the extended setting as well.   □

Given a regular extended Lagrangian $L : \mathbb{R} \times TQ \to \mathbb{R}$, define the *exact extended discrete Lagrangian* to be

$$L_d^E(t_0, q_0, t_1, q_1) = \int_{t_0}^{t_1} L(t, q_{0,1}(t), \dot{q}_{0,1}(t))\, dt, \qquad (4.6.1)$$

where $q_{0,1} : [t_0, t_1] \to Q$ is the unique solution of the Euler–Lagrange equations satisfying $q_{0,1}(t_0) = q_0$ and $q_{0,1}(t_1) = q_1$.

Lemma 4.6.2. A regular extended Lagrangian $L : \mathbb{R} \times TQ \to \mathbb{R}$ and the corresponding exact extended discrete Lagrangian $L_d^E : \bar{Q} \times \bar{Q} \to \mathbb{R}$ satisfy the relations

$$\bar{\mathbb{F}}^+ L_d^E(t_0, q_0, t_1, q_1) = \bar{\mathbb{F}}L(t_1, q_{0,1}(t_1), \dot{q}_{0,1}(t_1)),$$
$$E_{L_d^E}^+(t_0, q_0, t_1, q_1) = E_L(t_1, q_{0,1}(t_1), \dot{q}_{0,1}(t_1)),$$
$$\bar{\mathbb{F}}^- L_d^E(t_0, q_0, t_1, q_1) = \bar{\mathbb{F}}L(t_0, q_{0,1}(t_0), \dot{q}_{0,1}(t_0)),$$
$$E_{L_d^E}^-(t_0, q_0, t_1, q_1) = E_L(t_0, q_{0,1}(t_0), \dot{q}_{0,1}(t_0)),$$

for sufficiently close $q_0, q_1 \in Q$ and $t_0, t_1 \in \mathbb{R}$.

Proof. The calculation for the Legendre transforms is essentially the same as in the time-independent case. For the energies, we calculate

$$\frac{\partial L_d^E}{\partial t_0} = -L(t_0, q_{0,1}(t_0), \dot{q}_{0,1}(t_0)) + \int_{t_0}^{t_1} \left[\frac{\partial L}{\partial q} \cdot \frac{\partial q_{0,1}}{\partial t_0} + \frac{\partial L}{\partial \dot{q}} \cdot \frac{\partial \dot{q}_{0,1}}{\partial t_0} \right] dt$$

$$= -L(t_0, q_{0,1}(t_0), \dot{q}_{0,1}(t_0)) - \int_{t_0}^{t_1} \left[\frac{\partial L}{\partial q} - \frac{d}{dt}\left(\frac{\partial L}{\partial \dot{q}} \right) \right] \cdot \frac{\partial q_{0,1}}{\partial t_0} \, dt$$

$$+ \left[\frac{\partial L}{\partial \dot{q}}(t, q_{0,1}(t), \dot{q}_{0,1}(t)) \cdot \frac{\partial q_{0,1}}{\partial t_0}(t) \right]_{t_0}^{t_1},$$

using integration by parts. However, $q_{0,1}(t)$ is a solution of the Euler–Lagrange equations for L and so the middle term is zero. Now note that we have

$$\frac{\partial q_{0,1}(t; t_0, q_0, t_1, q_1)}{\partial t_0}\bigg|_{t=t_0} = \dot{q}_{0,1}(t_0; t_0, q_0, t_1, q_1),$$

$$\frac{\partial q_{0,1}(t; t_0, q_0, t_1, q_1)}{\partial t_0}\bigg|_{t=t_1} = 0,$$

and using this gives

$$\frac{\partial L_d^E}{\partial t_0} = -L(t_0, q_{0,1}(t_0), \dot{q}_{0,1}(t_0)) + \left[\frac{\partial L}{\partial \dot{q}}(t, q_{0,1}(t), \dot{q}_{0,1}(t)) \cdot \frac{\partial q_{0,1}}{\partial t_0}(t)\right]_{t_0}^{t_1}$$

$$= -L(t_0, q_{0,1}(t_0), \dot{q}_{0,1}(t_0)) + \frac{\partial L}{\partial \dot{q}}(t_0, q_{0,1}(t_0), \dot{q}_{0,1}(t_0)) \cdot \dot{q}_{0,1}(t_0)$$

$$= E_L(t_0, q_{0,1}(t_0), \dot{q}_{0,1}(t_0)).$$

The result for $D_4 L_d^E$ can be established by a similar calculation, using the fact that

$$\frac{\partial q_{0,1}(t; t_0, q_0, t_1, q_1)}{\partial t_1}\bigg|_{t=t_0} = 0,$$

$$\frac{\partial q_{0,1}(t; t_0, q_0, t_1, q_1)}{\partial t_0}\bigg|_{t=t_1} = -\dot{q}_{0,1}(t_1; t_0, q_0, t_1, q_1). \qquad \square$$

One interesting consequence of this lemma is that, for an exact extended discrete Lagrangian, the extended discrete Euler–Lagrange equations are always functionally dependent. Indeed, the second equation (4.3.7b) becomes a consequence of the first equation (4.3.7a), just as in the continuous case. Before we prove this, however, we give a theorem relating discrete and continuous solution curves.

Theorem 4.6.3. Given a regular extended Lagrangian $L : \mathbb{R} \times TQ \to \mathbb{R}$, let $L_d^E : \bar{Q} \times \bar{Q} \to \mathbb{R}$ be the associated exact extended discrete Lagrangian. Consider a solution $q : [t_0, t_N] \to Q$ of the extended Euler–Lagrange equations for L, and take any sequence $\{t_k\}_{k=0}^N \subset [0, T]$ with sufficiently small $|t_{k+1} - t_k|$. Setting $q_k = q(t_k)$, we now have that $\{(t_k, q_k)\}_{k=0}^N$ is a solution of the extended discrete Euler–Lagrange equations for L_d^E.

Conversely, given any solution $\{(t_k, q_k)\}_{k=0}^N$ of the extended discrete Euler–Lagrange equations for L_d^E, define a curve $q : [t_0, t_N] \to Q$ by $q(t) = q_{k,k+1}(t)$ for $t \in [t_k, t_{k+1}]$, where $q_{k,k+1} : [t_k, t_{k+1}] \to Q$ is the unique solution of the extended Euler–Lagrange equations for L satisfying $q_{k,k+1}(t_k) = q_k$ and $q_{k,k+1}(t_{k+1}) = q_{k+1}$. Then $q(t)$ is a solution of the extended Euler–Lagrange equations for L on $[t_0, t_N]$.

Proof. The proof of this theorem is essentially identical to that of Theorem 1.6.4. Forming $\{(t_k, q_k)\}_{k=0}^{N}$ from a given solution $q(t)$, we see that the discrete Euler–Lagrange equations, which are just a matching of discrete momenta and energies at each t_k, are satisfied because the discrete quantities are equal to the continuous ones.

In the reverse direction, the first part (4.3.7a) of the extended discrete Euler–Lagrange equations, together with Lemma 4.6.2, implies that the first part (4.2.9a) of the Euler–Lagrange equations is satisfied for $q(t)$. It can now be checked, in the same way as the proof of Theorem 1.6.4, that $q(t)$ is C^2. As the second part of the Euler–Lagrange equations is dependent on the first, this means that $q(t)$ automatically satisfies the full extended Euler–Lagrange equations for L. $\qquad\square$

Note that in the last part of the above proof the fact that the discrete curve also satisfied the second part (4.3.7b) of the extended discrete Euler–Lagrange equations was not used. This allows us to prove the following.

Corollary 4.6.4. Given a regular extended Lagrangian $L : \mathbb{R} \times TQ \to \mathbb{R}$, let $L_d^E : \bar{Q} \times \bar{Q} \to \mathbb{R}$ be the associated exact extended discrete Lagrangian. Then the second part (4.3.7b) of the extended discrete Euler–Lagrange equations for L_d^E is satisfied whenever the first part (4.3.7a) is satisfied.

Proof. Consider points $(t_0, q_0, t_1, q_1, t_2, q_2)$ which satisfy the first part of the extended discrete Euler–Lagrange equations for L_d^E. Now define $q_{0,1,2}(t) :$ $[t_0, t_2] \to Q$ by $q_{0,1,2}(t) = q_{k,k+1}(t)$ for $t \in [t_k, t_{k+1}]$, as in Theorem 4.6.3. From the proof of that theorem it is clear that $q(t)$ is a solution of the Euler–Lagrange equations for L on $[t_0, t_2]$, and thus

$$E_L(t_1, q_{0,1}(t_1), \dot{q}_{0,1}(t_1)) = E_L(t_1, q_{0,1,2}(t_1), \dot{q}_{0,1,2}(t_1))$$

$$= E_L(t_1, q_{1,2}(t_1), \dot{q}_{1,2}(t_1)).$$

By Lemma 4.6.2, however, the left- and right-hand parts of this expression give

$$E_{L_d^E}^{+}(t_0, q_0, t_1, q_1) = E_{L_d^E}^{-}(t_1, q_1, t_2, q_2)$$

which is exactly the second part of the extended discrete Euler–Lagrange equations for L_d^E. $\qquad\square$

Indeed, as we will see in Section 4.8, the above statement is actually both necessary and sufficient for L_d to be an exact extended discrete Lagrangian.

Note that this corollary means that the exact extended discrete Lagrangian is not discretely well-posed, and thus does not define an extended discrete Lagrangian map $\bar{F}_{L_d} : \bar{Q} \times \bar{Q} \to \bar{Q} \times \bar{Q}$. Indeed, any time-step forward will give a valid solution. This means that the statements about symplecticity and momentum conservation do not hold literally in this case. Instead, we must consider a generalized first variation interpretation of symplecticity or momentum conservation, which would hold for any tangent vectors to the set of solution curves. See Marsden *et al.* (1998) for the outline of this idea, although it is used there for a different reason.

As we have seen above, it is exactly because any time-step gives a solution of the extended discrete Euler–Lagrange equations for L_d^E that we cannot define $\bar{F}_{L_d}$ or $\tilde{F}_{L_d}$. For a given time-step h, however, we can define the map $\varphi^{E,h} : \mathbb{R} \times T^*Q \to \mathbb{R} \times T^*Q$ by the conditions

$$\varphi^{E,h} \circ \bar{\mathbb{F}}^- L_d^E = \bar{\mathbb{F}}^+ L_d^E,$$
$$t_1 = t_0 + h,$$

where $(t_1, q_1, p_1) = \varphi^{E,h}(t_0, q_0, p_0)$. This map plays the same role as the restriction of $\tilde{F}_{L_d}$ to φ^h defined in Section 4.5.4, although $\varphi^{E,h}$ is not the restriction of anything. Using now the fact that $\bar{F}_L^h(t_0, q_{0,1}(0), \dot{q}_{0,1}(0)) = (t_1, q_{0,1}(h), \dot{q}_{0,1}(h))$ and combining the definition of $\varphi^{E,h}$ with Lemma 4.6.2 shows that $\varphi^{E,h}$ satisfies the following commutative diagram.

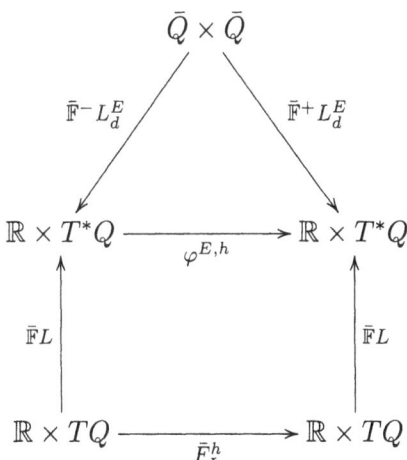

This proves the following theorem.

Theorem 4.6.5. Consider a regular extended Lagrangian $L : \mathbb{R} \times TQ$ with corresponding extended Hamiltonian $H : \mathbb{R} \times T^*Q$ and exact extended discrete Lagrangian $L_d^E : \bar{Q} \times \bar{Q} \to \mathbb{R}$. Then $\varphi^{E,h}$ defined as above is exactly the extended Hamiltonian flow map $\bar{F}_H^h$.

The above construction will be useful when we consider the relationship between exact extended discrete Lagrangians and extended Hamilton–Jacobi theory in Section 4.8.

4.7. Background: Extended Hamilton–Jacobi theory

4.7.1. Extended generating function for the flow

As we have already shown, the flow map $\bar{F}_H^h : (t_0, q_0, p_0) \mapsto (t_1, q_1, p_1)$ of an extended Hamiltonian system is an extended symplectic map for each fixed h. From the preceding theory, it must therefore have a representation in terms of some extended generating function $\bar{S}^h$. The extended Hamilton–Jacobi equation is a PDE that defines this generating function.

Considering that the flow map acts

$$\bar{F}_H : (\mathbb{R} \times T^*Q) \times \mathbb{R} \to (\mathbb{R} \times T^*Q)$$

we calculate the tangent map of this to be

$$T\bar{F}_H((t, q, p), h) \cdot ((\delta t, \delta q, \delta p), \delta h)$$

$$= T\bar{F}_H^h(t, q, p) \cdot (\delta t, \delta q, \delta p) + \frac{\partial}{\partial h} \bar{F}_H^h(t, q, p) \cdot \delta h$$

$$= T\bar{F}_H^h(t, q, p) \cdot (\delta t, \delta q, \delta p) + \bar{X}_H \circ \bar{F}_H^h(t, q, p) \cdot \delta h,$$

using the fact that $\bar{F}_H$ is the flow of the vector field $\bar{X}_H$. This then shows that

$$\bar{F}_H^*(\bar{\Omega}_H) = (\pi_{\mathbb{R} \times T^*Q})^*(\bar{\Omega}_H),$$

which follows from (4.4.3). Consider now the space $(\mathbb{R} \times T^*Q) \times \mathbb{R} \times (\mathbb{R} \times T^*Q)$ with projections

$$\pi_1 : (\mathbb{R} \times T^*Q) \times \mathbb{R} \times (\mathbb{R} \times T^*Q) \to (\mathbb{R} \times T^*Q) \times \mathbb{R},$$
$$\pi_2 : (\mathbb{R} \times T^*Q) \times \mathbb{R} \times (\mathbb{R} \times T^*Q) \to \mathbb{R} \times T^*Q$$

onto the first and second parts, respectively. Define the one-form

$$\hat{\Theta} = \pi_2^* \bar{\Theta}_H - \pi_1^* \pi_{\mathbb{R} \times T^*Q}^* \bar{\Theta}_H$$

and the corresponding two-form

$$\hat{\Omega} = -\mathbf{d}\hat{\Theta} = \pi_1^* \pi_{\mathbb{R} \times T^*Q}^* \bar{\Omega}_H - \pi_2^* \bar{\Omega}_H.$$

The graph of the flow map is a subset $\Gamma(\bar{F}_H) \subset (\mathbb{R} \times T^*Q) \times \mathbb{R} \times (\mathbb{R} \times T^*Q)$, and we denote the corresponding inclusion map by $i_{\bar{F}_H} : \Gamma(\bar{F}_H) \to (\mathbb{R} \times T^*Q) \times \mathbb{R} \times (\mathbb{R} \times T^*Q)$. Now note that

$$\pi_1 \circ i_{\bar{F}_H} = \pi_1|_{\Gamma(\bar{F}_H)},$$
$$\pi_2 \circ i_{\bar{F}_H} = \bar{F}_H \circ \pi_1 \text{ on } \Gamma(\bar{F}_H),$$

with which we calculate

$$
\begin{aligned}
(i_{\bar{F}_H})^*\hat{\Omega} &= (i_{\bar{F}_H})^*(\pi_1)^*(\pi_{\mathbb{R}\times T^*Q})^*(\bar{\Omega}_H) - (i_{\bar{F}_H})^*(\pi_2)^*(\bar{\Omega}_H) \\
&= (\pi_{\mathbb{R}\times T^*Q} \circ \pi_1 \circ i_{\bar{F}_H})^*(\bar{\Omega}_H) - (\pi_2 \circ i_{\bar{F}_H})^*(\bar{\Omega}_H) \\
&= (\pi_{\mathbb{R}\times T^*Q} \circ \pi_1 \circ i_{\bar{F}_H})^*(\bar{\Omega}_H) - (\bar{F}_H \circ \pi_1 \circ i_{\bar{F}_H})^*(\bar{\Omega}_H) \\
&= (\pi_{\mathbb{R}\times T^*Q} \circ \pi_1 \circ i_{\bar{F}_H})^*(\bar{\Omega}_H) - (\pi_{\mathbb{R}\times T^*Q} \circ \pi_1 \circ i_{\bar{F}_H})^*(\bar{\Omega}_H) \\
&= 0.
\end{aligned}
$$

This thus establishes that $\mathbf{d}(i^*_{\bar{F}_H}\hat{\Theta}) = 0$, and so there is locally a function $\bar{S} : \Gamma(\bar{F}_H) \to \mathbb{R}$ with $i^*_{\bar{F}_H}\hat{\Theta} = \mathbf{d}\bar{S}$.

4.7.2. Extended Hamilton–Jacobi equation

We now choose a particular set of coordinates on $\Gamma(\bar{F}_H)$ and derive a co-ordinate expression for $i^*_{\bar{F}_H}\hat{\Theta} = \mathbf{d}\bar{S}$. Taking coordinates on $(\mathbb{R} \times T^*Q) \times \mathbb{R} \times (\mathbb{R} \times T^*Q)$ of $((t_0, q_0, p_0), h, (t_1, q_1, p_1))$, we will take (t_0, q_0, t_1, q_1) as coordinates on $\Gamma(\bar{F}_H)$. The function $\bar{S}$ is thus a map $\bar{S} : \mathbb{R} \times Q \times \mathbb{R} \times Q \to \mathbb{R}$ and has differential

$$
\mathbf{d}\bar{S} = \frac{\partial \bar{S}}{\partial t_0}\mathbf{d}t_0 + \frac{\partial \bar{S}}{\partial q_0}\mathbf{d}q_0 + \frac{\partial \bar{S}}{\partial t_1}\mathbf{d}t_1 + \frac{\partial \bar{S}}{\partial q_1}\mathbf{d}q_1.
$$

In these coordinates we also have

$$
\hat{\Omega} = (p_1\mathbf{d}q_1 - H(t_1, q_1, p_1)\mathbf{d}t_1) - (p_0\mathbf{d}q_0 - H(t_0, q_0, p_0)\mathbf{d}t_0).
$$

Equating coefficients, we now see that the equation $i^*_{\bar{F}_H}\hat{\Theta} = \mathbf{d}\bar{S}$ is

$$
p_0 = -\frac{\partial \bar{S}}{\partial q_0}(t_0, q_0, t_1, q_1), \tag{4.7.1a}
$$

$$
H\left(t_0, q_0, -\frac{\partial \bar{S}}{\partial q_0}(t_0, q_0, t_1, q_1)\right) = \frac{\partial \bar{S}}{\partial t_0}(t_0, q_0, t_1, q_1), \tag{4.7.1b}
$$

$$
p_1 = \frac{\partial \bar{S}}{\partial q_1}(t_0, q_0, t_1, q_1), \tag{4.7.1c}
$$

$$
H\left(t_1, q_1, \frac{\partial \bar{S}}{\partial q_1}(t_0, q_0, t_1, q_1)\right) = -\frac{\partial \bar{S}}{\partial t_1}(t_0, q_0, t_1, q_1). \tag{4.7.1d}
$$

Of these four equations, (4.7.1a), (4.7.1c) and (4.7.1d) can be regarded as the generating function equations (4.4.7) for the map $\bar{F}_H$, while (4.7.1b) is the equation which must be satisfied if $\bar{F}_H$ is to be the Hamiltonian flow. Note that the first three equations do not actually specify $\bar{F}_H$ uniquely, as any time reparametrization of the flow will satisfy them. To ensure uniqueness, we must augment the above equations with the condition $t_1 = t_0 + h$.

The last equation which $\bar{S}$ must satisfy to be an extended generating function for the flow is known as the *extended Hamilton–Jacobi equation* and is given by

$$H\left(t_0, q_0, -\frac{\partial \bar{S}}{\partial q_0}(t_0, q_0, t_1, q_1)\right) = \frac{\partial \bar{S}}{\partial t_0}(t_0, q_0, t_1, q_1). \qquad (4.7.2)$$

As in the autonomous case, it is necessary to specify boundary conditions for this partial differential equation.

4.7.3. Jacobi's solution

The solution of the extended Hamilton–Jacobi equation can be written in terms of the action associated to the Lagrangian which is the Legendre transform of H. This is known as the *extended Jacobi's solution* and is

$$\bar{S}(t_0, q_0, t_1, q_1) = \int_{t_0}^{t_1} L(t, q(t), \dot{q}(t)) \, dt, \qquad (4.7.3)$$

where $q : [t_0, t_1] \to Q$ is the unique solution of the extended Euler–Lagrange equations for L satisfying $q(t_0) = q_0$ and $q(t_1) = q_1$. The proof that this is indeed a solution is essentially identical to the proof of Lemma 4.6.2.

4.8. Discrete variational mechanics: Hamilton–Jacobi viewpoint

For a fixed time-step h, we saw in Section 4.5.4 that the restriction of $\tilde{F}_{L_d}$ to the extended phase space $\mathbb{R} \times T^*Q$ is generated by the restriction of the discrete Lagrangian to the set $t_1 - t_0 = h$. We then showed that there is a particular choice of extended discrete Lagrangian, called the exact extended discrete Lagrangian, for a particular restricted map is exactly the extended Hamiltonian flow map.

In the preceding section we have seen that the generating function for the extended Hamiltonian flow map must satisfy the extended Hamilton–Jacobi equation. It is clear that this is the case for the exact extended discrete Lagrangian, as it is simply the standard Jacobi's solution to the extended Hamilton–Jacobi equation.

Note that the story is not as simple as in the autonomous case, however. The extended discrete Lagrangian in fact defines a map $\tilde{F}_{L_d}$ on the enlarged space $(\mathbb{R} \times T^*Q) \times \mathbb{R}$, and it is only upon restriction that we have a generating function construction. This restriction essentially discards the equation for updating the time-step.

The definition of $\tilde{F}_{L_d}$ relied upon L_d being discretely well-posed, which fails for the exact discrete Lagrangian. Nonetheless, we saw that by adding the condition $t_1 = t_0 + h$ then the *restriction* of $\tilde{F}_{L_d}$ still makes sense even when L_d is not discretely well-posed, and it is this map which is exactly the Hamiltonian flow. It is precisely this same extra equation which was also necessary to specify the map generated by the solution to the Hamilton–Jacobi equation.

Another way of viewing the situation is to consider a Hamiltonian system $H : \mathbb{R} \times T^*Q \to \mathbb{R}$ and to attempt to find a discrete Lagrangian for which $(\bar{\mathbb{F}}^\pm L_d)^* H = E_{L_d}^\pm$ or, equivalently, $(\bar{\mathbb{F}}^\pm L_d)^* \Theta_H = \Theta_{L_d}$. Such a discrete Lagrangian would have the appealing property that $\tilde{F}_{L_d}$ would exactly preserve the true energy, symplectic form and momentum maps.

Considering the statement $(\bar{\mathbb{F}}^+ L_d)^* H = E_{L_d}^+$, however, we see that this is exactly the extended Hamilton–Jacobi PDE: that is, the only extended discrete Lagrangian for which the pullback of the Hamiltonian is the discrete energy is the exact extended discrete Lagrangian. The same statement also holds with energy replaced by canonical one-forms.

Note that this does not say that the standard autonomous discrete Hamiltonian map will preserve H if and only if it exactly integrates the flow, as H could be preserved even though $E_{L_d}^\pm$ was not equal to $(\bar{\mathbb{F}}^\pm L_d)^* H$. Indeed, as we saw above, this would be the case for reparametrizations of the flow map, as discussed in Ge and Marsden (1988).

4.9. Time-dependent variational integrators

Just as we can regard autonomous discrete variational systems as integrators for a given autonomous Lagrangian system, so too can we consider an extended discrete Lagrangian system as an integrator for an extended Lagrangian system $L : \mathbb{R} \times TQ \to \mathbb{R}$. For this to be the case, we need that the discrete Lagrangian $L_d : \bar{Q} \times \bar{Q} \to \mathbb{R}$ satisfies

$$L_d(t_0, q(t_0), t_1, q(t_1)) = \int_{t_0}^{t_1} L\big(t, q(t), \dot{q}(t)\big)\, dt + \mathcal{O}(|t_1 - t_0|^{r+1}),$$

where $q : [t_0, t_1] \to Q$ is any solution of the extended Euler–Lagrange equations for L. Here r is known as the *order* of the discrete Lagrangian, and we require $r \geq 1$ for L_d to be *consistent*.

Having chosen an extended discrete Lagrangian of some order, we can then use the extended discrete Hamiltonian map $\tilde{F}_{L_d} : (\mathbb{R} \times T^*Q) \times \mathbb{R} \to (\mathbb{R} \times T^*Q) \times \mathbb{R}$ as an integrator for the Hamiltonian system associated to L. This maps from the point (t_k, q_k, p_k) and the time-step h_k to give a new point $(t_{k+1}, q_{k+1}, p_{k+1})$, as described in Section 4.5.3.

Although this map, based on updating time-steps, was useful for understanding the relationship of extended discrete variational mechanics to extended Hamilton–Jacobi theory, it turns out to be more convenient to implement an equivalent method which updates points in conjunction with energies, rather than time-steps. To construct this, we define the maps

$$\hat{\mathbb{F}}^+ L_d : (t_0, q_0, t_1, q_1) \mapsto (t_1, q_1, p_1 = D_4 L_d, E_1 = -D_3 L_d),$$
$$\hat{\mathbb{F}}^- L_d : (t_0, q_0, t_1, q_1) \mapsto (t_0, q_0, p_0 = -D_2 L_d, E_0 = D_1 L_d),$$

and define the alternative extended discrete Hamiltonian map $\hat{F}_{L_d} : \mathbb{R} \times T^* Q \times \mathbb{R} \to \mathbb{R} \times T^* Q \times \mathbb{R}$ by

$$\hat{F}_{L_d} = \hat{\mathbb{F}}^+ L_d \circ (\hat{\mathbb{F}}^- L_d)^{-1}.$$

Regarding $\hat{F}_{L_d}$ as an integrator, it maps

$$\hat{F}_{L_d} : (t_k, q_k, p_k, E_k) \mapsto (t_{k+1}, q_{k+1}, p_{k+1}, E_{k+1})$$

and is defined by the relations

$$p_k = -D_2 L_d(t_k, q_k, t_{k+1}, q_{k+1}), \qquad (4.9.1a)$$
$$E_k = D_1 L_d(t_k, q_k, t_{k+1}, q_{k+1}), \qquad (4.9.1b)$$
$$p_{k+1} = D_4 L_d(t_k, q_k, t_{k+1}, q_{k+1}), \qquad (4.9.1c)$$
$$E_{k+1} = -D_3 L_d(t_k, q_k, t_{k+1}, q_{k+1}), \qquad (4.9.1d)$$

together with the requirement that $t_{k+1} > t_k$, which is built in to the definition of the extended discrete path space. Computationally, the implicit equations (4.9.1a) and (4.9.1b) must be solved simultaneously for t_{k+1} and q_{k+1} under the restriction $t_{k+1} > t_k$, and then the explicit equations (4.9.1c) and (4.9.1d) must be evaluated to give p_{k+1} and E_{k+1}. Note that the calculations to produce a trajectory with $\tilde{F}_{L_d}$ are the same as those for calculating the same trajectory with $\hat{F}_{L_d}$, but the presentation is clearer in this form.

As we have seen in previous sections, the method defined by (4.9.1) preserves an extended discrete symplectic form and extended discrete momentum maps and it satisfies a discrete energy evolution equation which exactly conserves the discrete energy if the discrete Lagrangian is autonomous.

4.9.1. Initial conditions

To actually use $\hat{F}_{L_d}$ as an integrator, it is necessary to choose initial conditions. Given $(t_0, q_0, p_0) \in \mathbb{R} \times T^* Q$, this reduces to the question of choosing an initial energy $E_0 \in \mathbb{R}$.

While the natural choice may, at first, appear to be the Hamiltonian H evaluated at (t_0, q_0, p_0), this is not a feasible option. As we have seen, requiring that the Hamiltonian coincides with the discrete energies is equivalent to the extended Hamilton–Jacobi PDE, and so using this as an initial condition will generally make the first step of $\hat{F}_{L_d}$ ill-defined. Even when this equation is solvable, this approach provides no control over the size of the initial time-step.

Instead, a better choice is to select an initial time-step $h_0 \in \mathbb{R}$ and then to set $t_1 = t_0 + h_0$ and to solve the equation

$$p_0 = -D_2 L_d(t_0, q_0, t_1, q_1)$$

for q_1. We can then evaluate $E_0 = E_{L_d}^-(t_0, q_0, t_1, q_1)$ to obtain the initial condition for the discrete energy.

To see the difficulties encountered with choosing $E_h = H(t_0, q_0, p_0)$, consider the following example.

Example 4.9.1. Taking the discrete Lagrangian L_d^α from Example 2.3.2 with $\alpha = 1$ and extending it to the time-dependent case gives

$$L_d(t_0, q_0, t_1, q_1) = (t_1 - t_0)L\left(t_1, q_1, \frac{q_1 - q_0}{t_1 - t_0}\right),$$

and for the case of the autonomous Lagrangian $L(t, q, \dot{q}) = \frac{1}{2}\dot{q}^T M \dot{q} - V(q)$ we can calculate

$$\bar{\mathbb{F}}^- L_d(t_0, q_0, t_1, q_1) = \left(t_0, q_0, M\left(\frac{q_1 - q_0}{t_1 - t_0}\right)\right),$$

$$E_{L_d}^-(t_0, q_0, t_1, q_1) = \frac{1}{2}\left(\frac{q_1 - q_0}{t_1 - t_0}\right)^T M \left(\frac{q_1 - q_0}{t_1 - t_0}\right) + V(q_1),$$

and hence

$$\left((\mathbb{F}^- L_d)^* H\right)(t_0, q_0, t_1, q_1) = \frac{1}{2}\left(\frac{q_1 - q_0}{t_1 - t_0}\right)^T M \left(\frac{q_1 - q_0}{t_1 - t_0}\right) + V(q_0).$$

If we now take $E_0 = H(t_0, q_0, p_0)$ and calculate the first step of $\hat{F}_{L_d}$, then equations (4.9.1a) and (4.9.1b) are

$$(t_0, q_0, p_0) = \bar{\mathbb{F}}^- L_d(t_0, q_0, t_1, q_1),$$

$$\left((\mathbb{F}^- L_d)^* H\right)(t_0, q_0, t_1, q_1) = E_{L_d}^-(t_0, q_0, t_1, q_1),$$

which reduce to

$$q_1 = q_0 + (t_1 - t_0)M^{-1}p_0,$$

$$V(q_1) = V(q_0).$$

For a nontrivial potential function $V : Q \to \mathbb{R}$ these equations will not have a solution (t_1, q_1), except for special choices of p_0.   ◇

4.9.2. Order of accuracy and local errors

As we have seen while considering the choice of initial conditions, and as we will discuss in more detail below, existence of solutions for an extended discrete Lagrangian system can be problematic for their use as integrators. For this reason we will not discuss the order of accuracy of the extended discrete

Hamiltonian map $\tilde{\mathbb{F}}_{L_d}$ or the variant $\hat{\mathbb{F}}_{L_d}$ here, but instead we will consider the relationship between the order of accuracy of the extended discrete Lagrangian and its discrete energies and extended Legendre transforms.

Theorem 4.9.2. Consider an extended Lagrangian system $L : \mathbb{R} \times TQ$ and a corresponding extended discrete Lagrangian system $L_d : \bar{Q} \times \bar{Q} \to \mathbb{R}$ of order r. Then the extended discrete Legendre transforms $\bar{\mathbb{F}}^\pm L_d : \bar{Q} \times \bar{Q} \to \mathbb{R} \times T^*Q$ and the discrete energies $E_{L_d}^\pm : \bar{Q} \times \bar{Q} \to \mathbb{R}$ are also of order r, in the sense that

$$\bar{\mathbb{F}}^+ L_d(t_0, q(t_0), t_1, q(t_1)) = \bar{\mathbb{F}}L(t_1, q(t_1), \dot{q}(t_1)) + \mathcal{O}(|t_1 - t_0|^{r+1}),$$
$$E_{L_d}^+(t_0, q(t_0), t_1, q(t_1)) = E_L(t_1, q(t_1), \dot{q}(t_1)) + \mathcal{O}(|t_1 - t_0|^{r+1}),$$
$$\bar{\mathbb{F}}^- L_d(t_0, q(t_0), t_1, q(t_1)) = \bar{\mathbb{F}}L(t_0, q(t_0), \dot{q}(t_0)) + \mathcal{O}(|t_1 - t_0|^{r+1}),$$
$$E_{L_d}^-(t_0, q(t_0), t_1, q(t_1)) = E_L(t_0, q(t_0), \dot{q}(t_0)) + \mathcal{O}(|t_1 - t_0|^{r+1}),$$

for any solution $q : [t_0, t_1] \to Q$ of the extended Euler–Lagrange equations for L.

Proof. This can be proved in the same way as Theorem 2.3.1, using the results of Lemma 4.6.2.   □

4.9.3. Autonomous discrete Lagrangians

If we have an autonomous discrete Lagrangian $L'_d : Q \times Q \to \mathbb{R}$, then we can either treat it using the autonomous theory and obtain the integrator $\tilde{F}_{L'_d} : T^*Q \to T^*Q$, or we can form the corresponding extended discrete Lagrangian by

$$L_d(t_0, q_0, t_1, q_1) = L'_d(q_0, q_1, t_1 - t_0),$$

as was considered in Section 4.3.7. As we saw there, the time translation invariance of L_d means that the two discrete energies are equal, and using the map $\hat{F}_{L_d}$ will integrate the system while preserving the discrete energy. Indeed, this method is exactly the symplectic-energy-momentum-conserving integrator presented in Kane *et al.* (1999*a*), written in a slightly different form. That paper gives some simple specific examples that show the advantage in using these types of integrators.

   Although using the autonomous integrator $\tilde{F}_{L'_d}$ for this system will not exactly preserve the discrete energy or the Hamiltonian, backward error analysis techniques can be used to show that the Hamiltonian, and hence the discrete energies, will be nearly preserved for exponentially long times (see, for example, Hairer (1994), Hairer and Lubich (1997) and Reich (1999*a*)).

   For autonomous systems, it will thus generally be preferable to use a standard variational integrator, and not to enforce exact energy conservation. Even in this case, however, the theory developed above for non-autonomous systems gives much insight into the geometric structure.

4.9.4. Existence of solutions

The question of the existence of solutions to the extended discrete Euler–Lagrange equations is, in general, rather subtle. There are two main problems which may arise.

First, it may be that for certain choices of initial condition (t_0, q_0, p_0, E_0), or correspondingly (t_0, q_0, t_1, q_1), that the extended discrete Euler–Lagrange equations simply have no solutions.

For specific choices of discrete Lagrangian it is sometimes possible to analyse this situation in more detail. For example, in Kane *et al.* (1999a) it is proved that, for Lagrangians of the form $L(q, \dot q) = \frac{1}{2}\dot q^T M \dot q - V(q)$ and the discrete Lagrangian L_d^α from Example 2.3.2, then the extended discrete Euler–Lagrange equations have unique solutions away from points where the kinetic energy is near zero. This criterion of being away from turning points is illustrated in that paper with numerical examples.

A particular choice of discrete Lagrangian, albeit in a different setting, is also analysed in Lew, Marsden, Ortiz and West (2001), where the existence of solutions to a time-adaptive variational PDE integrator is investigated.

The second problem which may arise is that the solution times t_k may not be unbounded, that is, they may converge to some finite value $t_k \to T \in \mathbb{R}$. This is similar to the issue of completeness of the differential equations in the continuous case, where a solution may escape to infinity in finite time.

4.9.5. Examples of extended variational integrators

To actually construct an extended discrete Lagrangian given an extended Lagrangian, most of the methods discussed in Section 2.6 can be readily extended to the nonautonomous case. Typically the time-step h should be replaced by $t_{k+1} - t_k$ and the continuous Lagrangian should be evaluated at the appropriate interpolated times. We give here a number of examples of this.

Example 4.9.3. (Low-order methods) Consider the discrete Lagrangian L_d^α from Example 2.3.2, and extend it to a nonautonomous Lagrangian by

$$\bar L_d^\alpha(t_0, q_0, t_1, q_1) = (t_1 - t_0)L\left((1 - \alpha)t_0 + \alpha t_1, (1 - \alpha)q_0 + \alpha q_1, \frac{q_1 - q_0}{t_1 - t_0}\right).$$

Similarly we can extend $L_d^{\mathrm{sym},\alpha}$ to

$$\bar L_d^{\mathrm{sym},\alpha} = \frac{1}{2}\bar L_d^\alpha + \frac{1}{2}\bar L_d^{1-\alpha}.$$

It is straightforward to check that the order of these discrete Lagrangians is the same as in the autonomous case, namely $\bar L_d^\alpha$ is first-order unless $\alpha = \frac{1}{2}$, when it is second-order, and $\bar L_d^{\mathrm{sym},\alpha}$ is always second-order.

This is reflected in the expression for the discrete energy $E_{\bar{L}_d^1}^-$ given in Example 4.9.1 above, which is clearly a first-order approximation to the Hamiltonian, as implied by Theorem 4.9.2.   ◊

Example 4.9.4. (Symplectic partitioned Runge–Kutta methods)
The symplectic partitioned Runge–Kutta methods discussed in Section 2.6.5 have a standard extension to nonautonomous systems, although they will clearly not be symplectic in the normal sense on T^*Q in that case. Taking coefficients c_i, b_i, a_{ij}, $\tilde{c}_i$, $\tilde{b}_i$, $\tilde{a}_{ij}$ for $i, j = 1, \ldots, s$, then the corresponding partitioned Runge–Kutta method for a regular extended Lagrangian system $L : \mathbb{R} \times TQ \to \mathbb{R}$ is a map $(t_0, q_0, p_0, h) \mapsto (t_1 = t_0 + h, q_1, p_1)$ defined by

$$q_1 = q_0 + (t_1 - t_0) \sum_{j=1}^{s} b_j \dot{Q}_j, \quad p_1 = p_0 + (t_1 - t_0) \sum_{j=1}^{s} \tilde{b}_j \dot{P}_j, \tag{4.9.2a}$$

$$Q_i = q_0 + (t_1 - t_0) \sum_{j=1}^{s} a_{ij} \dot{Q}_j, \quad P_i = p_0 + (t_1 - t_0) \sum_{j=1}^{s} \tilde{a}_{ij} \dot{P}_j, \quad i = 1, \ldots, s,$$
$$\tag{4.9.2b}$$

$$P_i = \frac{\partial L}{\partial \dot{q}}(t_i, Q_i, \dot{Q}_i), \qquad \dot{P}_i = \frac{\partial L}{\partial q}(\tilde{t}_i, Q_i, \dot{Q}_i), \qquad i = 1, \ldots, s,$$
$$\tag{4.9.2c}$$

where $t_i = (1 - c_i)t_0 + c_i t_1$ and $\tilde{t}_i = (1 - \tilde{c}_i)t_0 + \tilde{c}_i t_1$ are the interpolated times. As in the autonomous case, we can regard (4.9.2) as defining p_0, p_1, Q_i, P_i, $\dot{Q}_i$ and $\dot{P}_i$ for $i = 1, \ldots, s$ as functions of (t_0, q_0, t_1, q_1). Assuming this, we define the extended discrete Lagrangian

$$L_d(t_0, q_0, t_1, q_1) = (t_1 - t_0) \sum_{i=1}^{s} b_i L(t_i, Q_i, \dot{Q}_i),$$

and, if the coefficients satisfy the extended conditions

$$b_i \tilde{a}_{ij} + \tilde{b}_j a_{ji} = b_i \tilde{b}_j, \qquad i, j = 1, \ldots, s, \tag{4.9.3a}$$
$$b_i = \tilde{b}_i, \qquad i = 1, \ldots, s, \tag{4.9.3b}$$
$$c_i = \tilde{c}_i, \qquad i = 1, \ldots, s, \tag{4.9.3c}$$

then it is clear from the result in Section 2.6.5 that the partitioned Runge–Kutta method is exactly the restriction φ^h of the extended discrete Hamiltonian map $\tilde{F}_{L_d}$ defined in Section 4.5.4.

If we instead use the full map $\tilde{F}_{L_d}$ or the alternative $\hat{F}_{L_d}$ then the theory developed above shows that we will have a symplectic integrator in the extended sense.   ◊

PART FIVE
Further topics

In this part we briefly discuss some additional topics and future directions that are related to those covered in this review. We cite relevant references in the literature for further information.

5.1. Discrete symmetry reduction

The theory and applications of reduction of mechanical systems with symmetry both from the Hamiltonian and Lagrangian perspectives have been an active area of investigation for quite some time. For a recent review, see Marsden, Ratiu and Scheurle (2000). It is natural to investigate the discrete counterpart of this theory and such an investigation has begun. The discrete analogue of the Euler–Poincaré equations and reduction theory (rigid body and fluid-type equations on Lie algebras) lead to the DEP (discrete Euler–Poincaré) equations: see Marsden *et al.* (1999*a*, 1999*b*), and Bobenko and Suris (1999*a*, 1999*b*). The latter references also make intriguing links with discrete integrable systems. Other intriguing links between discrete mechanics, rigid body systems and optimal control are given in Bloch, Crouch, Marsden and Ratiu (1998).

The DEP context assumes that the configuration space of the system is a Lie group. The development of a more general reduction theory for group actions on more general configuration manifolds has begun, with the case of abelian group actions given in Jalnapurkar, Leok, Marsden and West (2000). The general nonabelian case as well as discrete analogues of singular reduction are also of considerable interest.

It would also be of interest to combine these variational methods for systems with symmetry with general methods that preserve symmetry structure to take advantage of both approaches: see Iserles, McLachlan and Zanna (1999) and Iserles *et al.* (2000), and literature cited therein.

5.2. Multisymplectic integrators for PDEs

The basic extension of the methods of this paper to the context of PDEs was given in Marsden *et al.* (1998), and Marsden and Shkoller (1999). These papers lay down the variational discretization of PDEs in a multisymplectic context. The examples in these papers were fairly simple, while more interesting examples in the context of continuum mechanics are given in Marsden, Pekarsky, Shkoller and West (2001*a*, 2001*b*).

An important point about the variational methods for multisymplectic PDEs is that they do not require that the PDE and multisymplectic structure be discretized separately (as in Bridges and Reich (200x)). Instead, the discrete multisymplectic structure arises directly from the discrete variational principle, which also immediately gives a discrete multisymplectic form formula: that is, the PDE analogue of the symplectic nature of variational integrators.

This procedure is illustrated in West (2001), where the method of Suris is extended to construct discrete Lagrangians for product Runge–Kutta discretizations of multisymplectic PDEs. From the variational principle we derive the discrete multisymplectic structure, which turns out to coincide with that proposed in Reich (2000).

Lew *et al.* (2001) develop and apply the theory of asynchronous variational integrators (AVIs) in the context of shell dynamics. This is the PDE analogue of integrators that use adaptive time-steps to achieve exact energy and momentum balance as well as being multisymplectic. We expect that similar integrators can be developed for other problems in continuum mechanics, including fluids.

5.3. Open problems

5.3.1. Reversibility

It would be interesting to make closer links with methods that have been developed for reversible systems (see, for example, Cano and Sanz-Serna (1997), Barth, Leimkuhler and Reich (1999), McLachlan, Quispel and Turner (1998*b*) and references therein). We should keep in mind that there are many interesting systems that are Lagrangian but not time-reversible (such as gyroscopic systems, including particles in magnetic fields) as well as systems that are reversible but not Lagrangian. See McLachlan, Quispel and Robidoux (1998*a*) for a discussion of the general classification of these systems as well as those with a first integral, volume-preserving integrators, *etc.*

5.3.2. Variational backward error analysis

Given a symplectic integrator F for a Hamiltonian vector field, traditional backward error analysis finds a modified vector field $\bar{X}_H$ which is exactly integrated by F. It is then shown that this is in fact the Hamiltonian vector field of a modified Hamiltonian $\bar{H}$, so that $\bar{X}_H = X_{\bar{H}}$, which is therefore exactly preserved by F. As H and $\bar{H}$ are close, this shows that F almost preserves H for long times.

We could alternatively do such an analysis on the Lagrangian side. Given a discrete Lagrangian that approximates the action of a given Lagrangian system $L_d \approx \int L \, dt$, we seek a modified Lagrangian whose action is exactly equal to the discrete Lagrangian, so that $L_d = \int \bar{L} \, dt$.

To do this, follow the same idea as for traditional backward error analysis and expand $\bar{L} = L + hL^1 + h^2 L^2 + \dots$. Substituting this into $L_d = \int \bar{L} \, dt$, expanding both sides in h, and equating terms allows us to solve for the L^i. To make this rigorous we would need to investigate the convergence of the sum, which will undoubtedly require optimal truncation techniques.

The primary advantage of a Lagrangian backward error analysis is that it may be able to be extended to Lagrangian PDEs, via the techniques of variational multisymplectic integrators, and to situations where the Hamiltonian vector field is not defined, such as nonsmooth contact problems.

5.3.3. Discrete multisymplectic reduction

A natural extension of the discrete reduction theory for ODEs would be its multisymplectic counterpart. Some modest progress has been made in this direction, for instance in Castrillon Lopez, Ratiu and Shkoller (2000), but much remains to be done. The examples of Maxwell's equations (where reduction theory is understood from the infinite-dimensional function space perspective, as in Marsden and Weinstein (1982)) and fluids are challenging, but progress seems likely, both with the theory and, eventually, the associated numerics.

5.3.4. Splitting methods

If $L = \sum L^i$, what can be said about constructing a discrete Lagrangian for L given discrete Lagrangians L_d^i for the L^i? The answer to such a question would be an interesting analogue of the corresponding question for Hamiltonian systems, as in, for example, McLachlan and Scovel (1996) and references therein. What makes the Hamiltonian case in a sense easier is that the equations are linear in the Hamiltonian, whereas this is not the case with the Euler–Lagrange equations (when written in the form $\dot{x} = f(x)$). On the other hand, Hamilton's variational principle is linear in the Lagrangian, so we should be able to exploit that structure.

5.3.5. Evolution of conserved quantities for forced systems

When forcing or dissipation is added to a Lagrangian or Hamiltonian system, then the symplectic form, momentum maps and energy are no longer preserved by the flow. It is often of importance in applications, however, to be able to correctly simulate the amount by which these various quantities change over time.

The variational framework for discrete systems with forcing given in Section 3.2 offers a way in which this evolution can be studied, both for the discrete system and for the true Lagrangian system. Numerical results in Kane *et al.* (2000) for simple systems indicate that, for weakly damped systems, correctly estimating energy and momentum decay requires the use of conservative integrators. It may be possible to make these statements rigorous using the ideas of variational backward error analysis and variational integrators with forcing.

5.3.6. Nonsmooth mechanics

Although we have concentrated on mechanical systems which follow smooth trajectories, there are many physical situations which demand nonsmooth models, such as collision and fragmentation problems. In such cases it can be very profitable to directly use nonsmooth techniques, as described in Kane *et al.* (1999*b*), rather than try to take smooth approximations.

When dealing with truly nonsmooth mechanics, many of the conventional definitions and derivations no longer apply, relying as they do on standard calculus. The variational approach, and in particular discrete variational mechanics, can be extended to nonsmooth situations with little difficulty, however, and provides a useful tool for the analysis of nonsmooth systems and numerical integrators for them (see Kane, Ortiz and Marsden (1998) and Fetecau *et al.* (2001)).

5.3.7. Systems with nonholonomic constraints

A nonholonomic constraint is, loosely speaking, a constraint on both the position and velocity variables. This can be defined by a constraint function $\phi : TQ \to \mathbb{R}$, and we seek trajectories $q(t)$ so that $\phi(q(t), \dot{q}(t)) = 0$ for all t. The variational principle which gives the correct equations of motion in this case is the Lagrange–d'Alembert principle used in Section 3.1 to add forcing (Bloch, Krishnaprasad, Marsden and Murray 1996*a*).

We could extend the discrete variational mechanics to systems with nonholonomic constraints by using the same discrete Lagrange–d'Alembert principle as in Section 3.2. This would yield a constrained discrete Hamiltonian map which would approximate the continuous constrained system, and preserve the nonholonomic constraints.

As nonholonomic systems do not, in general, preserve the standard symplectic structure, we would not expect the discrete system to preserve the discrete symplectic structure. It would be interesting, however, to see whether the evolution was qualitatively correct, as discussed above for the energy decay of dissipative systems. See McLachlan and Perlmutter (200x) for steps in this direction.

5.3.8. Other Galerkin methods

In Section 2.6.6 we considered variational integrators derived by taking polynomial approximations to the trajectory segments, and we saw that this is equivalent to Galerkin projection onto the space of polynomials.

While polynomials are a good choice for general smooth mechanical systems, there are many problems for which some other choice of curves may be more appropriate. For example, in systems with highly oscillatory components, variational integrators based on sinusoidal trajectory approximations may have superior accuracy or stability properties.

Acknowledgements

We thank many colleagues for their explicit and implicit help in putting this article together, including Razvan Fetecau, Arieh Iserles, Sameer Jalnapurkar, Couro Kane, Melvin Leok, Adrian Lew, Ben Leimkuhler, Michael Ortiz, George Patrick, Sergey Pekarsky, Reinout Quispel, Sebastian Reich, Steve Shkoller, and Robert Skeel. This work was supported by the California Institute of Technology and NSF/KDI grant ATM-9873133, as well as NSF grant DMS-9874082.

REFERENCES

R. Abraham, J. E. Marsden and T. Ratiu (1988), *Manifolds, Tensor Analysis, and Applications*, Vol. 75 of *Applied Mathematical Sciences*, 2nd edn, Springer, New York.

M. P. Allen and D. J. Tildesley (1987), *Computer Simulation of Liquids*, Oxford University Press, New York.

H. Anderson (1983), 'RATTLE: A velocity version of the SHAKE algorithm for molecular dynamics calculations', *J. Comput. Phys.* **52**, 24–34.

U. M. Ascher and L. R. Petzold (1998), *Computer Methods for Ordinary Differential Equations and Differential-Algebraic Equations*, SIAM, Philadelphia, PA.

M. A. Austin, P. S. Krishnaprasad and L. S. Wang (1993), Almost Poisson integration of rigid body systems, *J. Comput. Phys.* **107**, 105–117.

J. C. Baez and J. W. Gilliam (1994), An algebraic approach to discrete mechanics, *Lett. Math. Phys.* **31**, 205–212.

E. Barth and B. Leimkuhler (1996), 'Symplectic methods for conservative multibody systems', in *Integration Algorithms and Classical Mechanics* (Toronto, ON, 1993), American Mathematical Society, Providence, RI, pp. 25–43.

E. Barth, B. Leimkuhler and S. Reich (1999), 'A time-reversible variable-stepsize integrator for constrained dynamics', *SIAM J. Sci. Comput.* **21**, 1027–1044 (electronic).

G. Benettin and A. Giorgilli (1994), 'On the Hamiltonian interpolation of near-to-the-identity symplectic mappings with application to symplectic integration algorithms', *J. Statist. Phys.* **74**, 1117–1143.

J. S. Berg, R. L. Warnock, R. D. Ruth and E. Forest (1994), 'Construction of symplectic maps for nonlinear motion of particles in accelerators', *Phys. Rev. E* **49**, 722–739.

A. M. Bloch, P. S. Krishnaprasad, J. E. Marsden and R. M. Murray (1996a), 'Nonholonomic mechanical systems with symmetry', *Arch. Rational Mech. Anal.* **136**, 21–99.

A. M. Bloch, P. S. Krishnaprasad, J. E. Marsden and T. Ratiu (1996*b*), 'The Euler–Poincaré equations and double bracket dissipation', *Comm. Math. Phys.* **175**, 1–42.

A. M. Bloch, P. Crouch, J. E. Marsden and T. S. Ratiu (1998), 'Discrete rigid body dynamics and optimal control', *Proc. CDC* **37**, 2249–2254. Longer manuscript in preparation.

A. I. Bobenko and Y. B. Suris (1999*a*), 'Discrete Lagrangian reduction, discrete Euler–Poincaré equations, and semidirect products', *Lett. Math. Phys.* **49**, 79–93.

A. I. Bobenko and Y. B. Suris (1999*b*), 'Discrete time Lagrangian mechanics on Lie groups, with an application to the Lagrange top', *Comm. Math. Phys.* **204**, 147–188.

V. Brasey and E. Hairer (1993), 'Symmetrized half-explicit methods for constrained mechanical systems', *Appl. Numer. Math.* **13**, 23–31.

T. J. Bridges and S. Reich (200x), Multi-symplectic integrators: Numerical schemes for Hamiltonian PDEs that conserve symplecticity, *Phys. Lett. A*, to appear.

C. J. Budd and A. Iserles (1999), 'Geometric integration: Numerical solution of differential equations on manifolds', *Philos. Trans. Royal Soc. London Ser. A, Math. Phys. Eng. Sci.* **357**, 945–956.

J. A. Cadzow (1970), 'Discrete calculus of variations', *Internat. J. Control.* **11**, 393–407.

J. A. Cadzow (1973), *Discrete-Time Systems: An Introduction with Interdisciplinary Applications*, Prentice-Hall.

B. Cano and J. Lewis (1998), A comparison of symplectic and Hamilton's principle algorithms for autonomous and non-autonomous systems of ordinary differential equations, Technical report, Departamento de Matemática Aplicada y Computación, Universidad de Valladolid.

B. Cano and J. M. Sanz-Serna (1997), 'Error growth in the numerical integration of periodic orbits, with application to Hamiltonian and reversible systems', *SIAM J. Numer. Anal.* **34**, 1391–1417.

M. Castrillón López, T. S. Ratiu and S. Shkoller (2000), Reduction in principal fiber bundles: Covariant Euler–Poincaré equations, *Proc. Amer. Math. Soc.* **128**, 2155–2164.

P. J. Channell and C. Scovel (1990), 'Symplectic integration of Hamiltonian systems', *Nonlinearity* **3**, 231–259.

R. De Vogelaére (1956), Methods of integration which preserve the contact transformation property of the Hamiltonian equations, University of Notre Dame preprint.

D. Estep and D. French (1994), 'Global error control for the continuous Galerkin finite element method for ordinary differential equations', *RAIRO Modelisation Mathématique et Analyse Numérique* **28**, 815–852

R. Fetecau, J. E. Marsden, M. Ortiz and M. West (2001), Nonsmooth Lagrangian mechanics. Preprint.

E. Forest and R. D. Ruth (1990), '4th-order symplectic integration', *Physica D* **43**, 105–117.

Z. Ge and J. E. Marsden (1988), 'Lie–Poisson integrators and Lie–Poisson Hamilton–Jacobi theory', *Phys. Lett. A* **133**, 134–139.

S. Geng (1995), 'Construction of high-order symplectic PRK methods', *J. Comput. Math.* **13**, 40–50.

S. Geng (2000), 'A simple way of constructing symplectic Runge–Kutta methods', *J. Comput. Math.* **18**, 61–68.

J. W. Gilliam (1996), Lagrangian and symplectic techniques in discrete mechanics, PhD thesis, University of California, Riverside, Department of Mathematics. Available from: `http://math.ucr.edu/home/baez`

R. Gillilan and K. Wilson (1992), 'Shadowing, rare events and rubber bands: A variational Verlet algorithm for molecular dynamics', *J. Chem. Phys.* **97**, 1757–1772.

H. Goldstein (1980), *Classical Mechanics*, 1950, 2nd edn 1980, Addison-Wesley.

O. Gonzalez (1996a), Design and analysis of conserving integrators for nonlinear Hamiltonian systems with symmetry, PhD thesis, Stanford University, Department of Mechanical Engineering.

O. Gonzalez (1996b), 'Time integration and discrete Hamiltonian systems', *J. Nonlin. Sci.* **6**, 449–467.

O. Gonzalez (1999), 'Mechanical systems subject to holonomic constraints: Differential-algebraic formulations and conservative integration', *Physica D* **132**, 165–174.

O. Gonzalez and J. C. Simo (1996), 'On the stability of symplectic and energy-momentum algorithms for nonlinear Hamiltonian systems with symmetry', *Comput. Meth. Appl. Mech. Eng.* **134**, 197–222.

O. Gonzalez, D. J. Higham and A. M. Stuart (1999), 'Qualitative properties of modified equations', *IMA J. Numer. Anal.* **19**, 169–190.

M. Gotay, J. Isenberg and J. E. Marsden (1997), Momentum maps and classical relativistic fields, Part I: Covariant field theory. Unpublished; available from: `http://www.cds.caltech.edu/ marsden/`

E. Hairer (1994), 'Backward analysis of numerical integrators and symplectic methods', *Ann. Numer. Math.* **1**, 107–132.

E. Hairer (1997), 'Variable time step integration with symplectic methods', *Appl. Numer. Math.* **25**, 219–227.

E. Hairer (1998), 'Numerical geometric integration'. Notes available from: `http://www.unige.ch/math/folks/hairer/polycop.html`.

E. Hairer (200x), Geometric integration of ordinary differential equations on manifolds, *BIT*, to appear.

E. Hairer and C. Lubich (1997), 'The life-span of backward error analysis for numerical integrators', *Numer. Math.* **76**, 441–462.

E. Hairer and C. Lubich (1999), 'Invariant tori of dissipatively perturbed Hamiltonian systems under symplectic discretization', *Appl. Numer. Math.* **29**, 57–71.

E. Hairer and C. Lubich (2000), 'Long-time energy conservation of numerical methods for oscillatory differential equations', *SIAM J. Numer. Anal.* **38**, 414–441.

E. Hairer and G. Wanner (1996), *Solving Ordinary Differential Equations II: Stiff and Differential-Algebraic Problems*, Vol. 14 of *Springer Series in Computational Mathematics*, 2nd edn, Springer.

E. Hairer, S. P. Nørsett and G. Wanner (1993), *Solving Ordinary Differential Equations I: Nonstiff Problems*, Vol. 8 of *Springer Series in Computational Mathematics*, 2nd edn, Springer.

W. R. Hamilton (1834), On a general method in dynamics, *Philos. Trans. Royal Soc. London* Part II, 247–308; Part I for 1835, pp. 95–144.

D. J. Hardy, D. I. Okunbor and R. Skeel (1999), 'Symplectic variable step size integration for N-body problems', in *Proceedings of the NSF/CBMS Regional Conference on Numerical Analysis of Hamiltonian Differential Equations* (Golden, CO, 1997), *Appl. Numer. Math.* **29**, 19–30.

T. J. R. Hughes (1987), *The Finite Element Method: Linear Static and Dynamic Finite Element Analysis*, Prentice-Hall.

B. Hulme (1972a), 'One-step piecewise polynomial Galerkin methods for initial value problems', *Math. Comput.* **26**, 415–426.

B. Hulme (1972b), 'Discrete Galerkin and related one-step methods for ordinary differential equations', *Math. Comput.* **26**, 881–891.

C. L. Hwang and L. T. Fan (1967), 'A discrete version of Pontryagin's maximum principle'. *Oper. Res.* **15**, 139–146.

A. Iserles, R. I. McLachlan and A. Zanna (1999), 'Approximately preserving symmetries in the numerical integration of ordinary differential equations', *Europ. J. Appl. Math.* **10**, 419–445.

A. Iserles, H. Munthe-Kaas and A. Zanna (2000), 'Lie-group methods', in *Acta Numerica*, Vol. 9, Cambridge University Press, pp. 215–365.

T. Itoh and K. Abe (1988), 'Hamiltonian-conserving discrete canonical equations based on variational difference equations', *J. Comput. Phys.* **77**, 85–102.

C. G. K. Jacobi (1866), *Vorlesungen über Dynamik*, Verlag G. Reimer.

S. M. Jalnapurkar and J. E. Marsden (200x), 'Discretization of Hamiltonian Systems'. In preparation.

S. M. Jalnapurkar, M. Leok, J. E. Marsden and M. West (2000), Discrete Routh reduction. Preprint.

G. Jaroszkiewicz and K. Norton (1997a), 'Principles of discrete time mechanics, I: Particle systems', *J. Phys. A* **30**, 3115–3144.

G. Jaroszkiewicz and K. Norton (1997b), 'Principles of discrete time mechanics, II: Classical field theory', *J. Phys. A* **30**, 3145–3163.

L. O. Jay (1996), 'Symplectic partitioned Runge–Kutta methods for constrained Hamiltonian systems', *SIAM J. Numer. Anal.* **33**, 368–387.

L. O. Jay (1999), 'Structure preservation for constrained dynamics with super partitioned additive Runge–Kutta methods', *SIAM J. Sci. Comput.* **20**, 416–446.

B. W. Jordan and E. Polak (1964), 'Theory of a class of discrete optimal control systems'. *J. Electron. Control* **17**, 697–711.

C. Kane, M. Ortiz and J. E. Marsden (1998), The convergence of collision algorithms. Notes.

C. Kane, J. E. Marsden and M. Ortiz (1999a), 'Symplectic energy-momentum integrators', *J. Math. Phys.* **40**, 3353–3371.

C. Kane, E. A. Repetto, M. Ortiz and J. E. Marsden (1999b), 'Finite element analysis of nonsmooth contact', *Comput. Meth. Appl. Mech. Eng.* **180**, 1–26.

C. Kane, J. E. Marsden, M. Ortiz and M. West (2000), 'Variational integrators and the Newmark algorithm for conservative and dissipative mechanical systems', *Internat. J. Numer. Math. Eng.* **49**, 1295–1325.

R. A. Labudde and D. Greenspan (1974), 'Discrete mechanics: A general treatment', *J. Comput. Phys.* **15**, 134–167.

R. A. Labudde and D. Greenspan (1976*a*), 'Energy and momentum conserving methods of arbitrary order for the numerical integration of equations of motion, I: Motion of a single particle', *Numer. Math.* **25**, 323–346.

R. A. Labudde and D. Greenspan (1976*b*), 'Energy and momentum conserving methods of arbitrary order for the numerical integration of equations of motion, II: Motion of a system of particles', *Numer. Math.* **26**, 1–16.

F. Lasagni (1988), 'Canonical Runge–Kutta methods', *ZAMP* **39**, 952–953.

T. D. Lee (1983), 'Can time be a discrete dynamical variable?', *Phys. Lett. B* **122**, 217–220.

T. D. Lee (1987), 'Difference equations and conservation laws', *J. Stat. Phys.* **46**, 843–860

B. Leimkuhler and G. Patrick (1996), Symplectic integration on Riemannian manifolds, *J. Nonlin. Sci.* **6**, 367–384.

B. Leimkuhler and S. Reich (1994), 'Symplectic integration of constrained Hamiltonian systems', *Math. Comput.* **63**, 589–605.

B. J. Leimkuhler and R. Skeel (1994), 'Symplectic numerical integrators in constrained Hamiltonian systems', *J. Comput. Phys.* **112**, 117–125.

A. Lew, J. E. Marsden, M. Ortiz and M. West (2001), 'Asynchronous variational integrators'. Preprint.

J. D. Logan (1973), 'First integrals in the discrete calculus of variations', *Aequationes Mathematicae* **9**, 210–220.

R. MacKay (1992), Some aspects of the dynamics of Hamiltonian systems, in *The Dynamics of Numerics and the Numerics of Dynamics* (D. Broomhead and A. Iserles, eds), Clarendon Press, Oxford, pp. 137–193.

R. I. McLachlan (1995), 'On the numerical integration of ordinary differential equations by symmetric composition methods', *SIAM J. Sci. Comput.* **16**, 151–168.

R. I. McLachlan and M. Perlmutter (200x), Geometric integration of nonholonomic mechanical systems. In preparation.

R. I. McLachlan and C. Scovel (1995), 'Equivariant constrained symplectic integration', *J. Nonlin. Sci.* **5**, 233–256.

R. I. McLachlan and C. Scovel (1996), 'A survey of open problems in symplectic integration', *Fields Inst. Commun.* **10**, 151–180.

R. I. McLachlan, G. R. W. Quispel and N. Robidoux (1998*a*), 'Unified approach to Hamiltonian systems, Poisson systems, gradient systems, and systems with Lyapunov functions or first integrals', *Phys. Rev. Lett.* **81**, 2399–2403.

R. I. McLachlan, G. R. W. Quispel and G. Turner (1998*b*), 'Numerical integrators that preserve symmetries and reversing symmetries', *SIAM J. Numer. Anal.* **35**, 586–599.

R. I. McLachlan, G. R. W. Quispel and N. Robidoux (1999), 'Geometric integration using discrete gradients', *Philos. Trans. Royal Soc. London Ser. A, Math. Phys. Eng. Sci.* **357**, 1021–1045.

S. Maeda (1980), 'Canonical structure and symmetries for discrete systems', *Math. Japonica* **25**, 405–420.

S. Maeda (1981a), 'Extension of discrete Noether theorem', *Math. Japonica* **26**, 85–90

S. Maeda (1981b), 'Lagrangian formulation of discrete systems and concept of difference space', *Math. Japonica* **27**, 345–356.

J. E. Marsden (1999), 'Park City lectures on mechanics, dynamics and symmetry', in *Symplectic Geometry and Topology* (Y. Eliashberg and L. Traynor, eds), American Mathematical Society, Providence, RI, Vol. 7 of *IAS/Park City Math. Ser.*, pp. 335–430.

J. E. Marsden and T. Ratiu (1999), *Introduction to Mechanics and Symmetry*, Vol. 17 of *Texts in Applied Mathematics*, 2nd edn, Springer.

J. E. Marsden and S. Shkoller (1999), 'Multisymplectic geometry, covariant Hamiltonians, and water waves', *Math. Proc. Cambridge Philos. Soc.* **125**, 553–575.

J. E. Marsden and A. Weinstein (1982), The Hamiltonian structure of the Maxwell–Vlasov equations, *Physica D* **4**, 394–406.

J. E. Marsden, G. W. Patrick and S. Shkoller (1998), 'Multisymplectic geometry, variational integrators, and nonlinear PDEs', *Comm. Math. Phys.* **199**, 351–395.

J. E. Marsden, S. Pekarsky and S. Shkoller (1999a), 'Discrete Euler–Poincaré and Lie–Poisson equations', *Nonlinearity* **12**, 1647–1662.

J. E. Marsden, S. Pekarsky and S. Shkoller (1999b), Symmetry reduction of discrete Lagrangian mechanics on Lie groups, *J. Geom. Phys.* **36**, 140–151.

J. E. Marsden, T. Ratiu and J. Scheurle (2000), Reduction theory and the Lagrange–Routh equations, *J. Math. Phys.* **41**, 3379–3429.

J. E. Marsden, S. Pekarsky, S. Shkoller and M. West (2001a), 'Variational methods, multisymplectic geometry and continuum mechanics', *J. Geom. Phys.*, to appear.

J. E. Marsden, S. Pekarsky, S. Shkoller and M. West (2001b), 'Multisymplectic continuum mechanics in Euclidean spaces and multisymplectic discretizations'. Preprint.

J. Moser and A. P. Veselov (1991), 'Discrete versions of some classical integrable systems and factorization of matrix polynomials', *Comm. Math. Phys.* **139**, 217–243.

A. Murua and J. M. Sanz-Serna (1999), 'Order conditions for numerical integrators obtained by composing simpler integrators', *Philos. Trans. Royal Soc. London Ser. A, Math. Phys. Eng. Sci.* **357**, 1079–1100.

U. Mutze (1998), Predicting classical motion directly from the action principle. Preprint.

N. N. Newmark (1959), 'A method of computation for structural dynamics', *ASCE J. Eng. Mech. Div.* **85**, 67–94.

E. Noether (1918), 'Invariante Variationsprobleme', *Kgl. Ges. Wiss. Nachr. Göttingen. Math. Physik* **2**, 235–257.

K. Norton and G. Jaroszkiewicz (1998), 'Principles of discrete time mechanics, III: Quantum field theory', *J. Phys. A* **31**, 977–1000.

W. Oevel and M. Sofroniou (1997), Symplectic Runge–Kutta schemes, II: Classification of symmetric methods. Preprint.

P. J. Oliver and J. Sivaloganathan (1988), The structure of null Lagrangians, *Nonlinearity* **1**, 389–398.

M. Ortiz (1986), 'A note on energy conservation and stability of nonlinear time-stepping algorithms', *Comput. Structures* **24**, 167–168.

M. Ortiz and L. Stainier (1999), 'The variational formulation of viscoplastic constitutive updates', *Comput. Meth. Appl. Mech. Eng.* **171**, 419–444.

M. Qin and W. J. Zhu (1992), 'Construction of higher order symplectic schemes by composition', *Computing* **27**, 309–321.

R. Radovitzky and M. Ortiz (1999), 'Error estimation and adaptive meshing in strongly nonlinear dynamic problems', *Comput. Meth. Appl. Mech. Eng.* **172**, 203–240.

S. Reich (1996), 'Symplectic integration of constrained Hamiltonian systems by composition methods', *SIAM J. Numer. Anal.* **33**, 475–491.

S. Reich (1997), 'On higher-order semi-explicit symplectic partitioned Runge–Kutta methods for constrained Hamiltonian systems', *Numer. Math.* **76**, 231–247.

S. Reich (1999a), 'Backward error analysis for numerical integrators', *SIAM J. Numer. Anal.* **36**, 1549–1570.

S. Reich (1999b), 'Preservation of adiabatic invariants under symplectic discretization', *Appl. Numer. Math.* **29**, 45–55.

S. Reich (2000), Multi-symplectic Runge–Kutta collocation methods for Hamiltonian wave equations, *J. Comput. Phys.* **157**, 473–499.

R. D. Ruth (1983), 'A canonical integration technique', *IEEE Trans. Nuclear Sci.* **30**, 2669–2671.

J. Ryckaert, G. Ciccotti and H. Berendsen (1977), 'Numerical integration of the Cartesian equations of motion of a system with constraints: Molecular dynamics of n-alkanes', *J. Comput. Phys.* **23**, 327–341.

J. M. Sanz-Serna (1988), 'Runge–Kutta schemes for Hamiltonian systems', *BIT* **28**, 877–883.

J. M. Sanz-Serna (1992a), The numerical integration of Hamiltonian systems, in *Computational Ordinary Differential Equations* (J. Cash and I. Gladwell, eds), Clarendon Press, Oxford, pp. 437–449.

J. M. Sanz-Serna (1992b), 'Symplectic integrators for Hamiltonian problems: An overview', in *Acta Numerica*, Vol. 1, Cambridge University Press, pp. 243–286.

J. M. Sanz-serna and M. P. Calvo (1994), *Numerical Hamiltonian Problems*, Chapman and Hall, London.

T. Schlick, R. Skeel *et al.* (1999), 'Algorithmic challenges in computational molecular biophysics', *J. Comput. Phys.* **151**, 9–48.

W. M. Seiler (1998a), 'Numerical analysis of constrained Hamiltonian systems and the formal theory of differential equations', *Math. Comput. Simulation* **45**, 561–576.

W. M. Seiler (1998b), 'Position versus momentum projections for constrained Hamiltonian systems', *Numer. Algorithms* **19**, 223–234.

W. M. Seiler (1999), 'Numerical integration of constrained Hamiltonian systems using Dirac brackets', *Math. Comput.* **68**, 661–681.

Y. Shibberu (1994), 'Time-discretization of Hamiltonian systems', *Comput. Math. Appl.* **28**, 123–145.

M. Shimada and H. Yoshida (1996), 'Long-term conservation of adiabatic invariants by using symplectic integrators', *Publ. Astronomical Soc. Japan* **48**, 147–155.

J. Simo and N. Tarnow (1992), 'The discrete energy momentum method: Conserving algorithms for nonlinear elastodynamics', *ZAMP* **43**, 757–792.

J. C. Simo, N. Tarnow and K. K. Wong (1992), 'Exact energy-momentum conserving algorithms and symplectic schemes for nonlinear dynamics', *Comput. Meth. Appl. Mech. Eng.* **100**, 63–116.

R. D. Skeel and K. Srinivas (2000), 'Nonlinear stability analysis of area-preserving integrators', *SIAM J. Numer. Anal.* **38**, 129–148.

R. D. Skeel, G. Zhang and T. Schlick (1997), 'A family of symplectic integrators: Stability, accuracy, and molecular dynamics applications', *SIAM J. Sci. Comput.* **18**, 203–222.

M. Sofroniou and W. Oevel (1997), 'Symplectic Runge–Kutta schemes, I: Order conditions', *SIAM J. Numer. Anal.* **34**, 2063–2086.

Y. B. Suris (1989), 'The canonicity of mappings generated by Runge–Kutta type methods when integrating the system $\ddot{x} = -\partial u/\partial x$', *USSR Comput. Math. Phys.* **29**, 138–144.

Y. B. Suris (1990), 'Hamiltonian methods of Runge–Kutta type and their variational interpretation', *Math. Simulation* **2**, 78–87.

W. C. Swope, H. C. Andersen, P. H. Berens and K. R. Wilson (1982), 'A computer-simulation method for the calculation of equilibrium-constants for the formation of physical clusters of molecules: Application to small water clusters', *J. Chem. Phys.* **76**, 637–649.

V. Thomée (1997), *Galerkin Finite Element Methods for Parabolic Problems*, Springer, New York.

L. Verlet (1967), 'Computer experiments on classical fluids', *Phys. Rev.* **159**, 98–103.

A. P. Veselov (1988), 'Integrable discrete-time systems and difference operators', *Funct. Anal. Appl.* **22**, 83–93.

A. P. Veselov (1991), 'Integrable Lagrangian correspondences and the factorization of matrix polynomials', *Funct. Anal. Appl.* **25**, 112–122.

R. L. Warnock and R. D. Ruth (1991), 'Stability of nonlinear Hamiltonian motion for a finite but very long-time', *Phys. Rev. Lett.* **66**, 990–993.

R. L. Warnock and R. D. Ruth (1992), 'Long-term bounds on nonlinear Hamiltonian motion', *Physica D* **56**, 188–215.

J. M. Wendlandt and J. E. Marsden (1997*a*), 'Mechanical integrators derived from a discrete variational principle', *Physica D* **106**, 223–246.

J. M. Wendlandt and J. E. Marsden (1997*b*), Mechanical systems with symmetry, variational principles and integration algorithms, in *Current and Future Directions in Applied Mathematics* (M. Alber, B. Hu and J. Rosenthal, eds), Birkhäuser, pp. 219–261.

M. West (2001), Variational Runge-Kutta methods for ODEs and PDEs. Preprint.

J. Wisdom, S. J. Peale and F. Mignard (1984), The chaotic rotation of Hyperion, *Icarus* **58**, 137–152.

H. Yoshida (1990), 'Construction of higher-order symplectic integrators', *Phys. Lett. A* **150**, 262–268.

Acta Numerica (2001), pp. 515–560

Semidefinite optimization

M. J. Todd*

School of Operations Research and Industrial Engineering,
Cornell University, Ithaca,
New York 14853, USA
E-mail: miketodd@cs.cornell.edu

Optimization problems in which the variable is not a vector but a symmetric matrix which is required to be positive semidefinite have been intensely studied in the last ten years. Part of the reason for the interest stems from the applicability of such problems to such diverse areas as designing the strongest column, checking the stability of a differential inclusion, and obtaining tight bounds for hard combinatorial optimization problems. Part also derives from great advances in our ability to solve such problems efficiently in theory and in practice (perhaps 'or' would be more appropriate: the most effective computational methods are not always provably efficient in theory, and *vice versa*). Here we describe this class of optimization problems, give a number of examples demonstrating its significance, outline its duality theory, and discuss algorithms for solving such problems.

CONTENTS

* Research supported in part by NSF through grant DMS-9805602 and ONR through grant N00014-96-1-0050.

1. Introduction

Semidefinite optimization is concerned with choosing a symmetric matrix to optimize a linear function subject to linear constraints and a further crucial constraint that the matrix be positive semidefinite. It thus arises from the well-known linear programming problem by replacing the vector of variables with a symmetric matrix and replacing the nonnegativity constraints with a positive semidefinite constraint. (An alternative way to write such a problem is in terms of a vector of variables, with a linear objective function and a constraint that some symmetric matrix that depends affinely on the variables be positive semidefinite.) This generalization nevertheless inherits several important properties from its vector counterpart: it is convex, has a rich duality theory (although not as strong as that of linear programming), and admits theoretically efficient solution procedures based on iterating interior points to either follow the central path or decrease a potential function. Here we will investigate this class of problems and survey the recent results and methods obtained.

While linear programming (LP) as a subject grew very fast during the 1950s and 1960s, due to the availability of Dantzig's very efficient simplex method, semidefinite optimization (also known as semidefinite programming or SDP, the term we shall use) was slower to attract as much attention. Partly this was because, since the feasible region is no longer polyhedral, the simplex method was not applicable, although related methods do exist. As soon as theoretically efficient (as well as practically useful) algorithms became available in the late 1980s and 1990s, research in the area exploded. The recent *Handbook of Semidefinite Programming* (Wolkowicz, Saigal and Vandenberghe 2000) lists 877 references, while the online bibliography on semidefinite programming collected by Wolkowicz (2001) lists 722, almost all since 1990.

The development of efficient algorithms was only one trigger of this explosive growth: another key motivation was the power of SDP to model problems arising in a very wide range of areas. We will describe some of these applications in Section 3, but these only cover part of the domain. The handbook of Wolkowicz *et al.* (2000) has chapters on applications in combinatorial optimization, on nonconvex quadratic programming, on eigenvalue and nonconvex optimization, on systems and control theory, on structural design, on matrix completion problems, and on problems in statistics.

Bellman and Fan (1963) seem to have been the first to formulate a semidefinite programming problem. Instead of considering a linear programming problem in vector form and replacing the vector variable with a matrix variable, they started with a scalar LP formulation and replaced each scalar variable with a matrix. The resulting problem (though equivalent to the general formulation) was somewhat cumbersome, but they derived a dual problem and established several key duality theorems, showing that additional regularity is needed in the SDP case to prove strong duality. However, the importance of constraints requiring that a certain matrix be positive (semi)definite had been recognized much earlier in control theory: Lyapunov's characterization of the stability of the solution of a linear differential equation in 1890 involved just such a constraint (called a linear matrix inequality, or LMI), and subsequent work of Luré, Postnikov, and Yakubovich in the Soviet Union in the 1940s, 1950s, and 1960s established the importance of LMIs in control theory (see Boyd, El Ghaoui, Feron and Balakrishnan (1994)). In the early 1970s, Donath and Hoffman (1973) and then Cullum, Donath and Wolfe (1975) showed that some hard graph-partitioning problems could be attacked by considering a related eigenvalue optimization problem – as we shall see, these are closely connected with SDP. Then Lovász (1979) formulated an SDP problem that provided a bound on the Shannon capacity of a graph and thereby found the capacity of the pentagon, solving a long-open conjecture. At that time, the most efficient method known for SDP problems was the ellipsoid method, and Grötschel, Lovász and Schrijver (1988) investigated in detail its application to combinatorial optimization problems by using it to approximate the solution of both LP and SDP relaxations. Lovász and Schrijver (1991) later showed how SDP problems can provide tighter relaxations of $(0, 1)$-programming problems than can LP.

Fletcher (1981, 1985) revived interest in SDP among nonlinear programmers in the 1980s, and this led to a series of papers by Overton, and Overton and Womersley; see Overton and Womersley (1993) and the references therein. The key contributions of Nesterov and Nemirovski (1992, 1994) and Alizadeh (1995) showed that the new generation of interior-point methods pioneered by Karmarkar (1984) for LP could be extended to SDP. In particular, Nesterov and Nemirovski established a general framework for solving nonlinear convex optimization problems in a theoretically efficient way using interior-point methods, by developing the powerful theory of self-concordant barrier functions. These works led to the huge recent interest in semidefinite programming, which was further increased by the result of Goemans and Williamson (1995) which showed that an SDP relaxation could provide a provably good approximation to the max-cut problem in combinatorial optimization.

Our coverage will necessarily be incomplete and biased. Let us therefore refer the reader to a survey paper by Vandenberghe and Boyd (1996) which discusses in particular a number of applications, especially in control theory; the book of Boyd *et al.* (1994), which describes the latter in much further detail and gives the history of SDP in control theory; the excellent paper of Lewis and Overton (1996) in this journal on the very closely related topic of eigenvalue optimization; and the aforementioned handbook edited by Wolkowicz *et al.* (2000). We also mention that SDP is both an extension of LP and a special case of more general conic optimization problems. Nesterov and Nemirovski (1992, 1994) consider general convex cones, with the sole proviso that a self-concordant barrier is known for the cone. Nesterov and Todd (1997, 1998) consider the subclass of self-scaled cones, which admit symmetric primal-dual algorithms (these cones turn out to coincide with symmetric (homogeneous self-dual) cones). Another viewpoint is that of Euclidean Jordan Algebras, developed by Faybusovich (1997*a*, 1997*b*) and now investigated by a number of authors: see Alizadeh and Schmieta (2000). Since the area is receiving so much attention, it is hard to keep abreast of recent developments, but this is immeasurably assisted by three websites, those of Helmberg (2001) and Alizadeh (2001) on semidefinite programming, and that of Wright (2001) on interior-point methods. The latter also allows one to sign up for the interior-point methods mailing list, where almost all papers addressing interior-point methods for SDP are announced.

The rest of the paper is organized as follows. In the next section, we define the SDP problem in both primal and dual form and introduce some useful notation for expressing it. We also establish weak duality. Then Section 3 gives nine examples of the application of SDP to diverse areas; along the way, we list a number of useful facts about symmetric matrices that allow this development. The following section is devoted to duality, and presents some examples demonstrating the anomalies that can occur; then conditions sufficient for strong duality are established. Section 5 introduces the very important logarithmic barrier function for the cone of positive semidefinite matrices, and uses it to define the central path and to derive some of its important properties. Then in Section 6 we consider path-following and potential-reduction algorithms, and also methods based on nonlinear programming reformulations of the SDP problem. Section 7 contains some concluding remarks.

Notation
Most matrices occurring in this paper will be real symmetric matrices of order n: we let $S\mathbb{R}^{n \times n}$ denote the space of such matrices. $U \bullet V$ denotes the inner product between two such matrices, defined by trace $(U^T V)$ (the transpose makes this valid for nonsymmetric and even nonsquare matrices also). The associated norm is the Frobenius norm, written $\|U\|_F := (U \bullet U)^{\frac{1}{2}}$ or just $\|U\|$, while $\|P\|_2$ denotes the L_2-operator norm of a matrix. Norms on vectors will always be Euclidean unless otherwise noted.

We write $U \succeq 0$ to mean that U is positive semidefinite. Similarly, $U \succ 0$ indicates that U is positive definite, and these terms always refer to symmetric matrices unless there is an explicit statement otherwise. We write $S\mathbb{R}_+^{n \times n}$ ($S\mathbb{R}_{++}^{n \times n}$) to denote the set of positive semidefinite (positive definite) symmetric matrices of order n. We use $U \preceq V$ or $V \succeq U$ to mean $V - U \succeq 0$, and $U \prec V$ and $V \succ U$ similarly mean $V - U \succ 0$. If $U \succeq 0$, we write $U^{\frac{1}{2}}$ for the (symmetric) positive semidefinite square root of U.

We write $\operatorname{diag}(U)$ for the vector of diagonal entries of $U \in S\mathbb{R}^{n \times n}$, and $\operatorname{Diag}(u)$ for the diagonal matrix with the vector $u \in \mathbb{R}^n$ on its diagonal. We extend this to general block diagonal matrices: if $U_1, U_2, \ldots, U_k$ are symmetric matrices, then $\operatorname{Diag}(U_1, U_2, \ldots, U_k)$ denotes the block diagonal matrix with the U_is down its diagonal.

As is customary, lower-case Roman letters usually denote vectors and upper-case letters $n \times n$ matrices; we reserve K, L, P, and Q (Q will usually be orthogonal) for not necessarily symmetric matrices, with all other letters denoting members of $S\mathbb{R}^{n \times n}$. We use lower-case Greek letters for scalars, and script letters for linear operators on (usually symmetric) matrices. We introduce the useful notation $P \odot Q$ for $n \times n$ matrices P and Q (usually P and Q are symmetric). This is an operator from $S\mathbb{R}^{n \times n}$ to itself defined by

$$(P \odot Q)U := \frac{1}{2}(PUQ^T + QUP^T). \tag{1.1}$$

2. Problems

The SDP problem in *primal standard form* is

$$
\begin{aligned}
\text{(P)}: \quad &\min_X \ C \bullet X, \\
&A_i \bullet X = b_i, \quad i = 1, \ldots, m, \\
&X \succeq 0,
\end{aligned}
$$

where all $A_i \in S\mathbb{R}^{n \times n}$, $b \in \mathbb{R}^m$, $C \in S\mathbb{R}^{n \times n}$ are given, and $X \in S\mathbb{R}^{n \times n}$ is the variable. We also consider SDP problems in *dual standard form*:

$$
\begin{aligned}
\text{(D)}: \quad &\max_{y, S} \ b^T y, \\
&\textstyle\sum_{i=1}^m y_i A_i + S = C, \\
&S \succeq 0,
\end{aligned}
$$

where $y \in \mathbb{R}^m$ and $S \in S\mathbb{R}^{n \times n}$ are the variables. This can also be written as

$$\max_y \ b^T y, \quad \sum_{i=1}^m y_i A_i \preceq C,$$

or

$$\max_y \ b^T y, \quad C - \sum_{i=1}^m y_i A_i \succeq 0,$$

but we shall see the benefit of having the 'slack matrix' S available when we discuss algorithms.

We should strictly write 'inf' and 'sup' instead of 'min' and 'max' above, not just because the problems might be unbounded, but also because even if the optimal values are finite they might not be attained. We stick to 'min' and 'max' both to highlight the fact that we are interested in optimal solutions, not just values, and because we shall often impose conditions that ensure that the optimal values are in fact attained where finite.

The last form of the problem in dual standard form shows that we are try-ing to optimize a linear function of several variables, subject to the constraint that a symmetric matrix that depends affinely on the variables is restric-ted to be positive semidefinite. (Henceforth, as is common in mathematical programming, we use 'linear' to mean 'affine' in most cases: however, linear operators will always be linear, not affine.) We will encounter several ex-amples of such problems, and will not see the need to express them explicitly in the form above, but it is straightforward to do so.

We have been somewhat coy in referring to the problems above as SDP problems in primal and dual form respectively. If they are defined by the same data A_i, $i = 1, \ldots, m$, b, and C, they are in fact dual problems, and have a beautiful theory that will be studied in Section 4. However, we find it useful to discuss some examples before we investigate duality in detail. Here we just note the following trivial but key fact.

Proposition 2.1. (Weak duality) If X is feasible in (P) and (y, S) in (D), then

$$C \bullet X - b^T y = X \bullet S \geq 0. \tag{2.1}$$

Proof. We find

$$C \bullet X - b^T y = \left(\sum_{i=1}^{m} y_i A_i + S \right) \bullet X - b^T y$$

$$= \sum_{i=1}^{m} (A_i \bullet X) y_i + S \bullet X - b^T y = S \bullet X = X \bullet S.$$

Moreover, since X is positive semidefinite, it has a square root $X^{\frac{1}{2}}$, and so $X \bullet S = \text{trace}\,(XS) = \text{trace}\,(X^{\frac{1}{2}} X^{\frac{1}{2}} S) = \text{trace}\,(X^{\frac{1}{2}} S X^{\frac{1}{2}}) \geq 0$. Here we used the facts that $\text{trace}\,(PQ) = \text{trace}\,(QP)$, that $X^{\frac{1}{2}} S X^{\frac{1}{2}}$ is positive semidefinite since S is (from the definition), and that the trace of a positive semidefinite matrix is nonnegative (as the sum of its nonnegative diagonal elements or the sum of its nonnegative eigenvalues). □

It is convenient to introduce some notation to make the problems above easier to state. We define the linear operator $\mathcal{A} : S\mathbb{R}^{n \times n} \to \mathbb{R}^m$ by

$$\mathcal{A}X := (A_i \bullet X)_{i=1}^{m} \in \mathbb{R}^m.$$

Note that, for any $X \in S\mathbb{R}^{n \times n}$ and $v \in \mathbb{R}^m$, $(\mathcal{A}X)^T v = \sum_{i=1}^{m} (A_i \bullet X) v_i =$

$(\sum_{i=1}^{m} v_i A_i) \bullet X$, so the adjoint of $\mathcal{A}$ is given by

$$\mathcal{A}^* v = \sum_{i=1}^{m} v_i A_i,$$

a mapping from $\mathbb{R}^m$ to $S\mathbb{R}^{n \times n}$. Using this notation, we can rewrite our problems as

$$(P): \quad \min C \bullet X, \quad \mathcal{A}X = b, \quad X \succeq 0,$$

and

$$(D): \quad \max b^T y, \quad \mathcal{A}^* y + S = C, \quad S \succeq 0.$$

The weak duality chain of equations can then be written as

$$C \bullet X - b^T y = (\mathcal{A}^* y + S) \bullet X - b^T y = (\mathcal{A}X)^T y + S \bullet X - b^T y = X \bullet S.$$

We call the difference between the optimal value of (P) and that of (D), which is always nonnegative by the result above, the *duality gap*. Strong duality is the assertion that the duality gap is zero and both problems attain their optima whenever both problems are feasible, but it does not always hold for SDP problems. We investigate this in detail in Section 4.

3. Examples

In this section we present a number of examples of SDP problems. In order to do so, we also introduce some simple facts about symmetric matrices. Here is our first example.

Example 1: Minimizing the maximum eigenvalue. This arises in stabilizing a differential equation, for instance. Suppose we have a symmetric matrix, say $M(z)$, depending linearly (affinely) on a vector z. We wish to choose z to minimize the maximum eigenvalue of $M(z)$. Note that $\lambda_{\max}(M(z)) \leq \eta$ iff $\lambda_{\max}(M(z) - \eta I) \leq 0$, or equivalently iff $\lambda_{\min}(\eta I - M(z)) \geq 0$. This holds iff $\eta I - M(z) \succeq 0$. So we get the SDP problem in dual form:

$$\max -\eta, \quad \eta I - M(z) \succeq 0, \tag{3.1}$$

where the variable is $y := (\eta; z)$.

To introduce other examples, we need to use a collection of very handy tools concerning symmetric matrices. We list these below (usually) without proof, but most are not hard to show.

Fact 1. If $P \in \mathbb{R}^{m \times n}$ and $Q \in \mathbb{R}^{n \times m}$, then trace $(PQ) =$ trace (QP).

Fact 2. $\mathcal{A}$ and $\mathcal{A}^*$ above are adjoints.

Fact 3. If $U, V \in S\mathbb{R}^{n \times n}$, and Q is orthogonal, then $U \bullet V = (Q^T U Q) \bullet (Q^T V Q)$. More generally, if P is nonsingular, $U \bullet V = (PUP^T) \bullet (P^{-T} V P^{-1})$.

Fact 4. Every $U \in S\mathbb{R}^{n \times n}$ can be written as $U = Q \Lambda Q^T$, where Q is orthogonal and Λ is diagonal. Then $UQ = Q\Lambda$, so the columns of Q are the eigenvectors, and the diagonal entries of Λ the corresponding eigenvalues of U. We write $Q(U) := Q$, $\Lambda(U) := \Lambda$ and $\lambda(U) := \mathrm{diag}(\Lambda)$. (Together with Fact 3, this means that we can often assume that one symmetric matrix under study is diagonal, which can simplify some proofs.)

Fact 5. The following are norms on $S\mathbb{R}^{n \times n}$: $\|\lambda(U)\|_2 = \|U\|_F$, $\|\lambda(U)\|_\infty = \|U\|_2$, and $\|\lambda(U)\|_1$. If $U \succeq 0$, $\|\lambda(U)\|_1 = \sum_j |\lambda_j(U)| = \sum_j \lambda_j(U) = I \bullet \Lambda(U) = \mathrm{trace}(U) = I \bullet U$.

Fact 6. For $U \in S\mathbb{R}^{n \times n}$, the following are equivalent:

(a) $U \succeq 0$ $(U \succ 0)$;
(b) $v^T U v \geq 0$ for all $v \in \mathbb{R}^n$ $(v^T U v > 0$ for nonzero $v \in \mathbb{R}^n)$;
(c) $\lambda(U) \geq 0$ $(\lambda(U) > 0)$; and
(d) $U = P^T P$ for some matrix P $(U = P^T P$ for some square nonsingular matrix $P)$.

Immediate corollaries are that $uu^T \succeq 0$ for all $u \in \mathbb{R}^n$, that every $U \succeq 0$ has a positive semidefinite square root $U^{\frac{1}{2}}$ (take $U^{\frac{1}{2}} = Q(U)\Lambda^{\frac{1}{2}}(U)Q^T(U)$, where $\Lambda^{\frac{1}{2}}(U)$ is the diagonal matrix whose diagonal contains the (nonnegative) square roots of the eigenvalues of U), and that if $U \succ 0$, then U is nonsingular, with $U^{-1} = Q(U)\Lambda^{-1}(U)Q^T(U)$. It also follows that $S\mathbb{R}_+^{n \times n}$ is a closed convex cone, pointed (i.e., $(S\mathbb{R}_+^{n \times n}) \cap (-S\mathbb{R}_+^{n \times n}) = \{0\}$) and with nonempty interior $S\mathbb{R}_{++}^{n \times n}$, an open convex cone. Finally, hence we get $\{(\eta; z) : \eta \geq \lambda_{\max}(M(z))\} = \{(\eta; z) : \eta I - M(z) \succeq 0\}$, as used above, and since this is a convex set, $\lambda_{\max}(M(\cdot))$ is a convex function.

Fact 7. If $U \succeq 0$, then each $u_{jj} \geq 0$, and if $u_{jj} = 0$, $u_{jk} = u_{kj} = 0$ for all k. Similarly, if $U \succ 0$, then each $u_{jj} > 0$.

Fact 8. If $U \succeq 0$, then $PUP^T \succeq 0$ for any P of appropriate column dimension. If P is square and nonsingular, then $U \succ 0$ if and only if $PUP^T \succ 0$.

Fact 9. If

$$U = \begin{pmatrix} U_{11} & U_{12} \\ U_{12}^T & U_{22} \end{pmatrix} \succeq 0 \quad (\succ 0),$$

then $U_{11} \succeq 0$ $(\succ 0)$. Using Fact 8 with P a permutation matrix, we see that every principal submatrix of a positive semidefinite (definite) matrix is also positive semidefinite (definite).

Fact 10. $U \succeq 0$ ($\succ 0$) iff every principal minor is nonnegative (positive). In fact, $U \succ 0$ iff every leading principal minor is positive. Also, $U \succ 0$ iff $U = LL^T$ for some nonsingular lower triangular matrix L (the Cholesky factorization).

We can prove Fact 10 using the preceding facts, induction, and the following very useful property.

Fact 11. Suppose

$$U = \begin{pmatrix} A & B \\ B^T & C \end{pmatrix}$$

with A and C symmetric and $A \succ 0$. Then

$$U \succeq 0 \quad (\succ 0) \qquad \text{if and only if} \quad C - B^T A^{-1} B \succeq 0 \quad (\succ 0).$$

The matrix $C - B^T A^{-1} B$ is called the *Schur complement* of A in U. This is easily proved using the factorization

$$\begin{pmatrix} A & B \\ B^T & C \end{pmatrix} = \begin{pmatrix} I & 0 \\ B^T A^{-1} & I \end{pmatrix} \begin{pmatrix} A & 0 \\ 0 & C - B^T A^{-1} B \end{pmatrix} \begin{pmatrix} I & A^{-1} B \\ 0 & I \end{pmatrix}.$$

Fact 12. (Representing quadratics) If $U \in S\mathbb{R}^{n \times n}$, then $x^T U x = U \bullet xx^T$.

Fact 13. (Self-duality) $S\mathbb{R}_+^{n \times n} = (S\mathbb{R}_+^{n \times n})^* := \{V : U \bullet V \geq 0 \text{ for all } U \in S\mathbb{R}_+^{n \times n}\}$.

Proof. (i) $S\mathbb{R}_+^{n \times n} \subseteq (S\mathbb{R}_+^{n \times n})^*$: We want to show that $U \bullet V \geq 0$ for all positive semidefinite U and V. We can show this directly using Facts 3 and 4 to assume that one is diagonal, or use Fact 6 to obtain a square root of U, and then note that $U \bullet V = \text{trace}\, UV = \text{trace}\, U^{\frac{1}{2}} V U^{\frac{1}{2}} \geq 0$ since $U^{\frac{1}{2}} V U^{\frac{1}{2}}$ is positive semidefinite.

(ii) $S\mathbb{R}_+^{n \times n} \subseteq (S\mathbb{R}_+^{n \times n})^*$: We show that if $U \notin S\mathbb{R}_+^{n \times n}$, then $U \notin (S\mathbb{R}_+^{n \times n})^*$. Indeed, in this case we have $v^T U v < 0$ for some $v \in \mathbb{R}^n$, and then $U \bullet vv^T < 0$ shows that U is not in $(S\mathbb{R}_+^{n \times n})^*$. □

Fact 14. If $U \succ 0$, then $U \bullet V > 0$ for every nonzero $V \succeq 0$, and $\{V \succeq 0 : U \bullet V \leq \beta\}$ is bounded for every positive β. Indeed, if $\lambda := \lambda_{\min}(U) > 0$, then $U \bullet V = (U - \lambda I) \bullet V + \lambda I \bullet V \geq \lambda I \bullet V = \lambda I \bullet \Lambda(V) = \lambda \| \lambda(V) \|_1 \geq \lambda \| V \|_F$ for $V \succeq 0$. This shows the first part directly, and the second since then any V in the set has Frobenius norm at most β / λ.

Fact 15. If $U, V \succeq 0$, then $U \bullet V = 0$ if and only if $UV = 0$. This is easy to show using the eigenvalue decomposition of U, and considering separately its positive and zero eigenvalues.

Fact 16. If $U, V \in S\mathbb{R}^{n \times n}$, then U and V commute iff UV is symmetric iff U and V can be simultaneously diagonalized (*i.e.*, they have eigenvalue decompositions with the same Q).

We can now return to considering other examples of semidefinite programming problems.

Example 2: Minimizing the L_2-operator norm of a matrix. By considering the two cases where $P \in \mathbb{R}^{m \times n}$ is zero and nonzero, and using Fact 11, we can easily see that

$$\eta \geq \|P\|_2 \quad \text{if and only if} \quad \begin{pmatrix} \eta I & P \\ P^T & \eta I \end{pmatrix} \succeq 0.$$

Hence we can solve the problem of minimizing $\|P(z)\|_2$, where $P(z)$ depends affinely on z, by solving the SDP problem

$$\max \; -\eta, \quad \begin{pmatrix} \eta I & P(z) \\ P(z)^T & \eta I \end{pmatrix} \succeq 0, \tag{3.2}$$

where the variable is $y := (\eta; z)$.

Example 3: LP. The linear programming $\max\{b^T y : A^T y \leq c\}$, where $A \in \mathbb{R}^{m \times n}$ and the vectors have appropriate dimensions, can be written as the SDP problem in dual form

$$\max \; b^T y, \quad \text{Diag}(c - A^T y) \succeq 0.$$

Here, in our standard notation, $C = \text{Diag}(c)$ and $A_i = \text{Diag}(a_i)$, with a_i the ith column of A^T. Note that its semidefinite dual problem involves a symmetric $n \times n$ matrix X, and hence seems to differ from the usual linear programming dual. We will discuss this further very shortly.

Example 4: A quasi-convex nonlinear programming problem. Now consider the problem

$$\min \; \frac{(b^T y)^2}{d^T y}, \quad A^T y \leq c,$$

where we assume that $d^T y > 0$ for all feasible y. If we note that the objective function can be written as $(b^T y)(d^T y)^{-1}(b^T y)$ we see the resemblance to the Schur complement, and then it is easy to check that (for feasible y),

$$\eta \geq \frac{(b^T y)^2}{d^T y} \quad \text{if and only if} \quad \begin{pmatrix} \eta & b^T y \\ b^T y & d^T y \end{pmatrix} \succeq 0.$$

It follows that our nonlinear programming problem can be written as

$$\max_{\eta, y} -\eta, \quad \mathrm{Diag}(c - A^T y) \succeq 0, \quad \begin{pmatrix} \eta & b^T y \\ b^T y & d^T y \end{pmatrix} \succeq 0.$$

This has two semidefinite constraints, but of course they can be combined into a single constraint:

$$\mathrm{Diag}\left(\mathrm{Diag}(c - A^T y), \begin{pmatrix} \eta & b^T y \\ b^T y & d^T y \end{pmatrix} \right) \succeq 0.$$

Here C and the A_is are all block diagonal, with m 1×1 blocks and one 2×2 block.

In the last two examples we have seen cases where the data matrices C and the A_is share the same block diagonal structure. Indeed, as in the last example, this arises whenever several semidefinite constraints are combined into a single constraint. Let $\mathcal{S}$ denote the space of block diagonal symmetric matrices of the form

$$M = \begin{pmatrix} M_{11} & 0 & \cdots & 0 \\ 0 & M_{22} & \cdots & 0 \\ \vdots & \vdots & \ddots & \vdots \\ 0 & 0 & \cdots & M_{kk} \end{pmatrix},$$

where $M_{jj} \in \mathbb{R}^{n_j \times n_j}$ for $j = 1, \ldots, k$. Let us suppose C and all A_is lie in $\mathcal{S}$. Then any feasible S in the dual problem, with $\mathcal{A}^* y + S = C$, also lies in $\mathcal{S}$. So (D) can alternatively be written as

$$\max_{(y,S) \in \mathbb{R}^m \times \mathcal{S}} b^T y, \quad \mathcal{A}^* y + S = C, \quad S \succeq 0.$$

Its dual is

$$\min_{X \in \mathbb{SR}^{n \times n}} C \bullet X, \quad \mathcal{A} X = b, \quad X \succeq 0;$$

can we restrict X also to $\mathcal{S}$? If so, then in the LP case, X will be block diagonal with 1×1 blocks, and thus we regain the usual LP dual. Consider any $X \in \mathbb{SR}^{n \times n}$, and partition it as M above:

$$X = \begin{pmatrix} X_{11} & X_{12} & \cdots & X_{1k} \\ X_{21} & X_{22} & \cdots & X_{2k} \\ \vdots & \vdots & \ddots & \vdots \\ X_{k1} & X_{k2} & \cdots & X_{kk} \end{pmatrix}.$$

Then, with obvious notation,

$$A_i \bullet X = A_{i11} \bullet X_{11} + \cdots + A_{ikk} \bullet X_{kk} \quad \text{for each } i,$$
$$C \bullet X = C_{11} \bullet X_{11} + \cdots + C_{kk} \bullet X_{kk}.$$

Also, if $X \succeq 0$, then $X_{jj} \succeq 0$ for $j = 1, \ldots, k$, and then

$$\tilde{X} := \begin{pmatrix} X_{11} & 0 & \cdots & 0 \\ 0 & X_{22} & \cdots & 0 \\ \vdots & \vdots & \ddots & \vdots \\ 0 & 0 & \cdots & X_{kk} \end{pmatrix} \succeq 0.$$

Hence, if X is feasible in (P), then so is $\tilde{X} \in \mathcal{S}$, and with the same objective value. It follows that we can restrict X to $\mathcal{S}$ without loss of generality.

It is important to realize that, if $\mathcal{S}$ denotes instead the set of symmetric matrices with a given sparsity structure and all A_is and C lie in $\mathcal{S}$, then any feasible S also lies in $\mathcal{S}$ but it is no longer the case that we can restrict feasible Xs to $\mathcal{S}$.

Let us mention two ways in which block diagonal structure arises in SDP problems in primal form. First, if we have inequality constraints like $A_i \bullet X \leq b_i, i = 1, \ldots, m, X \succeq 0$, then we can add slack variables to reach $A_i \bullet X + \xi_i = b_i, i = 1, \ldots, m, X \succeq 0, \xi \geq 0$. But these can be written as equality constraints in the positive semidefinite variable $\tilde{X} := \mathrm{Diag}(X, \mathrm{Diag}(\xi))$, and then all matrices have the same block diagonal structure with one $n \times n$ block followed by m 1×1 blocks.

Similarly, if we have several matrix variables and our problem is

$$\begin{aligned} \min \quad & C_{11} \bullet X_{11} + \cdots + C_{kk} \bullet X_{kk}, \\ & A_{i11} \bullet X_{11} + \cdots + A_{ikk} \bullet X_{kk} = b_i, \quad i = 1, \ldots, m, \\ & X_{11} \succeq 0, \ \ldots \qquad\qquad X_{kk} \succeq 0, \end{aligned}$$

then we can express this as an SDP problem involving just one positive semidefinite variable

$$X := \begin{pmatrix} X_{11} & 0 & \cdots & 0 \\ 0 & X_{22} & \cdots & 0 \\ \vdots & \vdots & \ddots & \vdots \\ 0 & 0 & \cdots & X_{kk} \end{pmatrix} \succeq 0,$$

and again we have common block diagonal structure in all the matrices.

All of our results below and all algorithms can exploit this block diagonal structure (and obviously must to be efficient), but for simplicity we write $S\mathbb{R}^{n \times n}$ as the matrix space henceforth.

Now we return to our examples of SDP problems.

Example 5: Convex quadratically constrained programming. Here we consider optimizing a linear function subject to convex quadratic constraints (we can easily convert the minimization of a convex quadratic constraint subject to similar constraints to this form). So we address

$$\max \ b^T y, \qquad f_i(y) \leq 0, \quad i = 1, \ldots, l,$$

where $f_i(y) := y^T C_i y - d_i^T y - \epsilon_i, C_i \succeq 0, i = 1, \ldots, l$. Let $C_i = G_i^T G_i$. Then $f_i(y) \leq 0$ can be written as

$$\begin{pmatrix} I & G_i y \\ (G_i y)^T & d_i^T y + \epsilon_i \end{pmatrix} \succeq 0$$

using Schur complements, or alternatively as

$$\begin{pmatrix} (1 + d_i^T y + \epsilon_i)I & \left(\dfrac{1 - d_i^T y - \epsilon_i}{2 G_i y} \right) \\ \left(\dfrac{1 - d_i^T y - \epsilon_i}{2 G_i y} \right)^T & 1 + d_i^T y + \epsilon_i \end{pmatrix} \succeq 0.$$

The advantage of the second formulation is that the semidefinite constraint

$$\begin{pmatrix} \alpha I & v \\ v^T & \alpha \end{pmatrix} \succeq 0$$

can be expressed as

$$\begin{pmatrix} \alpha \\ v \end{pmatrix} \in K_2 := \left\{ \begin{pmatrix} \beta \\ w \end{pmatrix} : \beta \geq \|w\|_2 \right\},$$

the second-order or Lorentz cone. This is a more efficient way to solve the problem – second-order cones are to be preferred to semidefinite cones in general: see Nesterov and Nemirovski (1992, 1994).

Example 6: Robust mathematical programming. Here we model uncertainty in the data of an optimization problem (or in the implementation of a solution) by requiring that the solution be feasible whatever the realization of the data (see Ben-Tal and Nemirovski (1998)). Without loss of generality we can assume that the objective function is deterministic. Let us consider robust LP with ellipsoidal uncertainty. The problem

$$\max b^T y, \qquad a_j^T y \leq c_j, \quad \text{for all } (a_j; c_j) \in \mathcal{E}_j, \quad j = 1, \ldots, n,$$

can be rewritten, after introducing an extra variable and changing notation, as

$$\max b^T y,$$
$$a_j^T y \leq 0, \quad \text{for all } a_j \in \mathcal{E}_j \text{ and } j = 1, \ldots, k,$$
$$a_j^T y \leq c_j, \quad \text{for } j = k + 1, \ldots, n.$$

Suppose $\mathcal{E}_j = \{\bar{a}_j + G_j u_j : \|u_j\|_2 \leq 1\}$. Then, for a given vector y, we have $a_j^T y \leq 0$ for all $a_j \in \mathcal{E}_j$ if and only if $\bar{a}_j^T y + (G_j u_j)^T y \leq 0$ for all u_j of norm at most one, which holds if and only if $\|G_j^T y\|_2 \leq -\bar{a}_j^T y$, or $(-\bar{a}_j^T y; G_j^T y) \in K_2$, the second-order cone. So we can model the robust LP above using second-order cones, or if we wish as an SDP problem. Ben-Tal and Nemirovski discuss a number of other robust mathematical programming problems: for instance, the robust version of the convex quadratically constrained programming problem above can be formulated as an SDP problem (see Ben-Tal and Nemirovski (1998)).

Example 7: Control theory. There are many applications of semidefinite programming (or the feasibility version, called a linear matrix inequality in the field) in control systems. We will describe a very simple case, leaving the discussion of more general and realistic situations to Vandenberghe and Boyd (1996) and Boyd *et al.* (1994).

Suppose $x = x(t)$ satisfies the differential inclusion

$$\dot{x} \in \mathrm{conv}\{A_1, \dots, A_m\}x, \quad x(0) = x_0,$$

where $A_1, \dots, A_m$ are given matrices in $\mathbb{R}^{n \times n}$. We want to determine whether $x(t)$ necessarily remains bounded.

This holds if and only if there is some $P \succ 0$ so that $v(x) := x^T P x$ remains uniformly bounded, and this certainly follows if v is nonincreasing. Such a function is called a Lyapunov function. Hence a sufficient condition for uniform boundedness is that

$$\frac{\mathrm{d}}{\mathrm{d}t}(x^T P x) = \dot{x}^T P x + x^T P \dot{x} \leq 0.$$

If x_0 is arbitrary, and $\dot{x}(0)$ can be anywhere in the appropriate convex set, then we need

$$A_i^T P + P A_i \preceq 0, \quad \text{for all } i = 1, \dots, m.$$

We also want $P \succ 0$, and, since the constraints above are homogeneous, we may require $P \succeq I$. If we seek a matrix P with, say, minimum condition number, we are then led to the SDP problem

$$\begin{aligned}
\max \ &-\eta, \\
&A_i^T P + P A_i \preceq 0, \quad \text{for all } i = 1, \dots, m, \\
&\eta I \succeq P \succeq I,
\end{aligned}$$

where the variables are η and the entries of the symmetric matrix P. Note that again we have block diagonal structure in this SDP problem.

Now we turn to applications of SDP in obtaining good relaxations of combinatorial optimization problems, where 'relaxation' means that the feasible region is enlarged to obtain a tractable problem whose optimal value provides a bound for that of the problem of interest (in some cases, the optimal solution of the relaxed problem is also of great use). We discuss two: Lovász's theta function and the max-cut problem.

Example 8: Lovász's theta function (Lovász 1979). Here we seek a bound on the Shannon capacity, or on the stability number, of an undirected graph $G = (N, E)$ with node set N and edge set E; we write ij instead of $\{i, j\}$ for an edge linking nodes i and j. We will assume that $N = \{1, \dots, n\}$.

A *stable* or *independent* set is a set of mutually nonadjacent nodes, and $\alpha(G)$ is the maximum size of a stable set: it is NP-hard to compute. A *clique* of G is a set of mutually adjacent nodes, and $\bar{\chi}(G)$ is the minimum cardinality of a collection of cliques that together include all the nodes of G (a clique cover): this is also NP-hard to compute. Note that $\bar{\chi}(G) = \chi(\bar{G})$, the chromatic number of the complement $\bar{G}$ of G in which two nodes are adjacent if and only if they are nonadjacent in G. Clearly, since each node in a stable set must be in a different clique in a clique cover,

$$\alpha(G) \leq \bar{\chi}(G).$$

Our aim is to find approximations to these numbers: in particular, we will define $\theta(G)$, which lies between $\alpha(G)$ and $\bar{\chi}(G)$ and is the optimal value of an SDP problem. If G is a so-called *perfect* graph, then $\alpha(G) = \bar{\chi}(G)$ and we can calculate these invariants of the graph exactly by computing $\theta(G)$.

We define

$$\theta(G) := \max\{ee^T \bullet X : I \bullet X = 1, \ x_{ij} = 0 \text{ if } ij \in E, \ X \succeq 0\}, \qquad (3.3)$$

where $e \in \mathbb{R}^n$ denotes the vector of ones. Clearly this is an SDP problem in primal form, but in maximization form. Its dual can be written as $\min\{\eta :$ $\eta I + \sum_{ij \in E} y_{ij} M_{ij} \succeq ee^T\}$, where M_{ij} is the symmetric matrix that is zero except for ones in the ijth and jith positions. The constraint on η can also be written as $\eta I \succeq V + ee^T$, where V is a symmetric matrix that is zero on the diagonal and in positions $ij \notin E$. As we shall see in the next section, strong duality holds for this pair of SDP problems, so we can also define

$$\theta(G) := \min\{\lambda_{\max}(V + ee^T) : \operatorname{diag}(V) = 0, \ v_{ij} = 0 \text{ for } ij \notin E, \ V \in \mathbb{SR}^{n \times n}\}.$$

It is also instructive to give another definition of $\theta(G)$. An *orthonormal representation* of G is a set $\{u_i : i \in N\}$ of unit vectors in $\mathbb{R}^n$ with u_i and u_j orthogonal if $ij \notin E$. Then $\theta(G)$ can also be defined as the minimum, over all orthonormal representations $\{u_i : i \in N\}$ of G and all unit vectors c, of

$$\max_{i \in N} \frac{1}{(c^T u_i)^2},$$

where $1/0$ is taken to be $+\infty$. To illustrate these three definitions, consider the square viewed as a graph on four nodes, with edges 12, 23, 34, and 14. Then $\alpha(G) = \bar{\chi}(G) = 2$, so $\theta(G) = 2$ also. For the first definition, an optimal X has $1/4$ in positions 11, 13, 22, 24, 31, 33, 42, and 44, with zeros elsewhere. An optimal $V + ee^T$ for the second definition is 4 times this matrix. And to get an optimal orthonormal representation, consider an umbrella with just four ribs, and imagine opening it up until nonadjacent ribs are orthogonal. Then the u_is are unit vectors along the ribs $((\pm 1; 0; -1)/\sqrt{2}$ and $(0; \pm 1; -1)/\sqrt{2})$ and the unit vector c is a unit vector along the handle: $(0; 0; -1)$. (A similar example for the pentagon uses a five-ribbed umbrella, and gives $\theta = \sqrt{5}$, while $\alpha = 2$ and $\bar{\chi} = 3$; $\sqrt{5}$ is also the Shannon capacity of the pentagon.) It is not immediately apparent that this last definition gives the same value as the previous ones: we refer to Grötschel *et al.* (1988) for a proof of this (and several other definitions of $\theta = \theta(G)$). See also the survey article of Goemans (1997). We just note here the relationship between positive semidefinite matrices and sets of vectors. If V gives an optimal solution to the problem defining θ as a maximum eigenvalue, then $\theta I - V - ee^T$ is positive semidefinite and hence can be factored as $W^T W$, and clearly we then know something about the inner products of the columns w_i of W. We obtain an orthonormal system for G by manipulating these vectors w_i.

We conclude our discussion of this application by showing that all three definitions give an upper bound on $\alpha(G)$. Let $K \subseteq N$ be a maximum-cardinality stable set of G, with cardinality $k = \alpha(G)$. For the first definition, choose for X the symmetric matrix that is all zeros, except that $x_{ij} = 1/k$ for all $i, j \in K$. It is clear that this is feasible, and it gives an objective value of k. The maximum value is thus at least as large. For the second definition, consider any feasible V, and note that the (K, K) principal submatrix of $V + ee^T$ consists of all ones, and hence has maximum eigenvalue equal to its order, k. Since the largest eigenvalue of any matrix is at least that of any principal submatrix (*e.g.*, from considering Rayleigh quotients), we conclude that the optimal value of the eigenvalue problem is at least k. Finally, let $\{u_i : i \in N\}$ be an orthonormal representation of G and c a unit vector. Then $\{u_i : i \in K\}$ is a set of orthonormal vectors (which can be completed to an orthonormal basis), and so

$$1 = \|c\|^2 \geq \sum_{j \in K} (c^T u_j)^2.$$

It follows that one of the summands is at most $1/k$, and hence

$$\max_{i \in N} \frac{1}{(c^T u_i)^2} \geq \max_{j \in K} \frac{1}{(c^T u_j)^2} \geq k.$$

Example 9: The max-cut problem. Once again we have an undirected graph $G = (N, E)$, and a nonnegative vector $w = (w_{ij}) \in \mathbb{R}_+^E$. For $K \subseteq N$, $\delta(K)$ denotes $\{ij \in E : i \in K, j \notin K\}$, the *cut* determined by K, with *weight* equal to $w(\delta(K)) := \sum_{ij \in \delta(K)} w_{ij}$. We want to find the cut of maximum weight. (This problem arises in VLSI and in finding the ground state of a spin glass; see Poljak and Tuza (1995).) We can assume that the graph is complete (each node is adjacent to all others) by setting $w_{ij} = 0$ for all non-edges ij; we also set $w_{ii} = 0$ for all i.

We start with two (nonconvex) quadratic programming formulations. We use $x \in \mathbb{R}^n$, with each $x_i = \pm 1$, to represent the cut $\delta(K)$, where $x_i = 1$ if and only if $i \in K$. Then clearly $x_i x_j$ is -1 if $ij \in \delta(K)$, $+1$ otherwise. Let us define $C \in S\mathbb{R}^{n \times n}$ by setting $c_{ij} = -w_{ij}/4$ for $i \neq j$ and $c_{ii} = \sum_j w_{ij}/4$ for all i. Then for the x above, we have

$$w(\delta(K)) = \frac{1}{2} \sum_{i<j} w_{ij}(1 - x_i x_j) = \frac{1}{4} \sum_i \sum_j w_{ij}(1 - x_i x_j) = x^T C x.$$

Since every $(+1, -1)$-vector corresponds to a cut, the max-cut problem can be written as the integer quadratic programming problem

$$\text{(IQP)}: \quad \max x^T C x, \quad x_i \in \{+1, -1\}, i \in N,$$

or as the nonconvex quadratically constrained quadratic problem

$$\text{(NQCQP)}: \quad \max x^T C x, \quad x_i^2 = 1, i \in N.$$

We now discuss three different ways to arrive at an SDP relaxation of this problem. First, we note that (NQCQP) is linear in the products $x_i x_j$, and these are the entries of the rank one matrix $X = xx^T$. Note that $X \in S\mathbb{R}^{n \times n}$, with $X_{ii} = 1$ for all i and $X \succeq 0$. (We write the entries of X as X_{ij} to avoid confusion with the vector x.) Conversely, it is easy to see that any such matrix that also has rank one is of the form xx^T for some $(+1, -1)$-vector x. Since $x^T C x = C \bullet (xx^T)$, we see that (IQP) is equivalent to

$$\max C \bullet X, \quad X_{ii} = 1, i \in N, \ X \succeq 0, \ X \text{ rank one.}$$

If we relax the last constraint, we get the SDP problem

$$\max C \bullet X, \quad X_{ii} = 1, i \in N, \ X \succeq 0. \tag{3.4}$$

Secondly, note that in (IQP) we associate a 1-dimensional unit vector x_i (± 1) with each node. As in the previous example, we now associate an n-dimensional unit vector p_i with each node, and let P be the matrix whose rows are these vectors. (P corresponds to the vector x, whose rows correspond to the 1-dimensional vectors.) We then replace the objective $C \bullet (xx^T)$ with $C \bullet (PP^T)$, and the constraints $x_i \in \{+1, -1\}$ by $\text{diag}(PP^T) = e$. Since PP^T is positive semidefinite, and every such matrix can be factored as PP^T, we see that our problem has become the SDP problem above. It is clearly a relaxation, since if we restrict each row of P to a multiple of a fixed unit vector (± 1), then we recover (IQP).

The third way to derive the SDP relaxation is by taking the dual twice. (This approach was apparently first considered by Shor (1990); see also Poljak, Rendl and Wolkowicz (1995).) Given any optimization problem $\max\{f(x) : g(x) = b, x \in \Xi\}$, where we have distinguished a certain set of m equality constraints and left the rest as an abstract set restriction, the Lagrangian dual obtained by dualizing the $g(x) = b$ constraints is defined to be

$$\min_{y \in \mathbb{R}^m} h(y), \quad \text{where } h(y) := \max_{x \in \Xi} [f(x) - y^T(g(x) - b)].$$

Note that h, as the pointwise maximum of a set of linear functions, is always convex. It is easy to see that the optimal value of this dual problem always provides an upper bound on that of the original.

We now apply this scheme to (NQCQP), dualizing the constraints $x_i^2 = 1$, $i = 1, \ldots, n$. The dual problem is to minimize over all $y \in \mathbb{R}^n$

$$h(y) := \max_{x \in \mathbb{R}^n} \left(x^T C x - \sum_i y_i(x_i^2 - 1) \right)$$

$$= e^T y - \min_{x \in \mathbb{R}^n} (x^T(\text{Diag}(y) - C)x).$$

The minimum here is 0 if $\text{Diag}(y) - C$ is positive semidefinite, and $-\infty$ otherwise. Hence there is an implicit semidefinite constraint, and the dual problem becomes

$$\min e^T y, \quad \text{Diag}(y) - C \succeq 0.$$

This is an SDP problem in dual form, and its dual is precisely the SDP problem above. Again, these dual problems satisfy the conditions of the next section guaranteeing strong duality, so either provides a relaxation of the original max-cut problem. These bounds on the value of a maximum weight cut were obtained by Delorme and Poljak (1993).

Since we have a relaxation, the optimal *value* of the SDP problem provides an upper bound on the value of the max cut. But in this case, we can also use the *solution* of the primal problem to generate a provably good cut, as was shown in a beautiful contribution of Goemans and Williamson (1995) (see also the survey article of Goemans (1997)). This uses the second derivation of the SDP problem above. So let us suppose the optimal solution is $X \succeq 0$, and then factor $X = PP^T$. Then the rows of P, p_i for each i, give unit vectors for each node. If these vectors were all collinear, then we could obtain a maximum weight cut by choosing the nodes whose vectors were equal to p_1 as K, and those with vectors equal to $-p_1$ as $N \setminus K$. In general, we proceed as follows. Choose a random vector v uniformly on the unit sphere, and set $K := \{i \in N : v^T p_i \geq 0\}$. Then we get a random cut, and it is not hard to show that its expected weight, $Ew(\delta(K))$, is at least 0.878 of the optimal value of the SDP problem, which is at least the value of a maximum weight cut. Hence we achieve at least this fraction of the best cut (on average) in this way. In fact, it is possible to derandomize this procedure, to achieve a deterministic cut that is provably close to maximum weight. For the pentagon (again!) with all weights equal to one, the ratio of the optimal values of the max-cut problem and its SDP relaxation is about 0.884, so the bound above is about the best one could hope for.

4. Duality

Now it is time to discuss the relation between (P) and (D). We have already shown weak duality, and here we will give conditions for strong duality to hold. But first, since we have discussed Lagrangian duality in Example 9, we show that each of these problems is the Lagrangian dual of the other, dualizing the equality constraints in each case. It is easy to see that (D) is the Lagrangian dual of (P) (of course, we have to switch max and min in our derivation of the dual). Let us show that (P) is the Lagrangian dual of (D), when we dualize the constraints $A^*y + S = C$. Since this is an equation between symmetric matrices, our dual variable will also be a symmetric matrix, and we shall denote it by X. Hence our dual problem is

$$\min_{X \in S\mathbb{R}^{n \times n}} h(X), \quad h(X) := \max_{y \in \mathbb{R}^m, S \succeq 0} [b^T y - (A^*y + S - C) \bullet X].$$

The maximum can be written as

$$C \bullet X - \min_{y \in \mathbb{R}^m} [(AX - b)^T y] - \min_{S \succeq 0} [S \bullet X].$$

Since y ranges over all of $\mathbb{R}^m$, this is $+\infty$ unless $AX - b = 0$. Also, by Fact 13 (self-duality of the cone of positive semidefinite matrices), it is $+\infty$ unless $X \succeq 0$. If X satisfies both these conditions, the maximum reduces to just $C \bullet X$, and we retrieve (P).

Next we present a number of examples, from Vandenberghe and Boyd (1996) and Luo, Sturm and Zhang (2000), showing how strong duality can fail. Further examples can be found in the latter reference.

Consider first

$$\max -y_1, \quad \begin{pmatrix} -1 & 0 \\ 0 & 0 \end{pmatrix} y_1 + \begin{pmatrix} 0 & 0 \\ 0 & -1 \end{pmatrix} y_2 \preceq \begin{pmatrix} 0 & 1 \\ 1 & 0 \end{pmatrix}.$$

Equivalently, we require that

$$\begin{pmatrix} y_1 & 1 \\ 1 & y_2 \end{pmatrix} \succeq 0.$$

It is easy to see that the feasible region is $\{(y_1; y_2) : y_1 > 0, y_2 > 0, y_1 y_2 \geq 1\}$. So the optimal value is 0, but it is not attained. (We can get arbitrarily close with solutions of the form $(\epsilon; 1/\epsilon)$ for arbitrarily small positive ϵ.) The dual of this problem is

$$\min \begin{pmatrix} 0 & 1 \\ 1 & 0 \end{pmatrix} \bullet X, \quad \begin{pmatrix} -1 & 0 \\ 0 & 0 \end{pmatrix} \bullet X = -1, \quad \begin{pmatrix} 0 & 0 \\ 0 & -1 \end{pmatrix} \bullet X = 0, \ X \succeq 0,$$

for which the only feasible (and hence optimal) solution is

$$X = \begin{pmatrix} 1 & 0 \\ 0 & 0 \end{pmatrix}$$

with optimal value 0. Here there is no duality gap, but one of the values is not attained.

Our next example is

$$\min \begin{pmatrix} 0 & 0 & 0 \\ 0 & 0 & 0 \\ 0 & 0 & 1 \end{pmatrix} \bullet X$$

$$\begin{pmatrix} 1 & 0 & 0 \\ 0 & 0 & 0 \\ 0 & 0 & 0 \end{pmatrix} \bullet X = 0,$$

$$\begin{pmatrix} 0 & 1 & 0 \\ 1 & 0 & 0 \\ 0 & 0 & 2 \end{pmatrix} \bullet X = 2,$$

$$X \succeq 0.$$

Any feasible X is of the form

$$\begin{pmatrix} 0 & \xi_1 & \xi_2 \\ \xi_1 & \xi_3 & \xi_4 \\ \xi_2 & \xi_4 & 1-\xi_1 \end{pmatrix},$$

and, since it must be positive semidefinite, in fact

$$\begin{pmatrix} 0 & 0 & 0 \\ 0 & \xi_3 & \xi_4 \\ 0 & \xi_4 & 1 \end{pmatrix}$$

for suitable ξ_is. It follows that an optimal X is

$$\begin{pmatrix} 0 & 0 & 0 \\ 0 & 0 & 0 \\ 0 & 0 & 1 \end{pmatrix},$$

with optimal value 1. The dual problem is

$$\max 2y_2, \quad \begin{pmatrix} 1 & 0 & 0 \\ 0 & 0 & 0 \\ 0 & 0 & 0 \end{pmatrix} y_1 + \begin{pmatrix} 0 & 1 & 0 \\ 1 & 0 & 0 \\ 0 & 0 & 2 \end{pmatrix} y_2 \preceq \begin{pmatrix} 0 & 0 & 0 \\ 0 & 0 & 0 \\ 0 & 0 & 1 \end{pmatrix}.$$

Equivalently, we require

$$S = \begin{pmatrix} -y_1 & -y_2 & 0 \\ -y_2 & 0 & 0 \\ 0 & 0 & 1-2y_2 \end{pmatrix} \succeq 0,$$

so y_2 must equal 0 and y_1 be nonpositive. Thus $y = (0;0)$ is optimal, with optimal value 0. Here both problems attain their optimal values, but there is a gap between them.

Note that in both primal and dual, a matrix that is required to be positive semidefinite has a zero on the diagonal, and this forces the off-diagonal entries in that row and column to be zero also. It is instructive to see what happens when this implicit constraint is removed by perturbing the diagonal entry. The reader may wish to check that, if b_1 (using the usual notation) is changed to $\epsilon > 0$, then both optimal values become 0, while if c_{22} is changed to $\epsilon > 0$, then both optimal values become 1. (If both changes are made, the optimal values again become equal, but now both are $3/4$.)

Our last example is

$$\min \begin{pmatrix} 0 & 0 \\ 0 & 0 \end{pmatrix} \bullet X$$

$$\begin{pmatrix} 1 & 0 \\ 0 & 0 \end{pmatrix} \bullet X = 0,$$

$$\begin{pmatrix} 0 & 1 \\ 1 & 0 \end{pmatrix} \bullet X = 2,$$

$$X \succeq 0.$$

Any feasible X must have $(1,1)$ entry 0 and $(1,2)$ entry 1, and such a matrix cannot be positive semidefinite, so the optimal value (using the usual convention) is $+\infty$. The dual problem is

$$\max 2y_2, \quad \begin{pmatrix} 1 & 0 \\ 0 & 0 \end{pmatrix} y_1 + \begin{pmatrix} 0 & 1 \\ 1 & 0 \end{pmatrix} y_2 \preceq \begin{pmatrix} 0 & 0 \\ 0 & 0 \end{pmatrix}.$$

Equivalently, we require

$$S = \begin{pmatrix} -y_1 & -y_2 \\ -y_2 & 0 \end{pmatrix} \succeq 0,$$

so y_2 must equal 0 and y_1 be nonpositive. Thus $y = (0; 0)$ is optimal, with optimal value 0. Here there is an infinite gap between the optimal values.

Having given examples of strong duality failing, we now turn to conditions ensuring that it holds. It turns out that a Slater (strict feasibility) condition suffices. Let us define

$$F(\mathrm{P}) := \{X \in S\mathbb{R}^{n \times n} : \mathcal{A}X = b, X \succeq 0\},$$
$$F^0(\mathrm{P}) := \{X \in F(\mathrm{P}) : X \succ 0\},$$
$$F(\mathrm{D}) := \{(y, S) \in \mathbb{R}^m \times S\mathbb{R}^{n \times n} : \mathcal{A}^*y + S = C, S \succeq 0\},$$
$$F^0(\mathrm{D}) := \{(y, S) \in F(\mathrm{D}) : S \succ 0\}.$$

Also, we say the *linear independence condition* holds if $A_1, \ldots, A_m$ are linearly independent in $S\mathbb{R}^{n \times n}$.

Theorem 4.1. (Strong duality) Suppose that $F(\mathrm{P})$ and $F^0(\mathrm{D})$ are nonempty. Then (P) has a nonempty compact set of optimal solutions, and the optimal values of (P) and (D) are equal.

Proof. Let $\hat{X} \in F(P)$ and $(\hat{y}, \hat{S}) \in F^0(D)$. Then we can add the constraint $C \bullet X \leq C \bullet \hat{X}$ to (P) without changing its optimal value or the set of its optimal solutions. But, using (2.1), this inequality is equivalent to $\hat{S} \bullet X = C \bullet X - b^T \hat{y} \leq C \bullet \hat{X} - b^T \hat{y} = \hat{S} \bullet \hat{X}$. So (P) has the same optimal value and set of optimal solutions as

$$(\text{P}') : \qquad \min C \bullet X, \qquad AX = b,\ \hat{S} \bullet X \leq \hat{S} \bullet \hat{X},\ X \succeq 0.$$

But by Fact 14, this problem has a compact feasible region since $\hat{S} \succ 0$. The objective function being continuous, this implies the first assertion of the theorem.

Now let ζ_* denote the optimal value of (P) and ϵ be positive. We want to show that there is a feasible solution of (D) with objective value at least $\zeta_* - \epsilon$. Consider the two sets $\mathcal{G}_1 := S\mathbb{R}_+^{n \times n}$ and $\mathcal{G}_2 := \{X \in S\mathbb{R}^{n \times n} : AX = b, C \bullet X \leq \zeta_* - \epsilon\}$. These two sets are closed, convex, and disjoint, and have no common direction of recession (any such would be a nonzero $X \succeq 0$ satisfying $C \bullet X = 0$, $AX = 0$, showing that the set of optimal solutions of (P) is unbounded, a contradiction). Hence, by a separating hyperplane theorem (Rockafellar (1970, Corollary 11.4.1)), there exist $S \in S\mathbb{R}^{n \times n}$ and $\sigma \in \mathbb{R}$ with

$$S \bullet X > \sigma \quad \text{for any } X \in \mathcal{G}_1, \qquad S \bullet X < \sigma \quad \text{for any } X \in \mathcal{G}_2.$$

Since $0 \in \mathcal{G}_1$, σ is negative. Since $\lambda uu^T \in \mathcal{G}_1$ for any positive λ and any $u \in \mathbb{R}^n$, it follows that $S \succeq 0$.

Next we have that $AX = b$, $C \bullet X \leq \zeta_* - \epsilon$ imply $S \bullet X \leq \sigma$. By a theorem of the alternative (for linear inequalities – there are no semidefinite constraints here), there exist $y \in \mathbb{R}^m$ and $\eta \geq 0$ with

$$C\eta - A^*y = S, \qquad (\zeta_* - \epsilon)\eta - b^T y \leq \sigma.$$

Suppose $\eta = 0$. Then $-b^T y \leq \sigma < 0$ and also $-b^T y = -(A\hat{X})^T y = \hat{X} \bullet (-A^*y) = \hat{X} \bullet S \geq 0$, a contradiction. Hence η is positive, and by scaling y, S, and σ we can assume that $\eta = 1$. But then $C - A^*y = S \succeq 0$ and $b^T y \geq \zeta_* - \epsilon - \sigma \geq \zeta_* - \epsilon$, and we have a feasible solution to (D) with value at least $\zeta_* - \epsilon$. Since ϵ was arbitrary, we have shown that there is no duality gap. $\qquad\square$

The result above is asymmetric between (P) and (D). We now make a few remarks concerning these two problems and their presentation. The gist is that each can be rewritten in the format of the other. For this, we assume the linear independence condition. (Our assumption is basically without loss of generality. If the A_is are linearly dependent, and $A^*y = 0$ implies $b^T y = 0$, then we can choose a basis, say A_i, $i = 1, \ldots, k$, for them and remove the last $m - k$ primal constraints and dual variables to get equivalent problems where the assumption holds. If we have $\mathcal{A}^*\hat{y} = 0$ and $b^T\hat{y} > 0$, then (P) is infeasible and (D) unbounded if it is feasible.)

Given the assumption, we choose $D \in S\mathbb{R}^{n \times n}$ satisfying $\mathcal{A}D = b$ and let $G_1, \ldots, G_k$ be a basis for the orthogonal complement of the span of the A_is in $S\mathbb{R}^{n \times n}$. Finally, let $h_j := C \bullet G_j$, $j = 1, \ldots, k$. Then it is not hard to see that (P) is equivalent to

$$C \bullet D - \max_{w \in \mathbb{R}^k, X \in S\mathbb{R}^{n \times n}} \left\{ h^T w : \sum_j w_j G_j + X = D, X \succeq 0 \right\},$$

an SDP problem in dual form. Similarly, (D) is equivalent to

$$C \bullet D - \min_{S \in S\mathbb{R}^{n \times n}} \{ D \bullet S : G_j \bullet S = h_j, j = 1, \ldots, k, S \succeq 0 \},$$

an SDP problem in primal form. We can use this construction for moving between the two forms. Let us note that, given the linear independence condition, for any S there is at most one y with (y, S) feasible for (D), which allows us to extend boundedness results from just S to the pair (y, S).

Applying this procedure to the previous theorem, we obtain the following result.

Corollary 4.1. Suppose the linear independence condition holds and that $F^0(P)$ and $F(D)$ are nonempty. Then (D) has a nonempty compact set of optimal solutions, and there is no duality gap.

We also find the following.

Corollary 4.2. Suppose the linear independence condition holds and that both (P) and (D) have strictly feasible solutions. Then each has a nonempty compact set of optimal solutions, and there is no duality gap.

We will give an alternative proof of the second corollary in the next section. We note the historical fact that Corollary 4.2 was proved by Bellman and Fan (1963) for the following pair of SDP problems:

$$\min \sum_j C_j \bullet X_j,$$
$$\sum_j (A_{ij} X_j + X_j A_{ij}^T) = B_i, \quad \text{for all } i,$$
$$X_j \succeq 0, \quad \text{for all } j,$$

$$\max \sum_i B_i \bullet Y_i,$$
$$\sum_i (Y_i A_{ij} + A_{ij}^T Y_i) \preceq C_j, \quad \text{for all } j.$$

Here all B_is and C_js, as well as the variables X_j and Y_i, are symmetric matrices of order n, while the A_{ij}s are not necessarily symmetric matrices of the same order; also j runs from 1 to k and i from 1 to m throughout. Clearly this form of the problems was inspired by systems arising in control theory, but no connections were made. It is possible to show that (P) and

(D) can be formulated as above, and that these problems can be formulated as (P) and (D).

Since the 'standard' dual (D) of (P) may lead to a positive duality gap, we can ask whether there is a perhaps more complicated dual problem for which strong duality always holds, without any additional regularity assumptions. The answer is in the affirmative: see Ramana (1997) and Ramana, Tunçel and Wolkowicz (1997).

Finally, if we assume that strong duality holds, then we have as necessary and sufficient optimality conditions the following:

$$\begin{aligned} \mathcal{A}^*y + S &= C, \quad S \succeq 0, \\ \mathcal{A}X \phantom{{}+S} &= b, \quad X \succeq 0, \\ XS &= 0. \end{aligned}$$

(Here the natural last condition stating that the duality gap is zero, $X \bullet S = 0$, has been replaced by the seemingly stronger condition that the matrix product is zero using Fact 15, since both matrices are positive semidefinite.)

5. The logarithmic barrier function and the central path

We define f on $S\mathbb{R}^{n \times n}$ by

$$f(X) := -\ln \det X \text{ if } X \succ 0, \quad f(X) := +\infty \text{ otherwise}.$$

and call it the *logarithmic barrier function* for the cone $S\mathbb{R}_+^{n \times n}$ of positive semidefinite matrices. For $n = 1$, we get the smooth function $-\ln x$, which is defined on the positive axis and tends to $+\infty$ as x approaches 0 from above. In the same way, f is defined on the positive definite matrices and tends to $+\infty$ as X approaches a matrix $\bar{X} \in \partial S\mathbb{R}_+^{n \times n}$ through positive definite values. We say that f *has the barrier property* for $S\mathbb{R}_+^{n \times n}$.

The idea, which we shall investigate in detail below, is to replace the problem (P) with the somewhat awkward constraint that X be positive semidefinite by the sequence of problems (barrier problems parametrized by $\nu > 0$):

$$\text{BP}(\nu): \quad \min C \bullet X + \nu f(X), \quad \mathcal{A}X = b \quad (X \succ 0),$$

where there is only a linear constraint, the implicit positive definite constraint being enforced by the barrier property of f.

Clearly f is smooth on the interior $S\mathbb{R}_{++}^{n \times n}$ of $S\mathbb{R}_+^{n \times n}$; we evaluate its first few derivatives. Let $X \succ 0$, $H \in S\mathbb{R}^{n \times n}$. Then

$$\begin{aligned} f(X + \alpha H) &= -\ln \det[X(I + \alpha X^{-1}H)] \\ &= -\ln \det X - \ln(1 + \alpha \operatorname{trace} X^{-1}H + O(\alpha^2)) \\ &= f(X) - \alpha X^{-1} \bullet H + O(\alpha^2), \end{aligned}$$

so that $f'(X) = -X^{-1}$ and $Df(X)[H] = -X^{-1} \bullet H$.

Similarly we find

$$f'(X + \alpha H) = -[X(I + \alpha X^{-1}H)]^{-1}$$
$$= -[I - \alpha X^{-1}H) + O(\alpha^2)]X^{-1}$$
$$= f'(X) + \alpha X^{-1}HX^{-1} + O(\alpha^2).$$

Hence $f''(X)[H] = X^{-1}HX^{-1}$ and $D^2 f(X)[H, J] = X^{-1}HX^{-1} \bullet J$. In the notation introduced in (1.1), $f''(X) = X^{-1} \odot X^{-1}$. It is easy to see that the adjoint of $P \odot Q$ is $P^T \odot Q^T$, so this operator is self-adjoint (*i.e.*, $[(P \odot Q)U] \bullet V = [(P \odot Q)V] \bullet U$) if P and Q are symmetric; further, it is positive definite (*i.e.*, $[(P \odot Q)U] \bullet U > 0$ if $U \in S\mathbb{R}^{n \times n}$ is nonzero) if $P \succ 0$, $Q \succ 0$. Hence $f''(X)$ is self-adjoint and positive definite. In the same way we find $f'''(X)[H, J] = -X^{-1}HX^{-1}JX^{-1} - X^{-1}JX^{-1}HX^{-1}$.

We now introduce the important notion of *self-concordance*, defined and developed in great detail by Nesterov and Nemirovski (1994). We know that f is convex if, for every $X \succ 0$ and every $H \in S\mathbb{R}^{n \times n}$, $\phi(\alpha) := f(X + \alpha H)$ is convex in α. We say that f is *self-concordant* if it is convex and 3-times differentiable and if, for every such X and H, ϕ defined as above satisfies

$$|\phi'''(0)| \leq 2[\phi'(0)]^{3/2}.$$

Finally, f is a θ-*normal barrier for* $S\mathbb{R}_{++}^{n \times n}$ (or for $S\mathbb{R}_{+}^{n \times n}$) if it is convex, self-concordant, has the barrier property, and is logarithmically homogeneous of degree θ:

$$f(\alpha X) = f(X) - \theta \ln \alpha, \quad \text{for all } X \succ 0, \; \alpha > 0.$$

We now have the following result.

Theorem 5.1. $f(X) := -\ln \det X$ is an n-normal barrier for $S\mathbb{R}_{+}^{n \times n}$.

Proof.   Define ϕ as above. Then it is finite on the convex set of α such that $X + \alpha H \succ 0$, and on this set

$$\phi''(\alpha) = D^2 f(\bar{X})[H, H] = (\bar{X}^{-1}H\bar{X}^{-1}) \bullet H,$$

where $\bar{X} := X + \alpha H$. Since this matrix is positive definite, so is $V := \bar{X}^{-\frac{1}{2}}$, and then

$$\phi''(\alpha) = V^2 H V^2 \bullet H = \text{trace}\,(V^2 H V^2 H)$$
$$= \text{trace}\,([VHV][VHV]) = \|VHV\|_F^2 \geq 0.$$

So ϕ is convex. Indeed, the quantity above is positive if H is nonzero, so in fact then ϕ and hence f is strictly convex. We have also shown that $f''(X)$ is a positive definite and hence nonsingular operator.

Setting $V := X^{-\frac{1}{2}} \succ 0$, we obtain $\phi''(0) = \text{trace}\,([VHV][VHV])$. If $\lambda := \lambda(VHV)$, then $\phi''(0) = \text{trace}\,(\text{Diag}(\lambda)\,\text{Diag}(\lambda)) = \|\lambda\|_2^2$. Next,

$$\phi'''(0) = -2(X^{-1}HX^{-1}HX^{-1}) \bullet H$$
$$= -2\,\text{trace}\,(V^2HV^2HV^2H) = -2\,\text{trace}\,([VHV][VHV][VHV])$$
$$= -2\sum \lambda_i^3.$$

So we conclude that

$$|\phi'''(0)| = 2|\sum \lambda_i^3| \le 2\sum |\lambda_i^3| = 2\|\lambda\|_3^3 \le 2\|\lambda\|_2^3 = 2[\phi''(0)]^{3/2}.$$

Finally, we have already checked the barrier property, and, for $\alpha > 0$ and $X \succ 0$,

$$f(\alpha X) = -\ln\det(\alpha X) = -\ln(\alpha^n \det X) = f(X) - n\ln\alpha,$$

so the proof is complete.   □

Having the positive definite operator $f''(X)$, we can define the X-*norm* of a symmetric matrix by

$$\|H\|_X := (f''(X)H \bullet H)^{\frac{1}{2}} = \|\lambda(X^{-\frac{1}{2}}HX^{-\frac{1}{2}})\|_2 = \|X^{-\frac{1}{2}}HX^{-\frac{1}{2}}\|_F$$

and the *dual X-norm* of a symmetric matrix by

$$\|J\|_X^* := ([f''(X)]^{-1}J \bullet J)^{\frac{1}{2}} = \|X^{\frac{1}{2}}JX^{\frac{1}{2}}\|_F.$$

Note that $|J \bullet H| \le \|J\|_X^* \|H\|_X$ as in the Cauchy–Schwarz inequality.

The following properties follow from our formulae, but can also be obtained directly by differentiating the equation for logarithmic homogeneity.

Proposition 5.1. For $\alpha > 0$, $X \succ 0$ of order n,

$$f'(\alpha X) = \alpha^{-1}f'(X), \qquad f''(\alpha X) = \alpha^{-2}f''(X);$$
$$f'(X) \bullet X = -n, \qquad f''(X)X = -f'(X);$$
$$\|X\|_X = \sqrt{n}, \qquad \|f'(X)\|_X^* = \sqrt{n}. \qquad □$$

The last line also states that the X-norm of the Newton step for minimizing f from X, $-[f''(X)]^{-1}f'(X)$, is exactly $\sqrt{n}$. This shows that f satisfies the original definition of Nesterov and Nemirovski (which applies also to functions that are not logarithmically homogeneous) to be an n-self-concordant barrier function.

We now return to the barrier problem mentioned at the beginning of this section, defining the primal and dual barrier problems (parametrized by $\nu > 0$) to be

$$\text{BP}(\nu): \qquad \min C \bullet X + \nu f(X), \qquad \mathcal{A}X = b \quad (X \succ 0),$$

and

$$\text{BD}(\nu): \qquad \max b^T y - \nu f(S), \qquad \mathcal{A}^* y + S = C \quad (S \succ 0).$$

It is not hard to check that each is in fact the Lagrangian dual of the other up to an additive constant.

Suppose $BP(\nu)$ has an optimal solution X. Then $X \in F^0(P)$ and, by Lagrange's theorem, for some $y \in \mathbb{R}^m$ we have

$$C - \nu X^{-1} - \mathcal{A}^* y = C + \nu f'(X) - \mathcal{A}^* y = 0.$$

Let us set $S := \nu X^{-1} \succ 0$. Then we see that $(y, S) \in F^0(D)$, and we have a solution to the set of equations

$$\text{CPE}(\nu): \quad \begin{matrix} \mathcal{A}^* y + S = C, & S \succ 0, \\ \mathcal{A} X \qquad\quad = b, & X \succ 0, \\ XS \qquad\quad = \nu I. \end{matrix} \qquad (5.1)$$

We call these the *central path equations* for reasons that will become clearer shortly. Note that, except for the final right-hand side, these equations coincide with the optimality conditions stated at the end of the previous section.

If $BD(\nu)$ has an optimal solution (y, S), a similar derivation shows that, for some X, the above equations again hold.

Theorem 5.2. Suppose $F^0(P)$ and $F^0(D)$ are nonempty and the linear independence assumption holds. Then for every positive ν, there is a unique solution $(X(\nu), y(\nu), S(\nu))$ to $\text{CPE}(\nu)$. Further, $X(\nu)$ is the unique solution to $BP(\nu)$ and $(y(\nu), S(\nu))$ to $BD(\nu)$. Finally, if the assumption of strict feasibility fails, then $\text{CPE}(\nu)$, $BP(\nu)$, and $BD(\nu)$ have no solution.

Proof. First we establish existence. Choose $\hat{X} \in F^0(P)$ and $(\hat{y}, \hat{S}) \in F^0(D)$, and consider $BP(\nu)$. Suppose $\sigma := \lambda_{\min}(\hat{S}) > 0$. Now $\hat{X}$ is feasible for $BP(\nu)$, and for feasible X, $C \bullet X$ differs by a constant from $\hat{S} \bullet X$ (2.1). Hence $BP(\nu)$ has the same set of optimal solutions as

$$BP'(\nu): \quad \min \hat{S} \bullet X + \nu f(X),$$
$$\mathcal{A} X = b, \ \hat{S} \bullet X + \nu f(X) \le \hat{S} \bullet \hat{X} + \nu f(\hat{X}) \quad (X \succ 0).$$

Our aim is to show that this amounts to the minimization of a continuous function on a compact set, yielding existence.

Suppose X is feasible in $BP'(\nu)$, and let $\lambda := \lambda(X)$ and $e \in \mathbb{R}^n$ be again a vector of ones. Then we have $\lambda > 0$ and $\sigma e^T \lambda - \nu \sum \ln \lambda_j = \sigma I \bullet X + \nu f(X) \le \hat{S} \bullet X + \nu f(X) \le \hat{S} \bullet \hat{X} + \nu f(\hat{X}) =: \alpha$, so

$$\sum_j (\sigma \lambda_j - \nu \ln \lambda_j) \le \alpha.$$

Now the function $\sigma \tau - \nu \ln \tau$ has a unique minimizer at $\tau_* = \nu/\sigma$ and goes to $+\infty$ as τ goes to either 0 or $+\infty$. Let the minimum value be β and suppose that $\sigma \tau - \nu \ln \tau > \alpha - (n-1)\beta$ for $\tau \in (0, \underline{\tau}]$ or $\tau \in [\bar{\tau}, +\infty)$. Then the inequality above implies that $\lambda_j \in [\underline{\tau}, \bar{\tau}]$ for all j, so $\|X\|_F = \|\lambda\|_2 \leq \sqrt{n}\bar{\tau}$. Hence we have a bounded feasible set. Moreover, $\lambda_j \geq \underline{\tau} > 0$ for all j implies that $\hat{S} \bullet X + \nu f(X)$ is continuous on this set, so it is also closed and hence compact. We have just seen that the objective function of $\mathrm{BP}'(\nu)$ is continuous on the feasible set, and hence existence of a minimizer for $\mathrm{BP}(\nu)$ follows. Now such a minimizer must satisfy the necessary conditions, and hence we see as above that we have a solution to $\mathrm{CPE}(\nu)$.

Since the barrier problem is convex, these conditions are also sufficient for optimality. So any solution to $\mathrm{CPE}(\nu)$ yields a minimizer for $\mathrm{BP}(\nu)$. Moreover, the objective here is strictly convex, so the minimizer is unique. The equations $XS = \nu I$ show that S is also unique, and then the equations $A^*y + S = C$ and the linear independence assumption imply that y is also unique. The equations $\mathrm{CPE}(\nu)$ also provide necessary and sufficient conditions for the dual barrier problem. Finally, if strict feasibility fails for (P), there is no solution yielding a finite value for the objective function of $\mathrm{BP}(\nu)$; there is no solution satisfying the necessary conditions for optimality in $\mathrm{BD}(\nu)$; and there is no solution to $\mathrm{CPE}(\nu)$, since the X-part would give a strictly feasible solution. A similar argument applies to the dual, and the proof is complete.   □

So far we have established the existence of a unique solution to $\mathrm{CPE}(\nu)$ for each positive ν, but not that these solutions form a smooth path. This will follow from the implicit function theorem if we show that the equations defining it are differentiable, with a derivative (with respect to (X, y, S)) that is square and nonsingular at points on the path. Unfortunately, while the equations of (5.1) are certainly differentiable, the derivative is not even square since the left-hand side maps $(X, y, S) \in \mathbb{SR}^{n \times n} \times \mathbb{R}^m \times \mathbb{SR}^{n \times n}$ to a point in $\mathbb{SR}^{n \times n} \times \mathbb{R}^m \times \mathbb{R}^{n \times n}$; XS is usually not symmetric even if X and S are. We therefore need to change the equations defining the central path. There are many possible approaches, which, as we shall see, lead to different search directions for our algorithms, but for now we choose a simple one: we replace $XS = \nu I$ by $-\nu X^{-1} + S = 0$. As in our discussion of the barrier function f, the function $X \to -\nu X^{-1}$ is differentiable at nonsingular symmetric matrices, with derivative $\nu(X^{-1} \odot X^{-1})$. So the central path is defined by the equations

$$\Phi_P(X, y, S; \nu) := \begin{pmatrix} A^*y + S \\ AX \\ -\nu X^{-1} \quad + S \end{pmatrix} = \begin{pmatrix} C \\ b \\ 0 \end{pmatrix}, \qquad (5.2)$$

whose derivative (with respect to (X, y, S)) is

$$\Phi'_P(X, y, S; \nu) := \begin{pmatrix} 0 & A^* & I \\ A & 0 & 0 \\ \nu(X^{-1} \odot X^{-1}) & 0 & I \end{pmatrix}, \qquad (5.3)$$

where $\mathcal{I}$ denotes the identity operator on $S\mathbb{R}^{n\times n}$. We have been rather loose in writing this in matrix form, since the blocks are operators rather than matrices, but the meaning is clear. We want to show that this derivative is nonsingular, and for this it suffices to prove that its null-space is trivial. Since similar equations will occur frequently, let us derive this from a more general result.

Theorem 5.3. Suppose the operators $\mathcal{E}$ and $\mathcal{F}$ map $S\mathbb{R}^{n\times n}$ to itself, and that $\mathcal{E}$ is nonsingular and $\mathcal{E}^{-1}\mathcal{F}$ is positive definite (but not necessarily self-adjoint). Assume that the linear independence condition holds. Then, for any P, $R \in S\mathbb{R}^{n\times n}$ and $q \in \mathbb{R}^m$, the solution to

$$
\begin{aligned}
\mathcal{A}^* v + W &= P, \\
\mathcal{A} U &= q, \\
\mathcal{E} U + \mathcal{F} W &= R
\end{aligned}
\tag{5.4}
$$

is uniquely given by

$$
\begin{aligned}
v &= (\mathcal{A}\mathcal{E}^{-1}\mathcal{F}\mathcal{A}^*)^{-1}(q - \mathcal{A}\mathcal{E}^{-1}(R - \mathcal{F}P)), \\
W &= P - \mathcal{A}^* v, \\
U &= \mathcal{E}^{-1}(R - \mathcal{F}W).
\end{aligned}
\tag{5.5}
$$

Proof. The formulae for W and U follow directly from the first and third equations. Now substituting for W in the formula for U, and inserting this in the second equation, we obtain after some manipulation

$$(\mathcal{A}\mathcal{E}^{-1}\mathcal{F}\mathcal{A}^*)v = q - \mathcal{A}\mathcal{E}^{-1}(R - \mathcal{F}P).$$

Since $\mathcal{E}^{-1}\mathcal{F}$ is positive definite and the A_is are linearly independent, the $m \times m$ matrix on the left is positive definite (but not necessarily symmetric) and hence nonsingular. This verifies that v is uniquely determined as given, and then so are W and U. Moreover, these values solve the equations. □

In our case, $\mathcal{F}$ is the identity, while $\mathcal{E}$ is $\nu(X^{-1} \odot X^{-1})$ with inverse $\nu^{-1}(X \odot X)$. This is easily seen to be positive definite, just as $f''(X)$ is. Hence the theorem applies, and so the derivative of the function Φ_P is nonsingular on the central path (and throughout $S\mathbb{R}^{n\times n}_{++} \times \mathbb{R}^m \times S\mathbb{R}^{n\times n}_{++}$); thus the central path is indeed a differentiable path.

By taking the trace of the last equation of (5.1), we obtain the last part of the following theorem, which summarizes what we have observed.

Theorem 5.4. Assume that both (P) and (D) have strictly feasible solutions and the linear independence condition holds. Then the set of solutions to (5.1) for all positive ν forms a nonempty differentiable path, called the central path. If $(X(\nu), y(\nu), S(\nu))$ solve these equations for a particular positive ν, then $X(\nu)$ is a strictly feasible solution to (P) and $(y(\nu), S(\nu))$

a strictly feasible solution to (D), with duality gap

$$C \bullet X(\nu) - b^T y(\nu) = X(\nu) \bullet S(\nu) = n\nu. \tag{5.6}$$

□

We now justify our earlier claim that we could use the central path to prove strong duality.

Theorem 5.5. The existence of strictly feasible solutions to (P) and (D) and the linear independence condition imply that both have bounded nonempty optimal solution sets, with zero duality gap.

Proof. The last part follows from the existence of the central path, since by (5.6) the duality gap associated to $X(\nu)$ and $(y(\nu), S(\nu))$ is $n\nu$, and this approaches zero as ν tends to zero. (In fact, the central path approaches optimal solutions to the primal and dual problems as ν decreases to zero (Luo, Sturm and Zhang 1998, Goldfarb and Scheinberg 1998), but we shall not prove this here.)

To show that (P) has a bounded nonempty set of optimal solutions, we proceed as in the proof of Theorem 5.2, again choosing $(\hat{y}, \hat{S}) \in F^0(D)$. Clearly, the set of optimal solutions is unchanged if we change the objective function of (P) to $\hat{S} \bullet X$ and add the constraint $\hat{S} \bullet X \leq \hat{S} \bullet \hat{X}$. But this latter constraint (for $X \in \mathcal{P}$) implies that all the eigenvalues of X are bounded by $(\hat{S} \bullet \hat{X})/\sigma$, where again $\sigma > 0$ denotes the smallest eigenvalue of $\hat{S}$. This shows that all optimal solutions of (P) (if any) lie in a compact set of feasible solutions; but the minimum of the continuous function $\hat{S} \bullet X$ over this compact set (containing $\hat{X}$) is attained, and so the set of optimal solutions is nonempty and bounded. The proof that the set of optimal dual solutions is bounded and nonempty is similar: we start by noting that the objective of maximizing $b^T y$ can be replaced by that of minimizing $\hat{X} \bullet S$ using (2.1).   □

6. Algorithms

In this section we will discuss three classes of algorithms for solving SDP problems: path-following methods, potential-reduction methods and algorithms based on smooth or nonsmooth nonlinear programming formulations. The first two classes consist of interior-point methods, while the last contains both interior-point and non-interior-point approaches. Interior-point methods for SDP were first introduced by Nesterov and Nemirovski (see Nesterov and Nemirovski (1994)) and independently by Alizadeh (1995). In all cases we shall concentrate on feasible methods, in which all iterates are (strictly in the first two cases) feasible; if we are using Newton steps, this implies that P and q in the system (5.4) will be zero, while R will depend on the method. One easy way to allow infeasible iterates (satisfying positive definiteness, but not the equality constraints) is just to let P and q be the negatives of the residuals in the dual and primal equality constraints, but some theoretical results then do not hold. Alternatively, the problems (P) and (D) can be embedded in a larger self-dual system that always has strictly feasible solutions at hand and whose solution gives the required information about the original problems: see de Klerk, Roos and Terlaky (1997), Luo et al. (2000), and Potra and Sheng (1998), based on the work of Ye, Todd and Mizuno (1994) for linear programming.

6.1. Path-following methods

These methods are motivated by Theorem 5.4, and attempt to track points on the central path as the parameter ν is decreased to zero. We mention first primal and dual versions, and then discuss primal-dual methods.

Primal and dual path-following methods conform to the general scheme of Nesterov and Nemirovski (1992, 1994), where they were first introduced and analysed. The basic strategy of the primal method is to take some Newton steps towards the minimizer of $\mathrm{BP}(\nu)$ for some parameter $\nu > 0$, and then decrease ν and repeat. It is easy to see that Newton steps for minimizers of $\mathrm{BP}(\nu)$ are just the X-part of Newton steps for the zeros of $\Phi_P(\cdot\,;\nu)$ in (5.2), and Theorem 5.3 shows how these may be computed. It is not necessary to maintain the S iterates, but the y iterates are useful to give a test for when the Newton steps can be terminated and ν reduced. We want the gradient of $\mathrm{BP}(\nu)$, modified by a Lagrangian term, to be sufficiently small, and since gradients 'live in dual space', we measure this using the dual X-norm. Hence our proximity criterion is

$$\|C - \nu X^{-1} - \mathcal{A}^* y\|_X^* \leq \tau \nu,$$

where $\tau \in (0, 1)$. This has two nice consequences. Suppose we set $S := C - \mathcal{A}^* y$. Then we have $\|S - \nu X^{-1}\|_X^* \leq \tau \nu$, so that $\|\nu^{-1} S - X^{-1}\|_{X^{-1}} \leq \tau$, and using the eigenvalue characterization of this norm we see that $\nu^{-1} X^{\frac{1}{2}} S X^{\frac{1}{2}}$ and hence S is positive definite, and so (y, S) strictly feasible for (D). Secondly, the duality gap is

$$X \bullet S = X \bullet (\nu X^{-1} + [S - \nu X^{-1}]) \leq \nu n + \|X\|_X \|S - \nu X^{-1}\|_X^* \leq \nu(n + \tau \sqrt{n})$$

so that we are provably close to optimality when ν is small. The algorithm then becomes:

Choose a strictly feasible X for (P), $y \in \mathbb{R}^m$, and $\nu > 0$. Perform damped Newton steps, maintaining X positive definite, until the proximity criterion is satisfied. Stop if ν is sufficiently small. Otherwise, replace ν by $\theta\nu$ for some $\theta \in (0, 1)$ and continue.

By damped Newton steps we mean that (X, y) is replaced by $(X_+, y_+) := (X + \alpha \Delta X, y + \alpha \Delta y)$ for some $\alpha \in (0, 1]$, where $(\Delta X, \Delta y)$ is the usual (full)

Newton step obtained by setting the linearization of $\Phi_P(\cdot, \nu)$ to zero, which will now be called the Newton direction. Using Theorem 5.3, it is not hard to see that this direction can be found by first computing the $m \times m$ matrix M with entries $m_{ij} := \nu^{-1} A_i \bullet (X A_j X)$, then solving

$$M \Delta y = -\mathcal{A}(X - \nu^{-1} X [C - \mathcal{A}^* y] X),$$

and finally setting $\Delta X = X - \nu^{-1} X [C - \mathcal{A}^*(y + \Delta y)] X$. Note that the proximity criterion is satisfied (for X and $y + \Delta y$) if and only if the Newton step for X is small: $\|\Delta X\|_X \leq \tau$.

The beautiful theory of self-concordant functions developed by Nesterov and Nemirovski enables them to establish a polynomial convergence result for this method. Suppose the initial (X, y, ν) are such that the proximity criterion is satisfied for $\tau = 0.1$ (so that the first action of the algorithm will be to reduce ν). Suppose also that ν is reduced each time by the factor $\theta = 1 - 0.1/\sqrt{n}$. Then at each iteration we can choose $\alpha = 1$ (we do not need to damp the Newton steps), the proximity criterion will be satisfied after a single Newton step, and, in $O(\sqrt{n} \ln(1/\epsilon))$ steps, the duality gap will be reduced to ϵ times its original value. (The occurrence of $\sqrt{n}$ in these results arises since f is an n-normal barrier for the positive semidefinite cone, and more particularly from the size of $f'(X)$ established in Proposition 5.1. This shows that ν can be reduced by the factor θ above while not losing too much proximity, so that one Newton step restores it.)

Next we discuss the dual method. This can be viewed as taking Newton steps for the minimizer of $BD(\nu)$, or, equivalently, for the zero of $\Phi_D(\cdot; \nu)$, defined as $\Phi_P(\cdot; \nu)$ but with $X - \nu S^{-1}$ replacing $-\nu X^{-1} + S$ as its last part. Here it is not necessary to maintain the X iterates. It is not hard to see that the Newton direction is computed as follows. First find the $m \times m$ matrix M with entries $m_{ij} := \nu A_i \bullet (S^{-1} A_j S^{-1})$, then solve

$$M \Delta y = b - \nu \mathcal{A} S^{-1}, \tag{6.1}$$

and finally set $\Delta S = -\mathcal{A}^* \Delta y$. Continue taking damped Newton steps until the following proximity criterion is satisfied:

$$\|\Delta S\|_S \leq \tau.$$

Then reduce ν and continue. Here, $X := \nu[S^{-1} + S^{-1}(\mathcal{A}^* \Delta y) S^{-1}]$ is strictly feasible in (P) when this criterion holds. The same theoretical results hold as in the primal case. One advantage of the dual method arises when C and the A_is share a sparsity pattern. Then S will have the same sparsity, while X may well be dense. Of course, S^{-1} is likely to be dense, but we may be able to perform operations cheaply with this matrix using a sparse Cholesky factorization of S. Recently, Fukuda, Kojima, Murota and Nakata (2000) have investigated ways in which the primal-dual methods discussed next can exploit this form of sparsity.

Now we turn to primal-dual path-following methods. Here we maintain (X, y, S), and our steps are determined by both the current primal and the current dual iterates. Apart from the sparsity issue above, this seems to be worthwhile computationally, and leads to fewer difficulties if an iterate gets close to the boundary of the positive semidefinite cone. In addition, the Newton step is based on a system more like $XS - \nu I = 0$, which is certainly smoother than one involving inverses, especially for near-singular iterates. The Newton step is then regarded as a search direction, and damped steps are taken (possibly with different damping in the primal and dual spaces) to get the next iterates. As discussed in the previous section, we cannot take Newton steps for the function whose last part is defined by $XS - \nu I$, so we have to symmetrize this somehow, but now we do this without using the inverse function. The first idea is to replace this condition with $(XS + SX)/2 - \nu I$, and this was proposed by Alizadeh, Haeberly and Overton (1998). Linearizing this system gives the equation (in addition to the feasibility equations)

$$\frac{1}{2}(\Delta X S + S \Delta X + X \Delta S + \Delta S X) = \nu I - \frac{1}{2}(XS + SX).$$

Thus the resulting Newton direction (called the AHO search direction) is defined by a system as in (5.4) with

$$\mathcal{E} = S \odot I, \quad \mathcal{F} = X \odot I.$$

One difficulty with this system is that we do not have an explicit form for the inverse of $\mathcal{E}$; instead, to find $\mathcal{E}^{-1}U$ we need to solve a Lyapunov system. Also, the sufficient conditions of Theorem 5.3 do not hold for this choice, and Todd, Toh and Tütüncü (1998) give an example where the Newton direction is not well-defined at a pair of strictly feasible solutions. (This does not seem to cause difficulties in practice.)

A more general approach is to apply a similarity to XS before symmetrizing it. This was discussed for a specific pair of similarities by Monteiro (1997), and then in general by Zhang (1998). So let P be nonsingular, and let us replace the last part of Φ_P by

$$\frac{1}{2}(PXSP^{-1} + P^{-T}SXP^T) - \nu I. \tag{6.2}$$

(Zhang showed that this is zero exactly when $XS = \nu I$ as long as X and S are symmetric.) An alternative way to view this is to scale (P) so that the variable X is replaced by $\hat{X} := PXP^T$ and (D) so that S is replaced by $\hat{S} := P^{-T}SP^{-1}$; then apply the Alizadeh–Haeberly–Overton approach in this scaled space. The resulting search directions form the *Monteiro–Zhang* family. Of course, with $P = I$, we retrieve the AHO direction.

Since the need for symmetrization occurs because X and S do not commute, it seems reasonable to choose P so that the scaled matrices do commute. Three ways to do this are: choose $P = S^{\frac{1}{2}}$ so that $\hat{S} = I$; choose $P = X^{-\frac{1}{2}}$ so that $\hat{X} = I$; and choose $P = W^{-\frac{1}{2}}$, where

$$W = X^{\frac{1}{2}}(X^{\frac{1}{2}}SX^{\frac{1}{2}})^{-\frac{1}{2}}X^{\frac{1}{2}} \qquad (6.3)$$

is the unique positive definite matrix with $WSW = X$, so that $\hat{X} = \hat{S}$. The resulting search directions are known as the HRVW/KSH/M, dual HRVW/KSH/M, and NT directions. The first was introduced by Helmberg, Rendl, Vanderbei and Wolkowicz (1996), and independently by Kojima, Shindoh and Hara (1997), using different motivations, and then rediscovered from the perspective above by Monteiro (1997). The second was also introduced by Kojima *et al.* (1997) and rediscovered by Monteiro; since it arises by switching the roles of X and S, it is called the dual of the first direction. The last was introduced by Nesterov and Todd (1997, 1998), from yet another motivation, and shown to be derivable in this form by Todd *et al.* (1998). These and several other search directions are discussed in Kojima, Shida and Shindoh (1999) and Todd (1999).

In the first case, the Newton direction can be obtained from the solution of a linear system as in (5.4) with

$$\mathcal{E} = \mathcal{I}, \quad \mathcal{F} = X \odot S^{-1};$$

in the second case with

$$\mathcal{E} = S \odot X^{-1}, \quad \mathcal{F} = \mathcal{I};$$

and in the third case with

$$\mathcal{E} = \mathcal{I}, \quad \mathcal{F} = W \odot W$$

(it is not immediate that this last corresponds to the Newton system for (6.2) with $P = W^{\frac{1}{2}}$; see Todd *et al.* (1998) for the analysis). In all cases, it is easy to see that $\mathcal{E}^{-1}\mathcal{F}$ is positive definite (and in fact also self-adjoint), so that the Newton direction is well defined. However, in the second, a Lyapunov system must again be solved to apply $\mathcal{E}^{-1}$ to a matrix. For the first case, we define M by setting $m_{ij} = A_i \bullet (XA_jS^{-1})$, while for the last, $m_{ij} = A_i \bullet (WA_jW)$. We then solve (6.1) for Δy, set $\Delta S = -\mathcal{A}^*\Delta y$, and then set

$$\Delta X = -X + \nu S^{-1} + \frac{1}{2}[X(\mathcal{A}^*\Delta y)S^{-1} + S^{-1}(\mathcal{A}^*\Delta y)X]$$

for the first case, and

$$\Delta X = -X + \nu S^{-1} + W(\mathcal{A}^*\Delta y)W$$

for the last. Again, damped steps are taken to preserve positive definiteness.

We still need a proximity criterion, and here two possibilities have been considered. In both, we let $\mu := \mu(X,S) := (X \bullet S)/n$. Then the narrow neighbourhood (parametrized by $\tau \in (0,1)$) is

$$\mathcal{N}_F(\tau) :=$$
$$\{(X,y,S) \in F^0(\mathrm{P}) \times F^0(\mathrm{D}) : \|X^{\frac{1}{2}} S X^{\frac{1}{2}} - \mu I\|_F = \|\lambda(XS - \mu I)\|_2 \leq \tau\mu\},$$

while the wide neighbourhood is

$$\mathcal{N}_{-\infty}(\tau) := \{(X,y,S) \in F^0(\mathrm{P}) \times F^0(\mathrm{D}) : \lambda_{\min}(XS) \geq (1-\tau)\mu\}.$$

Algorithms that maintain all iterates in a narrow neighbourhood are called short-step methods, while those that keep the iterates in a wide neighbourhood are termed long-step methods. In practice, algorithms frequently ignore such criteria and just take steps a proportion α (say 0.99) of the way to the boundary; different steps can be taken for the primal and dual iterates.

Here is a typical short-step primal-dual path-following algorithm. Assume given an initial strictly feasible point $(X,y,S) \in \mathcal{N}_F(\tau)$. Choose $\nu = \sigma\mu$ for some $\sigma \in (0,1)$, compute the search direction chosen from the AHO, HRVW/KSH/M, dual HRVW/KSH/M, and NT search directions, and take a full Newton step. Repeat.

Monteiro (1998) showed that such an algorithm, with $\tau = 0.1$ and $\sigma = 1-0.1/\sqrt{n}$, generates a sequence of iterates all in the narrow neighbourhood, and produces a strictly feasible point with duality gap at most ϵ times that of the original point in $O(\sqrt{n}\ln(1/\epsilon))$ steps. (Included in this is the result of Monteiro and Zanjacomo (1997) that the AHO search direction is well-defined within such a narrow neighbourhood.) Predictor-corrector methods, which alternate taking $\sigma = 0$ (with a line search) and $\sigma = 1$, and use two sizes of narrow neighbourhood, also have the same complexity. See also Monteiro and Todd (2000).

A typical long-step primal-dual path-following algorithm assumes given an initial strictly feasible point $(X,y,S) \in \mathcal{N}_{-\infty}(\tau)$. Choose $\nu = \sigma\mu$ for some $\sigma \in (0,1)$, compute the search direction chosen from the AHO, HRVW/KSH/M, dual HRVW/KSH/M, and NT search directions, and take the longest step that keeps the iterate in $\mathcal{N}_{-\infty}(\tau)$. Here it is not certain that the AHO search direction will be well-defined, so our theoretical results are for the other cases.

Monteiro and Zhang (1998) showed that such an algorithm, with any τ and σ in $(0,1)$ and independent of n, and using the NT search direction, generates a strictly feasible point with duality gap at most ϵ times that of the original point in $O(n\ln(1/\epsilon))$ steps; using the HRVW/KSH/M or dual HRVW/KSH/M search direction increases the bound to $O(n^{3/2}\ln(1/\epsilon))$ steps. Again, another reference for these results is Monteiro and Todd (2000).

6.2. Potential-reduction methods

The methods of the previous subsection were based on approximately solving the barrier problems BP(ν) and BD(ν), and the parameter ν had to be explicitly adjusted towards zero. Here we combine the objective function and the barrier function in a different way, and avoid the need to adjust a parameter. Such *potential functions* were first introduced by Karmarkar in his seminal work on interior-point methods for linear programming (Karmarkar 1984).

Consider the Tanabe–Todd–Ye primal-dual potential function

$$\Psi_\rho(X, y, S) := (n + \rho) \ln X \bullet S - \ln \det X - \ln \det S - n \ln n,$$

defined for strictly feasible points (X, y, S) (Tanabe 1988, Todd and Ye 1990). If $\lambda := \lambda(X^{\frac{1}{2}} S X^{\frac{1}{2}})$, then it is easy to see that $\Psi_0(X, y, S) = n \ln(e^T \lambda / n) - \ln(\Pi_j \lambda_j)$, so the arithmetic-geometric mean inequality shows that this is always nonnegative. In fact, it is zero if and only if all eigenvalues of $X^{\frac{1}{2}} S X^{\frac{1}{2}}$ are equal, or equivalently if the point is on the central path. $\Psi_\rho(X, y, S)$ increases the weight on the logarithm of the duality gap, and therefore pushes points towards the optimum. Our aim is to decrease this function by a constant at each iteration.

Theorem 6.1. Suppose $(X_0, y_0, S_0) \in F^0(\mathrm{P}) \times F^0(\mathrm{D})$ satisfies

$$\Psi_0(X_0, y_0, S_0) \le \rho \ln \frac{1}{\epsilon}$$

for some $\epsilon > 0$. Then, if we generate a sequence of strictly feasible points (X_k, y_k, S_k) with

$$\Psi_\rho(X_k, y_k, S_k) \le \Psi_\rho(X_{k-1}, y_{k-1}, S_{k-1}) - \delta$$

for some constant $\delta > 0$ and all $k \ge 1$, then, in $O(\rho \ln(1/\epsilon))$ steps, we will have a strictly feasible point (X_K, y_K, S_K) with duality gap at most ϵ times that of (X_0, y_0, S_0).

Proof. Let $K := 2\rho \ln(1/\epsilon)/\delta$. Then, using the fact above, we have

$$\rho \ln X_K \bullet S_K \le \rho \ln X_K \bullet S_K + \Psi_0(X_K, y_K, S_K)$$
$$= \Psi_\rho(X_K, y_K, S_K)$$
$$\le \Psi_\rho(X_0, y_0, S_0) - K\delta$$
$$= \rho \ln X_0 \bullet S_0 + \Psi_0(X_0, y_0, S_0) - K\delta$$
$$\le \rho \ln X_0 \bullet S_0 - \rho \ln \frac{1}{\epsilon}. \qquad \square$$

Notice that there is no need to control the proximity of the iterates to the central path, as long as the requisite decrease in the potential function can be obtained. It turns out that this is possible as long as $\rho \ge \sqrt{n}$.

A reasonable way to try to effect such a decrease is to move in the direction of steepest descent with respect to some norm.

Let us consider first a dual method. Suppose our current dual strictly feasible iterate is (y, S), and that we have available a primal strictly feasible solution X (in fact, initially it is only necessary to have an upper bound on the dual optimal value). Let $\nabla_S \Psi$ denote the derivative of Ψ_ρ with respect to S,

$$\nabla_S \Psi = \frac{n + \rho}{X \bullet S} X - S^{-1},$$

let U be positive definite, and consider

$$\min \nabla_S \Psi \bullet \Delta S + \frac{1}{2} \|\Delta S\|_U^2, \quad \mathcal{A}^* \Delta y + \Delta S = 0. \tag{6.4}$$

Of course, it is natural to take $U = S$, but we shall soon see the value of the generality we have allowed. For now, let us choose $U = S$ and see what the resulting direction is. If we let P denote the Lagrange multiplier for the constraint, we need to solve

$$\mathcal{A}P = 0, \quad \mathcal{A}^* \Delta y + \Delta S = 0, \quad P + S^{-1} \Delta S S^{-1} = -\nabla_S \Psi. \tag{6.5}$$

Let us set $\nu := (X \bullet S)/(n+\rho)$. Then the last equation above, multiplied by ν, becomes $(\nu P) + \nu S^{-1} \Delta S S^{-1} = -X + \nu S^{-1}$. It follows that $(\Delta y, \Delta S)$ is exactly the same as the search direction in the dual path-following algorithm – see the paragraph including (6.1) – for this value for ν. If the resulting $\|\Delta S\|_S$ is sufficiently large, then a suitable step is taken in the direction $(\Delta y, \Delta S)$ and one can show that the potential function is thus decreased by a constant (X is unchanged). If not, then the solution of the problem above suffices to generate an improved X, exactly as we found below (6.1), and then updating X while holding (y, S) unchanged also can be shown to give a constant decrease in the potential function. It follows that we can attain the iteration complexity bound given in Theorem 6.1. Details can be found in, for example, Benson, Ye and Zhang (2000), which describes why this method is attractive for SDP problems arising in combinatorial optimization problems and gives some excellent computational results.

Now let us consider a symmetric primal-dual method. Suppose we have a strictly feasible point (X, y, S). In addition to the dual direction-finding problem (6.4) above, we need a primal problem to determine ΔX. Let $\nabla_X \Psi$ denote the derivative of Ψ_ρ with respect to X,

$$\nabla_X \Psi = \frac{n + \rho}{X \bullet S} S - X^{-1},$$

let V be positive definite, and consider

$$\min \nabla_X \Psi \bullet \Delta X + \frac{1}{2} \|\Delta X\|_V^2, \quad \mathcal{A} \Delta X = 0. \tag{6.6}$$

Here it is natural to choose $V = X$, and this would lead to a primal potential-reduction method with the same iteration complexity. But we would like to get search directions for both primal and dual problems without solving two optimization subproblems. This can be achieved by using $V = W$ in (6.6) and $U = W^{-1}$ in (6.4), where W is the scaling matrix of (6.3). The dual direction then comes from equations like (6.5), with W replacing S^{-1} on the left-hand side of the last equation. The primal direction, if we use a Lagrange multiplier q for the constraint, comes from the solution to

$$\mathcal{A}\Delta X = 0, \quad W^{-1}\Delta X W^{-1} - \mathcal{A}^* q = -\nabla_X \Psi. \qquad (6.7)$$

If we write R for $-\mathcal{A}^* q$ and pre- and postmultiply the last equation by W (noting that $WSW = X$ and $WX^{-1}W = S^{-1}$), we get

$$\mathcal{A}\Delta X = 0, \quad \mathcal{A}^* q + R = 0, \quad \Delta X + WRW = -\nabla_S \Psi.$$

Comparing these two systems, we see that they are *identical* if we identify ΔX with P and $(\Delta y, \Delta S)$ with (q, R). It thus turns out that both search directions can be obtained simultaneously by solving one system of the form (5.4). In fact, the search directions are exactly (up to a scalar factor) those of the NT path-following method of the previous subsection, and we have already discussed how those can be computed. (Again, we need to take $\nu = X \bullet S/(n+\rho)$.) It turns out that, by taking a suitable step in these directions, we can again achieve a constant decrease in the potential function. The analysis is somewhat complicated, and the reader is referred to the original article of Nesterov and Todd (1997), the subsequent paper (Nesterov and Todd 1998) which gives a simplified proof for the key Theorem 5.2 in the first paper, and the paper of Tunçel (2000) which provides an easier analysis for the SDP case.

The important point again is that a constant decrease leads easily (via Theorem 6.1) to the best known complexity bound for the number of iterations, and that this is achieved without any concern for the iterates staying close to the central path, yielding great flexibility for the algorithms.

6.3. Nonlinear programming approaches

Finally we turn to methods that are based on nonsmooth or smooth optimization techniques for nonlinear programming formulations of (P) or (D). Some of these place restrictions on the SDP problems that can be handled.

First we discuss nonsmooth methods for minimizing the maximum eigenvalue of a matrix which depends affinely on some parameters. This was our first example in Section 3, but it is remarkably general. Suppose X is bounded for feasible solutions to (P). Then we can add an inequality on the trace of X, and by adding a slack variable and making a block diagonal matrix, we can assume that the trace of X is fixed at some positive value;

by scaling, we suppose this is 1. So we assume that $\text{trace}\,X = 1$ for all feasible X. Note that this holds for Examples 8 and (after scaling) 9. If we add this constraint explicitly, the dual problem then becomes to minimize $\lambda_{\max}(\mathcal{A}^*y - C) - b^T y$ over $y \in \mathbb{R}^m$ (we switched to a minimization problem by changing the sign of the objective). We can also assume that the linear objective $b^T y$ does not appear by incorporating it into the first term (each A_i is replaced by $A_i - b_i I$). Hence any such constant trace problem has a dual that is a maximum eigenvalue minimization problem.

We now have a convex but nonsmooth optimization problem, to which standard methods of nonlinear programming can be applied. One such is the bundle method, which builds up a cutting-plane model of the objective function by computing subgradients of the maximum eigenvalue function. Let us set $g(y) := \lambda_{\max}(\mathcal{A}^*y - C)$. A *subgradient* of g at y is a vector z with $g(y') \geq g(y) + z^T(y' - y)$ for all y'; and in our case, one can be found as $\mathcal{A}(vv^T)$, where v is a unit eigenvector of $\mathcal{A}^*y - C$ associated with its maximum eigenvalue. It is useful also to consider so-called ϵ-*subgradients* for $\epsilon > 0$: z is one such if

$$g(y') \geq g(y) + z^T(y' - y) - \epsilon$$

for all y', and the set of them all is called the ϵ-*subdifferential* $\partial g_\epsilon(y)$. In our case this turns out to be

$$\partial g_\epsilon(y) = \{\mathcal{A}W : (\mathcal{A}^*y - C) \bullet W \geq \lambda_{\max}(\mathcal{A}^*y - C) - \epsilon,\ \text{trace}\,W = 1, W \succeq 0\}.$$

Helmberg and Rendl (2000) develop a very efficient algorithm, the *spectral bundle method*, by modifying the classical bundle method to exploit this structure. From the result above, it is easy to see that

$$g(y') \geq (\mathcal{A}^*y - C) \bullet W + (\mathcal{A}W)^T(y' - y) = (\mathcal{A}^*y' - C) \bullet W$$

for any $W \succeq 0$ with trace 1 and any y'. Hence, if we choose any subset $\mathcal{W}$ of such matrices,

$$g(y') \geq \hat{g}_{\mathcal{W}}(y') := \max\{(\mathcal{A}^*y' - C) \bullet W : W \in \mathcal{W}\}.$$

At every iteration, Helmberg and Rendl generate a search direction d for the current iterate y by minimizing $\hat{g}_{\mathcal{W}}(y + d) + (u/2)d^T d$ for some regularizing parameter u and some $\mathcal{W}$. Let $P \in \mathbb{R}^{n \times k}$ have orthonormal columns (think of them as approximate eigenvectors corresponding to almost maximal eigenvalues of $\mathcal{A}^*y - C$), and let $\bar{W} \succeq 0$ have trace 1 (think of this as a matrix containing useful past information). Then the spectral bundle method chooses

$$\mathcal{W} := \{\alpha \bar{W} + PVP^T : \alpha + \text{trace}\,V = 1, \alpha \geq 0, V \succeq 0\}.$$

The dual of the direction-finding subproblem turns out to be an SDP problem with a quadratic objective function in a lower dimensional space (V is of order k). This problem is solved to yield the search direction d and a new y trial vector is computed. If there is a suitable improvement in the objective function, this new point replaces the old; otherwise we stay at the old point. In either case, an approximate eigenvector corresponding to the maximum eigenvalue of the trial $\mathcal{A}^*y - C$ is computed, and this is added as a column to the P matrix. If there are too many columns, old information is incorporated into the aggregate matrix $\bar{W}$, and the process continues. Many details have been omitted, but the rough idea of the method is as above; it can be thought of as providing an approximation by considering only a subset of feasible X matrices, using this to improve the dual solution y, and using this in turn to improve the subset of feasible solutions in the primal.

As a version of the bundle method, the algorithm above has good global convergence properties, but no iteration bounds as for the interior-point methods of the previous subsections are known. Nevertheless, excellent computational results have been obtained for problems that are inaccessible to the latter methods due to their size; see Helmberg and Rendl (2000).

It is known that, for smooth optimization problems, second-order methods are much more attractive than first-order techniques such as the spectral bundle method, but it is not clear how second-order information can be incorporated in nonsmooth optimization. However, for the maximum eigenvalue problem, this is possible: Oustry (1999, 2000) devises the so-called $\mathcal{U}$-Lagrangian of the maximum eigenvalue function, uses this to get a quadratic approximation to the latter along a manifold where the maximum eigenvalue has a fixed multiplicity, and then develops a second-order bundle method using these ideas. This method retains the global convergence of the first-order method, but also attains asymptotic quadratic convergence under suitable regularity conditions. These bundle methods are further discussed, and improved computational results given, in Helmberg and Oustry (2000).

Fukuda and Kojima (2000) have recently given an interior-point method for the same class of problems, working just in the space of y to avoid difficulties for large-scale problems. This paper also has an excellent discussion of recent attempts to solve such problems efficiently. Note that Vavasis (1999) has developed an efficient way to compute the barrier and its gradient for this dual formulation.

Now we turn to methods that generate nonconvex nonlinear programming problems in a lower dimension, and apply interior-point or other techniques for their solution. Suppose first that (P) includes constraints specifying the diagonal entries of X:

$$(\mathrm{P}): \quad \min\ C \bullet X, \quad \mathrm{diag}(X) = d,\ \mathcal{A}X = b,\ X \succeq 0,$$

with dual problem

$$(\mathrm{D}): \quad \max\ d^T z + b^T y, \quad \mathrm{Diag}(z) + \mathcal{A}^*y + S = C,\ S \succeq 0.$$

Burer, Monteiro and Zhang (1999*a*) suggest solving (D) by an equivalent nonlinear programming problem obtained by eliminating variables. In fact, they only consider strictly feasible solutions of (D). Their procedure is based on a theorem stating that, for each $(w, y) \in \mathbb{R}^n_{++} \times \mathbb{R}^m$, there is a unique strictly lower triangular matrix $\bar{L} = \bar{L}(w, y)$ and a unique $z = z(w, y) \in \mathbb{R}^n$ satisfying

$$C - \mathrm{Diag}(z) - \mathcal{A}^* y = (\mathrm{Diag}(w) + \bar{L})(\mathrm{Diag}(w) + \bar{L})^T,$$

and that $\bar{L}(w, y)$ and $z(w, y)$ are infinitely differentiable. This takes care of the constraint that S be positive definite implicitly by requiring it to have a nonsingular Cholesky factorization. (D) is then replaced by the smooth but nonconvex problem

$$(\mathrm{D}') : \qquad \max_{w,y} d^T z(w, y) + b^T y, \; w > 0.$$

The authors then suggest algorithms to solve this problem: a log-barrier method and a potential-reduction method. A subsequent paper relaxes the requirement that the diagonal of X be fixed. Instead, they require in Burer, Monteiro and Zhang (1999*b*) that the diagonal be bounded below, so the first constraint becomes $\mathrm{diag}(X) \geq d$. This constraint can be without loss of generality, since it holds for any positive semidefinite matrix if we choose the vector d to be zero. The corresponding change to (D) is that now z must be nonnegative, and so the constraint $z(w, y) > 0$ is added to (D') (as we noted, Burer *et al.* only consider strictly feasible solutions to (D)). Once again, they consider log-barrier and potential-reduction methods to solve (D'). Although the problem (D') is nonconvex, Burer, Monteiro, and Zhang prove global convergence of their methods, and have obtained some excellent computational results on large-scale problems.

Finally, we mention the approach of Vanderbei and Yurttan Benson (1999): the primal variable X is factored as $L(X) \mathrm{Diag}(d(X)) L(X)^T$, where $L(X)$ is unit lower triangular and $d(X) \in \mathbb{R}^n$, and the constraint that X be positive semidefinite is replaced with the requirement that $d(X)$ be a nonnegative vector. The authors show that d is a concave function, and give some computational results for this reformulation.

We should mention that research is very active in new methods to solve large sparse SDP problems. The reader is urged to consult the web pages of Helmberg (2001) and Wright (2001) to see the latest developments.

7. Concluding remarks

We have investigated semidefinite programming from several viewpoints, examining its applications, duality theory, and several algorithms for solving SDP problems. The area has a rich history, drawing from several fields, and recently powerful methods for solving small- and medium-scale problems have been developed. The interior-point methods we have discussed can solve most problems with up to about a thousand linear constraints and matrices of order up to a thousand or so. However, as problems get larger, it is not clear that this class of methods can successfully compete with special-purpose algorithms that better exploit sparsity, and we have also considered a number of these. The limitations of such methods are being reduced, and they have successfully solved problems with matrices of order 10,000 and more. One limitation is that these more efficient methods usually solve the dual problem, and if a primal near-optimal solution is required (as in the max-cut problem using the technique of Goemans and Williamson to generate a cut), they may not be as appropriate. The topic remains exciting and vibrant, and significant developments can be expected over the next several years.

Acknowledgements

I would like to thank Michael Overton, Jos Sturm, and Henry Wolkowicz for helpful comments on this paper.

REFERENCES

F. Alizadeh (1995), Interior point methods in semidefinite programming with applications to combinatorial optimization, *SIAM J. Optim.* **5**, 13–51.

F. Alizadeh (2001), Semidefinite programming home page:
   `http://karush.rutgers.edu/~alizadeh/Sdppage/index.html`

F. Alizadeh and S. Schmieta (2000), Symmetric cones, potential reduction methods and word-by-word extensions, in *Handbook of Semidefinite Programming* (H. Wolkowicz, R. Saigal, and L. Vandenberghe, eds), Kluwer Academic Publishers, Boston/Dordrecht/London, pp. 195–233.

F. Alizadeh, J.-P. A. Haeberly, and M. L. Overton (1994), A new primal-dual interior-point method for semidefinite programming, in *Proceedings of the 5th SIAM Conference on Applied Linear Algebra* (J. G. Lewis, ed.), SIAM, Philadelphia, USA, pp. 113–117.

F. Alizadeh, J.-P. A. Haeberly, and M. L. Overton (1998), Primal-dual interior-point methods for semidefinite programming: convergence rates, stability and numerical results, *SIAM J. Optim.* **8**, 746–768.

R. Bellman and K. Fan (1963), On systems of linear inequalities in Hermitian matrix variables, in *Convexity*, Vol. 7 of *Proceedings of Symposia in Pure Mathematics*, American Mathematical Society, Providence, RI, pp. 1–11.

A. Ben-Tal and A. Nemirovski (1998), Robust convex optimization, *Math. Oper. Res.* **23**, 769–805.

S. J. Benson, Y. Ye, and X. Zhang (2000), Solving large-scale sparse semidefinite programs for combinatorial optimization, *SIAM J. Optim.* **10**, 443–461.

S. Boyd, L. El Ghaoui, E. Feron, and V. Balakrishnan (1994), *Linear Matrix Inequalities in System and Control Theory*, SIAM Studies in Applied Mathematics, SIAM, Philadelphia, USA.

S. Burer, R. D. C. Monteiro, and Y. Zhang (1999*a*), Solving semidefinite programs via nonlinear programming II: Interior point methods for a subclass of SDPs, Technical Report TR99-23, Department of Computational and Applied Mathematics, Rice University, Houston, TX.

S. Burer, R. D. C. Monteiro, and Y. Zhang (1999*b*), Interior point algorithms for semidefinite programming based on a nonlinear programming formulation, Technical Report TR99-27, Department of Computational and Applied Mathematics, Rice University, Houston, TX.

J. Cullum, W. E. Donath, and P. Wolfe (1975), The minimization of certain nondifferentiable sums of eigenvalues of symmetric matrices, *Math. Programming Study* **3**, 35–55.

W. E. Donath and A. J. Hoffman (1973), Lower bounds for the partitioning of graphs, *IBM J. Research and Development* **17**, 420–425.

C. Delorme and S. Poljak (1993), Laplacian eigenvalues and the maximum cut problem, *Math. Programming* **62**, 557–574.

L. Faybusovich (1997*a*), Euclidean Jordan algebras and interior-point algorithms, *Positivity* **1**, 331–357.

L. Faybusovich (1997*b*), Linear systems in Jordan algebras and primal-dual interior-point algorithms, *J. Comput. Appl. Math.* **86**, 149–175.

R. Fletcher (1981), A nonlinear programming problem in statistics (educational testing), *SIAM J. Sci. Statist. Comput.* **2**, 257–267.

R. Fletcher (1985), Semi-definite matrix constraints in optimization, *SIAM J. Control Optim.* **23**, 493–513.

M. Fukuda and M. Kojima (2000), Interior-point methods for Lagrangian duals of semidefinite programs, Technical Report B-365, Department of Mathematical and Computing Sciences, Tokyo Institute of Technology, Tokyo.

M. Fukuda, M. Kojima, K. Murota, and K. Nakata (2000), Exploiting sparsity in semidefinite programming via matrix completions I: General framework, *SIAM J. Optim.* **11**, 647–674.

M. X. Goemans (1997), Semidefinite programming in combinatorial optimization, *Math. Programming* **79**, 143–161.

M. X. Goemans and D. P. Williamson (1995), Improved approximation algorithms for maximum cut and satisfiability problems using semidefinite programming, *J. Assoc. Comput. Mach.* **42**, 1115–1145.

D. Goldfarb and K. Scheinberg (1998), Interior point trajectories in semidefinite programming, *SIAM J. Optim.* **8**, 871–886.

M. Grötschel, L. Lovász, and A. Schrijver (1988), *Geometric Algorithms and Combinatorial Optimization*, Springer, Berlin.

C. Helmberg (2001), Semidefinite programming home page:
`http://www.zib.de/helmberg/semidef.html`

C. Helmberg and F. Oustry (2000), Bundle methods to minimize the maximum eigenvalue function, in *Handbook of Semidefinite Programming* (H. Wolkowicz, R. Saigal, and L. Vandenberghe, eds), Kluwer Academic Publishers, Boston/Dordrecht/London, pp. 307–337.

C. Helmberg and F. Rendl (2000), A spectral bundle method for semidefinite programming, *SIAM J. Optim.* **10**, 673–696.

C. Helmberg, F. Rendl, R. Vanderbei, and H. Wolkowicz (1996), An interior-point method for semidefinite programming, *SIAM J. Optim.* **6**, 342–361.

F. Jarre (1999), A QQP-minimization method for semidefinite and smooth nonconvex programs, Technical Report, Institut für Angewandte Mathematik und Statistik, Universität Würzburg, Germany.

N. K. Karmarkar (1984), A new polynomial-time algorithm for linear programming, *Combinatorica* **4**, 373–395.

E. de Klerk, C. Roos, and T. Terlaky (1997), Initialization in semidefinite programming via a self-dual skew-symmetric embedding, *Operations Research Letters* **20**, 213–221.

M. Kojima, S. Shindoh, and S. Hara (1997), Interior-point methods for the monotone semidefinite linear complementarity problem in symmetric matrices, *SIAM J. Optim.* **7**, 86–125.

M. Kojima, M. Shida, and S. Shindoh (1999), Search directions in the SDP and the monotone SDLCP: Generalization and inexact computation, *Math. Programming* **85**, 51–80.

A. S. Lewis and M. L. Overton (1996), Eigenvalue optimization, in *Acta Numerica*, Vol. 5, Cambridge University Press, pp. 149–190.

L. Lovász (1979), On the Shannon capacity of a graph, *IEEE Trans. Inform. Theory* **25**, 1–7.

L. Lovász and A. Schrijver (1991), Cones of matrices and set-functions and 0–1 optimization, *SIAM J. Optim.* **1**, 166–190.

Z.-Q. Luo, J. F. Sturm, and S. Zhang (1998), Superlinear convergence of a symmetric primal-dual path-following algorithm for semidefinite programming, *SIAM J. Optim.* **8**, 59–81.

Z.-Q. Luo, J. F. Sturm, and S. Zhang (2000), Conic convex programming and self-dual embedding, *Optim. Methods Software* **14**, 169–218.

R. D. C. Monteiro (1997), Primal-dual path-following algorithms for semidefinite programming, *SIAM J. Optim.* **7**, 663–678.

R. D. C. Monteiro (1998), Polynomial convergence of primal-dual algorithms for semidefinite programming based on the Monteiro and Zhang family of directions, *SIAM J. Optim.* **8**, 797–812.

R. D. C. Monteiro and M. J. Todd (2000), Path-following methods, in *Handbook of Semidefinite Programming* (H. Wolkowicz, R. Saigal, and L. Vandenberghe, eds), Kluwer Academic Publishers, Boston/Dordrecht/London, pp. 267–306.

R. D. C. Monteiro and P. R. Zanjacomo (1997), A note on the existence of the Alizadeh–Haeberly–Overton direction for semidefinite programming, *Math. Programming*, **78**, 393–396.

R. D. C. Monteiro and Y. Zhang (1998), A unified analysis for a class of long-step primal-dual path-following interior-point algorithms for semidefinite programming, *Math. Programming* **81**, 281–299.

Yu. E. Nesterov and A. S. Nemirovski (1992), Conic formulation of a convex programming problem and duality, *Optim. Methods Software* **1**, 95–115.

Yu. E. Nesterov and A. S. Nemirovski (1994), *Interior Point Polynomial Algorithms in Convex Programming*, SIAM Publications, SIAM, Philadelphia, USA.

Yu. E. Nesterov and M. J. Todd (1997), Self-scaled barriers and interior-point methods for convex programming, *Math. Oper. Res.* **22**, 1–42.

Yu. E. Nesterov and M. J. Todd (1998), Primal-dual interior-point methods for self-scaled cones, *SIAM J. Optim.* **8**, 324–364.

F. Oustry (1999), The $\mathcal{U}$-Lagrangian of the maximum eigenvalue function, *SIAM J. Optim.* **9**, 526–549.

F. Oustry (2000), A second-order bundle method to minimize the maximum eigenvalue function, *Math. Programming* **89**, 1–33.

M. L. Overton and R. S. Womersley (1993), Optimality conditions and duality theory for minimizing sums of the largest eigenvalues of symmetric matrices, *Math. Programming* **62**, 321–357.

S. Poljak and Z. Tuza (1995), Maximum cuts and largest bipartite subgraphs, in *Combinatorial Optimization* (W. Cook, L. Lovász, and P. Seymour, eds), American Mathematical Society, Providence, RI, pp. 181–244.

S. Poljak, F. Rendl, and H. Wolkowicz (1995), A recipe for semidefinite relaxation for $(0, 1)$-quadratic programming, *J. Global Optimization* **7**, 51–73.

F. Potra and R. Sheng (1998), On homogeneous interior-point algorithms for semidefinite programming, *Optim. Methods Software* **9**, 161–184.

M. Ramana (1997), An exact duality theory for semidefinite programming and its complexity implications, *Math. Programming* **77**, 129–162.

M. Ramana, L. Tunçel, and H. Wolkowicz (1997), Strong duality for semidefinite programming, *SIAM J. Optim.* **7**, 641–662.

R. T. Rockafellar (1970), *Convex Analysis*, Princeton University Press, Princeton, NJ.

N. Z. Shor (1990), Dual quadratic estimates in polynomial and Boolean programming, *Ann. Oper. Res.* **25**, 163–168.

K. Tanabe (1988), Centered Newton method for mathematical programming, in *System Modeling and Optimization*, Springer, NY, pp. 197–206.

M. J. Todd (1999), A study of search directions in interior-point methods for semidefinite programming, *Optim. Methods Software* **11&12**, 1–46.

M. J. Todd and Y. Ye (1990), A centered projective algorithm for linear programming, *Math. Oper. Res.* **15**, 508–529.

M. J. Todd, K.-C. Toh, and R. H. Tütüncü (1998), On the Nesterov–Todd direction in semidefinite programming, *SIAM J. Optim.* **8**, 769–796.

L. Tunçel (2000), Potential reduction and primal-dual methods, in *Handbook of Semidefinite Programming* (H. Wolkowicz, R. Saigal, and L. Vandenberghe, eds), Kluwer Academic Publishers, Boston/Dordrecht/London, pp. 235–265.

L. Vandenberghe and S. Boyd (1996), Semidefinite programming, *SIAM Rev.* **38**, 49–95.

R. J. Vanderbei and H. Yurttan Benson (1999), On formulating semidefinite programming problems as smooth convex nonlinear optimization problems, Report ORFE 99-01, Operations Research and Financial Engineering, Princeton, NJ.

S. A. Vavasis (1999), A note on efficient computation of the gradient in semidefinite programming, Technical Report, Department of Computer Science, Cornell University, Ithaca, NY.

H. Wolkowicz (2001), Bibliography on semidefinite programming:
`http:liinwww.ira.uka.de/bibliography/Math/psd.html`

H. Wolkowicz, R. Saigal, and L. Vandenberghe, eds (2000), *Handbook of Semidefinite Programming*, Kluwer Academic Publishers, Boston/Dordrecht/London.

S. J. Wright (2001), Interior-point methods online home page:
`http://www-unix.mcs.anl.gov/otc/InteriorPoint/`

Y. Ye, M. J. Todd, and S. Mizuno (1994), An $O(\sqrt{n}L)$-iteration homogeneous and self-dual linear programming algorithm, *Math. Oper. Res.* **19**, 53–67.

Y. Zhang (1998), On extending some primal-dual interior-point algorithms from linear programming to semidefinite programming, *SIAM J. Optim.* **8**, 365–386.

For EU product safety concerns, contact us at Calle de José Abascal, 56–1°,
28003 Madrid, Spain or eugpsr@cambridge.org.

www.ingramcontent.com/pod-product-compliance
Ingram Content Group UK Ltd.
Pitfield, Milton Keynes, MK11 3LW, UK
UKHW060310090126

466816UK00021B/419